# Geology of the Brent Group

Geological Society Special Publications

*Series Editor* J. Brooks

GEOLOGICAL SOCIETY SPECIAL PUBLICATION NO 61

# Geology of the Brent Group

EDITED BY

A. C. MORTON
British Geological Survey
Keyworth
Nottingham, UK

R. S. HASZELDINE
Department of Geology and Applied Geology
The University of Glasgow, UK

M. R. GILES
Koninklijke/Shell Exploratie en Produktie Laboratorium
Rijswijk, Netherlands

S. BROWN
The Petroleum Science and Technology Institute
Edinburgh, UK

1992

Published by

The Geological Society

London

# THE GEOLOGICAL SOCIETY

The Society was founded in 1807 as the Geological Society of London and is the oldest geological society in the world. It received its Royal Charter in 1825 for the purpose of 'investigating the mineral structure of the Earth'. The Society is Britain's national learned society for geology with a Fellowship exceeding 6000. It has countrywide coverage and approximately one quarter of its membership resides overseas. The Society is responsible for promoting all aspects of the geological sciences and will also embrace professional matters on the completion of the reunification with the Institution of Geologists. The Society has its own publishing house to produce its international journals, books and maps, and is the European distributor for materials published by the American Association of Petroleum Geologists.

Fellowship is open to those holding a recognized honours degree in geology or cognate subject and who have at least two years relevant postgraduate experience, or have not less than six years relevant experience in geology or a cognate subject. A Fellow who has not less than five years relevant postgraduate experience in the practice of Geology may apply for validation and be able to use the designatory letters C. Geol (Chartered Geologist).

Further information about the Society is available from the Membership Manager, Geological Society, Burlington House, London, W1V 0JU, UK.

Published by the Geological Society from:
The Geological Society Publishing House
Unit 7
Brassmill Enterprise Centre
Brassmill Lane
Bath
Avon BA1 3JN
UK
(*Orders*: Tel. 0225 445046)

First published 1992



**Distributors**
USA
AAPG Bookstore
PO Box 979
Tulsa
Oklahoma 74101–0979
USA
(*Orders*: Tel: (918)584–2555)

Australia
Australian Mineral Foundation
63 Conyngham St
Glenside
South Australia 5065
Australia
(*Orders*: Tel: (08)379–0444)

**A catalogue record for this book is available from the British Library**

ISBN 0–903317–68–0

# Contents

# Geology of the Brent Group: Introduction

A. C. MORTON[1], R. S. HASZELDINE[2], M. R. GILES[3] & S. BROWN[4]

[1] *British Geological Survey, Keyworth, Nottingham NG12 5GG, UK*

[2] *Department of Geology and Applied Geology, The University of Glasgow, Glasgow G12 8QQ, UK*

[3] *Koninklijke/Shell Exploratorie en Produktie Laboratorium, Volmerlaan 6, 2280 AB Rijswijk, Netherlands*

[4] *The Petroleum Science and Technology Institute, Research Park, Riccarton, Edinburgh EH14 4AS, UK*

The Middle Jurassic Brent Group sediments and their correlatives on the Norwegian shelf are, in economic terms, the most important hydrocarbon reservoir in NW Europe. In 1971 the Brent Field was discovered by Shell/Esso and tested in 1972 with 1.8 billion barrels of recoverable oil; nine major Brent sandstone fields were discovered by the end of 1973 (Brennand *et al.* 1990). In 1980 the northern North Sea (overwhelmingly comprising fields with Brent Group reservoirs) was ranked as the 13th largest petroleum province in the world, containing 1.6% of produced and recoverable oil equivalent reserves (Ivanhoe 1980). By 1988, discovered Brent hydrocarbons comprised some 49% of the UK's recoverable reserves, totalling 22.5 billion barrels of oil equivalent. Brent recoverable hydrocarbons currently known in the Norwegian sector add approximately 8 billion barrels of oil equivalent (Brennand *et al.* 1990). Now that the UK Brent Province has reached maturity in exploration terms, this book provides a timely review of the geology and petroleum geology of one of the worlds major petroleum reservoirs. The book provides a wide-ranging coverage of Brent Group geology, including exploration history, structural evolution, sequence stratigraphy, sedimentology, diagenesis, palynology, hydrocarbon generation and migration, and petrophysics. Accounts of the geology of individual Brent Group fields are not included, as these are available in the books of Spencer *et al.* (1986) and Abbotts (1991). The book shows that despite the long passage of time since the original discovery was made, over 20 years ago, and despite the subsequent drilling of several hundred exploration and development wells, major controversies still exist, particularly over the depositional environment and diagenetic models.

The book commences with **Bowen** drawing upon his records and personal experience to outline the depositional history of the Brent Province, and to discuss future exploration potential. Although the Brent Group was originally interpreted to have been deposited during a period of active rifting and basin subsidence, **Yielding *et al.*** show that the major phase of rifting occurred earlier, with the Brent deposited during a thermal subsidence phase.

Several papers deal with the sedimentological framework of the Brent Group, beginning with the review by **Richards**. Although the original interpretation of the sequence as 'deltaic' (Bowen 1975) is still broadly acceptable, the book illustrates the controversy over the precise setting of Brent Group deposition. **Cannon *et al.*** and **Helland-Hansen *et al.*** consider the sequence as a prograding wave-dominated delta, supported by the detailed work of **Scott** on the nearshore and coastal Rannoch–Etive successions. Alternatively, **Richards** invokes a significant estuarine component and **Alexander** makes an analogy with the coeval prograding coastal plain sequences of Yorkshire. Problems of interpretation exist largely because of the difficulty in establishing clear time lines through the sequence, making palaeogeographic reconstructions speculative. Three papers (**Cannon *et al.*, Helland-Hansen *et al.*, Mitchener *et al.***) have dealt with this problem by using a combination of sedimentology, sequence stratigraphy and palynology: the earlier paper by Eynon (1981) may be considered a forerunner of this approach. These have provided a series of palaeogeographical 'snapshots', rather than one individual palaeogeography. The differences between the models depend to a large extent upon the interpretation of palynological events, so that the discussions by **Williams** and **Whitaker *et al.*** are especially relevant.

Three papers cover the topic of provenance of Brent Group sediments, underplayed in previous sedimentological interpretations of the sequence. **Morton** and **Stattegger & Morton** concentrate on the recently-developed garnet

*From* Morton, A. C., Haszeldine, R. S., Giles, M. R. & Brown, S. (eds), 1992, *Geology of the Brent Group*. Geological Society Special Publication No. 61, pp. 1–2.

geochemical technique, and **Mearns** uses the equally novel samarium–neodymium model age approach. Both methods demonstrate the importance of laterally-fed sediment to the basin, and show that input from Norway and the Shetland Platform was considerably more important than axially-transported material.

After the depositional facies, the most important individual process affecting the quality of Brent reservoirs is diagenesis. Past and current research activity is reflected by the inclusion of seven papers on this topic, which again demonstrate a lack of concensus of opinion. **Giles *et al.*** and **Bjørlykke *et al.*** provide the first regional syntheses of Brent sandstone diagenesis, both of which make use of extensive data sets acquired from the British and Norwegian sectors respectively. Both these papers identify good correlations between cement abundances and present-day subsidence depth, implying that burial is the main control on diagenetic cementation and inferring that pore fluid movement is of minor importance. By contrast, the other papers all invoke a degree of pore fluid movement during cementation. **Hogg *et al.*** utilize new cathode luminescence instrumentation to identify growth zones within quartz overgrowths, the major porosity-reducing cement. **Haszeldine *et al.*** combine petrographic, isotopic and geological information to consider the importance of open and restricted porefluid movement in Brent diagenesis. **Hamilton *et al.*** critically examine the evidence for dating cementation by illite, the major permeability-reducing cement. The remainder concentrate on particular aspects or geographical areas. Both **Glasmann** and **Harris** examine feldspar diagenesis and its influence on present-day rock mineralogy.

One paper deals with the important aspect of petroleum generation and migration. **Larter & Horstad** discuss the secondary migration of hydrocarbons through very restricted pathways and carrier beds into the Gullfaks structure of the Norwegian sector. Finally, two papers bridge the gap between geologist and petroleum engineer: **Kantorowicz *et al.*** draw on experience of producing Brent fields to predict reservoir properties in the Pelican Field area, and **Moss** describes and interprets the petrophysical properties of Brent sandstones in a geological context.

## References


ABBOTTS, I. L. (ed.) 1991. *United Kingdom Oil and Gas Fields, 25 Years Commemorative Volume*. Geological Society, London, Memoir **14**.

BOWEN, J. M. 1975. The Brent oilfield. *In*: WOODLAND, A. W. (ed.) *Petroleum and the Continental Shelf of North-West Europe, Vol 1: Geology*. Applied Science Publishers, London, 353–362.

EYNON, G. 1981. Basin development and sedimentation in the Middle Jurassic of the northern North Sea. *In*: ILLING, L. V. & HOBSON, G. D. (eds) *Petroleum Geology of the Continental Shelf of North-West Europe*. Heyden & Son, London, 196–204.

BRENNAND, T. P., VAN HOORN, B. & JAMES, K. H. 1990. Historical review of North Sea exploration. *In*: GLENNIE, K. W. (ed.) *Introduction to the Petroleum Geology of the North Sea*. Blackwell, Oxford, 1–33.

IVANHOE, L. F. 1980. World's giant petroleum provinces. *Oil and Gas Journal*, **30 June**, 146–147.

SPENCER, A. M. *et al.* (eds) 1986. *Geology of the Norwegian oil and gas fields*. Graham & Trotman, London.

# Exploration of the Brent Province

J. M. BOWEN

*Enterprise Oil Plc, Grand Buildings, Trafalgar Square, London WC2N 5EJ, UK*

**Abstract:** The Brent Province is situated in the central part of the northern North Sea between 60°30′ and 62° North. A few licences were issued in 1965 (Round 2) and again in 1969 (Round 3). Seismic exploration from the mid-1960s onwards showed a thick Tertiary/Cretaceous sequence draped unconformably over tilted fault blocks of unknown age. In 1971, coincident with the opening of the 4th Round, Shell's 211/29–1 pioneered drilling in the area with a major oil discovery, the Brent Field in Middle Jurassic deltaic sandstones. The 4th Round resulted in allocation of almost all the remaining blocks. Between 1972 and 1978, 32 discoveries were made in the UK sector of which 17 are currently in production and a further 7 in the pre-development stage; thereafter only minor reserves have been added, the area now being relatively mature. The exploration history in Norway has been similar although delayed by a few years; it is also less mature as several blocks have still to be explored. Total reserves for the province now approximate to 15 billion barrels of oil equivalent.

## How it began

The year: 1971. The scene: a small, rather work-worn semi-submersible alone in the empty expanse of the northern North Sea, nearly 500 km NNE of Aberdeen, operating what was then the world's most northerly offshore exploration well, 200 km beyond the nearest previous North Sea drilling activity. On the pipe deck were four men hammering on a core barrel reluctant to disgorge its contents.

Eventually 60 ft of core, mainly sandstone and looking encouragingly oily, was quickly recovered, furtively boxed up, rushed by helicopter to Sumburgh and thence by chartered plane directly to London arriving in the small hours.

These events occurred on 19 June 1971. Interest in that core was intense for commercial as well as geological reasons. The rig involved was Shell's 'Staflo'. The well was 211/29–1 and the core provided the first sight of the Middle Jurassic reservoir soon to become famous worldwide as the Brent Sandstone.

In reviewing the exploration of the Brent Province in general and the Brent sandstone reservoirs in particular, it is necessary to define the area which, for the purpose of this paper, lies on the UK side of the North Sea between 60°30′N and 62°N and also takes in the immediately adjacent area in Norwegian waters. The term 'Brent', in a lithostratigraphic sense, is frequently used outside this area, but for this historical review I shall stick to the 'Brent Province' as thus defined.

At the time of the first UK licensing round in 1964 the UK sector of the North Sea was designated by the British government by setting out the quadrants and blocks with which we are so familiar today. However, the area designated then (Fig. 1) extended from the Dover Strait only to the north tip of Shetland (61°N) and excluded a major part of the Brent Province, as well as the blocks immediately adjacent to the present day median line.

In the First Round all designated blocks were put on offer, as far north as 61°N, but the northernmost blocks actually to be licensed were far to the south in quadrants 20, 21 and 22 (58°N) which were pioneered by Shell, Amoco and Texaco.

By the time of the Second Round in 1965 (Fig. 2), most of quadrants 210 and 211 had been designated as well as all the blocks along the median line. Once again all blocks were put on offer, but the only blocks taken up in the far north were 10 boundary blocks in quadrants 3 and 9, licensed to the Total group.

To the author's knowledge one of the earliest reconnaissance seismic surveys shot in the far north took place in 1966. Data quality was very poor by present-day standards but the edge of the Shetland Platform was clearly apparent, bounding a deeper basinal area to the east which we then called the Shetland Trough. Within the basin area itself the deepest mappable reflector showed a number of four-way dip closed structural highs, some of very large areal extent. The deep reflector appeared to coincide with a major regional unconformity; below it, there were few coherent reflections, except within one or two of the major highs

*From* MORTON, A. C., HASZELDINE, R. S., GILES, M. R. & BROWN, S. (eds), 1992, *Geology of the Brent Group*. Geological Society Special Publication No. 61, pp. 3–14.

**Fig. 1.** Blocks offered and awarded in the 1st UK Offshore Licensing Round, 1964.

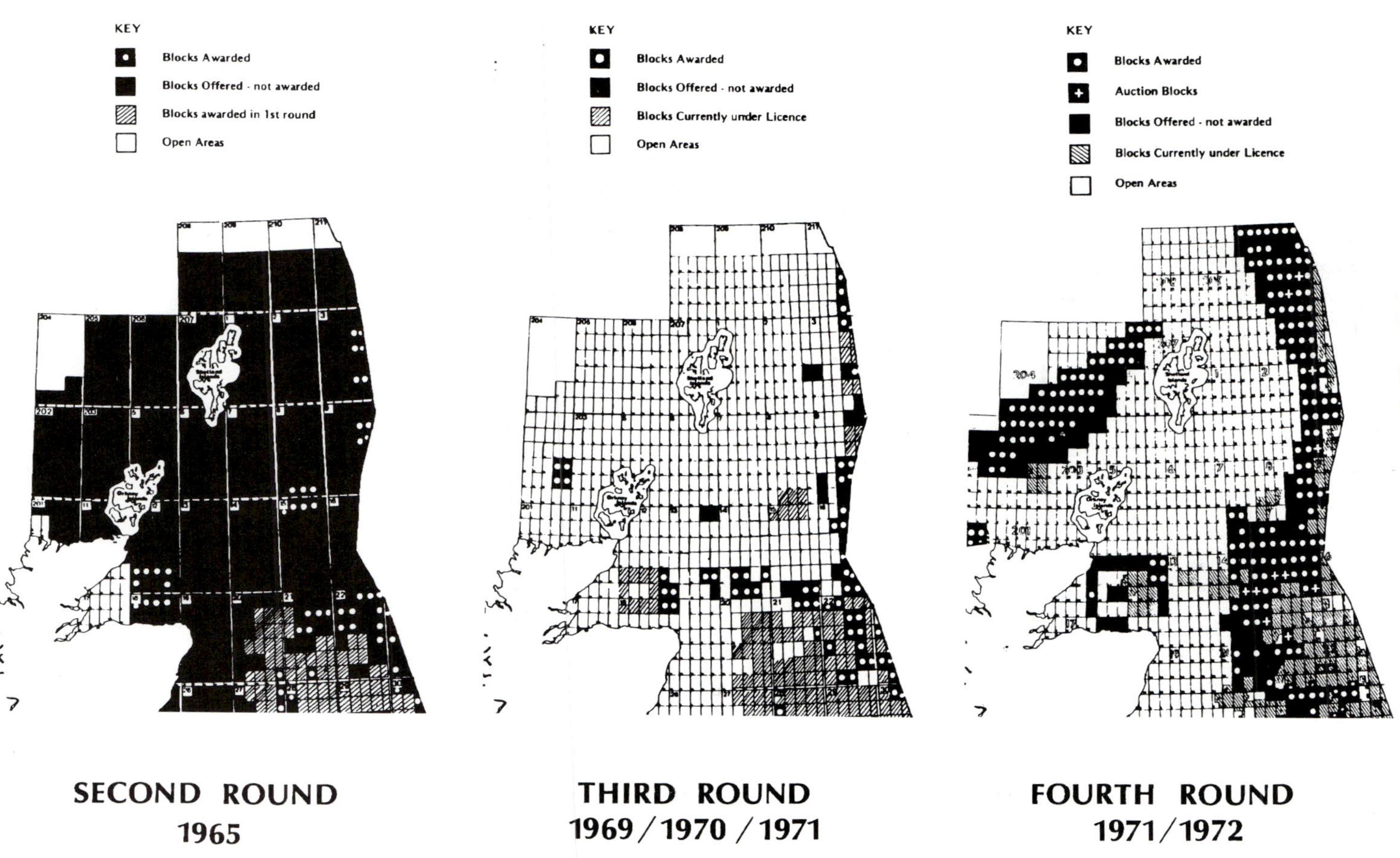


**Fig. 2.** Blocks offered and awarded in the 2nd, 3rd and 4th UK Offshore licensing rounds, Northern North Sea only.

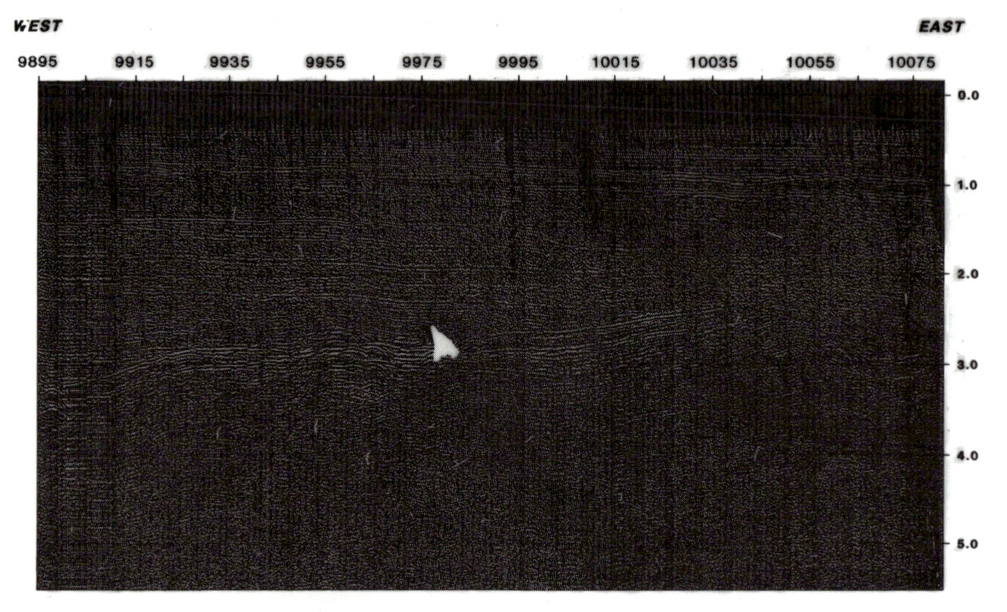


**Fig. 3.** A 1966 vintage seismic profile across the 'Shetland Trough'. The structural highs at SP9955 and SP10035 are the present day Hutton and Brent fields respectively.

trending roughly N–S, where westerly dipping sequences of unknown age and composition could just be discerned (Fig. 3).

Five years were to pass before the Third Round was announced in 1970. Only a few blocks were then offered in the north, mostly 'staking out' those parts of the median line not already licensed (Fig. 2). The seven northernmost blocks and part blocks were taken by Conoco (4), Shell (2) and Texaco (1), while Total picked up two more blocks further south.

The event which was finally, after seven years, to bring the spotlight to bear on this northern area was the UK Fourth Round in 1971. This was a major offering of 436 blocks and included virtually all the unlicensed prospective acreage south of 62°N, the limit of the then designated area. This round was the first offering since the recent discovery of giant fields such as Ekofisk and Forties in the Central Graben over 300 km to the south, so interest was expected to be intense.

Seismic shot in 1970 was of greatly improved quality and confirmed, at least in some areas, the presence of a stratified sequence below the regional unconformity.

## The first well

In the spring of 1971, Shell took the decision to drill the large structure in the Third Round block 211/29 in order to evaluate at least one of the known prospects in advance of the upcoming 4th licensing round. The structure as then mapped (Fig. 4) had an areal closure at the deep unconformity level, 'X', of 75 km$^2$, but no significant closure at the top Palaeocene (Balder Tuff) level. The Palaeocene sandstone play, exemplified by the recent major discovery at Forties, was therefore at best a secondary objective.

The big question to be answered was the age and nature of the rocks forming the tilted fault blocks below the regional unconformity. Shell's in-house well proposal, written in late 1970, states the primary objective of the well to be: 'the testing of a monoclinal dipping sequence below a pronounced regional unconformity at a depth of *c*. 8950 ft subsea in the Shetland Trough. Neither the time span of the unconformity nor the age of the rocks beneath is known ...' Elsewhere it said: 'stacking velocities of the sequence below Horizon 'X' indicate that sedimentary rocks can be expected.' On the subject of source rocks it said: 'the Shetland Trough probably contains a thick series of Jurassic source rocks and these have been sufficiently deeply buried to have generated oil, possibly since early to mid-Tertiary'.

Although the writers of the well proposal were understandably cagey when it came to sticking their necks out in the text regarding the age of the pre-unconformity sequence, Shell's prognosis sheet indicated the major uncon-

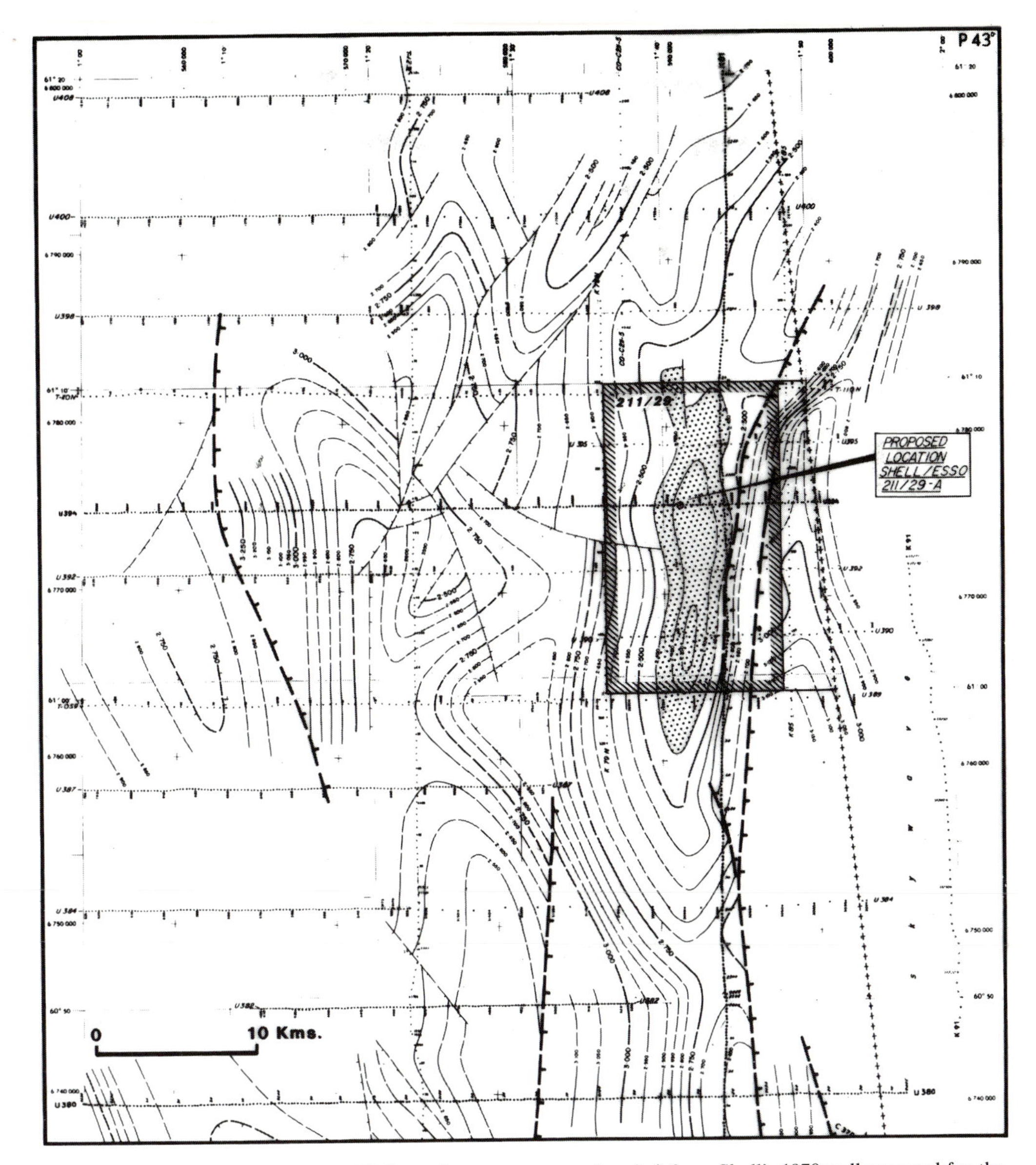


**Fig. 4.** Isochrone map at Horizon 'X' (base Cretaceous unconformity) from Shell's 1970 well proposal for the 211/29-A location.

formity as possibly intra-Jurassic separating early from mid-to-late Jurassic sequences; elsewhere in the text it was referred to as 'possibly Kimmerian'. The writer's recollection is that most of us favoured Jurassic to Triassic rocks below the unconformity although Palaeozoic or even Proterozoic (Torridonian) was given an outside chance! Whilst the primary objective lay below the unconformity, the possibility of the presence of onlapping sands above it led to the location being selected somewhat down dip on the west flank of the structure.

Perhaps the best indication that Jurassic sands might be present came from examination of what was then known as the 'Bullard fit' (Fig. 5), a contemporary pre-separation reconstruction of the north Atlantic area based on bathymetry. This indicated that the Shetland Trough occupied an analogous position to the south of the Mesozoic outcrop area of Scoresbysund in

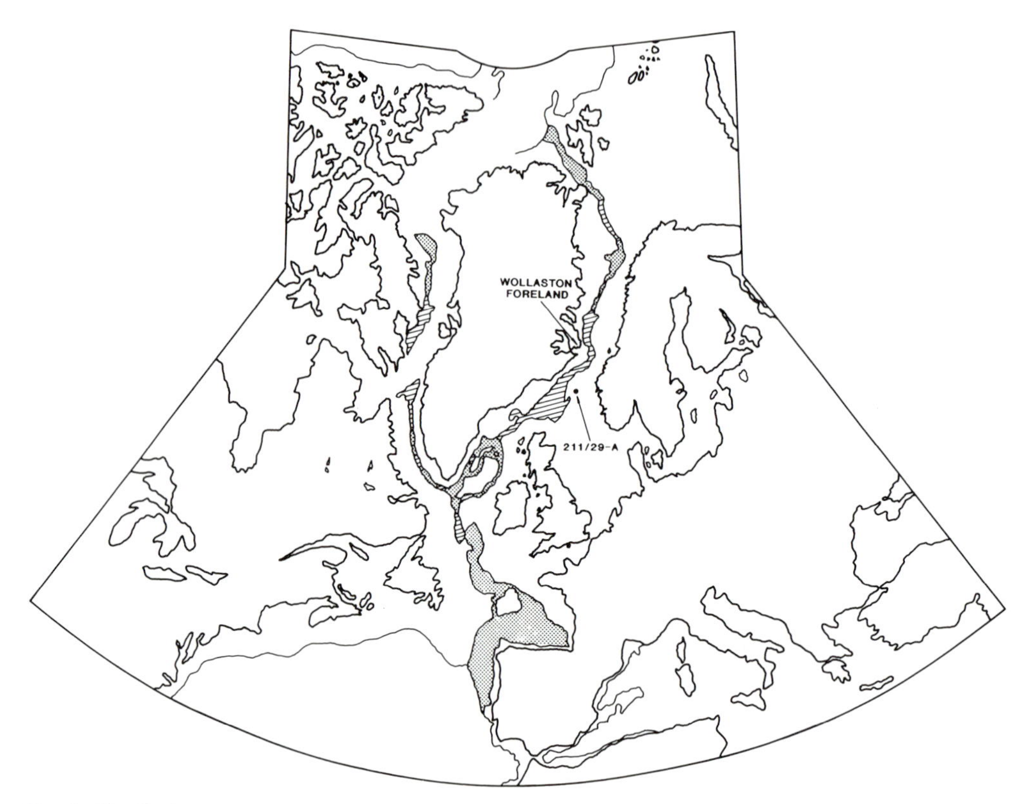


**Fig. 5.** The 'Bullard Fit' of the North Atlantic published in 1965. The stippled areas indicate gaps, while the cross hatched areas represent overlaps of the present day continental margin. After Bullard *et al.* (1965).

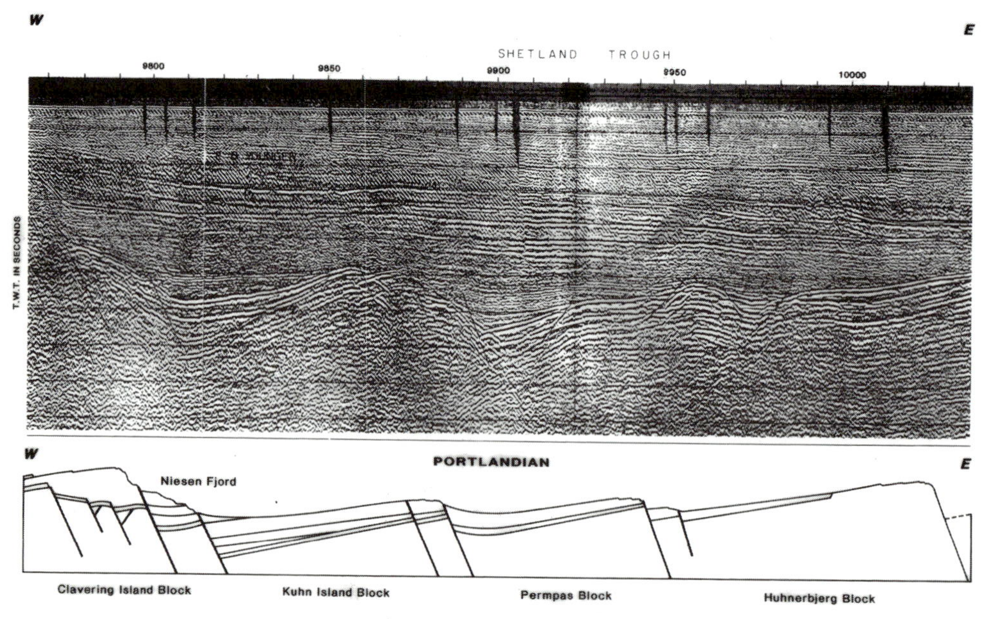


**Fig. 6.** A comparison of the structural styles of (above) a seismic profile from the Shetland Platform (left) to the 211/29 structure (right) with (below) a section across the Wollaston Foreland in Greenland representing the structural development in Portlandian times. After Haller (1971).

East Greenland and indeed, would once have been closer to it than the well-known Jurassic outcrop areas in the UK (Bullard *et al.* 1965). A section across the Wollaston Foreland in East Greenland showing the structural development during the Portlandian, then just published by Haller (1971), also showed a remarkable resemblance to the pre-unconformity structural style in the Shetland Trough (Fig. 6).

Because of the imminent Fourth Round and the large areas of prospective open acreage, the drilling of 211/29−1 had necessarily to be cloaked in secrecy. The nearest previous drilling was 200 km to the south at Balder in Norway and 300 km away in the UK sector, so it was obvious that the well would be closely watched by the competition.

No significant sands were encountered in the Tertiary below base Oligocene and drilling continued down to the vicinity of the regional unconformity where a core was taken. It consisted of black organic shale with pyritized fish scales and ammonities so that palaeontological help was scarcely needed to recognize it as Kimmeridge. Shortly after this a drilling break occurred and the well 'kicked' slightly bringing both sand and oil to the surface; this was only 18 ft above the prognosed depth for the top of the main objective section. Another core was immediately taken, as mentioned earlier, and secrecy became very much the order of the day until the well, having been deepened to 9912 ft, was logged, cased and suspended on 14 July 1971.

Logs showed 466 ft of net sandstone with excellent poroperm characteristics of which 140 ft net was oil-bearing; in view of the required security and the fact that one FIT recovered 38°API oil, DSTs were not carried out. An oil−water contact was inferred at 9129 ft (Fig. 7).

The sandstones in the core were described in the well summary as coarse, sorted, grading into fine sorted bioturbated sandstone thought

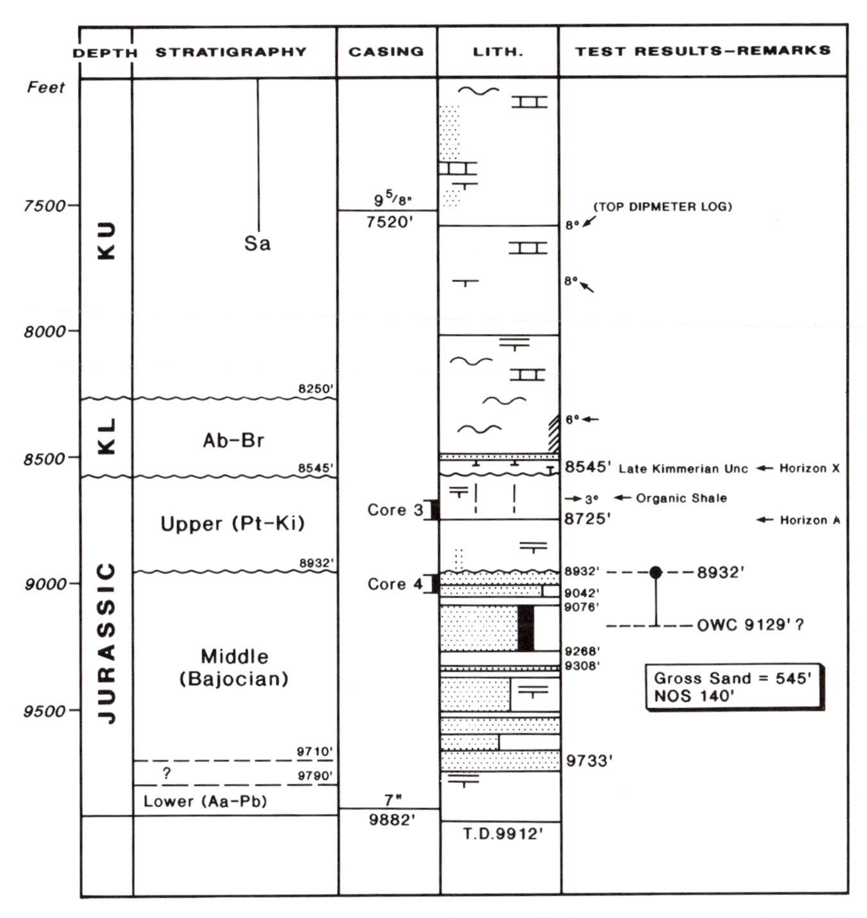


**Fig. 7.** Part of Shell's well summary sheet showing the deepest 3000 ft section encountered in the Brent discovery well 211/29−1.

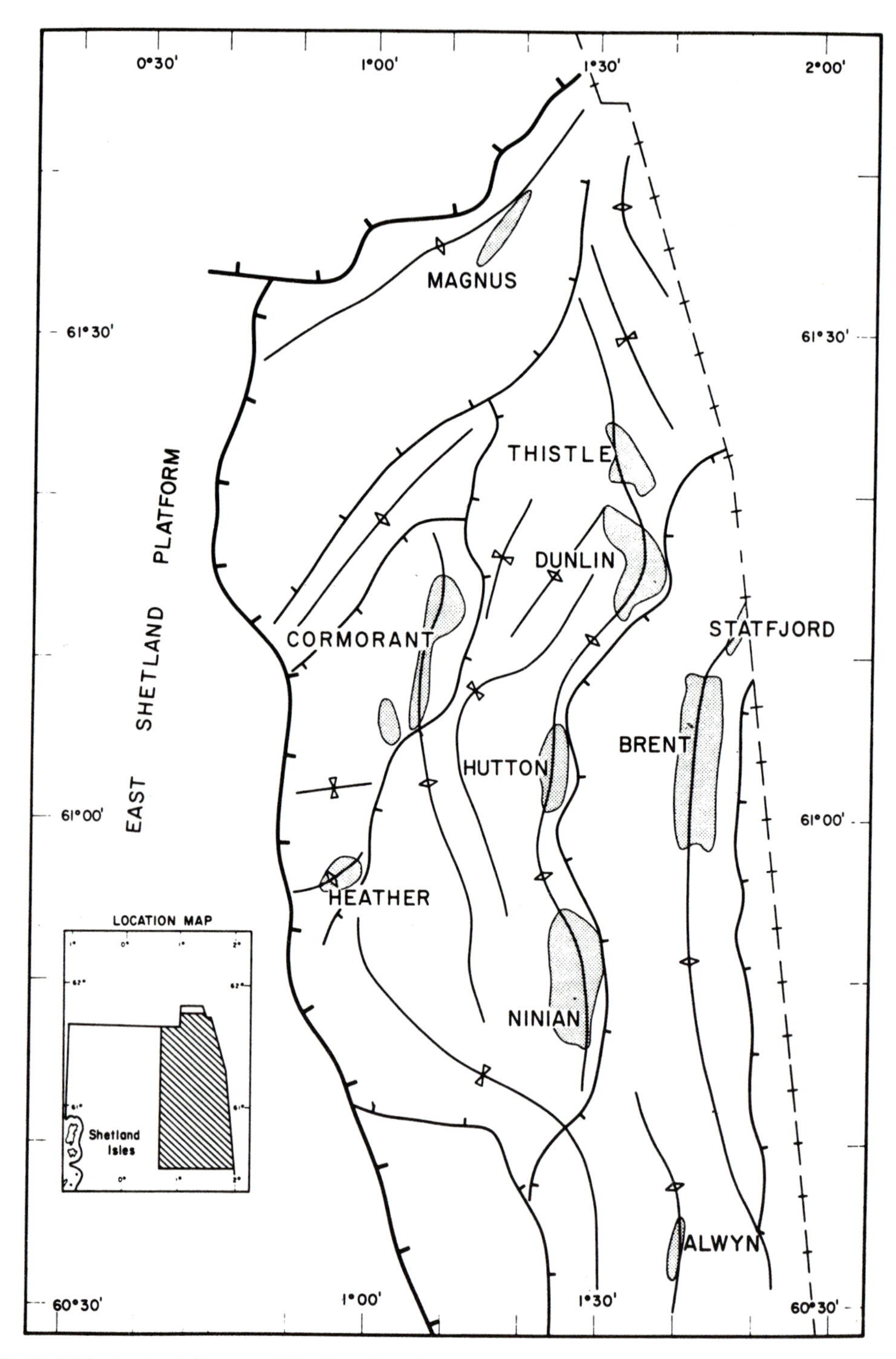


**Fig. 8.** Main structural features of the Brent Province as mapped in the early 1970s.

to have been deposited in a 'tidal coastal' environment. The uncored section, being barren of marine fauna and with many coals, was seen as shallow water, possibly deltaic. The well bottomed in a marine shale section dated as Lower Jurassic.

Thus the 211/29–1 well had resulted in the first major oil discovery in the reservoir to become known as the Brent sandstone. Preliminary estimates suggested the presence of a giant field with reserves in the order of 1 billion barrels. The main downside risk was that the water-bearing sands in the well might not be oil bearing up dip; the upside was that further sands, potentially hydrocarbon bearing, might be encountered deeper in the section and still within closure.

## The Fourth Round

The Fourth Round was announced on 25 June 1971, only 6 days after that first Brent core had been recovered. Applications and bids for the 421 discretionary blocks and the 15 bid blocks had to be submitted within 57 days! The Round was a resounding success with 46 out of 51 blocks offered in quadrants 210, 211, 2 and 3 being licensed (Fig. 2). One of the two blocks in the Brent area offered for cash bids was 211/21 which went to Shell/Esso for what seemed to most of the industry the incredible sum of £21 million, leaving almost £13 m 'on the table'.

The industry had now moved into the Brent area in a big way led by Shell, Amoco, BP, Signal, Burmah and Conoco. It has to be said however that many companies had gone north to continue the pursuit of the 'Forties sand play' that was going strong in the Central Graben at that time; structures with closure at top Palaeocene, such as Ninian and Magnus were therefore at a premium. Many of these coincided with the deeper structures at the late Kimmerian unconformity level as mapped by Shell at that time (Fig. 8).

## Exploration of the UK sector

Exploration drilling resumed in the summer of 1972 when Shell, Signal and Total drilled on blocks 211/26, 211/18 and 3/15 respectively. Shell also drilled a successful first appraisal well on Brent, further updip, which found a 545 ft hydrocarbon column in the Brent sands and bottomed in another massive sand, water-wet but with oil shows; this confirmed Brent as a giant field with still further upside potential. The exploration results that year, however, were not outstandingly successful. Shell found oil at 211/26–1 (Cormorant South), but only 123 ft net Brent sand above a thin water-bearing Triassic section, over basement; the reserve was calculated at only 90 million barrels. As this was next to the 'golden block' (211/21) there was considerable dismay in the Shell camp. Of the other two wells Signal on 211/18 found the small field not called Deveron but then named 'Thistle', while Total's well in 3/15 found only 'oil shows'.

The next season in 1973 was much more successful with six important discoveries in the Brent sandstones at Thistle, Dunlin, Alwyn (2), Hutton and Heather from only nine new field wildcats drilled. Only three failed to find commercial hydrocarbons, one of which was drilled by Shell on the 'golden block', 211/21, finding only a 21 ft oil column, while Conoco managed to drill a dry hole in the saddle between Brent and the mega-structure straddling the median line, now the Statfjord field.

The following five years were boom years in this area, with a further 59 wildcats being drilled in the UK sector to test Brent reservoirs. In November 1974, at the First Conference on the Northwest European Continental Shelf, the lid was lifted on the general geology of the Brent area, at least for people 'not in the know'. At the presentation on Brent (Bowen 1975), as for several other major new discoveries, there was standing room only, with an audience of over 1000. The stratigraphy then presented remains essentially intact today, although the Brent has been promoted from formation to group status and its former members, now more precisely defined, are now formations (Fig. 9).

The period 1974–78 saw a striking drop in the success rate from around 1 in 2 in 1974 when Ninian was discovered, to 1 in 7 in 1978 when Emerald was the only discovery. In fact, the last major Brent discoveries in this part of the UK sector were Murchison and North Alwyn in 1975, ending a 'bonanza' period of only 6 years. The subsequent 15 years have brought a number of relatively small independent discoveries and additions to existing fields, but none much in excess of 100 million barrels of oil or oil equivalent in gas.

## The Norwegian sector

In the Norwegian part of the area under discussion, the story of the exploration of the Brent has been much the same, only delayed by a few years and thereafter more extended, so that today it has not reached the same state of maturity as we see in the UK sector.

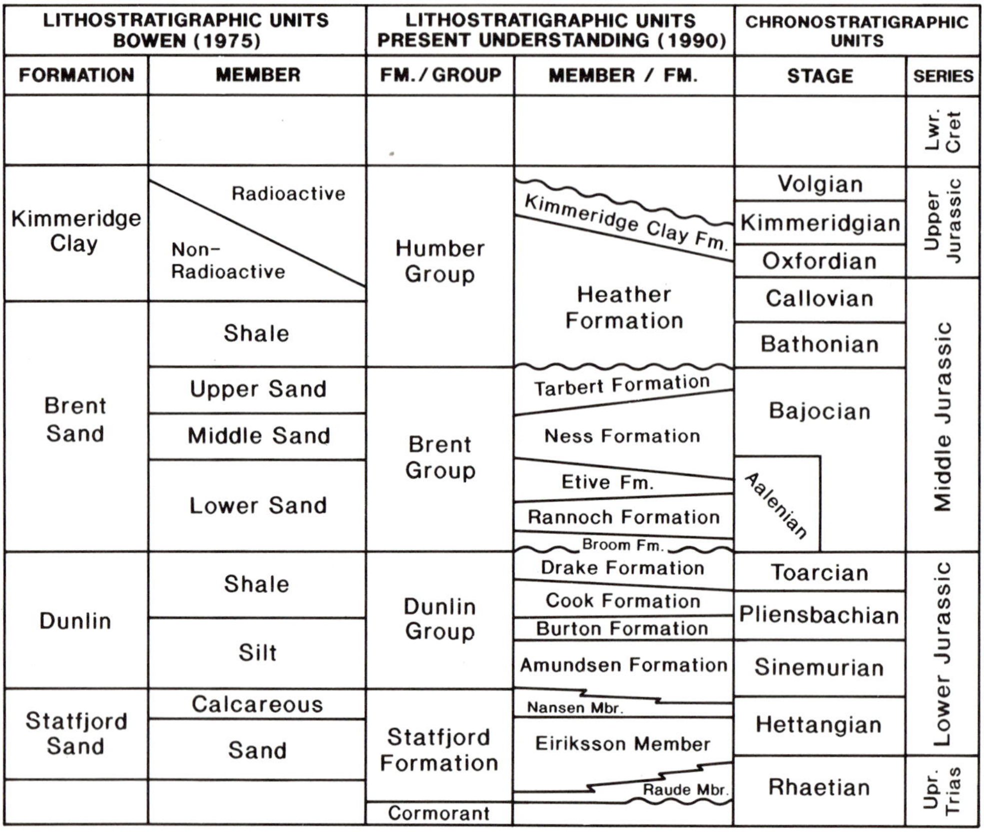

| LITHOSTRATIGRAPHIC UNITS BOWEN (1975) | | LITHOSTRATIGRAPHIC UNITS PRESENT UNDERSTANDING (1990) | | CHRONOSTRATIGRAPHIC UNITS | |
|---|---|---|---|---|---|
| FORMATION | MEMBER | FM./GROUP | MEMBER / FM. | STAGE | SERIES |
| | | | | | Lwr. Cret |
| Kimmeridge Clay | Radioactive | Humber Group | Kimmeridge Clay Fm. | Volgian | Upper Jurassic |
| | | | | Kimmeridgian | |
| | Non-Radioactive | | Heather Formation | Oxfordian | |
| | | | | Callovian | Middle Jurassic |
| Brent Sand | Shale | | | Bathonian | |
| | Upper Sand | Brent Group | Tarbert Formation | Bajocian | |
| | Middle Sand | | Ness Formation | | |
| | Lower Sand | | Etive Fm. | Aalenian | |
| | | | Rannoch Formation | | |
| | | | Broom Fm. | | |
| Dunlin | Shale | Dunlin Group | Drake Formation | Toarcian | Lower Jurassic |
| | | | Cook Formation | Pliensbachian | |
| | Silt | | Burton Formation | | |
| | | | Amundsen Formation | Sinemurian | |
| Statfjord Sand | Calcareous | Statfjord Formation | Nansen Mbr. | Hettangian | |
| | Sand | | Eiriksson Member | | |
| | | | Raude Mbr. | Rhaetian | Upr. Trias |
| | | Cormorant | | | |

**Fig. 9.** Jurassic stratigraphy of the Brent area as originally proposed compared with present usage.

As in the UK, all the acreage in question was offered to industry in the First Round in 1965 and again in the Second Round in 1968 (up to 61°N), but on neither occasion were licences awarded. A special award of blocks 33/9 and 33/12 was made in August 1973 following the UK Fourth Round excitement when the Brent and other major discoveries were made on the UK side. Other blocks were awarded subsequently in the 4th (1979) and later rounds.

Once again, as in the UK, the exploration started with a major discovery, the 2.8 billion barrel Statfjord field drilled by Mobil on block 33/12 in 1974. Two years were to pass before there was further exploration activity and it was not until 1978 that the second major discovery was made on Gullfaks by Statoil following another 'special award' earlier that year.

## The current status of exploration

The plots of reserves discovered against numbers of exploration wells in each sector bear an almost unbelievable resemblance (Fig. 10). If these two 'creaming' curves were to be believed, an ultimate reserve for the area under review would appear to be in the order of 17–19 billion barrels of oil equivalent, (including gas).

The classical shape of these curves, particularly the one for the UK sector, reflects the relative simplicity of the Brent play in both structural and stratigraphic terms. The major prospects were all, as we have seen, apparent even on the rather primitive seismic data obtained in the mid-1960s. By the early 1970s even quite small prospects could be defined in structurally less complex areas, so that the advance of seismic definition and the increasing availability of 3D has, while contributing essential information to the understanding of the fields themselves, played only a relatively minor role in the buildup of reserves by exploration.

Stratigraphically, although internally complex, the Brent does not appear to present opportunities for major stratigraphic trapping

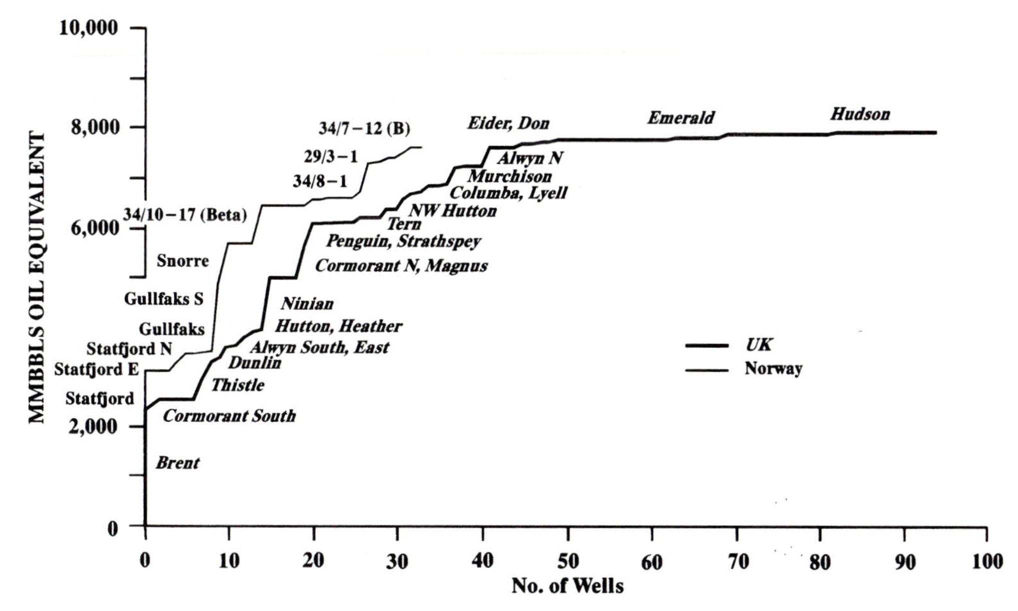


**Fig. 10.** 'Creaming' curves for the UK and Norwegian sectors for the area covered in this paper (N.B.: Whilst mostly in the Middle Jurassic Brent Sandstone, some Upper Jurassic and Lower Jurassic/Triassic reserves are also included in these figures).

within itself. Another, unfortunately negative characteristic, is the tendency for rapid deterioration of the reservoir with depth, as exemplified by a comparison of the excellent reservoir properties in the Brent field at 9000 ft with those in the deeper parts of Lyell or NW Hutton, only 4000 ft deeper. This certainly reduces prospectivity for oil in deeper more complex areas which might otherwise benefit from modern seismic exploration.

## The future

What of the future? On the UK side following the 1989 awards of 9 blocks in the 11th licensing round, there is currently very little unallocated acreage in the East Shetland Basin and Viking Graben, and by virtue of the UK licensing system, very little more will become available, except by voluntary relinquishment, in the short to medium term. Exploration will be stimulated by the recent awards and their consequential obligatory exploration drilling, but this increase in activity is unlikely to be sustained. Exploration will then be up to existing licencees and will probably be dominated by exploration of smaller prospects as possible satellites to existing infrastructure, although potential still exists for deeper gas-condensate plays in eastern part of Quadrant 3 and adjacent areas in Norwegian waters.

The density of wells in UK licenced acreage (10 per 1000 $km^2$) goes to demonstrate that this is a rather well-drilled, mature play. 3D surveys will almost certainly result in the definition of a number of high-risk exploration objectives such as isolated, perhaps downthrown fault traps, but oil discoveries are likely to be small, with anything in excess of 100 mmbbl a rarity. A very few larger discoveries, up to 250 mmbbl, cannot be entirely discounted, given the surprises we get used to in this business; anything much larger would, in the writer's opinion, be little short of miraculous!

On the other hand, in the Norwegian part of the area, exploration well density, even in the licensed acreage, is only half that seen on the UK side and large areas of unallocated and undrilled acreage remain. The rather mature looking creaming curve applies only to the licensed and hence explored acreage; consequently in this area the scope for further substantial discoveries must, from a purely statistical standpoint, be much greater. In particular there must be considerable remaining potential for gas.

The author's thanks are due to K. East of Enterprise Oil who helped in the preparation of this paper and also to Shell and Esso and particularly J. R. Parker of Shell Expro who very kindly made original material available, some of which is quoted and reproduced here.

## References

Bowen, J. M. 1975. The Brent Oil Field *In*: Woodland, A. W. (ed.) *Petroleum and the Continental Shelf of Northwest Europe*. Applied Science Publishers, London, 353–360.

Bullard, E. C., Everett, J. E. & Gilbert Smith, A. 1965. The fit of the continents around the Atlantic. *Philosophical Transactions of the Royal Society of London*, **A258**, 41–51.

Haller, J. 1971. *The Geology of the Greenland Caledonides*. Interscience Publishers, London.

# An introduction to the Brent Group: a literature review

P. C. RICHARDS

*British Geological Survey, 19 Grange Terrace, Edinburgh EH9 2LF, UK*

**Abstract:** There is a vast volume of published literature on the Brent Group, and consequently many controversies or differing points of view regarding the age, nature and palaeogeographic evolution of the Group. Brent Group lithostratigraphy is, however, relatively uncontroversial. A five-fold subdivision of the Brent Group into the Broom, Rannoch, Etive, Ness and Tarbert Formations is used by most authors. The Oseberg Formation is used to describe Broom Formation equivalent strata in Norwegian waters. There are, however, problems of applying this simple lithostratigraphy in some parts of the basin, where the artificial subdivision may unduly constrain depositional system correlation. Many companies use their own, field-specific, stratigraphic schemes for reservoir subdivision purposes, and some authors have applied a sequence stratigraphic framework to the Brent Group in preference to the lithostratigraphic scheme. There is, however, discord over the actual stratigraphic position of sequence boundaries and the time equivalence of strata. Brent Group biostratigraphy is less well documented in the literature than lithostratigraphy, and represents the area where most new work is needed if the evolution of the depositional system is to be better understood.

Analyses of the depositional environment of the Brent formations are numerous but more or less similar. The origin of the Broom Formation is the most controversial aspect, with a range of shallow to deep water interpretations; a fan delta interpretation is preferred in recent publications. The Rannoch and Etive Formations are generally interpreted as shoreface to foreshore/channel deposits, and the Ness Formation as delta top sediments. The Tarbert Formation represents an essentially transgressive unit. There are two main published models to explain the regional evolution of the Brent Group, and this is possibly the most active of the controversies in Brent Group geology. The first model considers the system as a northwards prograding delta with a southerly source and a concentric arrangement of facies belts across the basin. The second model envisages a dominantly transverse sediment supply to the basin, with subsequent localized northwards progradation within the basin.

The Brent Group is economically the most important succession in the North Sea. Most of Britain's oil reserves are found in these Middle Jurassic sandstones, trapped in the tilted fault blocks of the East Shetland Basin.

Brent Group sediments are recorded in the East Shetland Basin, the northern Viking Graben and over parts of the Horda Platform to the east of the Viking Graben (Fig. 1). Lithologically similar and possibly partly age equivalent strata have also been described from the Unst Basin (Johns & Andrews 1985) and the Beryl Embayment area of the central Viking Graben (Richards 1990, 1991). By definition, the term Brent Group is restricted to sediments north of about 60°N, and the somewhat similar sediments of the Beryl Embayment area are more properly referred to as the Sleipner and Hugin Formations of the Vestland Group. Brent Group strata pass laterally northwards (north of about 61°50′N) to age equivalent, un-named shales of marine origin (Eynon 1981).

Brent Group strata are absent over some fault crests in the East Shetland Basin, the Magnus Ridge to the north, most of the Shetland Platform, and probably also over the Transitional Shelf in the southwest of the basin. Possibly part age equivalent sediments (the Emerald Formation) have been recorded in the Transitional Shelf area by Wheatley *et al.* (1987).

The Brent Group is approximately 300 m thick in the East Shetland Basin. The succession is more or less complete across most of this area, but the Tarbert Formation is occasionally thin or missing over tilted fault block crests. Isopach maps in Brown *et al.* (1987) show variations in thickness of the Broom, Rannoch plus Etive, and Ness Formations in the UK sector due in part to deposition across differentially subsiding fault blocks and, especially in the case of the Ness Formation, to later erosion. Johnson & Stewart (1985) and Livera (1989) also emphasized the importance of synsedimentary faulting on Brent Group development. Thicknesses of up to 300 m for the Brent Group have also been mapped by Karlsson (1986) in the Norwegian sector. Karlsson emphasized the southwesterly thickening of the Brent Group in the Norwegian

*From* Morton, A. C., Haszeldine, R. S., Giles, M. R. & Brown, S. (eds), 1992, *Geology of the Brent Group*. Geological Society Special Publication No. 61, pp. 15–26.

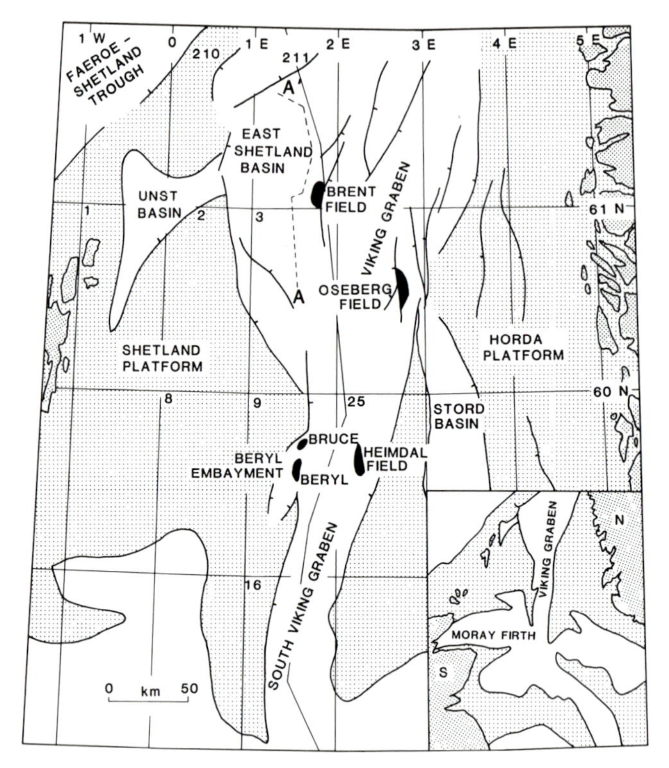


**Fig. 1.** Location map showing the main structural elements of the northern North Sea. S, Scotland; N, Norway. A–A' is the position of the correlation panel shown in Fig. 2.

area, suggesting sediment derivation from that direction.

Because of its vast economic importance, the Middle Jurassic succession in the northern North Sea has been extensively studied, and around 200 papers have been published on aspects of Brent Group stratigraphy, structure, sedimentology and oilfield geology. There is considerable controversy in this literature regarding virtually every aspect of Brent Group geology. For example, numerous sedimentological interpretations have been proposed for the Broom and Etive Formations. Much controversy occurs over the age of Brent Group units, and there are several differences of interpretation of the regional architecture of the depositional system. There is also controversy regarding the relative place of the Brent Group in the tectonic evolution of the Viking Graben.

This brief review cannot address all aspects of Brent Group geology covered in the vast volume of literature. The intention is to summarize only the key elements and controversies regarding the stratigraphy, depositional environments and depositional models that have been published previously.

## Stratigraphy

Brent Group lithostratigraphy is essentially simple. Bowen (1975) recognized a five-fold subdivision of the Brent Formation, which Deegan & Scull (1977) modified and formalized, giving the Brent succession group status, and the five subdivisions formation status. These five formations, from the base upwards are, Broom, Rannoch, Etive, Ness and Tarbert (Fig. 2). The five formation names form the acronym BRENT as an aide memoir.

Despite the controversy surrounding almost every other aspect of Brent Group geology, there has been surprisingly little dissent concerning the simple lithostratigraphic subdivision, although several authors have found it to be too restrictive, masking important lithological and stratigraphic variation. For example, Eynon (1981) abandoned the Deegan & Scull (1977) lithostratigraphy, stating that it is, 'both inadequate and misleading'. Eynon considered that from the point of sedimentology and facies variation, it is more useful to consider the Brent Group in terms of five depositional units, which he termed Basal Sand, Bajocian Delta Lobe,

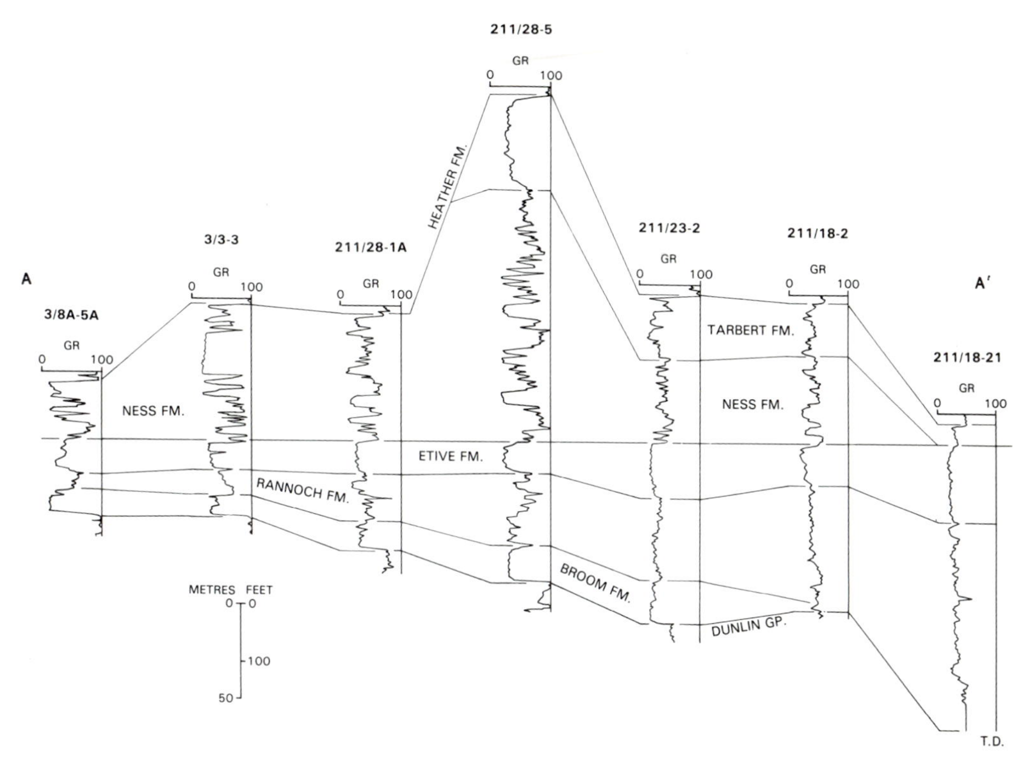


**Fig. 2.** North–south correlation panel of the Brent Group in the East Shetland Basin showing the five-fold subdivision of the Group and the general thickness trends. See Fig. 1 for location.

Bajocian–Bathonian marine transgression, Lower Bathonian Delta Lobe and Upper Bathonian Delta Lobe. Eynon's (1981) suggestions for stratigraphic revision have not, however, been adopted by any of the more recent authors, who essentially use the Deegan & Scull (1977) terminology.

Brown & Richards (1989) also recognized minor problems with the existing Brent Group lithostratigraphy near the northern limit of delta progradation in the East Shetland Basin. Here, the lithologically variable Ness Formation appears to be commonly absent, and massive sandstones more comparable to the Etive Formation occupy this stratigraphic interval. The massive sandstones are essentially Ness Formation deposits, but are developed in Etive Formation facies, leading to potential confusion in the application of the simple five-fold lithostratigraphic subdivision in the area. A comparable situation has also been described from the Oseberg Field by Hurst & Morton (1988). They noted that heavy mineral assemblages in one suspected Etive Formation succession were different to the Etive assemblages in adjacent wells, but concluded that they were similar to nearby Ness Formation sandstones, and that Ness channels can be locally difficult to distinguish from Etive channels. Hurst & Morton (1988) have also, therefore, shown that the apparently simple lithostratigraphic subdivision may not always work or be easy to apply.

Another significant difficulty arising from the lithostratigraphic scheme of Deegan & Scull (1977) is the recognition of the Tarbert Formation on well logs, especially in truncated Brent Group successions. Deegan & Scull (1977) placed the Ness–Tarbert boundary at the top of the uppermost, prominent shale unit. In their revision of Norwegian lithostratigraphy, Vollset & Doré (1984) placed the boundary at the top of the uppermost fining-upward unit, that is at the top of a shale bed or a coal. Given that post-Brent Group erosion can cut partly into a Ness-type sandstone and that Tarbert is now known to be locally heterolithic, such definitions are not wholly reliable. The availability of conventional core can allow the recognition of coarse-grained transgressive lag deposits above ravinement surfaces and/or the recognition of marine bioturbation in sandstones to assist in locating the base of the Tarbert Formation.

Graue *et al.* (1987) have improved the lithostratigraphy of the Brent Group by suggesting the incorporation of a new unit, the Oseberg Formation, to describe the basal sandstones in the Norwegian sector that are essentially equivalent to the Broom Formation deposits in UK waters.

Minor revisions to the Brent Group lithostratigraphy have also been proposed by other authors; for example Cannon *et al.* (this volume) present a three-fold subdivision of the Ness Formation, essentially formalizing the existing three-fold subdivision postulated by Budding & Inglin (1981). Significantly more complicated subdivisions of the Brent Group are employed within oil companies to subdivide and classify the Brent reservoir in individual oilfields; these are too numerous and localized to warrant further description here, although the reader is referred to the Brent Field paper by Livera (1989) for an excellent example of such a subdivision.

In an attempt to improve on the sometimes difficult to use lithostratigraphic zonation of the Middle Jurassic sediments, some authors have applied a sequence stratigraphic approach to the problem of subdivision. For example, Vail & Todd (1981) identified the base of the Broom Formation as a sequence boundary, and its top as a depositional hiatus. They also recognized a sequence boundary between the Ness and Tarbert Formations. This view is supported by Mitchener *et al.* (this volume), but Graué *et al.* (1987) and Falt *et al.* (1989) have proposed that the Ness and Tarbert Formations are in places coeval, suggesting the absence of a sequence stratigraphic boundary at this level.

### *Biostratigraphy*

The precise age of the Brent Group is controversial. Suggestions of its age range from late Toarcian to early Bathonian (Ryseth 1989) or entirely post-Aalenian (Helland-Hansen *et al.* 1989). Richards *et al.* (1990) have suggested that the bulk of the Group is probably of Aalenian to earliest Bajocian age, although the upper part of the Group may extend up into the Bathonian. Lithologically similar sediments in the central Viking Graben may range from Aalenian to Callovian (Fig. 3).

The Brent Group is probably conformable on the underlying Dunlin Group in the East Shetland Basin, although Hallett (1981) noted the presence of local unconformities. Further south, in the central Viking Graben, oil company logs often show an unconformity between the Lower and Middle Jurassic successions, while in the south Viking Graben the Middle Jurassic rests unconformably on Triassic (Larsen & Jaarvik 1981; Graué *et al.* 1987).

To date, published biostratigraphic data for the Brent Group are relatively sparse. However, the recent biostratigraphic determinations by Graué *et al.* (1987) and Falt *et al.* (1989) go some way to improving the definition of time lines within the Brent Group (see also Mitchener *et al.* this volume).

Using palynology, Graué *et al.* (1987) dated the Brent Group as Toarcian/Aalenian to early Bathonian (Fig. 3). Four biozones are recognized, namely biozone A (early Toarcian), B (late Toarcian–Aalenian), C (Aalenian–Bajocian) and D (early Bathonian). From the well sections figured by Graué *et al.* (1987, their figs 11 and 12) Dunlin Group strata are represented within biozones A, B and locally C (see Well G, fig. 11). The Oseberg Formation lies variously within biozone C or largely within biozone B, the Rannoch, Etive and Ness Formations also lie within biozone C, although upper Ness lies within biozone D in places, and the Tarbert Formation falls within biozone D, but locally in part within biozone C.

Biostratigraphic resolution of individual Brent Group formations and the establishment of the exact age relationships between the formations remains one of the most important aspects of Brent Group geology in need of further work. A substantially improved biostratigraphic framework is necessary in order to allow resolution of many of the existing controversies surrounding the regional depositional models that have been presented for the Group, and to establish a sequence stratigraphic scheme that is durable enough for general use.

## **Depositional environments**

The Brent Group has been described as a regressive–transgression wedge (Brown *et al.* 1987; Graué *et al.* 1987). The northwards-prograding, regressive part of the succession is usually considered to be the Broom, Rannoch, Etive and Lower Ness Formations, with the succeeding Upper Ness and Tarbert Formations deposited during retreat of the system in response to a relative sea level rise (Fig. 4).

### *Broom and Oseberg Formations*

The Broom Formation at the base of the Group ranges in thickness from about 48 m along the western margin of the basin and thins to the east and northeast, pinching out some 55 km to the east of the basin margin faults. Considerable

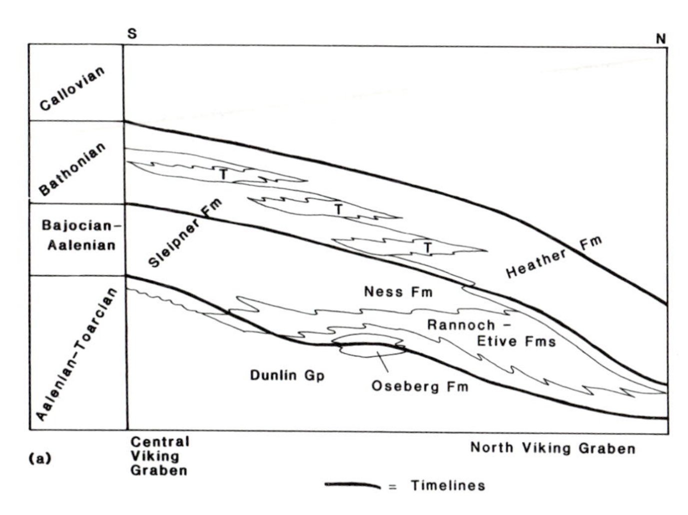


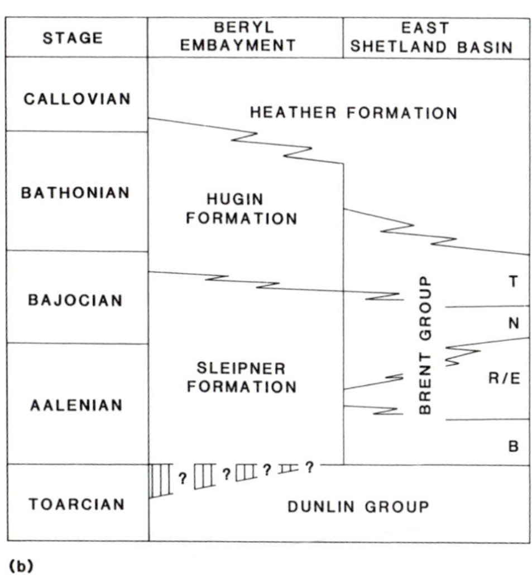


**Fig. 3.** Two of the recently published assessments of age of the Brent Group and similar strata in the central Viking Graben/Beryl Embayment area. (**A**) Modified from Graué *et al.* (1987) and shows a south–north schematic section illustrating the time equivalence of various depositional units. T, Tarbert Formation. (B) From Richards (1989) and Richards (1991). Both schemes record a time equivalence between the progradational parts of the Brent Group and the Sleipner Formation in the Beryl Embayment area, although there is some debate over the specific place of these formations within the Aalenian to Bajocian interval (see for example the discussion and reply debate between Falt & Steel 1990 and Richards *et al.* 1990).

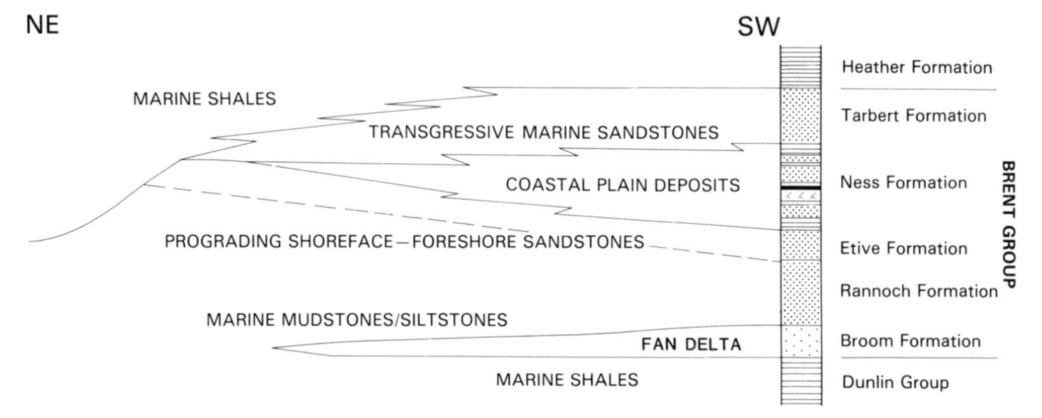


**Fig. 4.** A schematic southwest–northeast section of the Brent Group in the East Shetland Basin showing the relative geometries of the individual formations and their possible interpretation. (Modified after Brown 1990).

thickness variations occur across intra-basinal faults. Although the general sparseness of cores prevents the detailed identification of possible facies variations across these intra-basinal faults, large-scale facies variations are recorded across the basin.

In the Heather field area the Broom Formation dominantly comprises medium to coarse grained, poorly sorted, frequently carbonate cemented sandstones with bifurcated, sub-parallel to wavy, carbonaceous streaks in places. Mudstones with floating, coarse sand grains are also common. Burrows are frequent at some horizons, and include shallow water forms such as *Arenicola*. Slightly further east, in the NW Hutton field, the Formation is predominantly moderately well sorted, medium to coarse grained sandstones with metre-scale planar cross bedding and shallow marine burrows. Further east still, in well 3/9a–2, the Formation ranges in coarseness up to pebbly sandstones or conglomerates, and is interbedded with hummocky cross stratified shoreface sandstones of the Rannoch Formation. In contrast, the Broom Formation succession in the Dunlin oilfield area in the northeast of the basin comprises a thin unit of coarse, oolitic sandstones. The Formation here marks the transition from offshore, silty shales of the Dunlin Group, up to the fissile, highly micaceous, pro-deltaic siltstones of the lower part of the Rannoch Formation.

The Broom Formation was probably derived by erosion of the Shetland Platform area (Morton 1985; Richards 1990) and has been variously interpreted as transgressive tidal flat sands (Hay 1977), offshore sheet sands (Budding & Inglin 1981), a mass flow deposit (Parry *et al.* 1981), and a cliff base beach (Eynon 1981). However, the eastward thinning of the succession, coupled with the eastward change from shallow to deeper water deposits may suggest that the Formation was deposited as some form of fan delta system like the coeval but geographically separate Oseberg Formation on the other side of the Viking Graben (Richards & Brown 1987; Cannon *et al.* this volume).

Richards *et al.* (1986) used the interfingering relationship between the Broom Formation and the storm dominated Rannoch sandstones in the east of the East Shetland Basin to constrain their depositional model for the Broom Formation, and suggested that towards its eastern limits the Formation was deposited under storm-wave influenced shoreface conditions.

The generally coeval Oseberg Formation in Norwegian waters has been studied by Graué *et al.* (1987) and interpreted as the deposits of a westwardly-building fan delta. They recognized sediments deposited by marine gravity flows, and by sheet floods and shallow streams. Dipmeter data together with lithofacies information led Graué *et al.* (1987) to propose a Gilbert-type fan delta model with at least two depositional lobes active in the Oseberg area.

### *Rannoch and Etive Formations*

The Rannoch and Etive Formations overlying the Broom Formation can be considered together in terms of a single genetic package. According to conventional interpretations (e.g. Budding & Inglin 1981; Brown *et al.* 1987; Graué *et al.* 1987) these two formations represent the main marine to coastal, progradational phase of the Brent delta. The two formations attain a combined maximum thickness of about 154 m in the UK sector, and are thickest in a NW–SE-trending zone in the NE part of the basin. The formations form a variable, but generally coarsening upwards succession.

Over much of the East Shetland Basin the Rannoch Formation has a basal micaceous, siltstone dominated unit containing thin, sharp-based sands, and coarsens up to very fine-grained sandstones. These sandstones are micaceous, with the micas and carbonaceous detritus concentrated preferentially into low angle ($<10°$ of dip) laminae. The laminae are mostly planar, with occasional low angle truncations defining hummocky cross stratification (Richards & Brown 1986; Elliott & Buller 1987; Graué *et al.* 1987; Scott, this volume). Tops of beds are commonly bioturbated; the most common burrow types are *Rosselia* and *Planolites*, with other forms such as *Schaubycylindrichus* recorded in places. These sands were probably deposited within a storm wave influenced shoreface setting (Richards & Brown 1986; Brown *et al.* 1987; Graué *et al.* 1987).

In the southwestern part of the East Shetland Basin the Rannoch Formation is sometimes coarser grained than elsewhere, being dominantly fine rather than very fine grained. These fined-grained deposits are also probably of shoreface origin, and their coarser grain size may reflect a SW–NE proximal to distal progradational trend within the Formation (Richards *et al.* 1988).

The Etive Formation usually overlies the Rannoch Formation either as part of an upwards coarsening succession, or as a distinct fining upwards unit (or complex of fining up units) above a sharp base. The Etive Formation can also be missing locally, where sediments more typical of the overlying Ness Formation rest directly on the Rannoch Formation (Brown *et al.* 1987).

Where the Etive Formation occurs as part of a coarsening upward succession above the Rannoch Formation, it is usually composed dominantly of fine grained, moderately well sorted, usually structureless sandstones, often with a sharp base. Such successions are usually overlain by lagoonal siltstones or coals of the Ness Formation, and are generally interpreted as barrier bar/beach deposits formed by progradation of the coastline over the Rannoch Formation shoreface deposits (Budding & Inglin 1981; Johnson & Stewart 1985; Brown *et al.* 1987; Graué *et al.* 1987).

Sharp-based, fining-upwards variations of the Etive Formation frequently fine from medium or coarse to fine or very fine grained, and two or more such fining up beds are sometimes stacked vertically. These sandstones are often parallel laminated or planar cross bedded, and have been interpreted as channel deposits by Brown *et al.* (1987) and Brown & Richards (1989).

Elliott (1989) observed that channel sandstones are common in the lower part of the Etive Formation, where they form a multistorey, multilateral channel-belt sandstone body. He noted that this feature is not typical of modern wave-dominated deltas (which is the generally accepted interpretation of the Brent Delta), where distributary channels tend to be less numerous and/or more stable. Elliott argued for a sand-dominated braid plain occupying the lower delta plain of the Brent delta, but commented again that this would contrast with the character of modern deltas.

In some areas where the Etive Formation is missing or largely missing, the stratigraphic interval is occupied by Ness Formation deposits displaying facies types which are not observed above complete Etive Formation successions. For example, in well 211/27−4a in the NW Hutton field, a few tens of centimetres of Etive Formation sand is overlain by a lithologically complex and variable lower Ness unit of interbedded siltstones and sandstones. These Ness Formation sandstones display steeply inclined muddy laminae and scattered mud clasts, while the siltstones contain burrows and a marine microflora. Such successions have been interpreted by Brown *et al.* (1987) as tidal channel deposits cutting through the barrier bar. Although such successions are relatively rare, sufficient examples have been observed in oilfield areas to allow the delineation of possible NW−SE-trending tidal inlets.

Budding & Inglin (1981) suggested that thickness variations in the Rannoch and Etive Formations in the Cormorant Field are due to variations in the rate of progradation of the shoreface to foreshore barrier complex, and that the slower the progradation rate, the thicker would be the resulting depositional pile. Clearly, the relative effects of varying rates of basin subsidence and the creation of available depositional accommodation space are equally important in determining progradational thickness.

## *Ness Formation*

The Ness Formation is the most lithologically variable unit of the Brent Group, and occurs in successions up to about 180 m thick in the East Shetland Basin. The thickest succession is found in the structurally low area between the Hutton-Ninian and Brent fault trends, and the Formation thins towards the west.

In a detailed study of the Ness Formation in the Brent Field (UK Block 211/29), Livera (1989) characterized the Formation as a complex of deltaic and coastal-plain sediments. The deltas were relatively small, building into lagoons behind a protective barrier formed by the Rannoch−Etive deposits. Livera also noted the influence of wave processes on the lagoonal deposits and recorded the presence of hummocky cross-stratified sandstones in places. Wave influence is also recorded by Brown & Richards (1989) in the Ness Formation in fields farther north. An increase in marine influence is recorded by Livera (1989) at the top of the Ness in the Brent Field: a comparable increase in saline conditions is reported from other parts of the East Shetland Basin by Brown *et al.* (1987) and is compatible with the view of Graué *et al.* (1987) and Falt *et al.* (1989) that upper Ness deposition occurred behind a coastal complex during overall transgression.

Three informal units can be recognized in the Formation in the East Shetland Basin north of 61°N: a Lower Ness Unit; a Middle Ness Unit recognized as a correlatable siltstone horizon; and an Upper Ness Unit. Cannon *et al.* (this volume) have given these informal units member status.

The Lower Ness Unit is up to about 60 m thick and consists dominantly of: lagoonal siltstones with wavy bedding and a brackish to marine microflora; lagoonal shoal, beach and washover sandstones with wave ripples, parallel lamination and flaser bedding; mouth bar deposits forming minor, coarsening upward cycles; minor fluvial channel sands displaying fining upward profiles; and thin ($<1$ m) coal seams, sometimes with roots below. This Lower Unit is interpreted as a lagoonal or lower delta plain deposit by Budding & Inglin (1981), although Livera (1989) has also noted the local development of upper delta plain environments.

The Middle Unit, often informally termed the Mid Ness Shale, typically ranges from 4.5 m to 8 m in thickness, and reaches a maximum thickness of 18 m to the north of the Dunlin oilfield. It is found over an area of some 2100 square kilometres. The unit is predominantly composed of mid to dark grey, lenticular to wavy bedded siltstones, with some sharp-based, possibly wave ripple cross laminated sands. Budding & Inglin (1981) have shown that in the Cormorant oilfield area the unit consists of a vertical succession of facies similar to that seen in the underlying Lower Ness Unit. These facies formed in a large, high energy lagoon which developed when subsidence temporarily outpaced sediment supply.

The Upper Ness Unit usually ranges in thickness from 10 m to 15 m, but a thickness of up to 41 m has been recorded in the NW Hutton oilfield area. However, the top of the unit is often eroded, and this may mask the original thickness trends. For the most part of the Upper Ness Unit has a higher sand:shale ratio than the Lower Ness Unit, and has been interpreted as a dominantly upper delta plain deposit by Livera (1989).

## *Tarbert Formation*

The Formation base is often defined by a sharp-based, coarse- to very coarse-grained sandstone interpreted as a transgressive lag deposit developed above a ravinement surface (Brown *et al.* 1987). Overlying sandstones are occasionally fine grained and highly bioturbated, with rare, planar, dipping laminae developed in places. Some of the fine-grained units are highly bioturbated, and Livera (1989) has suggested they are similar to the Upper Jurassic shelf sandstones found in the central North Sea, and probably represent shelf ridge deposits. The finer, laminated sands sometimes pass up to micaceous, parallel laminated to hummocky cross stratified, shoreface sandstones identical to those in the Rannoch Formation (Brown *et al.* 1987).

The work of Ronning & Steel (1987), Graué *et al.* (1987) and Falt *et al.* (1989) has demonstrated that the Tarbert Formation consists of stacked transgressive–regressive couplets formed during the overall, southward retreat of Brent delta shorelines. Brown *et al.* (1987) and Graué *et al.* (1987) interpreted thin, coarse-grained deposits at several levels within the Formation as transgressive lags resting on shoreface ravinement surfaces formed during coastline retreat. Where distinct periods of regressive sedimentation can be observed between transgressive pulses, such as in the South Alwyn oilfield, the Formation still consists dominantly of marine sediments, although barrier washover, lagoon and coal swamp deposits of non-marine affinity are found at many levels (Ronning & Steel 1987).

Graué *et al.* (1987) suggested that a decrease in sediment supply and an over-extension of the deltaic system may have been sufficient, with background subsidence, to promote the drowning of the Brent delta. This view was modified

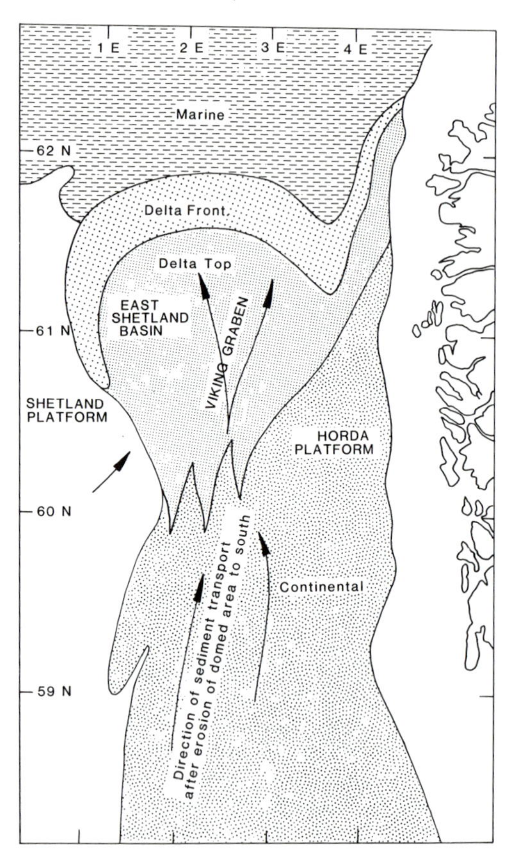


**Fig. 5.** Cartoon illustrating one generally held hypothesis regarding the deposition of the progradational part of the Brent Delta. Rivers carried the erosional products from a thermally upwarped area to the south axially along the Viking Graben, and deposited a wave-dominated delta with facies belts orientated concentrically across the Viking Graben and East Shetland Basin. (Modified after Eynon 1981).

by Helland-Hansen *et al.* (1989) who postulated, based on computer-modelling of the Brent Group succession, that Tarbert deposition was initiated by an increase in subsidence rates, and that the regressive pulses evident in the succession were promoted, at least in part, by eustatic sea-level falls.

Although the top (sometimes all) of the Tarbert Formation is eroded over the crests of some tilted fault blocks, a conformable transition up to Heather Formation mudstones is seen in the Brent field (Livera 1989).

## Depositional models and palaeogeography

There are two main views in the literature concerning the regional depositional models and palaeogeographic setting of the Brent Group. The older, more conventional model for the regressive phase of sedimentation has been documented by many authors, including Budding & Inglin (1981), Ziegler (1982), Johnson & Stewart (1985), Graué *et al.* (1987), Helland-Hansen *et al.* (1989), Falt *et al.* (1989) and Falt & Steel (1990). This model suggests that, following the latest early Jurassic or earliest mid-Jurassic growth of a thermally domed area to the south, erosional products were shed off the dome, and prograded northwards along the Viking Graben. This resulted in the deposition of the wave-dominated regressive deltaic succession in a system of facies belts arranged concentrically across the graben (Fig. 5). Johnson & Stewart (1985) re-enforced the hypothesis of northwards progradation of a single deltaic system with a concentric arrangement of facies belts across the graben by comparing a palaeogeographic time-slice of the Brent Group with the modern, wave-dominated Nile Delta (Fig. 6).

Eynon (1981) modified the conventional view of Brent Group progradation, but also envisaged the concentric deposition of facies belts across the graben (Fig. 5). However, Eynon's model has the deposition of a basal sand sheet (the Broom Formation), followed by delta lobe progradation, marine transgression, and then a further two phases of lobe progradation.

A number of authors (e.g. Morton & Humphreys 1983; Leeder 1983; Morton 1985; Hamilton *et al.* 1987) have, on the basis of the mineralogy of the sediments, disputed an origin for the bulk of the regressive succession by erosion from the south. These authors prefer a source on the adjacent Platforms. Many other authors (e.g. Eynon 1981; Ziegler 1982) have also indicated at least an element of sediment derivation from the basin margin areas.

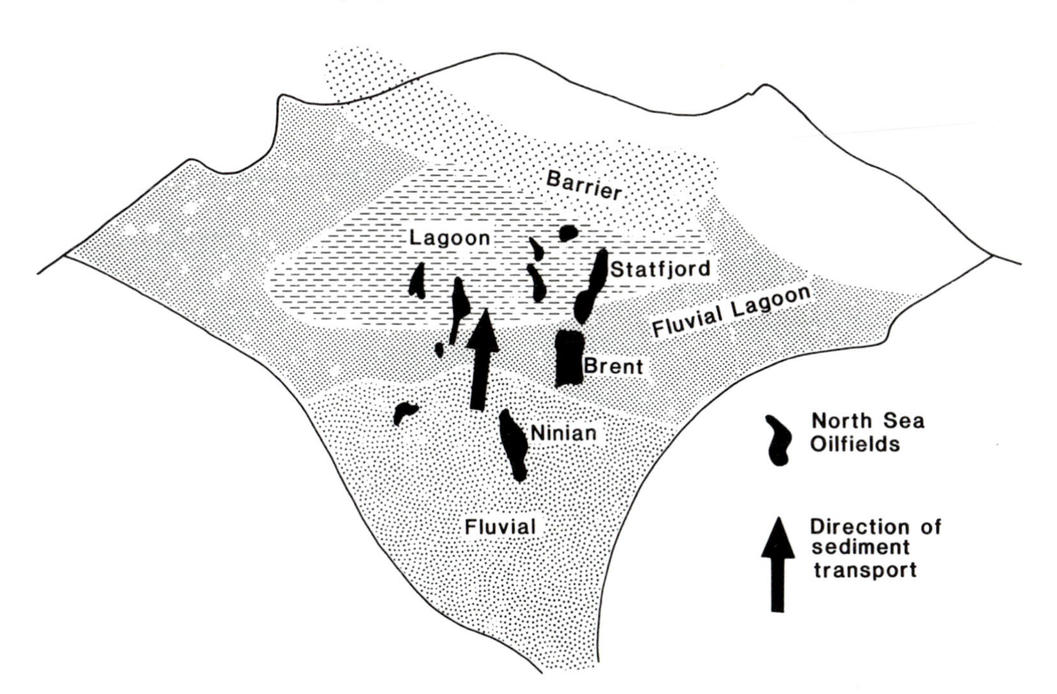


**Fig. 6.** Comparison of facies distributions in the Brent Delta and the modern, wave-dominated Nile Delta, illustrating the postulated concentric arrangement of facies belts in the Viking Graben. (Modified from Johnson & Stewart 1985).

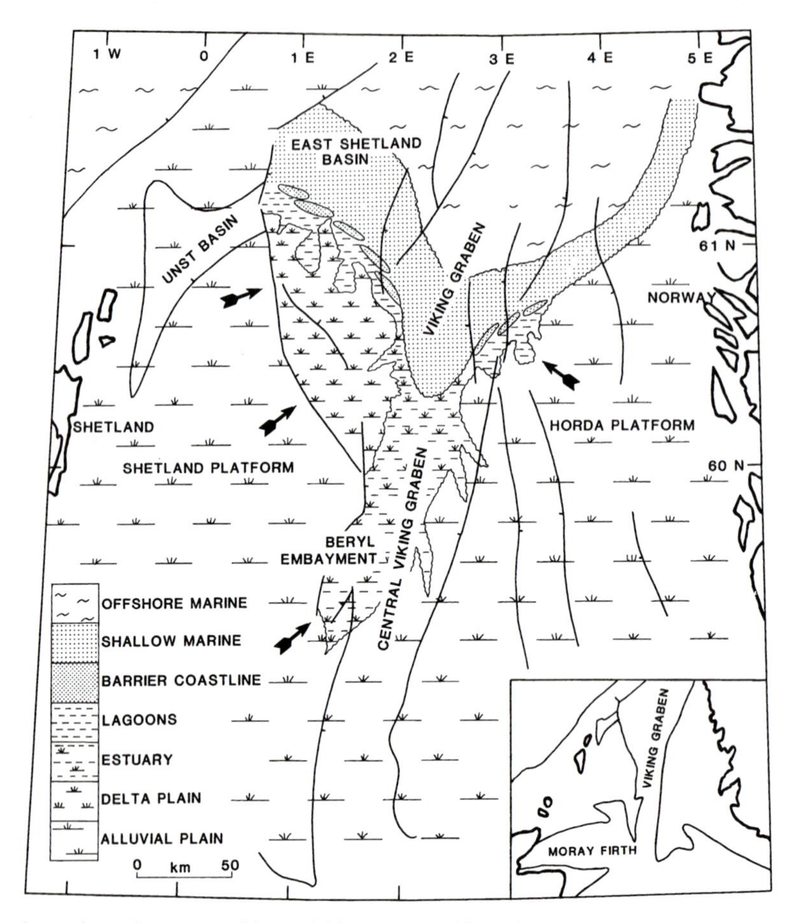


**Fig. 7.** An alternative palaeogeographic model for the deposition of the Brent Group postulated by Richards *et al.* (1988) and Richards (1990). This figure shows a possible reconstruction of the depositional environments in the Aalenian to earliest Bajocian, before the time of maximum progradation of the Brent Group. There is no concentric arrangement of facies belts postulated across the Viking Graben area in this model, which envisages dominantly transverse sediment supply to the basin followed by localized northwards progradation in response to local palaeoslopes. However, as shown in the figure, it is envisaged that substantial thicknesses of marine sands might be encountered in the axial area of the Viking Graben adjacent to the East Shetland Basin.

An alternative model (Fig. 7) has been postulated for the evolution of the succession by Richards *et al.* (1988) and Richards (1990) which accounts for this apparent derivation of material from adjacent platforms. This model envisages rapid sea-level fall and/or basin margin uplift in the latest Toarcian or earliest Aalenian which resulted in a rapid clastic input transversely into the graben. Richards *et al.* (1988) and Richards (1990) demonstrated that the Rannoch Formation thickens and decreases in grain size towards the northeast within the East Shetland Basin, suggesting derivation from the southwest, off the Shetland Platform. Falt & Steel (1990) also recognized a northwesterly component to sediment supply across the Horda Platform, and Hurst & Morton (1988) suggested on the basis of heavy mineral analyses that the Etive and Oseberg Formations in the Oseberg field share a common provenance to the east. Richards *et al.* (1990) have noted that, while most of the sediment was ultimately derived from the basin margins, local transportation may have occurred along the basin axis in response to the regional palaeoslope. At the same time, a marine connection is envisaged to have been maintained into the central Viking Graben area (Richards 1990), with the deposition of estuarine sediments there (Richards 1991) during the Aalenian to earliest Bajocian, before the time of maximum progradation of the Brent deltaic sediments.

There remains much controversy in the literature surrounding the favoured palaeogeographic models. It is hoped that the papers contained within this volume will stimulate further debate, and perhaps eventually lead to the development of a consensus view of Brent palaeogeographic evolution.

S. Brown and M. Dean are thanked for their invaluable discussions on the nature of the Brent Group through the many stages of its study over the years. M. Giles, H. Johnson, B. Levell and J. Marshall are all thanked for pointing out to the author the topics they thought should be covered in a brief review of the geology of the Brent Group. The UK Department of Energy is thanked for permission to publish, but any views presented are those of the author and do not necessarily represent those of the Department. Publication is by permission of the Director, British Geological Survey (NERC).

## References

BOWEN, J. M. 1975. The Brent oilfield. *In*: WOODLAND, A. W. (ed.) *Petroleum and the Continental Shelf of Northwest Europe*. Applied Science Publishers, London, 353–60.

BROWN, S. 1990. Jurassic. *In*: GLENNIE, K. W. (ed.) *Introduction to the Petroleum Geology of the North Sea*. Blackwell Scientific Publications, Oxford, 219–54.

—— & RICHARDS, P. C. 1989. Facies and development of the mid Jurassic Brent delta near the northern limit of its progradation, UK North Sea. *In*: WHATELEY, M. K. G. & PICKERING, K. T. (eds) *Deltas: Sites and traps for fossil fuels*. Geological Society, London, Special Publication, **41**, 253–67.

——, —— & THOMSON, A. R. 1987. Patterns in the deposition of the Brent Group (Middle Jurassic) UK North Sea. *In*: BROOKS, J. & GLENNIE, K. W. (eds) *Petroleum Geology of North West Europe*. Graham & Trotman, London, 899–913.

BUDDING, M. C. & INGLIN, H. F. 1981. A reservoir geological model of the Brent sands in Southern Cormorant. *In*: ILLING, L. V. & HOBSON, G. D. (eds) *Petroleum Geology of the Continental Shelf of North-West Europe*. Heyden, London, 326–34.

CANNON, S. J. C., GILES, M. R., WHITAKER, M. F., PLEASE, P. M. & MARTIN, S. V. 1992. A regional reassessment of the Brent Group, UK Sector, North Sea. *this volume*.

DEEGAN, C. E. & SCULL, B. J. 1977. *A standard lithostratigraphic nomenclature for the Central and Northern North Sea*. Report of the Institute of Geological Sciences, **77/25**.

ELLIOTT, T. 1989. Delta systems and their contribution to an understanding of basin-fill successions. *In*: WHATELEY, M. K. G. & PICKERING, K. T. (eds) *Deltas: sites and traps for fossil fuels*. Geological Society, London, Special Publication, **41**, 3–10.

—— & BULLER, A. T. 1987. *Deltaic sedimentation with reference to the North Sea Jurassic*. Joint Association for Petroleum Exploration Courses, UK, Course Notes No. **51**.

EYNON, G. 1981. Basin development and sedimentation in the Middle Jurassic of the northern North Sea. *In*: ILLING, L. V. & HOBSON, G. D. (eds) *Petroleum Geology of the Continental Shelf of North-West Europe*. Heyden, London, 196–204.

FALT, L. M. & STEEL, R. 1990. A new palaeogeographic reconstruction for the Middle Jurassic of the northern North Sea: a discussion. *Journal of the Geological Society, London*, **147**, 1085–90.

——, HELLAND, R., WIIK JACOBSON, V. & RENSHAW, D. 1989. Correlation of transgressive-regressive depositional sequences in the Middle Jurassic Brent/Viking Group megacycle, Viking Graben, Norwegian North Sea. *In*: COLLINSON, J. D. (ed.) *Correlation in Hydrocarbon Exploration*. Graham & Trotman, London, 191–201.

GRAUE, E., HELLAND-HANSEN, W., JOHNSON, J., LØMO, L., NOTTVEDT, A., RONNING, K., RYSETH, K. and STEEL, R. J. 1987. Advance and retreat of Brent Delta system, Norwegian North Sea. *In*: BROOKS, J. & GLENNIE, K. W. (eds) *Petroleum Geology of North West Europe*. Graham & Trotman, London, 915–37.

HALLETT, D. 1981. Refinement of the geological model of the Thistle Field. *In*: ILLING, L. V. & HOBSON, J. D. (eds) *Petroleum Geology of the Continental Shelf of North-West Europe*. Heyden, London, 315–25.

HAMILTON, P. J., FALLICK, A. E., MACINTYRE, R. M. & ELLIOTT, S. 1987. Isotope tracing of the provenance and diagenesis of Lower Brent Group sands, North Sea. *In*: BROOKS, J. & GLENNIE, K. W. (eds) *Petroleum Geology of North West Europe*. Graham & Trotman, London, 939–50.

HAY, J. T. C. 1978. Structural development in the northern North Sea. *Journal of Petroleum Geology*, **1**, 65–77.

HELLAND-HANSEN, W., STEEL, R., NAKAYAMA, K. and KENDALL, C. G. St. C. 1989. Review and computer modelling of the Brent Group stratigraphy. *In*: WHATELEY, M. K. G. & PICKERING, K. T. (eds) *Deltas: sites and traps for fossils fuels*. Geological Society, London, Special Publication, **41**, 237–252.

HURST, A. R. & MORTON, A. C. 1988. An application of heavy mineral analysis to lithostratigraphy and reservoir modelling in the Oseberg Field, northern North Sea. *Marine and Petroleum Geology*, **5**, 157–69.

JOHNS, C. R. & ANDREWS, I. J. 1985. The petroleum geology of the Unst Basin, North Sea. *Marine and Petroleum Geology*, **2**, 361–72.

JOHNSON, H. D. & STEWART, D. J. 1985. Role of clastic sedimentology in the exploration and production of oil and gas in the North Sea. *In*: BRENCHLEY, P. J. & WILLIAMS, B. P. J. (eds) *Sedimentology: recent developments and applied aspects*. Geological Society, London, Special Publication, **18**, 249–310.

KARLSSON, W. 1986. The Snorre, Statfjord and Gullfaks oilfields and the habitat of hydrocarbons on the Tampen Spur, offshore Norway. *In*: *Habitat of Hydrocarbons on the Norwegian Continental Shelf*. Graham & Trotman, London, 181–96.

Larsen, R. M. & Jaarvik, L. J. 1981. The geology of the Sleipner Field complex. *In*: *Norwegian Symposium on Exploration* Norwegian Petroleum Society, Geilo, 15/1–31.

Leeder, M. R. 1983. Lithostratigraphic stretching and North Sea Jurassic clastic sourcelands. *Nature*, **305**, 510–13.

Livera, S. E. 1989. Facies associations and sand body geometries in the Ness Formation of the Brent Group, Brent field. *In*: Whateley, M. K. G. & Pickering, K. T. (eds) *Deltas: sites and traps for fossil fuels*. Geological Society, London, Special Publication, **41**, 269–86.

Mitchener, B. C., Lawrence, D. A., Partington, M. A., Bowman, M. B. J. & Gluyas, J. 1992. Brent Group: sequence stratigraphy and regional implications. *This volume*.

Morton, A. C. 1985. A new approach to provenance studies: electron microprobe analysis of detrital garnets from Middle Jurassic sandstones of the northern North Sea. *Sedimentology*, **32**, 553–66.

—— & Humphreys, B. 1983. The petrology of the Middle Jurassic sandstones from the Murchison field, North Sea. *Journal of Petroleum Geology*, **5**, 245–60.

Parry, C. C., Whitley, P. K. J. & Simpson, R. D. H. 1981. Integration of palynological and sedimentological methods in facies analysis of the Brent Formation. *In*: Illing, L. V. & Hobson, G. D. (eds) *Petroleum Geology of the Continental Shelf of North-West Europe*. Heyden, London, 205–15.

Richards, P. C. 1989. *Lower and Middle Jurassic sedimentology of the Beryl Embayment, and implications for the evolution of the northern North Sea*. PhD thesis, University of Strathclyde.

——. 1990. The early to mid-Jurassic evolution of the northern North Sea. *In*: Hardman, R. F. P. & Brooks, J. (eds) *Tectonic events responsible for Britain's oil and gas reserves*. Geological Society, London, Special Publication, **55**, 191–205.

——. 1991. An estuarine facies model for the Middle Jurassic Sleipner Formation: Beryl Embayment, North Sea. *Journal of the Geological Society, London*, **148**, 459–71.

—— & Brown, S. 1986. Shoreface storm deposits in the Rannoch Formation (Middle Jurassic), North West Hutton oilfield. *Scottish Journal of Geology*, **22**, 367–75.

—— & —— 1987. *The nature of the Brent Delta, North Sea: a core workshop*. British Geological Survey, Open File Report **87/17**.

——, —— & Dean, J. M. 1986. Storm deposits in the Broom Formation (Aalenian), UK North Sea. *British Sedimentology Research Group, Annual Meeting, Nottingham* (Abstract).

——, ——, —— & Anderton, R. 1988. A new palaeogeographic reconstruction for the Middle Jurassic of the northern North Sea. *Journal of the Geological Society, London*, **145**, 883–6.

——, ——, —— & ——. 1990. A new palaeogeographic reconstruction for the Middle Jurassic of the northern North sea: a reply. *Journal of the Geological Society, London,* **147**, 1085–90.

Ronning, K. & Steel, R. J. 1987. Depositional sequences within a 'transgressive' reservoir sandstone unit: the Middle Jurassic Tarbert Formation. Hild area, northern North Sea. *In*: Kleppe, J. *et al.* (eds) *North Sea Oil and Gas Reservoirs*. Graham & Trotman, London, 169–76.

Ryseth, A. 1989. Correlation of depositional patterns in the Ness Formation. Oseberg area. *In*: Collinson, J. D. (eds) *Correlation in Hydrocarbon Exploration*. Graham & Trotman, London, 313–26.

Scott, E. 1992. The palaeoenvironments and dynamics of the Rannoch–Etive nearshore and coastal successions, Brent Group, northern North Sea. *This volume*.

Vail, P. R. & Todd, R. G. 1981. North Sea Jurassic unconformities, chronostratigraphy and sea-level changes from seismic stratigraphy. *In*: Illing, L. V. & Hobson, J. D. (eds) *Petroleum Geology of the Continental Shelf of North-West Europe*. Heyden, London, 216–35.

Vollset, J. & Doré, A. G. 1984. *A revised Triassic and Jurassic lithostratigraphic nomenclature for the Norwegian North Sea*. Norwegian Petroleum Directorate, Bulletin 3.

Wheatley, T. J., Biggins, D., Buckingham, J. & Holloway, N. H. 1987. The geology and exploration of the Transitional Shelf, an area to the west of the Viking Graben. *In*: Brooks, J. & Glennie, K. W. (eds) *Petroleum Geology of North West Europe*. Graham & Trotman, London, 979–90.

Ziegler, P. A. 1982. *Geological Atlas of Western and Central Europe*. Shell, The Hague.

# The structural evolution of the Brent Province

GRAHAM YIELDING, MICHAEL E. BADLEY & ALAN M. ROBERTS

*Badley, Ashton & Associates Ltd, Winceby House, Winceby, Horncastle, Lincolnshire LN9 6PB, UK*

**Abstract:** The structural evolution of the Viking Graben has been the fundamental control on the deposition of the Brent Group and on the development of trapping geometries. Major crustal extension in the early Triassic caused tilting of basement fault blocks, which can still be clearly seen at the basin margins. By the mid-Triassic a post-rift thermal-subsidence basin was established. Local fault control on Brent Group deposition indicates the onset of a second period of crustal extension, although the main control on subsidence at this time was still thermal relaxation following Triassic rifting. Extension during deposition of the Brent was of only *c*. 1% magnitude.

The second period of extension increased in magnitude following Brent deposition and peaked during the late Jurassic, creating the main structural traps for Brent Province oil and gas. Footwalls to major normal faults were uplifted and eroded. The amount of uplift on a given fault-block can be predicted using quantitative models (flexural and domino models). Subsidence in the adjacent half-graben generally outpaced sedimentation, leading to deep-water basins into which footwall material could collapse. On all but the largest fault-blocks it is likely that footwall uplift rates were low compared with erosion rates, and so footwall crests would have been degraded faster than they would have been uplifted above sea-level. Flushing by meteoric water during the late Jurassic is therefore expected only on the largest fault-blocks (e.g. Snorre, Gullfaks).

The 'Brent Province' of hydrocarbon discoveries is geographically coincident with the northern part of the Viking Graben and its flanking terraces. The structural evolution of the Viking Graben has been the fundamental control on the deposition of the Brent Group and on the development of trapping geometries. This paper aims briefly to review recent advances in knowledge of the history of the Viking Graben, with particular emphasis on:

(i) the evidence for major, early Triassic faulting;
(ii) post-Triassic thermal subsidence, which acted as the main control on patterns of Brent Group deposition;
(iii) the Late Jurassic fault movements that created the tilted fault-blocks in which most of the Brent Province oil and gas is trapped.

These processes affect the Brent Group at all scales, from regional cross-sections down to detailed well correlation within a single field.

The Viking Graben is an extensional basin. Deep seismic reflection profiling (Klemperer 1988) and gravity modelling (Donato & Tully 1981; Zervos 1987) show that the pre-Mesozoic basement has been thinned to as little as 12–15 km beneath the centre of the graben, compared to a thickness of *c*. 30 km beyond the limits of the basin. The overlying sedimentary pile shows a complementary increase in thickness towards the basin centre, reaching a maximum thickness of perhaps 10 km. These relationships can be seen in more detail in Fig. 1, which shows a set of full-crustal cross-sections across the basin (from Marsden *et al.* 1990). The sedimentary pile can divided into three major, structurally-controlled sequences.

First, the Triassic–Middle Jurassic sequence forms the syn-rift and initial thermal-subsidence fill of an early stretching episode. Basement fault-blocks underlying this fill can be seen most clearly at the basin margins, e.g. west of the Tern–Eider Ridge on Profile II, and the eastern end of Profiles I-IV (Fig. 1). In these areas the overprint by later Jurassic fault movement is minimal and therefore the earlier structures can still be recognized.

The second major sequence comprises the Upper Jurassic (and locally lowermost Cretaceous) rocks, which represent the syn-rift fill of the second stretching episode. The Upper Jurassic is characterized by wedge-shaped packages of sediment that partially infill a tilted-block topography. This topography was created by further movement on the major normal faults that bound the underlying Triassic half-graben.

The third major sequence comprises the Cretaceous–Tertiary rocks overlying the Jurassic tilted fault-blocks. This represents the thermal subsidence basin following Late Jurassic stretching. Initially, this subsidence would have been enhanced because the lithosphere was not

*From* Morton, A. C., Haszeldine, R. S., Giles, M. R. & Brown, S. (eds), 1992, *Geology of the Brent Group*. Geological Society Special Publication No. 61, pp. 27–43.

thermally re-equilibrated when the Late Jurassic stretching occurred (i.e. post-Triassic thermal subsidence was still occurring) (Giltner 1987). Sedimentation rates during the Cretaceous were low, and basin subsidence probably out-paced the sedimentation, leading to significant water depths (Bertram & Milton 1989; Barr 1991). Increased clastic input in the Tertiary then infilled the basin, giving a total Tertiary and Cretaceous thickness of about 3 km on the East Shetland Terraces and about 5 km in the axis of the Viking Graben. Fault activity during the Cretaceous and Tertiary rapidly diminished, becoming restricted to the major faults at the basin margins. A significant part of this late minor movement was probably caused by the continuing compaction of the Mesozoic sedimentary pile.

## The Triassic stretching episode

The clearest evidence for major Triassic fault activity in the north Viking Graben comes from seismic data across the Norwegian Horda Platform. It has been known for some years that wedge-shaped packages of Triassic rocks overlie a tilted-block basement topography (e.g. Eynon 1981; Badley *et al.* 1984, 1988). Only one well (N31/6–1) has reached basement in this area, and found Lower Triassic resting on gneiss (Lervik *et al.* 1989). However, the top basement reflection is relatively clear on seismic data, allowing the basement fault-blocks to be mapped. The major normal faults trend N–S and are typically about 15 km apart.

A number of these faults give rise to fault-plane reflections, which have been the subject of a detailed depth-migration study by Yielding *et al.* (1991). Figure 2 shows a cross-section based on their depth-migrated profiles. The eastern fault on the profile, the Øygarden Fault, lies just offshore from the Norwegian coast, and drops top basement from less than 1 km below sea-level to *c.* 5 km. The western fault on the profile, which now dips at only 25–30°, shows a displacement at top basement of *c.* 6 km. Displacement at the base of the Cretaceous is less than 1 km. We interpret these offsets to indicate major fault movement in the early Triassic, followed by minor reactivation in the earliest Cretaceous. The extension associated with the Triassic faulting would correspond to a stretching factor ($\beta$) of *c.* 1.3–1.4, whereas the later reactivation had a $\beta < 1.1$.

The Triassic fault movement created major half-graben in which were deposited continental red beds in monotonous sandstone-mudstone sequences (Fisher 1986; Lervik *et al.* 1989). The large-scale basement topography under these successions is important in controlling the compaction of the sedimentary pile (discussed below).

Triassic fault activity is also documented from the western margins of the north Viking Graben. The Unst Basin (Fig. 1, map) is a small half-graben within the East Shetland Platform that contains a wedge of Permo-Triassic red beds up to 3 km thick (Johns & Andrews 1985). The bounding fault to this half-graben, the Pobie Fault, shows negligible post-Triassic movement. However, the northeastwards continuation of the Pobie Fault forms the western bounding fault of the Tern–Eider Ridge (Fig. 1, map and Fig. 3). This fault shows increasing late Jurassic reactivation northeastwards past the Tern and Eider Fields (Speksnijder 1987). As with the faults on the Horda Platform, late Jurassic reactivation is relatively minor in comparison to the earlier Triassic movement (generally <1 km offset of Middle Jurassic compared to several kilometres offset of top basement).

Apart from the thick Triassic wedge against the Pobie–Tern–Eider fault, the Triassic is generally fairly thin along the western margin of north Viking Graben. At the western edge of the East Shetland Terraces, there is typically only *c.* 100 m of Triassic, resting on basement (Lervik *et al.* 1989). However, there is considerable basinwards thickening, so that >1900 m are present on the Brent–Statfjord–Snorre trend. Basement has not been drilled here, and so the total Triassic thickness is unknown. The basinwards thickening occurs in two distinct ways. Firstly, the upper part of the Triassic (post-Lomvi Fm) thickens continuously eastwards, within the present-day fault blocks. Due to the westwards dip imposed during late Jurassic block-tilting, the upper Triassic thickening is now in an up-dip direction. The Triassic as a whole, however, appears to show significant fault control, with abrupt thickness increases occurring over major basement faults. An example of this is given in Fig. 4, which shows a depth section across the Hutton Fault, separating the Hutton and Brent fault blocks. Top basement on the Hutton block is reasonably well-constrained (from wells along strike on Ninian), but that on the Brent block is more speculative. However, the fault plane itself gives rise to a reflection, and depth migration of this reflection constrains the fault geometry as approximately planar, dipping at *c.* 30–35° (Yielding *et al.* 1991). Displacement at top basement level is *c.* 6 km, compared to 1–2 km at Middle Jurassic levels. Again, significant ?early Triassic fault movement was followed by a

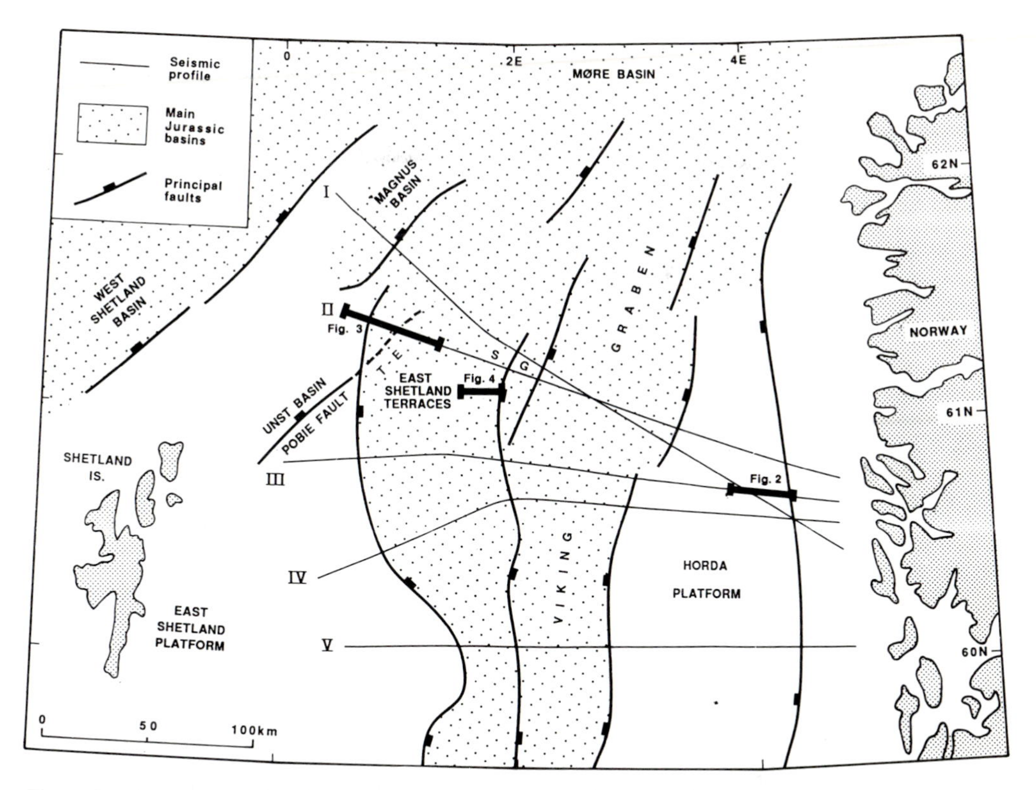


**Fig. 1.** Location map and (overleaf) full-crustal cross-sections of the North Viking Graben (from Marsden *et al.* 1990). On the location map, T-E is Tern-Eider Ridge, S is Statfjord and G is Gullfaks. The locations of Figs 2, 3 and 4 are also shown.

smaller amount of later (mostly late Jurassic) offset.

Nelson & Lamy (1987, their figs 4 & 5) show a section across the Snorre Fault very similar to our profile of the Hutton Fault. Their depth conversion gives a present fault dip (in basement) of *c.* 30°, with over 10 km of displacement at the base Triassic and *c.* 5 km at Middle Jurassic. In this case, both the early (Triassic) and later (late Jurassic) episodes of fault activity involved major displacement.

The general conclusion to be drawn from the above examples is that significant extensional faulting occurred in the north Viking Graben prior to the late Jurassic fault activity. This earlier stretching episode probably occurred during the early Triassic, given the general lack of Permian sediments (Lervik *et al.* 1989). On the basin margins, where the Triassic faulting is best-defined, the same major normal faults were active in both the early Triassic and late Jurassic stretching episodes. Fault movement during the Triassic appears to have been as great or greater than fault movement in the late Jurassic, corresponding to stretching factors ($\beta$) of up to *c.* 1.4. Marsden *et al.* (1990) infer a whole-basin Triassic extension of 38 km, based on observable fault offsets and sediment thicknesses; this is likely to be a conservative estimate.

In the graben axis, seismic data are unable to image the top of the basement. The geometry (or even existence) of Triassic fault blocks is therefore speculative. However, all the evidence from thickness trends on the graben terraces suggests that the present graben axis was also the axis of Triassic fault activity. The deep part of the Viking Graben is likely to contain the thickest Triassic sequences and the most-rotated basement fault blocks.

Following the early Triassic extension, the Viking Graben underwent a phase of post-rift (thermal) subsidence. Continuing fluvial sedimentation progressively covered the fault-block crests, producing a broad, saucer-shaped late Triassic basin. The late Triassic continental deposits then pass upwards into the Statfjord Formation, deposited in coastal environments (Røe & Steel 1985), followed by the marine

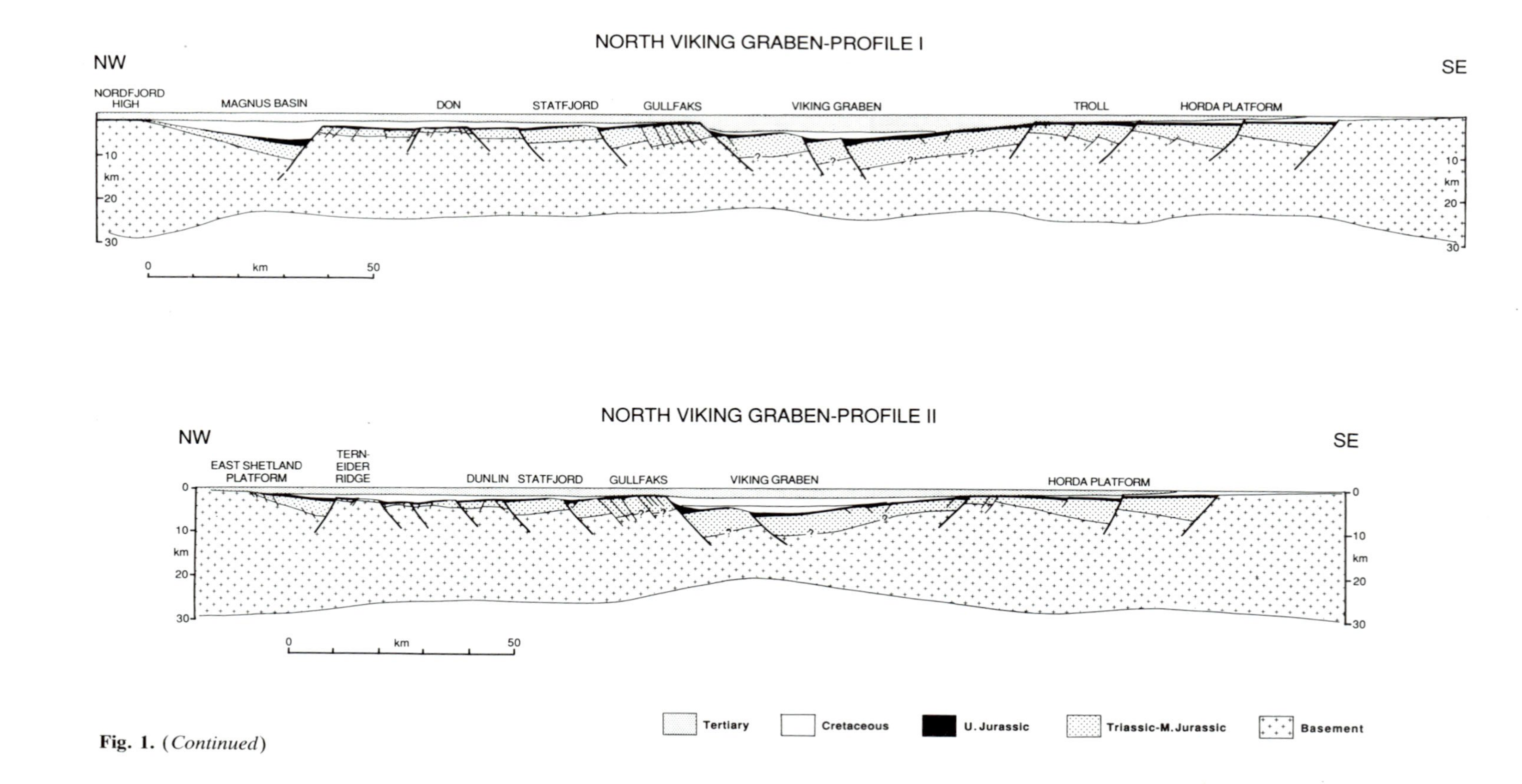


**Fig. 1.** (*Continued*)

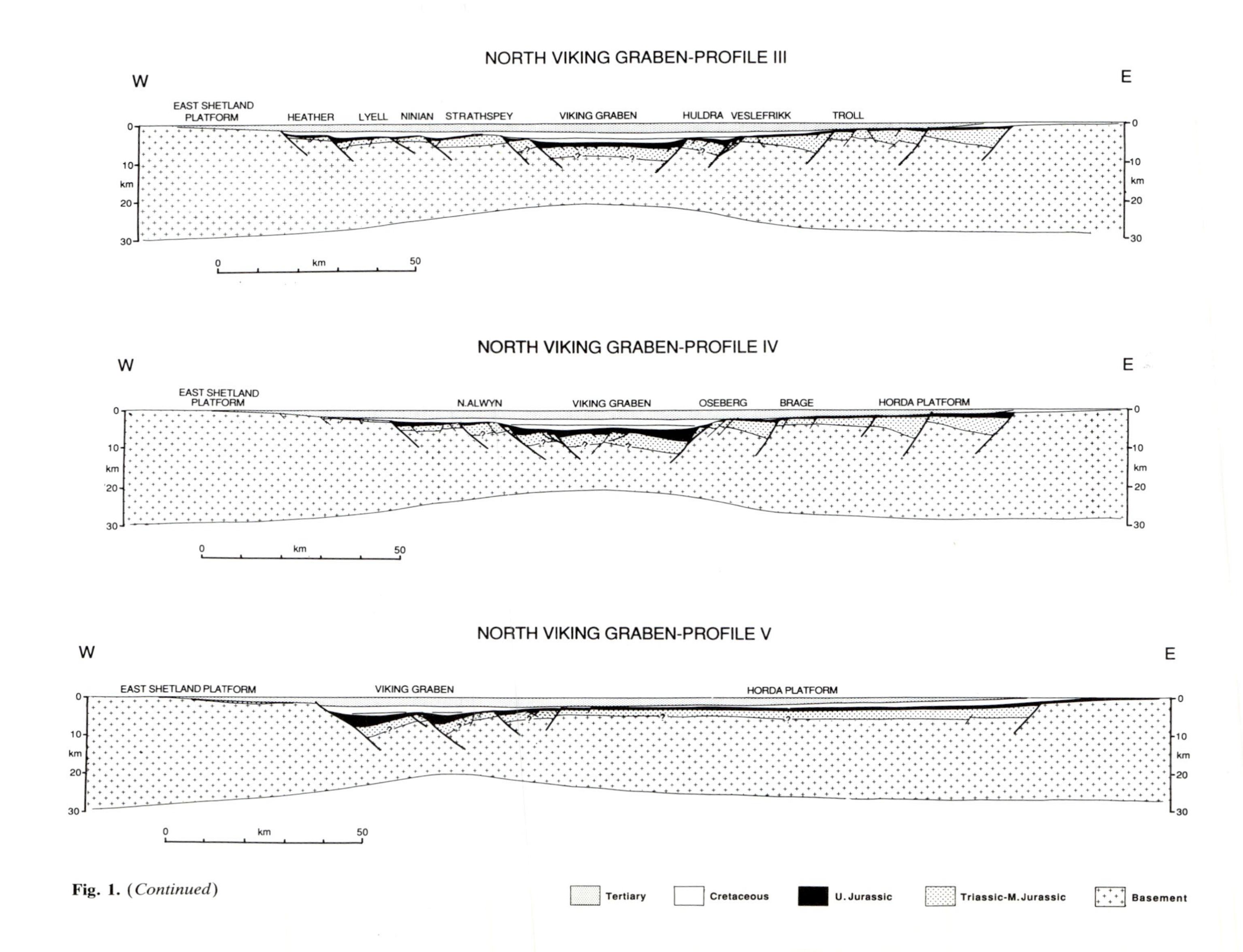


**Fig. 1.** (*Continued*)

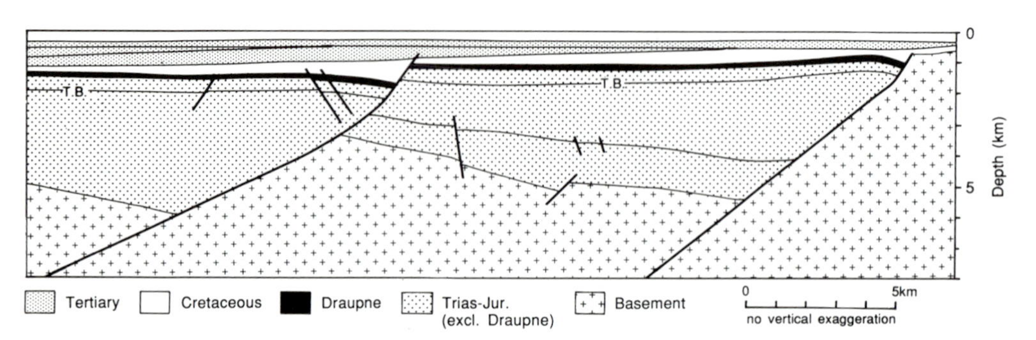


**Fig. 2.** W–E depth cross-section of the eastern margin of the Horda Platform (approximately along eastern end of Profile III in Fig. 1). Note the tilted basement fault-blocks, with Triassic half-graben fill. T. B. is Top Brent. Fault-plane positions are constrained by depth migration of fault-plane reflections. After Yielding *et al.* (1991).

shales of the Lower Jurassic Dunlin Group. This continental-to-marine transition implies that sedimentation was not keeping pace with the subsidence. However, the development of the Brent coastal plain in the Middle Jurassic reversed this trend, establishing the top of the sedimentary pile at or near sea-level.

The underlying reason for the establishment of the Brent delta-plain at sea-level is unclear. Waning thermal subsidence may have contributed by allowing sedimentation to 'catch up' with subsidence, but this does not explain the change in sediment type between the Dunlin and the Brent. The derivation of the sediments which comprise the Brent delta-plain has previously been ascribed to a Middle Jurassic 'prerift dome' within the Central North Sea (e.g. Ziegler 1988), which, it is suggested, shed detritus northwards into the Viking Graben. Throughout the North Sea and Norwegian shelf, however, the Middle Jurassic, wherever encountered, is sand-prone and interpreted as delta-plain or shallow marine in origin (e.g. Gjelberg *et al.* 1987; Berglund *et al.* 1986; Frandsen 1986; papers this volume). A more fundamental cause of sand derivation is therefore required than a local source in the North Sea. There is no evidence that the Brent, or any other Middle Jurassic sequence in the North Sea/Norwegian shelf area, was deposited coincidentally with fault activity of sufficient magnitude to generate extensive, uplifted source areas (see below). A lowering of sea-level at the beginning of the Middle Jurassic may therefore be the fundamental cause of widespread sand-prone deposition at this time.

Late Triassic, Lower Jurassic and Middle Jurassic units all thicken towards the basin

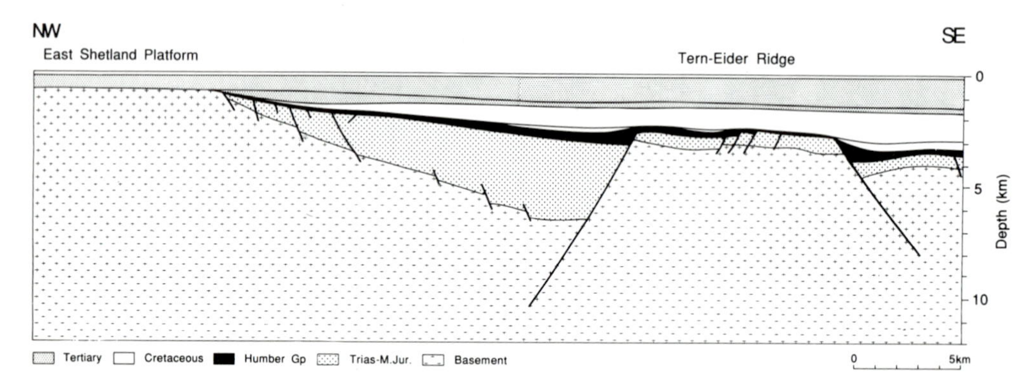


**Fig. 3.** Depth cross-section across the Tern-Eider Ridge (from W end of Profile II in Fig. 1). Note the major Triassic half-graben to the northwest of the Tern–Eider Ridge; only minor reactivation occurred here in the late Jurassic.

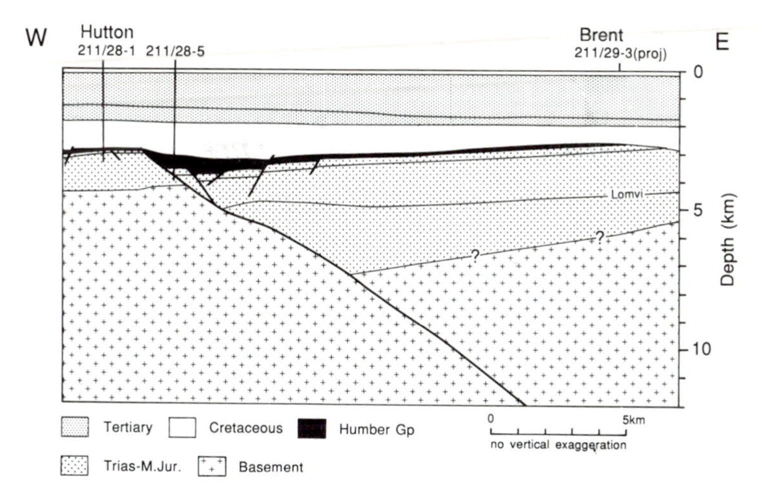


**Fig. 4.** Depth cross-section between the Hutton and Brent fields, East Shetland Terraces. Note the major Triassic growth on the Hutton Fault, with relatively minor late Jurassic reactivation. The position of the fault-plane is constrained by depth migration of a fault-plane reflection (Yielding *et al.* 1991). Note also the eastwards thickening of the post-Lomvi Triassic, in a present-day updip direction. The uppermost sub-division of the 'Trias-M.Jur.' unit is the Brent Group. See Fig. 1 for location.

centre. This depositional pattern was interpreted by Giltner (1987) and Badley *et al.* (1988) as a response to the thermal subsidence following early Triassic stretching; greater extension at the basin axis having produced greater thermal subsidence. As with the Triassic, thickness information for the Lower and Middle Jurassic is sparse in the graben axis. However, the combined Triassic and Lower–Middle Jurassic sequence is >3 km thick on the Brent–Statfjord–Snorre trend immediately west of the graben. It is therefore likely that this early syn- and post-rift sequence may be as much as 5 km thick at the basin centre, as shown in Fig. 1 (from Marsden *et al.* 1990).

## Structural controls on Brent Group deposition

The Brent Group was deposited prior to the zenith of renewed rifting which occurred in the Late Jurassic. The main structural control on Brent Group deposition was therefore the thermal subsidence following the early Triassic crustal stretching. It was this subsidence that created the trough into which the 'Brent delta' was able to prograde. However, superimposed on this first-order mechanism, the Brent Group also gives evidence of the beginnings of the second rift episode, i.e. renewed fault activity. The depositional patterns of the Brent Group result from the interplay of this renewed faulting with the ongoing background subsidence. Third-order effects, such as compaction of the underlying Triassic sequence and local footwall uplift, may also be recognisable.

### *Fault control*

Structural control on Brent Group thicknesses is clearly seen in E-W transects of the basin. Figure 5 (from Badley *et al.* 1988) shows representative Brent and Dunlin Group thicknesses between 59° and 60°N. More detailed maps of drilled thicknesses are available for smaller areas (e.g. Brown *et al.* 1987), but the important point to be made here is the basinwide variation. Also, the values in Fig. 5 have been decompacted, to show the thickness as it was immediately following deposition. There is a progressive increase in thickness from both basin margins to the basin axis, where the original uncompacted thickness of the Brent Group may have been *c.* 850 m. This basinwide pattern was ultimately controlled by the distribution of post-Triassic thermal subsidence.

In more detail, the basinwards thickening of the Brent Group is not uniform. Within each major fault-block, the Brent tends to be relatively tabular, but significant increases in thickness occur over the major boundary faults (see cross-sections, Figs 4 and 6). This pattern is particularly marked in stepping from the Ninian–Hutton block (decompacted thickness *c.* 180 m) to the Brent–Statfjord block (decompacted thickness *c.* 430 m), see Fig. 5. Since the top of the Brent Group was approximately flat,

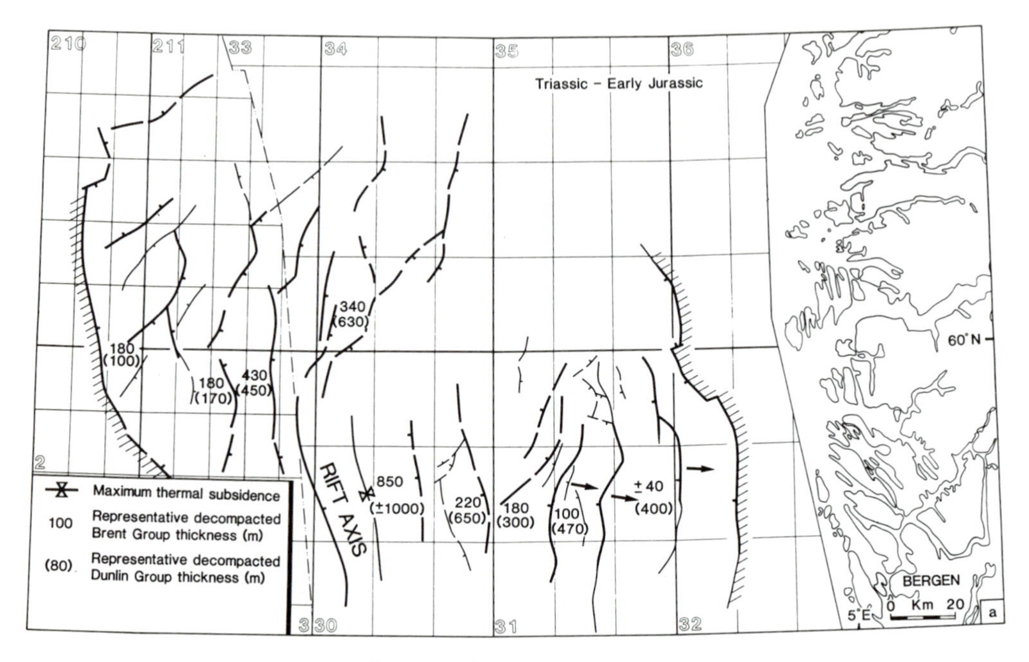


Triassic–Early Jurassic

**Fig. 5.** Tectonic sketch-map showing the basinwide variation in thickness of the Brent Group and Dunlin Group. Thicknesses have been decompacted using parameters of Sclater & Christie (1980). Note the thickening from both basin margins to the rift axis. This thickness variation is mainly the result of post-Triassic thermal subsidence (from Badley *et al.* 1988).

at or near sea-level, this fault-related thickening is good evidence for syn-Brent fault activity. However, because of the control by the ongoing thermal subsidence, the Brent would be expected to thicken progressively basinwards even in the absence of faulting. Thus the faulting *per se* was not causing enhanced subsidence in the basin centre. The simplest interpretation of the thickness pattern (tabular within a fault block but thickening across the faults) is that minor fault movement caused a westward-tilting of the fault blocks, approximately equal to the basinward tilt caused by differential thermal subsidence. Given the thickness gradient shown in Figs 5 and 6, this basinward tilt of the base Brent surface would have been *c.* 0.5°. Thus a 0.5° westward tilt on all of the fault blocks would produce a near-flat base Brent surface within each block. The regional thickening would then be accommodated by the individual major faults.

Assuming the top Brent to approximate to a sea-level datum, the thickness changes shown in Figs 5 and 6 indicate the syn-Brent vertical displacement (throw) on each of the major faults. Making a reasonable estimate of the fault dips (between 30° and 60°), the throw can be used to estimate the syn-Brent horizontal displacement (heave). Summing heaves along the whole E-W basin transect gives a (very approximate) estimate of the overall basin extension during the time of Brent deposition. Performing this exercise on the section of Fig. 6 gives an extension of *c.* 2 km, i.e. a $\beta$ of *c.* 1.01 over the 200 km profile. In terms of basin development this is a negligible amount of extension (1%), and approximately an order of magnitude less than conservative estimates of the total amount of Jurassic extension (Giltner 1987; Badley *et al.* 1988; Marsden *et al.* 1990). To this extent, description of the Brent Group as a 'syn-rift' unit is misleading. However, in terms of hydrocarbon prospectivity these fault-related thickness changes are obviously of immense importance, as they control reservoir thickness on individual fault blocks.

## *Compaction effects*

We have seen in the previous section that the Jurassic tilted fault-blocks are underlain by tilted basement, with great variation in the thickness

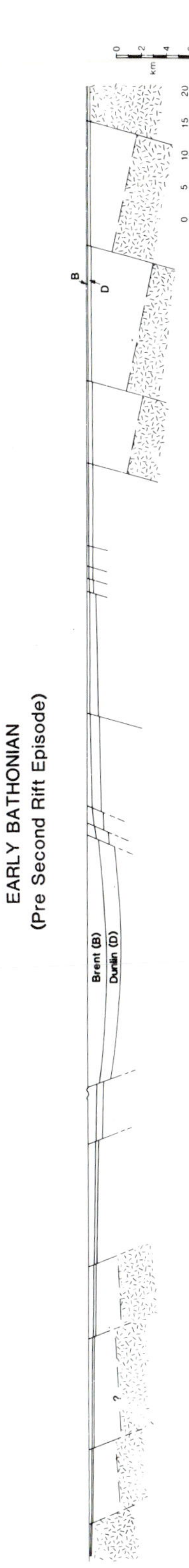


**Fig. 6.** Approximately E–W cross-section through the Viking Graben decompacted and schematically restored to end-Brent time: note the *c.* 2:1 vertical exaggeration. Both Brent and Dunlin intervals thicken regionally from the graben flanks into the graben axis. This regional trend is overprinted by more abrupt thickening across individual faults. The syn-Brent extension associated with these faults is *c.* 1%. The section location is approximately coincident with profile III (Fig. 1). (From Badley *et al.* 1988).

of the Triassic fill. As the Dunlin and Brent sediments were added to the basin fill, the underlying Triassic would have been continually compacting, making room for additional sediment. Clearly, more compaction of the sedimentary column can occur in areas where the basement is deep than in areas where basement is shallow. The basement structure can therefore modify the sedimentary thicknesses even when there is no actual fault movement. To estimate quantitatively the effect of differential compaction on Brent Group thicknesses, we have used the Hutton-Brent depth profile shown in Fig. 4. Figure 7a shows a simplified restoration using Top Dunlin as a horizontal datum. All older sediments have been decompacted using the shaley sand parameters of Sclater & Christie (1980). Up to 5 km of sedimentary fill is present in the hangingwall, compared to *c.* 2 km on the footwall (Hutton). Figure 7b shows a similar restoration taking Top Brent as a horizontal datum. The thickening across the Hutton boundary fault is constrained by wells on both sides of the fault (cf. Fig. 4). Figure 7c shows a forward model of predicted Brent thickness, derived by taking the restoration of Fig. 7a and adding an increment of subsidence to the basement. A basement subsidence step of 110 m, uniformly across the profile, was used. New sediment was allowed to infill the surficial hole, up to datum. However, this additional sediment causes compaction of the underlying sediment pile, which in turn allows deposition of a little more sediment. The resulting equilibrium sediment profile shown in Fig. 7c has *c.* 180 m of new sediment (Brent) on the footwall block and *c.* 220 m maximum in the hangingwall. This *c.* 25% change in Brent thickness has been produced solely by differential compaction of the underlying sediment pile. Thus we should expect to see significant changes in Brent thickness from footwall to hangingwall *even when there is no fault movement.*

However, the differential compaction cannot account for all of the observed thickness changes (cf. Fig. 7b). In particular, the observed thickening is greater than that attributable to differential compaction (*c.* 445 m *contra* *c.* 220 m). Also, the compaction is greatest above the deepest basement, whereas the observed thickening occurs abruptly across the fault at Brent levels. As discussed earlier, the position of the fault plane is well-constrained by depth migration (Yielding *et al.* 1991), and so the position of maximum compaction cannot be brought closer to the footwall at Brent level. Thus, although differential compaction can produce significant thickness changes, in this

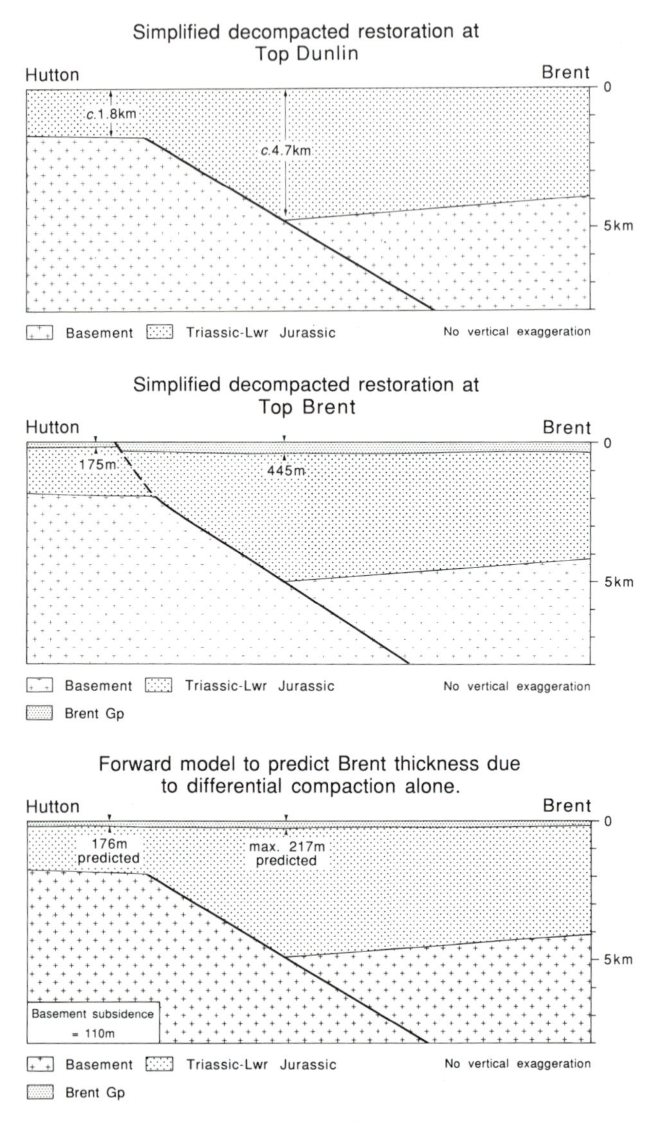


**Fig. 7.** (**a**) Simplified, decompacted restoration of the profile in Fig. 4 (between Hutton and Brent), at 'Top Dunlin' time. Note the thickness of the Triassic half-graben fill. (**b**) Simplified, decompacted restoration of the same profile at 'Top Brent' time. Note the doubling of Brent Group thickness over the Hutton Fault. (**c**) Forward model of predicted Brent Group thickness, starting from the restoration (a) and assuming *no* fault movement. Differential compaction of the Triassic half-graben fill produces thickness variations in the Brent of *c*. 25%, but this is not sufficient to match the restored observed profile (b).

particular case the fault-related thickness changes are even greater. Fault activity during Brent deposition is required to account for these increases in thickness.

## *Footwall uplift*

If the major basement faults *were* active during the Middle Jurassic, we might expect to see thinning or even erosion/non-deposition of the Brent Group on the fault-block crests due to footwall uplift. Quantitative structural models are now available to predict the vertical and lateral extent of footwall uplift associated with normal faulting. Such models can be used to predict where footwall uplift might be expected, given particular fault movements, or to constrain the fault movement, given direct observations of footwall thinning or erosion. Once confidence in a model or family of models is developed, detailed prediction of reservoir thickness can be made in areas beyond good well and seismic control.

We have found two models of extensional faulting to be particularly useful in matching fault-block geometries in the North Sea basins. First, elastic-dislocation models (e.g. Gibson *et al.* 1989) consider the volumetric strains around a fault embedded in an elastic medium. Secondly, flexural-isostatic models (e.g. Kusznir *et al.* 1991) incorporate the effects of isostasy when fault displacements become large enough to create topography or bathymetry within the basin. Both types of model predict subsidence of the hangingwall and uplift of the footwall, with both subsidence and uplift decaying approximately exponentially with distance from the fault. However, the wavelength of the deformation is different for the two models. Dislocation models produce deformation in a volume approximately the same diameter as the fault surface; in flexural models, the wavelength of uplift and subsidence is controlled by the strength of the upper crust. These differences are illustrated in Fig. 8, which shows the predicted base Brent profile around a planar normal fault, for both elastic and flexural models (see figure caption for detailed model parameters). If this fault movement is considered 'syn-Brent', the deformation at base Brent indicates the potential thickness variations in the Brent Group around the fault. The elastic-dislocation model (solid line in Fig. 8) would predict that significant thinning of the Brent Group is restricted to within *c.* 2–3 km of the fault. The flexural-isostatic model (dashed line), however, shows a much more gradual profile, with thickness changes on the footwall being spread over some 20 km.

The important point from Fig. 8 is that, whichever model is more applicable in a particular circumstance, syn-Brent footwall thinning is likely to be difficult to recognize. Wells are heavily concentrated at present structural crests. As discussed in the following section, the present crest of a tilted fault-block is typically a few kilometres from the major bounding fault, due to late Jurassic erosion/degradation of the fault scarp. Therefore, assuming an elastic-dislocation model, the narrow zone of syn-Brent footwall thinning is likely either to have been eroded away in the late Jurassic, or to be severely affected by low-angle slide surfaces (cf. Fig. 10a). Alternatively, assuming a flexural model to apply, thickness changes vary only gradually with distance from the fault, and

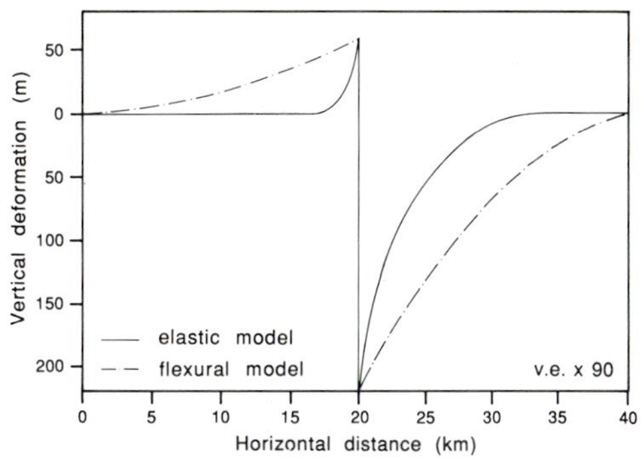


**Fig. 8.** Examples of predicted deformation profiles around the centre of a planar normal fault, 25 km in along-strike length, dipping at 50°. The solid line is a profile calculated using the elastic dislocation model of Gibson *et al.* (1989); the dashed line uses the flexural model of Jackson & McKenzie (1983). The profiles are intended to show the possible deformation of the 'base Brent' surface by syn-Brent fault movement. The Brent sediments are then assumed to infill this basin-floor topography. Both models would predict local thickening of the Brent on the downthrown side of the fault, and thinning on the upthrown side of the fault. However, the wavelength of the deformation is much smaller for the elastic dislocation model (solid line) than for the flexural model (dashed line).

therefore are unlikely to be reliably detected by a number of closely-spaced wells at the crest of the structure, particularly so given the dominance of post-Triassic thermal subsidence as an overprint on fault-related effects. Recognition of the complete thinning and thickening profile is generally only possible where there is good downflank well control (in both directions). Detailed biostratigraphic work is usually required to constrain the timing of any thinning within the Brent and to establish its relationship with adjacent fault movement (e.g. Helland Hansen *et al.* and Mitchener *et al.* this volume).

In conclusion, a number of structural controls on Brent Group deposition can be recognized.

(i) Thermal subsidence (following early Triassic extension) was the fundamental control, in providing a trough into which the Brent delta could prograde. This subsidence was greatest at the graben axis.

(ii) Superimposed on the post-Triassic subsidence, renewed fault activity was beginning to occur, heralding the onset of the late Jurassic extension. Syn-Brent extension on normal faults was only *c.* 1%, but locally this is manifest as a

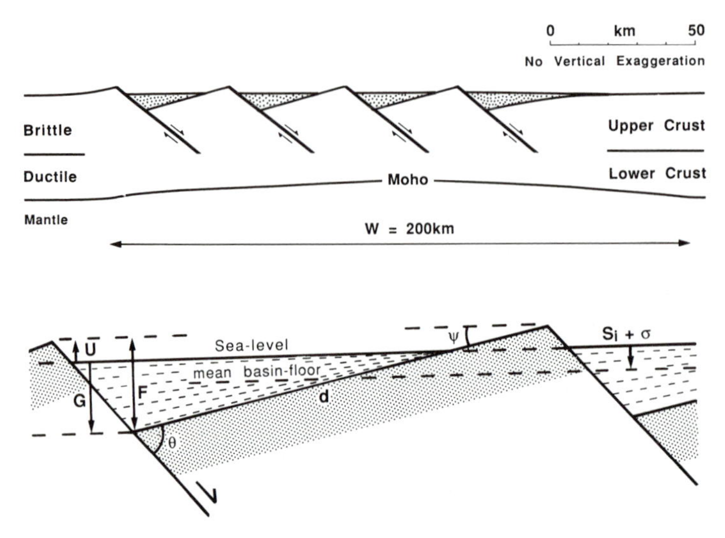


**Fig. 9.** (**a**) Example basin profile produced by the flexural cantilever model of Kusznir *et al.* (1991). Upper-crustal deformation is by simple shear on faults, whereas the lower crust and mantle deform by distributed pure shear. Note the local uplift of the footwalls to the major normal faults. (**b**) Basin geometry for the domino model. Rigid fault-blocks (and their planar bounding faults) rotate as extension proceeds. The fault throw (F) is partitioned into footwall uplift (U) and hangingwall subsidence (G) depending on the amounts of extension and sedimentary loading.

doubling of Brent thickness across a single major fault.

(iii) Differential compaction of the underlying Triassic sediment pile, over the tilted basement fault blocks, can produce thickness variations in the Brent of *c.* 25% across a major fault block.

(iv) Local thinning of the Brent in the footwalls to major faults has possibly occurred, but is difficult to observe because of late Jurassic erosion/degradation at the fault-block crest.

## Late Jurassic faulting and the creation of the Brent-Province traps

The second rift episode affected the north Viking Graben from the Bathonian to the Ryazanian, i.e. from the late Middle Jurassic to the earliest Cretaceous. From detailed stratigraphic modelling (Roberts *et al.* in press) we consider it likely that much of the fault activity was concentrated near the centre of this time span, in the Oxfordian–Kimmeridgian. Certainly the syn-Brent fault movement represents less than one-tenth of the total faulting, as discussed in the previous section.

The late Jurassic extension is characterised by wedge-shaped packages of sediment (Humber Group) that partially infill a tilted-block topography, created by further motion on the major Triassic normal faults (see Fig. 1). On the scale of a seismic reflection section, the upper surface of the tilted blocks is most conveniently taken as the top of the Brent Group, though as we have seen some fault motion had already occurred during Brent Group deposition. The onset of block tilting was not synchronous throughout the basin, occurring as late as Kimmeridgian on the northern Horda Platform (Badley *et al.* 1988). The overlying marine shales of the Heather Formation and Kimmeridge Clay (Draupne) Formation indicate rapidly-increasing water depths, but local uplift and erosion is the norm at footwall crests.

The tilted blocks of the north Viking Graben are the main hydrocarbon traps of the Brent Province (e.g. Spencer & Larsen 1991). Horizon dips within the blocks are typically about 5° and rarely more than 10°. Late Jurassic displacements on the major field-bounding faults are typically up to a few kilometres. However, the trapping geometry did not form simply by tilting followed by transgression of the late Jurassic seas. The 'east flanks' of many of the UKCS fields are notoriously difficult to interpret (both seismically and using wells). To unravel the complexity of these areas requires an understanding of continental extension at the crustal scale. Footwall uplift and half-graben water depths are key points in this discussion.

## *Basin models and footwall uplift*

At the largest scale, the formation of extensional basins can be described by the McKenzie (1978) model, which predicts the vertical movement of the basin floor through time. However, this model, (and its many derivatives), explicitly ignores the presence of upper crustal fault blocks and only addresses the problem of the 'average' basin floor. Kusznir & Egan (1989) and Kusznir *et al.* (1991) have formulated models in which the lower crust and mantle deform by pure shear (as in the McKenzie model), but where the upper crustal deformation is by simple shear on faults. Isostatic loads generated by this faulting (e.g. fault-block topography, sediment fill) are distributed flexurally. The resuting model (termed the 'flexural cantilever' model) is a generalized model of lithosphere extension that can be used to construct basin cross-sections. A schematic example is shown in Fig. 9a. Four major planar faults control the upper crustal deformation. In the lower crust and mantle, an equal amount of extension occurs by pure shear distributed over a region 200 km across. Each normal fault produces uplift in its footwall and subsidence in its hangingwall, and these movements decay with distance from the fault (see basin margins in Fig. 9a). The relative amounts of uplift and subsidence are strongly dependent upon the amount of sediment loading within the half-graben. For blocks bounded on each side by a major normal fault, the uplift at one fault and subsidence at the other cause a net rotation, giving the characteristic tilted-block profile.

Marsden *et al.* (1990) have used the flexural cantilever model to study the development of basin profile I in Fig. 1. They found that modest footwall uplift (a few hundred metres above sea-level) occurred during the Late Jurassic adjacent to the major faults. Late Jurassic sedimentation, however, was unable to keep pace with the rapid hangingwall subsidence, and therefore substantial water depths (*c.* 800 m average) rapidly developed in the half-graben. This is reflected in the thickness of the Humber Group, which typically does not fully infill the tilted-block topography (Figs 1, 3 & 4).

Where the major normal faults are all of similar displacement and are evenly-spaced, the flexural cantilever model can be approximated by a model of rigid-block rotation, the domino model (Barr 1987; Jackson *et al.* 1988). In the domino model, the upper-crustal fault blocks are assumed to be completely rigid, and all blocks and faults must rotate at the same rate as extension proceeds (Fig. 9b). The amount of extension accommodated by this block rotation is balanced at depth by distributed extension in the fashion of McKenzie (1978), so the domino model can be considered as a McKenzie model with upper-crustal fault blocks floating on the top. Amounts of footwall uplift and hangingwall subsidence can be calculated very simply (e.g. Barr 1987; Jackson *et al.* 1988).

The domino model has been applied to fault blocks in the Brent Province by Barr (1987), Yielding (1990), White (1990), and Yielding & Roberts (in press). Predicted uplift agrees with observed depth of erosion across a whole range of structures. Very large fault blocks, such as Snorre and Gullfaks, are typically bounded by large faults and show *c.* 1 km of erosion. Medium-sized fault blocks (e.g. Brent, Statfjord, Ninian) typically show a few hundred metres of erosion. Erosion is minimal on small blocks (e.g. Heather, Thistle), and generally absent when the fault spacing is <7 km; blocks of this size do not uplift when faulting occurs, but immediately subside even at their crest.

The detailed profiles of the field crests are usually one of two types (Fig. 10). Most fields, which experienced up to a few hundred metres of Late Jurassic footwall uplift, show a relatively rounded profile (Fig. 10a). The present structural crest occurs a few kilometres back from the major boundary fault. The reason for this is that the fault scarp, being composed of poorly-consolidated sands and shales, collapsed into the adjacent half-graben where there was deep water (hundreds of metres). Typical examples are Brent (Bowen 1975; Livera & Gdula 1990), Statfjord (Kirk 1980; Roberts *et al.* 1987), and Alwyn North (Johnson & Eyssautier 1987). The scarp degradation occurred by a range of processes, from detachment of large-scale slide blocks to complete disintegration and redeposition of sediment. Repeat Brent sequences, due to stacking of submarine slides, are known from a number of wells. Well correlation in such areas is fraught with difficulty, since it is not always clear how much of the section is in situ.

By contrast, the larger fault-blocks show a relatively flat profile, cf. Fig. 10b. The prime example is Snorre (Hollander 1987; Karlsson 1986), and to a lesser extent Gullfaks (Skarpnes *et al.* 1982) and N34/8 (Alhilali & Damuth 1987). These blocks, bounded by much larger faults, experienced significantly more late Jurassic footwall uplift (as much as 1400 m at Snorre). This uplift raised the upper Triassic rocks of the footwall to sea-level. Presumably these rocks were significantly better cemented and lithified than the much younger Brent Group sediments, and were less prone to col-

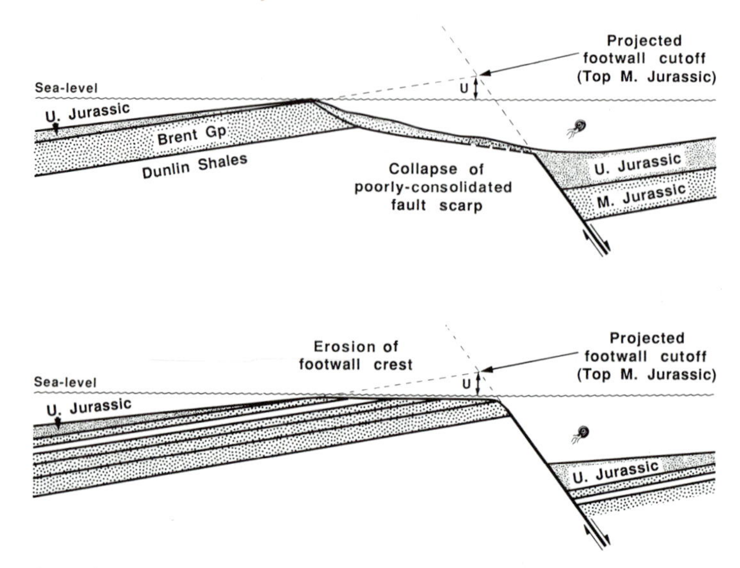


**Fig. 10.** Schematic erosion profiles across Brent Province fault blocks, with different degrees of footwall uplift (U). Profile (**a**) is typical of medium-sized fault-blocks, such as Brent, Ninian, Statfjord, where U is a few hundred metres. Poorly-consolidated Brent and Dunlin Group sediments collapsed into the adjacent deep-water half-graben, producing locally-complex sequences on the east flanks. Profile (**b**) is typical of the largest fault-blocks, e.g. Snorre, where U is about 1 km. Cemented Triassic sediments have been uplifted to sea-level and were less prone to submarine degradation; the bevelled profile suggests erosion at or near sea-level.

lapse into the adjacent half-graben. Instead of submarine degradation, erosion at or near wave-base was the prime process by which the footwall crest was removed, producing a relatively flat-topped structure. Sand eroded from Snorre in this way was redeposited downflank as the Munin member (Gradijan & Wiik 1987).

Fault blocks with a rounded profile like that in Fig. 10a are generally covered by a very thin veneer of Kimmeridge Clay Formation of Volgian/Ryazanian age. From application of a domino model to these fault blocks, it is clear that the footwall crests would have been rising, relative to sea-level, throughout the active rifting period (Barr 1987; Yielding 1990). Therefore, the presence of the condensed marine shale over the crests implies that uplift (and hence fault activity) was largely complete by the Volgian.

These relationships have important implications for the possible role of early diagenesis in the fault-block crests. From the amount and timing of the crestal erosion, we can make reasonable estimates of the uplift rate of a crestal area. For example, a medium-size block such as Brent experienced *c.* 400 m of uplift adjacent to its boundary fault (cf. U in Fig. 10a). If this footwall uplift occurred at a uniform rate between, say, mid-Bathonian and end-Ryazanian (a period of *c.* 34 Ma), that rate would have been *c.* 12 m/Ma, i.e. only about 1 cm per thousand years. Any Brent or Dunlin sediment at the footwall crest would easily have been eroded away faster than it could be uplifted significantly above sea-level. Even if all the fault-related uplift were concentrated in the Kimmeridgian (*c.* 6 Ma), the uplift rate would only be *c.* 7 cm per thousand years. It therefore seems very unlikely that 'footwall islands' would have been created during the faulting, but rather that any uplifted areas would be eroded/degraded faster than they could rise above sea-level. Substantial flushing by meteoric water is therefore not expected to be a significant process.

On the largest fault-blocks, however, the amount of uplift may have been great enough to outpace erosion. For example, if the 1400 m of uplift at Snorre were concentrated in the Kimmeridgian, the uplift rate would have been *c.* 230 m/Ma. Unlike the smaller fault blocks, Snorre and Gullfaks do not have a near-complete seal of Volgian–Ryazanian Kimmeridge Clay, and marine sedimentation was not resumed until the Early Cretaceous. The erosion profile (cf. Fig. 10b) suggests that a footwall island may have been formed, and then subsequently eroded down to sea-level. Meteoric flushing might be recognised below the erosional unconformity.

## Conclusions

Although it is apparently one of the simplest areas of the North Sea, the structural history of the Brent Province has involved processes at a range of scales. Major crustal extension occurred in the early Triassic and again in the late Jurassic. The early extension caused tilting of basement fault blocks, which were then buried under the post-rift thermal-subsidence basin. East–west thickness variations in the Brent relate to variations in the amount of thermal subsidence following the Triassic extension. However, the second (late Jurassic) faulting episode was beginning during Brent deposition, so that fault control is locally important. In terms of basin evolution, however, syn-Brent extension was negligible (*c.* 1%).

The late Jurassic faulting caused further block-tilting and created the main structural traps for the Brent Province oil and gas. Although overall the late Jurassic stretching caused basin subsidence, footwall crests were uplifted and eroded. The amount of erosion on a given fault-block can be predicted using quantitative models. On all but the largest fault-blocks it is likely that footwall uplift rates were far less than likely erosion rates, and so footwall crests would have collapsed into adjacent deep water faster than they could have been uplifted above sea-level. Flushing by meteoric water during the late Jurassic is therefore not expected to be common.

We would like to thank the Basin Modelling and Fault Analysis Groups at Liverpool University for many enlightening discussions on structural processes. We are also grateful to many colleagues throughout the industry who have influenced our ideas on the Brent Province, and to D. Blundell and J. Underhill for their reviews of the original manuscript.

## References

ALHILALI, K. A. & DAMUTH, J. E. 1987. Slide block (?) of Jurassic sandstone and submarine channels in the basal Upper Cretaceous of the Viking Graben: Norwegian North Sea. *Marine and Petroleum Geology*, **4**, 35–48.

BADLEY, M. E., EGEBERG, T. & NIPEN, O. 1984. Development of rift basins illustrated by the structural evolution of the Oseberg feature, Block 30/6, offshore Norway. *Journal of the Geological Society, London*, **41**, 639–49.

——, PRICE, J. D., RAMBECH DAHL, C. & AGDESTEIN, T. 1988. The structural evolution of the northern Viking Graben and its bearing upon extensional modes of basin formation. *Journal of the Geological Society, London*, **145**, 455–72.

BARR, D. 1987. Lithospheric stretching, detached normal faulting and footwall uplift. *In*: COWARD, M. P., DEWEY, J. F. & HANCOCK, P. L. (eds) *Continental Extensional Tectonics*. Geological Society, London, Special Publication, **28**, 75–94.

—— 1991. Subsidence and sedimentation in semi-starved half-graben: a model based on North Sea data. *In*: ROBERTS, A. M., YIELDING, G. & FREEMAN, B. (eds) *The Geometry of Normal Faults*. Geological Society, London, Special Publication, **56**, 17–28.

BERGLUND, L. T., AUGUSTSON, J., FÆRSETH, R., GJELBERG, J. & RAMBERG-MOE, H. 1986. The evolution of the Hammerfest Basin. *In*: SPENCER, A. M. *et al.* (eds) *Habitat of Hydrocarbons on the Norwegian Continental Shelf*. Graham & Trotman, London, 319–338.

BERTRAM, G. T. & MILTON, N. 1989. Reconstructing basin evolution from sedimentary thickness; the importance of palaeobathymetric control, with reference to the North Sea. *Basin Research*, **1**, 247–257.

BOWEN, J. M. 1975. The Brent Oil-field. *In*: WOODLAND, A. W. (ed.) *Petroleum and the Continental Shelf of Northwest Europe*. Applied Science Publishers, London, 353–60.

BROWN, S., RICHARDS, P. C. & THOMPSON, A. R. 1987. Patterns in the deposition of the Brent Group (Middle Jurassic) UK North Sea. *In*: BROOKS, J. & GLENNIE, K. W. (eds) *Petroleum Geology of North West Europe*. Graham & Trotman, London, 899–913.

DONATO, J. A. & TULLY, M. C. 1981. A regional interpretation of North Sea gravity data. *In*: ILLING, L. V. & HOBSON, G. D. (eds) *Petroleum Geology of the Continental Shelf of North-West Europe*. Heyden, London, 65–75.

EYNON, G. 1981. Basin development and sedimentation in the Middle Jurassic of the northern North Sea. *In*: ILLING, L. V. & HOBSON, G. D. (eds) *Petroleum Geology of the Continental Shelf of North West Europe*. Heyden, London, 196–204.

FRANDSEN, N. 1986. *Middle Jurassic deltaic and coastal deposits in the Lulu-1 well of the Danish Central Trough*. Danmarks Geologiske Undersøgelse. SerieA. Nr 9.

FISHER, M. J. 1986. Triassic. *In*: GLENNIE, K. W. (ed.) *Introduction to the Petroleum Geology of the North Sea*, Second edition. Blackwell Scientific Publications, Oxford, 113–132.

GIBSON, J. R., WALSH, J. J. & WATTERSON, J. 1989. Modelling of bed contours and cross-sections adjacent to planar normal faults. *Journal of Structural Geology*, **11**, 317–328.

GILTNER, J. P. 1987. Application of extensional models to the Northern Viking Graben. *Norsk Geologisk Tidsskrift*, **67**, 339–52.

GJELBERG, J., DREYER, T., HØIE, A., TJELLAND, T. & LILLENG, T. 1987. Late Triassic to Mid Jurassic sandbody development on the Barents and Mid-Norwegian shelf. *In*: BROOKS, J. & GLENNIE, K. W. (eds) *Petroleum Geology of North West Europe*. Graham & Trotman, London, 1105–1130.

GRADIJAN, S. J. & WIIK, M. 1987. Statfjord Nord. *In*: SPENCER, A. M. *et al.* (eds) *Geology of the Norwegian Oil and Gas Fields*. Graham & Trotman, London, 341–350.

HELLAND HANSEN, W., ASHTON, M., LOMO, L. & STEEL, R. 1992. Advance and retreat of the Brent delta: recent contributions to the depositional model. *This volume*.

HOLLANDER, N. B. 1987. Snorre. *In*: SPENCER, A. M. *et al.* (eds) *Geology of the Norwegian Oil and Gas Fields*. Graham & Trotman, London, 307–318.

JACKSON, J. & MCKENZIE, D. 1983. The geometrical evolution of normal fault systems. *Journal of Structural Geology*, **5**, 471–482.

——, WHITE, N. J., GARFUNKEL, Z. & ANDERSON, H. 1988. Relations between normal-fault geometry, tilting and vertical motions in extensional terrains, an example from the southern Gulf of Suez. *Journal of Structural Geology*, **10**, 155–70.

JOHNS, C. & ANDREWS, I. J. 1985. The petroleum geology of the Unst Basin, North Sea. *Marine and Petroleum Geology*, **2**, 361–72.

JOHNSON, A. & EYSSAUTIER, M. 1987. Alwyn North Field and its regional geological context. *In*: BROOKS, J. & GLENNIE, K. W. (eds) *Petroleum Geology of North West Europe*. Graham & Trotman, London, 963–977.

KARLSSON, W. 1986. The Snorre, Statfjord and Gullfaks oilfields and the habitat of hydrocarbons on the Tampen Spur, offshore Norway. *In*: SPENCER, A. M. *et al.* (eds) *Habitat of Hydrocarbons on the Norwegian Continental Shelf*. Graham & Trotman, London, 181–98.

KIRK, R. H. 1980. Statfjord Field: a North Sea giant. *In*: HALBOUTY, M. T. (ed.) *Giant Oil and Gas Fields of the Decade: 1968–1978*. Memoir of the American Association of Petroleum Geologists **30**, 95–116.

KLEMPERER, S. 1988. Crustal thinning and nature of extension in the northern North Sea from deep seismic reflection profiling. *Tectonics*, **7**, 803–822.

KUSZNIR, N. J. & EGAN, S. S. 1989. Simple-shear and pure-shear models of extensional sedimentary basin formation: application to the Jeanne d'Arc Basin, Grand Banks of Newfoundland. *In*: TANKARD, A. J. & BALKWILL, H. R. (eds) *Extensional Tectonics and Stratigraphy of the North Atlantic Margins*. American Association of Petroleum Geologists Memoir **46**, 305–322.

——, MARSDEN, G. & EGAN, S. S. 1991. A flexural cantilever simple-shear/pure-shear model of continental lithosphere extension: application to the Jeanne d'Arc Basin, Grand Banks and Viking Graben, North Sea. *In*: ROBERTS, A. M., YIELDING, G. & FREEMAN, B. (eds) *The Geometry of Normal Faults*. Geological Society, London, Special Publication, **56**, 41–60.

LERVIK, K. S., SPENCER, A. M. & WARRINGTON, G. 1989. Outline of the Triassic stratigraphy and structure in the central and northern North Sea. *In*: COLLINSON, J. D. (ed.) *Correlation in Hydrocarbon Exploration*. Graham & Trotman, London, 173–190.

LIVERA, S. E. & GDULA, J. E. 1990. Brent Oil Field. *In*: BEAUMONT, E. A. & FOSTER, N. H. (eds) *Structural Traps II, Traps Associated with Tectonic Faulting*. Atlas of Oil and Gas Fields, American Association of Petroleum Geologists, 21–63.

MARSDEN, G., YIELDING, G., ROBERTS, A. M. & KUSZNIR, N. J. 1990. Application of a flexural cantilever simple-shear/pure-shear model of continental lithosphere extension to the formation of the northern North Sea Basin. *In*: BLUNDELL, D. J. & GIBBS, A. D. (eds) *Tectonic evolution of the North Sea Rifts*. Oxford University Press, Oxford, 240–261.

MCKENZIE, D. 1978. Some remarks on the development of sedimentary basins. *Earth and Planetary Science Letters*, **40**, 25–32.

MITCHENER, B. C., LAWRENCE, D. A., PARTINGTON, M. A., BOWMAN, M. B. J. & GLUYAS, J. 1992. Brent Group: sequence stratigraphy and regional implications. *This volume*.

NELSON, P. H. H. & LAMY, J. M. 1987. The Møre/West Shetland area: a review. *In*: BROOKS, J. & GLENNIE, K. W. (eds) *Petroleum Geology of North West Europe*. Graham & Trotman, London, 775–84.

ROBERTS, A. M., YIELDING, G. & BADLEY, M. E. 1992. Tectonic and bathymetric controls on stratigraphic sequences within evolving half-graben. *In*: WILLIAMS, G. D. & DOBBS, A. (eds) *Tectonics and Seismic Sequence Stratigraphy*. Geological Society, London, Special Publication, In press.

ROBERTS, J. D., MATTHIESON, A. S., & HAMPSON, J. M. 1987. Statfjord. *In*: SPENCER, A. M. *et al.* (eds) *Geology of the Norwegian Oil and Gas Fields*. Graham & Trotman, London, 319–340.

RØE, S-L. & STEEL, R. 1985. Sedimentation, sea-level rise and tectonics at the Triassic-Jurassic boundary (Statfjord Formation), Tampen Spur, Northern North Sea. *Journal of Petroleum Geology*, **8**, 163–186.

SCLATER, J. G. & CHRISTIE, P. A. F. 1980. Continental stretching: an explanation of the post mid-Cretaceous subsidence of the Central North Sea Basin. *Journal of Geophysical Research*, **85**, 3711–39.

SKARPNES, O., BRISEID, E. & MILTON, D. I. 1982. The 34/10 Delta prospect of the Norwegian North Sea: exploration study of an unconformity trap. *In*: HALBOUTY, M. T. (ed.) *The deliberate search for the subtle trap*. American Association of Petroleum Geologists Memoir **32**, 297–216.

SPEKSNIJDER, A. 1987. The structural configuration of Cormorant Block IV in context of the northern Viking Graben structural framework. *Geologie en Mijnbouw*, **65**, 357–79.

SPENCER, A. M. & LARSEN, V. B. 1991. Fault traps in the northern North Sea. *In*: HARDMAN, R. F. P. & BROOKS, J. (eds) *Tectonic events responsible for Britain's oil and gas reserves*. Geological Society, London, Special Publication, **55**, 281–298.

WHITE, N. J. 1990. Does the Uniform Stretching Model work in the North Sea? *In*: BLUNDELL, D.

J. & GIBBS, A. D. (eds) *Tectonic Evolution of the North Sea Rifts*. Oxford University Press, Oxford, 217–239.

YIELDING, G. 1990. Footwall uplift associated with Late Jurassic normal faulting in the northern North Sea. *Journal of the Geological Society*, London, **147**, 219–222.

—— & ROBERTS, A. M. 1992. Footwall uplift during normal faulting — implications for structural geometries in the North Sea. *In*: LARSEN, R. M. *et al.* (eds) Structural and tectonic modelling and its application to petroleum geology. In press.

——, BADLEY, M. E. & FREEMAN, B. 1991. Seismic reflections from normal faults in the northern North Sea. *In*: ROBERTS, A. M., YIELDING, G. & FREEMAN, B. (eds) *The Geometry of Normal Faults*. Geological Society, London, Special Publication, **56**, 79–89.

ZERVOS, F. 1987. A compilation and regional interpretation of the northern North Sea gravity map. *In*: COWARD, M. P. *et al.* (eds). *Continental Extensional Tectonics*. Geological Society, London, Special Publication, **28**, 477–93.

ZIEGLER, P. A. 1988. *Evolution of the Arctic-North Atlantic and the Western Tethys*. American Association of Petroleum Geologists Memoir **43**.

# Brent Group: sequence stratigraphy and regional implications

B. C. MITCHENER, D. A. LAWRENCE, M. A. PARTINGTON, M. B. J. BOWMAN[1] & J. GLUYAS[2]

*BP Exploration, 301 St Vincent Street, Glasgow G2 5DD, UK*

*[1] Present address: BP Exploration, 4/5 Long Walk, Stockley Park, Uxbridge, Middlesex UB11 1BP, UK*

*[2] Present address: BP Exploration, Forusbeen 35, P.O. Box 197, 4033 Forus, Stavanger, Norway*

**Abstract:** Previous published models for the stratigraphic evolution of the Middle Jurassic (Aalenian–Bathonian) Brent Group and its equivalents rely largely upon lithostratigraphy, with few reliable time lines for correlation across the Northern North Sea basin. To rectify this a new, largely well-based, multi-disciplinary regional study was undertaken, combining detailed biostratigraphic and sedimentological data from more than 450 wells across the whole of the Northern North Sea basin.

A basin-wide sequence stratigraphy for the Middle Jurassic has been developed, comprising two tectono-stratigraphic units (J20, Aalenian–Late Bajocian and J30, latest Bajocian–Middle Callovian), subdivided into five sequences: J22 (Aalenian), J24 (Early Bajocian), J26 (Early–Late Bajocian), J32 (latest Bajocian–Late Bathonian) and J34 (latest Bathonian–Middle Callovian). The tectono-stratigraphic units developed in response to large scale (second order) tectonic processes: pre-rift thermal uplift (J20) and the onset of rifting (J30), whilst sequence development was controlled by a combination of changes in basin subsidence, sediment supply and eustatic sea level.

A series of palaeogeographic and isopach maps outline the basin evolution and sedimentary response for each sequence. They describe a progressive basinward shift of the J22 and J24 depocentres in response to the thermal uplift; by the end of J26, sediment supply could not keep pace with increased basin subsidence caused by the crustal extension. J32 and J34 represent the first periods where active rifting can be clearly identified in the North Sea basin, with the development of tilted fault block geometries in the East Shetland Basin and Bruce–Beryl Embayment. Continued extension and rifting in these areas led to the development of an overall retrogradational system which extends through to the latest Jurassic.

The study provides a model for North Sea basin evolution and a predictive sequence stratigraphic framework within which the local lithostratigraphy can be resolved and basin potential evaluated.

The Middle Jurassic Brent Group has formed a major exploration target in the North Sea since the discovery of the giant Brent and Ninian fields in the early seventies (Bowen, this volume). As such, its stratigraphy and sedimentology have been the focus of continual interest and analysis by a large number of geoscientists.

The broad palaeogeographic elements of the Brent Group were first described by Eynon (1981) and Skarpnes *et al.* (1980), who interpreted the succession in terms of a prograding deltaic complex. These, and subsequent refinements (e.g. Brown *et al.* 1987; Graue *et al.* 1987; Cannon *et al.* this volume) have been largely confined to reviewing the succession within a lithostratigraphic framework. Little detail has emerged of how the 'deltaic complex' and its related depositional systems evolved with time. This is largely a consequence of poor biostratigraphic control, particularly in the coastal plain-dominated sections which in places dominate large parts of the Brent succession. As a consequence, workers have relied on conceptual depositional models, developed largely from specific fields (e.g. Budding & Inglin 1981; Johnson & Stewart 1985; Livera & Caline 1990).

Only recently has there been some success in unravelling the time-stratigraphic evolution of what is traditionally viewed as a broadly regressive–transgressive megacycle (Helland-Hansen *et al.* 1989, this volume; Falt *et al.* 1989). To date no publications have succeeded in providing a synthesis which encompasses all of the Northern North Sea Basin. This paper attempts to rectify this, presenting the results of an integrated multi-disciplinary basin analysis carried out by BP Exploration (BP). It addresses both the temporal and spatial evolution of the Middle Jurassic at a basin scale. As such, it forms part of a much larger structural and

*From* MORTON, A. C., HASZELDINE, R. S., GILES, M. R. & BROWN, S. (eds), 1992, *Geology of the Brent Group*. Geological Society Special Publication No. 61, pp. 45–80.

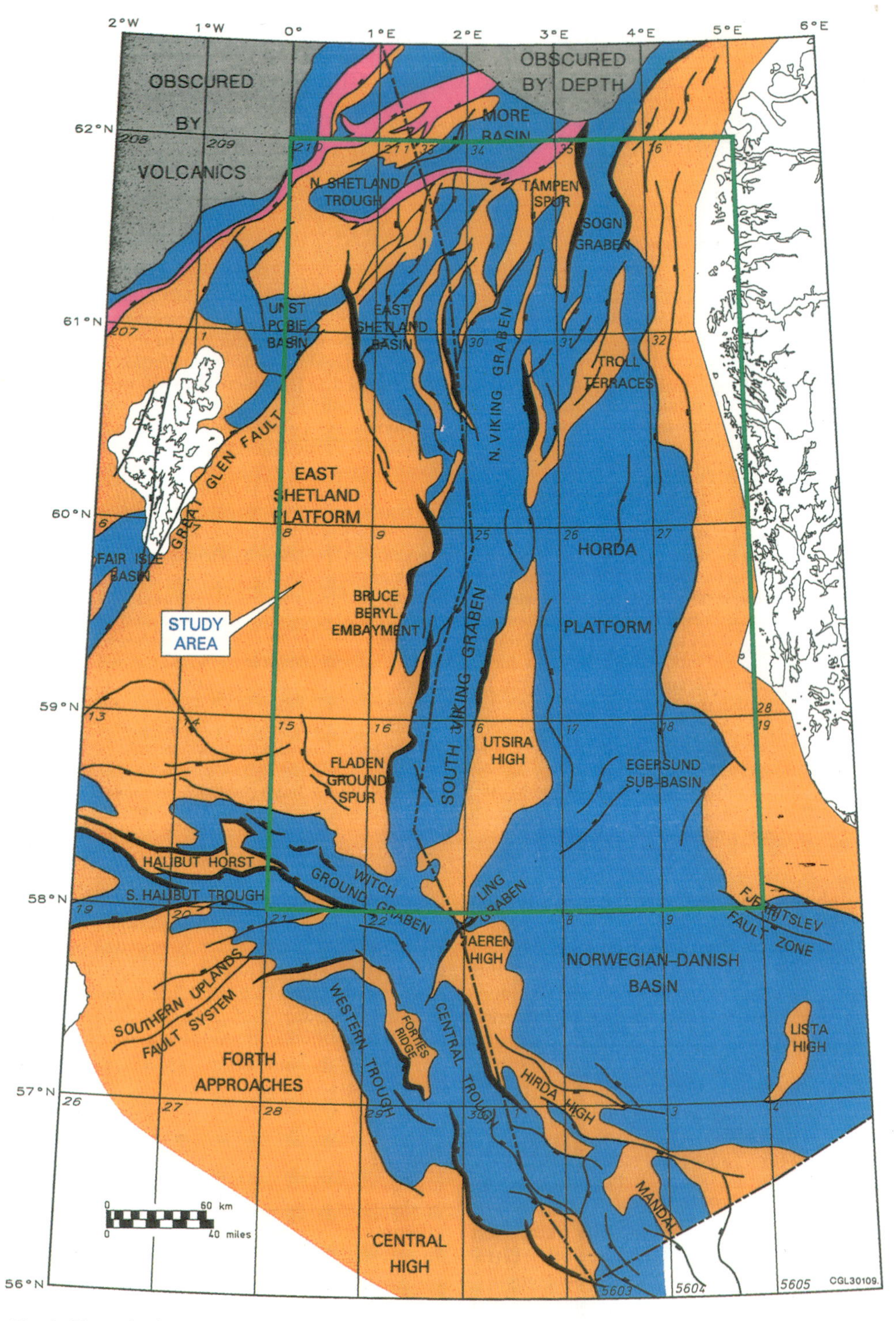


**Fig. 1.** Tectonic elements map of the UK continental shelf, showing location of study area.

stratigraphic analysis of the Jurassic rift system in the Central and Northern North Sea (Rattey & Hayward, in press).

Our approach has been to apply principles of sequence stratigraphy (van Wagoner *et al.* 1988) to unravel the Middle Jurassic palaeogeographic evolution over an area extending northwards from the South Viking Graben, including both UK and Norwegian sectors of the North Sea (Fig. 1). The key to this work has been a refined biostratigraphic framework, based initially on a scheme developed in the Bruce–Beryl embayment (UKCS Quadrant 9) and subsequently extended successfully on a regional scale. Our results clearly demonstrate a strong diachroneity of certain lithostratigraphic units; they further enable division of the succession into two meaningful tectono-stratigraphic units and five sequences. From the geometry and internal heterogeneities within the sequences, it has been possible to comment on the principal controls on their evolution.

## Lithostratigraphic background

The Middle Jurassic Brent Group and the correlative Vestland Group have been subdivided historically into a number of lithostratigraphic units (formations) each of which generally represents a particular gross depositional environment (Deegan & Scull 1977; Vollset & Dore 1984; review in Graue *et al.* 1987). They overlie an older Lower Jurassic (Hettangian to Toarcian) marine mudstone dominated succession (the Dunlin Group) and are succeeded by marine mudstones of the Bathonian to Late Jurassic, Humber Group.

Table 1 summarizes the main lithostratigraphic schemes developed by workers during exploration, appraisal and development of the Middle Jurassic; onshore outcrop schemes are also provided for comparison. From the table, it is clear that different operating petroleum companies have often used their own local lithostratigraphy.

Little has been achieved regarding the provision of a unified scheme, leading to confusion over the definition of variability of facies/facies associations and predictive palaeogeographies.

Within the context of a confused lithostratigraphy, facies associations within the Brent Group and its equivalents have been traditionally considered in terms of a megacycle, recording the northward advance and subsequent retreat of a wave-dominated deltaic complex (Brown *et al.* 1987; Graue *et al.* 1987). Recently, the basal Broom Formation and correlative Oseberg Formation (Norwegian Sector), have been regarded as a separate depositional episode from the main deltaic advance (Graue *et al.* 1987), on the basis of their coarser-grained nature, facies belt orientation (north–south) and provenance, which are all consistent with an early phase of basin-filling, with marginal sediment supply; this contrasts with overlying formations which are considered to have been supplied by major, largely axial, drainage systems. The Rannoch and Etive formations overly the Broom, recording the advance of a weather-dominated shoreline/shoreface system (see Scott, this volume), which forms an overall upward-coarsening unit where well developed (Richards & Brown 1986; Brown & Richards 1989). This is capped by a thick (up to 250 m) coastal plain succession, the Ness Formation, which shows considerable facies variability (e.g. Budding & Inglin 1981; Parry *et al.* 1981; Livera 1989). The final, retreat phase of the megacycle is traditionally considered to comprise shoreline-related and marine sandstones of the Tarbert and Hugin formations, backed by coastal plain associations assigned to the Ness and Sleipner formations.

## Database

The sequence stratigraphic analysis presented here is based largely upon well data. The internal configuration of the Middle Jurassic interval is rarely resolvable on seismic at 2 seconds or more TWT, where the gross interval appears mostly as a package of variably continuous high amplitude reflectors. Seismic data have only been used to constrain stratal and boundary geometries in the thicker parts of the basin (notably the Bruce–Beryl embayment and Viking Graben axis) as well as helping to define the basin outline and limits.

More than 450 wells have been used in this study (Fig. 2). Of these, more than 300 have been subject to detailed biostratigraphic evaluation and more than 3 km of core have been analysed sedimentologically. The database extends over all of the study area, including both UK and Norwegian sectors of the North Sea (Fig. 2). It has been supplemented wherever possible by existing publications together with unpublished production data from the numerous Brent fields, notably those in which BP has or had interests (e.g. Ninian, Thistle, Don, Bruce, Murchison and Statfjord).

## Approach

The principal aims of our analysis have been to develop a time framework for basin-wide strati-

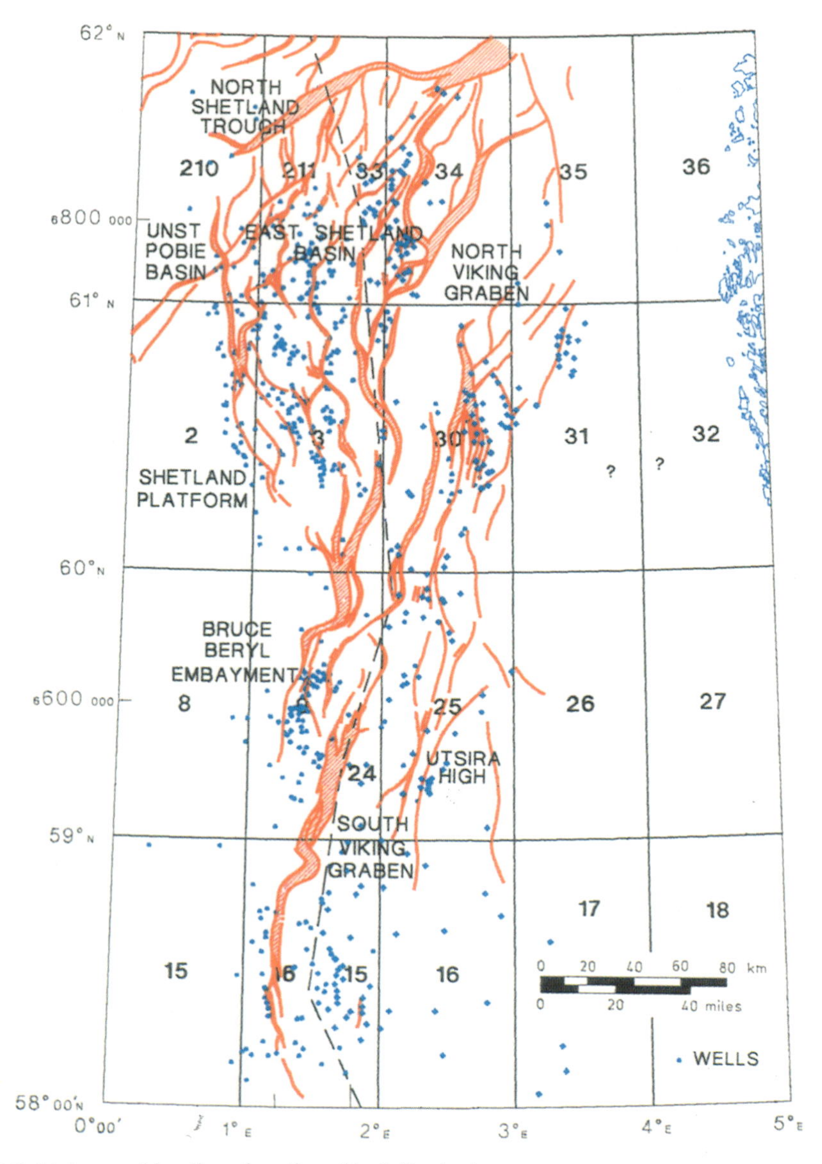


**Fig. 2.** Well database and location of northern North Sea basins.

graphic models in order to enhance understanding and prediction of reservoir character and distribution away from the well data points largely located over structural highs. An essential pre-requisite of any such analysis is a reliable and detailed biozonation scheme.

## *Biostratigraphy*

The relative chronology we have developed is based exclusively on palynological analysis. Through careful, high quality processing of core chips, cuttings and sidewall cores, 16 biozones have been defined across the study area, comprising 26 basinwide correlatable events and 9 subregional events of local correlative value. This is a considerable improvement on the biostratigraphic resolution of previous studies. The biozonation, detailed in Table 2, forms the framework for the sequence analysis. The strength of the scheme is that it is largely independent of depositional facies. This is mainly due to sophisticated proprietary palynological processing techniques (involving slow oxidation with KCl and $HNO_3$ followed by staining); these have allowed good marine dinoflagellate cyst assemblages to be consistently recovered in fluvial-dominated, coastal plain intervals where, in the past, marine taxa are thought to have been absent.

The scheme is based on extinctions and inceptions of marine microplankton. All of the biozones are either total range biozones, acme biozones or concurrent range biozones. Partial range biozones and miospore datums are given subregional event status. Whilst the authors accept that the subregional events based on miospore assemblages are prone to facies variations, they have proved useful in local field correlations where marine microplankton diversity is reduced (e.g. the 'Upper' Ness Formation). Such events provide a useful supplement to the regional biozonation. Palynofacies analysis is also possible within the biozones; this is particularly true for the coastal-plain Ness Formation (see Whitaker *et al.*, this volume).

Detailed discussion of the definition, assemblage characteristics and chronostratigraphic significance of each biozone and bioevents are beyond the scope of this paper and will instead form the basis of a separate publication. Table 2 provides a summary of the key characteristics of each biozone and their constituent events calibrated against chronostratigraphy, our sequence stratigraphy, ammonite zones and absolute age (Haq *et al.* 1987). In exploration wells, it is often possible to resolve chronostratigraphy beyond the conventional macrofossil dated ammonite zones for the Early Callovian, Late Bathonian and Toarcian, by reference to published onshore data (Woollam & Riding 1983) and BP's proprietary field data. The lack of accurately dated Late Bajocian (J26, see later) onshore sections using macrofossils prevent the microplankton biozones from being positively assigned to specific ammonite zones.

## *Sequence stratigraphy*

Combining the biostratigraphy with sedimentological data from core and wireline logs, and incorporating the existing lithostratigraphy has enabled definition of two tectono-stratigraphic units (J20, J30 of Rattey & Hayward, in press). Each defines a major phase of basin development containing consistent and predictable major structural and stratigraphic elements. The two tectono-stratigraphic units have been subdivided into five sequences (J22, J24, J26, J32 and J34). The sequences have been numbered following Rattey & Hayward's stratigraphic scheme (in press); this provides flexibility for additional sequences to be inserted if identified (e.g. J23) in subsequent work. Each sequence is defined according to van Wagoner *et al.* (1988) as 'a relatively conformable succession of genetically related strata bounded at its top and base by unconformities and their correlative conformities'. Definition of lower order sequence stratigraphic elements (systems tracts, flooding surfaces and parasequences) is also contained in van Wagoner *et al.* (1988). It was realized early in our studies that rigid adherence to sequence definitions and concepts without an appreciation of the basin tectonics, could lead to difficulties in identifying genetically and temporally related units of rocks.

In areas of active rifting as exemplified by the Mid- to Late Jurassic of the North Sea Basin, unconformity bound packages, and therefore identifiable sequence boundaries will often be confined to a relatively narrow zone around the basin margins. Here, uplift of the rift shoulders and development of tilted fault block topographies will produce stacked unconformities (sequence boundaries). Elsewhere, (most notably in the rift axis), a sympathetic increase in hanging-wall and general basinal subsidence rates will often lead to transgression and the development of a flooding surface. Without an appreciation of the basin margin uplift and its effects, it would be easy to produce a stratigraphic succession characterized by stacked flooding surfaces without any sequence boundaries, clearly an error!

Another problem posed by the Middle Jurassic succession in the Northern North Sea is that we are dealing with a remnant of an original shallow ramp-type basin without a distinct shelf/slope break and its consequent (i.e. clinoform) geometries. This makes identification of major sequence boundaries less easy, as shifts of facies belts are rarely dramatic, subtle facies changes being the norm. On this basis, we believe that it is highly likely that further analyses will lead to identification of additional sequences which are more subtly manifest in the log and core data.

As discussed by Rattey & Hayward (in press), the interpretations and predictive capability of such studies is the result of a multi-disciplinary, integrated approach to basin analysis. The remaining sections of the paper deal further with the depositional settings recognized from core and log data.

These form the basis of a suite of palaeogeographic maps for each sequence which are described in turn. The paper is concluded with a discussion of the implications of our results for the tectono-stratigraphic evolution of the North Sea basin during the Jurassic.

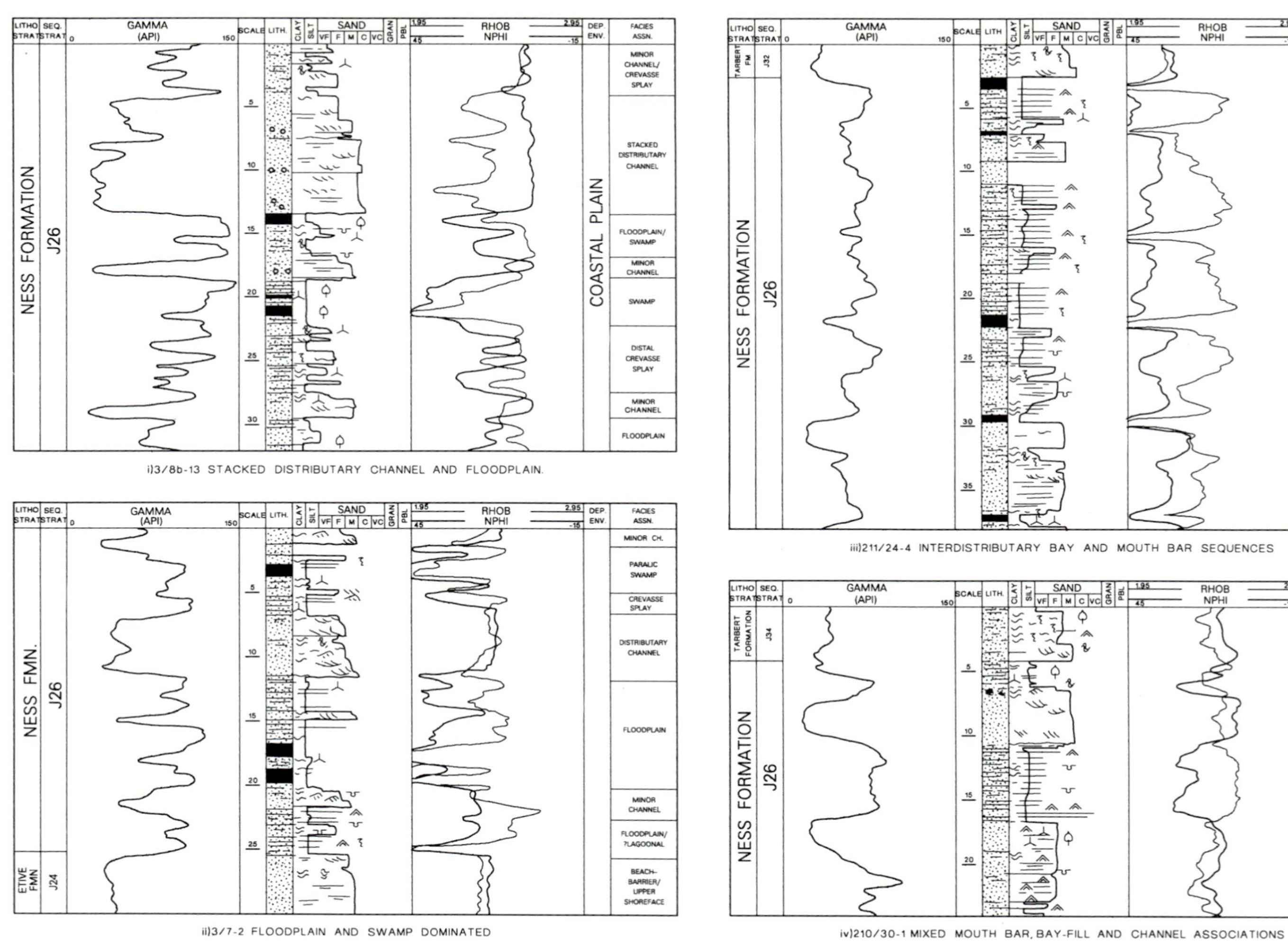


**Fig. 3.** Coastal plain facies associations: summary graphical and wireline logs.

## Sedimentology

A hierarchical approach has been used to characterize depositional setting within sequences. Thirty wells (*c.* 3 km core) have been logged to define the range of variability at the facies level. From these, a suite of facies associations have been defined which are recognizable on wireline logs; each association is typical of a particular sub-environment within a gross depositional setting. The facies association scheme was refined by incorporating proprietary and published core descriptions from a further 150 wells (*c.* 5 km core).

Facies associations are often mappable at field level (e.g. Livera 1989; Marshall *et al.* pers. comm.) as part of this study groups of associations (e.g. non-channel v. channel) have also been mapped to define broad regional trends within sequences. The facies associations are the building blocks for defining gross depositional settings. They are described in detail as an appendix to this paper. The principal facies associations identified within each sequence, are illustrated in Figs 3–5. Mapping out facies associations within the different sequences has formed the basis for recognition of major palaeogeographic basin elements (depositional systems *sensu* Galloway 1989).

## Sequence stratigraphy

The following sections present a summary description of the five sequences J22–J34, drawing on the biostratigraphic, lithostratigraphic and sedimentological framework to provide an overall basin synthesis. Each sequence is discussed in turn, around a suite of maps and cross sections which illustrate the basin's palaeogeography during maximum regressive/progradational phases. Figure 6 illustrates the sequence stratigraphy, biozonation and sedimentological breakdown for two separate well sections from the East Shetland Basin (3/8b–10) and Bruce Beryl embayment (9/9b–3). A similar subdivision has been used for each of the study wells and forms the basis for the sequence paleogeographic maps described in the ensuing sections.

## Structural setting

Detailed description of the pre-Middle Jurassic structural framework of the Northern North Sea is beyond the scope of this paper. Rattey & Hayward (in press) eloquently argue that the structural setting and palaeogeographic configuration at the onset of Middle Jurassic deposition owe much to earlier Devonian and Permo Triassic rifting episodes. Such an interpretation is entirely consistent with the observations of structural grain and principal depocentre for the pre-Aalenian and Aalenian stratigraphy in the study area.

### *Sequence J22* (Fig. 7)

*Age.* Aalenian.

*Lithostratigraphy.* Broom Formation and lower Rannoch Formation (Rannoch Shale, East Shetland Basin); Oseberg Formation (Norwegian North Sea); Bruce 'C' Sands, B-C Coal and Coaly Facies (Bruce–Beryl embayment).

*Boundary conditions.* The base of J22 is defined by a basinward shift of facies belts beyond the early Jurassic shoreline, following an early Jurassic marine transgression which inundated the Triassic rift-basin. This basinward shift was in response to a major relative fall in sea level. It resulted in a significant unconformity around the basin margins with Middle Jurassic strata resting largely on older Jurassic and Triassic sections.

Hinterland rejuvenation associated with the unconformity resulted in a substantial increase in coarse clastic input to the basin margins. In the East Shetland Basin this caused a change from largely anoxic (sediment starved) conditions to more oxygenated waters. Further south, in the Bruce–Beryl embayment, there is an abrupt upward transition from shallow marine muds into coastal plain sedimentation.

A relative rise of sea level at the end of J22 caused a landward shift of facies belts, to a narrow zone around the basin margin, and culminated in the development of a marine flooding surface over the East Shetland and Norwegian marine settings, and a laterally extensive thick (5–10 m) coal (Bruce B-C Coal) across the Bruce Beryl Embayment coastal plain setting.

J22 therefore comprises two distinct units, a lower progradational and an upper transgressive unit. The J22 palaeogeographic map represents the maximum progradation of the thicker lower unit. The later marine transgression was very rapid and resulted in the deposition of a relatively very thin interval.

*Basin forming processes.* The favoured mechanism for the relative drop in sea level and associated hinterland uplift in early J22, is pre-rift thermal uplift in the triple junction area of the Central North Sea. Ziegler (1982, 1988) considers that this thermal event was of sufficient

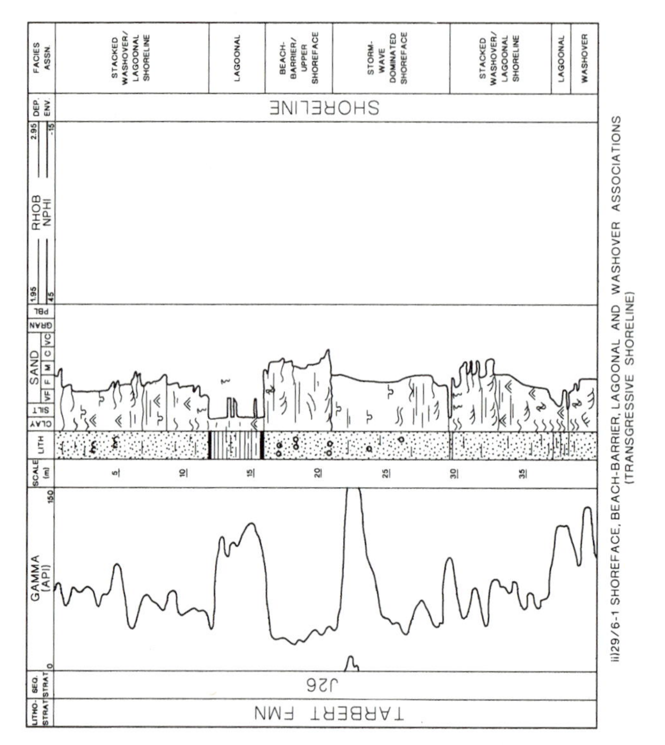


ii)29/6-1 SHOREFACE, BEACH-BARRIER, LAGOONAL AND WASHOVER ASSOCIATIONS (TRANSGRESSIVE SHORELINE)

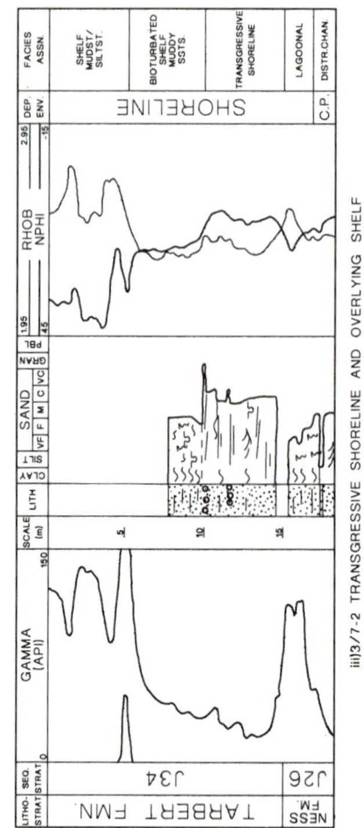


iii)3/7-2 TRANSGRESSIVE SHORELINE AND OVERLYING SHELF

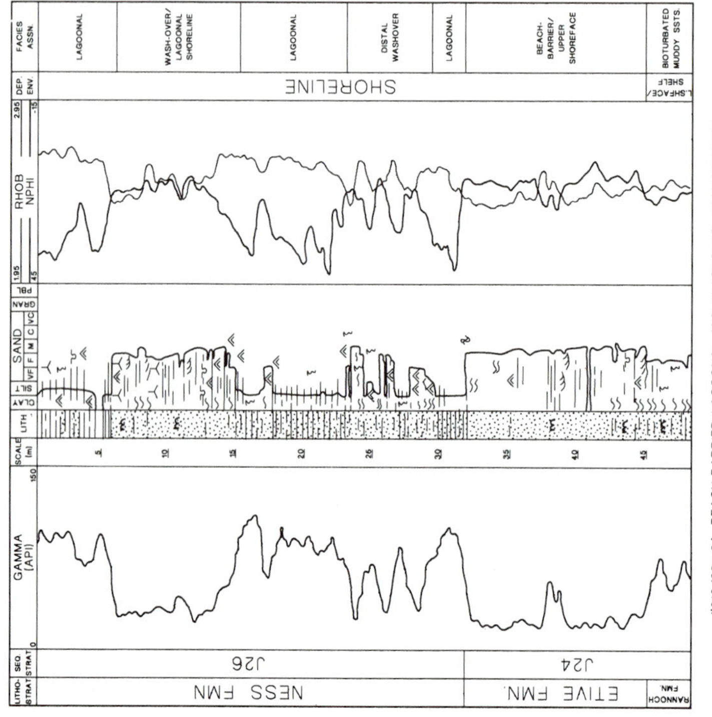


i)210/25c-6A BEACH-BARRIER, LAGOONAL AND WASHOVER ASSOCIATIONS (PROGRADATIONAL SHORELINE)

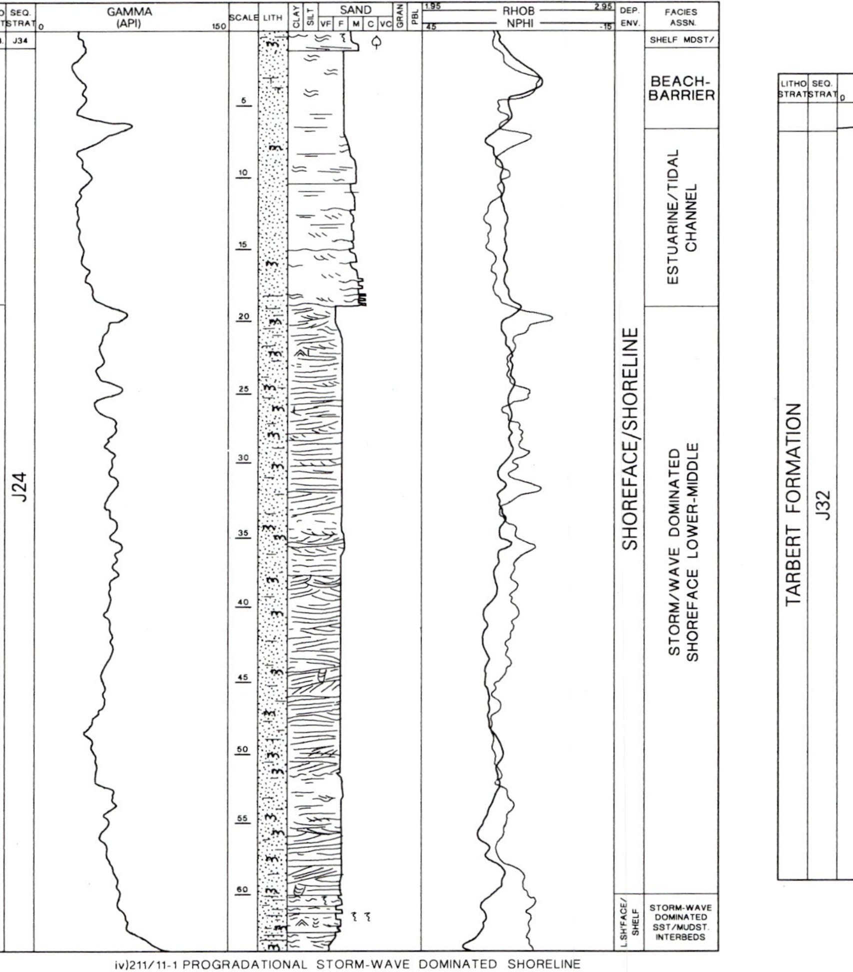

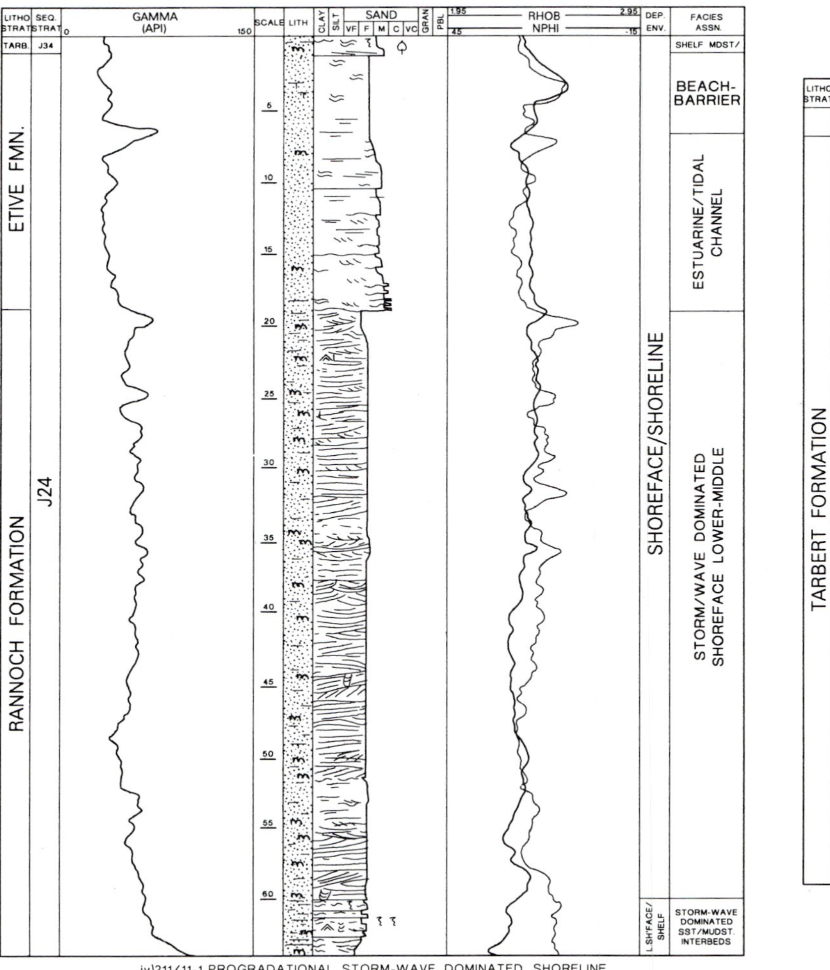


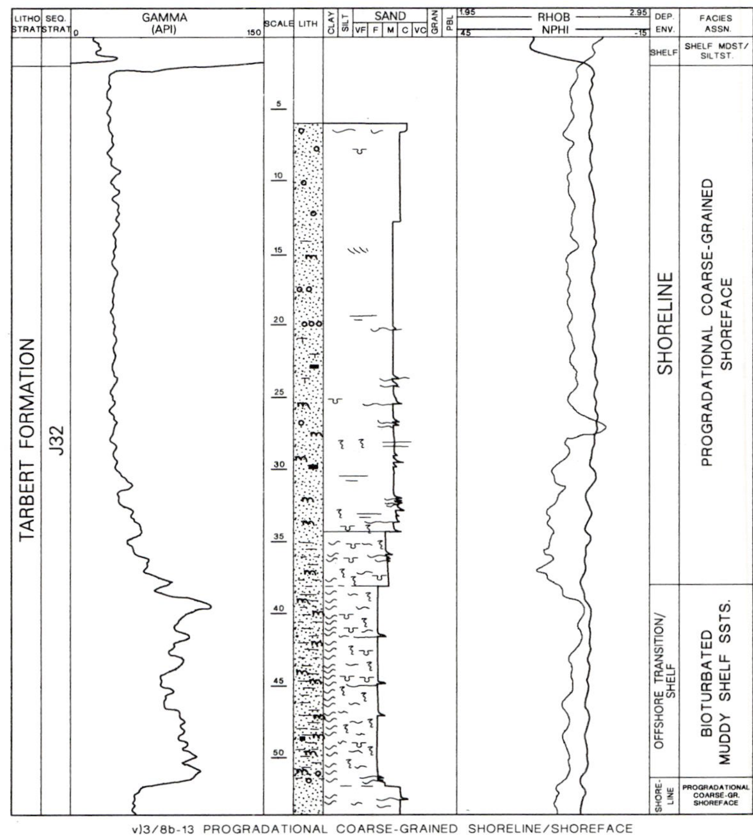


**Fig. 4.** Shoreline facies associations: summary graphical and wireline logs.

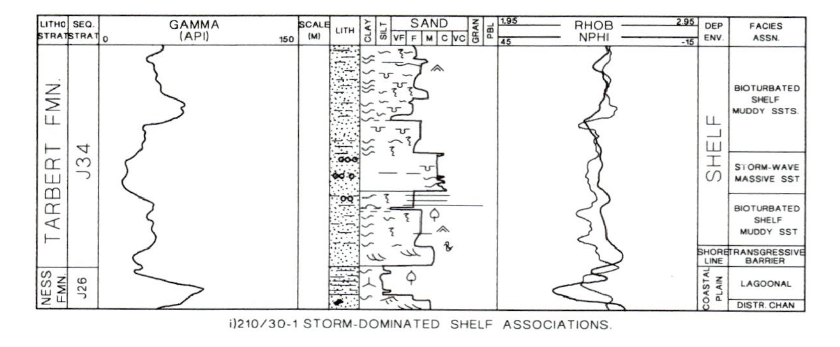

i) 210/30-1 STORM-DOMINATED SHELF ASSOCIATIONS.

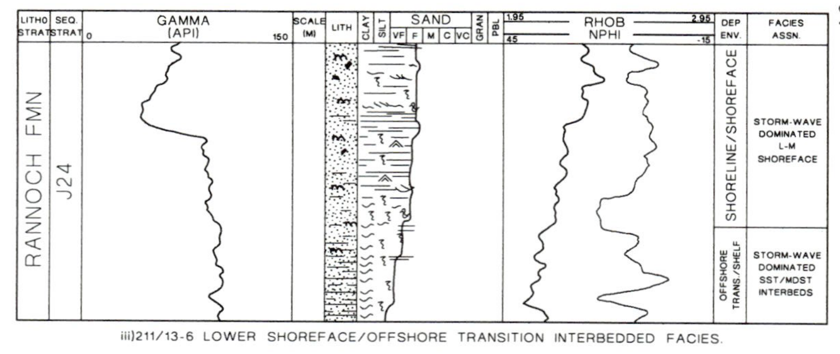

iii) 211/13-6 LOWER SHOREFACE/OFFSHORE TRANSITION INTERBEDDED FACIES.

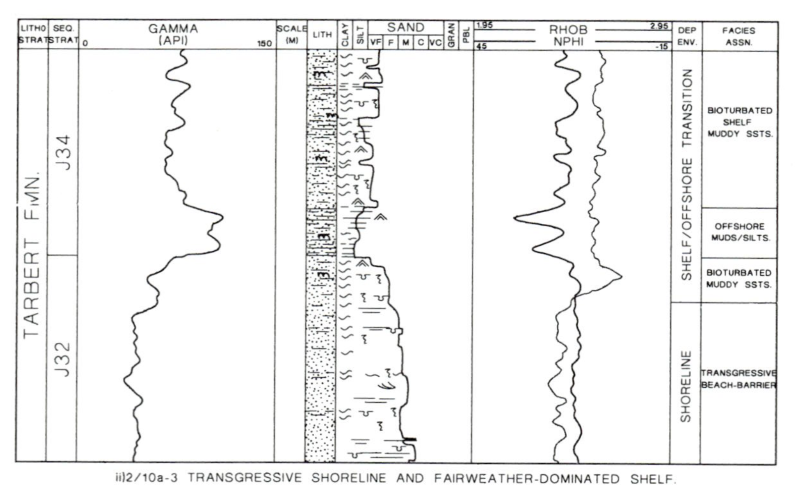

ii) 2/10a-3 TRANSGRESSIVE SHORELINE AND FAIRWEATHER-DOMINATED SHELF.

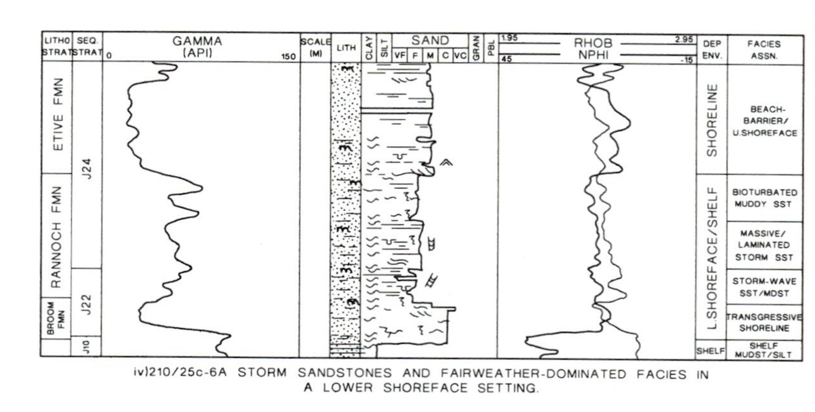

iv) 210/25c-6A STORM SANDSTONES AND FAIRWEATHER-DOMINATED FACIES IN A LOWER SHOREFACE SETTING.

**Fig. 5.** Shelfal plain facies associations: summary graphical and wireline logs.

a

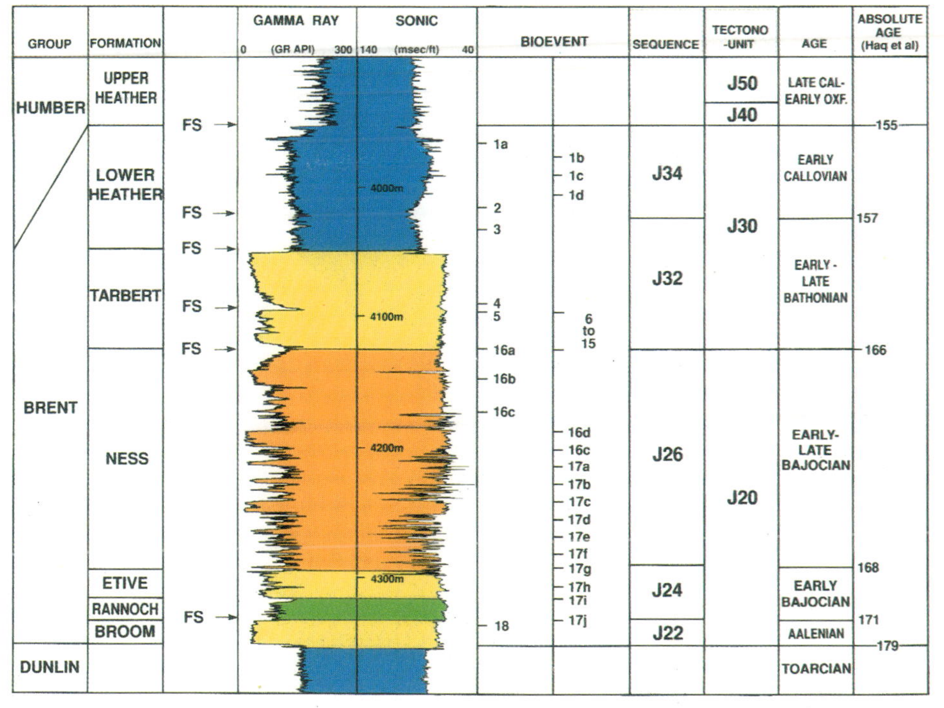


b

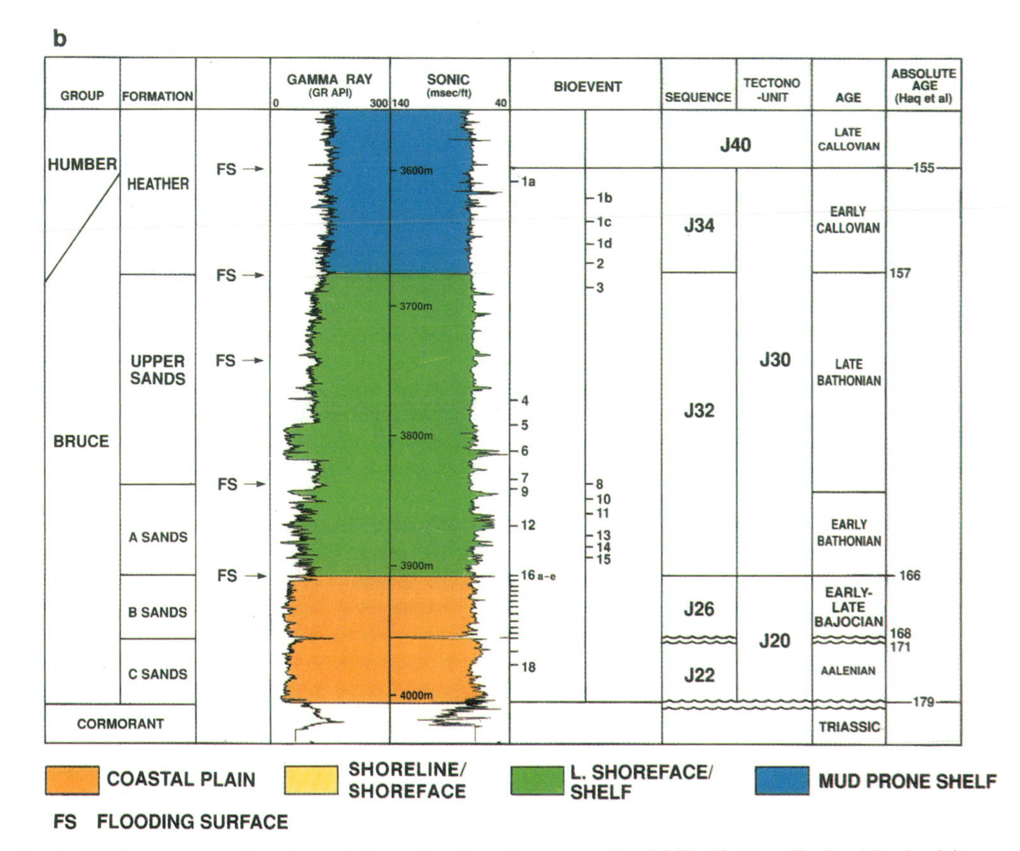


**Fig. 6.** (**a**) Summary stratigraphy and depositional environments, Well 3/8b-10, East Shetland Basin. (**b**) Summary stratigraphy and depositional environments, Well 9/9b-3, Bruce–Beryl Embayment.

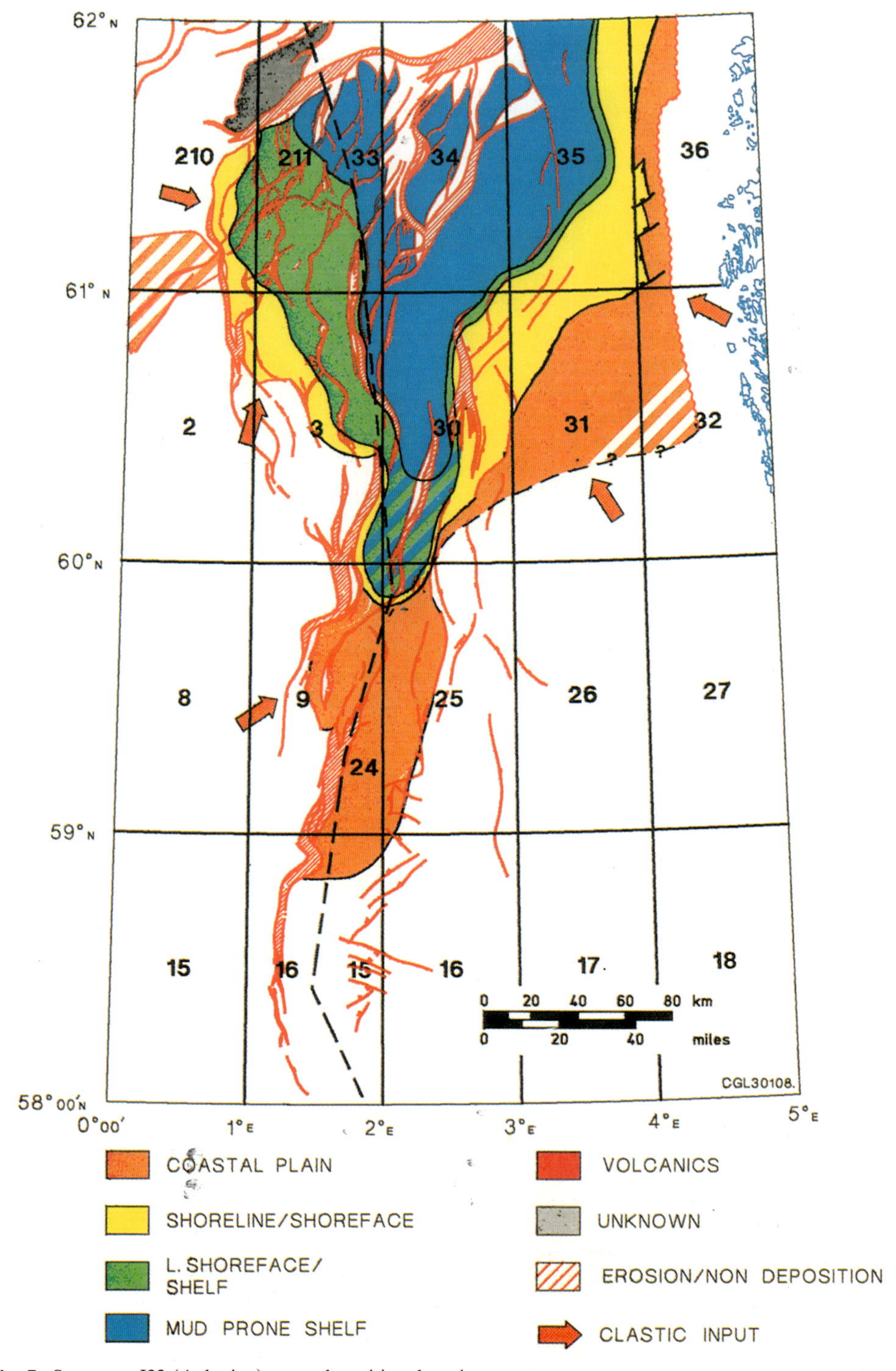


**Fig. 7.** Sequence J22 (Aalenian) gross depositional environments.

magnitude to affect much of the North West European area. It is, however, also possible that the sea level fall and resulting stratigraphy is attributable to eustatic changes, superimposed on a remnant Triassic bathymetry (Rattey & Hayward, in press).

In general the isopach of J22 suggests passive infill of an older remnant 'Triassic' rift topography over most of the basin, with no clearly active intrabasinal faults. This fact together with the apparent rapidity of the marine transgression in late J22 suggest that the cause of the base level rise may be eustatic.

*Sedimentary response*. In early J22, alluvial fans, fan deltas and related depositional systems dominate basin margins, infilling and prograding across pre-existing fault-block topographies (cf. Graue *et al*. 1987). In the Oseberg area (Graue *et al*. 1987), thickness changes in the order of 100 m across fault blocks reflect local, minor extension associated with the basin margin uplift. Outwith marginal areas, there are no major facies changes within the basin.

The rise of base level in Late J22 led to reworking of these marginal sediments and the deposition of transgressive shoreline/shelf sands (Upper Broom) around the basin margin and lower shoreface to shelf muds toward the basin centre (Rannoch Shale) over the Norwegian North Sea and East Shetland Basin areas. The coastal plain was drowned in the Bruce Beryl Embayment, with backstepping alluvial fan and braidplain systems, depositing a laterally extensive coal (Bruce B-C Coal).

### *Sequence J24* (Fig. 8)

*Age*. Early Bajocian.

*Lithostratigraphy*. Rannoch Formation; Etive Formation; lower Ness Formation (East Shetland Basin and Norwegian North Sea);

*Boundary conditions*. J24 commences with a major basinward shift of facies belts beyond the late J22 shoreline, and change in polarity of the main depocentres and facies belts. Biostratigraphic data suggest that J24 is either missing or forms part of a condensed section in the Bruce–Beryl embayment. The base of J24 is conformable in basinal areas where the basal boundary is marked by anoxic mudstones with a high gamma response representing a late J22 marine flooding episode.

A relative rise of sea level in late J24 initially caused aggradation, as sedimentation kept apace with sea level rise, followed by a rapid landward shift of facies belts to a narrow zone around the basin margin culminating in the development of a marine flooding surface (base 'Mid Ness Shale').

J24 comprises a lower progradational unit, a middle aggradational unit and an upper transgressive unit. The J24 palaeogeographic map represents the maximum extent of these progradational and aggradational units. The later marine transgression was very rapid and resulted in the deposition of a relatively very thin interval.

*Basin forming process*. Hinterland rejuvenation associated with continued thermal uplift of the Central North Sea and the development of a proto-triple point, is favoured as the trigger for re-activation of sediment supply to the basin in early J24. However, major eustatic changes in sea level cannot be discounted as at least a contributory factor to the basinal shift of facies belts.

Sequence isopachs (Fig. 8a and b) reflect a largely E to W depositional pattern of facies belts with no clear evidence of intrabasinal fault activity. Local isopach variations are attributable to the final infill of the relict Triassic rift topography. The rise of base level in Late J24 therefore may not result from tectonic subsidence but a combination of reduced clastic supply and more importantly a possible eustatic rise of sea level.

*Sedimentary response*. A storm-wave dominated shoreline system prograded rapidly northwards at the beginning of J24. The northerly limit of this shoreline system is shown in Fig. 8c. Lower/middle shoreface and upper shoreface/beach-barrier sediments are represented by the Rannoch and Etive formations respectively. Trends of isopach thickness, and the distribution of coarser-grained facies in the southern part of UKCS Quadrant 3, suggest a strong element of lateral sediment supply to the basin (cf. Richards *et al*. 1988) from the southern part of the East Shetland Basin (via transfer faults in the former Triassic fault system) and possibly from the Bruce–Beryl embayment. In contrast, J24 thins considerably onto the Horda Platform, in the Norwegian sector.

At its northerly limit (e.g. Don/Thistle area), stacked shoreface sequences are indicative of aggradation with sediment supply keeping pace with relative sea level rise following initial rapid progradation. Here extensive lagoonal and back-barrier associations accumulated (lower Ness Formation below the 'Mid Ness Shale') behind the shoreline.

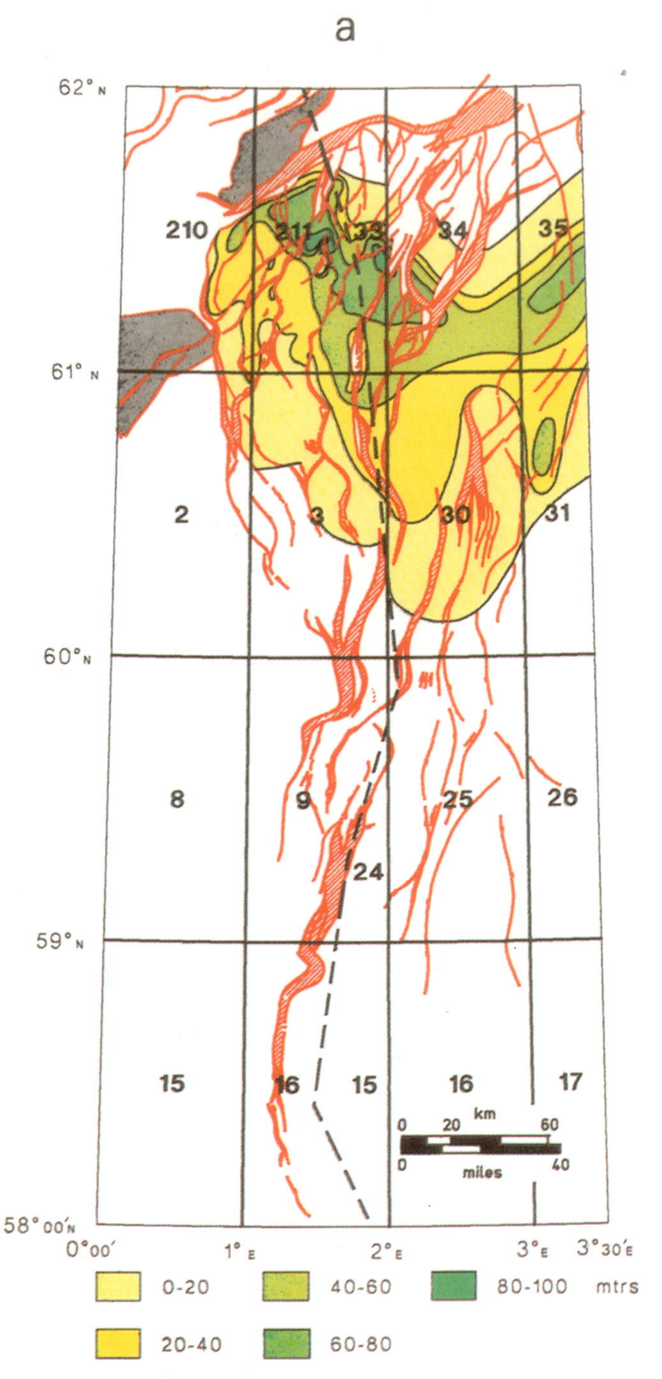


**Fig. 8.** Sequence J24 (Early Bajocian): (**a**) Mid–lower shoreface isopach; (**b**) upper shoreface/foreshore isopach; (**c**) gross depositional environments.

In late J24 transgression ensued as sediment supply was unable to keep pace with the relative rise of sea-level. This produced a major flooding surface which drowned the former lagoonal-barrier island coastline and parts of the back barrier coastal plain. This flooding surface is

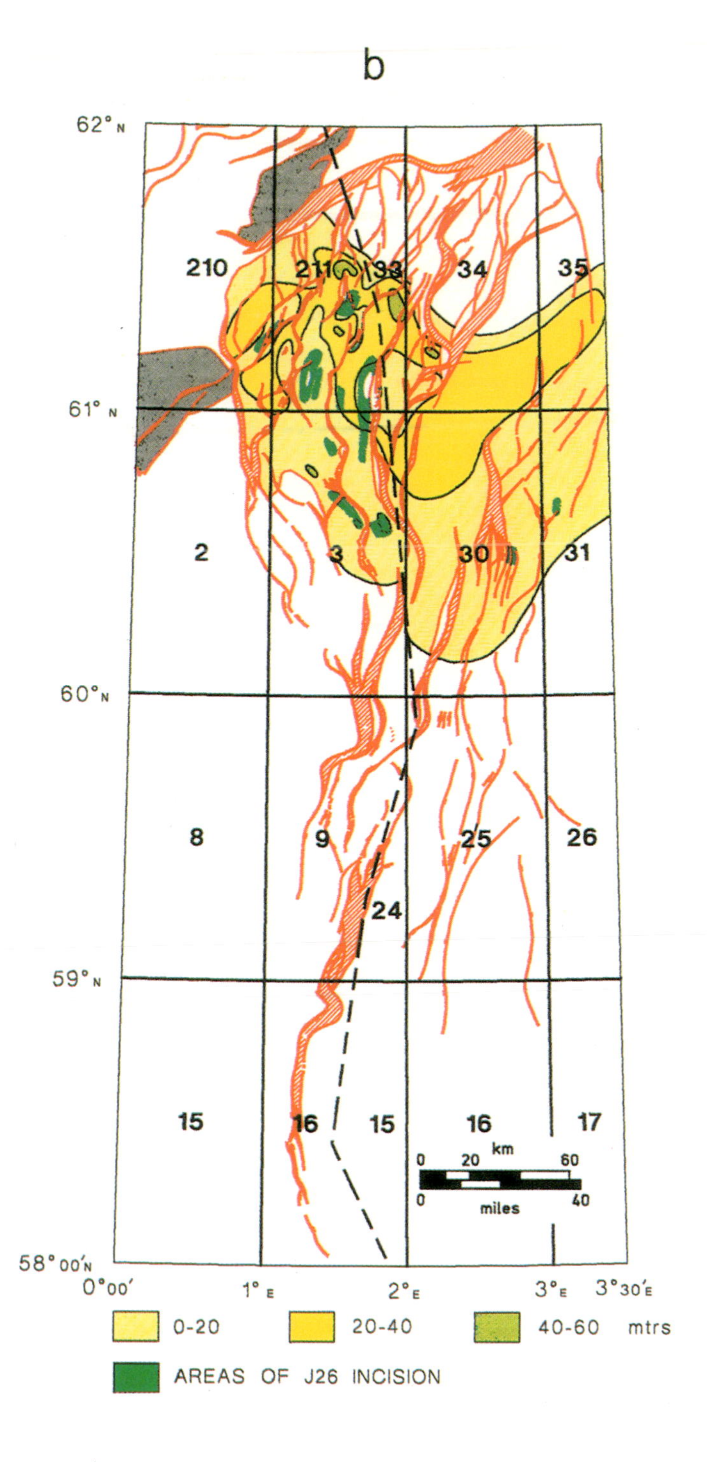


represented by the *base* of the 'Mid Ness Shale'. Little transgressive reworking is evident during this initial drowning of the shoreline system. Whilst the base of the 'Mid Ness Shale' is isochronous, the unit itself extends over *approximately 10 bio-events* in some parts of the basin. It is often eroded or strongly diachronous at its top. It follows that the Mid Ness Shale

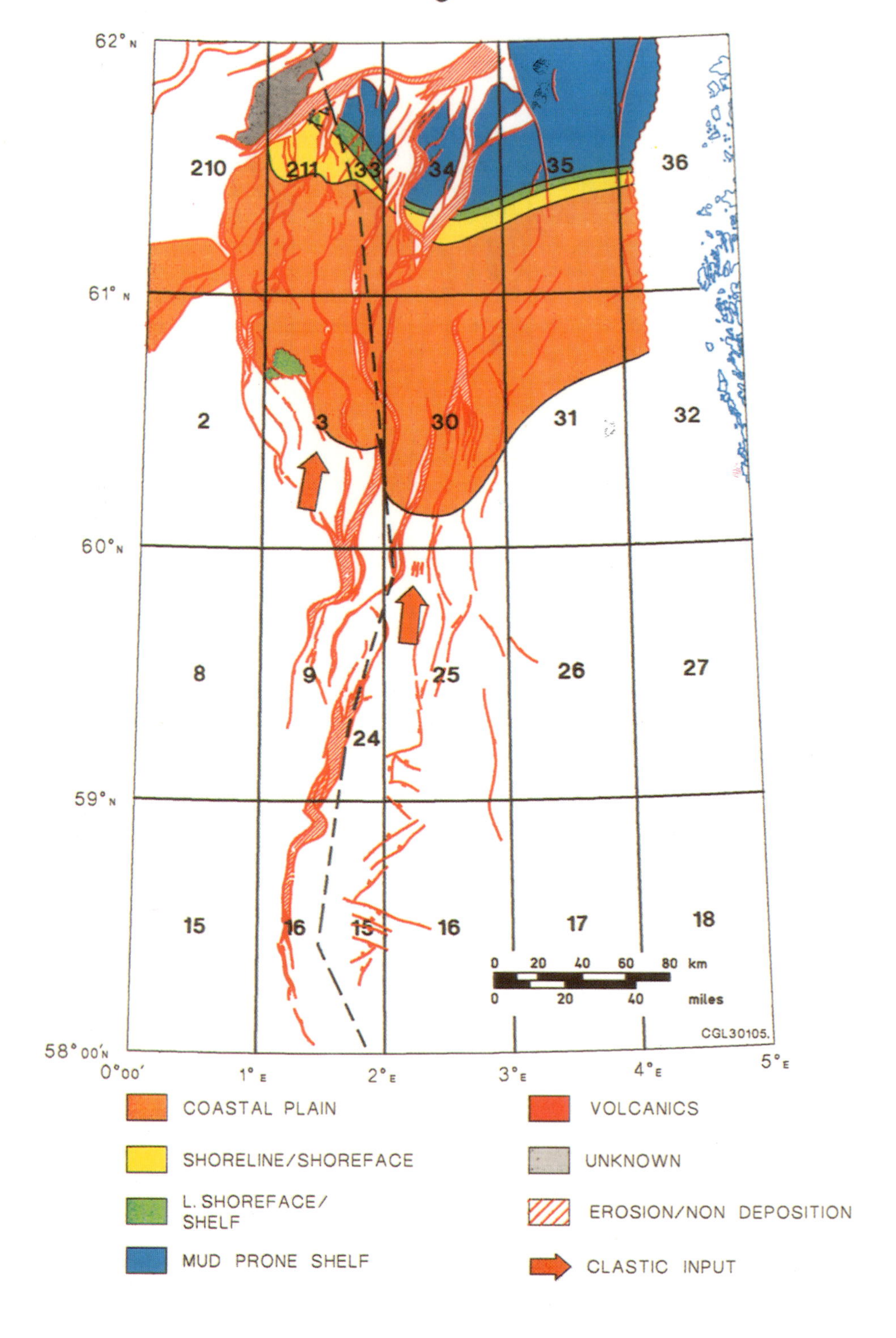
C
62° N
61° N
60° N
59° N
58° 00'N
0°00'
1° E
2° E
3° E
4° E
5° E
210
211
33
34
35
36
2
3
30
31
32
8
9
25
26
27
24
15
16
15
16
17
18
0 20 40 60 80 km
0 20 40 miles
CGL30105.
COASTAL PLAIN
SHORELINE/SHOREFACE
L. SHOREFACE/ SHELF
MUD PRONE SHELF
VOLCANICS
UNKNOWN
EROSION/NON DEPOSITION
CLASTIC INPUT

should not be used as a reliable unit for regional correlation (cf. Budding & Inglin 1981; Richards *et al.* 1988).

### *Sequence J26* (Fig. 9)

*Age.* Early to late Bajocian.

*Lithostratigraphy.* Ness Formation (East Shetland Basin, Norwegian North Sea); Bruce B Sands and Coaly Facies (Bruce–Beryl embayment/South Viking Graben).

*Boundary conditions.* Following the Mid Ness Shale transgression, coarse clastic deposition returned to the Bruce–Beryl embayment and South Viking Graben. The locus of sedimentation shifted rapidly northwards and re-established itself behind the former J24 shoreface/shoreline.

With continued rise of sea level in late J26, the sedimentary system retrograded and finally drowned culminating in a marine flooding episode. The J26 palaeogeographic map represents the maximum progradational limit of the early J26 shore-line. Within the constraints of the biostratigraphic framework, it is not possible to describe the progressive landward shift of the shoreline in late J26.

It is possible to argue that J26 is not a separate sequence, but forms part of the J24 transgressive systems tract. Our current database is inadequate to fully resolve this problem. However, the favoured interpretation is that J26 represents a separate basinal sedimentation episode as there is clear evidence of an increase in the rate of basin subsidence and coarse clastic supply.

*Basin forming processes.* Sequence isopachs (Fig. 9a) show substantial thickness variations across some of the major N–S-trending tectonic elements (e.g. Ninian–Hutton fault trend). There was, however, no associated fault-scarp topography as there is no evidence of erosion or significant facies changes across fault-blocks. Major channel belts are preferentially developed in hanging wall areas but their axes are offset from the faulted margin (Fig. 9b). These data are considered to indicate that the thickness changes reflect differential compaction and subsidence rates but no clear indication of active extensional faulting has been observed.

This differential compaction is accompanied by the resumption of sedimentation in the Bruce Beryl Embayment and South Viking Graben and indicate an increase in rate of basin subsidence. Sedimentation is able however to keep pace with increased basin subsidence, and there is a dramatic increase in the volume of sediment supply to the basin from J24 to J26.

The increased rate of sediment supply is interpreted to represent major thermal uplift of the hinterland, which has also caused stretching of the crust inducing basin subsidence along old Triassic fractures, increasing the available accommodation space in the Northern North Sea basins.

*Sedimentary response.* Following the 'Mid Ness Shale' marine incursion, coastal plain sedimentation resumed with a rapid shore-line advance to a position very similar to J24 (Fig. 9c). Differential subsidence along N–S fault trends results in the marked thickness changes discussed earlier, with coastal plain facies varying from 20–250 m (East Shetland Basin).

Sediment supply in lower J26 can be related to two principal point sources, in the south of UKCS Quadrant 3, and in the Bruce–Beryl embayment, both transfers in the old Triassic fault system. In both areas the proportion of channel and related facies increases to 80–90% in hangingwall settings. In contrast, whilst the thickness of J26 coastal plain facies association increases into the axis of the Viking Graben, the proportion of channel-related facies decreases, suggesting that the major channel belts were preferentially located in fault terraces on the west flank of the basin. Around the Bruce–Beryl embayment J26 is characterized by the development of a linear belt of alluvial fans which attest to significant topography along that basin margin.

The upper part of J26 is distinguished by an increasing frequency of minor marine flooding events throughout the section. Whilst the onset of retrogradation cannot be accurately timed, there is clear evidence in the northern part of the East Shetland Basin for the breakdown and repeated southward shift of the shoreline system. Here the upper part of J26 is marked by a prominant ravinement surface overlain by a transgressive lag. As the rate of clastic supply was gradually out-paced by tectonic subsidence, the coastal plain was finally drowned and the coarser-grained deposits related to the Late J26 shoreline are largely confined to the basin margins.

### *Sequence J32* (Fig. 10)

*Age.* Latest Bajocian–late Bathonian

*Lithostratigraphy.* Tarbert Formation and lower Heather Formation (East Shetland Basin, Norwegian North Sea); Bruce A and Upper

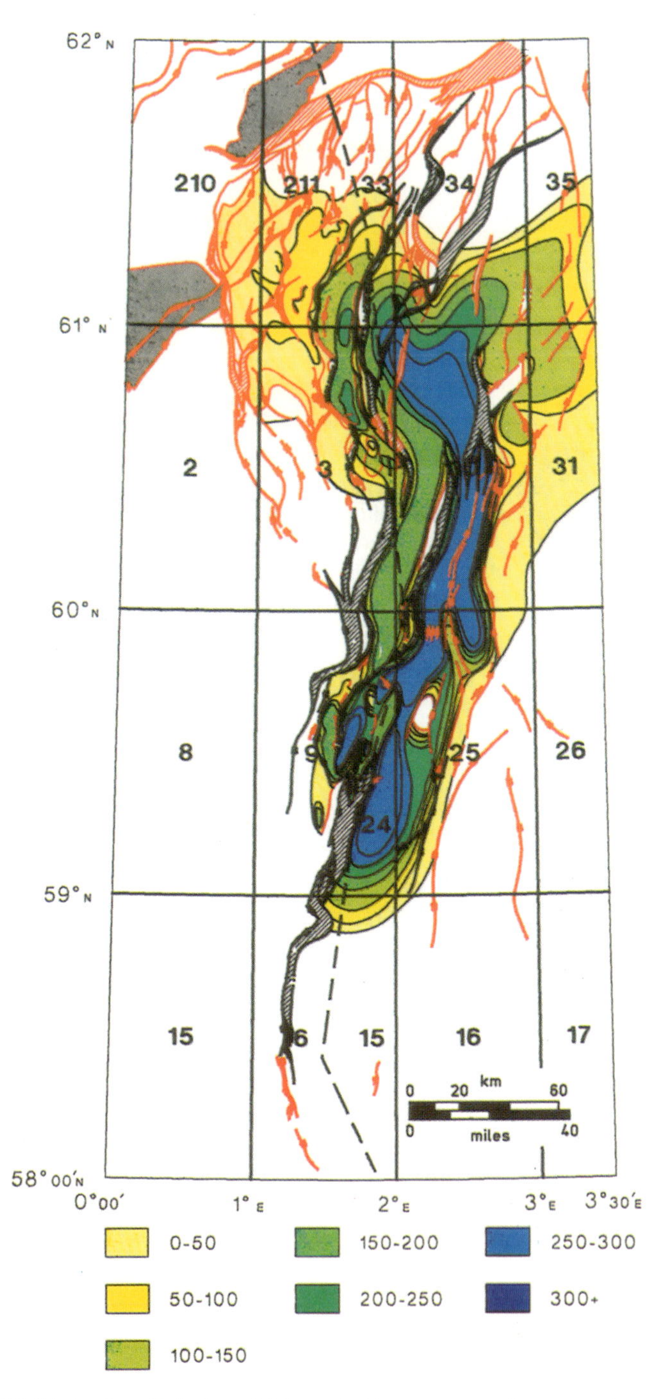


**Fig. 9.** Sequence J26 (Early–Late Bajocian): (**a**) isopach; (**b**) percentage channel sandstones; (**c**) gross depositional environments.

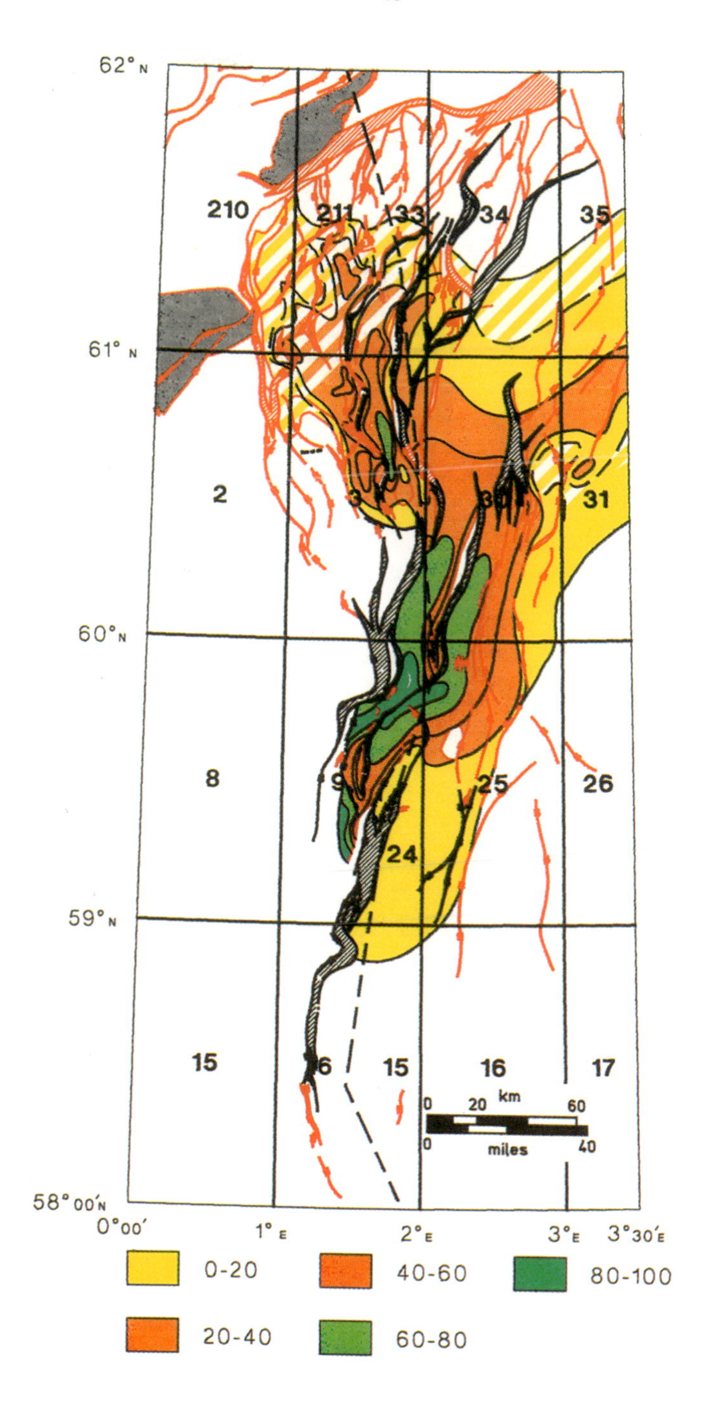
b
62° N
61° N
60° N
59° N
58° 00′ N
0° 00′
1° E
2° E
3° E
3° 30′ E
210
211
33
34
35
2
3
30
31
8
9
25
26
24
15
16
15
16
17
km
0
20
60
miles
0
40
0-20
40-60
80-100
20-40
60-80

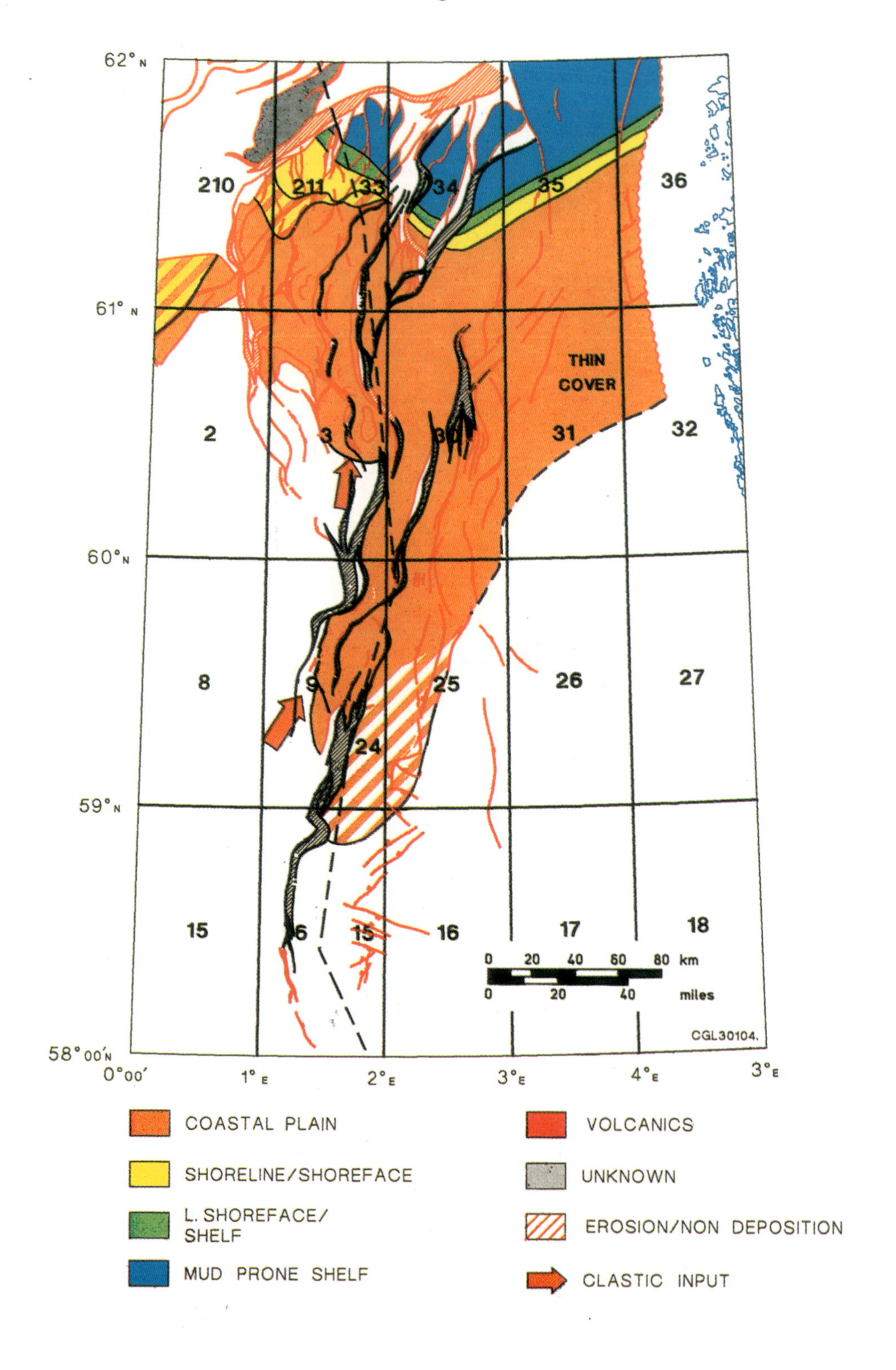
C
62°N
61°N
60°N
59°N
58°00'N
0°00'
1°E
2°E
3°E
4°E
3°E
210
211
33
34
35
36
THIN COVER
2
3
30
31
32
8
9
25
26
27
24
15
6
15
16
17
18
0 20 40 60 80 km
0 20 40 miles
CGL30104.
COASTAL PLAIN
SHORELINE/SHOREFACE
L. SHOREFACE/ SHELF
MUD PRONE SHELF
VOLCANICS
UNKNOWN
EROSION/NON DEPOSITION
CLASTIC INPUT

Sands and the Massive sandstone units (Bruce–Beryl embayment); Sleipner Formation and Rattray Volcanics (South Viking Graben);

*Boundary conditions*. J32 provides the first evidence of active rifting within the basin with the development of intrabasinal tilted fault blocks and uplift along the western footwall margins (Ninian–Hutton fault trend) of the major East Shetland Basin half graben. The characteristics of the sequence boundary varies according to tectonic setting. Clear evidence of an unconformity is confined to areas of uplift around rift shoulders and footwall crests.

In the East Shetland Basin, biostratigraphic data indicate that an area of non-deposition and sediment bypass lay to the west of the Ninian–Hutton fault trend (Figs 10b and 11). Erosion and emergence is only demonstrable in isolated wells at footwall crests, whilst stacked shoreline sequences accumulated on the hangingwall. By contrast, the Viking Graben axis was subject to increased subsidence and a local rise in base level. This led to the development of a major flooding surface which within the limits of the biostratigraphy is apparently coeval with the marginal uplift.

Further south, in the Bruce–Beryl embayment, the sequence boundary is seismically resolvable, with downlap and onlap of seismic events onto a rift topography. The western boundary fault to the embayment delineates the J32 depositional basin, being an area of stacked alluvial fan deltas.

It is clear from our analysis of J32, that a feature of sequence boundary development during active rifting will be coeval uplift in marginal areas and increased subsidence within the axial zone. This means that clear evidence of an unconformity bounded unit (sequence *sensu* van Wagoner *et al*. 1988) will be confined to a narrow zone around the rift margins and at footwall crests. Elsewhere, the boundary will appear as a flooding surface with a likely landward shift in facies belts along the rift axis; without evidence from the basin margins, it would be easy to miss the sequence boundary.

Major rift subsidence in earliest J34 led to further landward shift of the shoreline and the development of a marine flooding surface in hanging walls and along the rift axis. Unconformities developed over footwall crests and around the basin margin.

J32 comprises at least three retrogradational parasequences and the J32 palaeogeographic map represents the maximum progradation of the earliest parasequence. Over the East Shetland basin the map does however document the progressive landward shift of the shoreline from the east to the west of the Ninian–Hutton fault trend. The final drowning of J32 along hanging walls and the rift axis was very rapid and led to the deposition of a relatively very thin interval.

*Basin forming processes*. There are a number of lines of evidence to support the suggestion that J32 records the first indication of active rifting caused by breakup of the Central North Sea thermal dome:

(a) facies belts switch polarity from broadly E–W to N–S in response to the development of tilted fault block topography;
(b) resolvable time gaps with associated unconformities occur over footwall crests and along the rift shoulders (Fig. 11);
(c) major isopach thickening, with development of fan deltas in hanging walls of the major structural lineaments (Fig. 10);
(d) the extrusion of basaltic magmas (Rattray Volcanics) associated with crustal rupture at the triple point;
(e) seismic downlap and onlap of rift topography in the Bruce–Beryl embayment.

From this evidence, J32 is taken to mark the base of the Syn-rift mega-sequence in the Northern North Sea. Further south in the Moray Firth and Central Graben, J32 is characterized by a landward shift in coastal onlap (Rattey & Hayward, in press). In these areas, onset of active rifting appears to be later than the Northern North Sea.

*Sedimentary response*. The pattern of sedimentation within J32 is a result of a complex interplay between early rifting (local uplift and hinterland rejuvenation), and increased subsidence in axial areas which initiated transgression. The facies and systems tract development are described separately for each of the main geographic elements:

*(i) East Shetland Basin*. Where preserved in hanging wall areas, J32 comprises at least three retrogradational parasequence sets (Figs 11 and 12). In response to the early rifting, shorelines switched polarity and shifted into the hanging wall of the Ninian–Hutton fault trend. Facies associations point to a series of coalesced fan-deltas, fed from the west via transfers in the Ninian–Hutton fault trend. Textural and mineralogical data indicate that these J32 shoreline sandstones are distinct from the underlying J26 coastal plain section. Petrographically, they are more comparable with the Broom Formation (Morton, this volume). It is considered that these coarse-grained shoreline systems are more likely to have been derived from a hinterland to the west and south of the East Shetland

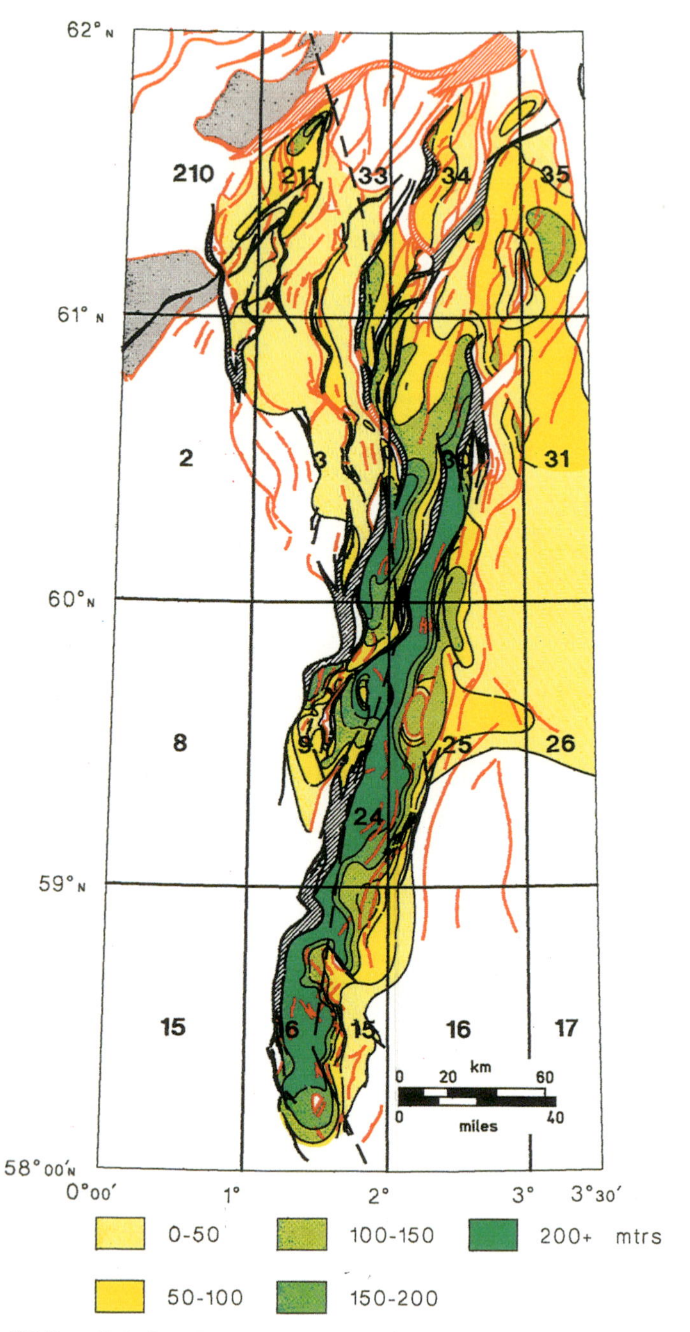


**Fig. 10.** Sequence J32 (Late Bajocian–Late Bathonian): (**a**) J30 isopach; (**b**) J32 gross depositional environments.

Basin via the transitional shelf (a transfer system to the former Triassic graben). The shoreline facies pass southward, parallel to the Ninian–Hutton trend into lagoonal/washover and back barrier, coastal plain facies (e.g. Hild Field, Ronning & Steel 1987).

The Early J32 shoreline system gradually drowned and the sea spread over the northern part of the East Shetland Basin and west of the Ninian–Hutton fault trend (Figs 13 and 14); a

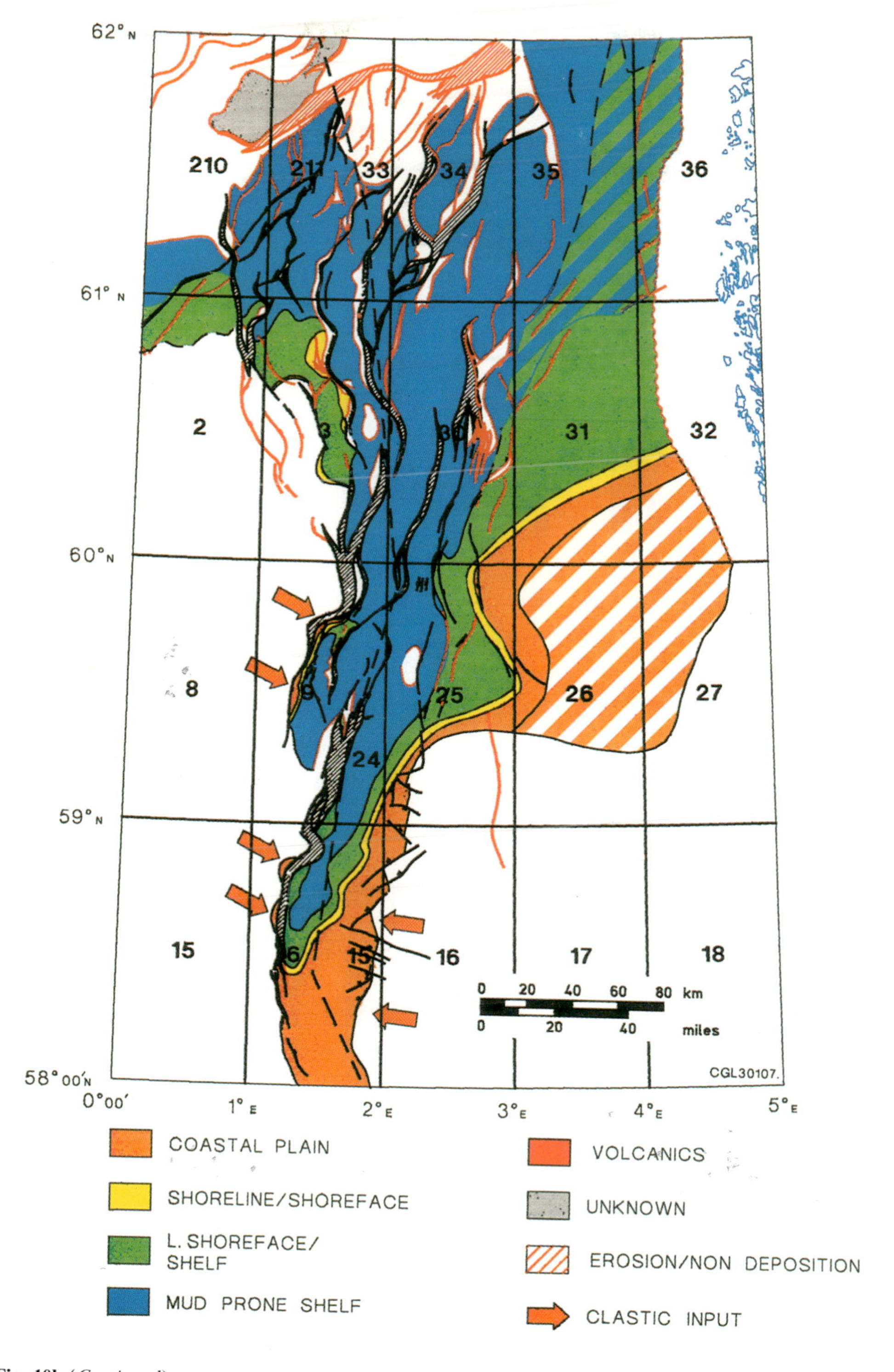


**Fig. 10b** (*Continued*)

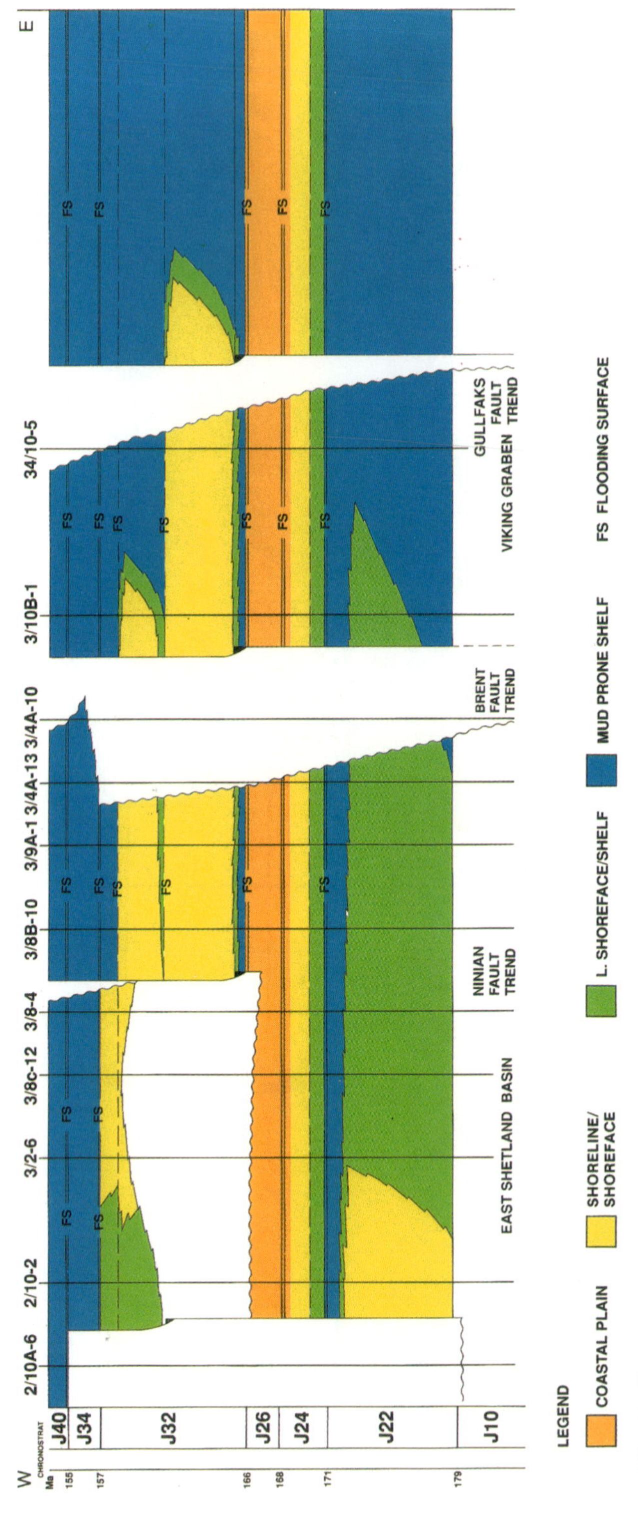


**Fig. 11.** East–west chronostratigraphic summary, Brent Province.

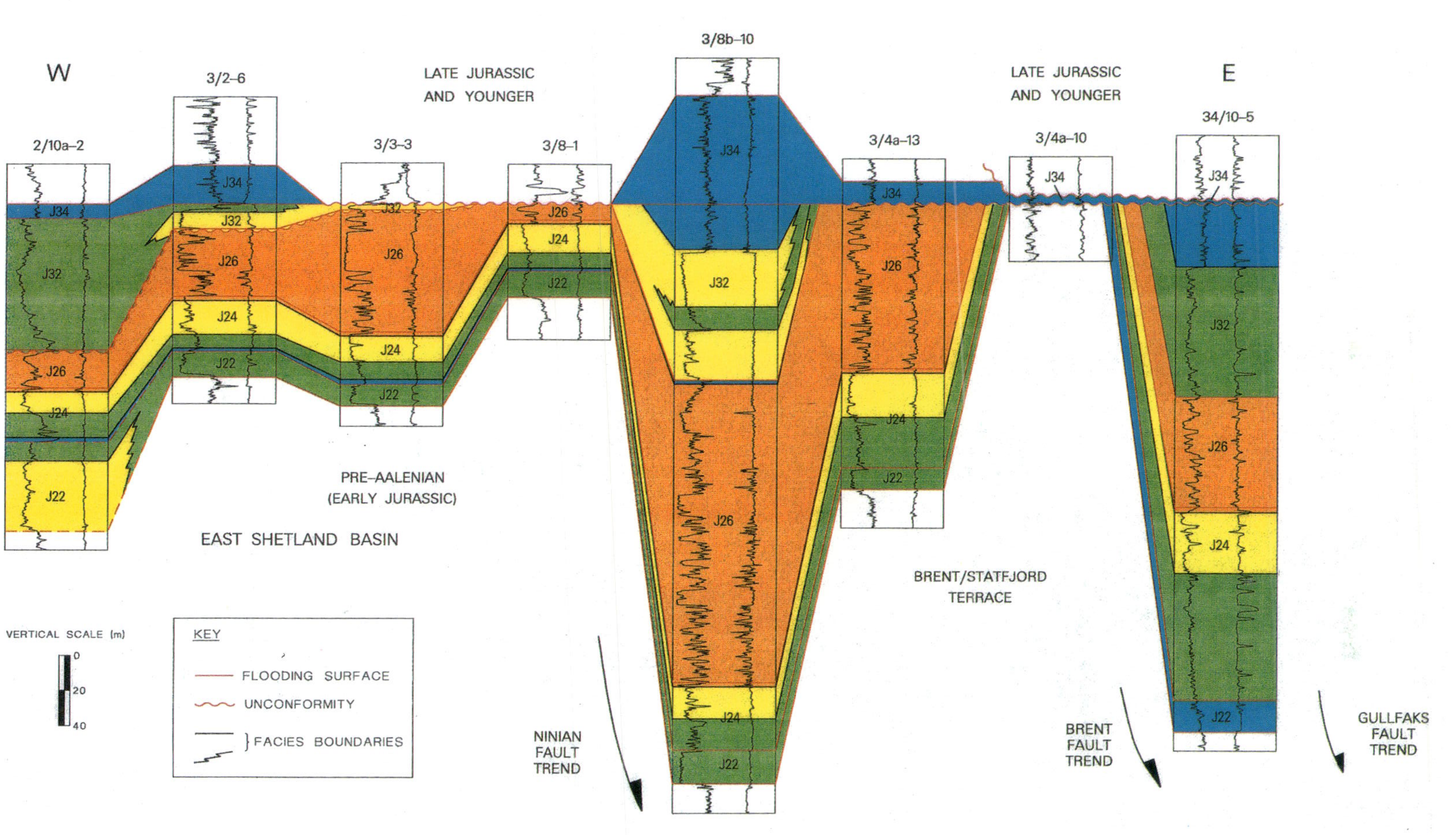


**Fig. 12.** East–west summary well correlation panel, Brent Province, (see Fig. 7 for key to gross depositional environments). Well spacing not to scale; approximate length of panel is 150 km.

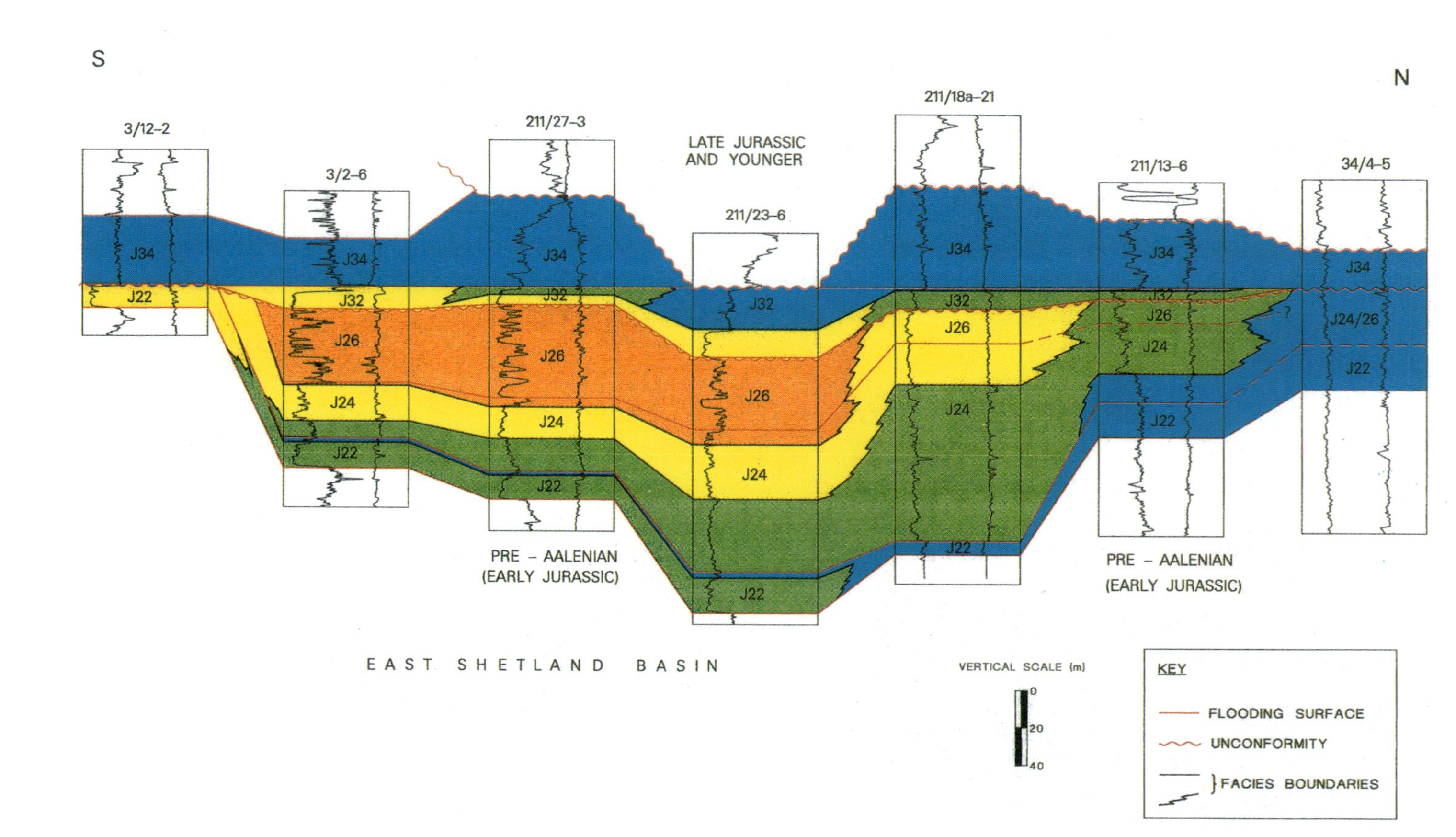


**Fig. 13.** North–South summary well correlation panel, Brent Province (see Fig. 7 for key to gross depositional environments). Length of panel is approximately 130 km.

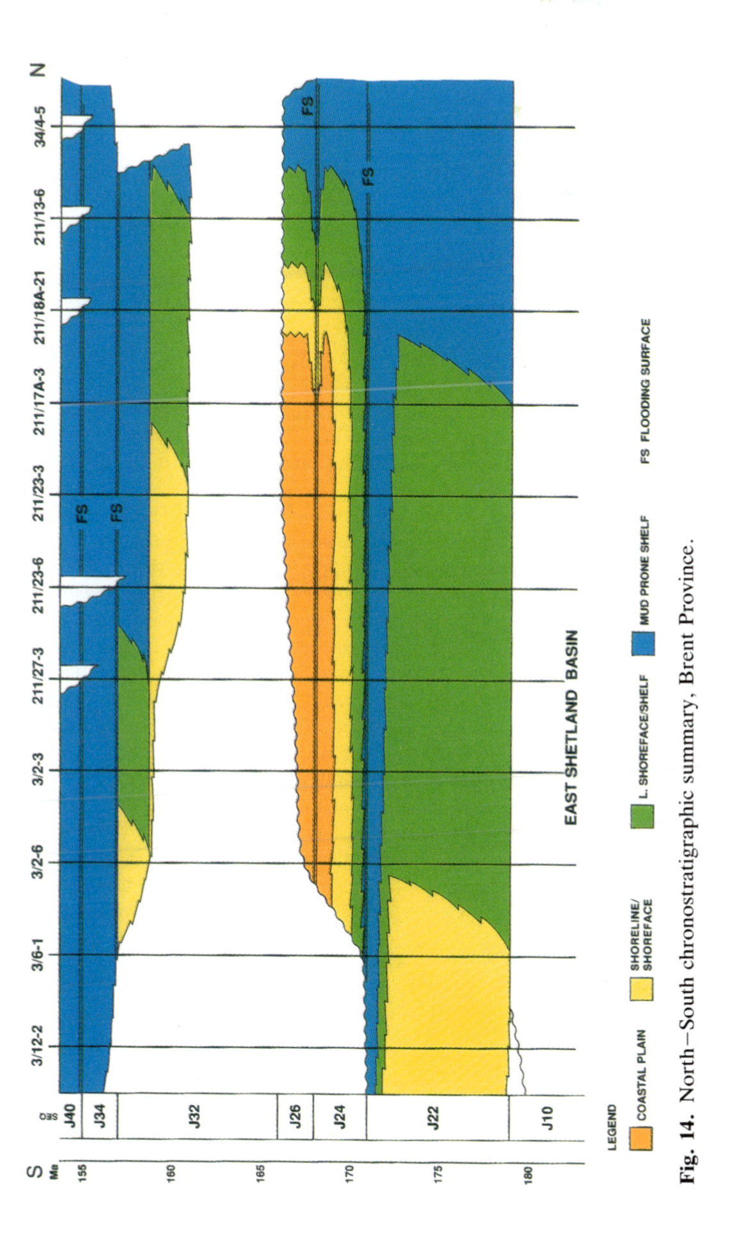


**Fig. 14.** North–South chronostratigraphic summary, Brent Province.

basal transgressive shoreline sand (e.g. Cormorant Field, Livera & Caline 1990) is associated with this late J32 transgressive event. This unit often overlies eroded J26 coastal plain facies; it passes up from a lower 10–20 m thick unit with granule lags to fully marine, bioturbated, fine-grained shelf sandstone and mudstone. Locally, around the Don/Thistle area (UKCS Blocks 211/8, 211/13), higher energy, storm-dominated shelf facies are developed (cf. Brown & Richards 1989).

By the end of J32, transgression had advanced southward behind the Ninian fault trend with seas covering the transitional shelf and clastic source areas had become increasingly reduced.

*(ii) Bruce–Beryl embayment.* To the west, the J26 alluvial fans were partially drowned by a marine transgression caused by an abrupt increase in the rate of basin subsidence. Following transgression, the initial stages of J32 deposition are distinguished by the development of fan deltas passing basinwards into storm dominated shelf sands. In contrast to the East Shetland Basin, sediment supply rates were sufficient to enable progradation and infilling of the newly developed tilted fault block topography. The fan delta shoreline system is overall retrogradational, so that with time, the facies belts shift landward, overstepping Palaeozoic and older basement.

*(iii) Viking Graben/Norwegian North Sea.* Within the basin axis there was a southward shift of the shoreline system. This was accompanied by a gradual eastward expansion of the shelf over much of the Norwegian North Sea which forms the rollover to the main basin half-graben. The coastal plain system persisted behind the J32 shoreline, but is much finer grained than J26 and reached a thickness of *c.* 500 m adjacent to the Viking Graben boundary fault. Further south, in the triple junction area, coastal plain facies associations interdigitate with volcaniclastics and volcanics (Rattray Formation). The coastal plain was drowned in late J32 and led to the deposition of a thick (5–10 m) laterally extensive coal which is approximately equivalent to the marine flooding surface and unconformities that developed at the base of J34.

## *Sequence J34* (Fig. 15)

*Age.* Late Bathonian to Middle Callovian.

*Lithostratigraphy.* Tarbert Formation and lower Heather Formation (East Shetland Basin, North Viking Graben); uppermost Sleipner Formation and lower Hugin Formation (South Viking Graben, Norwegian North Sea); Bruce Upper Sands (Bruce–Beryl embayment).

*Boundary conditions.* J34 developed in response to a further, more significant phase of rifting within the basin. Again, evidence for the sequence boundary unconformity is restricted to marginal uplifted areas and localized footwall crests. This is particularly clear in the South Viking Graben where significant topography was developed. Elsewhere, the sequence is marked by a continuation in the expansion and progressive deepening of the basin together with a gradual landward shift of facies belts.

In the East Shetland Basin exposure and erosion along isolated footwall crests attest to a relative fall in base level (Figs 11 and 12). Further south, in the Bruce–Beryl embayment, the locus of major clastic input supplying fan deltas became confined to the northern part of the area where seismic data again indicate downlap and onlap onto a fault block topography.

In other parts of the basin, J34 is dominated by shelfal facies associations. As with J32, the basal sequence boundary is taken at a flooding surface which developed in response to an increase in the rate basin subsidence and renewed extensional faulting with uplift around the basin margins. The final stages of deposition within J34 are characterized by widespread shelf mud deposition with minimal coarse clastic input. Clearly, by this time the sediment hinterland had been substantially reduced with sediment input being mainly confined to local areas around transfer zones.

The upper boundary of J34 is constrained by the development of a condensed section over much of the basin (particularly the East Shetland Basin). This phase of condensation is widespread and probably of plate-wide significance, implying some element of eustatic change in sea level. Rattey & Hayward (in press) described the overlying J40 tectono-stratigraphic unit as a tectonically quiescent period in the rift basin development and only record true onlap over the crests of a few intra-basinal fault blocks and rollovers which become emergent at the end of J34.

*Basin forming processes.* J34 is considered to represent a major pulse of rifting in the basin. As the crests of fault blocks and the basin margins were sites of active uplift, developing multiple stacked unconformities in response to rifting through the Late Jurassic, clear evidence of rifting is difficult to find. The chronostratigraphic diagram in Fig. 11 does however offer a

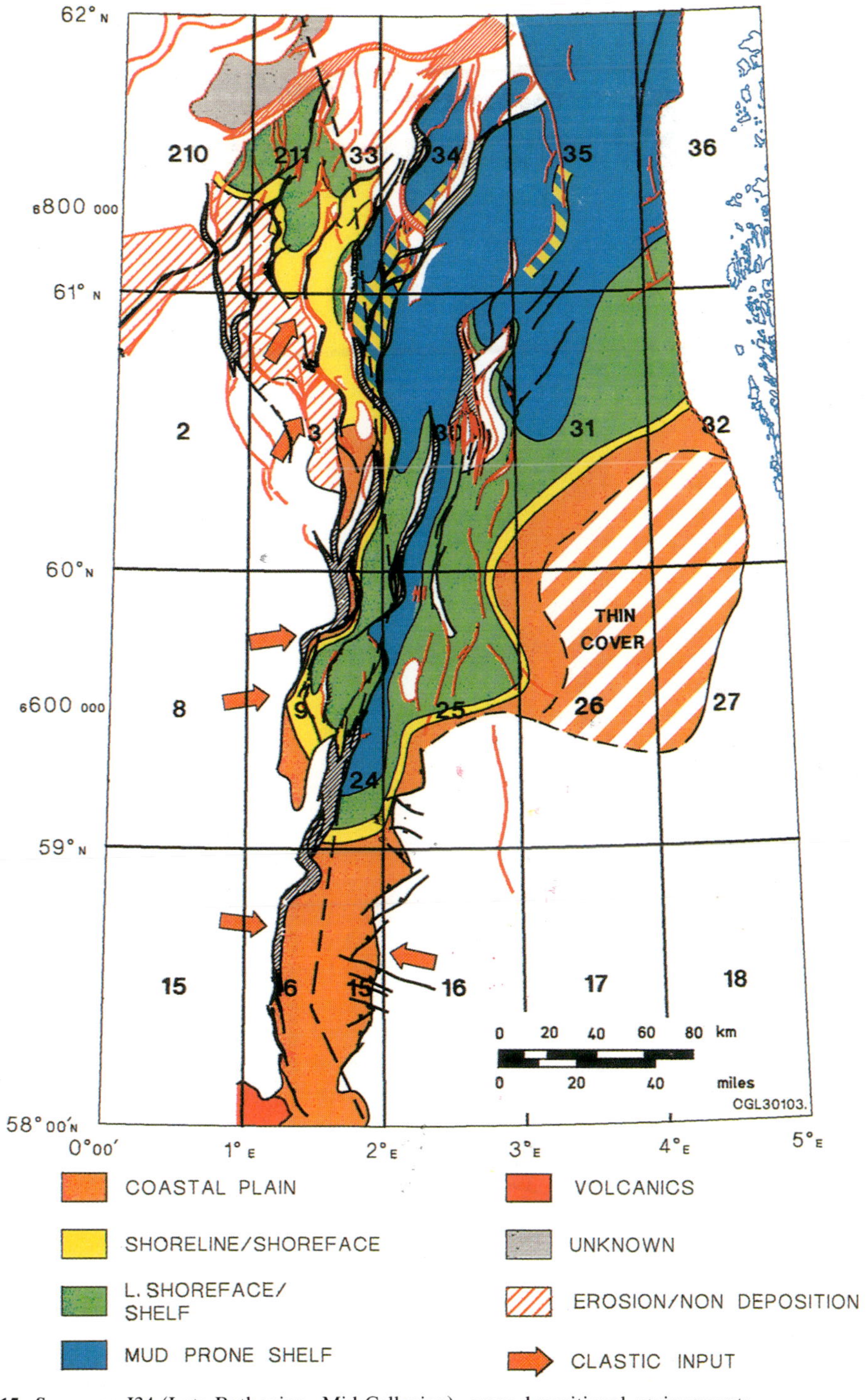


**Fig. 15.** Sequence J34 (Late Bathonian–Mid-Callovian), gross depositional environments.

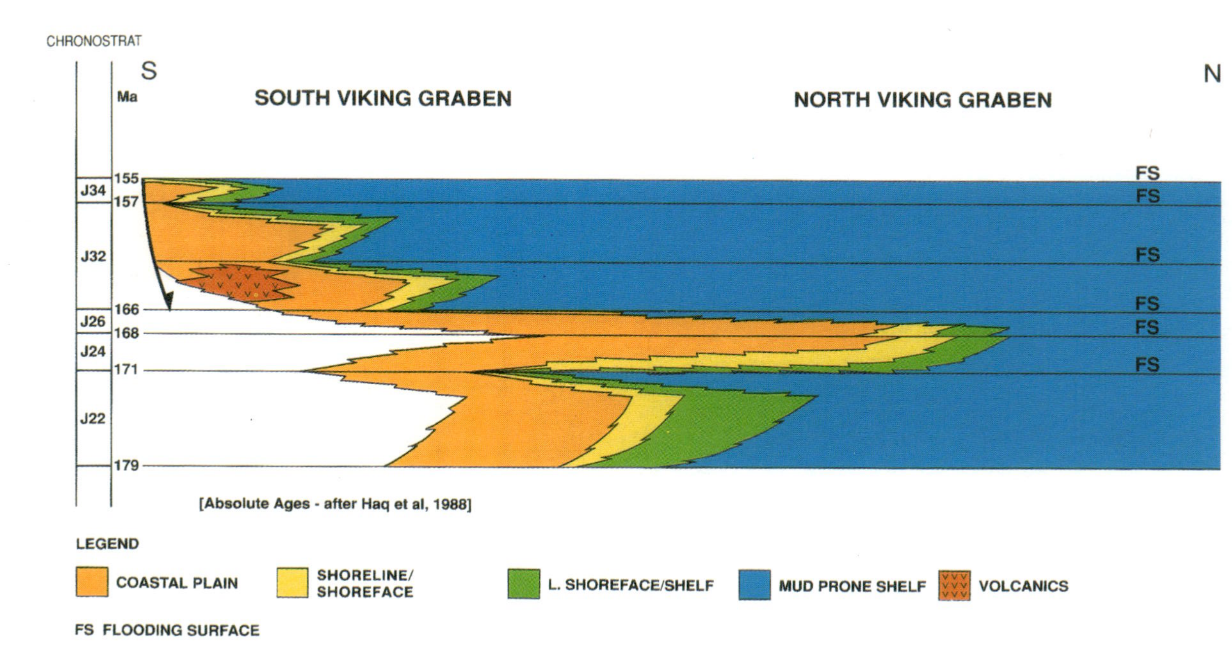


**Fig. 16.** Generalized north–south chronostratigraphic summary for the northern North Sea Basin.

possible candidate across the N. Alwyn–Brent fault trend.

Once again because of continued high sedimentation rates in the Bruce–Beryl embayment, the base of J34 is seismically resolvable, with downlap and onlap of seismic events onto a rift topography.

Passive subsidence followed the rifting, leading to a resumption in the overall expansion of the basin.

*Sedimentary response.* J34 records a series of retrogradational events within an overall deepening phase of the basin's history.

In the East Shetland Basin, shorelines fringed small islands and archipelagos, where crests of the Ninian–Hutton and similar trending fault blocks became emergent. The main shoreline trend remained largely static around the western boundary fault to the basin. High energy storm sands were driven offshore from these areas, forming stacked sheets; these pass gradually northwards into bioturbated, lower energy distal storm sandstones and shelf mudstones.

A similar situation prevailed in the Bruce–Beryl embayment with marginal shoreline systems passing offshore into stacked storm sandsheets. Coarse-grained alluvial fan-delta sedimentation was more confined than J32, with major transfer-zone input points remaining most active. South of the Bruce–Beryl area, J34 is characterized by a rapid southward shift in coastal-plain and shoreline systems, in response to overall basin expansion. The shoreline was fed by alluvial tracts directed axially along the graben. In early J34, the development of thick coals (c. 10 m thick) at the top of the coastal plain succession, attests to a reduction in coarse sediment input to the basin suggesting a much reduced hinterland to the south and west.

## Conclusions (Fig. 16)

The major implications of this study for the stratigraphic evolution of the Northern North Sea Basin during the Middle Jurassic are as follows.

(i) The tectono-stratigraphic unit development is considered to be controlled by large scale tectonic processes: thermal uplift (J20) and rifting (J30). Superimposed on these are five sequences which developed in response to changes in basin subsidence, sediment supply and eustatic sea level.

(ii) J22–26 developed in response to pre-rift thermal doming centred over the triple junction of the Viking, Witch Ground and Central Grabens. The uplift was punctuated by relative sea level rises caused by a combination of: eustacy and/or a reduction rate of sediment supply (J22, J24); and increased basin subsidence (J26).

The J22 and J24 depocentres progressively shifted basinward until in J26 the depositional systems began to aggrade, and retrograde as sediment supply eventually could not keep pace with increased basin subsidence in response to stretching of the crust by the thermal uplift.

(iii) J32 represents the first phase of active rifting which can be clearly identified in the North Sea Basin with the development of tilted fault blocks along the Ninian–Hutton trend and Bruce–Beryl boundary fault. Continued rifting led to a progressive landward shift of the shoreline and the development of a retrogradational system extending through the Late Jurassic.

The results of this evaluation provide a clear picture of the Middle Jurassic evolution and stratigraphy of the North Sea basin. It forms a framework within which local lithostratigraphy can be resolved, reservoir distribution viewed and basin potential evaluated. Its success is a consequence of an integrated multidisciplinary approach to basin analysis and most notably the availability of a refined biostratigraphic framework.

The work is published with the permission of BP Exploration. Particular thanks are due to P. Rattey, N. Milton, D. Ewen and D. Bailey for their contribution and encouragement with additional support from a large number of geoscientists in BP's Norwegian and UK-based exploration, development and production teams.

## Appendix: depositional environments and facies associations

A discussion of the approach and database used to characterize depositional setting within sequences is contained in the main part of the text. In this appendix, detailed descriptions of facies associations typical of each depositional environment are presented. For each facies association its description, wireline log response and stratigraphic occurrence are discussed. The principal wireline logs used to characterize facies associations in uncored sections were the gamma, sonic, formation density (RHOB) and neutron porosity (NPHI) logs. Examples of the more common facies associations are illustrated in Figs 3–5.

### *Depositional environments*

For the purposes of mapping palaeogeographic elements, five gross depositional environments were defined; namely (i) *continental*, (ii) *coastal plain*, (iii) *shoreline* (iv) *offshore transition/shelf*, (v) *basinal*. True continental and basinal environments were not recognized, though they may be present outwith the study area. Further refinement of depositional environment maps by breaking out groups of facies associations (e.g. percent channel facies associations in J26) was carried out to enhance understanding of palaeogeographic elements and evolution. A discussion of depositional environments and facies associations follows.

*Coastal plain*. Facies associations characteristic of this depositional environment are present in all sequences discussed (J32-J34), but are most typical of the Ness Formation (J24-J26) and Sleipner Formation (J32-J34 sequences).

(i) *Fluvial distributary channel association*. This comprises isolated (2–6 m thick) to vertically stacked (several tens of metres thick) erosively-based fining upward sequences dominated by planar and trough cross-bedding, parallel lamination, and current ripple cross-lamination (Fig. 3i). Sandstones vary from fine to very coarse grained and contain exotic pebble, coal and mud clasts. Gamma log response shows a sharp inflection at the base of channel sequences, with typically upward increasing gamma (20–40 API), increasing sonic, and sand cross-over on NPHI/RHOB.

(ii) *Minor channel association*. Isolated erosively-based small scale channel fills (1–2 m thick) characterize this association, and are generally medium-fine grained, dominated by small-scale planar cross-sets to current ripple lamination and may contain irregular or inclined mud-drapes in their upper parts (Fig. 3ii). Log response is similar to isolated channel-fills in (i), though increased clay content may lead to higher gamma response, particularly in the upper parts of channel fills.

(iii) *Crevasse splay/mouth bar association*. This association comprises various types of upward-coarsening sequences, usually 1–7 m thick. Crevasse splay sequences typically show a transition from structureless mudstone to current-ripple or parallel laminated to structureless sandstones and may be capped by a minor channel sandstone or rootlet horizon.

Mouth bar sequences are thicker, and may be dominated by ripple drift or wave ripple cross-lamination (Fig. 3iii). These sequences are capped by erosive distributary channel, or beach/spit associations. Evidence of emergence (rootlets, beach or ?aeolian facies) is common at the tops of thicker upward-coarsening sequences (e.g. Brent and Statfjord fields).

Log response is typified by a serrated often upward decreasing gamma response and shale cross-over on the NPHI/RHOB logs. Thicker mouth bar associations may show log responses very similar to progradational shoreface sequences.

(iv) *Floodplain/interdistributary bay association*. Siltstone and mudstone characterizes this association, with occasional thin sandstones (current or wave ripple laminated), coal, rootlet beds and common pyrite/siderite (Fig. 3i & ii). A transition is recognised into a paralic swamp association.

(v) *Paralic swamp association*. Dominated by siltstones/mudstones with seatearths, and single or composite coal horizons (Fig. 3i & ii).

A variable high gamma response (up to 150 API), and constant shale cross-over on FDC/CNL is typical of floodplain/interdistributary bay mudstones. Increase in carbonaceous mudstone and coals results in lower gamma, markedly increased sonic velocity and low RHOB/high NPHI.

*Coastal plain: interpretation*. Rapid vertical and lateral facies variability is common in coastal plain facies associations, recording the complex interplay of channel, crevasse splay/mouth-bar and floodplain environments. Two major channel belts are recognised along the eastern edge of the basin, in the Bruce-Beryl Embayment, and between the Ninian/Hutton and Brent fault trends (Fig. 9b). The position of these seems to be related to areas of increased subsidence in the hanging wall of re-activated faults.

Minor channels may be associated with crevasse splay sequences (crevasse channel origin), or as separate mixed to suspended load inter-distributary channel networks. The latter occurrence is more common in upper coastal plain areas (e.g. J26 sequence, Crawford Field, 9/28).

In general, the scale of major mouth bar sequences can be related to dimensions of distributary channel systems, and depth of water into which they prograde. Larger scale mouth bar sequences, often wave reworked into beach-spits, are interpreted from the Brent/Statfjord area. Livera (1989) describes complex lobate wave-reworked sand aprons fronting distributary channels entering wave-dominated lagoons. Smaller scale mouth bar sequences prograding into shallow wave-dominated lagoons are present to the west of the Ninian/Hutton fault (e.g. Dunlin and Hutton fields).

*Shoreline/shoreface*. This depositional environment includes a range of facies associations recording back-shore (washover/lagoonal), foreshore and shoreface deposition. Shoreline facies associations have been recognized from all sequences, occurring within the Broom/Oseberg, Rannoch, Etive, Ness and Tarbert formations in the northern part of the study area, and parts of the Hugin Formation in the South Viking Graben.

(i) *Lagoonal association*. The distinction between lagoonal and interdistributary bay associations is difficult, relying on subtle differences in the characteristics of minor sand bodies and extent of marine/wave influence. Lagoonal intervals generally comprise laminated to massive/bioturbated mudstone and silt-stone with thin silty to sandy wave ripple laminated lenses (Fig. 4i). Small scale hummocky and swaley cross-stratification is occasionally observed; coals are rare, but where present seem to be thin and laterally extensive.

Gamma log response is high (80–100 API), with low gamma spikes where thinly bedded sandstones are present. Gamma response may decrease upwards, with NPHI/RHOB converging from a predominantly shale cross-over, where lagoonal sequences are in-filled by thin sands.

(ii) *Washover/lagoonal shoreline association*.

Occurring together with beach and lagoonal associations, typical facies sequences comprise small scale (1–2 m) upward-coarsening thinly interbedded sandstone/mudstone beds passing up into current and wave ripple laminated fine grained sandstone (Fig. 4i). They may be capped by beach-barrier sands (e.g. J26 sequence, Cormorant area).

This association is difficult to recognise from wireline log response alone. A 'funnel-shaped' though often serrated gamma/sonic log response is typical, with RHOB/NPHI converging upwards from largely shale cross-over at the base.

(iii) *Beach-barrier/upper shoreface association*. This is characteristic of the Etive, and parts of the Ness and Tarbert formations (J24-J32). It consists of medium to fine grained well sorted sandstones which may coarsen or fine upwards, but more often show little grain size variation (Fig. 4i & ii). Sequences are 10–15 m thick, and internally massive or poorly laminated. In the J24 sequence, an abrupt erosion surface overlain by coarser small scale cross-bedded sandstone may separate this association from underlying laminated micaceous shoreface sediments (Fig. 4i).

The upper parts of this association may comprise backshore to aeolian facies (finer grained laminated sediments); these are often capped by a rootlet bed and thin lagoonal coal. In transgressive settings the top is truncated by a ravinement surface overlain by a heavy mineral lag or bioturbated sandstone/siltstone.

Constant low gamma (20–25 API), even sonic and consistant sand separation on NPHI/RHOB is very common in this association. Concentrations of mica (particularly at the base of upper shoreface sequences), and heavy mineral lags cause marked high gamma spikes.

(iv) *Transgressive barrier association*. This facies association is most commonly developed in the Broom Formation (J22 sequence), but has also been recognized from the Bruce 'B' sands (J26) and Tarbert Formation (J34, crestal Ninian/Alwyn South wells). It generally comprises blocky, coarse to very coarse grained erosive-based upward-fining sandstone 10–20 m thick (Fig. 4iii). Internally, sandstones are massive to poorly laminated (occasional low angle lamination or cross-bedding); bioturbation and irregular argillaceous/carbonaceous laminae are common especially towards the top.

A very blocky gamma log response is typical, with a sharp deflection at the base often decreasing upwards from *c*. 30 API to 40–50 API (increase in carbonaceous, micaceous and bioturbated facies). Both sonic and NPHI/RHOB are also constant and reflect high sand content. In the absence of core, the log response may be confused with isolated channel fill facies. This is a particular problem where J34 transgressive shoreline sands overlie channel-dominated J26 coastal plain facies, and can usually be resolved by detailed correlation.

(v) *Estuarine/tidal inlet association*. Erosively based sequences of medium/coarse-grained sandstone are typical of this association, erosively cutting down into foreshore or upper shoreface facies (Fig. 4iv). Single (2–3 m thick) or stacked upward-fining sequences containing small scale planar cross-sets, irregular mud drapes and vertical burrows are typical. This facies association has been recognised mainly in progradational shoreline settings (e.g. J24 sequence, Etive Formation of the Cormorant area).

Wireline log response is very similar to distributary channel associations in the coastal plain environment, with sharp gamma deflections at erosive bases, and upward increasing gamma profiles. Response may vary depending on degree of bioturbation and mud drapes, particularly in the upper parts of channel fills.

(vi) *Progradational coarse-grained shoreface association*. This association has not been widely recognized in literature on the Brent Group, and occurs commonly in the J32 sequence (Tarbert Formation) east of the Ninian/Hutton fault. Large-scale coarsening upward sequences (30–40 m thick) are medium to coarse grained with common granule/pebble lags (Fig. 4v). Sequences are essentially massive, bioturbated and argillaceous towards the base; pebble lined scours, graded laminae and rare small-scale cross-bedding occur in the upper parts. A transition probably exists between this and a fan-delta association described below.

Gamma log response reflects the overall coarsening upward nature, decreasing gradually from 40–60 API to 30–40 API. The sand-rich nature results in a constant to widening sand crossover on NPHI/RHOB logs.

(vii) *Fan delta association*. This has been described within the J22 sequence from the Oseberg Formation (Graue *et al*. 1987) and may also occur in more proximal facies of the Broom Formation (Livera & Caline 1990). Up to 50 m thick blocky to coarsening upward units comprise decimetre to centimetre scale graded massive beds at the base, passing upward into graded inclined fan delta foreset beds. Fan delta topset beds could equally suitably be grouped with channelized associations in the shoreline depositional environment.

A very blocky low gamma response with sharp base and top, and constant sand cross-over on NPHI/RHOB is typical of this association. Graue *et al*. (1987) have differentiated distinct fan-delta foreset packages using dipmeter profiles.

(viii) *Storm-wave dominated shoreface association*. Though characteristic of the Rannoch Formation (J24 sequence), examples are also recognised from J32 shoreline sandstones in the Thistle, Don and Dunlin areas. In prograding sections, it often comprises a single coarsening-upward sequence 40–50 m thick consisting of finely laminated micaceous medium to fine grained sandstone (Fig. 4iv). Structures include horizontal and low angle lamination, hummocky lamination, and rare scour-and-fill/high angle foresets. Bioturbation is rare and usually restricted to the base of the association. A more detailed description of this association is provided by Scott (this volume).

Gamma response is often variable, reflecting concentration of mica-rich laminae, but decreases upwards from *c*. 60–40 API. Sonic velocity is relatively constant (*c*. 80 μs/ft), and RHOB/NPHI logs may show a shale cross-over (more micaceous intervals), or sand cross-over (cleaner, or carbonate cemented intervals).

*Shoreline/shoreface: interpretation*. The range of

facies associations developed in the shoreline setting is highly variable; it depends largely on the interaction between sediment supply and reworking processes (mainly waves and storms). Differences in facies composition and stacking patterns between progradational and transgressive shoreline sequences are discussed below.

The back-barrier environment commonly comprises open wave-dominated lagoons supplied by minor washover and inlet mouth shoal sands. The importance of tidal processes in modifying beach and backshore/lagoonal sediments during the progradational phase of the Brent system (J24 sequence) is often debated (e.g. Richards & Brown 1986) but probably overestimated. In many cases, evidence for tidal processes affecting channel fill sequences is equivocal. It is also difficult to distinguish between fluvial channels cutting into relict shoreline sands, and true temporal equivalence of channel and shoreline systems (though see Hurst & Morton 1988 where this has been achieved).

Strongly progradational shoreline associations are recognised from two sequences in particular. Within the J24 sequence (upper Rannoch, Etive and lower Ness Formations) they are interpreted as deposits of a prograding barred wave-dominated shoreface/beach barrier system in a microtidal regime (Richards & Brown 1986; Scott, this volume). In the J32 sequence (lower Tarbert Formation) coarse-grained progradational shoreline and fan–delta sequences were probably dominated by flood-generated sediment input, and gravitational processes. Examples from the Oseberg Formation (J22 sequence) are documented in Graue *et al.* (1987), where bottomset/subaqueous fan, steeply dipping foreset and sheetflood-dominated topsets facies are interpreted.

Transgressive shoreline deposits are typical of the lower part of the Tarbert Formation (J34 sequence) over the northern part of the East Shetland Basin, and the dip slope of the Ninian/Hutton fault trend. These comprise thin coarse-grained sands with transgressive lags (stacked ravinement surfaces), passing abruptly upwards into shelfal muddy sandstones (e.g. Livira & Caline 1990). In areas where subsidence rates are higher, and sediment supply maintained, facies sequences record punctuated beach-barrier progradation interrupted by periods of transgression (Ronning & Steel 1987). During transgression, back-barrier lagoons are in-filled by washover sequences and succeeded by shoreface remnant and beach-barrier facies.

*Offshore transition/shelf.* A degree of overlap exists between facies associations recorded from the lower part of the shoreface, and true shelf sequences. Facies typical of this depositional environment are recorded from J22 (upper part of the Broom Fm), the lower part of J24 (mainly lower Rannoch Fm), and extensively in the J30 sequence (parts of the Tarbert and Lower Heather Formations of the East Shetland Basin, the Upper/'A' and 'Massive Sands' of the Bruce–Beryl Embayment, and Hugin Formation of the South Viking Graben).

(i) *Storm-wave dominated massive sandstone.* This association comprises coarse-to fine-grained sandstones 1–3 m thick, often erosively based with an overlying pebble lag (Fig. 5i). Internally beds may coarsen or fine upwards, contain faint parallel to hummocky lamination, and in places are amalgamated into units up to 10–15 m thick.

The generally coarse grained and clean nature of these sandstones produces a characteristic blocky low gamma response (*c.* 30 API). Sonic response is constant, and NPHI/RHOB logs show a clear sand crossover. In stacked examples, micaceous sandstones or mudstones produce high gamma spikes.

(ii) *Storm-wave dominated sandstone/mudstone interbeds.* This association has been recognized in the lower part of the J24 sequence (Rannoch Formation) and the J34 sequence (upper Tarbert/Lower Heather formations). It is typified by repeated thin (<1 m) erosively based beds of medium/fine grained sandstone interbedded with siltstone/mudstone (Fig. 5ii & iii). Structures include horizontal to low angle lamination, and small scale hummocky cross-stratification; beds may be capped by wave ripples and bioturbated siltstones.

The highly interbedded nature of this association produces a variable spikey and generally high gamma response (50–75 API), and constant shale cross-over on NPHI/RHOB logs.

(iii) *Bioturbated muddy shelf sandstones.* Facies of this type are common in the lower parts of the Rannoch Formation (J24 sequence) and much of the upper Tarbert Formation in the East Shetland Basin (J34 sequence). It is typified by heavily bioturbated medium to fine grained sandstones with irregular micaceous and carbonaceous laminae (Fig. 5iii & iv). Diffuse horizontal and rare wave ripple lamination are preserved.

This association occurs in both transgressive and progradational shelf settings. In the former case, it forms repeated small scale (2–3 m) sharp based upward-fining sequences with mud drapes and bioturbation increasing upwards, and may cap a transgressive shoreline sequence (Fig. 4iii). In the latter case, this association may form 20 m+ coarsening upward sequences succeeded by upper shoreface/beach-barrier sequences.

Log response of this association is highly variable. Gamma response is generally high (50– 60 API), though may vary considerably with clay content. NPHI/RHOB logs may show a relatively constant shale or narrow sand crossover.

(iv) *Shelf mudstones/siltstones.* Massive to laminated fully marine mudstones/siltstones are only recorded from the J22/24 boundary in the northern part of the Viking Graben ('Rannoch Shale') and the upper parts of the J34 sequence (Lower Heather Formation). Only minor, thin sandstone laminae are present, and sedimentary structures include wave rippled lenses, common slumping, scattered molluscan shell debris and pyrite nodules (Fig. 5ii).

*Offshore transition/shelf interpretation.* Thickly bedded sandstones show many characteristics of classical storm-generated beds though in places are clearly amalgamated (thin micaceous/mud drapes). A complete transition probably exists between amalgamated beds which may develop at the inner shelf/shoreface transition, and thinner beds representing more distal storm event below fairweather wave-base.

Bioturbated muddy sandstones formed in open shelf settings above fairweather wave base, with substantial in-faunal reworking. Examples of thick transgressive shelf sequences are uncommon in the literature, though examples of thinning and fining upward sequences have been described. Similar facies also formed in more protected progradational shoreface environments where bioturbation is more thorough, reflecting lower intensity and frequency of storms. The latter case includes examples from the J24 sequence (Rannoch Formation in the SE sector of Q211, and most of Q210) and the lower parts of progradational shoreline sequences in J32 east of the Ninian/Hutton fault trend.

Laminated mudstone/siltstone facies accumulated in open shelf settings, below fair-weather wave base affected by occasional distal storm.

## References


BOWEN, J. M. 1992. Exploration of the Brent Province. *This volume.*

BROWN, S. & RICHARDS, P. C. 1989. Facies and development of the Middle Jurassic Brent Delta near the northern limit of its progradation, UK North Sea. *In*: WHATELEY, M. K. G. & PICKERING, K. T. (eds) *Deltas: Sites and Traps for Fossil Fuels* Geological Society London, Special Publication, **41**, 253–267.

——, RICHARDS, P. C. & THOMSON, A. R. 1987. Patterns in the deposition of the Brent Group (Middle Jurassic) U.K. North Sea. *In*: BROOKS, J. & GLENNIE, K. (eds) *Petroleum Geology of NW Europe*. Graham & Trotman, London, 899–913.

BUDDING, M. C. & INGLIN, H. F. 1981. A reservoir geological model of the Brent sands in Southern Cormorant. *In*: ILLING, V. & HOBSON, G. D. (eds) *Petroleum Geology of the Continental Shelf of N. W. Europe*. Heyden, London, 326–34.

CANNON, S. J. C., GILLES, M. R., WHITAKER, M. J., PLEASE, P. M. & MARTIN, S. V. 1992. A regional reassessment of the Brent Group, UK Sector, North Sea. *This volume.*

DEEGAN, C. E. & SCULL, B. J. 1977. *A proposed standard lithostratigraphic nomenclature for the central and northern North Sea*. Institute of Geological Sciences, Report, No. **77/25**; Norwegian Petroleum Directorate Bulletin, no. **1**.

EYNON, G. 1981. Basin development and sedimentation in the Middle Jurassic of the Northern North Sea. *In*: ILLING, V. & HOBSON, G. D. (eds), *Petroleum Geology of the Continental Shelf of N. W. Europe*. Heyden, London, 98–103.

FALT, L. M., HELLAND, R., WIIK JACOBSEN, V. & RENSHAW, D. 1989. Correlation of transgressive-regressive depositional sequences in the Middle Jurassic Brent/Vestland Group megacycle, Viking Graben, Norwegian North Sea. *In*: COLLINSON, J. D. (ed.) *Correlation in Hydrocarbon Exploration*. Graham & Trotman, London, 191–200.

GALLOWAY, W. E. 1989. Genetic stratigraphic sequences in basin analysis I: architecture and genesis of flooding-surface bounded depositional units: *American Association of Petroleum Geologists Bulletin*, **73**, 125–142.

GRAUE, E., HELLEND-HANSEN, W., JOHNSEN, J., LOMO, L., NOTTVEDT, A., RONNING, K., RYSETH, A. & STEEL, R. 1987. Advance and retreat of the Brent delta system, Norwegian North Sea. *In*: BROOKS, J. & GLENNIE, K. W. (eds) *Petroleum Geology of N.W. Europe*, Graham & Trotman, London, 915–937.

HAQ, B. U., HARDENBOL, J. & VAIL, P. R. 1987. Chronology of fluctuating sea levels since the Triassic. *Science*, **235**, 1156–1167.

HELLAND-HANSEN, W., STEEL, R., NAKAYAMA, K. & KENDALL, C. G. St. C. 1989. Review and computer modelling of Brent Group stratigraphy. *In*: WHATELEY, M. K. G. & PICKERING, K. T. (eds), *Deltas: Sites and Traps for Fossil Fuels*. Geological Society, London, Special Publication, **41**, 237–252.

——, ASHTON, M., LOMO, L. & STEEL, R. 1992. Advance and retreat of the Brent delta: recent contributions to the depositional model. *This volume.*

HURST, A. & MORTON, A. C. 1988. An application of heavy mineral analysis to lithostratigraphy and reservoir modelling in the Oseberg Field, Northern North Sea. *Marine and Petroleum Geology*, **5**, 157–170.

JOHNSON, H. D. & STEWART, D. J. 1985. Role of clastic sedimentology in the exploration of oil and gas in the North Sea. *In*: BRENCHLEY, P. J. & WILLIAMS, B. P. J. (eds)., *Sedimentology: Recent Developments and Applied Aspects*. Geological Society, London, Special Publication, **18**, 249–310.

LIVERA, S. E. 1989. Facies associations and sand-body geometries in the Ness Formation of the Brent Group, Brent Field. *In*: WHATELEY, M. K. G. & PICKERING, K. T. (eds), *Deltas: Sites and Traps for Fossil Fuels*. Geological Society, London, Special Publication, **41**, 269–286.

—— & CALINE, B. 1990. The sedimentology of the Brent Group in the Cormorant Block IV Oilfield. *Journal of Petroleum Geology*, **13**, 367–396.

MORTON, A. C. 1992. Provenance of Brent Group sandstones: heavy mineral constraints. *This volume.*

PARRY, C. C., WHITLEY, P. J. K. & SIMPSON, R. D. H. 1981. Integration of palynological and sedimentological methods in facies analysis of the Brent Formation. *In*: ILLING, V. & HOBSON, G. D. (eds), *Petroleum Geology of the Continental Shelf of North-West Europe*. Heyden, London, 205–215.

RATTEY, R. P. & HAYWARD, A. H. (in press). Stratigraphy and structure of a failed rift system. The Middle Jurassic to early Cretaceous Basin Evolution of the Central and Northern North Sea. *In*: WILLIAMS, G. D. (ed.) *Tectonics and Seismic Sequence Stratigraphy*. Geological Society, London, Special Publication.

RICHARDS, P. C. & BROWN, S. 1986. Shoreface storm deposits in the Rannoch Formation (Middle Jurassic), North-West Hutton Oilfield. *Scottish*

*Journal of Geology*, **22**, 367–375.

Richards, P. C., Brown, S., Dean, J. M. & Anderton, R. 1988. A new palaeogeographic reconstruction for the Middle Jurassic of the northern North Sea. *Journal of the Geological Society, London*, **145**, 883–886.

Ronning, K. & Steel, R. J. 1987. Depositional sequences within a 'transgressive' reservoir sandstone unit: the Middle Jurassic Tarbert Formation, Hild area, Northern North Sea. *In*: Kleppe, J. *et al.* (eds) *North Sea Oil and Gas Reservoirs*. Graham & Trotman, London, 169–176.

Scott, E. 1992. The palaeoenvironments and dynamics of the Rannoch–Etive nearshore and coastal successions, Brent Group, Northern North Sea. *This volume*.

Skarpnes, O., Hamear, G. P., Jakobsson, K. H. & Ormaasen, D. E. 1980. Regional Jurassic setting of the North Sea north of the Central Highs. *In*: *The Sedimentation of the North Sea reservoir rocks*. Proceedings, Norwegian Petroleum Society, Paper XIII, 1–8.

Van Wagoner, J. C., Posamentier, H. W., Mitchum, R. M., Vail, P. R., Sarg, J. F., Loutit, T. S. & Hardenbol, J. 1988. An overview of sequence stratigraphy and key definitions. *In*: Wilgus, C. W. *et al.* (eds) *Sea level changes: an integrated approach*. Society of Economic Paleontologists and Mineralogists Special Publication, **42**, 39–45.

Vollset, J. & Dore, A. G. (eds) 1984. *A revised Triassic and Jurassic lithogratigraphic nomenclature for the Norwegian North Sea*. Norwegian Petroleum Directorate Bulletin No. **3**.

Whittaker, M. F., Giles, M. R. & Cannon, S. J. C. 1992. Palynological review of the Brent Group, northern North Sea. *This volume*.

Woollam, R. & Riding, J. B. 1983. *Dinoflagellate cyst zonation of the English Jurassic*. Institute of Geological Science Report **83/2**.

Ziegler, P. A. 1982. *Geological Atlas of Western and Central Europe*. Shell, the Hague.

—— 1988. Post Hercynian plate reconstruction in the Tethys and Arctic–North Atlantic domains. *In*: Manspeizer, X. (ed.) *Triassic–Jurassic rifting: continental breakup and the origin of the Atlantic Ocean and passive margins. Part B*. Elsevier, Amsterdam, 711–755.

# A regional reassessment of the Brent Group, UK sector, North Sea

S. J. C. CANNON[1], M. R. GILES, M. F. WHITAKER[1], P. M. PLEASE[2] & S. V. MARTIN

*Shell UK Exploration and Production, Shell-Mex House, Strand, London, UK*

[1] *Present address: The Geochem Group Limited, Chester Street, Chester CH48RD, UK*

[2] *Present address: Department of Applied Geology, University of New South Wales, Sydney, NSW 2033, Australia*

**Abstract:** The deltaic Brent Group sequence consists of three formations. The Rannoch is a shoreface section; the Etive is a barrier facies. The Ness contains three members: the Enrick, representing a marginal marine back barrier environment; the Oich, an organic-rich open lagoonal shale and the Foyers, a largely fluvially-dominated delta-top or coastal plain deposit. During the Bajocian, facies belts rapidly prograded northwards until the northern part of the area was reached. Here progradation stalled and thick delta front sediments were deposited. Shortly before the delta began to retreat, a temporary relative sea-level rise pushed the delta front southwards resulting in the deposition of an intra-Ness barrier. The progradational sequence is underlain by the Aalenian to lowermost Bajocian Broom Formation fan delta and overlain by the transgressive Tarbert sequence of largely Bathonian age.

The Brent Group of the UK sector of the North Sea represents a period of marginal marine to deltaic sedimentation during the Middle Jurassic Bajocian–earliest Bathonian stages. As part of the 11th UK Licensing Round, Shell (UK) Exploration & Production Ltd and their partners Esso (UK) Exploration & Production Ltd commissioned an internal review covering the sedimentology, stratigraphy and delta development (this paper and Whitaker *et al.* this volume) and diagenesis (Giles *et al.* and Hamilton *et al.* this volume; Samways & Marshall, pers. comm.) of the Group in the East Shetland Basin (Fig. 1). The study initially involved sedimentological logging of some 9000 ft of cores from the interval. The cores were logged at 1:50 scale and used to calibrate uncored sections at 1:200. A generalized lithofacies scheme was developed as part of the study (Fig. 2) which is broadly applicable to all sequences encountered in the Brent Group. Facies associations depicting different depositional environments can also be developed for specific formations and members of the Brent Group. Eighteen basin-wide correlation panels were constructed and lithostratigraphic correlation was carried out on the wells included in the study. Thickness and facies distribution maps were prepared and integrated with palynological data (Whitaker *et al.* this volume); a progradational model for the sequence was developed based on these data.

## Sedimentology and depositional environments

The Brent Group in the East Shetland Basin is divided into five formations (Deegan & Scull 1977). The earliest infill into the basin is represented by the Broom Formation of latest Toarcian-early Bajocian age, followed by the Rannoch, Etive and Ness Formations together making up the main Bajocian progradational phase of the delta. Finally the very latest Bajocian–early Bathonian Tarbert Formation was deposited which represents the retreat, possibly pulsed, and destruction of the delta. Each formation is discussed below in terms of their gross sedimentological characteristics, depositional environment and distribution.

### *Broom Formation*

The Broom Formation represents sediment infill from uplifted rift margins and is typified by coarse to granule grade arkosic sandstones which exhibit a variety of primary and secondary sedimentary/biogenic features. Four characteristic vertical sections are identified both from core material and wireline log pattern (Fig. 3) which relate to the distribution and thickness of sequences across the basin. The nature of the contact with the underlying Lower Jurassic Dunlin Group of marine shales may also be of significance in defining the nature of the Broom

*From* Morton, A. C. Haszeldine, R. S. Giles, M. R. & Brown, S. (eds), 1992, *Geology of the Brent Group*. Geological Society Special Publication No. 61, pp. 81–107.

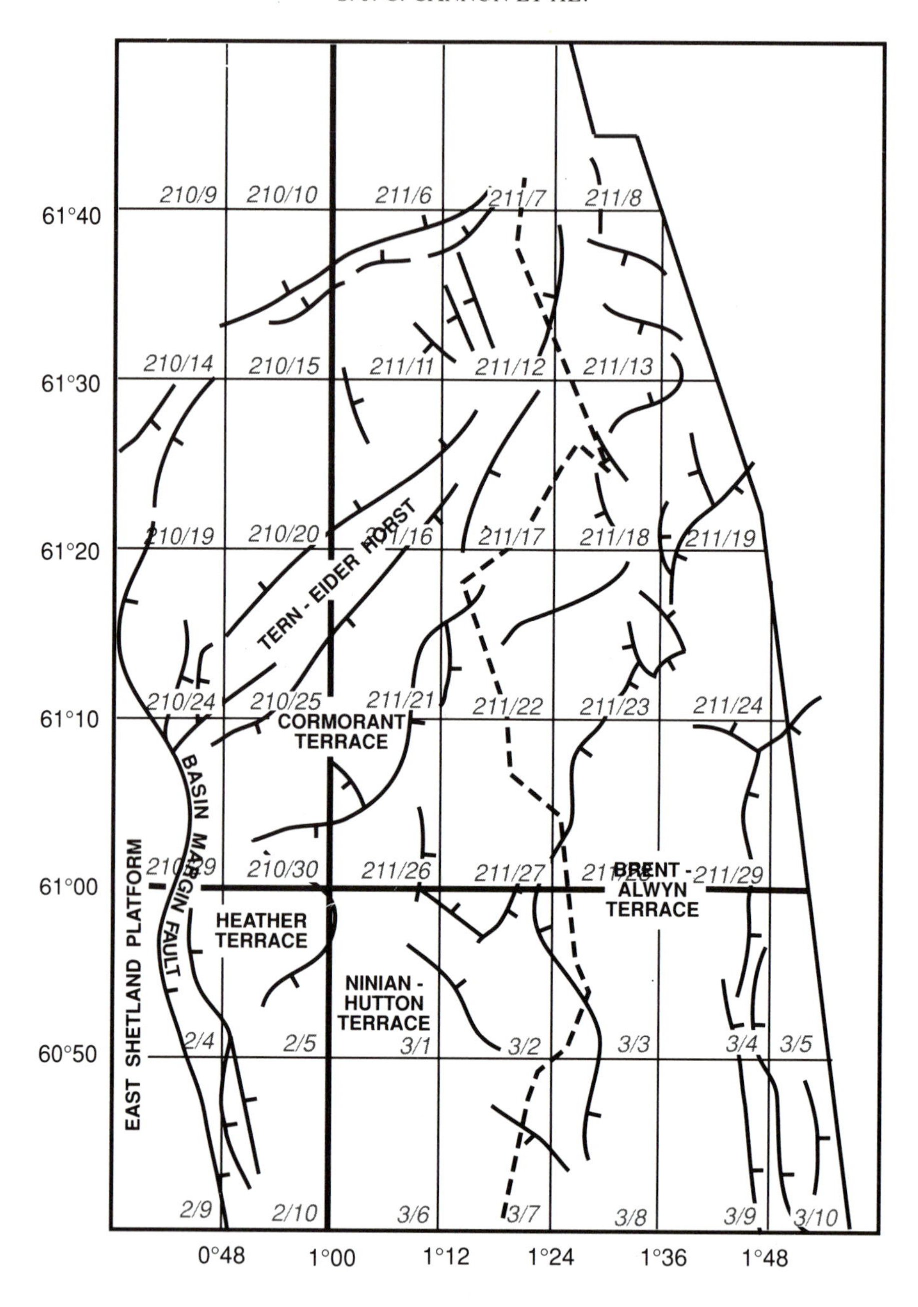


- - - N-S CORRELATION LINE (see Figure No. 16)

**Fig. 1.** Location map and main structural elements of the East Shetlands Basin.

Formation depositional environment and mode of emplacement.

*Distal sequences*. In distal sequences, the transition from the underlying Dunlin Group is marked by a gradual increase in fine sand and silt laminae, frequently bioturbated or rippled, which appear as upward coarsening cycles up to 3 ft thick. This grain-size change is reflected in the funnel shaped gamma-ray response (Fig. 3a). A sharp grain-size increase occurs, marking the base of the Broom Formation proper and

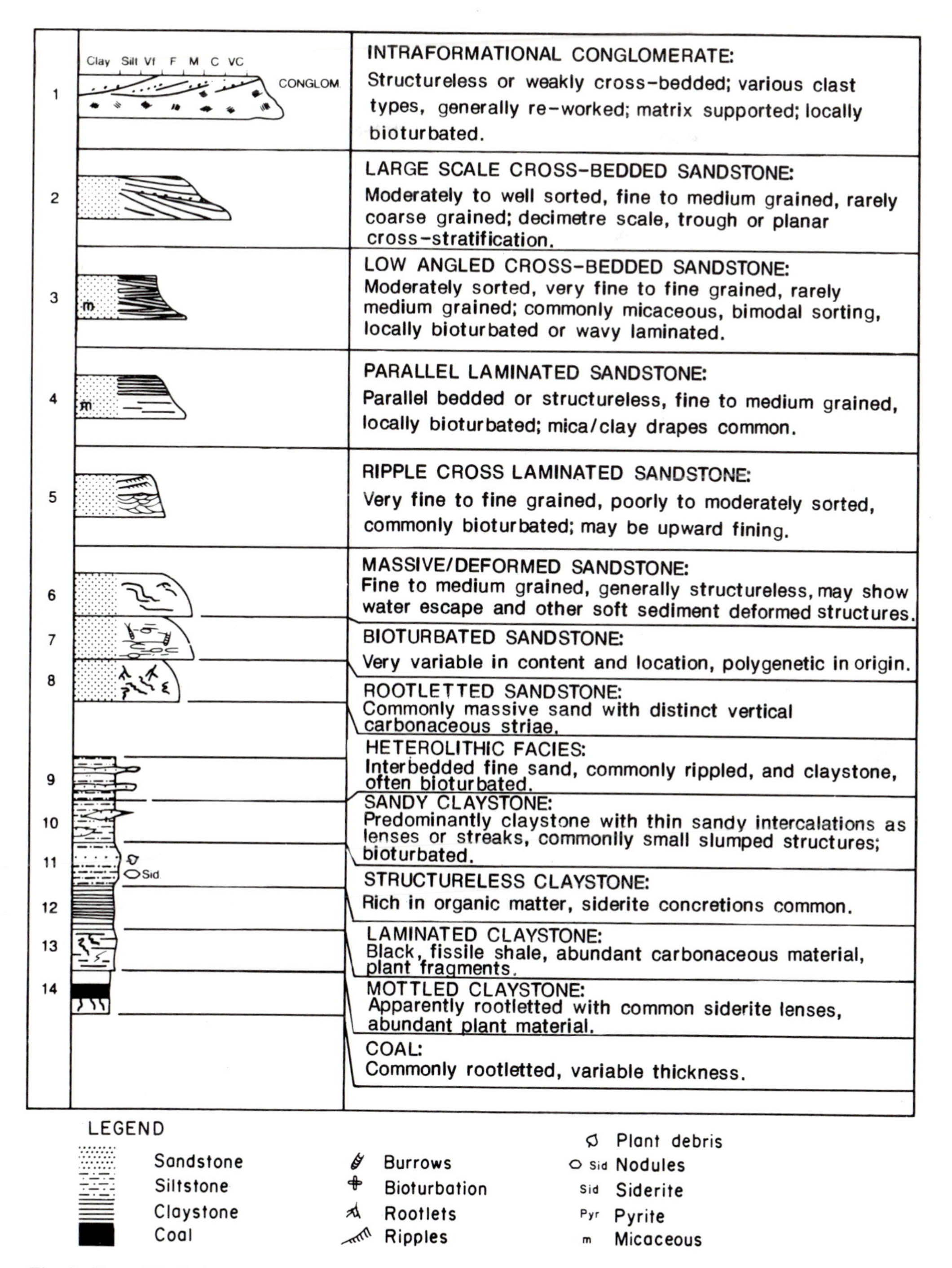


**Fig. 2.** Brent lithofacies scheme.

consists of a thin coarse sand to granule lag which is commonly structureless or may be weakly laminated towards the top. Above, the sediments are intensely bioturbated with clay lined horizontal burrows (?*Ophiomorpha*) being dominant. Distal sequences are generally

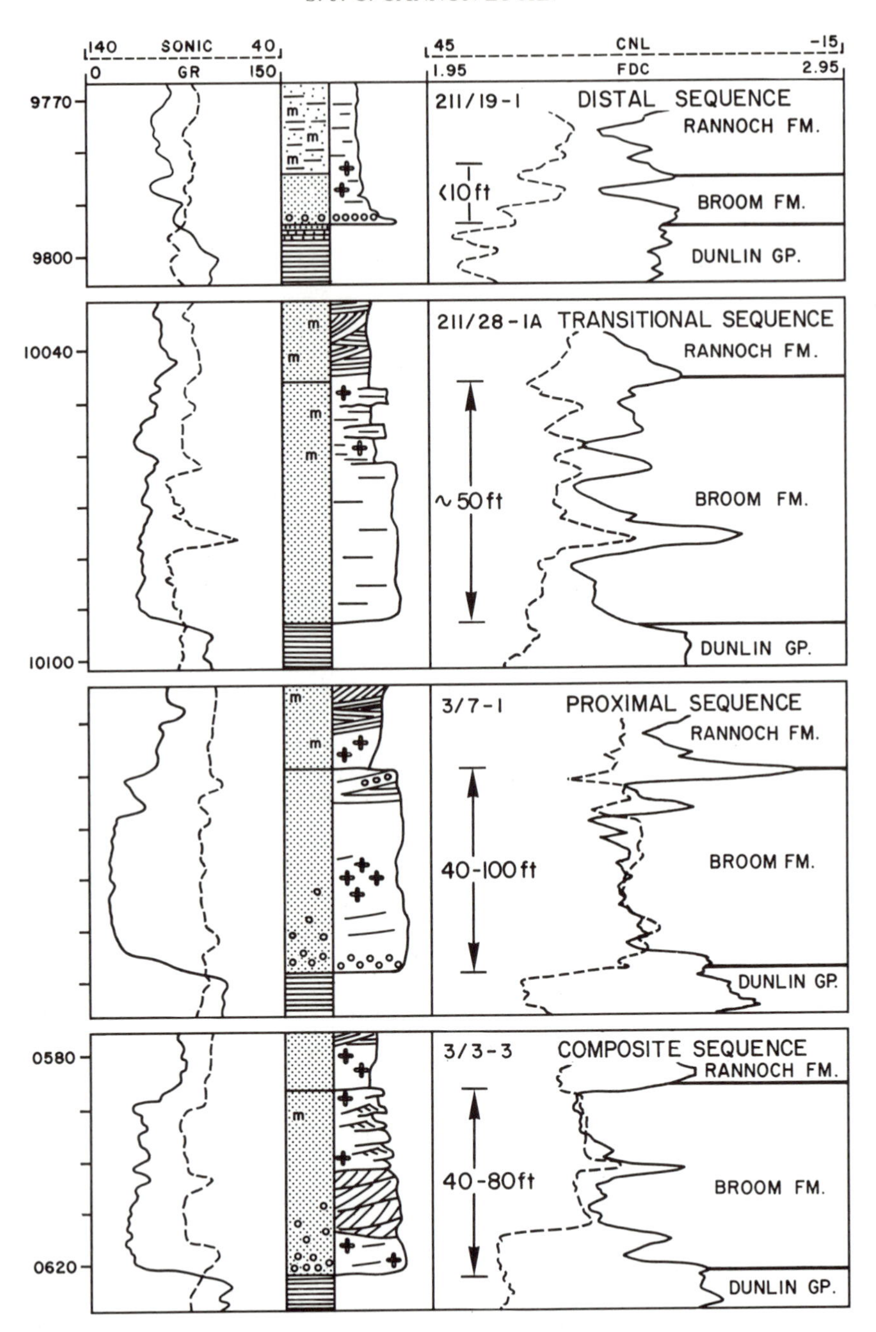


**Fig. 3.** Typical Broom Formation sequences.

thin, being less than 10 ft thick. In very distal locations the Broom Formation may be represented by very thin (<1 ft) layers of chamositic ooliths and floating quartz granules within marine shales. The presence of these ooliths in such relatively deep water suggests local reworking and emplacement by a high energy process such as periodic storm activity.

*Transitional sequences.* In transitional sequences, the boundary between the Broom Formation and the Dunlin Group is abrupt

(Fig. 3b). The basal part of the Broom Formation is argillaceous but rapidly coarsens upwards. The majority of the overlying section comprises vaguely laminated coarse sandstones which become massive towards the top. Occasional coarse granule grade layers are observed which help to pick out any lamination. Towards the top, the transitional nature of the sequence is seen by the presence of thin (2 ins), fine sand/mica laminae which herald the incoming of the Rannoch Formation and represent possible reworking or mixing of the two systems. The sequence as a whole fines upwards which suggests a progressive decrease in depositional energy. These sequences are commonly up to 30 ft in thickness.

*Proximal sequences.* Sharp upper and lower boundaries to the Broom Formation are seen in proximal sequences which give the familiar blocky gamma ray profile on wireline logs (Fig. 3c). The sequence comprises massive and bioturbated coarse sandstones commonly with a granule/pebble basal lag. The sequence may be capped by low angle cross-bedded units towards the top, each with a thin granule lag at the base. Proximal sequences tend to be the thickest representatives of the Broom Formation being 80–100 ft thick.

*Composite sequences.* The composite sequence comprises a sharply based proximal unit, which may coarsen upwards and is commonly bioturbated towards the top and base but is generally cross bedded in the centre. This unit is capped by a series of upward fining sandstones each with a rippled or bioturbated top. These upper sandstone units are finer and do not contain thin granule lags common in the lower proximal unit. Argillaceous wisps are commonly incorporated in the upper parts of units. The thickness of these sequences is of the order of 50–80 ft. The upper contact with the Rannoch Formation is commonly abrupt (Fig. 3d).

*Discussion.* The distribution and thickness of these four associations have been mapped out (Fig. 4) and show that the Broom Formation generally thins from west (southwest) to east in marked contrast with much of the Brent Group. Locally the interval thickens dramatically across faults. There is no evidence to suggest that thinning on the upthrown side of fault blocks is related to erosion. Within the sequence there is a proximal-distal relationship in facies associations also from west to east (northeast). The controlling influence of faults on facies distribution is especially marked on the western boundary of the basin.

Graue *et al.* (1987) propose that the Broom Formation may be equivalent to the Oseberg Formation in the Norwegian sector of the North Sea which has been interpreted as a fan-delta sequence. Some of the features recorded by this study would support this interpretation. The presence of intensely bioturbated and homogenized sandstones indicates a diverse fauna and suggests deposition in a shallow marine environment as does the position of the Broom Formation sandwiched between two other marine formations. The lack of traction generated features in general, indicates deposition by gravitational mass flow processes rather than fluvial or wave driven processes. The much greater thickness of the Oseberg Formation leads to a more complex hierarchical breakdown of lithofacies, something which is not as satisfactory for the Broom Formation. However, some comparisons may be drawn between the two formations. The presence of vague coarsening upwards profiles within units is suggestive of the prograding nature of the sequence (cf. sequences 1–3 of Graue *et al.*). Coarse-grained erosional lags seen in the proximal and lower parts of composite sequences may represent weak channel bases formed by shallow braided channel systems on a fan top (cf. sequence 4 of Graue *et al.*).

From the above it can be seen that in terms of distribution, sediment dispersal patterns and gross genetic attributes, the Broom Formation is fundamentally different from the overlying Rannoch–Etive–Ness package. The predominant Toarcian age attributed to the sediments is also at variance with the overlying Bajocian formations with which it is grouped. The authors suggest that it is arguable whether the Broom Formation should be re-assigned, possibly to within the Dunlin Group.

## *Rannoch Formation*

The Rannoch Formation is characterized by very fine- to fine-grained micaceous and argillaceous sandstones. Commonly an interval is seen to comprise 3 or 4 distinct facies which are genetically related to water depth and energy regimes. Two typical vertical sequences are shown (Fig. 5) which represent end members of the Rannoch Formation. The overall upward coarsening nature of the lower part of sequence can be masked on gamma-ray logs by the high mica content near the base. The upper boundary with the Etive Formation is generally quite distinct being marked by a significant grain-size

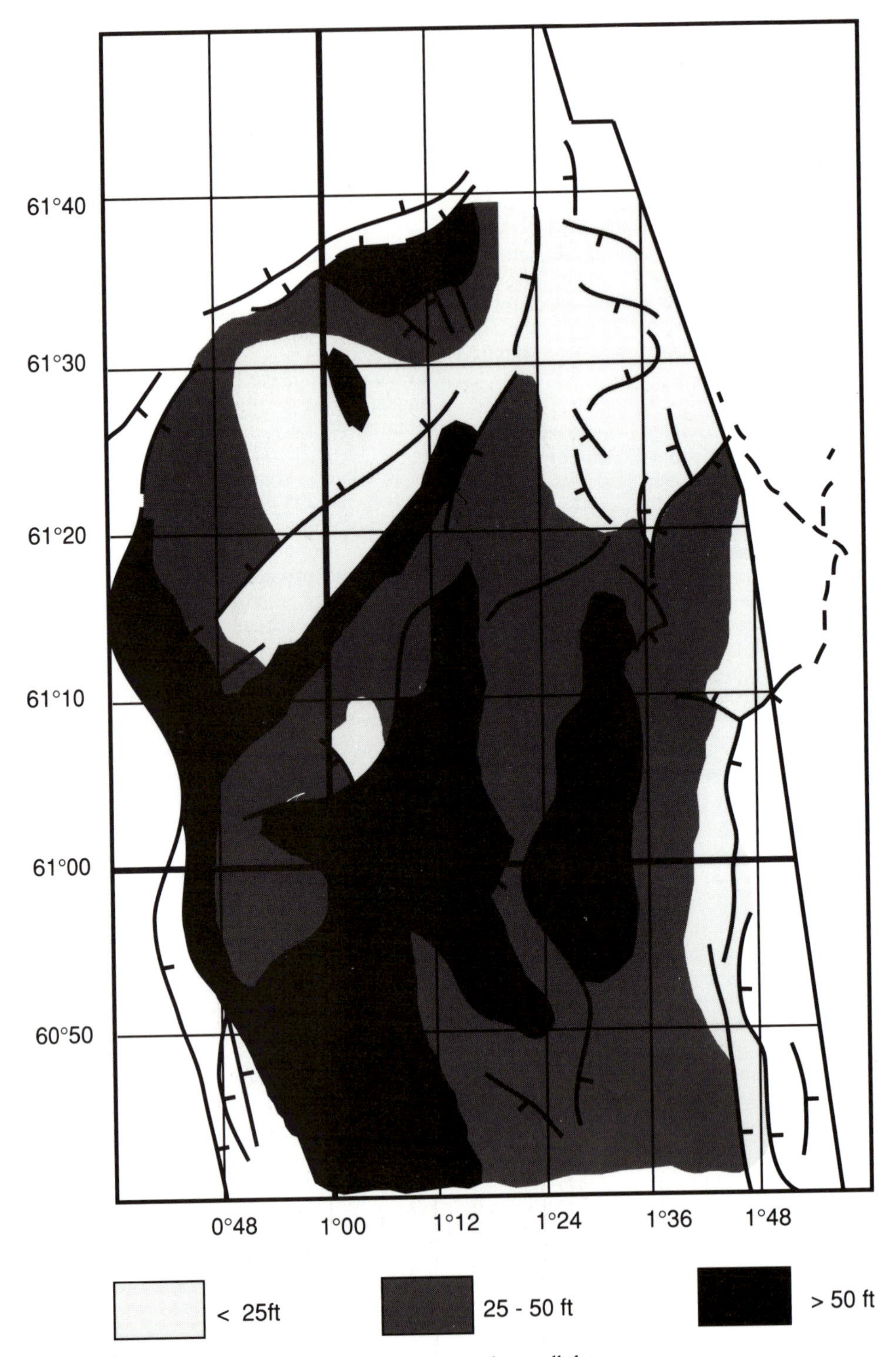


**Fig. 4.** Broom Formation thickness distribution map based on well data.

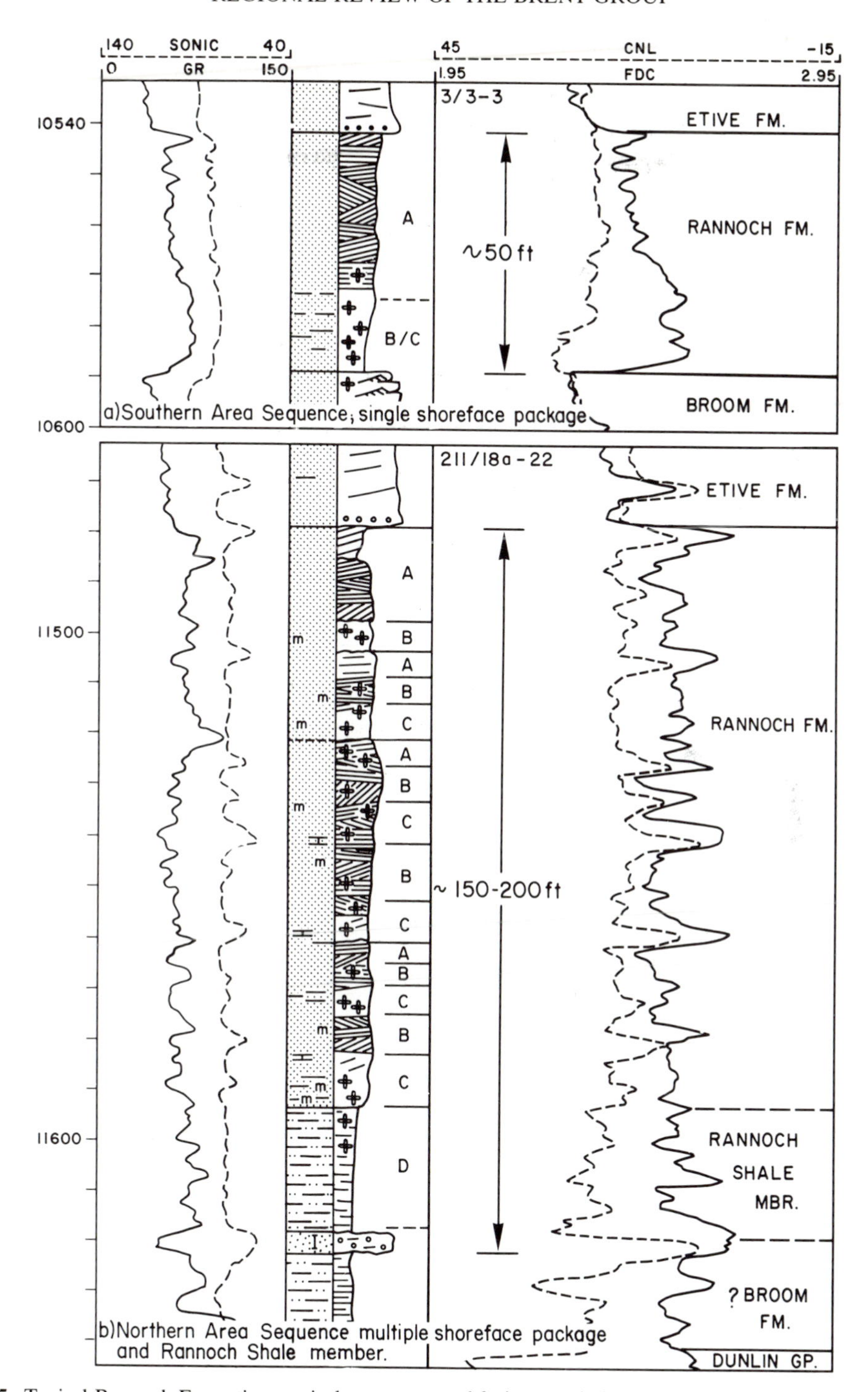


**Fig. 5.** Typical Rannoch Formation vertical sequences and facies associations.

increase. Figure 5 should be referred to during the following discussion.

Typically the base of the Rannoch Formation comprises a markedly argillaceous or micaceous unit (C) in which the individual beds show a clear bipartite separation between mica-rich and mica-poor laminae. Towards the base bioturbation is frequently intense, destroying all primary structure. This type of facies may be related to a regional drowning event associated with a sea-level high stand at or near the start of the Bajocian. Above, the sandstones are less micaceous and are made up of low-angle and planar laminations (B) which commonly show discrete scour and fill type features. Individual packages of sediment commonly have bioturbated tops, and where stacked, may represent individual storm emplaced units. The upper part (A) of a typical Rannoch sequence is characterized by better developed low angle cross-lamination and commonly higher angled scour surfaces. These sediments are rarely bioturbated.

In the northern part of the basin, the base of the Rannoch is marked by a variably thick laminated silty shale (Rannoch Shale Member-D) representing the most distal sediments. This facies may be indistinguishable from the underlying Dunlin Group unless a Broom Formation package is present. Detailed core descriptions have indicated the presence of a limited turbidite facies occurring within more northerly, distal sequences, where the boundary between the Rannoch Shale and main sandstones is gradational. The turbidites are recognised by small (1–3 ft) upward fining sequences with finely rippled tops and clay drapes. They are thought to represent delta-slope deposits preserved by later progradation of the main shoreface sequence.

Another significant feature of the Rannoch Formation is the frequent occurrence of carbonate cemented zones. Their size and frequency within a sequence is highly variable but they tend to occur at or near the top of individual units and may represent periods of non-deposition. Two distinct types of carbonate zone have been potentially identified which probably represent either conventional concretionary 'doggers' or those formed during periods of subsea cementation.

*Discussion.* The Rannoch Formation is interpreted as a full shoreface sequence which prograded in front of the Brent delta. The vertical sequences indicate the gradual shallowing of water from lower to upper shoreface and a corresponding increase in energy. Richards & Brown (1986) report the presence of hummocky cross stratification in certain Rannoch sequences and argue for storm induced deposition in the shoreface. Although such structures have not been positively identified in this study the presence of storm emplaced beds in the lower parts of individual Rannoch units have been recorded. In such instances the tops of individual sand units show common escape burrows indicating the rapid nature of sedimentation. In general the very sandy nature of the sequence as a whole suggests storm domination of the delta front.

The thickness distribution map of the Rannoch Formation (Fig. 6) shows an overall thickening trend to the east and northeast. In blocks 211/11, 211/12, 211/13 and 211/18 the sequence is at its thickest (250–300 ft) and appears to represent stacking of whole shoreface units. This may indicate the maximum progradation of the delta. It is also in this more northerly area that the Rannoch Shale Member becomes widespread, representing the most distal component of the sequence. The possibility that delta progradation effectively 'stalls' in the area is supported by the presence of coarser and cleaner sandstone units typical of the Etive Formation which interfinger with the normal sequence. This suggests that the delicate balance of the depositonal system was periodically disturbed by local fluctuations in relative sea-level. In the most northerly localities the Rannoch may be overlain directly by Tarbert Formation sandstones or marine shales attributed to the Heather Formation, indicating that progradation of the delta front had ceased.

### *Etive Formation*

The Etive Formation comprises a generally coarsening-upwards sequence of medium-grained sandstones. Minor facies variations are observed representing a variety of different environments in what is assumed to be an overall barrier system (Budding & Inglin 1981; Brown *et al.* 1987).

The dominant lithofacies present is that of massive, structureless sandstones which appear to be homogenized by dewatering or fluidization. Occasionally, vague parallel lamination is preserved representing distinct foreshore lamination. The basal part of the sequence commonly has a coarse grained, possibly erosional lag which may be overlain by decimetre scale planar and trough cross-laminated sandstones. The nature of the lag deposit is somewhat ambiguous, possibly representing a series of laterally migrating channel belts or a ravinement surface.

Similar units, generally 10–30 ft thick, are sometimes seen within the sequence and are thought to indicate possible channelized flow through the barrier. Bidirectional cross-lamination has been rarely observed in such

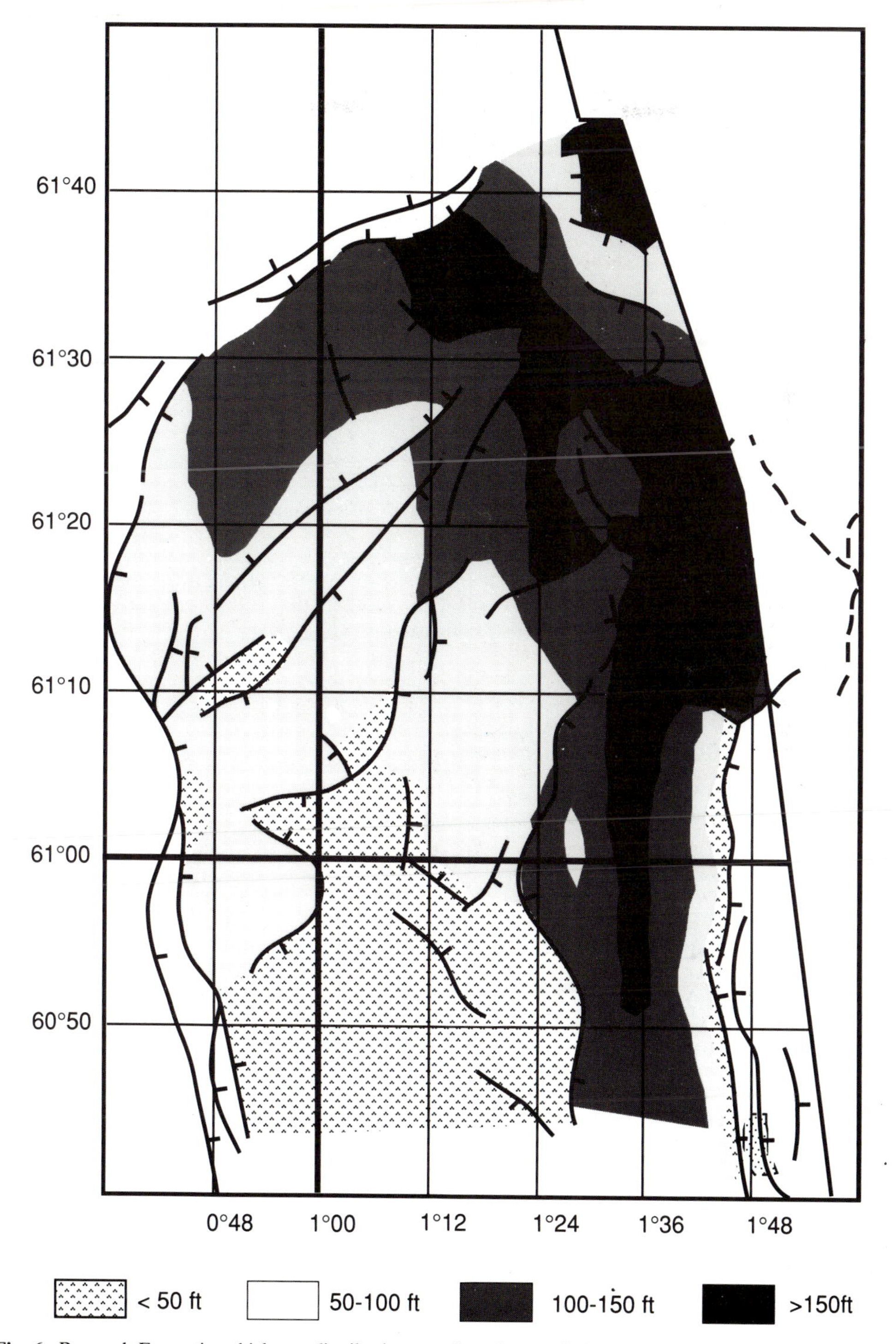


**Fig. 6.** Rannoch Formation thickness distribution map based on well data.

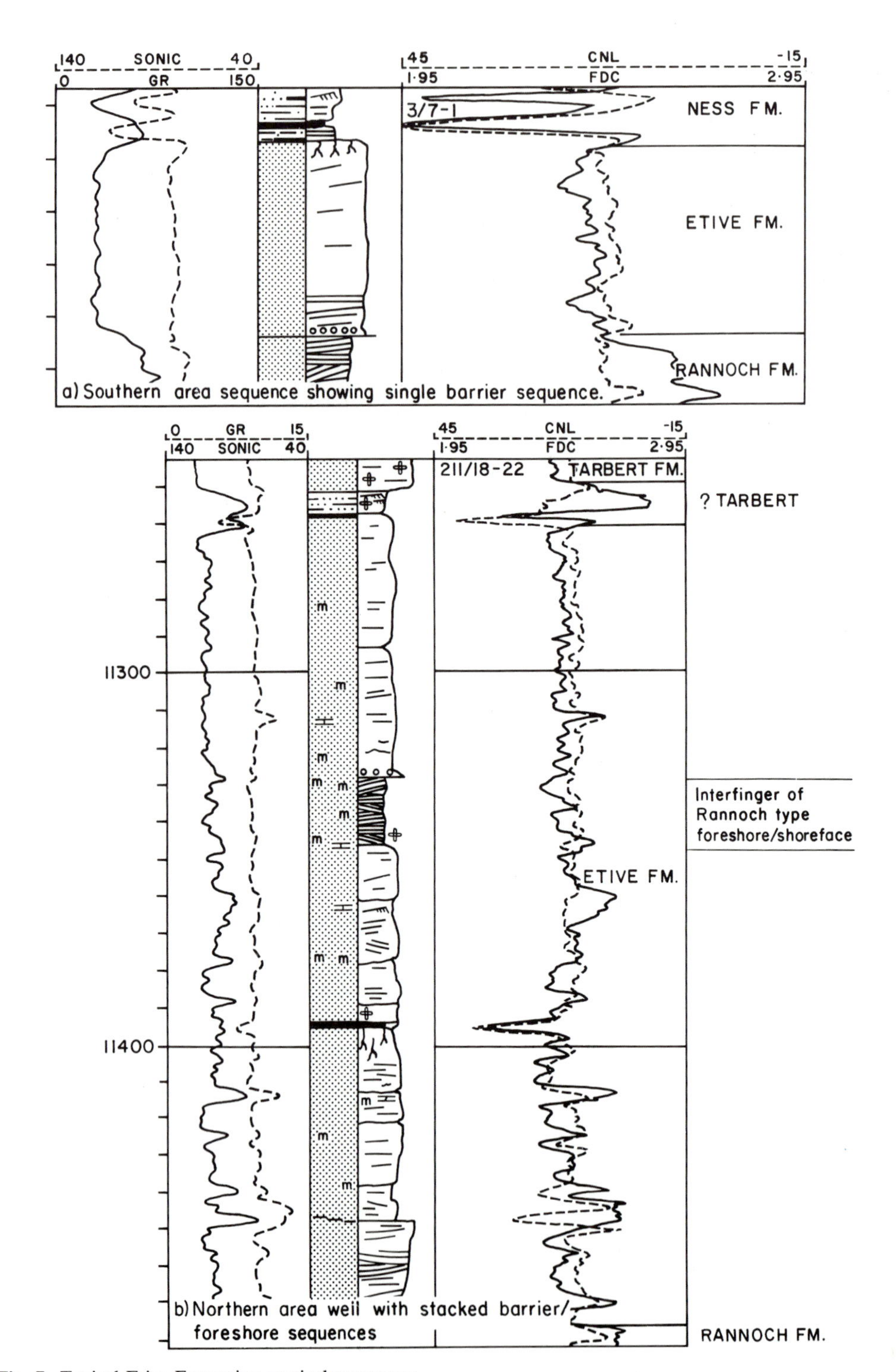


**Fig. 7.** Typical Etive Formation vertical sequences.

units which suggests a possible tidal influence in some channels. Low angle cross lamination with grainflow and grainfall structures have been recorded, usually towards the top of an interval, which suggests possible aeolian reworking of the barrier top. A variety of mottled sandstones have also been observed commonly occurring towards the top of an individual sequence. These include weakly bioturbated, wavy laminated and rootletted horizons representing a number of barrier top environments.

The Etive Formation exhibits a very distinctive wire-line log response with a strong positive separation of FDC/CNL traces and low gamma-ray indicating clean, homogeneous sandstones, in strong contrast to the underlying Rannoch Formation. The upper boundary is commonly marked by an in situ coal development normally included within the overlying Ness Formation. The absence of this marker due to erosion by overlying Ness channels can complicate the identification of the boundary. Occasionally, within thicker sections of the Etive a series of upward fining and coarsening units are observed suggesting repeated cycles of barrier and channel development. Cycle boundaries may be marked by further coal horizons or heavy mineral concentrations. Generally, however, a typical sequence comprises channelized beach plain deposits at the base passing upwards into barrier and back-barrier deposits with evidence of plant colonization (Fig. 7).

*Discussion.* The variety and combination of lithofacies observed in Etive sequences indicates the polygenetic origin of the deposits in an overall marginal marine environment. The lateral extent, apparent connectivity, and vertical relationship with the Rannoch and Ness Formations supports the barrier-bar complex proposed by Budding & Inglin (1981). This interpretation was based on the Cormorant Field and has subsequently been applied to other field areas.

A thickness distribution map for the interval between the Rannoch and Ness Formations (Fig. 8) shows a general thickening towards the north and east. A marked increase in the interval thickness is observed in the area of block 211/13 and 211/18 usually at the expense of the Ness sediments. The thickening is coincident with that observed in the Rannoch sequence and is thought to pinpoint the maximum edge of delta progradation and stalling of the barrier. It is also in this area that stacked distributary channel sequences are described by Brown & Richards (1988). This interpretation is not confirmed by this study; rather, we suggest that a complex stacking of barrier/foreshore sequences is occurring. This is to some extent supported by detailed core description and regional correlation which has identified a marked repetition of Etive-like sediments within the Ness Formation in areas north of block 211/22 (i.e. 211/11, 211/16, 211/17, 211/18, 211/19). These will be more fully discussed in the following section.

### *Ness Formation*

By convention, the Ness Formation has been divided in three units (Budding & Inglin 1981); Upper and Lower Ness, and the Mid-Ness Shale or Mid-reservoir Shale. This tripartite division represents the progradation of coastal plain and lagoonal sediments to upper delta plain fluvial dominated sequences. The basis of this subdivision has been satisfactory in most cases during the study; however, certain anomalies have been recorded with respect to the middle unit which make a revision of the stratigraphy desirable. The Mid-Ness Shale is a misnomer; in a number of more northerly areas e.g. Tern, no Lower Ness is recognized and in most cases the defined unit is not centrally located in the sequence. In addition, the unit has gained the distinction of being a time correlatable horizon representing a major flooding event (Livera 1989) or delta lobe abandonment (Eynon 1981). This view can no longer be supported as a result of new palynological evidence (Whitaker *et al.* this volume, and see discussion). It is proposed, therefore to adopt the following nomenclature.

| | *Present* | *Proposed* |
|---|---|---|
| | Upper Ness | Foyers Member |
| Ness | Mid-Ness Shale | Oich Member |
| Formation | Lower Ness | Enrick Member |

The new member names are taken from three rivers which flow into Loch Ness and were chosen for this reason. Each member will be discussed in further detail in the following sections. Environmental determination of individual facies within each member has been greatly enhanced by palynofacies investigations reported elsewhere in this volume (Whitaker *et al.*). Figures 9 and 10 represent typical vertical sequences through the Ness Formation from different areas which show a marked variation in development, especially within the Foyers Member.

*Enrick Member (formerly Lower Ness Member).* The lowermost unit of the Ness Formation, the

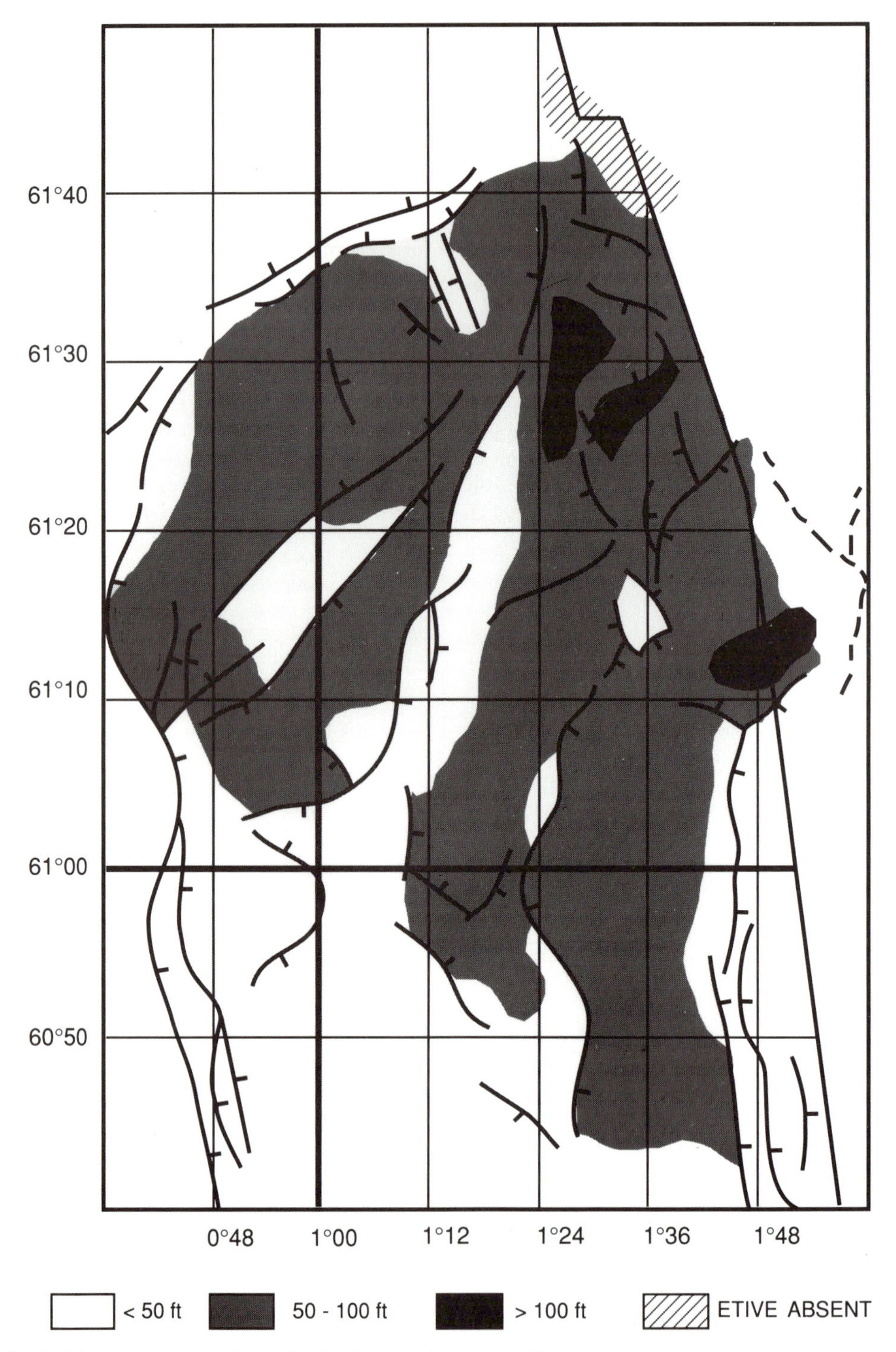


**Fig. 8.** Etive Formation thickness distribution map based on well data.

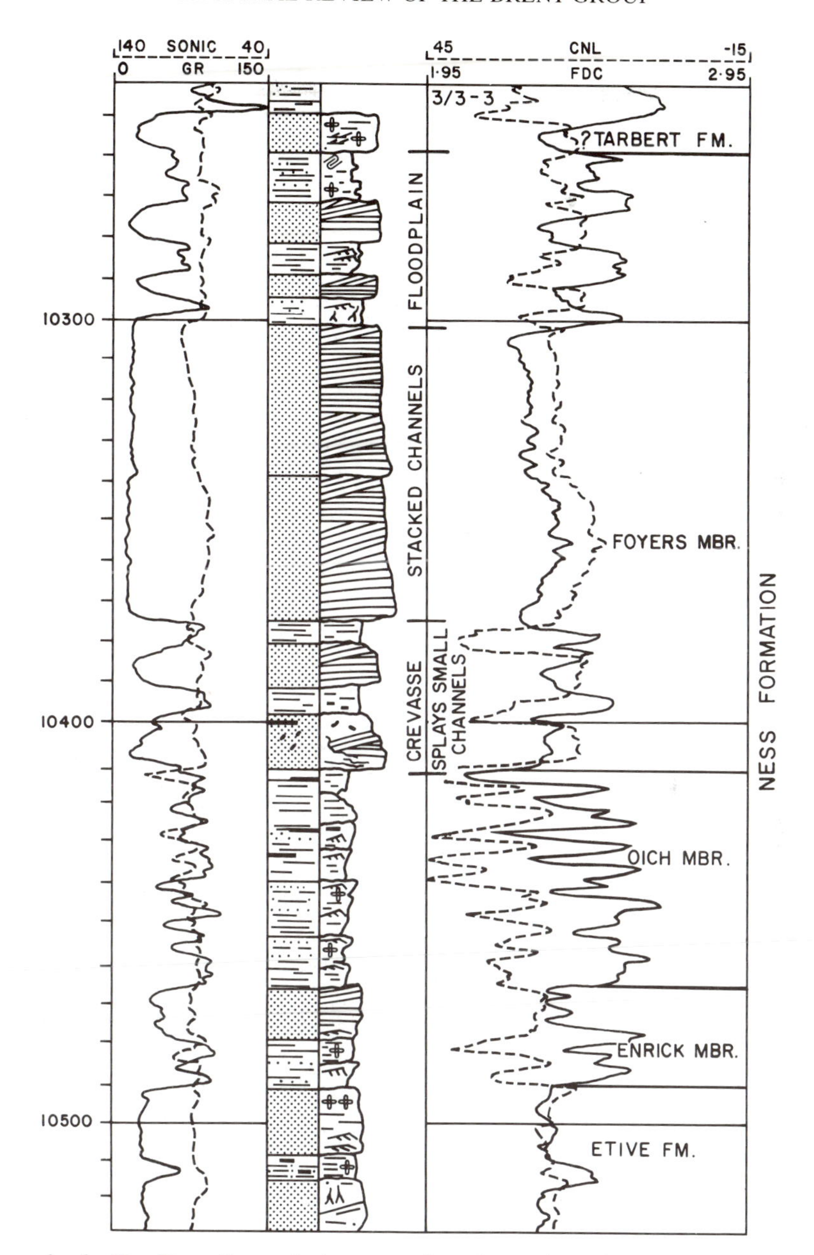


**Fig. 9.** Example of a Ness Formation vertical sequence from the southern 'fluvial fairway'. The stacked channel sequence in the Foyers Member is the thickest penetrated in the area. Nearby wells commonly show only one of these major channels. The lower part of the interval is dominated by lagoonal and minor mouth bar sediments. In the Enrick Member possible washover fans and channels are seen. The basal coal has been cut out by such a channel.

Enrick Member, represents the marginal marine back-barrier deposits, which overlie the Etive Formation in most cases. The common presence of marine microplankton within the sequence gives some indication of the degree of marine influence. The base of the Enrick Member is taken commonly at a rootletted coal horizon, identifiable except where it has been cut out by channelling. The upper boundary is less obvious because it is generally gradational into open

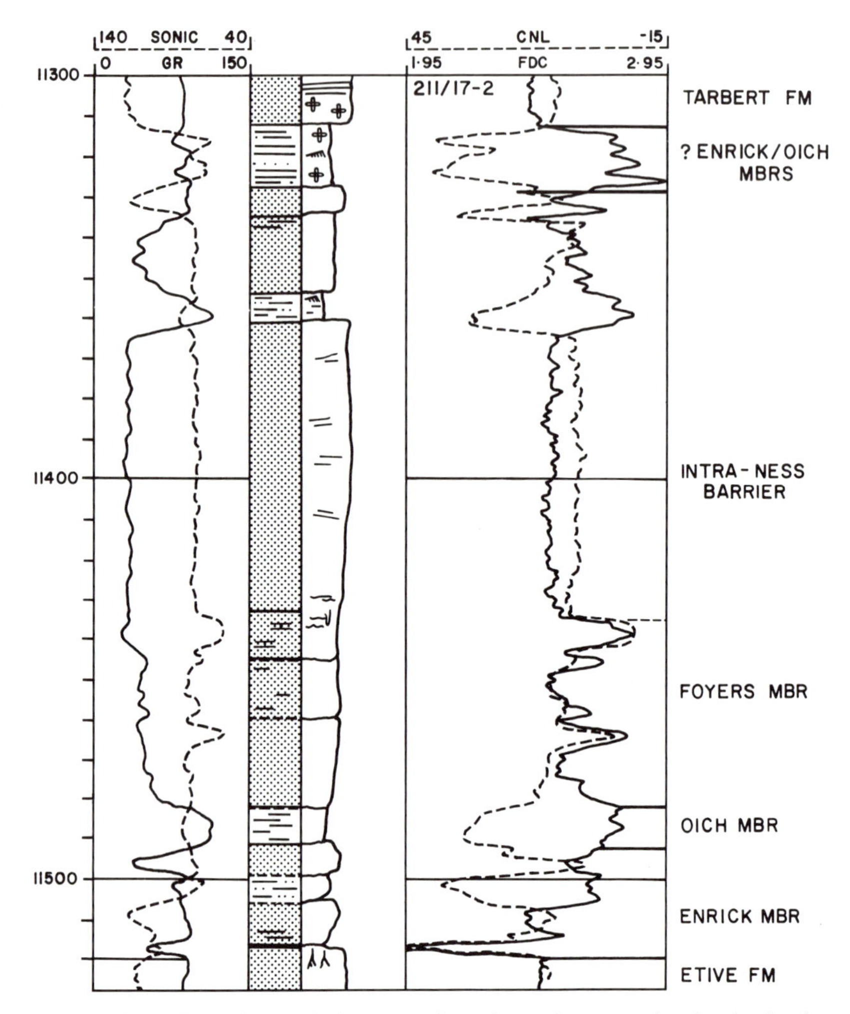


**Fig. 10.** Example of Ness Formation vertical sequence from the northern area showing the development of an Intra-Ness barrier sequence within the normally fluvially dominated Foyers Member. Above the barrier, back barrier/lagoonal deposits equivalent to the Enrick and Oich members re-occur.

lagoonal deposits of the Oich Member. The main facies associations within the Enrick Member are identified as follows.

(a) *Back barrier*. This facies association comprises predominantly very fine- to medium-grained sandstone commonly arranged in coarsening upward sequences above lagoonal shales. The sandstones exhibit small scale (millimetre–centimetre) cross lamination and are commonly rippled at the top. Minor bioturbation and rootletting is also recorded. These deposits are interpreted as wash-over fans and flood-tidal deltas reworked from the Etive barrier. Individual delta lobes may be abandoned and thereafter become vegetated. Occasionally small upward fining units are seen which may show bi-directional cross bedding and are thought to be tidal channels. Within the Cormorant Field (Marshall *et al.* pers. comm.) and the Brent Field (Livera 1989) distributary channels and minor mouthbars have been identified. Local channel switching driven by compactional autocyclic controls have been suggested for such features.

(b) *Stagnant swamp*. Periodic isolation of the back barrier resulted in the formation of small stagnant swamps. These deposits are typified by thin coals and finely laminated carbonaceous shales. Rarely, simple vertical burrows are recorded. The mechanism by which these swamps developed is probably related to autocyclic switching of small distributaries as discussed above.

(c) *Lacustrine*. Gradual influx of fresh water in response to further channel switching enlarged the swamps and developed small brackish lakes. This change is recorded by the presence of more heterolithic facies which commonly include coarsening upwards units. Local variations in spore content indicate a rapidly changing environment as new flora became established on lake margins (see Whitaker *et al.* this volume for discussion).

Distribution of the Enrick Member proper appears to be limited to the southern parts of the basin with a maximum northward extent in the lower half of blocks 211/27, 211/28. The Enrick Member is interpreted as an initial phase of progradation of immature delta top sediments, which were later drowned by a major flooding event represented by the overlying Oich Member in the more southerly parts of the basin.

*Oich Member (formerly Mid-Ness Shale)*. The Oich Member is generally characterized by a widespread organic-rich shale, which, based on palynology, suggests an open lagoon environment. The shales are commonly burrowed and are locally strongly bioturbated. Occasionally thin sandy intervals break up the package and may represent reworking of the barrier during storms and/or the more distal deposits of minor mouth-bars.

Within a field area the Oich Member is readily correlatable and is of great importance in reservoir development. However, within the Brent Province as a whole the sequence can be shown to be highly diachronous and to represent the progradation, temporary retreat and maximum development of a series of lagoon dominated environments (Whitaker *et al.* this volume).

*Foyers Member*. The Foyers Member comprises a variety of delta top and coastal plain environments ranging from south to north across the basin. The lower boundary is generally gradational with the underlying Oich Member and the upper boundary is frequently truncated or reworked by the Tarbert transgressive phase. Three main depositional environments have been recognized with additional local variations also seen.

(a) *Fluvial dominated delta top*. This environment comprises three main facies associations. A typical vertical log sequence is presented as Fig. 9 demonstrating their relationship.

(i) *Fluvial channel sandstones* comprise fine sand- to granule-grade material, are generally clean and dominated by decimetre-scale cross-bedded, upward-fining sequences. Channel bases are commonly sharp and erosive with a pebbly lag. Stacking of channel bodies is seen in the most sand prone areas e.g. block 3/3. The tops of abandoned channel sequence are commonly overlain by coals. The thickness, and therefore connectivity, of individual sands is highly variable ranging from 5–30 ft in the Brent Field (Livera 1989) to 30–90 ft further south (Lyell Area).

(ii) *Crevasse splay and overbank sandstones* are generally very fine to fine grained and show parallel and ripple cross lamination. Individual units are thin, less than 5 ft, but may be stacked into overall coarsening upwards sequences. Rare burrows and minor rootlet development are recorded towards the tops of individual and stacked units.

(iii) *Deltaplain (floodplain) shales and coals* make up the major part of the Foyers Member. Such deposits are poorly bedded or slumped and associated with minor sandstone lenses producing a heterolithic facies. The sequences are typical of vegetated, poorly drained floodplain. Coals are of varying thickness and may be of value as local correlatable horizons. These deposits are seen as the tops of major fining upwards sequences possibly representing gradual local relative sea level rises or more widespread flooding surfaces on the delta plain.

(b) *Coastal plain*. Marginal sediments especially towards the upper part of the Foyers Member record the transition from the fluvial dominated delta plain to a more low lying, lower energy coastal plain. These sediments are typified by very fine to medium grained sandstones with wispy mud intercalations (?flasers) and pervasive bioturbation. Distinct burrow types such as *Scoyenia* and *Planolites* attest to the marine influence. The presence of rare *in situ* coals and occasional thick (up to 10 ft) bioturbated shale packages indicate the development of local marine embayments with vegetated margins. These marginal environments are distinguished by palynofacies analysis as well, but their local high energy input cannot be clearly attributed to fluvial/distributary channel origin or marine reworking.

(c) *Intra-Ness barrier*. A third facies association has been identified in the more northerly areas (211/13, 211/17, 211/18, 210/25) which is thought to represent a barrier development (Fig. 10). In a number of wells in the area, above the defined Oich Member a thick (50–80 ft), homogeneous, massive fine-grained

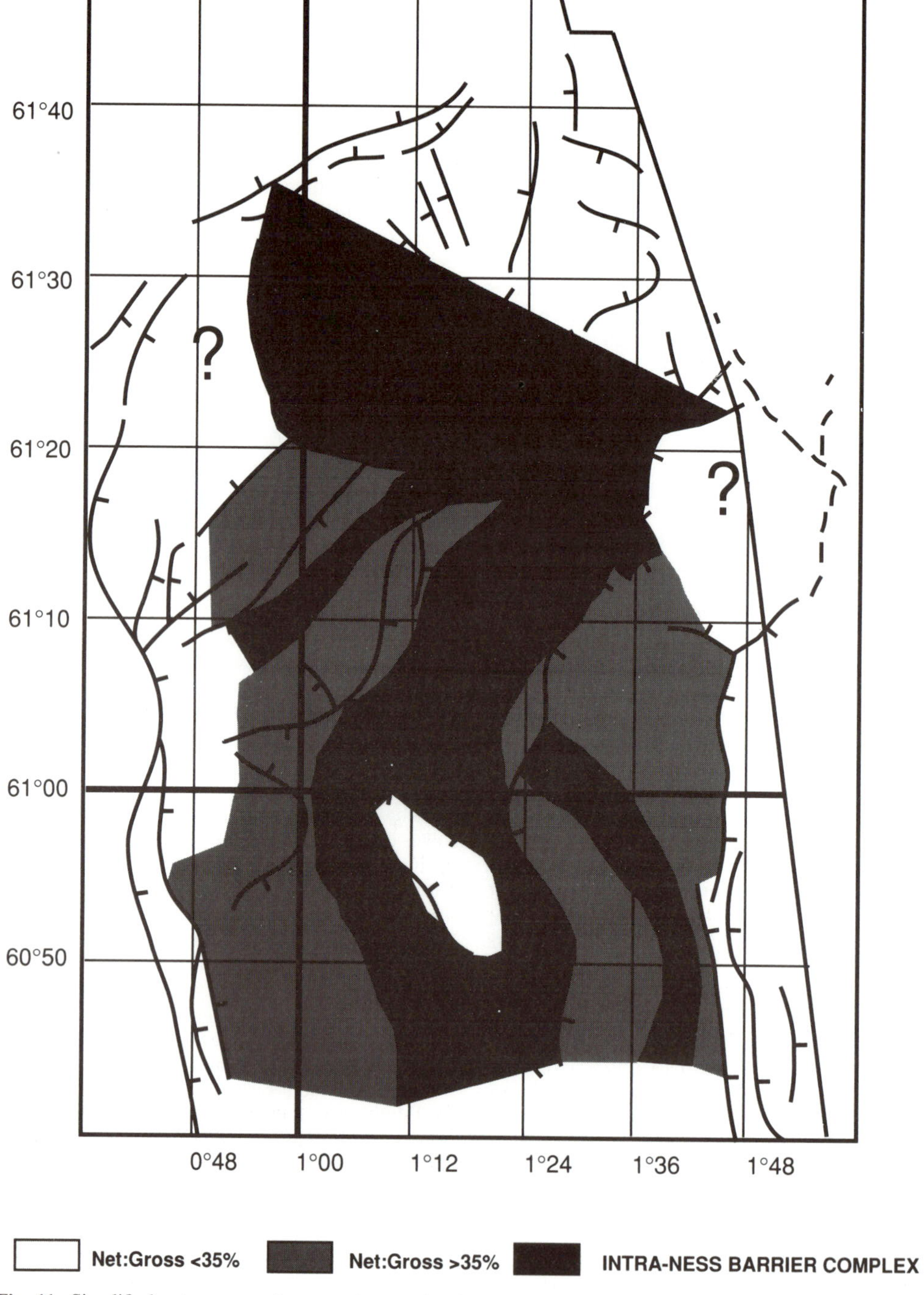


**Fig. 11.** Simplified net:gross sandstone ratio map for the Foyers Member.

sandstone unit is observed. This sequence commonly exhibits dewatering structures, weak bioturbation and local carbonate cementation. The interval has been correlated on regional panels and mapped out (see Fig. 11) as a widespread, presumably laterally persistent, sheet-like unit. The unit may lie directly above a thin Oich Member or be underlain by a thin Foyers Member. Commonly, above the sandstone unit, open lagoonal shales, similar to the Oich Member, are seen. In view of the previously discussed interfingering of Etive/Rannoch sequences in the same area and together with the observed lithofacies and distribution it is suggested that this unit represents a temporary retreat of the delta front and the formation of a second barrier sequence.

### *Controls on sedimentation*

Controls on sedimentation within the Ness Formation may be demonstrated by the distribution of the depositional environments of the Foyers Member and their relationship to local and regional structures. Figure 11 is a generalized net:gross ratio map of the Foyers Member based on a cut off of 35% net sand. Four areas of higher sand content are seen, including the almost 100% N:G of the intra-Ness Barrier. These are related to some of the main structural features of the East Shetland Basin. Thickness distributions of the whole Ness interval also show similar structural controls (Fig. 12).

*Brent–Alwyn Terrace*. This easterly fault terrace is dominated by a thick sequence of vegetated flood plain sediments with important correlatable coals within the Foyers Member. The area can be shown to have been actively subsiding throughout the Early and mid-Jurassic, as all formations of the Brent Group and also the Dunlin Group (Fig. 13) are commonly considerably thicker in this area than elsewhere. During the development of the Foyers Member this area was predominantly a slowly subsiding swamp dissected by laterally confined fluvial systems. The concentration of channel sands picked out by the N:G ratio map indicates a gradual migration of major sand bodies down the hanging wall of the fault terrace.

*Ninian–Hutton Terrace*. The next fault terrace to the west, this area comprises major fluvial sand bodies throughout the Foyers Member. It is here that the thickest channel sequences are recorded. A stacked channel body up to 90 ft thick is seen in well 3/3–3 in the southern part of this fault terrace. Local correlation shows the preservation of only single large channel sequences either side of this major channel indicating the switching and stacking of a fluvial system. The continuation of this sand prone trend picked out by N:G ratios runs to the north and northwest presumably down the palaeoslope of the terrace hanging wall before swinging eastward possibly in response to the presence of the Tern–Eider Horst. A separate sand prone trend crosses the Cormorant Terrace again deflected by this horst. More detailed mapping in the future of individual channel belts may give additional information regarding the relationship between fault activity and fluvial architecture. Towards the northern part of this area the major fluvial systems begin to breakdown, forming smaller, poorly connected channel systems. This occurs as progradation is nearing a maximum. Coastal plain facies rather than purely fluvial dominated delta top conditions become dominant.

*Tern–Eider Horst*. This structural feature is thought to have been a relative high throughout the progradation of the whole Brent Group; all depositional units are attenuated over it. The control exerted on fluvial development has already been mentioned with respect to the Foyers Member. The role played by the Tern–Eider Horst on more northerly development related for instance to the Middle Jurassic deltaic deposition in the Unst Basin (Johns & Andrews 1985) can only be speculated upon.

### *Tarbert Formation*

A detailed study of the Tarbert Formation is difficult to pursue due to the limited available information. The sequence is frequently absent through erosion and truncation and even when present, is seldom fully cored. Identification of the Tarbert is therefore often made using wireline-logs only; a risky technique as the sequence is effectively a heavily reworked interval of Rannoch/Etive or Ness sediments consequently showing little petrophysical contrast. It should therefore be stressed that without reliable core or biostratigraphic data the Tarbert Formation may be wrongly identified. The nature of the prograding system is such that in the north of the basin, where Ness deposition was restricted, the Tarbert Formation may directly overlie Etive and then Rannoch sequences making identification more problematical.

Where core material is available a typical vertical sequence can be described (Fig. 14) as

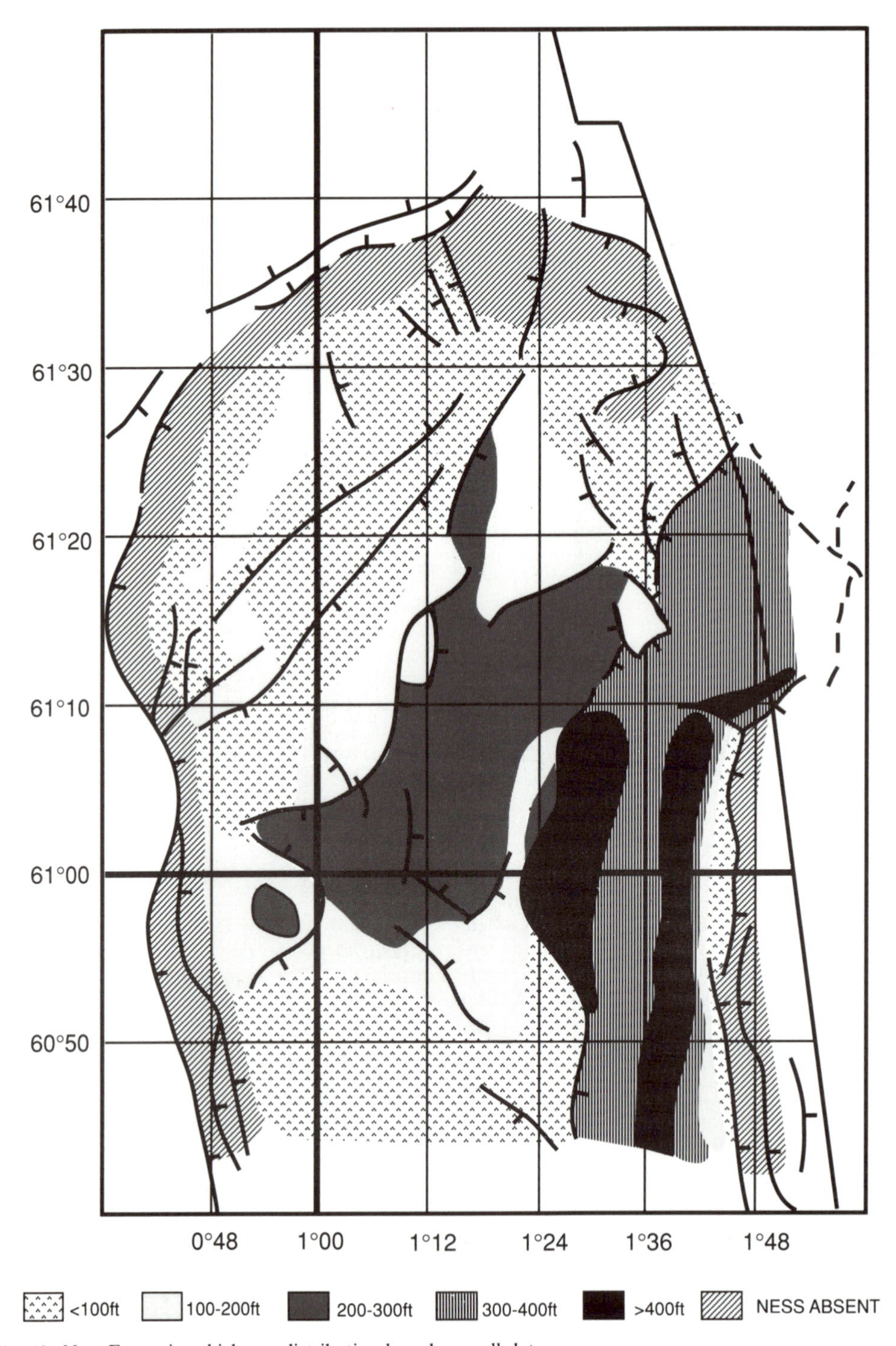


**Fig. 12.** Ness Formation thickness distribution based on well data.

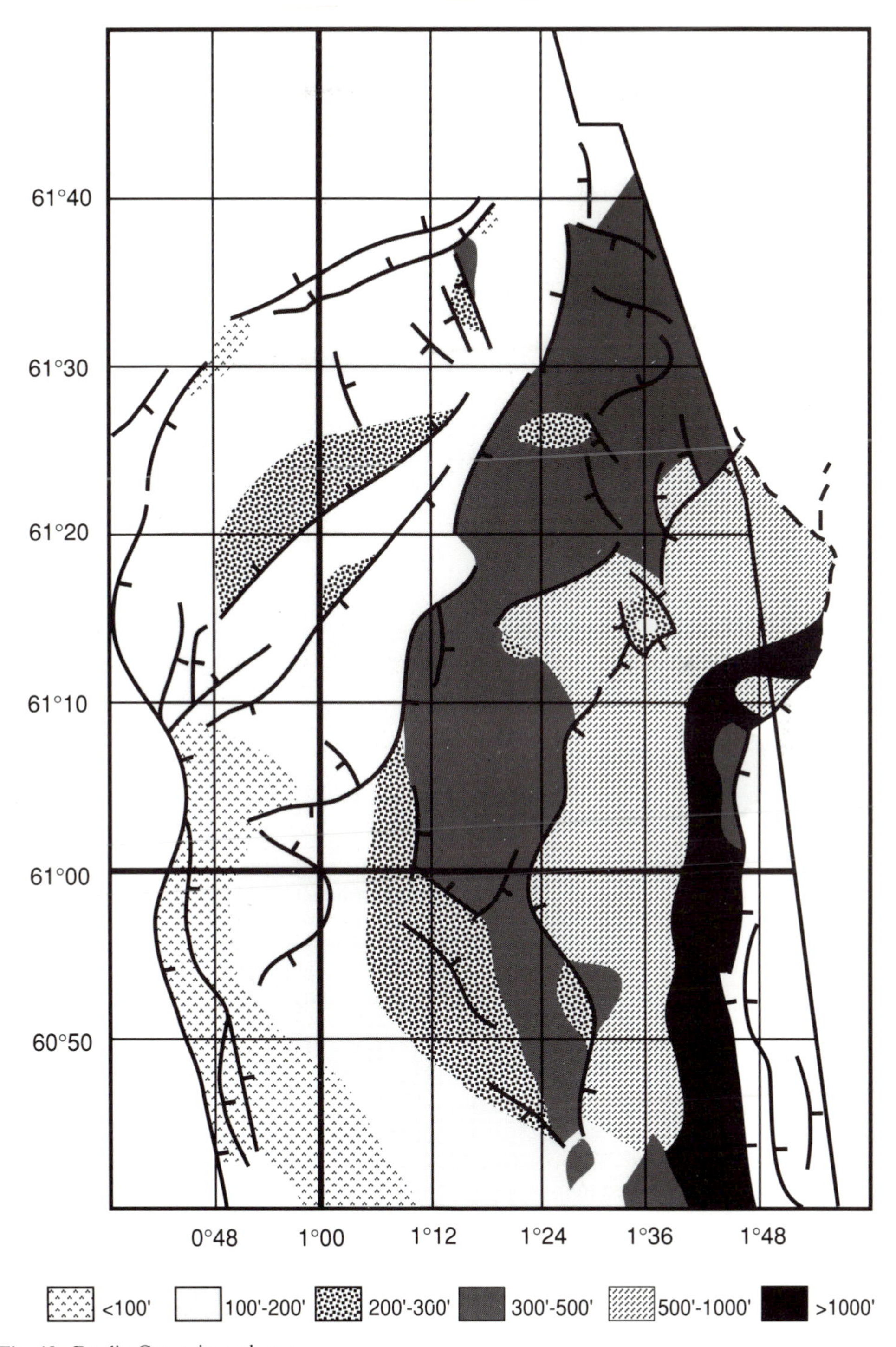


**Fig. 13.** Dunlin Group isopach map.

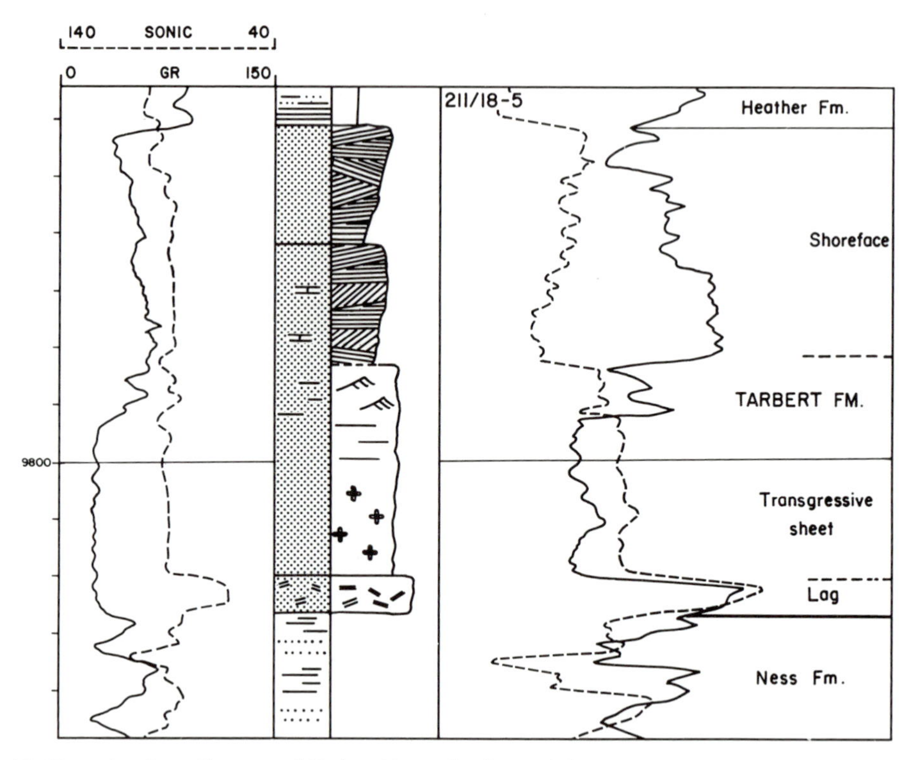


**Fig. 14.** Example of a well preserved Tarbert Formation interval showing basal lag with reworked Ness Formation shale and coal; massive and bioturbated transgressive sandsheet and stakced regressive (?) shoreface sequences.

having a coarse to granule grade erosional base with overlying fine to medium grained sandstones which are intensely bioturbated. The upper part of the sequence where preserved commonly comprises very fine- to fine-grained parallel-laminated sandstones similar to the Rannoch Formation and representing a shoreface environment. This later shoreface sequence is thought to be regressive in nature, occurring as part of an overall transgressive regime. This transgressive event is locally variable especially when viewed from north to south. This is particularly apparent when taken in relation to the underlying formation be it Ness, Etive or Rannoch.

In the northern parts of the basin (211/13, 211/14, 211/19 in part) the Tarbert transgression reworks Etive or Rannoch deposits. This is represented by thin granule lags attributed to ravinement surfaces (Ronning & Steel 1987) or minor channelling events (Brown *et al.* 1987). Further south (211/21) peculiar reddened zones (<1 ft) with coarse sand grains and small ooliths floating in a hematitic matrix have been recorded. These occur at both the base of the Tarbert Formation and higher in the interval and are thought to represent condensed sequences where the transgression may have temporarily by-passed a local high. Further evidence for the gradual nature of the transgression is recorded where repeated bioturbated units are separated by minor rootletted horizons (211/29–3) which indicate periods of emergence in the overall transgression.

The present-day distribution of the Tarbert Formation is patchy, largely due to later tectonically induced erosive events. Figure 15 shows the potential thickness of Tarbert Formation remnants and records the thickening of the interval to the south and east indicating continued fault controlled subsidence in this area.

Whitaker *et al.* (this volume) have dated all confirmed Tarbert Formation intervals studied in this review as no older than earliest Bathonian or very latest Bajocian in age, making correlation with the Hugin or Sleipner Formations unlikely (cf. Graue *et al.* 1987) as these are considered to be generally younger, i.e. Upper Bathonian–Callovian in age.

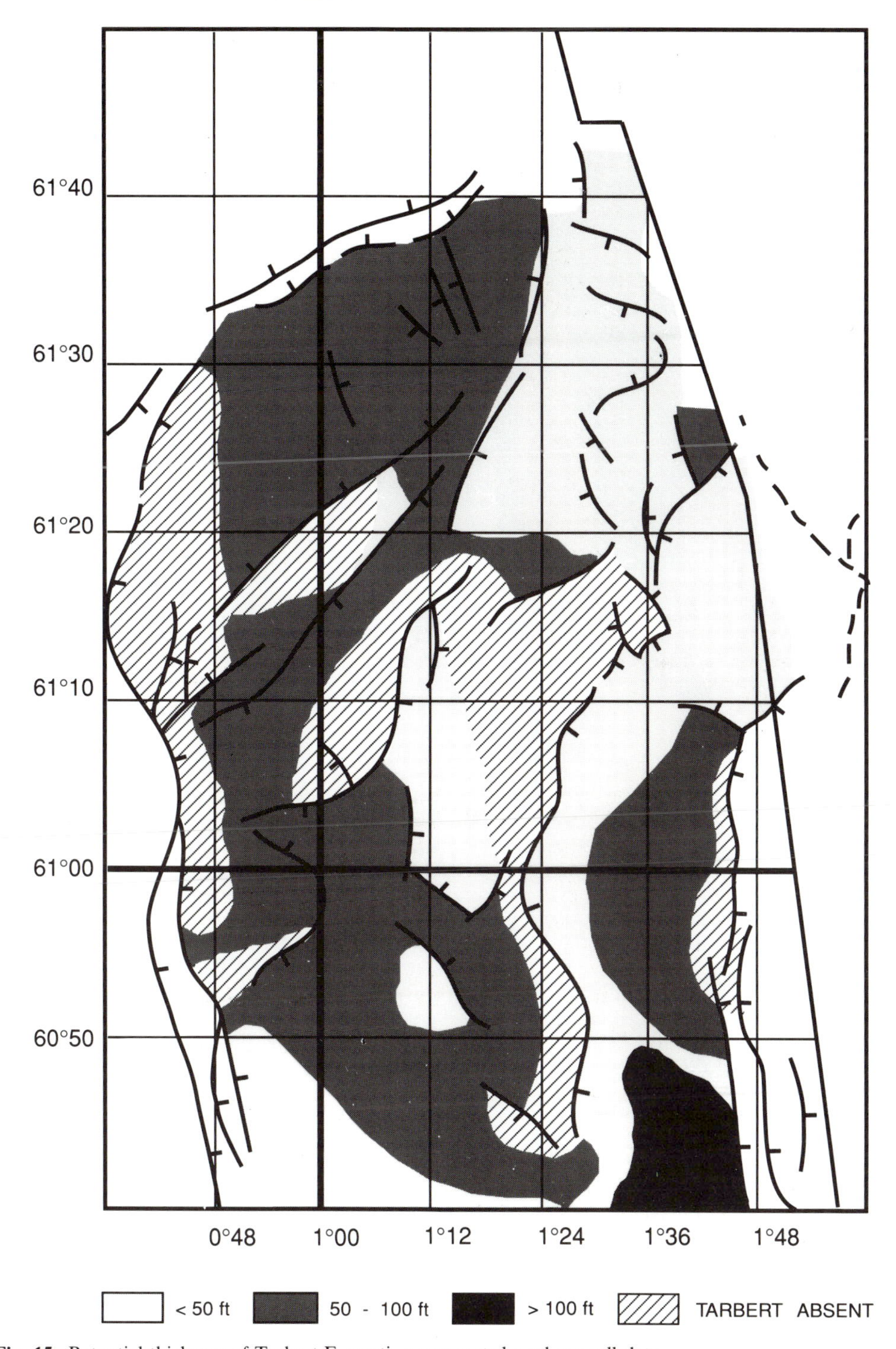


**Fig. 15.** Potential thickness of Tarbert Formation remnants based on well data.

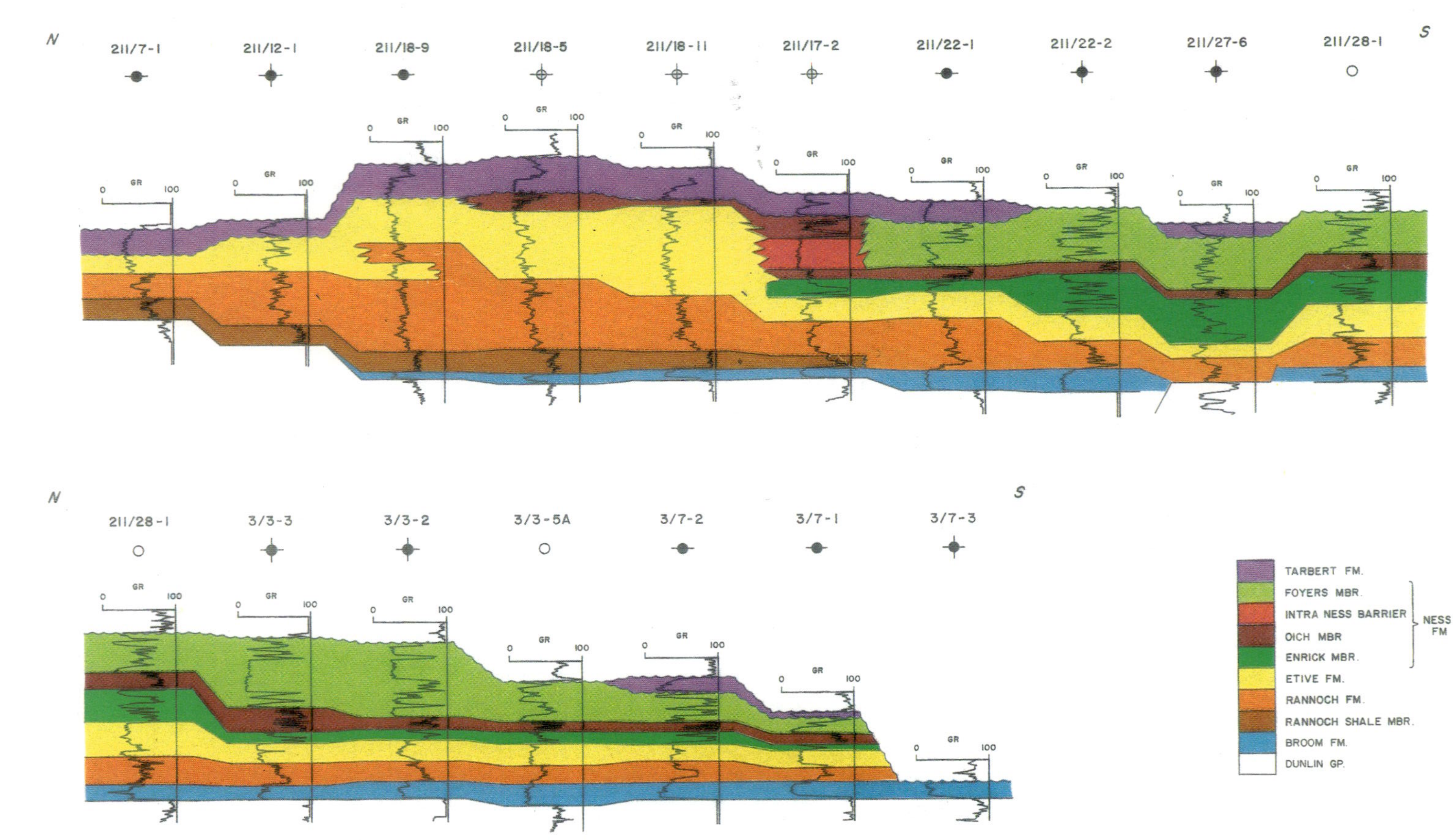


**Fig. 16.** N–S regional correlation panel through Brent delta. See Fig. 1 for location. Vertical scale 1:2500.

## Geological history

The Brent deltaic province has been the focus of detailed study by many authors (e.g. Budding & Inglin 1981; Johnson & Stewart 1985; Moiola *et al.* 1985; Brown *et al.* 1987; Graue *et al.* 1987; Mitchener *et al.* this volume). Attempts to distil the mountain of data into all encompassing models which can be related to modern day analogues are unlikely to succeed as realistically no single model is going to give 'the answer'. Complications abound because of the Holocene sea-level rise affecting shorelines and mouths of modern deltas. An example of one of the north–south correlation panels is included as Fig. 16, which show some of the lateral thickness variations and the relationships between different units.

During the Late Toarcian–Sinemurian the Northern North Sea was an area of substantial marine fine-grained deposition (Dunlin Shale). This period of quiescence was disturbed by the reactivation of marginal fault systems brought about by early rift activity (Ziegler 1981) in response to lithospheric stretching or post rift thermal subsidence from earlier (? Permian) rifting (Badley *et al.* 1984). In the west and east locally sourced fan delta deposition occurred (Broom and Oseberg Formations) in partial response to the uplifted rift margins. Heavy mineral analysis indicates separate sources for each fan delta system independent of later sources (Morton, this volume). The development of the fan deltas was strongly influenced by local fault activity and resulting topography producing distinct wedge shaped units. Infilling of tilted blocks by Broom sediments acted to smooth the sea-bed topography at the time and the subsequent palaeoslope. Further or continued uplift or thermal doming to the south during the early Bajocian produced an alternative source of sediment. In a period of overall sea-level high (Vail & Todd 1981) sediment input was sufficient to produce a rapidly prograding coastal/delta system.

By using the pseudo-timelines recorded by Whitaker *et al.* (this volume) it is now possible to produce schematic palaeographic reconstructions of the development of the Brent deltaic depositional system up to its maximum stage of progradation (Fig. 16). The later destructive phase of the Tarbert transgression is still ambiguous in its various stages.

Initially the lowermost part of the depositional system pushed northwards (Fig. 17.1) along the Brent–Alwyn Terrace as an active shoreface/barrier sequence (Rannoch/Etive). The effects of wave/storm activity were dominant in building out the barrier. Possible longshore currents aided sediment input from the east. Behind the barrier open lagoons existed, fed by a major axial drainage system, while on the floodplain small marginal lakes and coal swamps developed (Fig. 17.2).

At some stage in the early development of the system a major flooding event occurred destroying the immature delta top sediments (Enrick Mb.) in the south of the basin. This is represented by the widespread lagoonal sediments of the Oich Member in these areas (Fig. 17.3). This event was effectively synchronous but only occurred in the south. The cause of this event may be related to a breakdown in the protective barrier brought about by tectonic or isostatic re-adjustment of local fault blocks. These events may also have produced the next stage of progradation as discussed below.

During the next stage of progradation fluvial input switched from the east to the west with a major fluvial fairway located to the north along the Ninian–Hutton Terrace (Fig. 17.4). This had the effect of feeding sediment into the protected bay to the west of the 'nose' and smoothing the profile of the delta front. While fluvial dominated deposition continued to the west, the Brent–Alwyn Terrace maintained a slow but steady rate of subsidence. A large coastal (?) coal swamp developed with only minor fluvial systems crossing the area.

Westerly sourced peripheral fluvial input also occurred leading to delta plain sedimentation across the Cormorant Terrace, Tern–Eider Horst and possibly the Unst Basin. Throughout, the barrier and shoreface sequences continued to push north and eastward in a large concave arc to some extent controlled by the Tern–Eider Horst, drainage systems being deflected by this relative high (Fig. 17.5). Behind the delta front sediments, the lagoonal sequences also prograded but probably at different rates. This phase of progradation represented by the Oich Member is therefore diachronous and not related to the earlier drowning event recognized in the south. The open lagoon shows a period of contraction possibly in response to increased sediment input from the south (Fig. 17.6). However, soon after, progradation of the barrier outstripped fluvial input, possibly in response to additional longshore input allowing the lagoon to expand. It is represented locally by an extended Oich Member.

A temporary retreat of the barrier sequence occurred when the delta had prograded further north producing the intra-Ness barrier (Fig. 17.7). Whether this was a rapid or gradual event cannot be determined but the likelihood

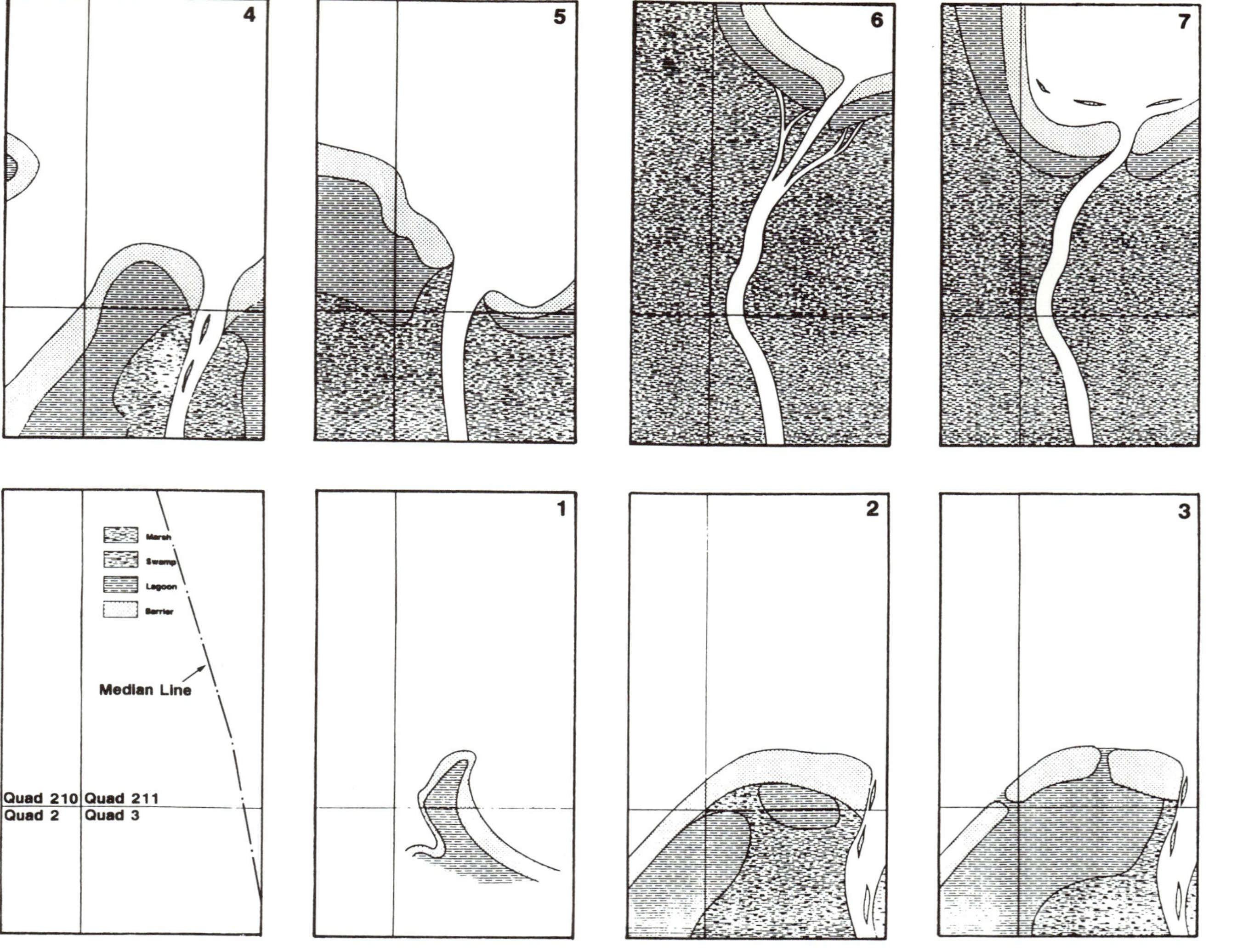


**Fig. 17.** Palaeographic reconstructions of Brent delta progradation (after Whitaker *et al.* this volume).

is that the delicate balance of the system was disturbed by eustatic or tectonic factors rather than sediment input falling off. The short lived nature of this event is reflected in the return to progradation of the system as a whole; however, little fluvial sedimentation occurred further north. A low lying coastal plain, probably already cut by tidally influenced channels and major distributaries developed as barrier and shoreface stalled. Sediment input to the barrier system could now have been even more strongly influenced by longshore currents (Morton this volume). Figure 18 presents a palaeogeographic reconstruction of the maximum phase of progradation of the different Brent facies belts.

The start of the Tarbert Formation marine transgression in the north is thought to have been gradual with the drowning of the barrier and retrogradation of the coastal plain. Reworking of the underlying sediments into a series regressive shorefaces also occurred. The cause of the transgression may be linked to many factors including eustatic or tectonic forces or by an over-extension of the whole system with an increase in sediment capture on the alluvial plain (Graue *et al.* 1987). The transgression gathered pace southwards becoming more erosive although local periods of quiescence occurred as areas were temporarily bypassed.

## Discussion

The Brent Delta has been compared with many modern day deltaic systems in search for a satisfactory analogue; Niger and Grijalva (Budding & Inglin 1987), Nile (Johnson & Stewart 1985) and the Lafourche lobe of the Mississippi Delta (Moiola *et al.* 1985). To some extent certain attributes of each of these example may be applicable to the Brent at various stages in its development.

Using the triangular classification scheme of Galloway (1975) and the prognostic approach of Coleman & Wright (1975) a series of delta types may be proposed for the evolution of the Brent. In its earliest stages the delta prograded through a fairly narrow basin or series of basins perpendicular to the main marine opening to the north. The rapid progradation of a sand-rich delta front indicates the dominance of fluvial input into a high energy environment. The development of a 'nose' of sediment suggests an elongate profile to the system at least during its earliest phase. Only when the delta has prograded further northwards can a more lobate profile be predicted, indicating the increasing influence of wave activity and possible longshore currents. This in itself may be related to a widening of the basin mouth. The eventual domination of wave energy built up a protective barrier or beach plain, behind which the fluvial and lagoonal environments developed.

As the rate of sediment input from fluvial sources waned modification of the delta front probably took place, resulting in a more concave profile. At the maximum phase of progradation little fluvial input to the delta front occurred, most sediment being trapped on the delta-plain; the delta effectively stalls and continued progradation is halted by an anticipated increase in water depth allowing greater space for delta front sediments to accumulate. These latest sediments may be derived from lateral sources, either longshore drift or additional fluvial input from the west.

It is at this stage in the development that wave and tidal influences exert complimentary effects, re-working the delta front and to a lesser degree the back-barrier and coastal plain sediments. The gradual progression from wave to tidal influence marks the initial drowning phase of the Brent delta as typified by the sequences in the north of the basin. Tidal domination of the whole delta was probably never achieved; however locally the effects of tides may have been significant. As the delta was transgressed southwards the likelihood is that wave driven processes dominated, producing thick regressive shoreface sequences. The low preservation potential of this phase of delta evolution in the East Shetland Basin means than an unequivocal understanding is not possible.

## Conclusions

This study has resulted in three important conclusions which may be listed as follows.

(1) Palynofacies studies are of significant value in the modelling of delta evolution (Whitaker *et al.* this volume).

(2) The identification of a phase of delta retreat and barrier development within the Ness Formation, prior to the main Tarbert transgression, greatly increasing reservoir connectivity in the north of the Brent Province.

(3) Potential tectonic and eustatic controls on sedimentation and distribution of Brent Group formations have been identified through detailed sedimentological core logging and regional correlation.

The Brent Group in the UK Sector of the North Sea is now a super-mature hydrocarbon province with little true exploration potential remaining. It is by detailed facies mapping, and sandstone distribution and connectivity studies

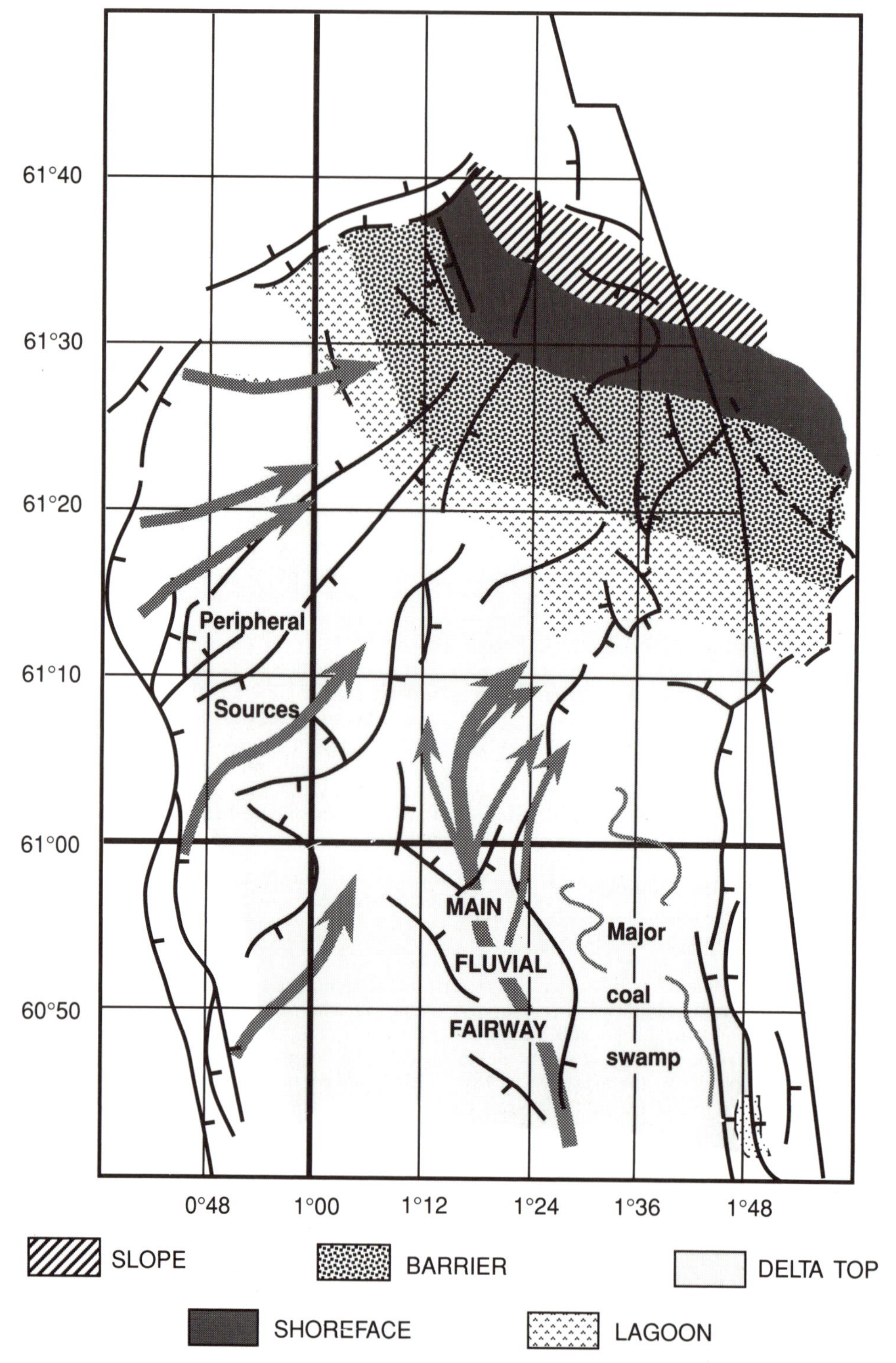


**Fig. 18.** Palaeogeographical representation of the Brent delta showing main facies domains at the time of maximum progradation.

that hydrocarbon production and reserves will be increased. It is believed that by undertaking such a broad regional approach to the study in the first place that the framework required for such further work has been established. The proposed changes to the stratigraphy also form part of the framework on which further detailed studies can be made.

The authors would like to thank Shell UK Exploration and Production Limited and their partners Esso UK Exploration and Production Limited for permission to publish the results of this largely internal project. The many operating companies in the area who gave access to cores and sample material are also gratefully acknowledged. Thanks are also given to the various colleagues both past and present who gave advice and encouragement during the project. Finally we are indebted to Shell Expro's drafting department and in particular to B. Hancock.

## References

Badley, M. E., Egeberg, T. & Nipen, O. 1984. Development of rift basins illustrated by the structural evolution of the Oseberg feature, Block 30/6, offshore Norway. *Journal of the Geological Society, London*, **141**, 639–649.

Brown, S. & Richards, P. C. 1988. Facies and development of the mid-Jurassic Brent delta near the northern limit of its progradation, UK North Sea. *In*: Pickering, K. J. & Whateley, M. (eds) *Deltas; sites and traps for fossil fuels*. Geological Society, London, Special Publication, **41**, 253–267.

——, —— & Thomson, A. R. 1987. Patterns in the deposition of the Brent Group (Middle Jurassic) UK North Sea. *In*: Brooks, J. & Glennie, K. (eds) *Petroleum Geology of North West Europe*. Graham & Trotman, London, 899–414.

Budding, M. C. & Inglin, H. F. 1981. A reservoir geological model of the Brent sands in Southern Cormorant. *In*: Illing, L. V. & Hobson, G. D. (eds) *Petroleum Geology of the Continental Shelf of North-West Europe*. Heyden, London, 326–334.

Coleman, J. M. & Wright, L. D. 1975. Modern river deltas: variability of processes and sand bodies. *In*: Broussard, M. L. (ed.) *Deltas*. Houston Geological Society, 99–149.

Deegan, S. E. & Scull, B. J. 1977. *A standard lithostratigraphic nomenclature for the Central and Northern North Sea*. Institute of Geological Sciences Report **77/25**.

Eynon, G. 1981. Basin development and sedimentation in the Middle Jurassic of the Northern North Sea. *In*: Illing, L. V. & Hobson, G. D. (eds) *Petroleum Geology of the Continental Shelf of North-West Europe*. Heyden, London, 196–204.

Graue, E., Helland-Hansen, W., Johnsen, J., Lomo, L., Nottveld, A., Ronning, K., Ryseth, A. & Steel, R. 1987. Advance and retreat of the Brent Delta System, Norwegian North Sea. *In*: Brooks, J. & Glennie, K. (eds) *Petroleum Geology of North West Europe*. Graham & Trotman, London, 915–938.

Galloway, W. E. 1975. Process framework for describing the morphological and stratigraphic evolution of deltaic depositional systems. *In*: Broussard, M. L. (ed.) *Deltas*. Houston Geological Society, 87–98.

Hamilton, P. J., Giles, M. R. & Ainsworth, P. 1992. K-Ar dating of illites in Brent Group reservoirs: a regional perspective. *This volume*.

Johns, C. R. & Andrews, I. J. 1985. The petroleum geology of the Unst Basin, North Sea. *Marine and Petroleum Geology*, **2**, 000–000.

Johnson, H. D. & Stewart, D. J. 1985. The role of clastic sedimentology in the exploration and production of oil and gas in the North Sea. *In*: Brenchly, P. J. & Williams, B. J. P. (eds) *Sedimentology; recent developments and applied aspects*. Geological Society, London, Special Publication, **18**, 249–310.

Livera, S. E. 1989. Facies associations and sand-body geometries in the Ness Formation of the Brent Group, Brent Field. *In*: Pickering, K. J. & Whateley, M. (eds) *Deltas; sites and traps for fossil fuels*. Geological Society, London, Special Publication, **41**, 269–288.

Mitchener, B. C., Lawrence, D. A., Partington, M. A., Bowman, M. B. J. & Gluyas, J. C. 1992. Brent Group: sequence stratigraphy and regional implications. *This volume*.

Moiola, R. J., Jones, E. L. & Shanmugan, G. 1985. Sedimentology and diagenesis of the Brent Group (Middle Jurassic), Statfjord field, Norway–United Kingdom. *In*: *Habitat of Hydrocarbons Norwegian Oil and Gas Finds*, abstract volume.

Morton, A. C. 1992. Provenance of Brent Group sandstones: heavy mineral constraints. *This volume*.

Richards, P. C. & Brown, S. 1986. Shoreface storm deposits in the Rannoch Formation (Middle Jurassic), North West Hutton oilfield. *Scottish Journal of Geology*, **22**, 367–375.

Ronning, K. & Steel, R. 1987. Depositional sequences within a 'Transgressive' reservoir sandstone unit: the Middle Jurassic Tarbert Formation, Hild area, Northern North Sea. *In*: Kleppe, J. *et al.* (eds) *North Sea Oil and Gas Reservoirs*. Graham & Trotman, London, 169–176.

Vail, P. R. & Todd, R. G. 1981. Northern North Sea Jurassic unconformities, chronostratigraphy and sea-level changes from seismic stratigraphy. *In*: Illing, I. V. & Hobson, G. D. (eds) *Petroleum Geology of North West Europe*. Graham & Trotman, London, 216–235.

Whitaker, M. F., Giles, M. R. & Cannon, S. J. C. 1990. Palynostratigraphical review of the Brent Group, UK sector North Sea. *This volume*.

Ziegler, P. A. 1981. Evolution of sedimentary basins in North-West Europe. *In*: Illing, L. V. & Hobson, G. D. (eds) *Petroleum Geology of the Continental Shelf of North-West Europe*. Heyden, London, 3–42.

# Advance and retreat of the Brent delta: recent contributions to the depositional model

W. HELLAND-HANSEN[1], M. ASHTON[2], L. LØMO[1] & R. STEEL[3]

[1] *Norsk Hydro Research Centre, PO Box 4313, N-5028 Bergen, Norway*

[2] *Badley, Ashton & Associates, Winceby, Horncastle, Lincs LN9 6PB, UK*

[3] *Norsk Hydro, PO Box 200, N-1321 Stabekk, Norway*

**Abstract:** The behaviour and the development of the Brent Group in the Northern North Sea is documented by cross-sections and paleogeographic maps. Although this work is based on a previously published Norsk Hydro study, we present recent progress in the understanding of the Brent Group which has been achieved by a better timing of depositional events, by integrating new well information, and by generating new models based on the new information.

The early (Aalenian) lateral infill of the Oseberg Formation was deposited as a response to a relative fall of sea-level, probably generated by a tectonic uplift. Fan deltaic sediments built out towards the west and northwest and backfilled the previously emergent areas during the subsequent relative sea-level rise.

The progradation of the Brent Delta (Rannoch, Etive and lower Ness Formations) took place in Late Aalenian to Early Bajocian. The thickness of the Etive relative to the Rannoch Formation indicates thickening due to bathymetric deepening in the area south of the maximum regression whereas subsidence controlled the thickening in the area of the maximum extent of the delta.

During the retrogradational part of the delta development (the Tarbert and upper Ness Formations, Early Bajocian to Early Bathonian), increasing tectonic activity can be documented, both in terms of enhanced differential subsidence across faults and by early rotational uplift on some fault blocks.

The Brent Group of Middle Jurassic age is a succession of sandstones, siltstones, shales and coals up to 600 m thick. The group comprises the most significant hydrocarbon-bearing level in the Northern North Sea and constitutes major reservoirs in fields such as Oseberg, Gullfaks, Statfjord (Norwegian sector) and Brent, Ninian, Cormorant and Murchison (UK, sector).

The Brent Group includes the Broom, Rannoch, Etive, Ness and the Tarbert Formations in addition to the Oseberg Formation which has been defined as the Broom Formation equivalent on the Norwegian side (Graue *et al.* 1987). The Broom and Oseberg Formations represent early lateral infill of the basin whereas the remaining formations, constituting the bulk volume of the Brent Group, comprise a major regressive (Rannoch, Etive and Ness Formations) to transgressive (Ness and Tarbert Formations) clastic wedge. This clastic wedge was early recognized to be be of deltaic origin, and a northerly direction of progradation was postulated (e.g. Bowen 1975; Deegan & Scull 1977; Proctor 1980; Skarpnes *et al.* 1980; Eynon 1981). Later, more information covering aspects ranging from depositional processes to palaeogeographic evolution have been published (e.g. Brown *et al.* 1987; Graue *et al.* 1987; Richards *et al.* 1988; Brown & Richards 1989; Fält *et al.* 1989; Livera 1989; Ryseth 1989).

The present paper attempts to elaborate upon the previously published Norsk Hydro work on the Brent Group (Graue *et al.* 1987, study area indicated in Fig. 1). There we put forward a depositional model for the Brent Group including an interpretation of the Oseberg Formation in terms of fan-deltaic deposition; the Tarbert Formation as being an offset backstepping unit; and the upper part of the Ness Formation being deposited during the retrogradational phase. We also emphasized the contrasting nature of the marine sands of the early lateral infill (Oseberg and Broom Formations) relative to those of the main Brent delta, the former being gravitationally emplaced whereas the latter indicate strong wave influence.

Since then, our understanding of the Brent Group has improved in several respects: the biostratigraphical dating is more precise than before and more time-lines have been established; the model for the Oseberg Formation has been refined; and our knowledge of the nature of both the progradational and retrogradational phases of the main delta has increased. Moreover, early tectonic uplift is demonstrated to be important particularly during

*From* Morton, A. C., Haszeldine, R. S., Giles, M. R. & Broom, S. (eds), 1992, *Geology of the Brent Group*. Geological Society Special Publication No. 61, pp. 109–127.

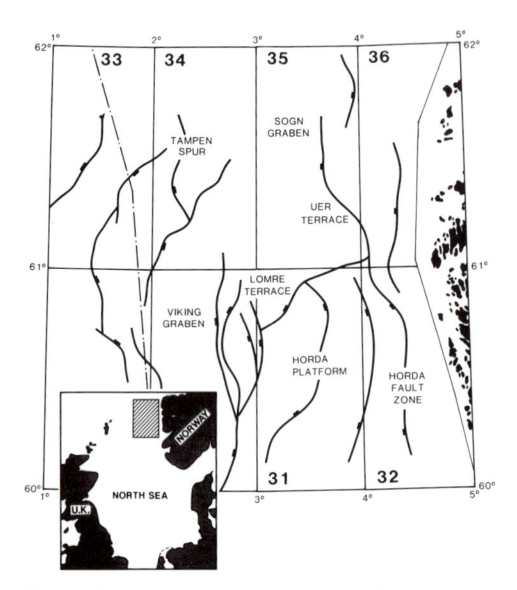


**Fig. 1.** Study area with major faults and physiographic provinces referred to in text.

the early lateral infill (Oseberg Formation), and locally during the retrogradational stage of the delta development. We also still maintain that both axial and lateral infill trends characterized Brent delta behaviour.

## Time-lines within the Brent Group

The establishment of time-lines within the Brent Group has been handicapped by the absence of a published and commonly accepted biostratigraphical framework for the Middle Jurassic. There is generally good agreement regarding the succession of palynological events, but their age interpretation varies among palynologists. For correlation purposes, a consistent biostratigraphy is more important than the accurate age determination. The palynological data have therefore been synthesized into a standard format developed for the northern North Sea by Norsk Hydro.

Due to their potentially high stratigraphical resolution, the occurrence of marine microplankton has been regarded as the best tool for establishing a regional palynostratigraphy through the Lower and Middle Jurassic. However, in a marginal marine setting, such as much of the Brent Group, the recovery of dinocysts is very low. Their first and last occurrences are closely related to the degree of marine influence and are therefore facies dependent.

In an attempt to get better stratigraphical resolution, the dinocyst stratigraphy has been combined with the relative distribution of land-derived palynomorphs. Detailed, quantitative analysis has been carried out for approximately 30 wells in the northern North Sea. The distribution of pollen and spores reflects the local and regional paleoflora. Some acmes of hinterland species have been consistently recorded and are probably related to climatic variations. These events cross formation boundaries and are therefore regarded as less dependent on local facies changes. They are interpreted as isochronous events, at least in a subregional context, and have been used in the definition of time-lines through the Ness and Tarbert formations.

**Table 1.** *Definition of time-lines*

| TIME-LINE | EVENT | AGE |
|---|---|---|
| 10 | ↲• C. hyalina | EARLY CALLOVIAN |
| 9 | ↰• CYSTE B/C | EARLY BATHONIAN |
| 8 | ↲ CYSTE B/C | EARLY BATHONIAN |
| 7 | ↰• A. australis<br>↲• ESCHARISPHAERIDIA spp. | LATE BAJOCIAN |
| 6 | ↲• C. macroverrucosus<br>↲ ESCHARISPHAERIDIA spp. | EARLY-LATE BAJOCIAN |
| 5 | ↰• COROLLINA spp. | EARLY BAJOCIAN |
| 4 | ↲• COROLLINA spp. | EARLY BAJOCIAN |
| 3 | ↰• NANNOCERATOPSIS spp. | LATE AALENIAN |
| 2 | ↰• SPHAEROMORPHS | EARLY TOARCIAN |
| 1 | ↰• SPHAEROMORPHS<br>↲ L. spinosa | LATEST PLIENS-BACHIAN |

↰ LAST STRATIGRAPHICAL OCCURRENCE

↲ FIRST STRATIGRAPHICAL OCCURRENCE

• COMMON - ABUNDANT OCCURRENCE

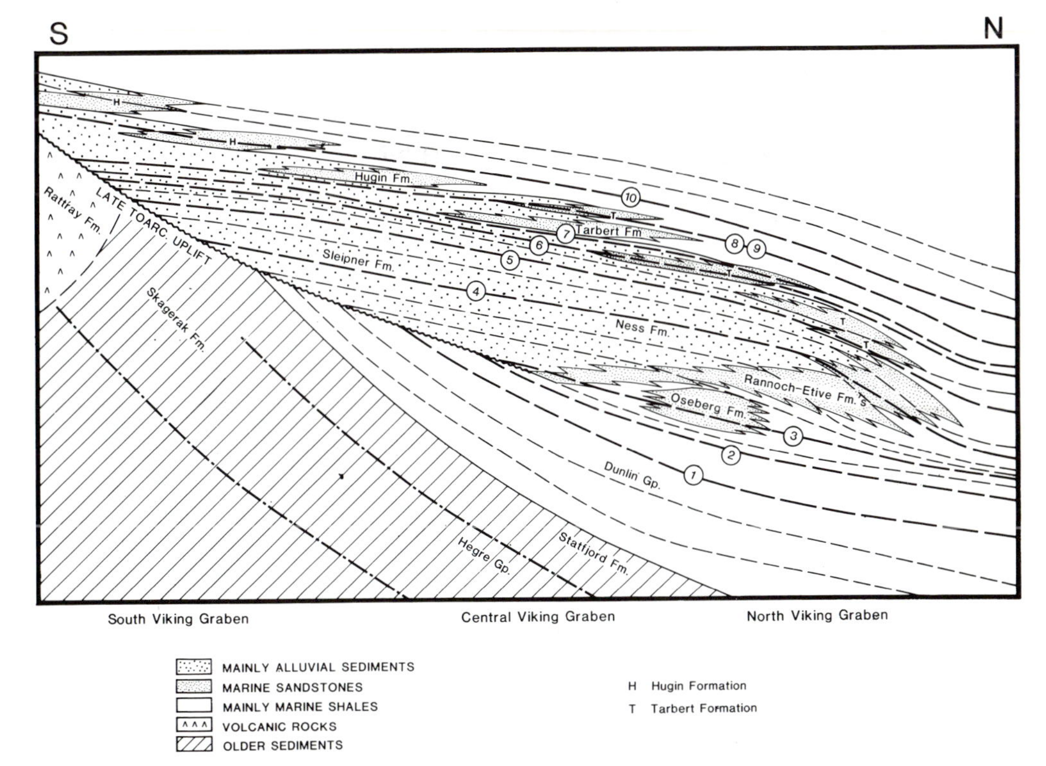


**Fig. 2.** Schematic south–north section through the Brent and Vestland Groups, showing formations and timelines within the megasequence (revised from Graue *et al.* 1987).

The succession of time-lines, their boundary criteria and their ages are summarized in Table 1. Figure 2 shows an updated version of Fig. 15 from the Graue *et al.* (1987) paper showing positions of timelines within the megasequence.

## *Definition of time-lines*

Time-line 1 (latest Pliensbachian) corresponds to the first stratigraphical occurrence of abundant sphaeromorph acritarchs in the early Jurassic. The bloom of sphaeromorphs is a widespread acme, which is believed to be caused by restricted water mass circulation in the North Sea. Additionally, line 1 is defined by the uppermost occurrence of *Luehndea spinosa*.

Time-line 2 (Early Toarcian) is defined by the last stratigraphical occurrence of abundant sphaeromorphs in the Early Jurassic.

Time-line 3 (Late Aalenian) corresponds to the last stratigraphical occurrence of common-abundant *Nannoceratopsis gracilis/senex*.

The time-lines 4 and 5 are defined by an acme of *Corollina torosus/meyeriana* during Early Bajocian times. This acme is interpreted as a response to a temporary increase in paleotemperature (cf. Vakhrameev 1970, 1978), and is regarded as an isochronous event in the northern North Sea. Line 4 corresponds to the base of this acme, while line 5 is recognized by its stratigraphical top.

Time-line 6 (Early–Late Bajocian) corresponds to the first stratigraphical occurrence of consistent *Escharisphaeridia* spp. and/or an increase in the amount of *Cerebropollenites macroverrucosus*.

Time-line 7 (Late Bajocian) is defined by the lowermost occurrence of common-abundant *Escharisphaeridia* spp. and/or a decrease in the influx of *Araucariacites australis*.

In the northern North Sea, time-line 8 and 9 are recognized by the acme of Cyste B/C (De Vains 1980), which may result in almost monotypic microplankton assemblages during Early Bathonian times. Line 8 corresponds to the base of this acme and is additionally characterized by the first stratigraphical occurrence of *Pareodinia evittii* and *P. brachythelis*. Line 9 is defined at the top of the Cyste B/C acme or by the lowermost occurrence of *Sirmiodinium grossii*.

Time-line 10 (Early Callovian) corresponds to the first stratigraphical occurrence of common *Chytreiosphaeridia hyalina*.

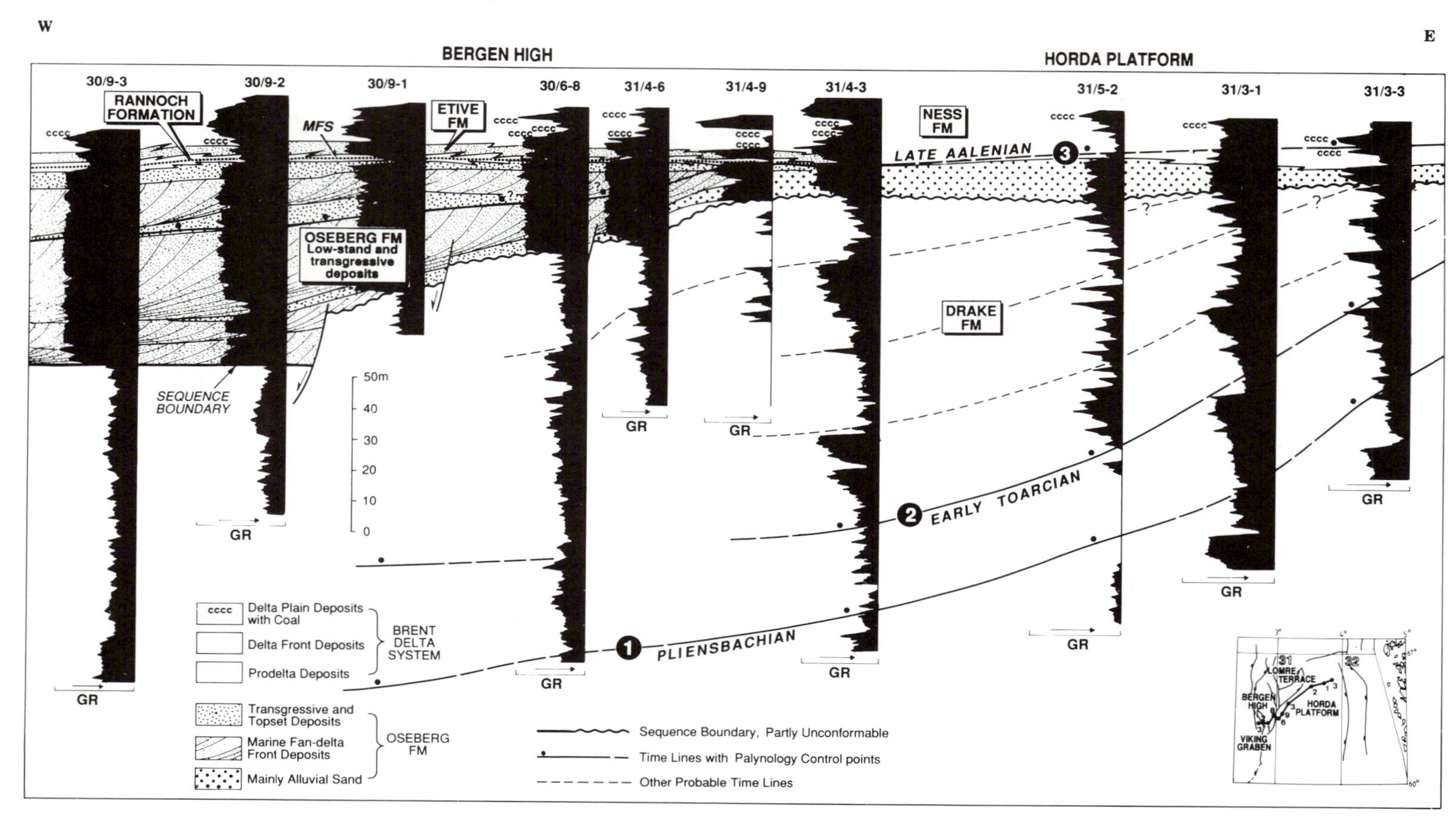


**Fig. 3.** East–west correlation of the Oseberg Formation, from Horda Platform towards Viking Graben.

## Early lateral infill of the basin

### *Pre-Aalenian infill*

During the Toarcian there had been repeated episodes of lateral infill of the northern North Sea basin from the Norwegian hinterland. This resulted in a series of sand wedges (within the Drake Formation) which tend to pinch-out before reaching the edge of the Horda Platform. (cf. Fig. 3). These sand wedges are the latest in a series of early Jurassic, basin margin-attached sand lobes which failed to reach or establish themselves in the axial parts of the basin. Since the Hettangian/Sinemurian, and prior to the establishment of Brent delta proper in latest Aalenian, there had been no well-developed, axially directed sediment transport in the northern North Sea basin (Steel & Gjelberg 1989; Gabrielsen *et al.* 1990).

### *Aalenian infill*

In the Aalenian there was a significant basinward shift in the locus of sedimentation, so that a younger series of sand lobes (Oseberg Formation) were now deposited beyond the edge of the Horda Platform, to the west and northwest and possibly locally to the southwest. Like the early Jurassic sand wedges, the Aalenian sands are basin margin-attached. However, they have quite a different internal geometry, showing steeply inclined progradational surfaces (Graue *et al.* 1987), indicative of relatively rapid outbuilding of coarse-grained sediments into shallow water. The Horda Platform itself (as well as adjacent terraces to the north and south) was largely emergent at this time and suffered active erosion along its eastern reaches. However it did accommodate, along its western and northern reaches, a thin, condensed alluvial unit which is in lateral continuity with the thicker marine lobes, which built out from the platform fringes.

### *Aalenian uplift and relative sea-level*

This basinward (westward and northwestward) shift of the Aalenian depocentres, the appearance of shallow marine and alluvial deposits above the Toarcian marine shales and sands, and the erosion and truncation of the older tilted deposits along the eastern reaches of the platform by the Aalenian sequence (see angular relationship between timelines 1, 2 and 3 in Figs 3 and 4), suggest that there was a relative fall of sea-level which generated the Oseberg Formation. The tilting indicates that the fall in large part was driven by tectonic uplift along the eastern edges of the older basin and along the Norwegian hinterland. The textural and mineralogical immaturity of the Oseberg Formation deposits further underlines this conclusion.

The Aalenian sedimentary system is thus quite distinct from the later Brent delta complex. The Oseberg Formation is seen as a product of a low relative sea-level stand and the subsequent sea-level rise, whereas the overlying Bajocian deltaic outbuilding occurred during the continued rising and subsequent high stand of sea-level.

### *Other Aalenian transverse-infill deposits*

The Broom Formation forms the 'mirror image' of the Oseberg Formation, along the western margin of the northern North Sea basin, in the East Shetland Basin (between 60°30′N and 61° 30′N) (Graue *et al.* 1987). Although generally thinner than the Oseberg Formation, the Broom Formation, like its eastern counterpart shows a fault-controlled, lobe-like development (Brown *et al.* 1987) and has also recently been interpreted in terms of fan-delta deposition (Stow & Kelman pers. comm.).

Along the southern closure of the Aalenian northern North Sea basin there is also likely to occur time-equivalent deposits as indicated by Mitchener *et al.* (this volume). They report alluvial deposits of the same age from the Bruce/Beryl embayment of the southern Viking Graben.

## Oseberg Formation interpretation

### *Depositional environment*

The Oseberg Formation was formally defined and its component facies were described by Graue *et al.* (1987). The formation tends to be dominated by medium or coarse grained, marine sandstones, either in bioturbated fairly flat-lying units which alternate with micaceous siltstones/mudstones or in rarely bioturbated, large-scale (up to 25 m) unidirectional, cross-stratified sets which dip internally at 10–30° (Figs 3 and 4). Rare units of coarse- to very coarse-grained, pebbly sandstones were interpreted as fluvial components in an otherwise marine-dominated association of façies (Graue *et al.* 1987). The vertical and lateral organization of these facies, together with dipmeter data which confirm the presence of large-scale (up to 45 m thick), single or multistorey, progradational sandbodies, led to an interpretation of Oseberg Formation in

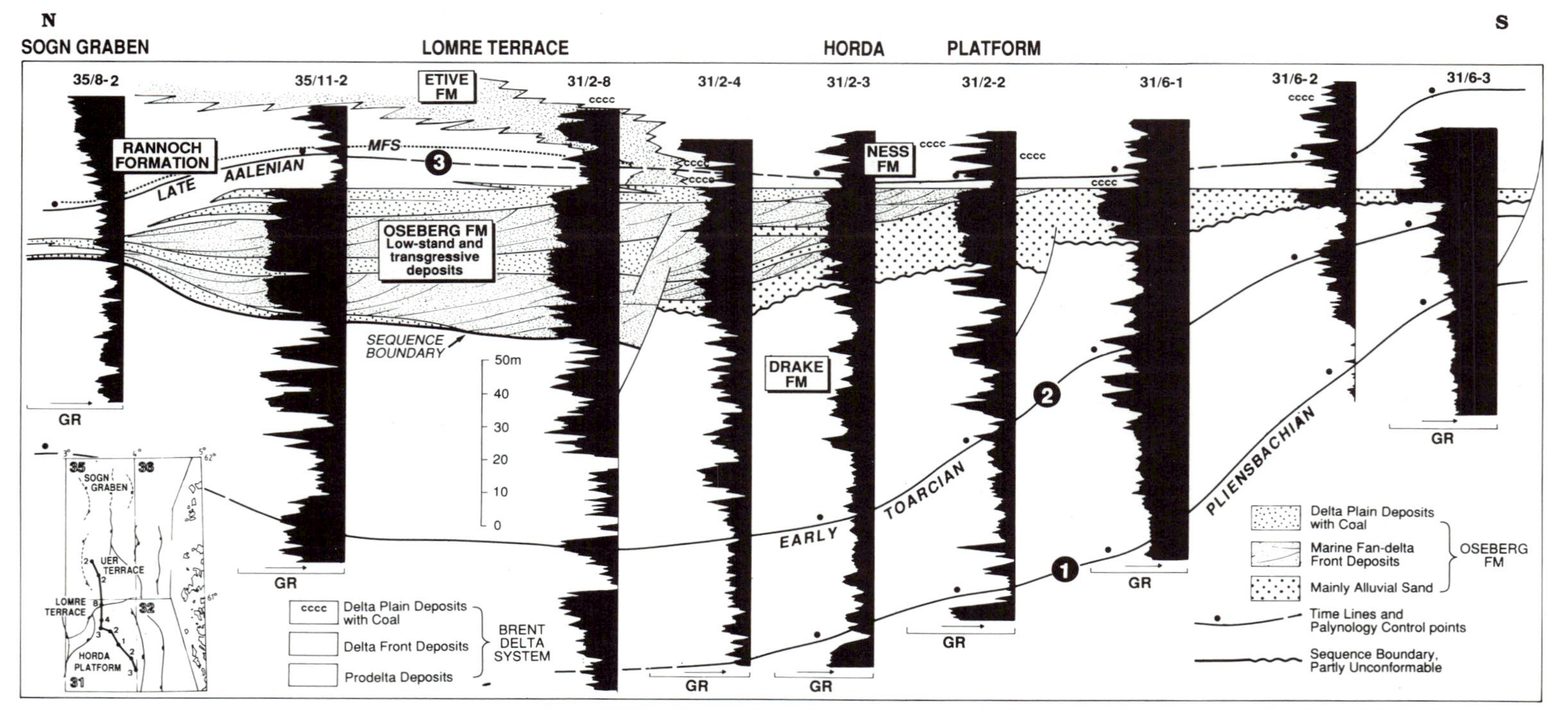


**Fig. 4.** North–south correlation of the Oseberg Formation from Horda Platform towards Sogn Graben.

terms of 'bottomset-modified' Gilbert-type fan deltas (Graue *et al.* 1987). Other general features which reinforce this interpretation are:

(1) the textural and mineralogical immaturity of these coarse-grained deposits;
(2) the fault-controlled development of sand-body thickness, as seen by 'growth' across some of the most important fault lineaments;
(3) the lateral connection between the marine sand lobes along and off the western and northwestern edge of the Horda Platform and a thin unit of alluvial deposits on the platform itself (Figs 3 and 4); the latter are commonly expressed by a 'blocky' gamma-ray pattern on the Horda Platform interpreted here in terms of braided alluvium of the fan-delta plain.

These features, together with the dominance of sediment gravity flow processes and the very large-scale, foresetted units make an interpretation in terms of tidally generated sandbodies (Bray *et al.* pers. comm.) rather unlikely.

## Oseberg sandbody development

### *Horda Platform – Bergen High area*

Figure 3 illustrates the development of the Oseberg Formation in a section from the Horda Platform west and southwestwards towards the Bergen high and the Viking Graben. Note the eastwards 'rising' of the pre-Oseberg time-lines, important evidence for the truncation of pre-Oseberg Formation strata due to uplift of the eastern reaches of the Horda Platform. Because of this truncation, the sharp base of the formation, and the implied abrupt shallowing of the depositional environment across the Oseberg/Drake Formation boundary, the latter is assigned 'sequence boundary' significance.

Fault control of the thickening of the formation from less than 15 m on the platform to more than 60 m thick beyond the western edge of the platform, as well as the component large-scale, cross-stratified units can be seen on Fig. 3.

The eastwards transition from mainly marine fan-delta front deposits to alluvial deposits on the platform is broadly determined by the easternmost occurrence and pinch-out of the maximum marine flooding surface (MFS in Fig. 3) at the top of Oseberg Formation. The latter surface represents the culmination of marine transgression (initiated already within the Oseberg Formation) onto the Horda Platform. This implies that the Oseberg Formation east of this pinch-out line will be capped by alluvial deposits. Moreover, judging from the thinness of the Oseberg Formation east of this line it is assumed that only alluvial deposits are present there. This is in accordance with the assumption of an initial Aalenian fall in relative sea-level (sequence boundary) to near the western edge, of Horda Platform and a late Aalenian relative sea-level high stand (near base Rannoch Formation) which did not flood the region east of the above-mentioned pinch-out.

### *Horda Platform – Lomre Terrace*

Figure 4 illustrates Oseberg Formation development from the Horda Platform northwestwards out towards the Sogn Graben. Many of the features discussed for Fig. 3 can also be seen here, but the thickening of the Oseberg Formation off the platform appears to be more gradual, suggesting that there may have been a slight northerly dip to the early Aalenian uplift of the Horda Platform. This interpretation implies that northwesterly sediment transport dominated the Aalenian lateral infill of the basin in this region.

Comparison of Figs 3 and 4 shows that deposition of Oseberg lobes continued later (post time-line 3) along the western edge of the platform than along its northerly fringes.

### *Palaeogeography*

Latest Toarcian and Aalenian palaeogeography along the eastern reaches of the northernmost North Sea basin is illustrated in Fig. 5. An early stage of easterly uplift, already discussed above, gave rise to easterly emergence, a relative fall of sea-level, a westerly and northwesterly shift of shoreline and depositional area, and a supply of coarse-grained sediment from the eroded platform and the Norwegian hinterland (Fig. 5).

During the time-interval subsequent to the sea-level lowstand and up to the time of maximum marine flooding on the Horda Platform in the Aalenian, fan-deltas built out into the basin in a transverse manner along the eastern reaches between about 60°N and 62°N. Deposition, although dominantly progradational at times, was overall transgressive and is inferred to have had a 'backfilling' tendency across the previously eroded hinterland. During latest Aalenian times the northerly fringe of the Horda Platform had been transgressed and was receiving fine-grained pro-delta shales. Deposition of sandy Oseberg Formation continued along the fault-bounded western edge of the platform. Trans-

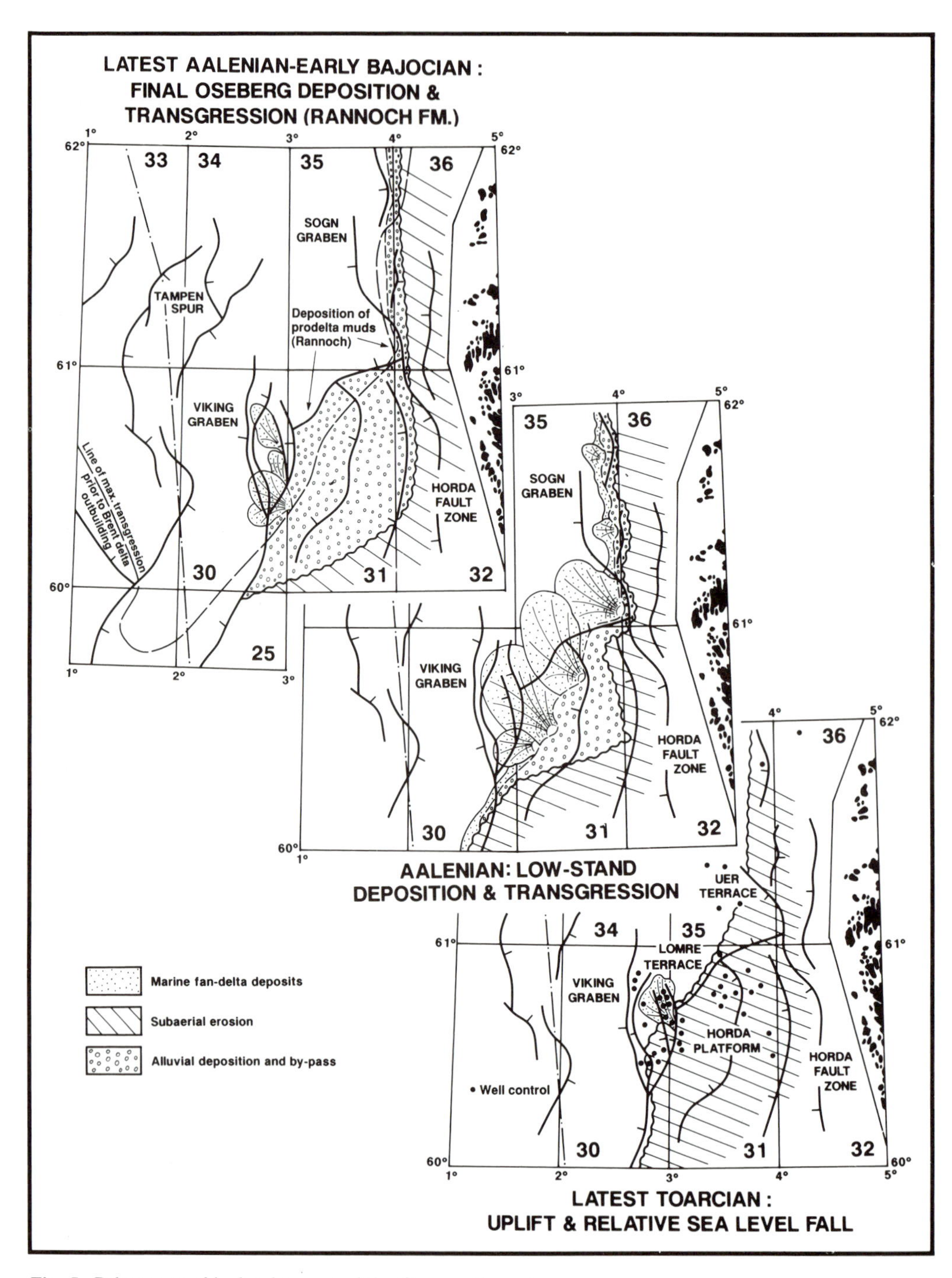


**Fig. 5.** Paleogeographic development of the Oseberg Formation.

gression culminated in latest Aalenian/earliest Bajocian times, and the landward extent of maximum marine flooding along the eastern side of the basin is shown in Fig. 5. This level of maximum marine flooding (near base Rannoch Formation) denotes the onset of progradation of Brent delta proper in the basin.

## Progradation of the Brent Delta

In this section the main progradational components of the Brent delta, which consists of the Rannoch–Etive Formations and the fluvial-dominated lower part of the Ness Formation, are discussed. Each of these units will be characterized sedimentologically as a prelude to discussion of the delta's progradational evolution, which will be addressed in terms of basin configuration, patterns of progradation and controls on deposition.

### *Sedimentological characterization*

Although the sedimentary character, thickness and sequence make-up of the Rannoch–Etive–Ness progradational wedge vary across the basin, the following simplified characterization of the formations serves as a reference for the subsequent discussion.

*Rannoch–Etive Formations.* Together these units form a variably thick, storm-wave-dominated, delta-front or barrier/shoreface, coarsening-up sequence, characterized by an upward-decreasing, gamma-ray profile. Ideally, the depositional couplet is composed of:

> mudrocks, which grade rapidly up into highly micaceous, very fine- to fine-grained, well-sorted sandstones, in which bioturbation is usually very subordinate to physical sedimentary structures, most notably flat- and wave-ripple-laminated units, together with hummocky cross-stratified sets (Richards & Brown 1986); alternating mica-rich and mica-poor, mm-scale lamination is characteristic of these *Rannoch Formation* deposits, which either grade up into or are sharply/erosively overlain by
>
> coarser grained, more poorly sorted, mica-poor sandstones of variable character, ranging from massive, to low-angle-laminated, to trough- and planar-cross-stratified; burrowing is rare. This lithotype variability is matched by a variation in the nature of vertical sequences (cf. Brown & Richards 1989), prompting many authors to envisage a polygenetic origin for these *Etive Formation* deposits (e.g. Brown *et al.* 1987; Graue *et al.* 1987).

There is a growing consensus that, in general, the Rannoch deposits represent lower to middle shoreface or lower delta front deposits, while the Etive sandstones reflect deposition in the upper shoreface/foreshore (barrier bar or upper delta front) realm (Budding & Inglin 1981; Brown *et al.* 1987; Graue *et al.* 1987; Helland-Hansen *et al.* 1989). The degree of fluvial influence in these shallower water facies is still a matter of some debate (e.g. Brown & Richards 1989) and there are some few wells which probably contain mouth bar and distributary channel facies. However, the predominance of a simple (not vertically interfingering) transition from marine delta-front (Rannoch–Etive) to continental delta-plain (Ness) environments in most wells clearly favours the progradation of a simple, straight shoreline for the Rannoch-Etive deposits, and by implication, one in which wave processes were dominant over fluvial across most of the deltaic coastline (Graue *et al.* 1987).

*Ness Formation.* This comprises a variably thick and heterolithic interval of delta-plain deposits, that are the terrestrial equivalents of the Rannoch–Etive delta-front facies. The mixed sandstone, mudrock and coal sequences reflect fluvial channel/mouth-bar, overbank, interdistributary bay and lagoonal subenvironments of the delta-plain (Graue *et al.* 1987; Helland-Hansen *et al.* 1989). However, it is important to note that while such lithological heterogeneity is typical of the entire Ness Formation, the progradational part discussed here tends to be dominated by fining-up fluvial sandbodies and allied flood-generated facies.

The fluvial dominance but overall heterogeneity of the Ness Formation is well illustrated by the gamma-ray profiles, which are both complex and characteristically represented by upwards-diminishing patterns over the major sandbodies (= channel fills).

### *Progradational history*

The marine flooding, in latest Aalenian/earliest Bajocian times, across the Oseberg and Broom sand systems produced an extensive marine basin, opening to the north, into which the Rannoch–Etive–Ness deltaic system built out. Although locally variable the basin's initial morphology can be generalised as crudely horseshoe-shaped with a shoreline developed to the south-east, south and southwest (Fig. 6); this southerly limit, defined by the maximum flooding during the latest Aalenian/earliest Bajocian, is recognized by the southerly loss of Rannoch–Etive facies (Fig. 7). While the existence of this shoreline across the western reaches of the Horda Platform area is readily acceptable (Fig. 7), its development across the Viking Graben is also important (Figs 6 and 8). This has a strong bearing on our non-acceptance of the Richards *et al.* (1988) model (their fig. 1),

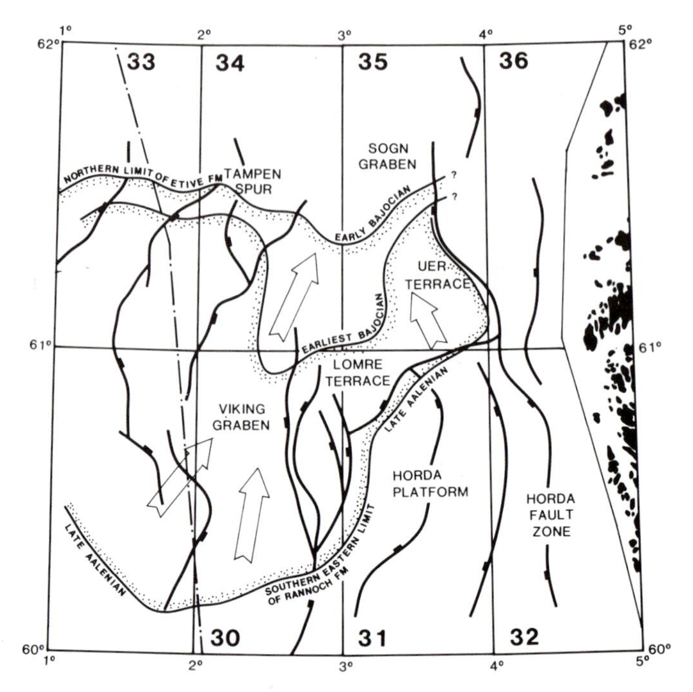


**Fig. 6.** Positions of the Rannoch-Etive shoreline at three stages of the Brent Delta progradation.

which states that there was a permanent marine connection from the north into the central Viking Graben during the build-up of the Brent delta. Northwards of the initial and southernmost shoreline the marine waters were shallow, particularly where the underlying Oseberg Formation is thickly developed, but deeper waters existed to the north and in the northern part of the Viking Graben (Graue *et al.* 1987).

The generation of the Late Aalenian transgression, and indeed the primary control of the basin's morphology at that time, is due to sediment starvation in a basin undergoing post-rift, thermal subsidence, following Triassic extension and crustal thinning across the North Viking Graben (Badley *et al.* 1988; Marsden *et al.* 1990; Yielding *et al.* this volume). Superimposed on this was a longer term, eustatic, sea-level rise (Haq *et al.* 1987). Direct tectonic influence, in the form of faulting, was of relatively minor importance, particularly as much of the pre-existing fault topography has been 'smoothed' by the deposition of the fan-delta lobes of the Oseberg Formation (see earlier section and Graue *et al.* 1987). We envisage, therefore, in Late Aalenian times, a relatively low gradient basin, deepening northwards and into the North Viking Graben area, where both the greater water depth and enhanced (thermal) subsidence provided greater potential for sediment accumulation and preservation.

## *Patterns of progradation*

Although well density varies significantly across the Brent-receiving basin, a reasonably coherent model can be proposed for the entire system. This differs significantly from that proposed by Richards *et al.* (1988). Here emphasis is given to the Norwegian sector but the basin as a whole is discussed.

The main progradation of the Brent delta can, within the limits of the biostratigraphical framework discussed earlier, be allocated to two discrete time intervals: late Aalenian–earliest Bajocian (i.e. approximately timelines 3–4) and earliest Bajocian–early Bajocian (i.e. approximately timelines 4–5).

This 'timing' of the out-building of the Brent delta demonstrates two main points: (1) the progradation proceeded south to north relatively rapidly, at least initially (Figs 6 and 7) and that (2) delta-front deposits persisted longer in the far north of the delta's development (Figs 6 and 9); a similar time-span encompasses Rannoch–Etive–Ness deposits in the south (timelines 3 to 4) but only Rannoch–Etive deposits in the far north (time-lines 4 to 5, cf. Figs 7 and 9).

However, from a purely depositional standpoint, a more refined understanding of the Brent's out-building can be elucidated by reference to three areas: East Shetland Terraces/Tampen Spur; Horda Platform/Uer Terrace; Bergen High/North Viking Graben.

*East Shetland Terraces/Tampen Spur.* This, the most highly drilled of the three areas, contains many of the larger North Sea oilfields, and lies predominantly in the UK sector. There is a developing consensus (Karlsson 1986; Brown *et al.* 1987; Richards *et al.* 1988) that in the UK sector the Brent delta prograded SW–NE across the Shetland Terraces and Tampen Spur area.

In the southwest, thin Rannoch–Etive couplets (<35 m) comprising near-equal proportions of Rannoch and Etive deposits, arranged as simple coarsening-upward sequences, represent deposition on stable terraces. There is little fault-related thickening (cf. Brown *et al.* 1987) and variations in the Rannoch:Etive ratio appear to be related to the presence/absence of Broom sandbodies, ie. it was bathymetrically controlled (e.g. Bertram & Milton 1988).

Northwards, the Rannoch–Etive couplet thickens (doubling its thickness in some cases) and the proportion of Etive decreases to constitute commonly <35% of the couplet, which tends to maintain a simple, cleaning-up gamma-ray motif. This change is probably best explained by enhanced basin accomodation, which manifested itself both in terms of an initial northwards increasing bathymetry, as it is across this area that the Broom thins significantly (cf. Brown *et al.* 1987 figs 5 and 6), and greater subsidence. In this area too there is more sign of inter-terrace thickening.

Further north, towards the limit of the delta's progradation, the Rannoch–Etive thickens to its maximum (>150 m; cf. Brown *et al.* 1987, fig. 6) and commonly develops 'depocentres' in which the very thickest sequences are characterised by high proportions of Etive (commonly 40–60% of the whole couplet). These features suggest that the delta-front had now built out into deeper water where progradation had been retarded, resulting, because of continued, high sand supply and continued subsidence, in overall vertical aggradation possibly with smaller shoreline oscillations (Figs 7 and 9). Northwards, beyond this line first the Ness and then, progressively, the Etive and Rannoch are lost as the delta-front over-extends and is replaced seaward by marine mudrocks. This pattern of 'shaling out' is in some East Shetland Basin wells obscured to some extent by the subsequent erosive activity of the transgressive Tarbert deposits, and by post-Brent erosion of the uplifted footwalls to major, normal faults (Yielding 1990).

*Horda Platform/Sogn Graben.* By comparison with the UK sector this area of the Brent province is poorly drilled and therefore the conclusions drawn from it are more speculative. However, sufficient data exist to demonstrate (Fig. 7) that the Brent delta prograded from the northern tip of the Horda Platform northwards to the Sogn Graben, a traverse that, in some respects, provides a mirror image of the Shetland Terraces as the delta builts out from a stable platform area, across a previously-deposited, sandbody complex (Oseberg Fm) of the Lomre Terrace, towards the deeper waters of the Sogn Graben. However, as this system is more proximal (i.e. closer to the stable platform), and probably more influenced by the thickness of the underlying Oseberg sand-pile than that of the Shetland Terraces, the progressive patterns of sedimentary change are not directly comparable.

Seawards from the Horda Platform the Brent delta-front deposits thicken progressively to a maximum of *c.* 80 m, with the bulk of the increase being taken up in the Rannoch Formation, particularly so beyond the northern limit of the Oseberg Formation (Fig. 7). As the thickness of the Etive Formation does not appear to change significantly across the area, the overall thickness changes are probably best explained in terms of similar progradation rates into a progressively more accommodating basin, characterized by deeper water and enhanced subsidence, particularly to the north, in the Sogn Graben. Although no wells have been drilled through the distal parts of the Brent delta in the Sogn Graben, the potential for repeated coarsening-upwards sequences, as seen in the Tampen Spur area, also exists for this area (Fig. 7).

*Bergen High/Northern Viking Graben.* Westward of the Horda Platform lies the Bergen High over which the Rannoch–Etive couplet is also thin (<30 m), largely because it rests on a thick Oseberg Formation; the implication is of limited basin accommodation at that time (Fig. 8). However, once this Oseberg sand-pile had been traversed, the enhanced basin accomodation of the North Viking Graben provided greater space for a thickened, delta-front sequence. This relationship is well illustrated by the transition from wells 30/2–2 to 34/10–23

(Fig. 8). Perhaps significantly, here too the thickening can be seen to affect both Rannoch and Etive, with the Etive:Rannoch ratio being high. This is similar to the pattern seen in the north of the Shetland Terraces and raises the possibility of another Brent depocentre in the north of the Viking Graben.

The exact nature of the Brent's development south to north along the Viking Graben is less clearly understood because of the paucity of wells. However, from the known data some critical observations can be made.

(1) The Rannoch–Etive couplet is not developed in the Viking Graben as far south as 60°N, but its presence on the graben's flanks, in the Hild and Oseberg Fields, together with its development in 34/10–23, strongly suggest that delta-front facies are present along most of the North Viking Graben (Fig. 6).

(2) The biostratigraphical evidence indicating relatively rapid progradation of the delta-front and the subdued impact of fault-related thickening during late Aalenian–early Bajocian times favours a progressive, rather than rapid, thickening of the Rannoch–Etive couplet northwards along the graben. This model is in sharp contrast with that proposed by Richards *et al.* (1988).

## *Northerly termination of the delta*

With the notable exception of Brown & Richards (1989), little has been published on the nature of the northern limit of the Brent delta's progradation, and it is therefore perhaps worth itemizing some of the more general characteristics. These largely relate to the area astride the Norwegian/UK divide, from the Tampen Spur to the northern extremity of the East Shetland Terraces. The most important features are as follows.

(1) As the northern limit of the delta's progradation is reached there is a progressive loss of first the delta-plain Ness facies, and then the Etive and finally the Rannoch facies; locally all may be overlain by transgressive deposits of the Tarbert, with the contact commonly showing signs of reworking.

(2) In the areas where the Ness is thinning out, the delta-front deposits commonly attain their maximum development, both in terms of absolute thicknesses and in enhanced Etive: Rannoch ratios. This appears to be a function of retarded progradation resulting in vertical aggradation as the delta meets deeper water. This event, which covers a significant time period (Fig. 9; note that time markers 4 and 5 are cut-out at the Tarbert–Etive interface in the northern wells 211/13–7 and 33/9–11), reflects the competition for supremacy between sand supply and marine reworking, as the delta's progradation reaches a standstill before finally retreating.

(3) The internal make-up of the Rannoch–Etive couplet is varied at its northern extremity, both in terms of lithofacies make-up (Brown & Richards 1989) and in terms of its sequence composition; repeated cleaning/coarsening-upward sequences are recognized (Figs 7 and 9), and the possibility of repeated delta-front units, separated by marine mudrocks, also exists (Fig. 7). The ready differentiation of the Rannoch and Etive components of the delta-front couplets on the gamma-ray log can also be obscured in these northern areas.

## *Progradation pathways*

Richards *et al.* (1988) and Richards (1990) have suggested a new palaeogeography for the Brent basin, including the progradational phase discussed here, in which transverse rather than axial progradational pathways are postulated together with a persistent marine connection from the Boreal Ocean into the central Viking Graben during the built-up of the Brent delta in the East Shetland Basin (Richards *et al.* 1988 p. 883 and see also fig. 1).

This theme has been expanded to rule out thermal up-doming of the mid North Sea High as a source for Brent deposits. An array of evidence has been cited to support this contention, the most notable of which are; the sourcelands for detrital garnets (cf. Morton 1985; Hurst & Morton 1988), the absence of volcanic detritus in the Brent (Leeder 1983), the inappropriate age of the 'Forties' volcanics, which are supposedly related to the doming (Latin *et al.* 1990), and the recognition of E–W orientated Ness channels in the Brent Field (Livera 1989).

While not disagreeing with some of the new proposals, nor with the specifics of the evidence forwarded by the original authors, we are sceptical to the complete replacement of axial sediment pathways by transverse ones, and by the assumption of a persistent marine link along the Viking Graben. In the earlier sections of this paper we have demonstrated the following.

(1) The Oseberg–Broom and Rannoch–Etive–Ness depositional systems are distinct entities; the former are clearly transverse in nature and limited in extent, while the main Brent system is significantly larger and basin pervasive. It will therefore have transverse and axial components.

(2) There is good evidence to favour the development of Rannoch–Etive delta-front

deposits in the northern part of the axial Viking Graben, although they do not in all probability extend as far south as 60°N. This suggest two things: (i) that delta-front progradation did indeed occur northwards up the axis of the Viking Graben, and (ii) that the absence of such facies further south implies that the late Aalenian transgression did not penetrate into that area. Both factors mitigate against the persistent seaway along the Viking Graben.

Such a view does not presume a thermal dome to have been the source area for the Brent, but merely emphasizes the fact that if any clastic system is large enough to extend across a basin it will generate an axial component that will preferentially 'flow' down the dominant palaeoslope.

### *Tectonic influences on deposition*

Biostratigraphical analysis has shown that the main Brent delta progradation was relatively rapid and occurred during the late Aalenian to early Bajocian (see especially Figs 7–9). During this time-span thickness changes were related primarily to bathymetry accommodation and subsidence pattern; the latter was principally controlled by post-rift thermal relaxation. However, towards the end of this time interval (i.e. post time line 4), some fault-related thickening/thinning patterns were beginning to manifest themselves in the far north between individual terraces of the East Shetland Basin. At this time similar tectonic influences were making an impact on Ness thicknesses in the Viking Graben (Fig. 8), and it is from this stage hence that renewed fault activity, contemporaneous with sedimentation, started to influence thickness patterns in the Brent. Thus we have a clear two-fold division in terms of thickness distributions:

(1) a northern province, dominated by the progradational Rannoch–Etive–Ness deposits, where the thickness zones are predominantly aligned E–W and relate to depositional rather than tectonic controls;
(2) a southern province, dominated by retrogradational Ness–Tarbert deposits, where synsedimentary faulting is the dominant control on thickness and the isopachs are more strongly aligned N–S, especially in the immediate vicinity of the Viking Graben (Brown *et al.* 1987, fig. 7).

## Retreat of Brent delta

In this section the retrogradational stage of the delta development is further discussed. The overall retreat started in the early Bajocian and continued into the Oxfordian, and is represented by the marine deposits of the Tarbert (of the Brent Group) and Hugin (of the Vestland Group) Formations and their time-equivalent continental deposits of the Ness (Brent Group) and Sleipner Formations (Vestland Group) (Graue *et al.* 1987, Fält *et al.* 1989).

It has been argued that the retreat of the Brent delta in the Norwegian sector took place by the offset backstepping of progradational prisms as evidenced by the predominance of coarsening-upward motifs (progradation) and interfingering with continental deposits (Graue *et al.* 1987). Recent studies on the Brent Group in the East Shetland Basin in the UK sector demonstrates less well-developed progradational trends in the Tarbert formation, but also here most sediments are likely to have been deposited during regressive depositional phases with shoreface erosion during intervening transgressions (Brown *et al.* 1987).

### *Tarbert Formation*

The Tarbert Formation is recognized by the first appearance of shoreline sediments (delta-front or shoreface/foreshore) in the upper part of the Brent Group, above the continental deposits of the Ness Formation. Moreover, the formation underlies, sometimes unconformably, the offshore shales and muds of the Heather Formation. This latter boundary is easy to define whereas the determination of the base sometimes can be a matter of discussion. Vollset & Dore (1984) defined the base of the Tarbert Formation at the top of the last fining-upward in the Ness Formation (i.e. top of an argillaceous bed or coal bed). Rønning & Steel (1987) put it at the base of the first marine sandstone above the Ness Formation delta-plain deposits.

In wells where the upper part of the Brent Group consists of one or several progradational shoreline packages separated or capped by continental 'Ness-like' deposits, the above definitions are inadequate and the question arises as to which of these deposits belong to the Ness Formation and which belong to the Tarbert Formation.

In our opinion the Tarbert Formation principally comprises the shoreline sediments (not merely marine sediments) associated with the overall retreat of the Brent delta, but also includes thinner units of continental deposits (laid down behind the shoreline) at the top of these largely progradational shoreline prisms. However, when these intervening or capping continental deposits have thicknesses more than about 10 m, an interfingering relationship to the Ness Formation with vertical repetition of formations can be established.

*Regressive and transgressive elements*. The offset backstepping model for the Tarbert Formation implies that the formation can consist of both retrogradational and progradational elements, but that the latter contributes the largest volume of sediments.

The transgressive phase is often thin or absent due to rapid transgression across a low-gradient and fine-grained substrate on the deltaic/coastal plain. However, when transgression takes place across a sand-dominated substrate (such as the Etive Formation) the transgressive phase tends to expand, particularly if the transgression is slow, giving enhanced shoreface erosion. Such sediments penetrated by wells close to the northern termination of the delta (e.g. 211/12–1, 211/13–7 and 33/9–11, cf. Fig. 9) are very fine- and fine-grained argillaceous and micaceous with extensive bioturbation, sometimes with weakly fining-upward (metre-scale) trends.

The progradational phase is typically composed of coarsening and cleaning-upwards sequences with thicknesses generally greater than 10 m. These sequences commonly resemble the Rannoch–Etive Formations, having in the lower part micaceous sand and sedimentary structures pointing to wave dominance, overlain by coarser, cleaner sands with a more 'blocky' log character exhibiting horizontal to low-angle, planar or trough stratification or no internal structures. Typical examples of progradational patterns are seen in the Norwegian wells 30/11–4, 30/9–1, 30/9–2 (Fig. 8), 35/8–1, 35/8–2 (Fig. 7).

In addition to regressive and transgressive developments, a few wells indicate vertical aggradation of marine sediments.

### Ness Formation

As noted earlier, significant amounts of sediments were also deposited in continental ('protected') environments during the overall retreat of the delta. This upper part of the delta may be difficult to separate from the lower, regressive phase. Sometimes, lagoonal or bay sediments are present below the Tarbert Formation, presaging the final transgression.

In this context it should be noted that coarsening-upward sequences in the Ness Formation of interdistributary bay origin with marine or brackish–marine bioturbation and wave-generated sedimentary structures can easily be mis-interpreted as (unprotected) shoreline deposits. However, the former sequences tend to be thinner than the shoreline progradational sequences (shallower water) and have normally finer grain sizes than both the progradational and retrogradational shoreline sequences.

### Regional patterns

In Fig. 10 we illustrate three phases of offset backstepping ranging in age from late early Bajocian to early Bathonian. These maps also show where progradational and retrogradational (or vertical aggradational) elements within each phase can be identified. The map has been constructed from a number of wells, some of which are represented on the correlation panels in Figs 7–9.

The earliest (late early Bajocian) and northernmost phase in the Tampen Spur area consists of several smaller regressive-transgressive cycles (cf. Fig. 7). In the East Shetland basin, in wells basinward of the pinchout of the Ness Formation, the transgressive character is more prominent (Fig. 9). The shoreline sediments of the late Bajocian phase are probably spread over a wider area including the Bergen High, Sogn Graben and Southern Tampen Spur areas. Also during this phase transgressive development can be identified in the East Shetland basin. The Bathonian phase, covering the southernmost and most flanking positions in the studied part of the basin, also shows regressive and vertical aggradational or transgressive elements.

Fält *et al.* (1989) have recently published a series of marine sandstone distribution maps for the Tarbert/Hugin formations. They identify at least six offset backstepping progradational prisms in the area as far south as 59°N and being as young as Late Callovian.

### Early rotational uplift

Detailed correlations at the uppermost Ness and Tarbert Formation level in the Horda Platform/Bergen High area (Fig. 11) reveal that in late Bajocian and mid-late Bathonian there was already rotation and uplift on some fault blocks. This is indicated by the truncations of timelines 5 and 8–9, respectively. Note also the relatively uniform development of the shoreline sequence (late Bajocian) (simplified palaeogeography in Fig. 12.1) and the tendency for the whole sequence to thicken structurally downflanks as a result of the combined effect of sedimentary expansion and differential erosion. As noted above, Bathonian shoreface sands in

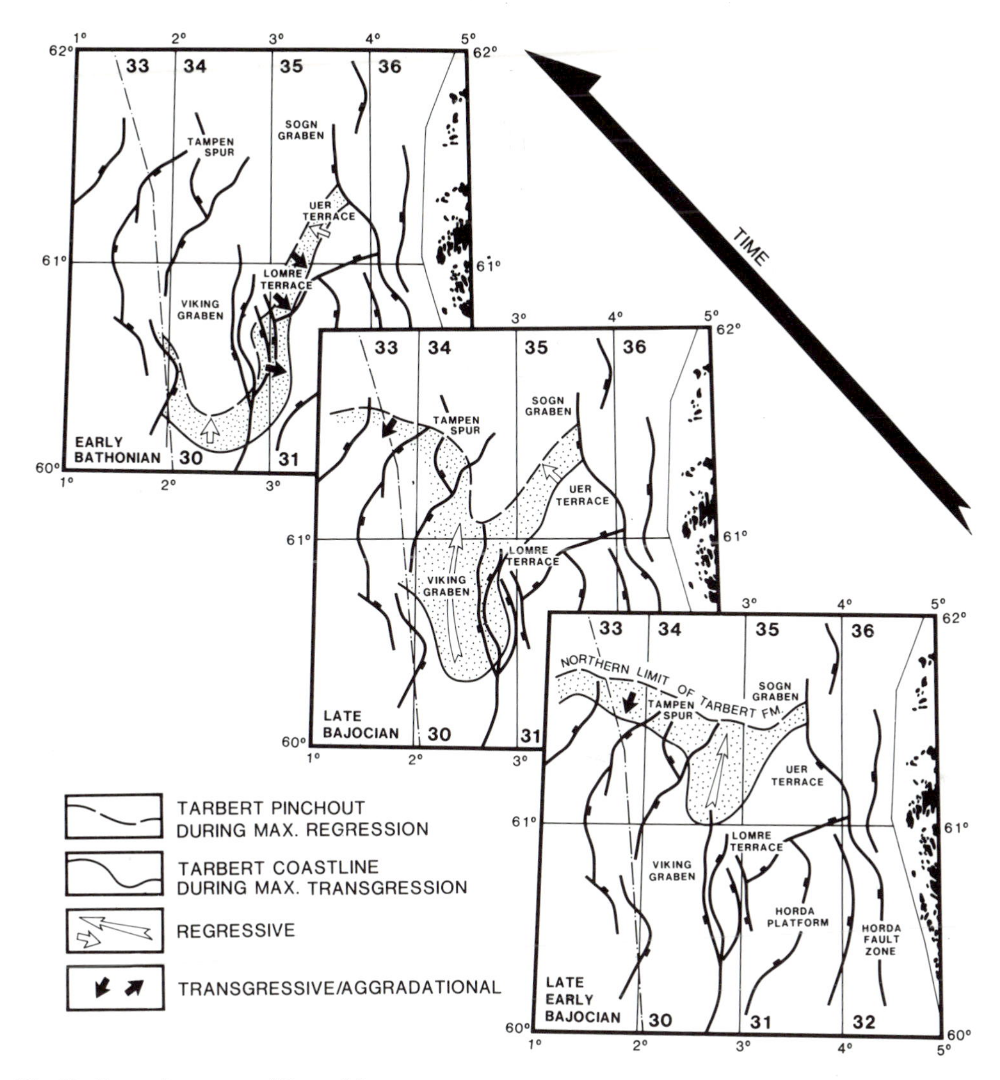


**Fig. 10.** Successive stages of Brent delta retreat.

structurally low positions (e.g. 30/6–8) display a vertical aggradational trend. These sediments may be products of upflank erosion or distal tongues of progradational prisms farther south. The former scenario is illustrated in Fig. 12.2–3.

## Conclusions

The Brent Group can be subdivided into three main phases: the first can be related to early lateral infill of the basin (Oseberg Fm, Broom Fm); the next records the advance of the delta (Rannoch, Etive and lower part of Ness Formation); and finally the last phase was a response to the drowning of the group (Tarbert Formation and upper part of the Ness Formation).

The Oseberg Formation was deposited as a result of Aalenian uplift due to a relative sea-level fall accompanied by a basinward shift of the depocenter. Fan-deltaic sand lobes backed by alluvial sediments built out towards the west and northwest. During the subsequent transgression, marine sediments 'backfilled' the previous emergent areas.

The main delta prograded towards its maximum position in Late Aalenian/Early Bajocian. There the shoreline position oscillated for some

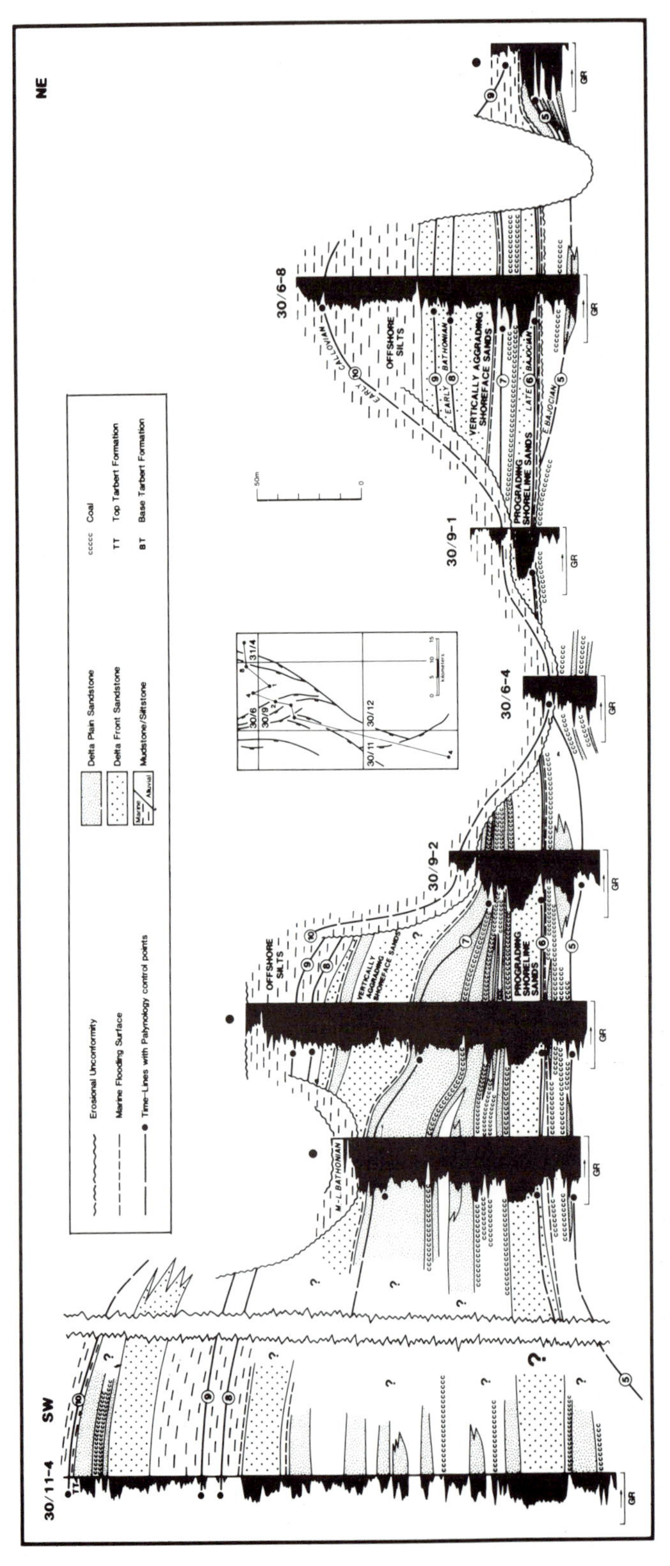


**Fig. 11.** Northeast–southwest correlation of the Tarbert Formation from the Horda Platform towards the Viking Graben.

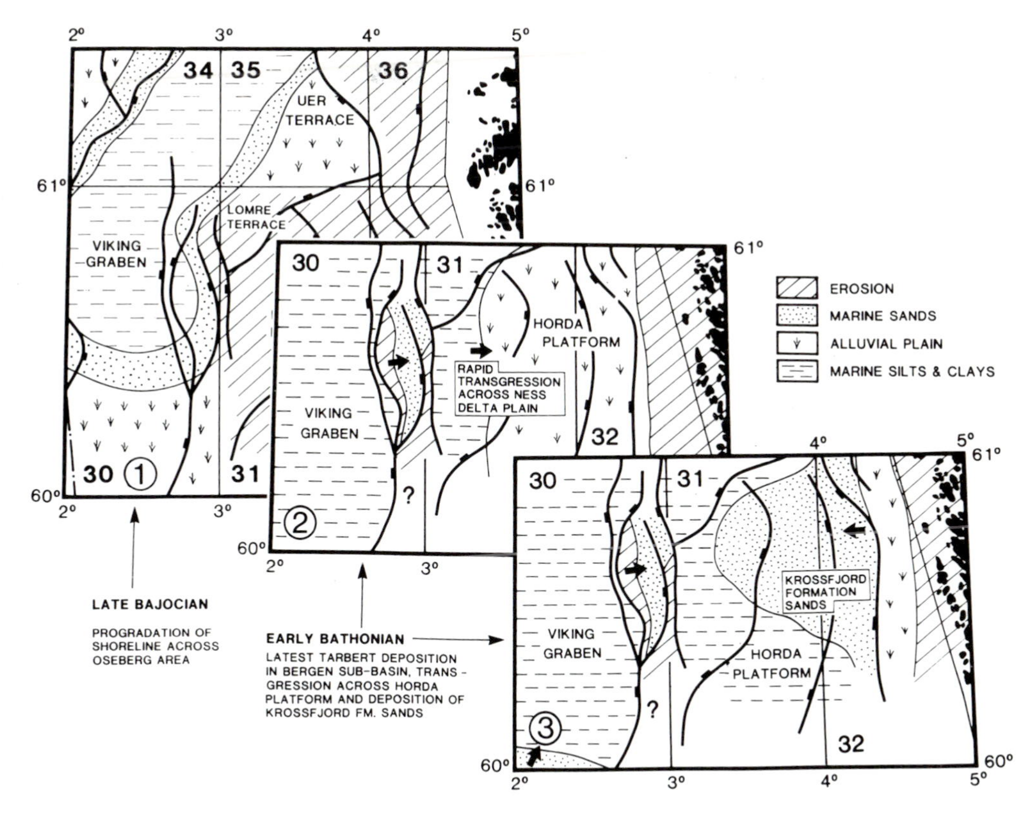


**Fig. 12.** Schematic palaeogeographic maps showing different stages of deposition of the Tarbert Formation in Quadrants 30 and 31.

time (early Bajocian) before the retreat of the delta took place from late early Bajocian. The thickness distribution of the combined Rannoch and Etive formations suggests that their northwards thickening was primarily a function of basin accomodation with bathymetry being important to the south of the area of maximum regression while retardation of the progradation rate in the northernmost positions gave more time forsubsidence to be a major control of thicknesses.

The Tarbert Formation, being deposited essentially as a series of offset backstepped progradational prisms, can locally, and particularly in the northernmost parts of the delta, show evidence of transgressive reworking. Early rotational uplift can locally be demonstrated to have influenced Tarbert Formation deposition, heralding the Late Jurassic rifting phase.

## References


Badley, M. E., Price, J. D., Rambech Dahl, C. & Agdestein, T. 1988. The structural evolution of the northern Viking Graben and its bearing upon extensional modes of basin formation. *Journal of the Geological Society, London*, **145**, 455–472.

Bertram, G. T. & Milton, N. 1988. Reconstructing basin evolution from sedimentary thickness; the importance of palaeobathymetric control, with reference to the North Sea. *Basin Research*, **1**, 247–257.

Bowen, J. M. 1975. The Brent oil-field. *In*: Woodland, A. W. (ed.) *Petroleum and the Continental Shelf of North-West Europe*. Applied Science Publishers, Barking, 353–362.

Brown, S. & Richards, P. C. 1989. Facies and development of the Middle Jurassic Brent Delta near the northern limit of its progradation, UK North Sea. *In*: Whateley, M. K. G. & Pickering, K. T. (eds) *Deltas: Sites and Traps for Fossil Fuels*. Geological Society, London, Special Publication, **41**, 253–267.

——, —— & Thompson, A. R. 1987. Patterns in the deposition of the Brent Group (Middle Jurassic) UK North Sea. *In*: Brooks, J. & Glennie, K. W. (eds) *Petroleum Geology of North West Europe*. Graham & Trotman, 899–913.

Budding, M. C. & Inglin, H. F. 1981. A reservoir geological model for the Brent Sands in Southern Cormorant. *In*: Illing, L. V. & Hobson, G. D.

(eds) *Petroleum Geology of the Continental Shelf of Northwest Europe*. Heyden, London, 326–334.

DEEGAN, C. E. & SCULL, B. J. 1977. *A proposed standard lithostratigraphic nomenclature for the Central and Northern North Sea*. Report of the Institute of Geological Sciences **77/25**.

EYNON, G. 1981. Basin development and sedimentation in the Middle Jurassic of the Northern North Sea. *In*: ILLING, L. V. & HOBSON, G. D. (eds) *Petroleum Geology of the Continental Shelf of Northwest Europe*. Heyden, London, 196–204.

FÄLT, L. M., HELLAND, R., JACOBSEN, V. W. & RENSHAW, D. 1989. Correlation of depositional sequences in the middle Jurassic Brent/Vestland Group megacycle, Viking Graben, Norwegian North Sea, *In*: COLLINSON, J. D. (ed.) *Correlation in Hydrocarbon Exploration*. Graham & Trotman, London, 191–200.

GABRIELSEN, R. H., FÆRSETH, R. B., STEEL, R. J., IDIL, S. and KLØVJAN, O. S. 1989. Architectural styles of basin fill in the northern Viking Graben, *In*: BLUNDELL, D. J. & GIBBS, A. D. (eds) *Tectonic Evolution of the North Sea Rifts*. Oxford University Press, 158–179.

GRAUE, E., HELLAND-HANSEN, W., JOHNSEN, J., LØMO, L., NØTTVEDT, A., RØNNING, K., RYSETH, A. & STEEL, R., 1987. Advance and retreat of Brent Delta System, Norwegian North Sea. *In*: BROOKS, J. & GLENNIE, K. (eds) *Petroleum Geology of North West Europe*. Graham & Trotman, 915–937.

HAQ, B.U., HARDENBOL, J. and VAIL, P. R. 1987. Chronology of fluctuating sea-levels since the Triassic. *Science*, **235**, 1156–1166.

HELLAND-HANSEN, W., STEEL, R., NAKAYAMA, K. and KENDALL, C. G. St. C. 1989. Review and computer modelling of the Brent Group stratigraphy. *In*: WHATELEY, M. K. G. & PICKERING, K. T. (eds) *Deltas: Sites and Traps for Fossil Fuels*. Geological Society, London, Special Publication, **41**, 237–252.

HURST, A. & MORTON, A. C. 1988. An application of heavy-minerals analysis to lithostratigraphy and reservoir modelling in the Oseberg Field, Northern North Sea. *Marine and Petroleum Geology*, **5**, 157–169.

KARLSSON, W. 1986. The Snorre, Statfjord and Gullfaks oilfields and the habitat of hydrocarbons on the Tampen Spur, offshore Norway. *In*: SPENCER, A. M. *et al*. (eds) *Habitat of hydrocarbons on the Norwegian Continental Shelf*. Graham & Trotman, London, 181–198.

LATIN, D. M., DIXON, J. E. & FITTON, J. G. 1989. Rift-related magmatism in the North Sea Basin. *In*: BLUNDELL, D. J. & GIBBS, A. D. (eds) *Tectonic evolution of the North Sea Rifts*. Oxford University Press, 101–144.

LEEDER, M. R. 1983. Lithospheric stretching and North Sea Jurassic clastic sourceland. *Nature*, **305**, 510–513.

LIVERA, S. E. 1989. Facies associations and sandbody geometries in the Ness Formation of the Brent Group, Brent field. *In*: WHATELEY, M. K. G. & PICKERING, K. J. (eds) *Deltas: Sites and Traps for Fossil Fuels*. Geological Society, London, Special Publication, **41**, 269–286.

MARSDEN, G., YIELDING, G., ROBERTS, A. M. and KUSZNIR, N. J. 1990. Application of a flexural continental simple-shear/pure-shear model of continental lithosphere extension to the formation of the northern North Sea Basin. *In*: BLUNDELL, D. J. & GIBBS, A. D. (eds) *Tectonic evolution of the North Sea Rifts*. Oxford University Press, 240–261.

MITCHENER, B. C., LAWRENCE, D. A., PARTINGTON, M. A., BOWMAN, M. B. J. & GLUYAS, J. 1992. Brent Group: sequence stratigraphy and regional implications. *This volume*.

MORTON, A. C. 1985. A new approach to provenance studies: electron microprobe analysis of detrital garnets from Middle Jurassic sandstones of the northern North Sea. *Sedimentology*, **32**, 553–566.

PROCTOR, C. V. 1980. Distribution of Middle Jurassic facies in the East Shetland Basin and their control on reservoir capability. *In*: *The sedimentation of the North Sea Reservoir Rocks, Geilo*. Norwegian Petroleum Society 15/1–22.

RICHARDS, P. C. 1990. The early to mid-Jurassic evolution of the northern North Sea. *In*: HARDMAN, R. F. P. & BROOKS, J. (eds). *Tectonic Events Responsible for Britain's Oil and Gas Reserves*. Geological Society, London, Special Publication, **55**, 191–205.

—— & BROWN, S. 1986. Shoreface storm deposits in the Rannoch Formation (Middle Jurassic), North West Hutton oilfield. *Scottish Journal of Geology*, **22**, 367–375.

——, ——, DEAN, J. M. & ANDERTON, R. 1988. A new palaeogeographic reconstruction for the Middle Jurassic of the northern North Sea. *Journal of the Geological Society, London*, **45**, 883–886.

RYSETH, A. 1989. Correlation of depositional patterns in the Ness Formation, Oseberg area. *In*: COLLINSON, J. D. (ed.) *Correlation in Hydrocarbon Exploration*. Graham & Trotman, London, 313–326.

RØNNING, K. & STEEL, R. J. 1987. Depositional sequences within a 'transgressive' reservoir sandstone unit: the middle Jurassic Tarbert formation, Hild area, northern North Sea, *In*: KLEPPE, J., BERG, E. W., BULLER, A. T., HJELMELAND, O. & TORSÆTER, O. (eds) *North Sea Oil and Gas Reservoirs*. Graham & Trotman, London, 169–176.

SKARPNES, O., HAMAR, G. P., JACOBSSON, K. H. & ORMAASEN, D. E. 1980. Regional Jurassic setting of the North Sea north of the central highs. *In*: *The sedimentation of the North Sea Reservoir Rocks, Geilo*. Norwegian Petroleum Society 13/1–8.

STEEL, R. J. & GJELBERG, J. 1989. Reservoir sand and sequence development on the Norwegian shelf (abs.), European Association of Petroleum Geologists, Berlin.

VAINS, R. DE 1980. *Etude du microplancton du Jurassique Moyen (Bathonien–Callovien) dans le Graben de Viking (Mer du Nord). Systematique et biostratigraphie*. These 3 cycle, Univ. P. Sabatier de Toulouse.

VAKHRAMEEV, V. A. 1970. Range and palaeoecology of Mesozoic conifers, the Cheirolepidiaceae. *Paleontologicheskii Zhurnal*, **1**, 19–34.

—— 1978. The climate of the Northern Hemisphere in the Cretaceous in the light of palaeobotanical data. *Paleontologicheslkii Zhurnal*, **2**, 3–17.

VOLLSET, J. & DORE, A. G. 1984. *A revised Triassic and Jurassic lithostratigraphical nomenclature for the Norwegian North Sea*. Bulletin of the Norwegian Petroleum Directorate **3**.

YIELDING, G. 1990. Footwall uplift associated with Late Jurassic normal faulting in the northern North Sea *Journal of the Geological Society, London*, **147**, 219–222.

——, BADLEY, M. E. & ROBERTS, A. M. 1992. The structural evolution of the Brent Province. *This volume*.

# The palaeoenvironments and dynamics of the Rannoch–Etive nearshore and coastal succession, Brent Group, northern North Sea

ELAINE S. SCOTT

*Department of Earth Sciences, University of Oxford, Parks Road, Oxford, OX1 3PR*

**Abstract:** This paper investigates the palaeoenvironments and dynamics of the Rannoch Formation in the East Shetland Basin and the sedimentology of the Etive Formation in one particular part of the basin, the southern Cormorant area. The study utilizes core and well data from the North Sea as well as comparative outcrop data from Utah, western USA.

Rannoch Formation facies, coupled with a slight coarsening and cleaning upward of sand-dominated profiles above offshore shales and below channelized delta top sediments, are indicative of a prograding shoreface. The middle shoreface is dominated by hummocky cross-stratified fine sand characterised by horizontal to very low angle, planar to undulatory laminations, often cut by steeper scour features.

The Etive Formation is characterized in the southern Cormorant area by development of a microtidal, wave-dominated barrier cut by associated inlet channels.

These associations are consistent with an interpretation of the Rannoch Formation as a high energy, storm-dominated environment, with strong wave oscillatory motion accompanied by weak translatory currents as the most likely depositional setting. In the overlying Etive Formation, shoreline characteristics are those of a dissipative coastal system with a complex three-dimensional inshore topography and at least one nearshore bar system developed during fairweather periods.

The East Shetland Basin lies on the western flank of the north Viking Graben in the northern North Sea (Fig. 1a). During Mid-Jurassic times this basin was the western part of the Brent Group depocentre which extended eastwards over the Horda Platform in the Norwegian sector (Fig. 1b) (Helland-Hansen *et al.* 1989). The Brent Group is a single regressive–transgressive system. It is subdivided into five lithostratigraphic units with the Rannoch, Etive and lower Ness Formations representing the northwards progradation of a storm-dominated shoreline (for example Budding & Inglin 1981; Richards & Brown 1986; Brown & Richards 1987; Helland-Hansen *et al.* 1989). In the UK Sector, the Rannoch and Etive Formations thicken to the northeast with the locus of maximum progradation lying approximately at the northern edge of the East Shetland Basin (Eynon 1981; Brown & Richards 1987).

This paper examines the sedimentology of the Rannoch Formation shoreface sediments and identifies the environmental associations in the overlying coastal succession of the Etive Formation. The Etive Formation is examined in the southern Cormorant Field where a coastal barrier system is developed. The study incorporates core and well data from the Brent Group in conjunction with a comparative study of outcrops from the Blackhawk Formation in east-central Utah, western USA. Subsurface data from 23 wells forms the basis of this work in addition to detailed examination of 180 m of core from six of the wells. The aims are to identify the dominant processes which acted during deposition and the type of shoreline that developed along the Brent delta coast.

## Rannoch Formation sedimentology and palaeoenvironments

The Rannoch Formation (Fig. 2) shows the coarsening-up, cleaning-up profile of the prograding shoreface of the Brent Group. The Rannoch Formation represents the lower and middle shoreface, dominated by horizontal to low-angle and undulatory lamination. In deviated core (Fig. 3a & b) individual laminae are seen to be <1 mm to 3 mm thick and are composed of very fine to fine sand. The dark–pale colour alternations are controlled by mica concentrations. Structures display upward curvature of laminations, low angle truncations, long wavelength, low relief forms with lamina dips of less than 15°.

This style of lamination has been described as hummocky cross-stratification (HCS), as proposed formally by Harms *et al.* (1975) and later by Walker (1982, Fig. 4) and assigned to a shallow marine origin (for example, Dott & Bourgeois 1982; Swift *et al.* 1983). Many authors believe that hummocky cross-stratification is

*From* Morton, A. C., Haszeldine, R. S., Giles, M. R. & Brown, S. (eds), 1992, *Geology of the Brent Group*. Geological Society Special Publication No. 61, pp. 129–147.

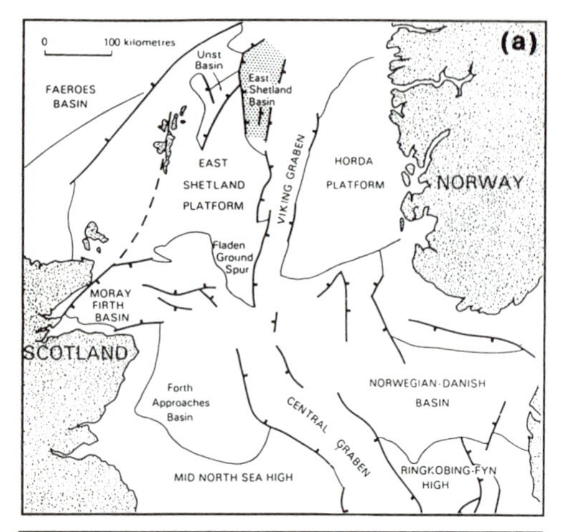


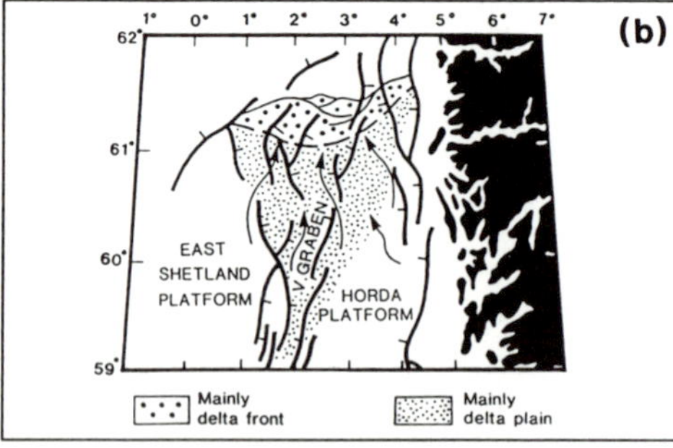


**Fig. 1.** (**a**) Location map showing East Shetland Basin, UK, Sector, northern North Sea. (**b**) Palaeoenvironment distribution near the northern limit of progradation of the Brent system in Bathonian times (from Helland-Hansen *et al.* 1989).

commonly formed under storm waves on the shoreface, although the precise contribution of purely oscillatory flow and/or a translatory component to the formation of the structure remains a matter of debate (for example, Dott & Bourgeois 1982; Allen 1985; Duke 1985; Nøttvedt & Kreisa, 1987; Allen & Underhill 1989; Southard *et al.* 1990; Arnott & Southard 1990).

However, some features of Rannoch Formation lamination apparently cannot be accounted for by assigning it to HCS. Observations from another core (Fig. 5) show some of the sequence to be composed of millimetre-scale parallel to low angle laminated intervals, with low angle truncations, which are punctuated at intervals of 10–60 cm by steep sided scours filled with onlapping laminae.

These scours present a problem as their steeper, shorter wavelength morphologies seem to preclude their inclusion as HCS. In fact, in most non-deviated cores, the unequivocal identification of HCS in the Rannoch Formation is, at best, difficult. A very low angle structure with wavelengths of 1–5 m in a core 10 cm wide will merely look like parallel lamination. However, deviated cores, for example the core in Fig. 3a, do allow identification of HCS structures, and in some cases these structures are associated with steeper scours. Clearly, given the limitations of vertical cores, any attempt to interpret the structures requires comparison with similar outcrop examples. This is especially important if any interpretation of the processes of formation is to be attempted.

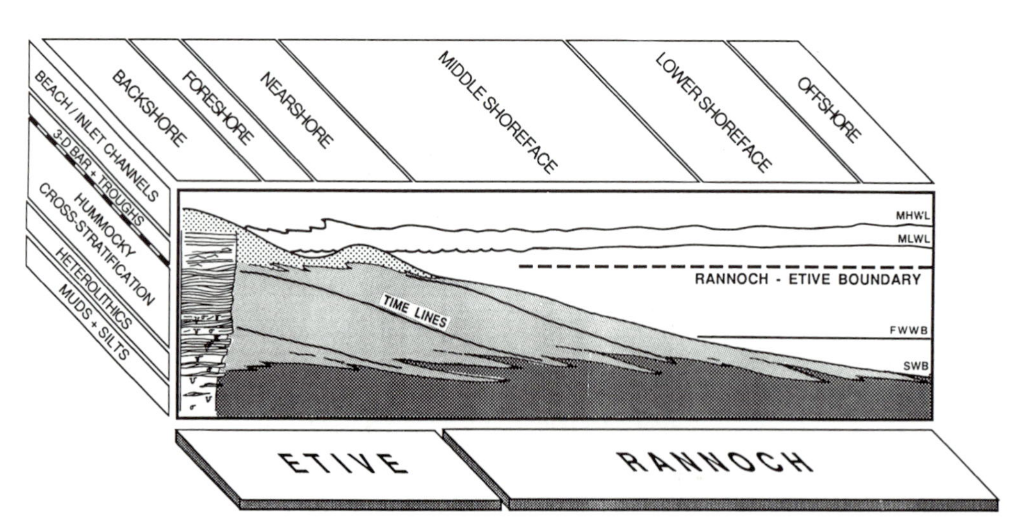


**Fig. 2.** Summary log and environmental interpretation of the Rannoch and Etive Formations.

**Fig. 3.** Deviated core from the Rannoch Formation displays long wavelength, low relief forms with upward curvature of laminations, low angle truncations, and lamina dips of less than 15°. Well deviation angle is 60° with way up to the top left. Scale bars are 5 cm. (**a**) Middle shoreface with amalgamated HCS and (**b**) lower shoreface with individual HCS beds surrounded by bioturbated heterolithics.

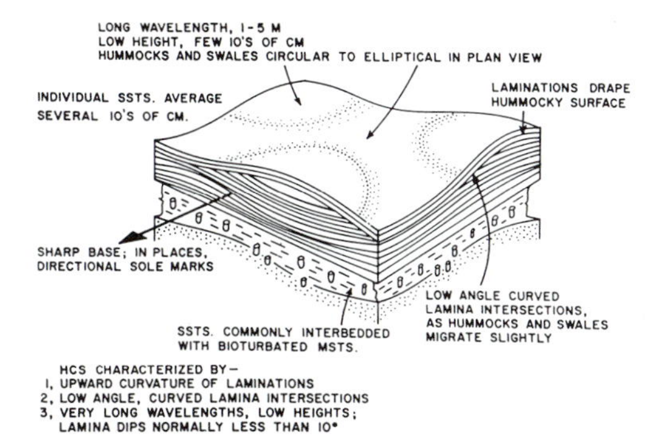


**Fig. 4.** Morphological features of hummocky cross-stratification (from Walker 1982).

**Fig. 5.** Deviated core from the Rannoch Formation shows some of the sequence composed of mm-scale parallel to low angle laminated intervals, with low angle truncations, punctuated at intervals of 10–60 cm by steep sided scours filled with onlapping laminae. Well deviation angle is 51°; scale bar is 5 cm.

## Field observations and measurements from outcrop examples

Outcrop examples of hummocky cross-stratification were studied from the shoreface of the Upper Cretaceous Blackhawk Formation which outcrops in east-central Utah, western USA. They form the nearshore deposits in the foreland basin which developed east of the Sevier orogenic belt (Fig. 6).

Several HCS beds from the shoreface of the Blackhawk Formation were studied in detail to determine their geometry, internal bounding surface relationships and lamination structures.

The simplest geometries found (Fig. 7) are sequential long wavelength, shallow (low amplitude) troughs which migrate laterally, truncating previous troughs and preferentially locating in topographic lows. Examining details of the internal geometry reveals that internal laminations generally onlap or *downlap* basal bounding surfaces. Draped bounding surfaces also occur, but less frequently. While draping lamination may suggest oscillatory depositional processes, the presence of downlapping laminae suggests lateral migration of the bedform. Some of the internal laminations have also been modifield by shorter wavelength, wave-formed structures.

Also developed are small, steep trough or scoop shaped scours, frequently stacked or offset vertically. A range of scour scales and types occur and some, with onlapping lamination, are very similar to those seen in Rannoch Formation core (Fig. 5).

Complex accretionary bedforms are found but, rather than simple draping of the topography, the laminae show subtle discontinuities due to slight lateral shifting of the bedforms.

Within the set, vertical burrows are truncated by internal bed bounding surfaces. These features suggest multiple scour events resulting in amalgamation of individual beds to form a set.

## *Palaeocurrent data*

Palaeocurrent data of maximum dip/dip direction measurements from bed bounding surfaces and associated internal laminations were collected for beds occurring at a similar stratigraphic level. The data points cluster around the edge of the stereoplot and generally fall within the 15° circle representing maximum expected dip for HCS (Fig. 8a). However, some of the steeper scour and fill structures show higher angle dips than generally expected for accretionary or scour and drape HCS. The points are contoured (Fig. 8b) and although there is a wide spread of directions, two nodes are evident, concentrating around WNW and ESE. These represent preferred dip/dip directions of trough limbs with the trough axes aligned between these nodes, orientated towards the north and south.

I suggest that the trough axes represent a weak mean flow direction of a unidirectional current superimposed on wave action. This would produce elongate hummocks with a weak directional preference to internal laminations, bounding surfaces and trough axes (in the direction of the unidirectional component).

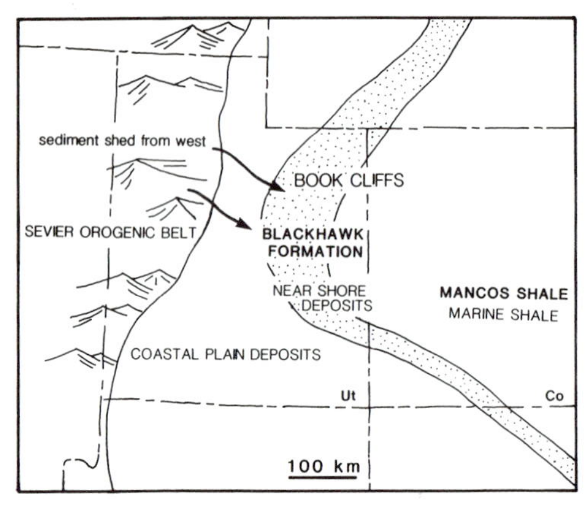


**Fig. 6.** Palaeoenvironments of the Blackhawk Formation during the Campanian (modified from McGookey *et al.* 1972).

## *Thin-section method*

Unfortunately, the palaeocurrent and bedform geometry analysis carried out on the Blackhawk Formation sediments is impossible from core. A more rigorous means of comparison with core was needed. A study of the internal fabric of the lamination was conducted to investigate if a lamination type could be determined and a comparison made. Results are encouraging with HCS from both the Rannoch and Blackhawk Formations showing a recurring lamination style with stacked, 1–2 mm couplets of a coarsening–upwards unit succeeded by a fining–upwards unit (Fig. 9). Hence, the subsurface Rannoch can provisionally be compared with surface HCS on the basis of internal lamination fabric. These relationships can be confirmed with additional analysis and comparison with other occurrences of HCS in outcrop.

The significance, with regard to the flow conditions of the depositional process, of these coarsening-up/fining-up couplets is still under investigation and could be due to one, or a combination of several, hydrodynamic mechanisms. However, although the precise hydrodynamic conditions remain unclear, it is possible to at least eliminate mechanisms which could not have formed the laminae textures found. To do this, consideration is given to the five possible depositional mechanisms and processes which include high energy flow conditions, with abundant sediment in suspension, and are therefore capable of producing appropriate, millimetre-scale, fining-up and/or coarsening-up laminations.

With increasing flow velocities, the *dune–plane bed transition* under sub-critical unidirec-

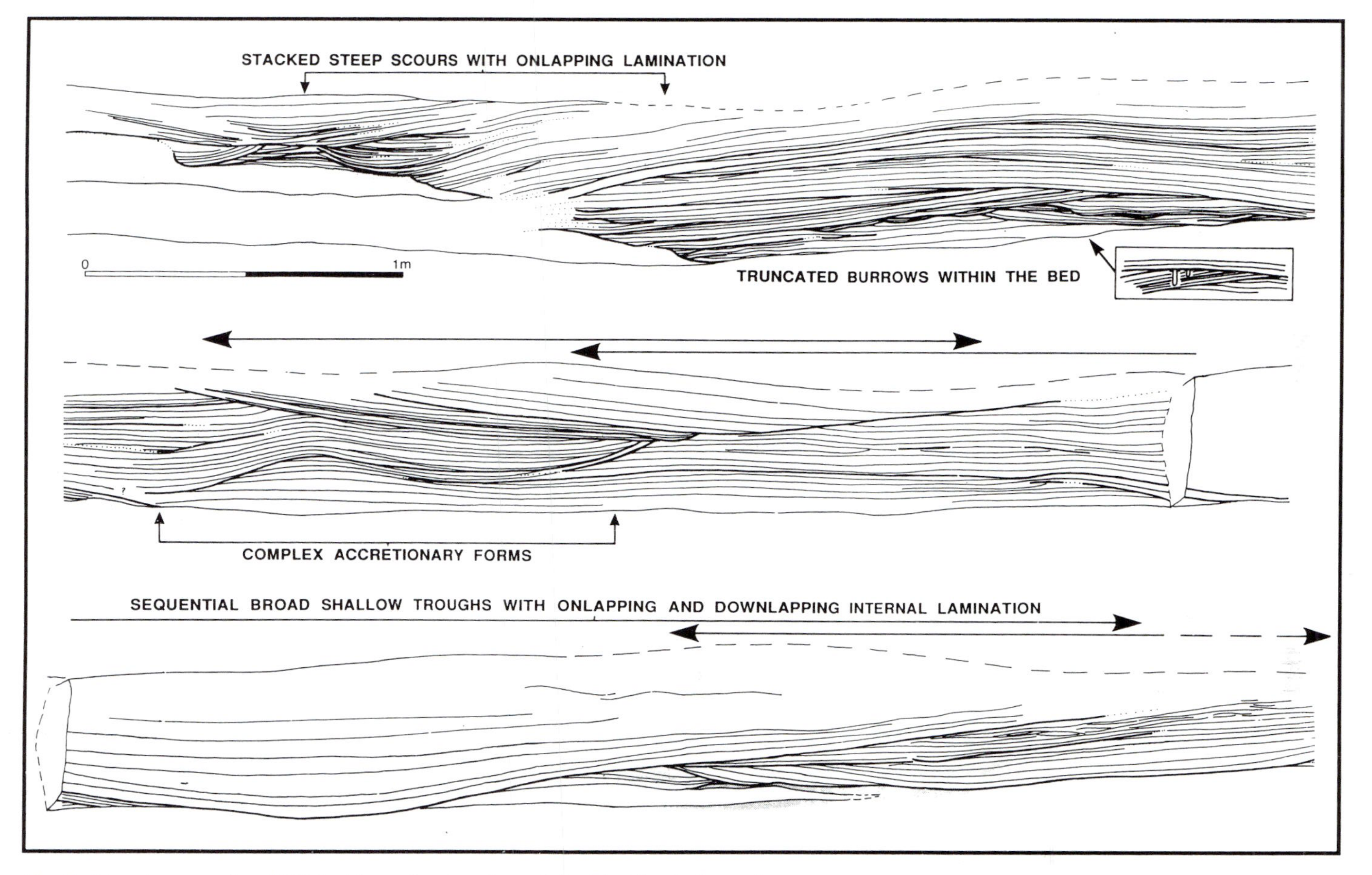


**Fig. 7.** Line drawing of an HCS bed from the shoreface of the Blackhawk Formation.

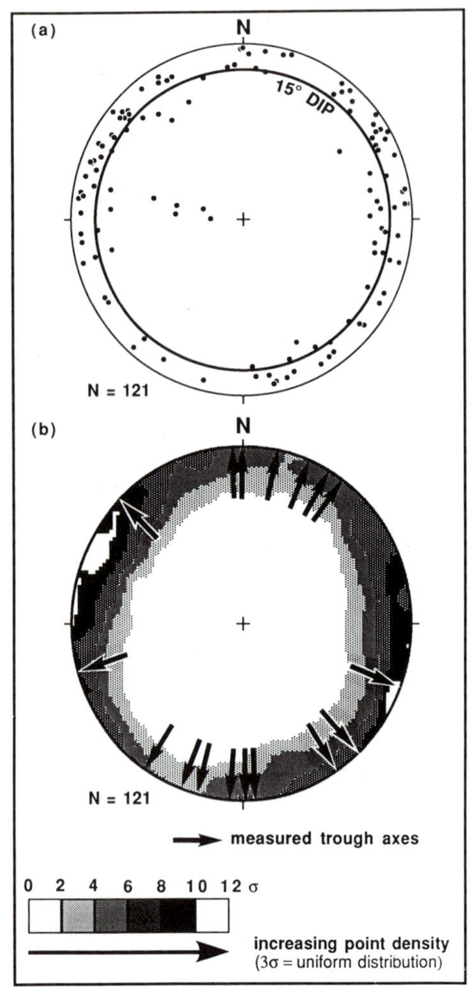


**Fig. 8.** (**a**) Equal area stereoplot with data points showing maximum dip/dip direction of bounding surfaces and internal laminations. Note that most dips are less than 15° (**b**) Contour stereoplot (Kamb 1959) of bounding surface and internal lamination data. Darker tones represent greater concentration of data points (compare with (**a**). Arrows representing measured trough axis directions plot between nodes showing preferred trough limb orientation (WNW–ESE).

tional flow gives way to the stability field of *upper stage plane beds with superimposed very low amplitude bedforms*. Similar mechanisms are responsible for the resulting lamination. The migration of these bed waves with amplitudes of a few grain diameters gives rise to laterally extensive, millimetre-thick, planar laminae with a dominantly fining-upwards internal structure (Bridge & Best 1988). This can be modified (Paola *et al.* 1989; Cheel 1990) by superimposed, higher frequency, turbulence-induced burst–sweep cycles. Bridge (1978) suggests that vertical sorting within the laminae reflects this turbulence with high velocity inrushes to the bed (sweeps) allowing deposition of coarsest grains at the top of the laminae, while the removal of fluid and localized decrease in shear stress (bursts) would disperse the finer particles prior to suspension deposition of the remainder of the fining-up laminae. Cheel & Middleton (1986) and Cheel (1990) suggest a burst–sweep mechanism whereby bursts deposit fining-up laminae and sweeps produce coarsening-up laminae through dispersive pressure effects. Burst–sweep cycles alone are unlikely to have formed the couplets seen as they have an irregular periodicity of a few seconds. This is probably too erratic and on too rapid a time scale, with only small lateral extents of the bed influenced, to explain the couplets.

*Fall out from suspension*, as a mechanism in its own right, would mean that settling velocity effects would be dominant, so producing a fining-up unit. In this case, any mica platelets present would be the last to be deposited on to the bed as they are hydraulically equivalent to silt or very fine sand (Doyle *et al.* 1983). However, the coarsening-up laminae cannot be explained by settling effects alone.

*High concentration bedload layers* (basal shear layers) are dominated by grain interactions as dispersive pressure works against gravity. As additional grains enter this layer from above, a plug of grains forms at the top where the shear gradient is lowest (Lowe 1988). Continued suspended-load fallout results in increasing plug thickness and eventual collapse and freezing of the entire layer as flow velocities decrease. A reversely graded unit results. Hence, the alternation of bedload layer deposition and fallout from suspension would explain the couplets.

*Intense oscillatory flow under gravity waves* is another process capable of producing the required scale of lamination. Allen (1985) argued that if HCS is formed under intense oscillatory flows then the dimensions of the hummocks should be related to flow conditions. Production of HCS purely under progressive storm waves is incompatible with observed hummock spacings in the ancient as predicted wavelengths are much too small. In direct conflict with the theoretical predictions of Allen (1985), Southard *et al.*'s (1990) experiments suggest that HCS can form under pure oscillatory flows at long periods and high orbital velocities, during a strong but

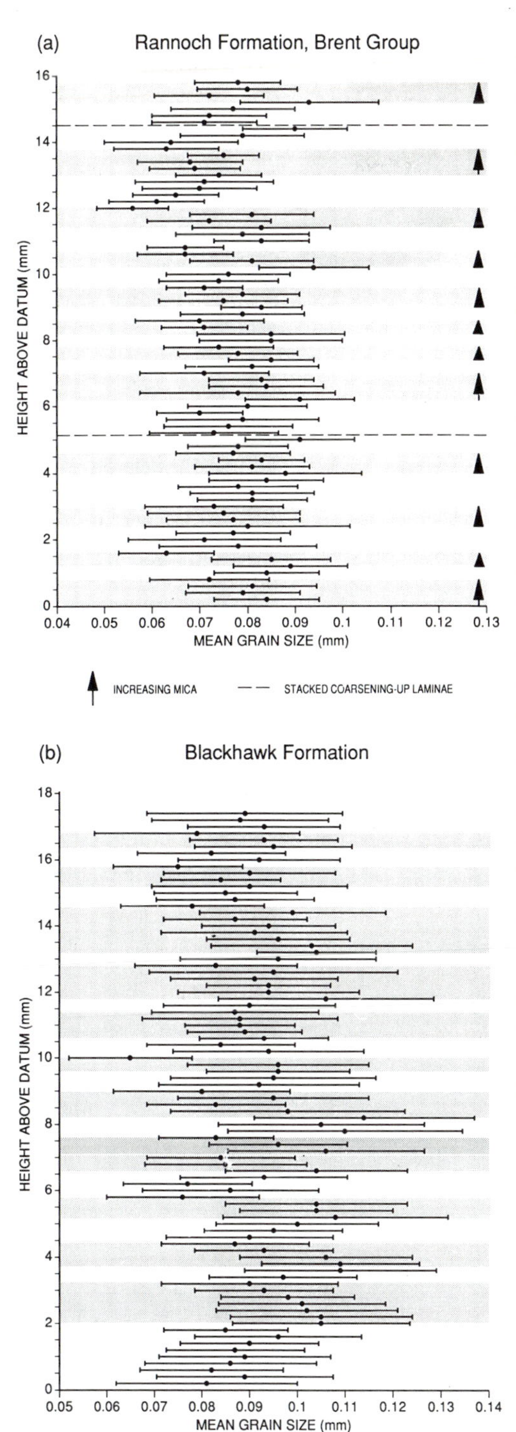


**Fig. 9.** Graphs of the internal lamination fabric of the Rannoch Formation (**a**) and the Blackhawk Formation (**b**).

waning flow which started under plane-bed conditions. However, this does not explain the presence of coarsening-up/fining-up couplets.

Any proposed mechanism would need to produce stacked, millimetre-scale couplets of a coarsening-up unit followed by a fining-up unit, and one which concentrates mica at the top of the fining-up unit. Therefore, I suggest that the most likely combination of depositional mechanisms would be one in which dispersive pressure maintains a high density shear layer near the bed, which acts to concentrate the coarsest grains at the top of the flow. If the flow velocity decreases below the threshold velocity for these coarse grains, the bedload freezes as a coarsening-up layer and finer sediment falls from suspension onto the bed producing a fining-up unit in which mica would be the last to be deposited on to the bed.

However, the exact nature of the process which combines these mechanisms during deposition to produce suitable bed conditions remains speculative. Clearly, none of these unidirectional or oscillatory flow mechanisms and processes seem, in isolation, capable of producing the repetitive fluctuations to produce the couplet-style of lamination found. In view of this, and the evidence from palaeocurrent data, it seems likely that a process involving a combined flow needs to be invoked. A detailed discussion on the nature of this combined flow and its required proportion of oscillatory and unidirectional components is outside the scope of this paper. However, recent flume experiments (Southard *et al.* 1990; Arnott & Southard 1990) have added ammunition to the lively debate on the merits of combined flows and the origins of HCS.

## Storm genesis of HCS

Characteristics of the HCS beds described include long wavelength, low amplitude bedforms with erosional set, and internal bed, boundaries. Internal laminations drape, onlap and downlap bed boundaries. These long wavelength structures are interpreted as suggesting a storm-wave origin. However, the associated steeper scours, with shorter wavelength internal structures, would require modified storm-wave processes to produce them in such intimate relation with the long wavelength structures. Palaeocurrent data suggest at least a component of unidirectional flow to produce the relationships seen. Truncated burrows within sets suggest multiple events, or at least a 'pulsed' single event, to allow sufficient decrease in current strength near the bed to allow colonis-

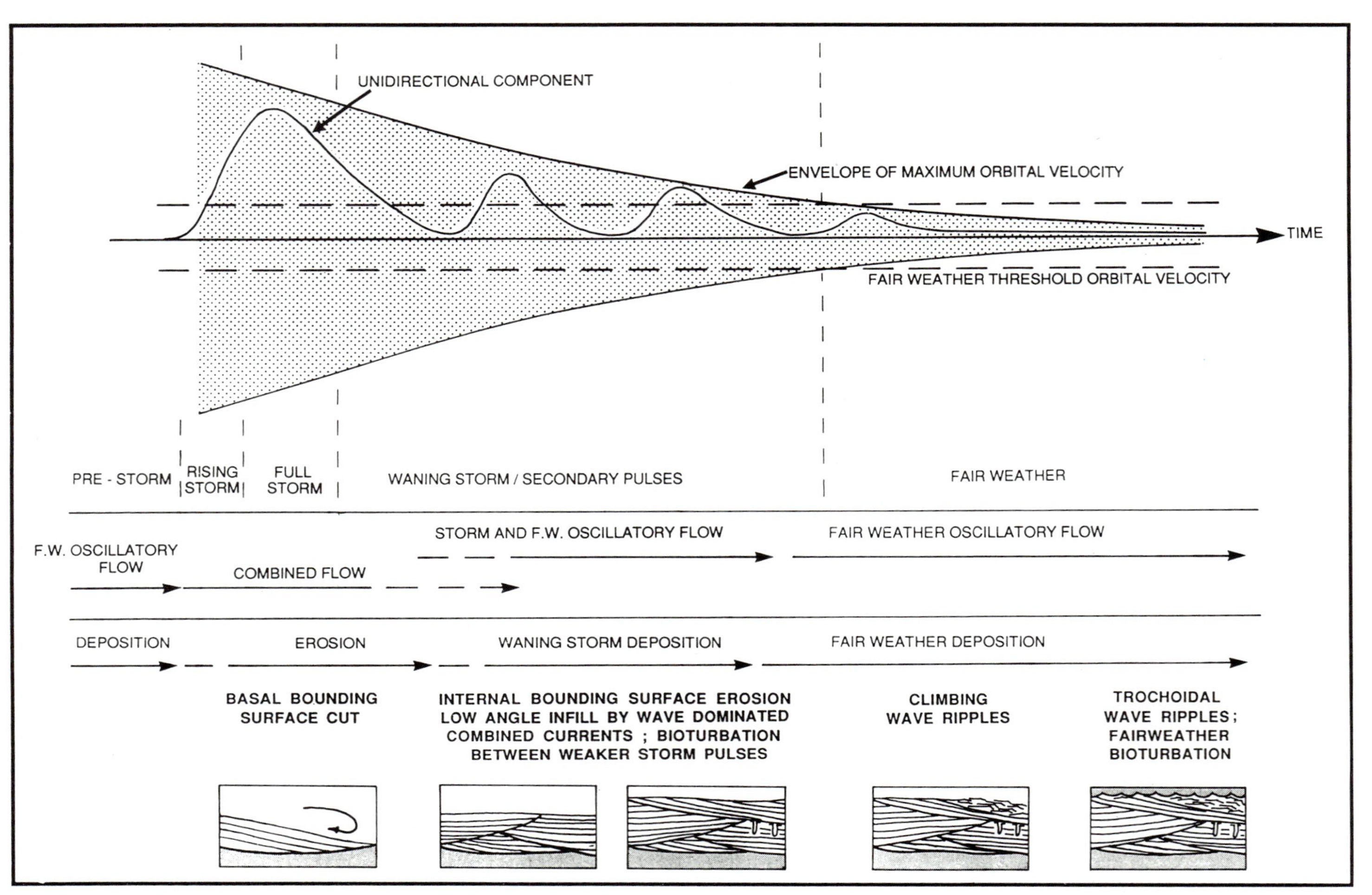


**Fig. 10.** Schematic sequence in the formation of storm generated hummocky cross-stratification (compare with Allen 1984 (page 487); Cheel 1991).

ation by burrowing organisms. Resumed higher energy conditions would truncate the burrows and the surrounding bed to produce an amalgamated set.

This evidence suggests that the particular occurrences of HCS found in the Rannoch and Blackhawk Formations are formed by storm activity (Fig. 10).

Figure 10 shows a schematic evolutionary diagram that is consistent with the field data. The envelope of maximum orbital velocity (shaded wedge) during the storm event(s) is dominant over the storm's unidirectional component, represented by the sinusoidal curve. Storm erosion or deposition is dominant over fair weather deposition while the storm oscillatory flow is greater than the fair weather threshold orbital velocity (dashed line, Fig. 10). Fair weather deposition is interrupted as the storm rises quickly to a full storm and a strong unidirectional component combines with storm oscillatory flow to erode a basal bounding surface (or an amalgamation plane). As the storm wanes the component of unidirectional flow decreases so that wave-dominated combined currents produce the low angle infill above the bounding surfaces. If storms are closely spaced, or there are secondary pulses to a main storm, then organisms may burrow the sediment if the flow energy is suppressed sufficiently during these inter-storm periods. Increase in storm activity would then cut internal bounding surfaces and truncate the burrows. As the storm finishes, the storm orbital velocity falls below the fairweather threshold. Fairweather oscillatory flow dominates, with final stage climbing wave ripples and finally trochoidal wave ripples and fairweather sedimentation and bioturbation. Resumption of storm activity would either cut an internal bounding surface to cause further amalgamation and addition of further beds to the set, or, if storms were infrequent and allowed sufficient time for fairweather heterolithics to accumulate, another basal set boundary.

During Rannoch Formation deposition storms were frequent enough and large enough over the middle shoreface to form a completely amalgamated sequence (Fig. 3a). Each event eroded far enough into the previous bed or beds to completely remove any evidence of fairweather processes and near-bed energy was hardly ever low enough to allow burrowing. In the lower shoreface, perhaps only the larger storms had an influence and individual beds and sets are found encased in bioturbated heterolithics (Fig. 3b).

## Etive Formation sedimentology and palaeoenvironments

In the Etive Formation of southern Cormorant (Fig. 11), two sequences are commonly found

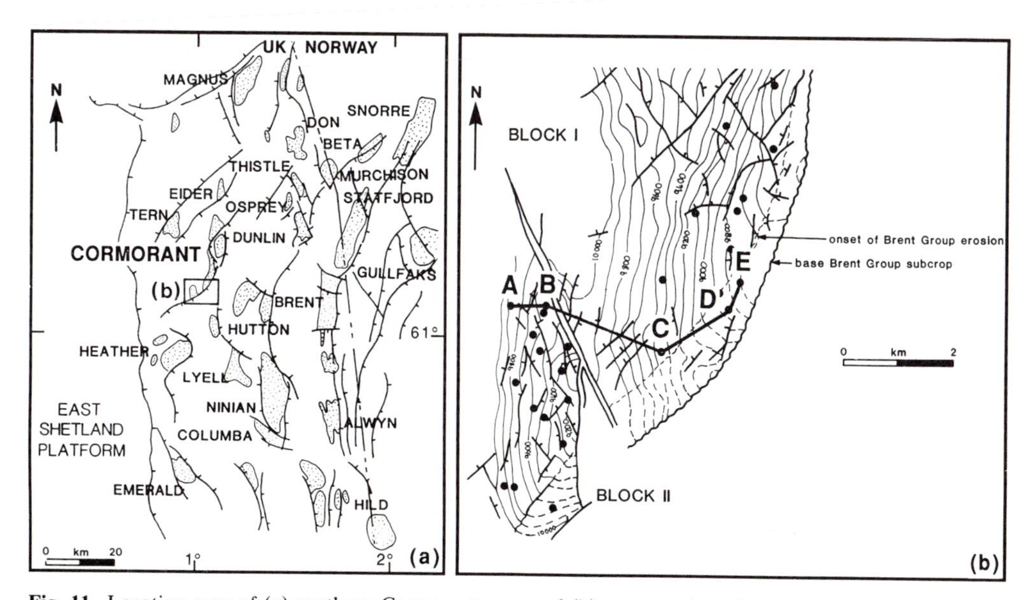


**Fig. 11.** Location map of (**a**) southern Cormorant area and (**b**) cross section of Fig. 12.

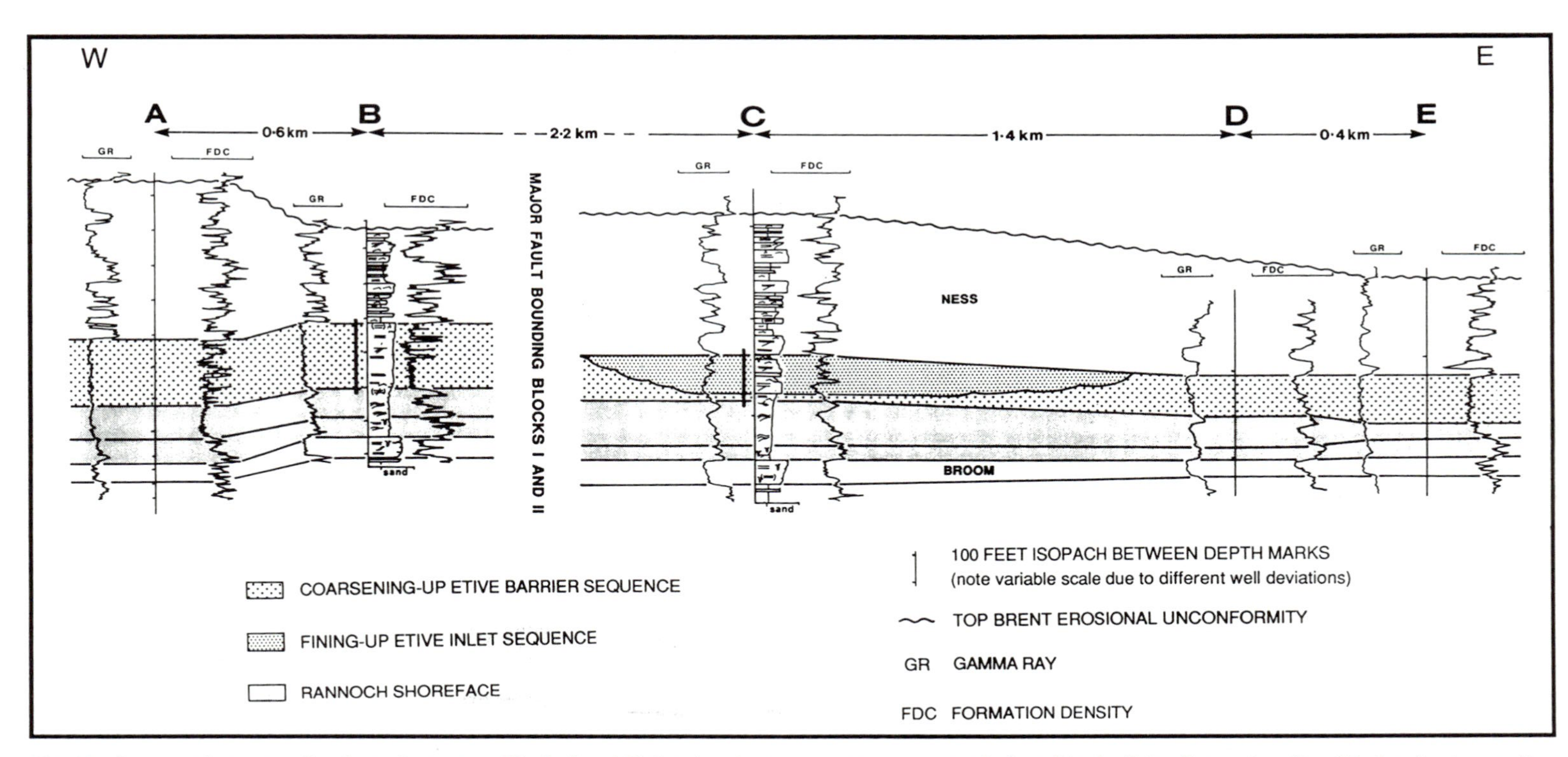


**Fig. 12.** Cross section across Southern Cormorant Blocks I and II showing the two sequences commonly found in the Etive Formation. Simplified grain size profiles of the two cored sections B and C are shown. Section B coarsens upward (Fig. 14 shows a log of the marked section) while section C fines upward (Fig. 18 shows a log of the marked section) and cuts the coarsening-up sequence. Comparison of log profile shapes suggests that sections A, D and E also coarsen upwards.

(see also Budding & Inglin 1981). One sequence (Fig. 12, section B) coarsens gradationally upwards from the underlying middle shoreface. It is dominated by parallel and low-angle lamination. The second sequence (Fig. 12, section C) fines upwards and is characterized by sharp-based, cross-bedded sandstones. In some cases the cross-bedded sandstones have mud rip-up clasts and diffuse muddy laminations.

Comparison of well log profiles from uncored sections with section B and C shows that sections A, D and E show similar profiles to section B. Sections A, D and E are therefore inferred to gradually coarsen up from the middle shoreface and show a similar facies development. Correlation of the logs in Fig. 11 shows that the fining upwards Etive Formation sequence (section C) cuts the coarsening upwards sequences (sections A, B, D and E).

Examination of these sequences reveals the nature of the Etive Formation coastline which can then be compared with the classification of Wright *et al.* (1979).

### *Comparative framework*

Shorelines may be either dissipative or reflective (Wright *et al.* 1979). The terminology was introduced during studies of the New South Wales coastline in southeast Australia. This coastline is a microtidal, wave-dominated, high energy coast and beaches along it show a spectrum of characteristics between the dissipative (Fig. 13a) and reflective (Fig. 13d) end members.

Reflective coastlines have steep, linear beach faces with a simple inshore topography of beach berms and cusps. Incident wave energy is reflected from the beach face (Fig. 13c and d). Low energy reflective coastlines only fully develop after extended periods of low swell. Dissipative coastlines, at the opposite end of the spectrum, have a smaller grain size, an increased sediment supply (abundant inshore sediments), a higher wave energy (especially during or immediately after major storms) and/or an increased width of inshore slope (Fig. 13a and b). In the high energy, dissipative case, wide, flat surf zones on exposed, open coasts result in turbulent dissipation of much of the wave energy before it reaches the beach. A complex three-dimensional inshore topography is developed. Longshore bars, which migrate landward during fairweather periods, are associated with longshore troughs and rip channels (Fig. 13a and b).

### *Coarsening-upwards sequence*

The coarsening upwards sequence is interpreted as a beach–barrier succession (Fig. 14). At the base, gradationally above the middle shoreface of the Rannoch Formation, rippled sandstones occur immediately below planar laminated sandstones. The ripples are in very fine–fine sand and have discontinuous, muddy and silty drapes or lenses. These muddy drapes are concentrated in the troughs and thin, frequently pinching-out, over the rounded, washed-out crests of the ripples where the erosional energy was higher (Fig. 15a). These ripples are associated with planar bedded, slightly coarser, sandstone above. It is probable that the facies is the remnant of a nearshore bar where upper flow regime conditions prevailed in the shallower water over the bar crest. Flow separation at the top of the bar slip face allowed sediment to avalanche down the slip face resulting in landward migration of the bar over the preceding trough. If the velocity of the water over the bar crest dropped then ripples may also have formed there, though their preservation potential, if upper flow regime conditions resumed, would have been low. Even so, occasional, poorly developed muddy ripples are seen in core associated with the parallel lamination (Fig. 15b). Hence, these two facies are closely related and represent the remnants of a nearshore bar and trough system.

A possible modern comparison is shown in Fig. 16 (Hayes 1974). This photograph of Kiawah Island, South Carolina shows a wide, flat, dissipative bar and trough (ridge and runnel) system seaward of the foreshore with a second system seaward of the first. Parallel lamination (P) is preserved on the bar crest and in this case antidunes (A) are preserved on the slipface of the bar, but lower flow regime ripples (R), as in the Etive Formation core, are seen in the trough.

Returning to the Etive, above the nearshore bar and trough system, several metres of clean, well sorted, fine sand with low angle, planar laminations dominate the sequence (Fig. 15c). This swash–backwash lamination can be weakly to moderately disturbed into convolute laminations or replaced by ripples (Fig. 15c). The disruptions are probably due to partial dewatering of water-logged sand on the foreshore. The ripples were possibly formed by small ephemeral 'streams' draining back to the sea or due to wind agitation of water in shallow pools on the foreshore. The 'active' foreshore zone passes gradationally into the 'inactive' backshore where wave activity is almost absent

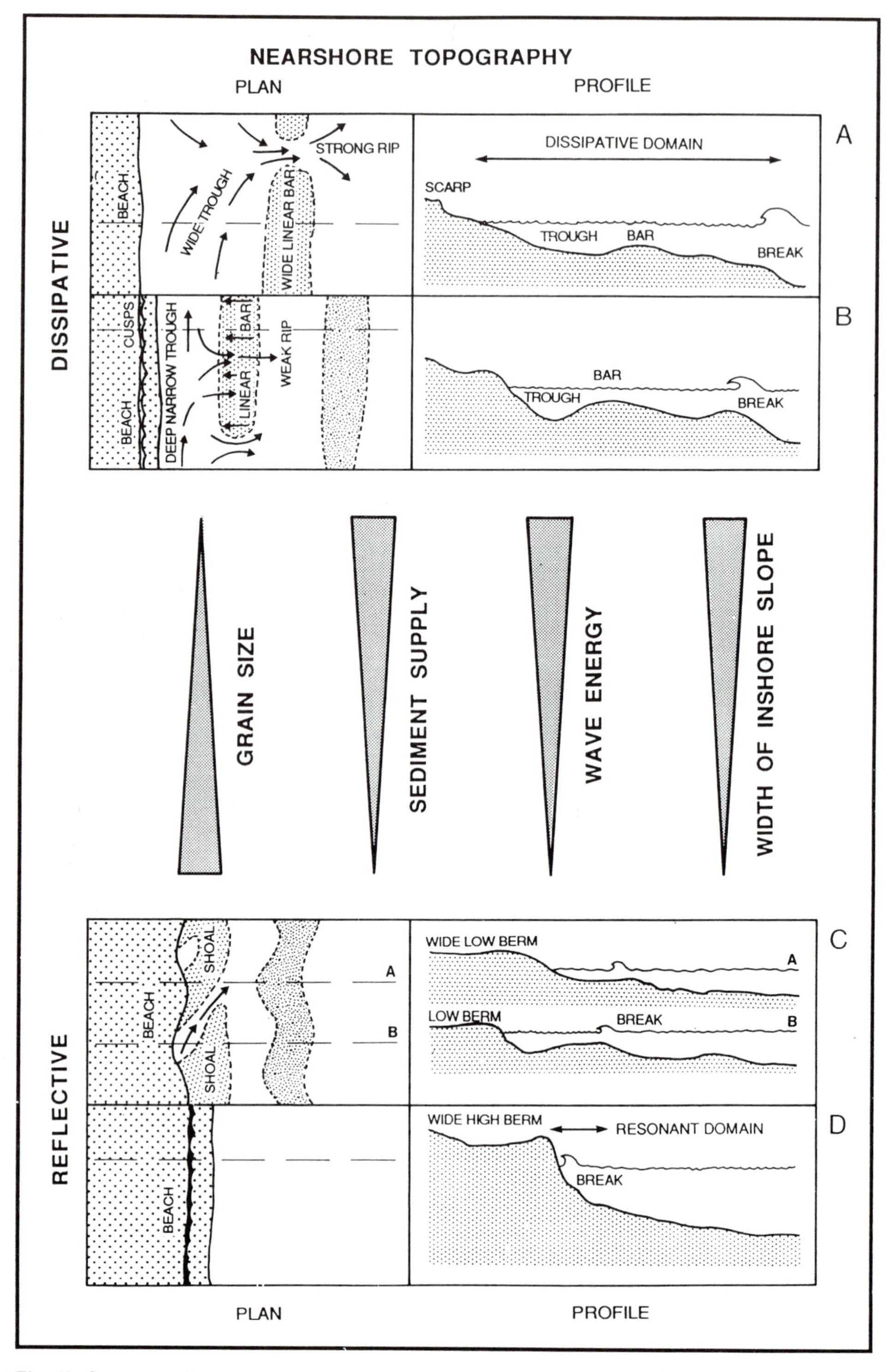


**Fig. 13.** Spectrum of nearshore topography, from dissipative (**a**) to reflective (**d**) coastlines and their controlling features (modified from Wright *et al.* 1979).

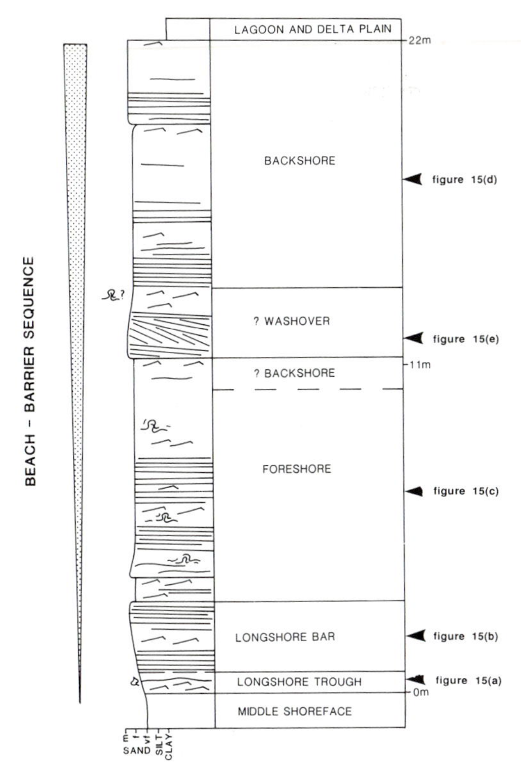


**Fig. 14.** Log from Section B (Fig. 12) core shows the coarsening-up beach–barrier sequence.

during fairweather periods. The internal structures become less well defined and any silt present becomes dispersed through the sand as destratification proceeds, perhaps due to the effects of the roots of sparse vegetation cover (Fig. 15d).

This gently coarsening-up succession is interrupted by a coarser sandstone with a sharp base which fines upward into the backshore above (Fig. 14). This is dominated by higher angle, tabular cross-bedding (Fig. 15e) which, with decreasing flow energy, wanes into ripple lamination above. It is interpreted to be a storm event which washed sand across the foreshore and cut erosively into the backshore.

The sequence shown in Fig. 14 shows a prograding barrier beach. During fairweather periods at least one, and possibly more than one, nearshore bar was developed creating a relatively flat, complex surf zone with a three-dimensional bar and trough topography. This shallow region would dissipate much of the wave energy before it reached the beach. However, the sequence shows evidence for storm activity which would periodically swamp and modify the dissipative coast. This activity is seen by the washover event and by the fact that, even though there may have been multiple bars in the fairweather system, there is only one bar preserved in the succession.

The modern Fripp Island (Fig. 17, Hayes, 1979) is a regressive barrier island which shows features similar to the Etive Formation in the southern Cormorant area. The long, straight barrier has multiple beach ridges and the intertidal zone has a bar (B) and trough (T) system similar to that discussed for the Etive Formation. Rip channels (C) associated with bar and trough development are also seen, and though not found in all wells due to their limited lateral extent, they are occasionally preserved (Fig. 18). Tidal channels wind over the contemporaneous delta or coastal plain (DP) and drain into the sea via inlet channels (IC) which cut the barrier.

## *Fining-upwards sequence*

The normal beach–barrier development is periodically cut by a fining-upwards sequence (Fig. 12, section C). A typical fining-up profile (Fig. 18) has a truncated coarsening-up sequence at the base. A bar and trough system with associated parallel (P) and ripple (R) lamination (Fig. 19a) is found, as in the barrier succession discussed above. Well-developed trough cross-bedding (Fig. 19b), preserved immediately below this bar–trough system is interpreted to be a closely associated rip channel. This drains high tide water quickly back to sea and generally cuts erosively into the shoreface below. Above this bar–trough system a weakly stratified section, with occasional parallel lamination, represents the shoreface–foreshore transition.

This coarsening-up profile is truncated by a sharp-based sandstone with angular mud clasts and carbonaceous debris immediately above the basal erosion surface (Figs 18 and 19c). The sandstone fines upwards from medium to fine grained and displays high angle, tabular cross-bedding (TCB) succeeded by climbing ripple lamination (RL) (Fig. 19d). Weakly developed, diffuse silty drapes are found on the bedding planes and ripple surfaces.

The fining-up sections are small channels in the barrier and may have been of either fluvial or tidal origin. Unequivocal evidence of tidal processes is absent but the presence of weak silty drapes, the similarity of the sand to the marine sands below and their position with respect to the barrier and mud flat environments suggests that these channels represent tidal

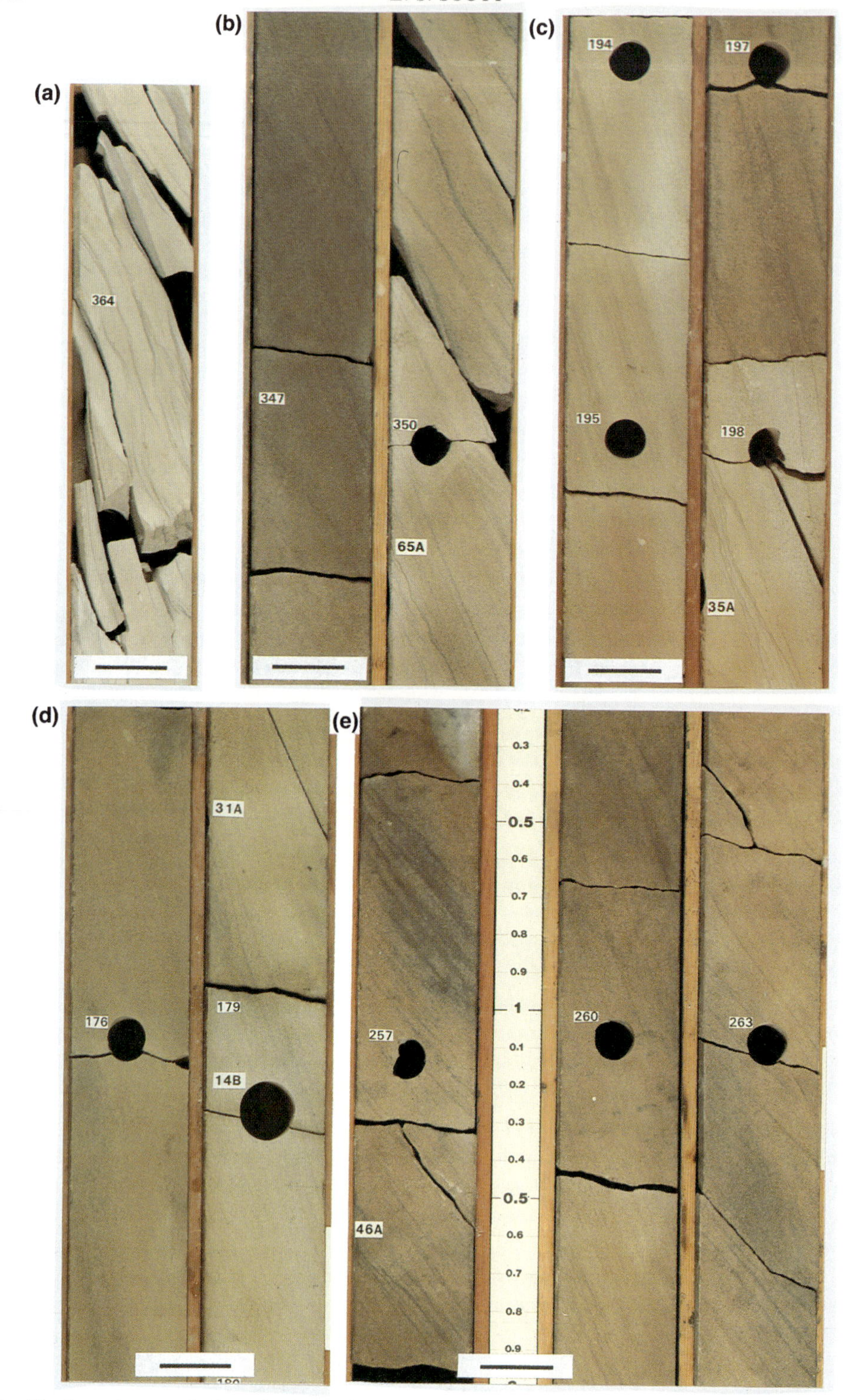


**Fig. 15.** Core photographs from Section B (Fig. 12) illustrating nearshore and beach–barrier sedimentary structures. Scale bars are 6 cm. (**a**) Muddy ripples deposited in a nearshore trough. (**b**) Parallel lamination and poorly developed ripples of a shallow nearshore bar. (**c**) Parallel swash–backwash lamination of the foreshore. (**d**) Destratified backshore sandstone with weak parallel lamination. (**e**) High angle trough cross-bedding of a storm washover sandstone.

**Fig. 16 (above).** Bar and trough (ridge and runnel) system developed in the nearshore of Kiawah Island, southern Carolina. A second system is developed seaward of the first. Parallel lamination (P) is preserved on the bar crest and in this case antidunes (A) are preserved on the slipface of the bar. Lower flow regime ripples (R) are found in the trough. Photograph by M. O. Hayes, 1974.

**Fig. 17 (left).** Fripp Island, southern Carolina is a regressive barrier island. The long, straight barrier has multiple beach ridges and the intertidal zone has a bar (B) and trough (T) system similar to that discussed for the Etive Formation. Rip channels (C) associated with bar and trough development are also found. Tidal channels wind over the contemporaneous delta or coastal plain (DP) and drain into the sea via inlet channels (IC) which cut the barrier. North is towards the top of the picture. Photograph by M. O. Hayes, 1979.

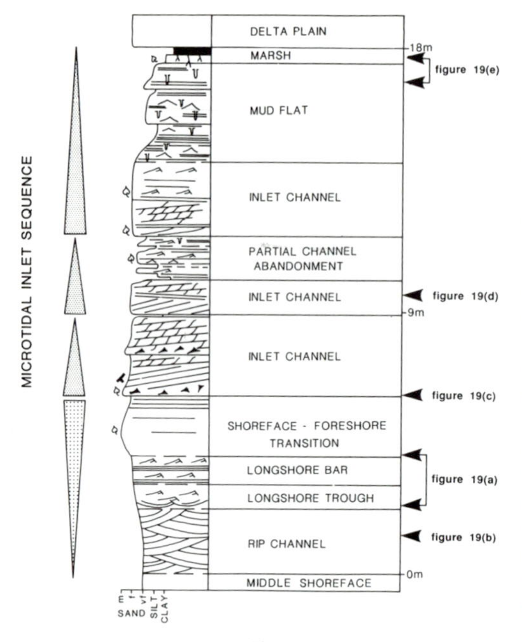


**Fig. 18.** Log from Section C (Fig. 12) core showing the fining-up microtidal inlet sequence.

inlets where any tidal currents acting along the coast would have been concentrated. Four small microtidal inlet channels are stacked and sequentially truncate each other, perhaps due to slight migration of the channel axis or changes in the intensity of channel activity.

The paucity of tidal features in these sediments and the only occasional development of inlet facies (e.g. only in section C of Fig. 12) in a continuous barrier coast (Fig. 12, sections A, B, D and E; compare with Fig. 17) suggests that the Etive Formation in southern Cormorant was a microtidal coast (Davies 1964; Hayes & Kana 1976) with a tidal range much less than 2 m.

The third channel partially abandons, probably due to slight migration, with the development of rippled and laminated fine sands and silts (H) (Fig. 19d). Organic fragments are abundant and the sediments are bioturbated. Channel activity is resumed with higher energy structures which pass gradationally upwards into heterolithics (H) as inlet activity is abandoned and the channel silts up. The sequence fines upwards, punctuated by thicker rippled sands, into a silty rooted horizon (RH) and finally a coal (C) (Fig. 19e). Above these mud flat and marsh horizons, interbedded sands and muds of the delta plain dominate.

Some of these features are seen in Fripp Island (Fig. 17). The inlet channel (IC) at the bottom of the photograph cuts into the normal barrier sequence as it migrates alongshore in a southerly direction. The delta plain (DP) lies landward of the barrier and is cut by small channels which silt up as they become inactive.

## Conclusions

The nearshore of the Brent was a storm-dominated, microtidal system with a barred, dissipative coastline (Fig. 20).

The Rannoch Formation forms the middle shoreface of a progradational sequence and is dominated by fine sand with horizontal to very low angle, planar and undulatory laminations, often cut by steeper scour features. Comparison with hummocky cross-stratification occurrences in Blackhawk Formation outcrop showed similar stratification styles and bedform geometries with low angle internal laminations showing draped, onlap and downlap relations to bed topography.

Outcrop palaeocurrent measurements show a wide dispersion but with a weak unidirectional component. This, coupled with bedforms showing transitional low angle planar–undulatory and wave-formed structures, indicates the presence of combined unidirectional and oscillatory flows.

There is strong evidence that these occurrences of hummocky cross-stratification were deposited in a high energy, storm-dominated shoreface environment, with dominant wave oscillatory motion accompanied by weak translatory currents.

The overlying Etive Formation sequences discussed show the development of nearshore bar and trough systems, with associated rip channels, passing upwards into a foreshore environment. The bars would migrate landward during fairweather periods, welding to the beachface, but would be wholly or partially washed out during storm activity. They would then reform once more during the next fairweather period. If storm activity was dominant or frequent with respect to the time required for the development and stabilisation of a nearshore bar system then only one or two bars would be developed or preserved after storm erosion. The shallow gradient, three-dimensional inshore topography developed was cut in some areas by inlet channels displaying a fining-up profile.

I thank Shell UK Exploration and Production and Esso Exploration and Production UK for funding this research as part of a DPhil. project at the University of Oxford, and for permission to publish the Rannoch

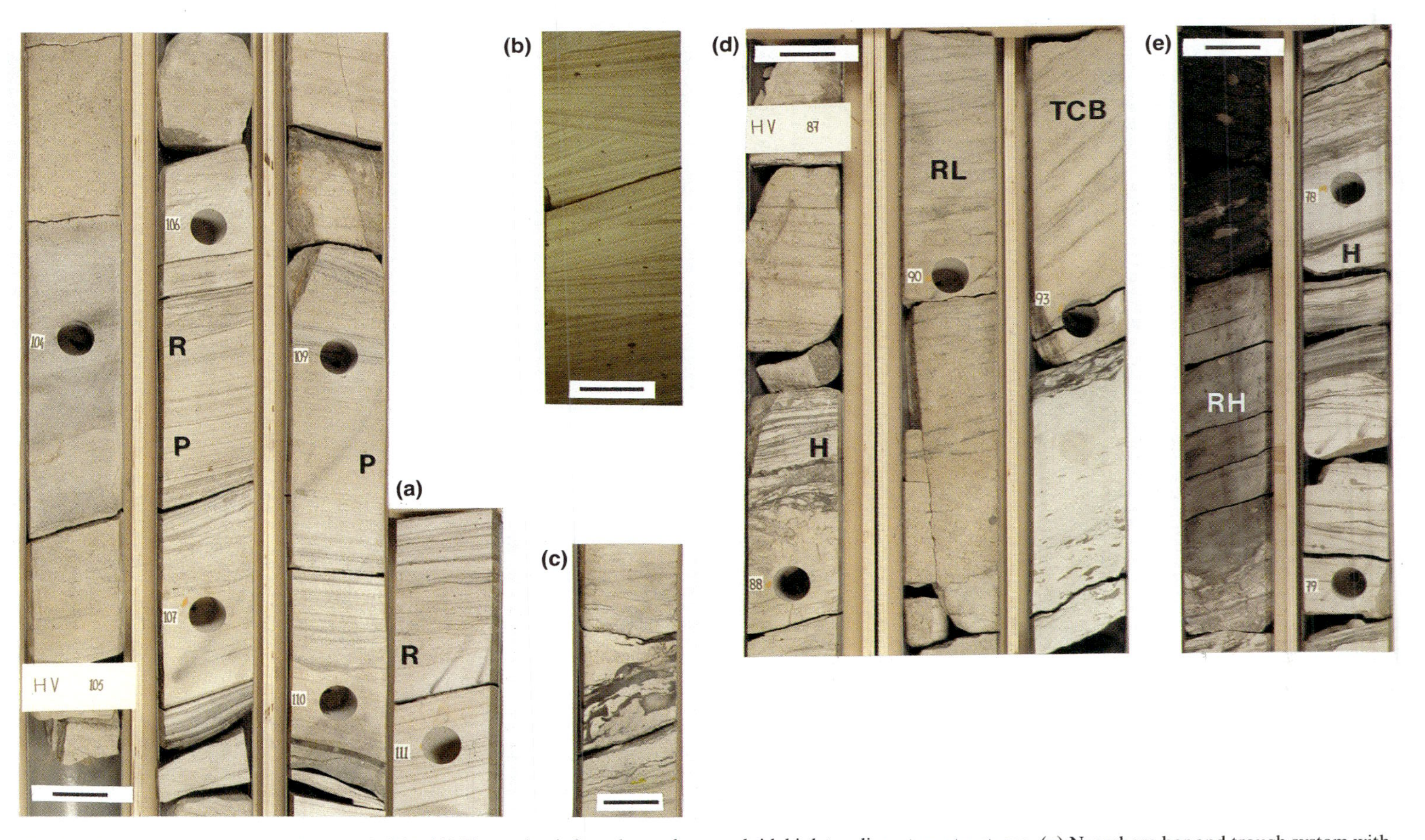


**Fig. 19.** Core photographs from Section C (Fig. 12) illustrating inferred nearshore and tidal inlet sedimentary structures. (**a**) Nearshore bar and trough system with parallel lamination (P) of the bar crest interbedded with rippled facies (R) of the trough. Scale bar is 4 cm. (**b**) Trough cross-bedding of a rip channel. Scale bar is 2.5 cm. (**c**) Angular mud clasts and carbonaceous debris in the erosional base of an inlet channel. Scale bar is 4 cm. (**d**) Fining-up microtidal inlet channel fill with high angle, tabular cross-bedding (TCB) succeeded by climbing ripple lamination (RL). Silt is found on the bedding planes and ripple surfaces, but unequivocal evidence of tidal processes is absent. Scale bar is 4 cm. (**e**) Higher energy structures of the active channel pass gradationally upwards into heterolithics (**H**) as inlet activity is abandoned and the channel silts up. The sequence fines upwards, punctuated by thicker rippled sands, into a silty rooted horizon (RH) and finally a coal (C). Scale bar is 4 cm.

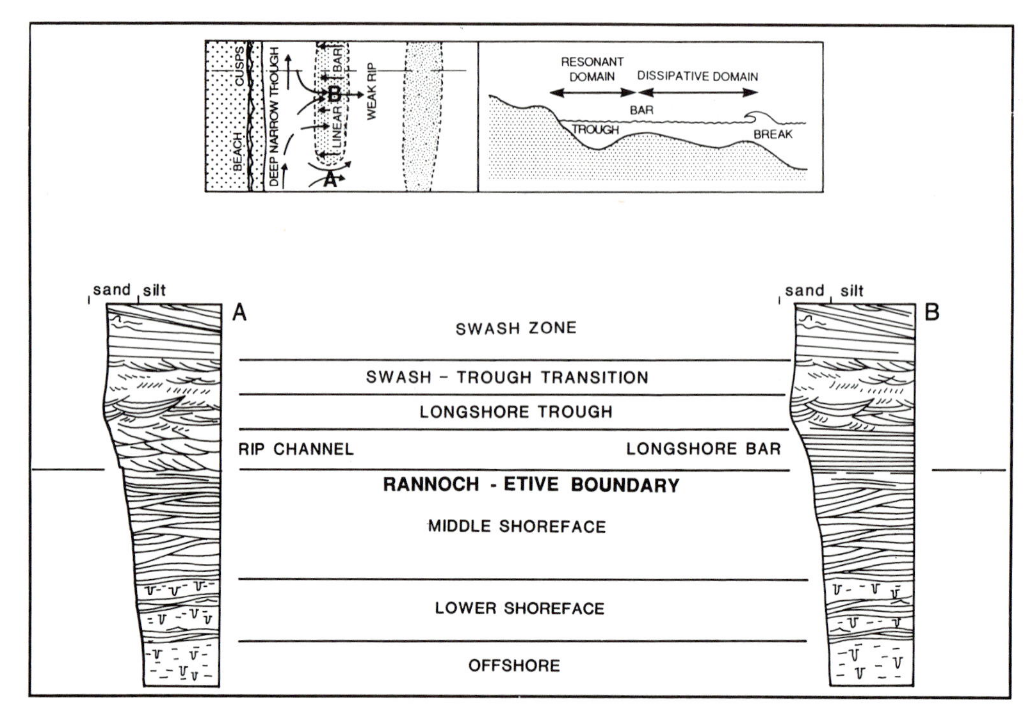


**Fig. 20.** The Rannoch and Etive show features representative of a progradational, storm-dominated, barred, dissipative coastline (modified from Wright *et al.* 1979; Shipp 1984; Hunter *et al.* 1979).

and Etive data. The Reservoir Geology Group at Shell are also thanked for ongoing help and costs to print the colour photographs. I am grateful to H. D. Johnson for a constructive and helpful review which led to improvements in the paper. Supervision and encouragement from P. A. Allen and H. G. Reading is appreciated. I would also like to thank G. Wach for useful discussion and S. Davey for assistance in the field.

## References

ALLEN, J. R. L. 1984. Sedimentary Structures: their character and physical basis. *Developments in Sedimentology*, **30**, Elsevier, New York.

ALLEN, P. A. 1985. Hummocky cross-stratification is not produced purely under progressive gravity waves. *Nature*, **313**, 562–564.

—— & UNDERHILL, J. R. 1989. Swaley cross-stratification produced by unidirectional flows, Bencliff Grit (Upper Jurassic), Dorset, UK. *Journal of the Geological Society, London*, **146**, 241–252.

ARNOTT, R. W. & SOUTHARD, J. B. 1990. Exploratory flow-duct experiments on combined-flow bed configurations and some implications for interpreting storm event stratification. *Journal of Sedimentary Petrology*, **60**, 211–219.

BRIDGE, J. S. 1978. Origin of horizontal lamination under a turbulent boundary layer. *Sedimentary Geology*, **20**, 1–16.

—— & BEST, J. L. 1988. Flow, sediment transport and bedform dynamics over the transition from dunes to upper-stage plane beds: implications for the formation of planar laminae. *Sedimentology*, **35**, 735–763.

BROWN, S. & RICHARDS, P. C. 1987. Patterns in the deposition of the Brent Group (Middle Jurassic) UK North Sea. *In*: BROOKS, J. & GLENNIE, K. (eds) *Petroleum Geology of North West Europe*. Graham & Trotman, London, 899–913.

BUDDING, M. C. & INGLIN, H. F. 1981. A reservoir geological model of the Brent Sands in southern Cormorant. *In*: ILLING, L. V. & HOBSON, G. D. (eds) *Petroleum Geology of the Continental Shelf of North-west Europe*. Heyden, London, 326–334.

CHEEL, R. J. 1990. Horizontal lamination and the sequence of bed phases and stratification under upper flow regime conditions. *Sedimentology*, **37**, 517–530.

—— 1991. Grain fabric in hummocky cross-stratified storm beds: genetic implications. *Journal of Sedimentary Petrology*, **61**, 102–110.

—— & MIDDLETON, G. V. 1986. Horizontal laminae formed under upper flow regime plane bed conditions. *Journal of Geology*, **94**, 489–504.

DAVIES, J. L. 1964. A morphogenic approach to

world shorelines. *Zeitschrift fur Geomorphologie*, **8**, 127–142.

DOTT, R. H. & BOURGEOIS, J. 1982. Hummocky stratification: Significance of its variable bedding sequences. *Geological Society of America Bulletin*, **93**, 663–680.

DOYLE, L. J., CARDER, K. L. & STEWARD, R. G. 1983. The hydraulic equivalence of mica. *Journal of Sedimentary Petrology*, **53**, 643–648.

DUKE, W. L. 1985. Hummocky cross-stratification, tropical hurricanes and intense winter storms. *Sedimentology*, **32**, 67–194.

EYNON, G. 1981. Basin development and sedimentation in the Middle Jurassic of the northern North Sea. *In*: ILLING, L. V. & HOBSON, G. D. (eds) *Petroleum Geology of the Continental Shelf of North-west Europe*. Heyden, London, 326–334.

HARMS, J. C., SOUTHARD, J. B., SPEARING, D. R. & WALKER, R. G. 1975. *Depositional environments as interpreted from primary sedimentary structures and stratification sequences*. Society of Economic Paleontologists and Mineralogists Short Course 2.

HAYES, M. O. 1974/1979. *Clastic Depositional Systems Slide Set*. Research Planning Institute Inc., Columbia, South Carolina.

—— & KANA, T. W. 1976. Terrigenous Clastic Depositional Environments — some modern examples. *Technical Report* 11-CRD. Coastal Research Division, Univ. of Southern Carolina, I-131, II-184.

HELLAND-HANSEN, W., STEEL, R., NAKAYAMA, K. & KENDALL, C. G. St. C. 1989. Review and computer modelling of the Brent Group stratigraphy. *In*: WHATELEY, M. K. G. & PICKERING, K. T. (eds) *Deltas: Sites and Traps for Fossil Fuels*. Geological Society, London, Special Publication, **41**, 253–267.

HUNTER, R. E., CLIFTON, H. E. & PHILLIPS, R. L. 1979. Depositional processes, sedimentary structures and predicted vertical sequences in barred nearshore systems, southern Oregon coast. *Journal of Sedimentary Petrology*, **49**, 711–726.

KAMB, W. B. 1959. Ice petrofabric observations from Blue Glacier, Washington, in relation to theory and experiment. *Journal of Geophysical Research*, **64**, 1891–1909.

LOWE, D. R. 1988. Suspended-load fallout rate as an independent variable in the analysis of current structures. *Sedimentology*, **35**, 765–776.

MCGOOKEY, D. P., HAUN, J. D., HALE, L. A., GOODELL, H. G., MCCUBBIN, D. G., WEIMER, R. J. & WULF, G. R. 1972. Cretaceous System. *In*: *Geologic Atlas of the Rocky Mountain Region*. Rocky Mountain Association of Geologists, Denver, Colorado, 190–229.

NØTTVEDT, A. & KREISA, R. D. 1987. Model for the combined-flow origin of hummocky cross-stratification. *Geology*, **15**, 357–361.

PAOLA, C., WIELE, S. M. & REINHART, M. A. 1989. Upper-regime parallel lamination as the result of turbulent sediment transport and low-amplitude bedforms. *Sedimentology*, **36**, 47–60.

RICHARDS, P. C. & BROWN, S. 1986. Shoreface storm deposits in the Rannoch Formation (Middle Jurassic), North West Hutton oilfield. *Scottish Journal of Geology*, **22**, 367–375.

SHIPP, R. C. 1984. Bedforms and depositional sedimentary structures of a barred nearshore system, eastern Long Island, New York. *Marine Geology*, **60**, 235–259.

SOUTHARD, J. B., LAMBIE, J. M., FREDERICO, D. C., PILE, H. T. & WEIDMAN, C. R. 1990. Experiments on bed configurations in fine sands under bidirectional purely oscillatory flow, and the origin of hummocky cross-stratification. *Journal of Sedimentary Petrology*, **60**, 1–17.

SWIFT, D. J. P., FIGUEIREDO, JR. A. G., FREELAND, G. L. & OERTEL, G. F. 1983. Hummocky cross-stratification and megaripples: a geological double standard? *Journal of Sedimentary Petrology*, **53**, 1295–1317.

WALKER, R. G. 1982. Hummocky and swaley cross-stratification. *In*: WALKER, R. G. (ed.) *Clastic units of the Front Ranges, Foothills and Plains in the area between Field B. C. and Drumheller, Alberta*. Guidebook for Excursion 21A, IAS, 11th International Congress on Sedimentology, 22–30.

WRIGHT, L. D., CHAPPELL, I., THOM, B. G. BRADSHAW, M. P. & COWELL, M. P. 1979 Morphodynamics of reflective and dissipative beach and inshore systems, southeastern Australia. *Marine Geology*, **32**, 105–140.

# A discussion of alluvial sandstone body characteristics related to variations in marine influence, Middle Jurassic of the Cleveland Basin, UK, and the implications for analogous Brent Group strata in the North Sea Basin

JAN ALEXANDER

*Geology Department, University of Wales College of Cardiff, PO Box 914, Cardiff CF1 3YE, UK*

**Abstract:** The pattern of deposition in the Middle Jurassic Cleveland and North Sea basins was complicated by spatial and temporal variations in the amount of marine influence. These variations ranged from major changes in shore-line position resulting in the alternating deposition of marine and non-marine formations to minor shore-line movements, incursions, tidal back-up, saline wedge intrusion and storm/tide/rain-induced flood events. The fluctuating extent of marine influence resulted from the interplay of factors including sedimentation rate, compaction, local tectonic activity and eustatic variations. The extent of marine influence on the alluvial environment affected the nature and architecture of sandstone bodies and the nature of early diagenesis.

It is probable that within the North Sea area local and regional relative sea-level fluctuations during Ness Formation deposition would have resulted in a range of alluvial sandstone body types similar to those of the Middle Jurassic of Yorkshire seen in outcrop. Major changes in sea-level relative to surface subsidence resulted in changing sandstone body architectural styles, while minor changes in marine influence resulted in significant changes in internal sandstone body characteristics.

Tectonic extension and related thermal subsidence together with sea-level changes, variations in sedimentation rate and sediment consolidation produced complex successions of marine and non-marine sediments in the Jurassic basins of North West Europe (Eynon 1981; Barton & Wood 1984; Badley *et al.* 1988; and others). The detailed basin history and resulting basin-fill architecture varies greatly between areas and detailed correlation of Middle Jurassic successions between basins is difficult. However, similar depositional conditions prevailed in several different basins and the sedimentary facies may be comparable.

The Middle Jurassic Ravenscar Group of the Cleveland Basin is frequently used as an illustrative analogue for the Brent Group. The analogue is far from perfect but, as the climate and sedimentary processes operating in the two areas were similar, many of the facies and facies patterns seen in the Yorkshire exposures illustrate possible variation in the Brent Group (Ravenne *et al.* 1987; Kantorowicz 1984; Mjøs & Walderhaug 1989; Mjøs *et al.* 1990; and others). The most useful analogues in the Ravenscar Group of Yorkshire are probably the Saltwick and Scalby Formations as they are exposed in extensive marine cliffs and the Saltwick Formation in particular is also well exposed inland (Fig. 1). By consideration of bedform, grain size and type, channel sandstone storey size distribution, environmental and palaeogeographical reconstructions (Knox 1969; Nami 1976*a*; Budding & Inglin 1981; Livera 1981; 1989; Ziegler 1982; Johnson & Stewart 1985; Alexander 1986*a*; Brown *et al.* 1987; Brown & Richards 1989; Helland-Hansen *et al.* 1989; and others), it seems likely that the Ness Formation of the Brent Group and the Scalby and Saltwick Formations were deposited by channels with hinterlands of a similar size distribution, character and climate. In addition, deposition was influenced by active tectonic surface deformation and sea-level fluctuations (Alexander 1986*b*; Brown *et al.* 1987; Graue *et al.* 1987; Harris & Fowler 1987; Johnson & Eyssautier 1987; Milsom & Rawson 1989). Within the Ness, Scalby and Saltwick Formations, fluctuations in the extent of marine influence occurred within the systems tracts that are generally considered to be dominantly alluvial (Knox 1969; Nami 1976*a*; Livera 1981, 1989; and others).

Alluvial sediments are generally divided into tidally and non-tidally influenced (fluvial) deposits on the basis of infauna, style and extent of bioturbation, facies and palaeocurrent patterns (Barwis 1978; de Raaf & Boersma 1971;

*From* Morton, A. C., Haszeldine, R. S., Giles, M. R. & Brown, S. (eds), 1992, *Geology of the Brent Group*. Geological Society Special Publication No. 61, pp. 149–167.

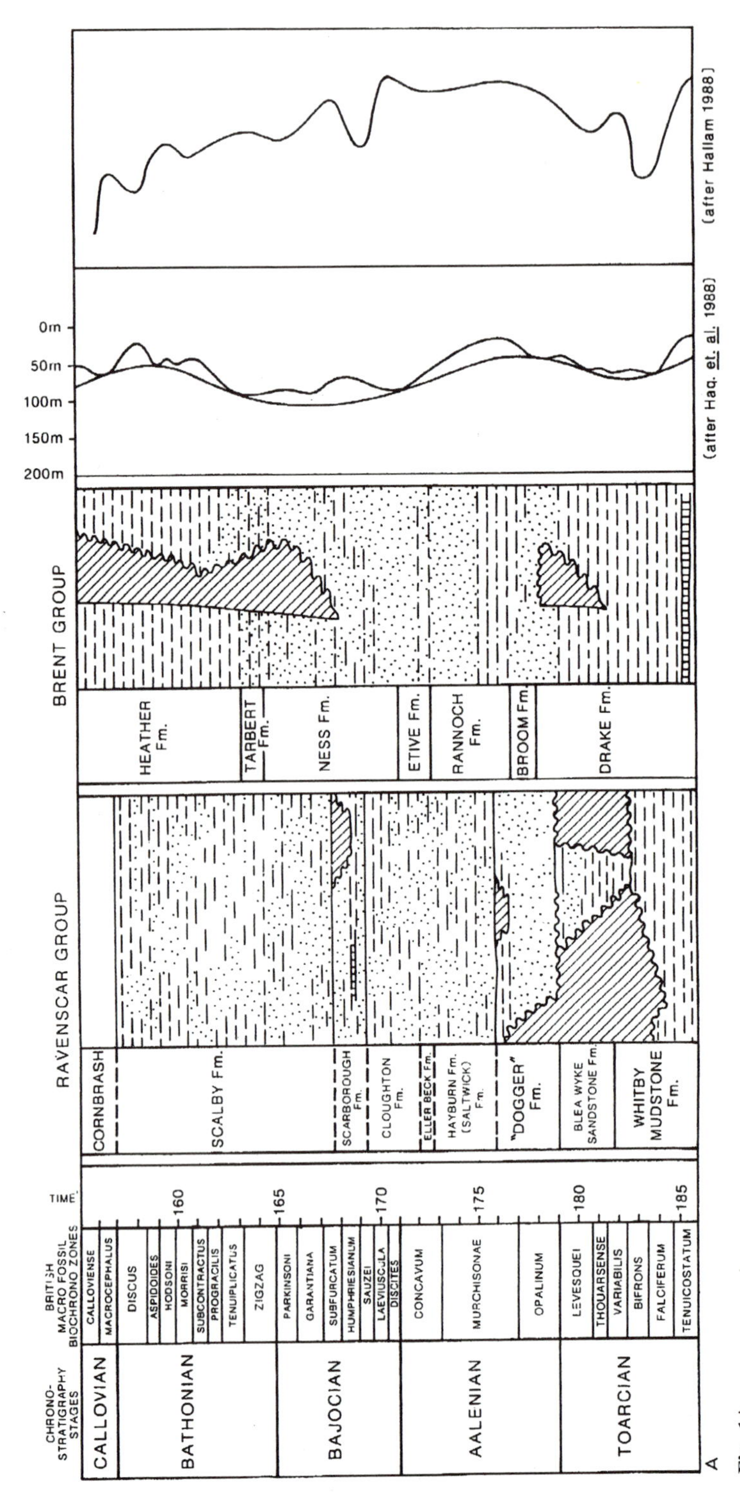

CHRONO-STRATIGRAPHY STAGES
BRITISH MACRO FOSSIL BIOCHRONO ZONES
TIME
CALLOVIAN
BATHONIAN
BAJOCIAN
AALENIAN
TOARCIAN
CALLOVIENSE
MACROCEPHALUS
DISCUS
ASPIDOIDES
HODSONI
MORRISI
SUBCONTRACTUS
PROGRACILIS
TENUIPLICATUS
ZIGZAG
PARKINSONI
GARANTIANA
SUBFURCATUM
HUMPHRIESIANUM
SAUZEI
LAEVIUSCULA
DISCITES
CONCAVUM
MURCHISONAE
OPALINUM
LEVESQUEI
THOUARSENSE
VARIABILIS
BIFRONS
FALCIFERUM
TENUICOSTATUM
160
165
170
175
180
185
RAVENSCAR GROUP
CORNBRASH
SCALBY Fm.
SCARBOROUGH Fm.
CLOUGHTON Fm.
ELLER BECK Fm.
HAYBURN Fm. (SALTWICK) Fm.
"DOGGER" Fm.
BLEA WYKE SANDSTONE Fm.
WHITBY MUDSTONE Fm.
BRENT GROUP
HEATHER Fm.
TARBERT Fm.
NESS Fm.
ETIVE Fm.
RANNOCH Fm.
BROOM Fm.
DRAKE Fm.
0m
50m
100m
150m
200m
(after Haq. et. al. 1988)
(after Hallam 1988)
A


Fig. 1A.

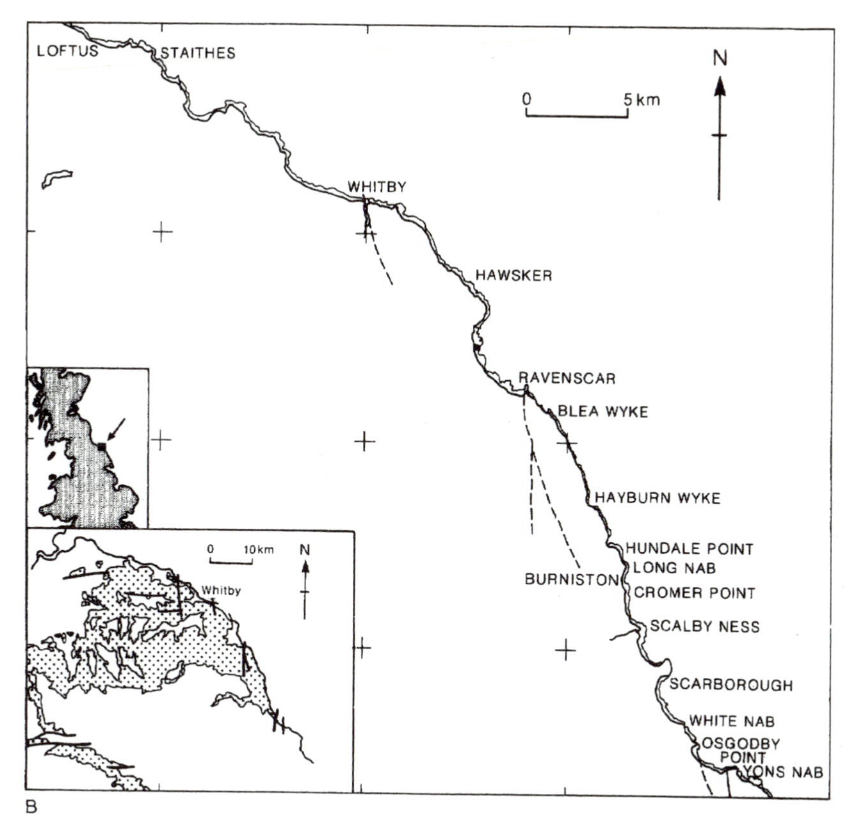


**Fig. 1.** (**A**) Stratigraphic nomenclature of the Middle Jurassic of Yorkshire and the Northern North Sea (the Brent Group stratigraphy is a generalization and not representative of any one field; after Deegan & Scull 1977; and others) with theoretical sea-level curves after Haq *et al.* (1988) and Hallam (1988). (**B**) Location map for the Cleveland Basin with an inset map showing the area over which Middle Jurassic rocks crop out.

Smith 1987; Terwindt 1988; van Straaten 1954; and others). The presence of diagnostic features depends on a wide range of environmental factors and the presence or absence of any particular characteristic is not a totally reliable diagnostic tool (Terwindt 1988). In the rock record the division is often arbitrary and frequently too ardently applied. It is clear that the features present and preserved depend on the climatic setting (Terwindt 1988) and this is frequently not sufficiently acknowledged in the rock record.

A division should be made between marine influenced alluvial deposits and deposits with no marine influence. This division will commonly not coincide with the boundary between tidally influenced and non-tidal influence. The application of a classification based on marine influence rather than tidal influence is of particular importance in micro and meso tidal regimes such as dominated the Middle Jurassic of Yorkshire and the northern North Sea.

## Humid subtropical delta and coastal plain environments

In a humid, subtropical, coastal plain or delta environment such as prevailed during Ness, Scalby and Saltwick Formation deposition, a wide range of channel sand body shapes can be generated, as a result of variations in channel form, migration and aggradation patterns. Avulsion and channel migration will move the locus of sand deposition over the depositional area in a non-random and non-chaotic fashion with time. The changing position of channels may result in lateral amalgamation of channel sandstone bodies, in a variety of cross-cutting or paralleling patterns and, with aggradation, will cause vertical stacking or channel sandstone body isolation.

Changes in relative sea-level influence facies characteristics and distribution in a variety of ways on the low gradient surface of delta or coastal plains. High rates of relative sea-level change strongly affect architectural patterns, while slight changes in the extent of marine influence may dramatically change facies characteristics. This paper aims to discuss the possible variations in sandstone body shape and characteristics in alluvial dominated deposits with varying extent of marine influence.

A notable characteristic of humid, subtropical, marine influenced deposits is the more common occurrence of fine sediment fractions in the channel deposits. This effect is due to high proportions of fine sediment in the river systems and flocculation at lower salinities than in cooler climates (Gibbs *et al.* 1989).

Very slight changes in salinity (for example 0.3‰ in the Delaware Bay, Gibbs *et al.* 1983; and 0.1‰ in the Gironde River, Gibbs *et al.* 1989) result in clay flocculation and increased probability of clay deposition in the channel sandbody. In a steady state system, or an instantaneous situation in a river/estuary system, it is possible to define the areas of fluvial sand-dominated channel deposition, of coagulation and flocculated clay incorporation in channel sandbodies and of clean tidal/estuary sand bodies where clay is 'winnowed out' by turbulence effects. In empirical studies of this type it is clear that the facies (environment) belts are variable in extent depending on climate, channel gradient, fresh water discharge and marine tide, storm and salinity conditions. In many cases the onset of clay flocculation is tens of kilometres into the fluvial environment and well inland of major facies changes that would be recognisable in the rock record.

The fossil microfauna and flora distribution in sedimentary rocks does not correlate directly with facies changes related to varying extent of marine influence. The reasons for this are that their distribution in life may be controlled by different threshold conditions than the flocculation and deposition of clay and that they are easily entrained in the water currents and are commonly deposited outside their life habitats. The salinity tolerance levels of most modern species are unknown (Ireland 1987) that of extinct forms can only be estimated. Marine microfauna and flora have been recorded hundreds of kilometres inland in some modern rivers. In Holocene and modern environments diatoms are frequently used as salinity indicators although the salinity tolerance of only about 15 species is known (Ireland 1987). Bakker & de Pauw (1974) considered that turbulence was the major factor controlling diatom distribution and salinity had a less exact effect. They should not therefore be used to indicate salinity variations and should be expected to have a different distribution to flocculated clay. Pollen analysis is also used to indicate changing climate and changing salinity conditions but few empirical studies exist that document changes in subtropical lowland areas with the exception of mangroves. Mangrove type plant remains have great potential for interpreting palaeoenvironments and environmental change in the Middle Jurassic (C. Hill pers. comm.).

Minor marine influence that cannot be reliably detected by changes in microfossils, may cause significant changes in sand body characteristics within the 'fluvial' deposits, and fluvial processes strongly affect the characteristics of the 'tidal' sediments. Studies of modern and ancient alluvial sediments have resulted in a variety of empirical facies models for channel sediment bodies with varying extent of marine influence (Terwindt 1988).

The first lateral accretion structures described were tidal in origin (van Straaten 1954) and Smith (1987) described mesotidally-influenced point bar facies that were very similar to Jackson's (1978) lithofacies class 1 formed in muddy fine-grained streams. Barwis (1978) observed that 'tidal creek and fluvial floodplain lithosomes are distinguishable not only on the basis of the relative importance of channel versus overbank deposits but on the basis of faunal diversity, bioturbation structures, orientation of cross-strata and characteristics of vertical point bar sequences that reflect tidal current stagnation'. The distinction in the case of warm, humid, microtidal to mesotidal systems, such as the Ness, Scalby and Saltwick Formations, is not as straightforward as this statement would suggest. Partly as factors other than the extent of tidal influence may affect the relative importance of channel versus overbank deposits, faunal diversity, bioturbation structures, orientation of cross-strata and characteristics of vertical point bar sequences. In addition fluctuations in the extent of marine influence may occur during the time taken to deposit a sandstone body. It is possible to make informed guesses on the extent of marine influence on sandstone bodies seen in extensive outcrops but applying any of these models to subsurface data is even more difficult.

No natural system is (or was) steady state; fluctuations in river discharge, temperature, tidal variations, and storm surge are some of the major factors that result in major shifts in facies belts on a variety of time scales. These changes are more extreme where they are superimposed on a progressive change resulting from eustatic, tectonic or other prolonged relative sea-level change.

## The Middle Jurassic of the Cleveland Basin

The Middle Jurassic Ravenscar Group (Fig. 1) crops out over an area greater than 2000 km$^2$ and is particularly well exposed along the Yorkshire coast and inland along 'sandstone edges' of the North Yorkshire Moors. The laterally extensive exposures allow the overbank and channel deposits to be examined in detail over large areas and the geometry of facies bodies to be recorded.

The extent of marine influence on deposition in the Scalby Formation is problematical (Hancock & Fisher 1981; Fisher & Hancock 1985; Leeder & Alexander 1985; Alexander 1986*b*, 1989). The problem results from the difficulties in quantifying the effects of lateral and temporal changes in the extent of marine influence on alluvial sedimentation. During the deposition of the Ravenscar Group, marine conditions persisted to the south of the Market Weighton Block thus the Cleveland Basin was never far removed from the coast. The alluvial plain was probably never more than about 100 km from the shoreline. Consequently a marine influence would have been present throughout Scalby and Saltwick Formation deposition even though the characteristics of the deposits remained dominantly alluvial.

A facies association classification is presented here, based on the variations seen in outcrop. This classification system stresses the variation in sandstone body characteristics to be expected in Middle Jurassic coastal or delta plain deposits in NW Europe. Differentiating between facies associations in the subsurface is frequently difficult, but a knowledge of the possible variations is important. Relative changes in marine influence caused lateral and vertical variations in facies associations and alluvial architectural style in the Ness, Scalby and Saltwick Formations.

### *Facies associations in outcrop*

A facies may be defined as a body of rock with specific characteristics, defined on the basis of colour, bedding, composition, texture, fossil content and sedimentary structures. It is possible on this basis to define over 40 facies and subfacies in the Scalby and Saltwick Formations and these are summarized in Table 1. These facies may be described from outcrop or core. This division of alluvial sediments into facies and subfacies, however, is insufficient for architectural interpretation or facies prediction. Facies associations based on facies and sediment body shape allow some interpretation of genesis and development of predictive models. The use of facies associations also has limitations as it is difficult to indicate the extent of sandstone variation between and within individual sandstone bodies. Differentiating between facies associations in the subsurface is difficult but facies association approach does allow the definition of end member types.

*Channel facies associations.* Channel facies associations are erosive-based deposits and include the coarsest sediments in the Ravenscar Group. They are classified on the basis of morphology, internal structure and sedimentary pattern as the deposits of alluvial channels. The channel deposits can be subdivided as below on a descriptive basis and these subdivisions bear some relation to the sedimentary processes responsible for their formation.

Geometrical classification of alluvial coarse members has been attempted as part of this paper. Potter & Blakeley (1967) defined a sandbody as 'any single interconnected mappable body of sand'. Friend *et al.* (1979) produced the first major descriptive classification of alluvial coarse members purely on the basis of their overall shape. Atkinson (1983) recorded seven main recurring alluvial sediment body geometries in the South Pyrenees Basin based on overall cross-section shape, width:depth ratio and the vertical arrangement of facies and grain size. Livera (1981) used a genetic facies classification for the Saltwick Formation. A modified combination of these approaches is used here.

The three main facies association divisions are: (1) simple channel deposits, (2) tabular channel deposits (sandstone body nomenclature after Friend 1978) and (3) complex channel deposits. These three architectural types are subdivided on the basis of internal characteristics. The possible channel sediment body architectures are diagramatically represented in Fig. 2 and these are largely independent of channel sandbody size or internal characteristics.

*Facies Association 1.* Simple channel deposits; solitary concave based ribbon deposits (width (w):thickness (t) ratio of less than 25:1 after Atkinson 1983). These are subdivided on the basis of lithology.

*1a. Homogeneous ribbon sandstone deposit.* This facies association shows no structures related to lateral accretion and appears to be formed by channel cut and fill. The sedimentary structures are generally trough and tabular cross-stratification and small-scale cross-lamination. Fossil logs and mudstone conglomerates are common near the basal scour. These channel deposits have been observed only in

**Table 1.** *Facies in the Saltwick and Scalby Formations of the Cleveland Basin*

| | Plant material | | | | | Bioturbation | | | | | | | | Fossils | | | | Fracture | |
|---|---|---|---|---|---|---|---|---|---|---|---|---|---|---|---|---|---|---|---|
| | Amorphous | Stems | Logs | Leaves | Roots | General | Recognizable burrows | Mudstone clasts | Ironstone clasts | Iron stained | Iron mottled | Siderite nodules | Siderite sphaerulites | Fresh water bivalves | Marine shell debris | Marine/ brackish microfossils | Dinosaur footprints | Blocky | Fissure |
| *Conglomerates* | | | | | | | | | | | | | | | | | | | |
| intraformational (mudstone) | x | | x | | x | | | x | | | | | | | | | | | |
| intraformational (ironstone) | x | | x | | x | | | x | x | | | | | | | | | | |
| interformational | x | | x | | x | | | x | x | | | | | | x | | | | |
| *Coarse sandstone* | | | | | | | | | | | | | | | | | | | |
| massive | x | | x | | | | | x | x | | | | | | | | | | |
| *Medium sandstone* | | | | | | | | | | | | | | | | | | | |
| massive | x | | x | | x | | | x | x | | | | | | | | | | |
| flat laminated | x | x | x | | x | | | x | x | | | | | | | | | | |
| planar cross-bedded | x | x | x | | x | | | x | x | | | | | | | | | | |
| trough cross-bedded | x | x | x | | x | | | x | x | | | | | | | | | | |
| mudstone lenses and laminae | x | x | x | x | x | | | | | | | | | | | | | | |
| wavy laminated | x | x | x | x | x | | | | | | | | | | | | | | |
| *Fine sandstone* | | | | | | | | | | | | | | | | | | | |
| massive | x | | x | | x | x | x | x | x | | | | | | | | | | |
| planar laminated | x | x | x | | x | x | x | x | x | | | | | | | x | x | | |
| planar cross-bedded | x | x | x | | x | | | x | x | | | | | | | | | | |
| trough cross-bedded | x | x | x | | x | | | x | x | | | | | | | | | | |
| ripple cross-laminated | x | x | x | | x | x | x | | | | | | | | | | x | | |
| wavy laminated | x | x | x | x | x | x | x | | | | | x | x | x | | x | | | |
| silt lenses and laminae | x | x | x | x | x | x | x | | | | | x | x | | | | | | |
| mud laminae | x | x | x | x | x | x | x | | | | | x | x | | | | x | | |
| thinly interbedded fine sandstone and mudstone | x | x | x | x | x | x | x | | | | | x | x | x | | x | x | | |
| *Siltstone* | | | | | | | | | | | | | | | | | | | |
| massive pale yellow | x | | | | x | x | x | | | x | x | x | x | | | | | | |
| laminated pale grey | x | x | | x | x | x | x | | | | | x | x | x | | | | | |
| pale green grey | x | x | | x | x | x | x | | | | | | | | | | | | |
| *Silty mudstone* | | | | | | | | | | | | | | | | | | | |
| green grey | x | | | x | x | x | x | | | x | x | x | x | | | | | x | x |
| green grey with sandstone lenses | x | | | x | x | x | x | | | | x | x | x | | | | | x | x |
| dark green grey | x | | | x | x | x | x | | | x | x | x | x | | | | | x | x |
| medium grey | x | x | | x | x | x | x | | | x | x | x | x | | | | | | x |
| medium grey with sandstone lenses | x | x | | x | x | x | x | | | | x | x | x | | | | | | x |
| *Mudstone* | | | | | | | | | | | | | | | | | | | |
| dark grey | x | x | | x | x | | | | | x | x | x | x | x | | | | | x |
| purple grey | x | x | | x | x | | | | | | | x | x | x | | | | | x |
| coaly | x | x | | x | x | | | | | | | x | x | x | | | | | x |
| *Coal* | x | x | x | x | x | | | | | | | | | | | | | | |

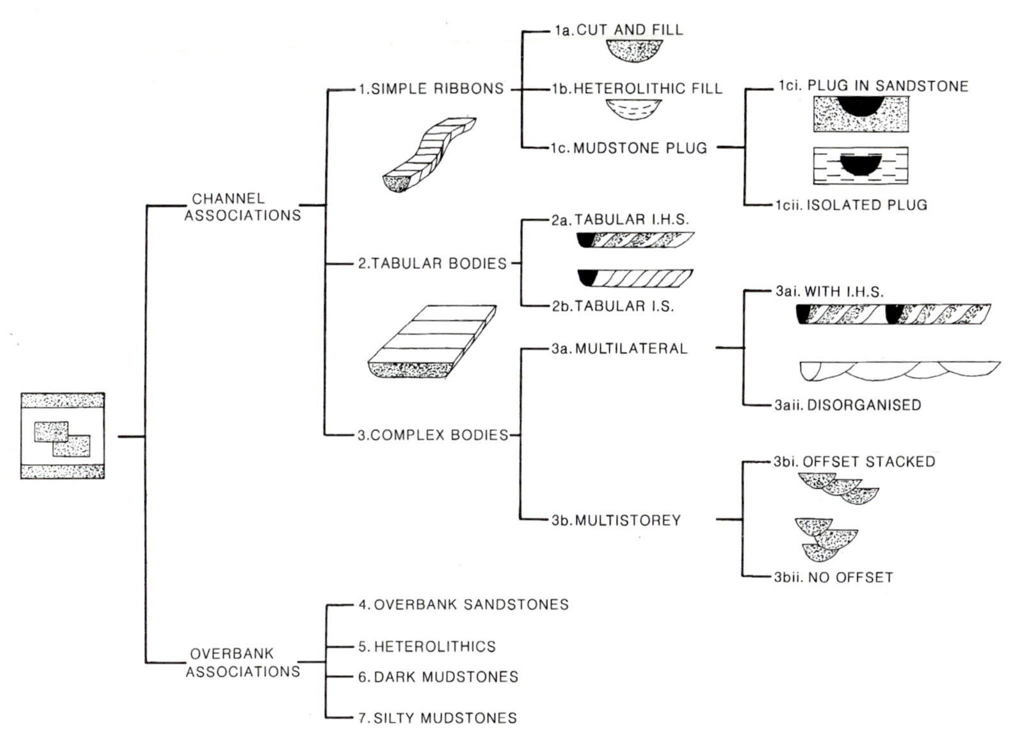


**Fig. 2.** Facies association classification diagram. The channel association architecture is independent of size. The facies associations are end members in a continuum of possibilities. Inclined homolithic stratification (I.S.) and inclined heterolithic stratification (I.H.S.) follow the usage of Thomas *et al.* (1987).

cross-section in the cliffs. It is unwise to infer the channel plan form from limited two dimensional exposure (Bridge, 1985), although classically the sediments would be interpreted as the deposits of low sinuosity channels. Some examples of these types of channel deposits appear to relate to incision events. An example of this type is seen in cliff exposures of the Scalby Formation north of Scalby Ness (Grid Ref. TA030920). The high w:t ratio and lack of channel migration is partly explained in this case by the greater resistance to lateral bank erosion during a period of incision.

*1b. Mixed sandstone and siltstone deposits.* These channel deposits are generally smaller than the 1a types and are often associated with recognizable flood sediments. The interbedded sandstone and siltstone are commonly deformed as the result of slumping off the channel banks and by post-depositional pedogenic activity, including root penetration. They show limited evidence of lateral migration, this may not necessarily imply low sinuosity but may be related to restricted channel occupation time or slow meander migration rate. An example of this type may be seen in the cliff south of Crook Ness (Grid Ref. TA029936).

*1c. Mudstone plugs.* Mudstone dominated channel deposits may be recognised within (i) over bank deposits and (ii) associated with channel deposits of another type. The latter situation represents the fine infill of the channel following a change in hydrological regime, while the former is considered the result of intrabasinal stream activity. The nature of the base of these deposits varies from a sharp erosional surface to a gradual upward fining sequence from the underlying sandstone dominated deposits. This variation relates to the difference in the rate of hydrological change associated with the change from channel forming to channel filling conditions. An upward fining sequence can be produced by a gradual meander cut-off, anabranch abandonment, slow processes of avulsion or rapid avulsion with the abandoned channel remaining in a proximal position to the coarse sediment supply.

Mudstone dominated channel-fills that are not associated with sandstone deposits are more difficult to recognise in the field. Detailed examination of the fines-dominated sequences in the Saltwick and Scalby Formations, however, commonly reveals the existence of such channels. A small example can be seen easily where a channel cut through an overbank sheet sandstone in the Scalby Formation near Cromer Point (Grid Ref. TA029927). It would be difficult if not impossible to identify such channel deposits in core or restricted outcrop, as the composition and fabrics will be the same as deposits of other floodplain depressions that may have considerably greater lateral extent. The fine channel sediments may have been deposited in minor floodplain channels that had no coarse sediment supply although they had the power to erode into the floodplain sediments. Such channels frequently occur as intrabasinal streams on modern-day coastal and delta plains, and act to drain runoff from the areas between the main channels and may also act as conduits in times of floods. In some cases they will follow stretches of abandoned major channels. The minor channels become more pronounced when their trunk streams undergo a period of incision.

The environment of deposition for the Middle Jurassic of NW Europe is thought to have been subtropical and humid, and therefore moderate drainage network densities may be expected as the amount and distribution of rainfall, plant cover and substrate strongly influences drainage density. By comparison with published empirical correlations (Abrahams 1972; Gregory & Gardiner 1975) a density of the order of 5 $km^{-1}$ seems reasonable. By consideration of channel network structure (Richards 1982), the majority of these channels would be relatively small floodplain draining channels that would result in deposits of facies type 1c and 1b, major channel deposits should be less in number. In the rock record deposits of these low order channels are rarely distinguished.

*Facies Association 2*. Tabular sandstone deposits (sandstone body nomenclature after Friend 1978). In cross section these deposits are not always easy to distinguish from types 1a and 1b, as at any given point along their course some of the channels may have had a low sinuosity or low rate of migration.

*2a. Tabular, sandstone-dominated bodies with inclined heterolithic stratification (terminology after Thomas* et al. *1987).*
These deposits frequently show classical features of epsilon cross bedding (cf. Allen 1963), with basal scour and associated intraformational lag conglomerates, passing upwards in to cross stratified sandstone that gives palaeocurrent readings approximately perpendicular to the dip of the major lateral accretion surfaces (Allen 1963). Examples of this type of facies association may be seen in the Whitby East Cliff (Grid Ref. NZ905114) and in the Scalby Formation on the south side of Yons Nab (Grid Ref. TA085840 Nami 1976*a*).

The 2a type facies association is commonly found associated with a 1c channel plug, the upper parts of which may show pedogenic alteration. These deposits were formed by a single meandering channel between avulsions. The channels had a low inter-avulsion period compared with the channel migration rate such that insufficient time was available for a mature meander-belt to develop.

*2b. Tabular, sandstone-dominated bodies with inclined homolithic stratification (terminology) after Thomas* et al. *1987).*
These deposits are similar in some ways to those of types 1a and 1b but show a stepwise movement that may be related to periodic migration of the river bank. These erosive events produced a stepped form to the basal scour but the deposits do not contain all the features typical of the classical lateral accretion units of type 2a. The palaeocurrent directions are approximately perpendicular to the dip of the erosion surfaces that pass through the deposits. They tend to be lithologically more homogeneous than types 1b or 2a deposits and relate to streams with a higher sand fraction in the load and or more constant discharge.

An example of type 2b is seen in the Saltwick Formation at Hawsker Bottoms (Grid Ref. NZ947080; Alexander 1987). This example shows clearly the dipping erosion surfaces which are often seen associated with soft-sediment deformation and intraformational conglomerates.

*Facies Association 3*. Complex channel sandstone deposits. This facies association comprises all the major sandstone bodies within the alluvial formations. Most of these are the result of amalgamation of channel deposits over a considerable time period.

The controls on channel size are complex but a major control is that of hinterland size; major modern rivers (Amazon, Mississippi, Niger, Nile etc.) are seen to drain very large continental areas (Stoddart 1971). In comparison the hinterland for the Jurassic rivers of Yorkshire and North Sea areas are likely to have been relatively small, considering palaeogeographic reconstructions (Ziegler 1982). The size of channel sandstone bodies and abandoned channel fills seen in outcrop and core tends to support the theory of relatively small channels (channel depths up to 5–10 m compared to 25 m for the

Brahmaputra; Bristow 1987). It is thought therefore that major sandstone bodies (several tens of metres thick or several kilometres wide) are not the result of very large rivers but of prolonged occupation by smaller channels.

*3a. Multilateral channel deposits*. A multilateral channel deposit can develop in a number of ways: mature meander-belt development, lateral migration of meander-belts, mature braid-plain development and amalgamation of channels of differing ages due to low aggradation rates or fluctuating incision and deposition. The resulting sandstone bodies can not always be distinguished from each other in the rock record. Multilateral sandstone bodies in the Ravenscar Group can be subdivided on the basis of their internal structures and these may be used to infer their mode of origin.

*3ai. Multilateral, sigmoidally bedded, sandstone-dominated deposits*. This type of deposit is best seen in the foreshore between Long Nab (Grid Ref. TA030940) and Scalby (Grid Ref. TA034913). This section was described as an exhumed meander-belt by Nami (1976*b*) and in detail is highly complex (Alexander 1992). An examination of lithology, internal structures and plan of lateral accretion units suggests that the sandstone body was not formed as a single meander-belt (Alexander 1992). It is not easy, however, to differentiate between the units of differing age. The deposit may have been the result of one trunk-system with its tributaries and distributaries occupying one site for a long period of time, by a system avulsing repeatedly back to the site or by superposition of different channels in one area through time. The foreshore between Long Nab and Scalby is made up predominantly of point bar inclined heterolithic stratification with minor amounts of channel plug, sheet flood and counterpoint bar deposits (Alexander 1992).

*3aii. Multilateral, sandstone-dominated deposits with no obvious migration pattern and minor sigmoidal bedding*. These sandstone deposits are made up of a number of type 2b deposits in lateral contact, as seen near the base of the Scalby Formation at Black Rocks (Grid Ref. TA050870) or may be of a more complex nature as seen at the base of the Scalby Formation around Hundale Point (Grid Ref. TA026949).

At Hundale Point (Grid Ref. TA026949) the complex sandstone body at the base of the Scalby Formation (the Moor Grit) is made up of large bed forms with complex interrelationships (Ravenne *et al.* 1987). At this site it is difficult to establish where the margins of the channel(s) were at any one time. This complex, laterally extensive sandstone body is demonstrably the result of lateral coalescence of a variety of channel sediment body types.

*3b. Multistorey stacked sandstone deposits*. Multistorey stacked sandstone deposits can be subdivided on the nature of their vertical stacking pattern. This division relates to the relative extent of restriction on channel position.

*3bi. Multistorey sandstone-dominated body with offset channel deposits*. Offset stacked channel sandstone bodies are difficult to identify in outcrop due to their size; however, they can be seen in the cliff to the south of Whitby. This facies association can be formed where channels repeatedly re-occupy the same site. The offset is a result of differential compaction of sandstone and overbank fines in the shallow subsurface (Anderson 1989). To produce an offset stacked channel body the floodplain aggradation rate must be relatively low to allow channels to cut down into the earlier sandstone bodies. This is best seen in the Saltwick Formation north of Sandsend (Grid Ref. NZ860138).

*3bii. Stacked sandstone body with no obvious offset of channel deposits*. The best examples of this type is seen in Whitby West Cliff (Grid Ref. NZ897115) and at the Loftus alum quarries (Grid Ref. NZ740200). In the Loftus example the existence of stacked deposits can be demonstrated from one dimensional logs as there is an abrupt change in palaeocurrent direction across a surface, although the sandstone itself appears the same above and below that surface except in the youngest sandstone body where there is a slight petrographic change. In limited outcrop or in core the multistorey nature of this sandstone body would be difficult to establish without the aid of palaeocurrent indicators. In multistorey sandstone bodies where there is little variation in the palaeocurrent direction, there is a greater difficulty in distinguishing storeys. This would be the case, for example if the stacked channel sandstone body that is seen in Whitby West Cliff were only recorded in core. In the Whitby example the multistorey nature of the deposits is demonstrated by the pattern of erosion surfaces within the outcrop. Locally the sandstone body is interrupted by fine channel fill deposits (type 1b and 1c). The channels repeatedly re-occupied the same site and maintained a similar orientation.

*Overbank facies associations*. These are volumetrically the most important rocks of the Ravenscar Group. Variation in the overbank environment resulting from minor changes in topography, depositional mechanism, sediment supply, water table depth and extent of marine influence caused the complex distribution of facies that is observed in outcrop. The overbank sediments were deposited from moving flood-water, standing water, minor marine incursions and aeolian dust. Most, if not all of these deposits have been altered by pedogenic activity. The nature of the facies variation was examined in detail at Burniston Wyke (TA 028935 Alexander 1986*a*), where lithological logs were measured at *c*. 25 m intervals over a distance of *c*. 2 km and at selected other sites where exposure was easily accessible.

*Facies association 4*. Sharp-based sheet sandstones. Sharp-based sheet sandstones form about 20% of the thickness of the Scalby or Saltwick Formations. The distribution is variable laterally and vertically through both formations. The sandstone is generally well sorted and fine to very fine grained. It is petrographically similar to the channel sandstone deposits, but often more tightly cemented (Kantorowicz 1985). Some of these sheet sandstones may be traced over several kilometres and they are locally associated with minor channelling. The sheet sandstones may have been deposited as crevasse splay deposits as described from the Mississippi by Fisk (1947) or, more likely, as sheet flood sandstones as described from the Brahmaputra by Bristow (1987). Some of the sheet sandstones show a reverse grading similar to that observed in modern flood deposits of the Brahmaputra.

Facies association 4 can be subdivided on the basis of the nature of flood event into three types: (a) extensive marine influenced sheet flood sandstone, (b) extensive sheet flood sandstone with limited or no marine influence and (c) laterally restricted sheet flood sandstone. In reality there is probably a complete continuum between these end members. In limited outcrop or core the three sub-facies would be difficult to distinguish as the processes of fluid flow and sediment deposition are the same on a local scale. Within one flood basin the grain size distribution and grading may vary with flood type but insufficient data on modern flood deposits exists to make definite distinctions on this basis. Type (a) may be distinguished in some cases by the presents of marine palynomorphs as is the case in the Burniston Food Print Bed (Fisher & Hancock 1985).

*Facies association 5*. Thinly interbedded sandstone and mudstone. This facies association makes up a small percentage of the Scalby and Saltwick Formations. These deposits are generally thin and laterally impersistent, and pass laterally into sandstone of facies association 3 or the mudstones of facies association 6. Wave ripple lamination, planar lamination, bioturbation and occasional body fossils (non-marine bivalves) indicate that some of these sediments were deposited in floodplain lakes. Locally brackish marine fossils are recorded and lagoonal or imbayment conditions occurred, notably at the base of the Saltwick Formation (Alexander & Hill, in prep).

*Facies association 6*. Dark humic mudstones. These mudstones with well preserved plant material occur locally throughout the Ravenscar Group. These sediments were deposited in shallow water lakes and lagoons on the coastal plain and had a low coarse clastic input. Plant material falling into the fine sediment of the stagnating lake waters had a high preservation potential. The salinity of the lakes varied from fresh to brackish on the basis of contained macrofossils.

*Facies association 7*. Silty mudstones deposited in the overbank environment. Overbank silty mudstones dominate many of the exposed coastal sections of the Scalby and Saltwick Formations. Many features in these sediments relate to pedogenic and diagenetic processes rather than to original sedimentary structure. Notable pedogenic features include colour mottling, curtains, roots and sphaerosiderite which indicate variation in the maturity and water saturation levels in the soils. Both fining up and coarsening upward sequences are seen in this facies association.

## *Marine influence, architectural style and sandstone body character*

***Controls on relative sea-level change***. During the mid-Jurassic, in the North Sea and Cleveland basins, as in any coastal setting, relative sea-level would have been controlled by: eustatic sea-level, tectonic subsidence, isostatic effects, sediment accumulation/sediment supply rate, sediment pile consolidation and inherited topography.

The extent of marine influence on alluvial channel deposition depended on: distance to river mouth, tidal range, discharge characteristics, gradient, storm activity, temperature and dissolved organics.

Hallam (1988), Vail *et al*. (1984), Haq *et al*. (1988) and others suggests major changes in eustatic sea-level through the Jurassic, although there is some disagreement on the exact timing and rate of these eustatic variations (J. Cope pers. comm.). If the sea-level curves are anything like correct, then several eustatic changes of tens of metres occurred during the mid-Jurassic (Fig. 1). The rates of topographic change related to tectonic activity and sedimentation may have been similar in the short term to the proposed sea-level changes, but the overall magnitude of the resulting topographic change may not have been so extreme.

Small and local changes in the extent of marine influence on the alluvial environment would have related to the interplay of eustatic, tectonic and sedimentary factors with storm and tidal effects. Minor effects, below the resolution of eustatic sea-level curves (and complicating the data on which they are based), are of major importance in controlling facies characteristics of channel sediment bodies on low gradient alluvial surfaces.

In the Cleveland Basin relative sea-level change occurred on a range of scales. For example the depth of water during Scarborough Formation deposition although variable (Taylor, pers. comm.; Gowland 1987; Livera 1981) reached maximum depths that were greater than 30 m. The land area became emergent soon after with deposition of the Scalby Formation. In contrast fluctuations of 1 m or less would be sufficient to explain some of the changes in facies seen higher in the Scalby Formation. In the northern North Sea relative sea-level changes explain features such as the nature and distribution of the Mid Ness Shale.

***Major changes in relative sea-level and architectural style.*** High rates of change of relative sea-level strongly affect the sandstone body architectural style. These changes relate to eustatics, tectonics and major changes in sediment aggradation. In the extreme, on a low lying delta or coastal plain, they will result in major transgression or regression and consequent complete change in sedimentary style. For example on an alluvial surface with a mean seaward slope of 1 in 10 000 (a figure estimated from modern deltas and coastal plains) a one metre change in sea-level, in the simplest case, would result in a horizontal movement of the shore line by 10 km. The eustatic fluctuations of tens of metres postulated for the Middle Jurassic would result in complete emergence or complete inundation of the alluvial surface assuming no other influence on topography. In reality the major eustatic sea-level changes never affect shore-line position in isolation. Other factors (listed above) result in long term and short term fluctuations that may at times reinforce eustatic effects but often oppose them, so that any eustatic change results in a complex number of shore-line fluctuations.

Given the above comments, it is generally true to say that major relative sea-level changes between these two extreme situations control the rate of change of channel gradient and therefore the hydrodynamic characteristics of the channels. Empirical and theoretical studies suggest (Bridge & Leeder 1979; Crane 1982; Fisk 1947; Ouchi 1985) that an increase in alluvial plain gradient may result in a straightening of the channel and a tendency towards incisions, while a lowering of the channel gradient may increase the tendency to meander and anastamose. In parallel to the changing channel forms, there are also changes in the lateral stability of channel position, channel aggradation, periodicity of avulsion, sand preservation potential (net to gross) and consequently type of channel facies association even when the character of the channel load does not change.

High rates of relative sea-level fall will result in incision and this in turn restricts the position of the channels (Fig. 3C). This is a factor ignored in theoretical alluvial architectural models (Bridge & Leeder 1979, Bridge & Mackey 1989). With increased sediment accommodation (rising relative sea-level) the incised channel must fill before the channel position can change, unless stream capture occurs. This process of incised channel fill will produce either an architectural style with narrow thick sandstone bodies (Facies association 1a or 1b, as seen in the Scalby Formation south of Cromer Point, Grid Ref. TA030920), or, if more time were available for enlargement of the incised channel, a vertically stacked channel sandstone system (Facies association 3bii; Fig. 3D).

In conditions where relative sea-level fluctuates by small amounts around a constant or slowly changing mean value, channels have time to migrate or avulse across the alluvial plain, assuming there is minimum lateral control by differential subsidence. As the channels migrate and when they avulse they erode into the older sediments by an amount dependent on the channel discharge and slope, and to a lesser extent on substrate. With near constant mean sea-level, successive channel bases will erode down to a similar level if there is little net aggradation. In this way a laterally extensive, complex alluvial sandstone body (Facies association 3ai or 3aii) may develop (Fig. 3A). This is seen, for example, at the base of the Scalby Formation where channel sandstone bodies with varying amounts of marine influence are juxtaposed to form a complex alluvial sheet sandstone that extends tens of kilometres laterally and varies in thickness up to about 20 m as discussed by Nami & Leeder (1978). A similar but less extreme case is seen in intermediate levels in

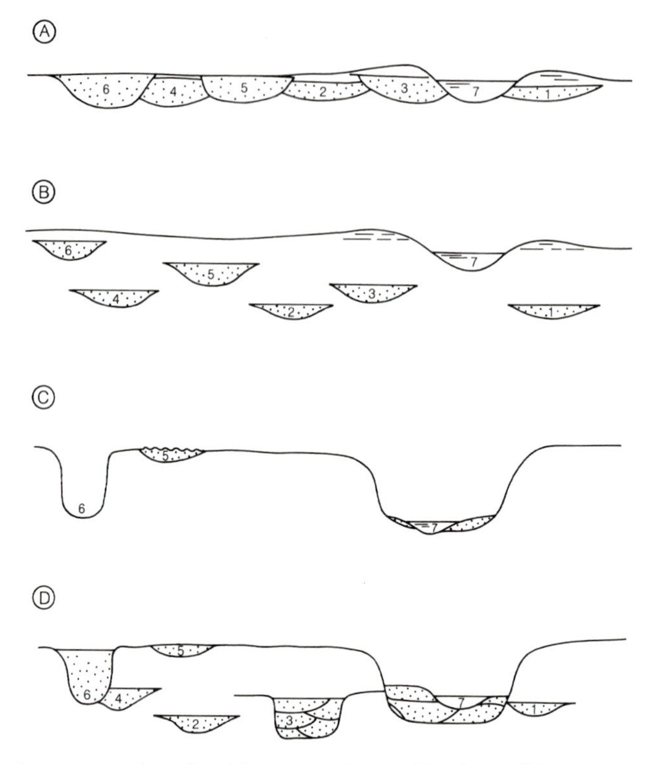


**Fig. 3.** Diagrammatic representation of architectural styles resulting from different rates of relative sea-level change and aggradation. The architectural styles presented are a result of changes in accommodation space and do not require changes in the nature or quantity of sediment supplied. The percentage volume of fines preserved in the sequence is a reflection of selective preservation and is not correlatable with percentage of fines in the channel load.
(**a**) At low aggradation rate and stable to slow changing relative sea-level the architecture will tend towards a domination of facies association 3aii.
(**b**) With rising mean sea-level and aggradation the channel deposits tend towards isolation (Facies associations 1 and 2). The channel deposits may locally come in contact where there is a lateral bias to channel position for a prolonged time period, for example, in a growth syncline.
(**c**) In conditions where base level (in most cases relative sea-level) drops rapidly, major incision results. Incision will continue until the relative sea-level becomes stable or starts to rise. Channels are confined to the incised valleys unless there is stream capture by headward erosion of another channel or the valley is filled. The width of the incised valleys is dependent on the rate of lateral erosion and incision.
(**d**) In most coastal plain or delta settings alternations of incision and aggradation occur as a result of varying, stream hydraulics, relative sea-level and aggradation. This alternation results in channel bodies with very low w:d ratios but increased probability of channel body interconnectedness.

the Saltwick Formation as suggested by Knox (1969). Another example that may be explained in this way is the lateral complex of stacked 'low sinuosity' channels in the Brent Group of the Bruce Field described by Ketter (pers. comm.). This complex sandstone body he interprets as being tidally influenced and the amount of marine influence increased with time.

If there is net aggradation during rising relative sea-level, sedimentation keeping pace with rising base level, the channels have maximum lateral freedom of movement and following avulsions or with migration successive channels erode to progressively younger stratigraphic levels. This results in architectural styles with isolated channel sandstone bodies (Fig. 3B) or, if there is a lateral bias to channel position resulting, for example, from differential tectonic or compactional subsidence, small clusters of sandstone bodies. The sandstone bodies will tend to be thinner than those where there have been periods of incision as, in any aggrading alluvial system, there is a threshold of channel bed aggradation that results in channel avulsion.

In most alluvial formations these architectural styles develop in different places through time and so the resulting formation architecture is invariably a combination of the types. An extreme example of this is seen in the Scalby Formation, where a laterally extensive complex alluvial sandstone body passes up into a mudstone dominated deposit with isolated channel deposits and evidence of periodic incision diagrammatically represented in Fig. 4.

Although alluvial architecture is controlled by relative base-level change (sea-level in this case) and aggradation the nature of the channel sediment bodies does not correlate with the architectural style. In humid, subtropical coastal plain and delta plain environments the nature of channel deposits depends on the river load characteristics, the hydraulic nature of the flow and the extent of marine influence, particularly minor changes in salinity.

***Minor changes in the extent of marine influence and sandstone body character.*** Minor changes in the relative sea-level are topographically restricted and shore line migration or facies belt migration will be irregular over the coastal area. Minor shore-line shifts may be recorded in the rock record but significant small variations may be masked by overall trends. Changes in the extent of marine influence are still more difficult to distinguish. From a reservoir geology viewpoint the mud content of a channel sandstone body is of major importance. In a coastal plain or delta setting, major changes in detrital clay content (mud drapes or dispersed clay) occur as a result of changes in salinity of the order of 0.1‰ (Gibbs *et al.* 1989; and others) and turbulence effects. In low gradient settings (as for the Ness Formation and the Ravenscar Group) with waters containing a high dissolved organic content (typical of humid subtropical systems) and micro or meso tidal regimes, flocculation may commence several tens of kilometres inland.

Minor changes in freshwater discharge, marine water influx (tidal or storm induced) or gradient will move the onset of flocculation and clay deposition kilometres or tens of kilometres. This in turn can dramatically change the reservoir characteristics of the channel sandstone body. In this respect, most of the channel sandstones in the Ravenscar Group and Ness Formation over much of the northern North Sea, suffered fluctuating extents of marine influence. There is no guarantee that an increase in clay flocculation and settling would correlate with a change in any preserved microfossil assemblage, as the salinity fluctuations may be less than the tolerance levels of the fauna and the time periods involved may be in the order of hours or days and therefore not develop a stable biota.

Sandstone body character can change laterally

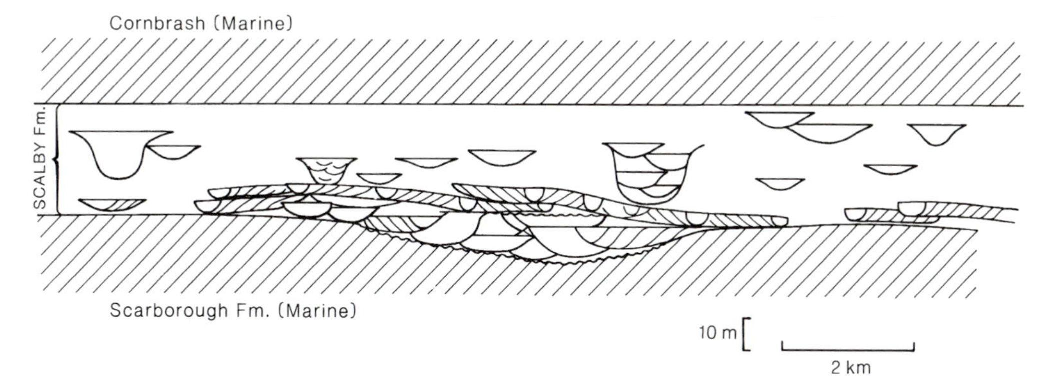


**Fig. 4.** Diagrammatic representation of the architecture of the Scalby Formation. The onset of Scalby Formation deposition marked a change from the shallow, sheltered marine conditions of the Scarborough Formation to a predominantly alluvial environment with variable amount of marine influence. The apparently sudden change in depositional environment and the significant topography on erosional base of the formation is taken to indicate an abrupt fall in relative sea-level resulting from either eustatic variation or from local tectonic uplift (Alexander 1986*a*, *b*, 1989; Eschard *et al.* 1989). Brackish marine microfossils have been recorded from the unit above the meander belt near Hundale Point (Fisher & Hancock 1985) and also from some of the sheet-flood sandstones. The thickness variation correlates with variations in tectonic subsidence over the area; it is not a result of localised major incision. The thickness of the Scarborough Formation varies with that of the Scalby Formation. The thickness of the Scarborough Formation and Cornbrash are *not* represented on this figure.

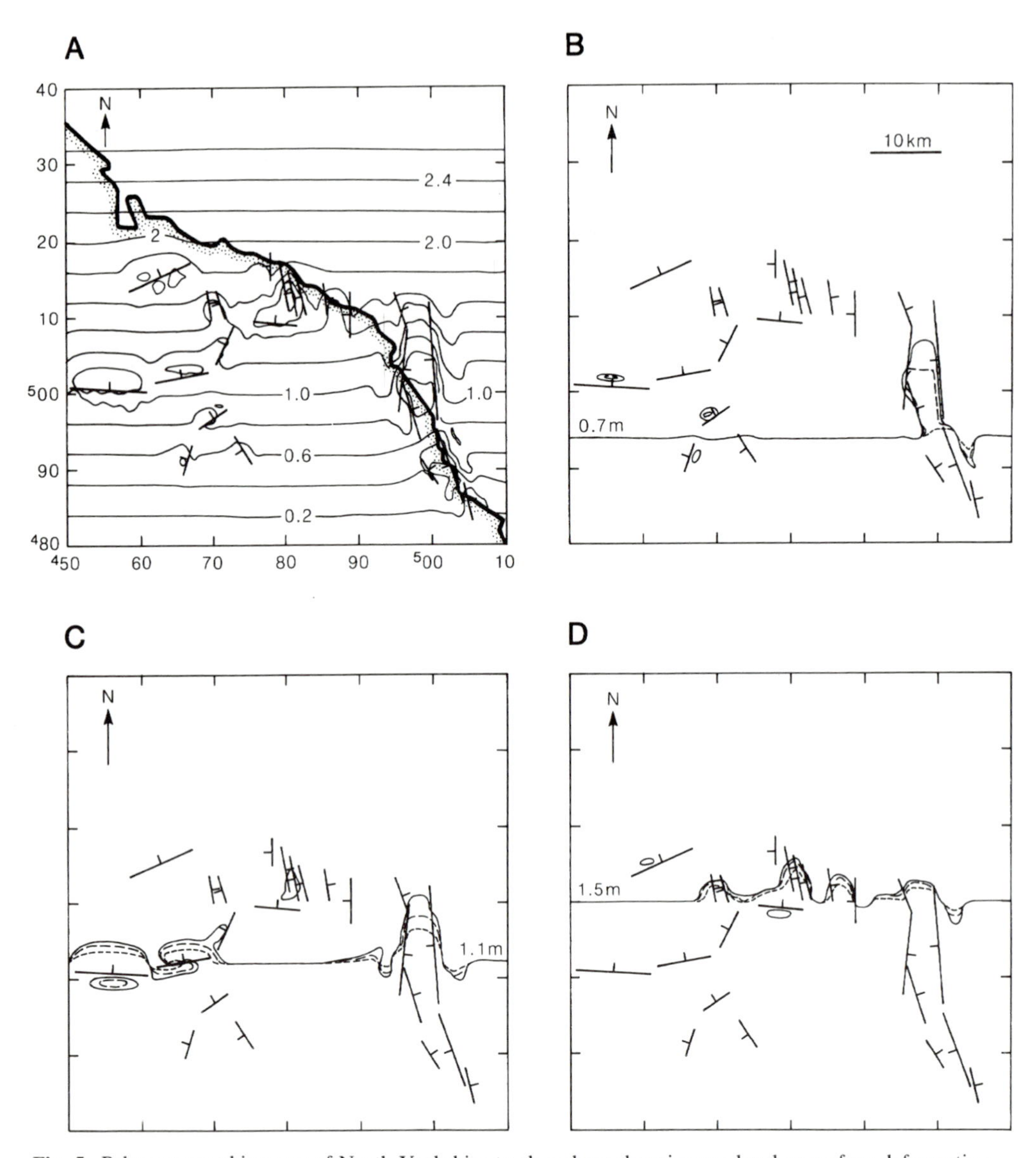


**Fig. 5.** Palaeogeographic maps of North Yorkshire to show how changing sea-level or surface deformation caused by tectonic activity would result in changes in the pattern and extent of marine influence.
(**a**) Theoretical topographic map of mid-Jurassic coastal plain in the Cleveland Basin. The gradient is 1 in 20000 and contours are in metres. Faults thought to have been active in the Jurassic are marked and surface deformation associated with movement on these faults is calculated by using Liverpool Fault Group displacement model (Gibson *et al.* 1989). The present day coast line is marked in bold.
(**b**) Palaeogeography with mean sea-level at 0.7 m above the lowest point of the area modelled. The solid line represents the shore-line with surface deformation resulting from a maximum fault displacement of 0.75 m. When the amount of fault displacement is reduced to 0.50 m or to 0.25 m the shore-line positions change as marked in the dashed lines (0.50 m, long dashes; 0.25 m, short dashes).
(**c**) Palaeogeography with mean sea-level at 1.1 m above the lowest poin. in the modelled area. Changes in shore-line resulting from changing the magnitude of maximum fault displacements are marked (0.25 m maximum fault displacement, short dashes; 0.50 m, long dashes; 0.75 m, solid line).
(**d**) Palaeogeography with mean sea-level at 1.5 m above the lowest point in the modelled area. Changes in shore-line resulting from changing the magnitude of maximum fault displacements are marked (0.25 m maximum fault displacement, short dashes; 0.50 m, long dashes; 0.75 m, solid line).

both down dip and along strike as a result of variations in the extent of marine influence. These effects are illustrated by a consideration of tectonic effects (below) but apply in a similar way to other local relative sea-level control. Minor changes in the extent of marine influence may occur during the deposition of single bar forms and examples of this may be seen in the Scalby and Saltwick Formations where the nature of the sandstone body varies through a lateral accretion unit, changing from inclined homolithic to inclined heterolithic stratification and resulting in a sandstone body hybrid between types 2a and 2b.

***Tectonic (topographic) depressions affecting the extent of marine influence.*** In areas where there is a lateral change in gradient across a delta or coastal plain, the extent of marine influence will vary in intensity along strike. The nature of the variation will depend on the range of topographic variation, position relative to coastline and the extent to which it causes increased channelling. The computer simulated models of the palaeogeography of North Yorkshire presented in Fig. 5 show how the change in shore-line position or inland extent of marine influence are controlled by magnitude of differential subsidence and relative sea-level. In this simple model the shore line position is controlled by altitude only.

In cases where there is significant lateral control of channel position the increased discharge (runoff capture) in the topographic depression may result in a seawards shift in the onset of flocculation due to the locally increased volume of freshwater discharge. An extreme example of this sort is seen in the Amazon River where the salinity change causing flocculation of the fluvially derived clay is near and often seaward of the channel mouth (Gibbs & Konwar 1986). A seawards deflexion of the area of marine influence may explain the clean fluvial sandstone bodies found in several of the synsedimentary tectonic depressions, where these sandstone bodies are stratigraphically equivalent to marine influenced deposits on the adjacent tectonic 'highs' (Fig. 6B). This may have been the case,

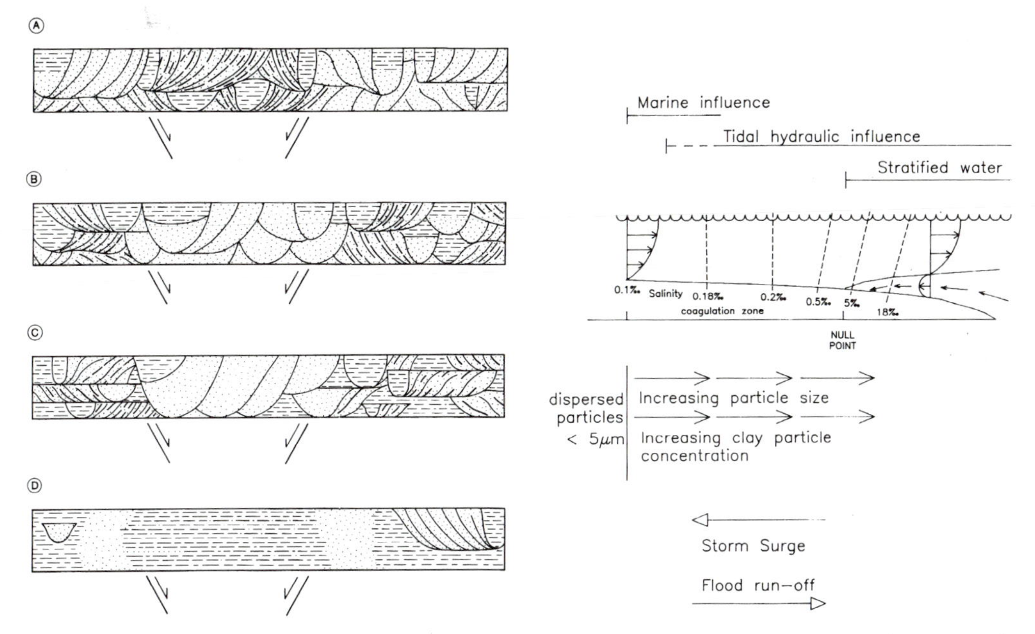


**Fig. 6.** Variation in sediment body architecture resulting from changing channel form and varying extent of marine influence.
**(a)** Local area of subsidence resulting in increased minor marine influence and a greater probability of detrital clay deposition within the channel sandstone body. Dashes represent mud and silt; dots represent sand.
**(b)** Local subsidence resulting in increased freshwater runoff in the depression and cleaner sandstone body.
**(c)** Local subsidence resulting in a change in the location of intertidal conditions and the formation of cleaner sandstone body in the depression and supratidal mud deposits on the flanks of the depression.
**(d)** Local subsidence resulting in marine incursion and lagoon or interdistributary bay deposition.
**(e)** Definition diagram for areas and nature of marine influence (After Gibbs *et al.* 1989).

for example, in the Whitby area (Grid Ref. NZ898115) for some of the time of Saltwick Formation deposition. Such an interpretation could also be applied to the Moor Grit at the base of the Scalby Formation near Hundale Point (Grid Ref. TA026949) (Fig. 4).

In areas where the topographic depression does not result in increased discharge of fresh water, the result of the depression may be to shift the area of minor marine influence (onset of clay flocculation, for example) inland. If this were the case the sandstone bodies in the depression would tend to have greater proportions of detrital clay than those on the adjacent highs, as illustrated in Fig. 6A.

Variations in the extent of minor marine influence may explain, for example, the changes seen in the Scalby Formation 'exhumed meander-belt' (Nami 1976*b*; Leeder & Nami 1979; Alexander 1989; 1992). The variation in the extent of bioturbation seen in different lateral accretion units of the same sandstone body in the foreshore between Hundale Point (Grid Ref. TA026949) and Scalby (Grid Ref. TA034913) may be indicative of a variation in the extent of marine influence on the channels, although variations in discharge and sedimentation characteristics may also influence the extent of bioturbation. The point bar deposits south of White Nab (Grid Ref. TA059861) are similar to those in the 'exhumed meander-belt' and are at the same lithostratigraphic level in the Scalby Formation (palaeostratigraphy is not accurate enough to judge if they are truly time equivalent). However the deposits are less extensively deformed or bioturbated, strongly suggesting that they were deposited in channels with less marine influence. These channel deposits seen at White Nab and between Hundale Point and Scalby Ness may be a continuous sandstone body with characteristics changing laterally as suggested in Fig. 6A. The lateral change in the extent of marine influence correlates with the tectonic depression of the Peak Trough (Milsom & Rawson 1989) and it seems likely that differential subsidence may have changed the range of marine influence without significantly altering the channel size, as suggested by the computer model presented in Fig. 5.

## Conclusions

Sandstone architecture in coastal or delta plain sediments may relate to changing climate and drainage density, changing channel form and lateral stability, or to sea-level by way of periodic incision, changing aggradation rate or channel load characteristics. Incision notably is a common process that strongly affects the sandstone body architecture and sandstone percentage even in net aggradational systems. Growing topographic features change the shore-line position and influence the inland extent of marine influence on the alluvial system. A growing depression, such as is caused by synsedimentary tectonic activity, may have three possible responses:

(a) increase the inland extent of marine influence on the alluvial system along the depression;
(b) force a major local environmental change in the depression from alluvial to marine bay or estuary; or,
(c) localize surface run off into the depression, increasing fresh water influence and moving the inland influence of marine processes towards the shore.

Variations in clay content in the form of clay drapes and dispersed detrital clay can result from very small changes in salinity (in the order of 0.1–1‰) and turbulence conditions. The amplitude and period of changes in salinity that result in flocculation may be within the tolerance limits of most fresh water fauna and flora and therefore the changes in the occurrence of detrital clay in channel sandstones may not correlate with changes in the preserved microfossil assemblage.

Divisions should be made between marine influenced and non-marine influenced alluvial deposits in addition to between tidally and non-tidally influenced alluvial deposits.

The shaping of the ideas presented in this paper where greatly influenced by the few hardy students who braved gale-force winds and torrential rain on a field trip (February 1990) to examine modern marine-influenced channel sediment and by discussions with J. Cope, J. Getliff, A. Hartley, C. Hill, D. Taylor, T. Young. Thanks to L. Thompson, M. Miller and P. Weston for help in preparing the manuscript.

## References

Abrahams, A. D. 1972. Drainage densities and sediment yields in E Australia. *Australian Geographical Studies*, **10**, 19–41.

Alexander, J. 1986*a*. *Sedimentary and tectonic controls of river facies distribution, Middle Jurassic Yorkshire and North Sea Basins and Holocene of SW Montana, USA*. PhD Thesis, University of Leeds.

—— 1986*b*. Idealised flow models to predict alluvial sandstone body distribution in the Middle Jurassic Yorkshire Basin. *Marine and Petroleum*

*Geology*, **3**, 298–305.
—— 1987. Syn-sedimentary and burial related deformation in the Middle Jurassic non-marine formations of the Yorkshire Basin. *In*: JONES, M. E. & PRESTON, R. M. F. (eds) *Deformation of Sediments and Sedimentary Rocks*. Geological Society, London, Special Publication, **29**, 315–324.
—— 1989. Delta or coastal plain? With an example of the controversy from the Middle Jurassic of Yorkshire. *In*: WHATELEY, M. K. G. & PICKERING, K. P. (eds) *Deltas: Sites and Traps for Fossil Fuels*. Geological Society, London, Special Publication, **41**, 11–19.
—— 1992. The nature of a laterally extensive alluvial sandstone body in the Middle Jurassic Scalby Formation. *Journal of the Geological Society, London*, in press.
ALLEN, J. R. L. 1963. The classification of cross-stratification units with notes on their origin. *Sedimentology*, **2**, 93–114.
ANDERSON, S. 1989. Differential compaction (Middle Jurassic, Yorkshire, U.K.) *4th International Conference on Fluvial Sedimentology abstract volume*, 63.
ATKINSON, C. D. 1983. *Comparative sequences of ancient fluviatile deposits in the Tertiary, South Pyrenean Basin Northern Spain*. PhD Thesis, University of Wales.
BADLEY, M. E., PRICE, J. D., RAMBECH DAHL, C. & AGDESTEIN, T. 1988. The structural evolution of the northern Viking Graben and its bearing upon extensional models of basin formation. *Journal of the Geological Society, London*, **145**, 455–472.
BAKKER, C. & DE PAUW, N. 1974. Comparison of brackish water plankton assemblages in an estuarine tidal (Westerschelde) and stagnant (Lake Vere) environment (S.W. Netherlands) I. Phytoplankton. *Hydrobiology Bulletin*, **8**, 197–189.
BARTON, P. & WOOD, R. 1984. Tectonic evolution of the North Sea Basin: crustal stretching and subsidence. *Geophysical Journal of the Royal Astronomical Society*, **79**, 987–1022.
BARWIS, J. H. 1978. Sedimentology of some South Carolina tidal-creek point bars, and a comparison with their fluvial counterparts. *In*: MIALL, A. D. (ed.) *Fluvial Sedimentology*. Canadian Society of Petroleum Geologists, Memoir, **5**, 129–160.
BRIDGE, J. S. 1985. Paleochannel patterns inferred from alluvial deposits: a critical evaluation. *Journal of Sedimentary Petrology*, **55**, 579–589.
—— & LEEDER, M. R. 1979. A simulation model of alluvial stratigraphy. *Sedimentology*, **26**, 617–644.
—— & MACKEY, S. D. 1989. Alluvial stratigraphy models. *4th International Conference on Fluvial Sedimentology abstract volume*, 85.
BRISTOW, C. S. 1987. Brahmaputra River: channel migration and deposition. *In*: ETHRIDGE, F. G., FLORES, R. M. & HARVEY, M. D. (eds) *Recent Developments in Fluvial Sedimentology*. Special Publication of the Society of Economic Paleontologists and Mineralogists, **39**, 63–74.
BROWN, S. & RICHARDS, P. C. 1989. Facies and development of the Middle Jurassic Brent delta near the northern limit of its progradation, UK North Sea, *In*: WHATELEY, M. K. G. & PICKERING, K. P. (eds) *Deltas: Sites and Traps for Fossil Fuels*. Geological Society, London, Special Publication, **41**, 253–267.
——, —— & THOMSON, A. R. 1987. Patterns in the deposition of the Brent Group (Middle Jurassic) UK North Sea. *In*: BROOKS, J. & GLENNIE, K. (eds) *Petroleum Geology of North West Europe*. Graham & Trotman, London, 899–913.
BUDDING, M. C. & INGLIN, H. F. 1981. A reservoir geological model of the Brent sands in Southern Cormorant. *In*: ILLING, L. V. & HOBSON, G. D. (eds) *Petroleum Geology of the Continental Shelf of NW Europe*. Hayden, London, 326–334.
CRANE, R. L. 1982. *A computer model for the architecture of avulsion-çontrolled alluvial suites*. PhD thesis, University of Reading.
DEEGAN, C. E. & SCULL, B. J. 1977. *A standard lithostratigraphic nomenclature for the Central and Northern North Sea*. Institute of Geological Science Report **77/25**.
ESCHARD, R. 1989. *Geometrie et Dynamique de Sequences de Depots dans un Systeme Deltaique (Jurassique moyen, Bassin de Cleveland, Angleterre) Implications sur l'architecture tridimensionelle des corps sedimentaires*. Thesis, Universite Louis Pasteur de Strasbourg.
——, HOUEL, P., RUDKIEWICZ, J. L. 1989. Three dimensional architecture and geostatistic simulation of a fluvial to estuarine valley fill complex (Middle Jurassic, Cleveland Basin). *International Conference on Fluvial Sedimentology abstract volume*, 125.
EYNON, G. 1981. Basin development and sedimentation in the Middle Jurassic of the Northern North Sea. *In*: ILLING, L. V. & HOBSON, G. D. (eds) *Petroleum Geology of the Continental Shelf of NW Europe*. Hayden, London, 196–204.
FISHER, M. J. & HANCOCK, N. J. 1985. The Scalby Formation (Middle Jurassic, Ravenscar Group) of Yorkshire: reassessment of age and depositional environment. *Proceedings of the Yorkshire Geological Society*, **45**, 293–298.
FISK, H. N. 1947. *Fine grained alluvial deposits and their effects on Mississippi River activity*. Mississippi River Commission, Vicksburg, Miss.
FRIEND, P. F. 1978. Distinctive features of some ancient river systems. *In*: MIALL, A. D. (ed.) *Fluvial sedimentology*. Canadian Society of Petroleum Geologists, Memoir, **5**, 531–542.
FRIEND, P. F., SLATER, M. J. & WILLIAMS, R. C. 1979. Vertical and lateral building of river sandstone bodies, Ebro Basin, Spain. *Journal of the Geological Society, London*, **136**, 39–46.
GIBBS, R. S. & KONWAR, L. 1986. Coagulation and settling of amazon River suspended sediment. *Continental Self Research*, **6**, 127–149.
——, —— & TERCHUNIAN, A. 1983. Size of flocs suspended in Delaware Bay. *Canadian Journal of Fish and Aquatic Science*, **40**, 102–104.
——, TSHUDY, D. M., KONWAR, L. & MARTIN, J. M. 1989. Coagulation and transport of sediments

in the Gironde Estuary. *Sedimentology*, **36**, 987–1000.

Gibson, J. R., Walsh, J. J. & Watterson, J. 1989. Modelling the bed contours and cross-sections adjacent to plane normal faults. *Journal of Structural Geology*, **11**, 317–328.

Gowland, S. 1987. *Facies analysis of three members of the Scarborough Formation (Middle Jurassic: Lower Bajocian) in the Cleveland Basin, northeast England: Blea Wyke, Byland Limestone and Crinoid Grit Members*. PhD Thesis, University of Hull.

Graue, E., Helland-Hansen, W., Johnsen, J., Lømo, L., Nøttvedt, A., Rønning, K., Ryseth, A. & Steel, R. 1987. Advance and retreat of Brent Delta system, Norwegian North Sea, *In*: Brooks, J. & Glennie, K. (eds) *Petroleum Geology of North West Europe*. Graham & Trotman, London, 915–937.

Gregory, K. J. & Gardiner, V. 1975. Drainage density and climate, *Zeitschrift fur Geomorphologie*,. **19**, 287–298.

Hallam, A. 1988. A reevaluation of Jurassic eustasy in the light of new data and the revised Exxon curve. *In*: Wilgus, C. K., Hastings, B. S., Kendall, C. G. St. C., Postamentier, H. W., Ross, C. A. & Van Wagoner, J. C. (eds) *Sea-level changes: an integrated approach*. SEPM Special Publication, **42**, 261–274.

Hancock, N. J. & Fisher, M. J. 1981. Middle Jurassic North Sea deltas with particular reference to Yorkshire. *In*: Illing, L. V. & Hobson, G. D. (eds) *Petroleum Geology of the Continental Shelf of NW Europe*, Heyden, London 86–195.

Haq, B. U., Hardenbol, J. & Vail, P. R. 1988. Mesozoic and Cenozoic chronostratigraphy and eustatic cycles. *In*: Wilgus, C. K., Hastings, B. S., Kendall, C. G. St. C., Postamentier, H. W., Ross, C. A. & Van Wagoner, J. C. (eds) *Sea-level changes: an integrated approach*. SEPM Special Publication, **42**, 71–108.

Harris, J. P. & Fowler, R. M. 1987. Enhanced prospectivity of the Mid-Late Jurassic sediments of the South Viking Graben, northern North Sea. *In*: Brooks, J. & Glennie, K. (eds) *Petroleum Geology of North West Europe*. Graham & Trotman, London, 879–898.

Helland-Hansen, W., Steel, R., Nakayama, K. & Kendall, C. G. St. C., 1989. Review and computer modelling of the Brent Group stratigraphy. *In*: Whateley, M. K. G. & Pickering, K. T. (eds) *Deltas: Sites and Traps of Fossil Fuels*. Geological Society, London, Special Publication, **41**, 237–252.

Ireland, S. 1987. The Holocene sedimentary history of coastal lagoons of the Rio de Janeiro State, Brazil. *In*: Tooley, M. J. & Shennan, I. (eds) *Sea-Level Changes*. Basil Blackwell, 1–24.

Jackson, R. G. II. 1978. Preliminary evaluation of lithofacies models for meandering alluvial streams. *In*: Miall, A. D. (ed.) *Fluvial sedimentology*, Canadian Society of Petroleum Geologist, Memoir, **5**, 543–576.

Johnson, A. & Eyssautier, M. 1987. Alwyn North Field and its regional geological context, *In*: Brooks, J. & Glennie, K. (eds) *Petroleum Geology of North West Europe*. Graham & Trotman, London, 963–977.

Johnson, H. D. & Stewart, D. J. 1985. Role of clastic sedimentology in the exploration and production of oil and gas in the North Sea. *In*: Brenchley, P. J. & Williams, B. P. J. (eds) *Sedimentology: Recent Developments and Applied Aspects*. Geological Society, London, Special Publication, **18**, 249–310.

Kantorowicz, J. D. 1984. The nature, origin and distribution of authigenic clay minerals from Middle Jurassic Ravenscar and Brent Group sandstones. *Clay Minerals*, **19**, 359–375.

—— 1985. The petrology and diagenesis of Middle Jurassic clastic sediments, Ravenscar Group, Yorkshire. *Sedimentology*, **32**, 833–853.

Knox, R. W. O'B. 1969. *Sedimentary studies of the Eller Beck Bed and Lower Deltaic Series in North-East Yorkshire*. PhD thesis, Newcastle University.

Leeder, M. R. & Alexander, J. 1985. Discussion of Fisher, M. J. and Hancock, N. J. The Scalby Formation (Middle Jurassic, Ravenscar Group) of Yorkshire: reassessment of age and depositional environment. *Proceedings of the Yorkshire Geological Society*, **45**, 297–298.

—— & Nami, M. 1979. Sedimentary models for the non-marine Scalby Formation (Middle Jurassic) and evidence for Late Bajocian/Bathonian uplift of the Yorkshire Basin. *Proceedings of the Yorkshire Geological Society*, **42**, 461–482.

Livera, S. E. 1981. *Sedimentology of the Bajocian Rocks from the Ravenscar Group of Yorkshire*. PhD thesis, University of Leeds.

—— 1989 Facies associations and sand-body geometries in the Ness Formation of the Brent Group, Brent Field. *In*: Whateley, M. K. G. & Pickering, K. T. (eds) *Deltas: sites and traps for fossil fuels* Geological Society, London, Special Publication, **4**, 269–286.

Milsom, J. & Rawson, P. F. 1989. The Peak Trough — a major control on the geology of the North Yorkshire coast. *Geological Magazine*, **126**, 699–705.

Mjøs, R. & Walderhaug, O. 1989. Spatial organisation and geometries of channels and crevasse splay sandstones. Ravenscar Group, Yorkshire and Ness Formation, Oseberg Field. *In*: *4th International conference on fluvial sedimentology, abstract volume*, 186.

——, Walderhaug, O. & Prestholm, E. 1990. Sandbody geometry and heterogeneity in the fluviodeltaic Ravenscar Group of Yorkshire. *In*: *Advances in reservoir geology, abstract volume*.

Nami, M. 1976*a*. *Sedimentology of the Scalby Formation (Upper Deltaic Series) in Yorkshire*. PhD thesis, University of Leeds.

—— 1976*b*. An exhumed Jurassic meander belt from Yorkshire, England. *Geological Magazine*, **113**, 47–52.

—— & Leeder, M. R. 1978. Changing channel morphology and magnitude in the Scalby Formation

(Mid Jurassic) of Yorkshire, England. *In*: MIALL, A. D. (ed.) *Fluvial sedimentology* Canadian Society of Petroleum Geologists, Memoir, **5**, 431–440.

OUCHI, S. 1985. Response of alluvial rivers to slow active tectonic movement. *Geological Society of America Bulletin*, **96**, 504–515.

POTTER, P. E. & BLAKELEY, R. F. 1967. Generation of a synthetic profile of a fluvial sandstone body. *Journal of Petroleum Technology*, **7**, 243–251.

RAAF, J. F. M., DE & BOERSMA, J. R. 1971. Tidal deposits and their sedimentary structures. *Geologie en Mijnbouw*, **50**, 470–504.

RAVENNE, C., ESCHARD, R., GALLI, A., MATHIEU, Y., MONTADERT, L. & RUDKIEWICZ, J.-L. 1987. Heterogeneities and geometry of sedimentary bodies in a fluvio-deltaic reservoir. *Society of Petroleum Engineers, Paper* **16752**, 115–122.

RICHARDS, K. 1982. *Rivers: form and process in alluvial channels*, Methuen & Co. Ltd.

SMITH, D. G. 1987. Meandering river point bar lithofacies models: modern and ancient examples compared. *In*: ETHRIDGE, F. G., FLORES, R. M. & HARVEY, M. D. (eds) *Recent Developments in fluvial Sedimentology*, Society of Economic Paleontologists and Mineralogists, Special Publication, **39**, 83–91.

STODDART, D. R. 1971. World erosion and sedimentation. *In*: CHORLEY, R. J. (ed.) *Introduction to fluvial processes*, Methuen, London, 8–29.

TERWINDT, J. H. J. 1988. Palaeo-tidal reconstructions of inshore tidal depositional environments. *In*: DE BOER, P. L., VAN GELDER, A. & NIO, S. D. (eds) *Tide-influenced sedimentary environments and facies* D Reidel, 233–263.

THOMAS, R. G., SMITH, D. G., WOOD, J. M., VISSER, J., CALVERLEY-RANGE, E. A. & KOSTER, E. H. 1987. Inclined Heterolithic Stratification — Terminology, description, interpretation and significance. *Sedimentary Geology*, **53**, 123–179.

VAIL, P. R., HARDENBOL, J. & TODD, R. G. 1984. Jurassic unconformities, chronostratigraphy, and sea-level changes from seismic stratigraphy and biostratigraphy. *American Association of Petroleum Geologists, Memoir*, **36**, 129–144.

VAN STRAATEN, L. M. J. U. 1954. *Composition and structure of recent marine sediments in the Netherlands*. Leidse Geologische Mededelingen **19**.

ZIEGLER, P. A. 1982. *Geological atlas of Western and Central Europe*. Shell Internationale Petroleum Maatschappij B.V., The Hague.

# Palynological review of the Brent Group, UK sector, north sea

M. F. WHITAKER[1], M. R. GILES[2] & S. J. C. CANNON[1]

[1] *The Geochem Group Ltd, Chester Street, Chester CH4 8RD, UK*

[2] *Shell UK Exploration and Production, Shell-Mex House, Strand, London, UK*

**Abstract:** A biostratigraphical zonation of the Brent Group, applicable within local oil field areas, has been established. It comprises 11 palynozones and subzones, eight of which are confined to the Ness Formation. Palynomorph associations and palynofacies types are shown to broadly correspond to specific palaeoenvironments. A regional 'event' scheme is outlined which establishes chronostratigraphical correlations and suggests temporal relationships between deltaic and delta front sediments. Integrated palynofacies data have allowed the development of deltaic sub-palaeoenvironments and lithofacies to be examined with respect to time and, in this way, to chart the possible progress of the delta as it prograded. Similarly, this approach enables possible areas of concentrated fluvial channel activity to be identified and the likely migration of these fluvial fairways with time. The implications of this depositional model for reservoir development and performance is discussed, showing the varying orientation of the delta front and barrier sands as it changed from a predominantly N–S trend during its early stages to WNW–ESE in its final stages.

In order to improve and refine the understanding of depositional processes controlling sediment distribution within the delta complex of the Brent Group a detailed palynological study of 44 wells was carried out in the East Shetland Basin area (Table 1). The study was initiated as a part of the work carried out by Shell Exploration and Production and supported lithological and sedimentological information presented by Cannon *et al.* (this volume).

The geographical distribution of the 44 wells in the study area is shown in Fig. 1. The sequence studied is from the uppermost Drake Formation to the lowermost Heather Formation, incorporating the Broom, Rannoch, Etive, Ness and Tarbert Formations of the standard Brent Group sequence (Deegan & Scull 1977, see Fig. 2). The Ness Formation is further subdivided into 3 members based on the occurrence of a finer grained interval towards the middle referred to as the Oich Member, separating an upper and lower Ness unit termed the Foyers and Enrick members respectively (see Fig. 2) (Cannon *et al.* this volume). Analytical work has allowed a detailed local, facies related, zonation scheme and a regional, time related, event scheme to be established.

**Table 1.** *Wells employed for this study*

| | | |
|---|---|---|
| 2/5–1 | 3/8–12 | 211/19–6 |
| 2/5–12A | 210/15–2 | 211/22–1 |
| 3/1–1 | 210/25–2 | 211/22–2 |
| 3/1–2 | 210/25–3b | 211/23–A32 |
| 3/2–1A | 210/25–5 | 211/23–A36 |
| 3/2–2 | 211/7–1 | 211/24–2 |
| 3/2–3 | 211/8–1 | 211/24–4 |
| 3/2–4 | 211/11–1 | 211/26–4 |
| 3/3–2 | 211/11–3 | 211/26a–12 |
| 3/3–3 | 211/12–1 | 211/27–6 |
| 3/3–5A | 211/13–6 | 211/27–11 |
| 3/4–1 | 211/13–7 | 211/28–1A |
| 3/7–1 | 211/18a–21 | 211/29–3 |
| 3/7–3 | 211/18–22 | 211/29–A16 |
| 3/8–4 | 211/19–1 | |

The primary purpose of the study, however, has been the detailed documentation of palynofacies distributions and their palaeoenvironmental significance. Based on an integration of the zonation schemes and palynofacies data, it has been possible to examine the development of deltaic sub-palaeoenvironments with respect to lithofacies distribution and to suggest their likely temporal relationships within the study area. In addition, the development of a depositional model together with palaeogeographical reconstructions based on selected palyno-events has provided a foundation for predicting the possible spatial distribution of lithofacies between well control points.

## Approach to palynofacies analysis

Analysis of the palynodebris and palynomorph content in their stratigraphical context has allowed a detailed understanding of the processes controlling deposition and provides a basis for palaeogeographical reconstruction. Palynofacies represents a geological interpretation of sedimentary dispersed organic material,

*From* MORTON, A. C., HASZELDINE, R. S., GILES, M. R. & BROWN, S. (eds), 1992, *Geology of the Brent Group*. Geological Society Special Publication No. 61, pp. 169–202.

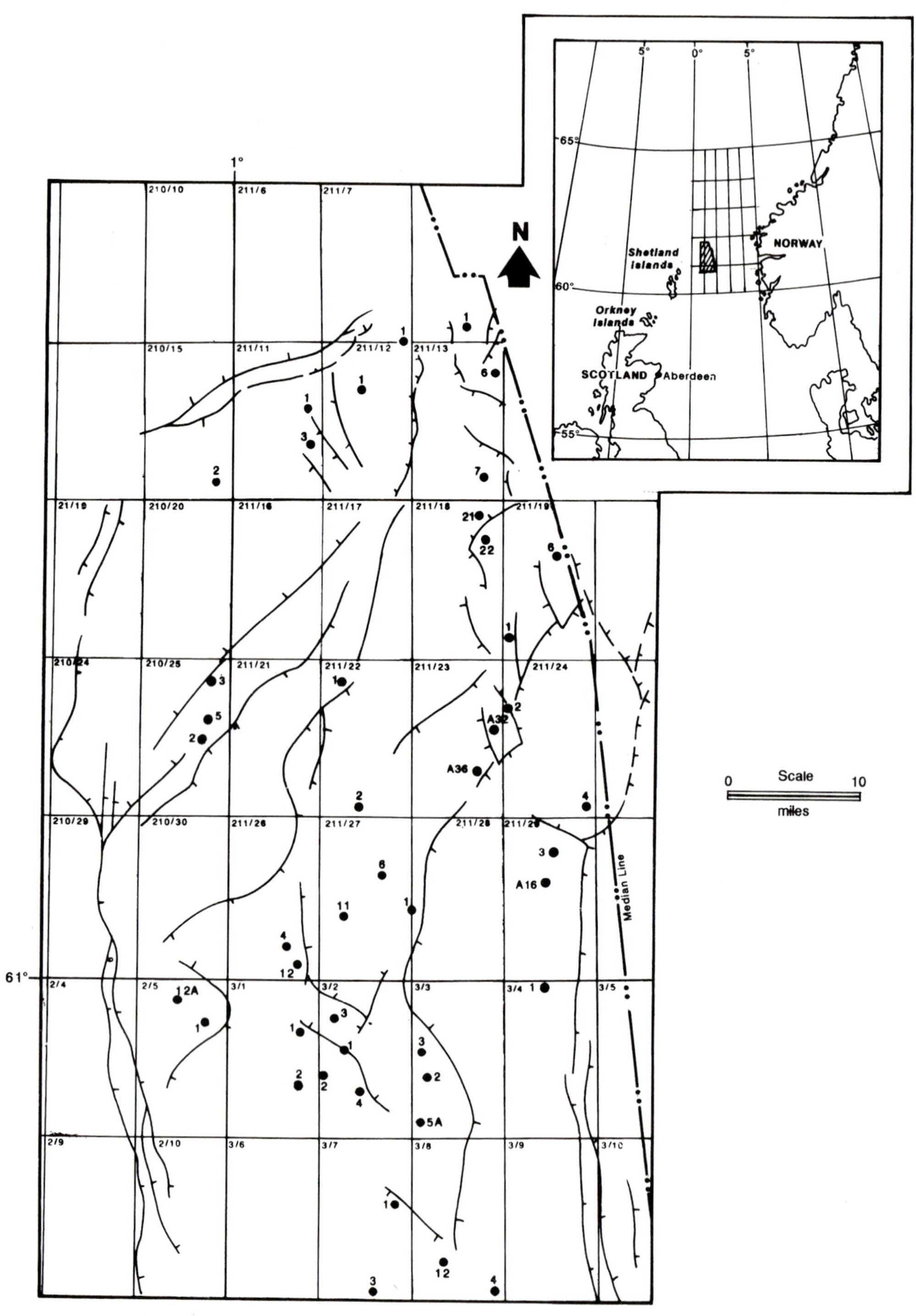


**Fig. 1.** Well location map.

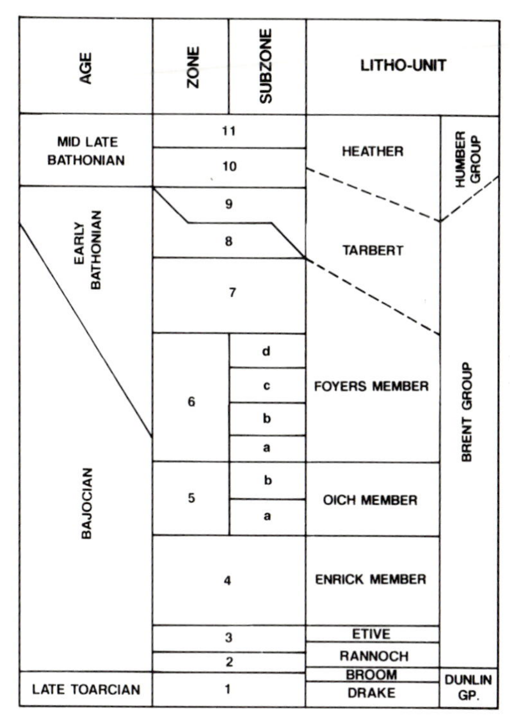


**Fig. 2.** Palynozonation.

which together with lithofacies studies can provide the key to identification of depositional environments. Each lithofacies has individual physical characteristics related to factors such as grain size and sorting variations, which are largely dependant upon depositional processes, such as energy conditions and sediment supply. However, similar depositional processes may not always be limited to one palaeoenvironment and can be found, in some cases, in contrasting types of depositional settings. Thus, from lithofacies alone, it is not always possible to reconstruct precisely the palaeoenvironment. In this respect, a palynofacies approach in conjunction with sedimentology can enable an accurate palaeoenvironmental appraisal and an enhanced analysis of sediment distribution within reservoir sequences.

Previous palynofacies work in the Brent Group was introduced by Denison & Fowler (1980) in a study which demonstrated this technique for palaeoenvironmental interpretation. Parry *et al.* (1981) provided a formal approach, classifying specific palynological constituents and associations which could be tentatively related to particular depositional settings. The present study is a refinement of this technique. In addition palynofacies studies have been carried out on the Middle Jurassic Deltaic Series of Yorkshire by Fisher (1980), Hancock & Fisher (1981), and Fisher & Hancock (1985). This work describes palynofacies analyses from outcrop sections which are well documented in terms of sedimentological and macrofossil characteristics. The results provide an excellent model for many of the different coastal and delta top settings encountered within the Brent Group. However, despite the remarkable similarity, comparisons are limited by the lack of information with regard to palynomorph content.

The various palynological constituents (macerals) which are employed in this study are summarized in Fig. 3 and their likely behavioural characteristics, with respect to the depositional environment are illustrated in Fig. 4. The macerals can be subdivided into a terrestrial input of woody debris and sporomorphs, and a marine content of dinoflagellate cysts (subsequently termed dinocysts), acritarchs, marine algae and foraminiferal test linings. In certain depositional settings a bacterial biomass is preserved, referred to as structureless (sapropelic) organic matter (S.O.M.). The woody debris may be subdivided further into macerals 1–4 according to their buoyancy and degradation characteristics. Although each palynomaceral may be of heterogeneous composition and origin, they can be generally equated to biological and coal petrographic types as shown in Fig. 4. By applying basic sedimentological principles to the terrestrial input and palaeontological principles to the marine content, depositional processes can be delineated and a palaeogeographic picture reconstructed. Examples of this approach as applied to other sequences are also documented by Van der Zwan (1989) from the Middle and Late Jurassic of the Troll and Draugen Field, Whitaker (1984) from the Troll Field and Bryant *et al.* (1988) from the Upper Lias sands of southern England. A more detailed discussion of palynofacies constituents their classification and likely behavioural characteristics is given by Tyson (1990). A palynofacies study is a comparative study and thus demands a consistent and controllable preparation method. The standard preparation method, applied to all samples in the present study, is outlined in Table 2.

The proportions of the different palynological constituents were estimated semi-quantitatively using a purely visual assessment. Their proportions were considered in transmitted light using a non-oxidized slide sieved at 15 $\mu$m.

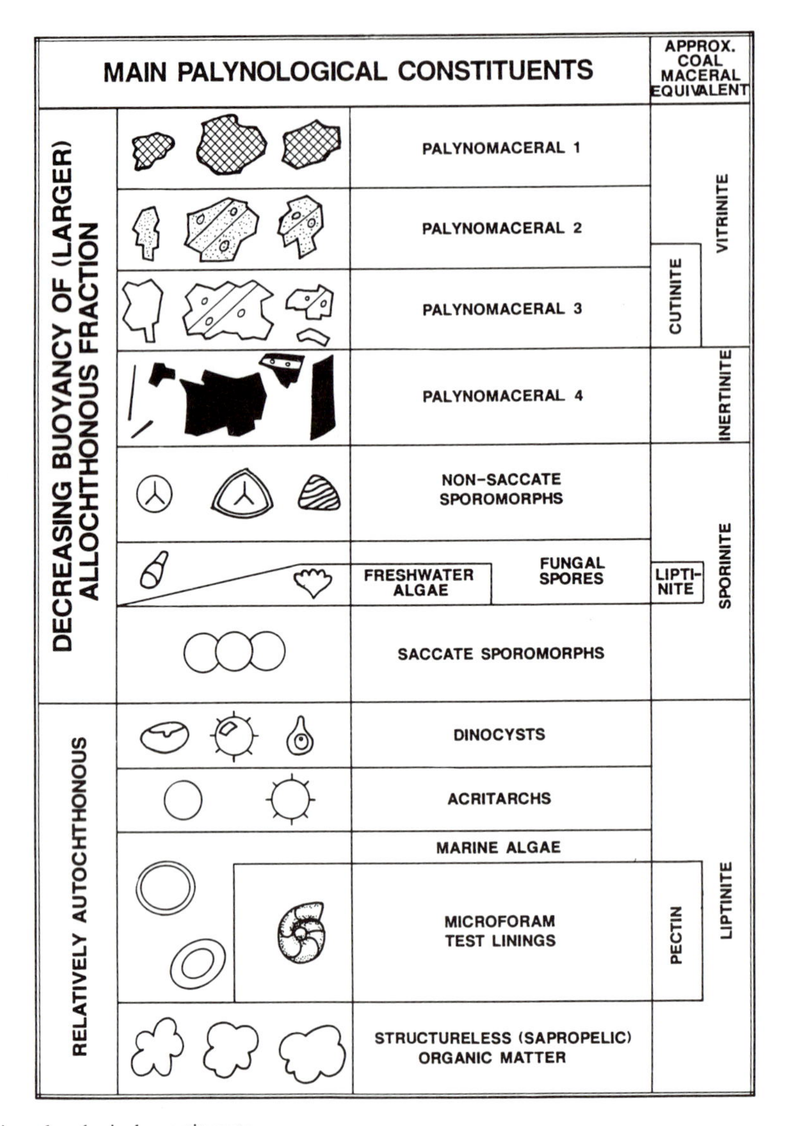


**Fig. 3.** Main palynological constituents.

Slides made of residues prior to sieving proved impractical to count, since the dominance of fine organic and mineral particles effectively diluted the coarser (structured) fraction. The palynomorphs were counted in detail using the oxidized and swirled slide.

## *Classification of palynological constituents*

The classification relates to the hydrodynamic properties of the palynological constituents and is based on their transmitted light characteristics, with additional support from ultra-violet reflected light studies.

*Palynomaceral 1.* Orange-brown or dark brown structured or structureless material, irregular in shape, variable in preservation and of dense appearance. In thin rock section it may occur as large tissues up to 500 $\mu$m, and may be dispersed or concentrated in layers. Generally, however, it is moderate in size in palynological macerations and is dispersed. It may include structured plant debris (mainly resinous cortex material), humic gel-like substances, resinous substances, and algal detritus (mainly degraded *Botryococcus*). It is considered to be of low buoyancy because of its frequently large size and higher specific gravity whilst its spongy cortex nature

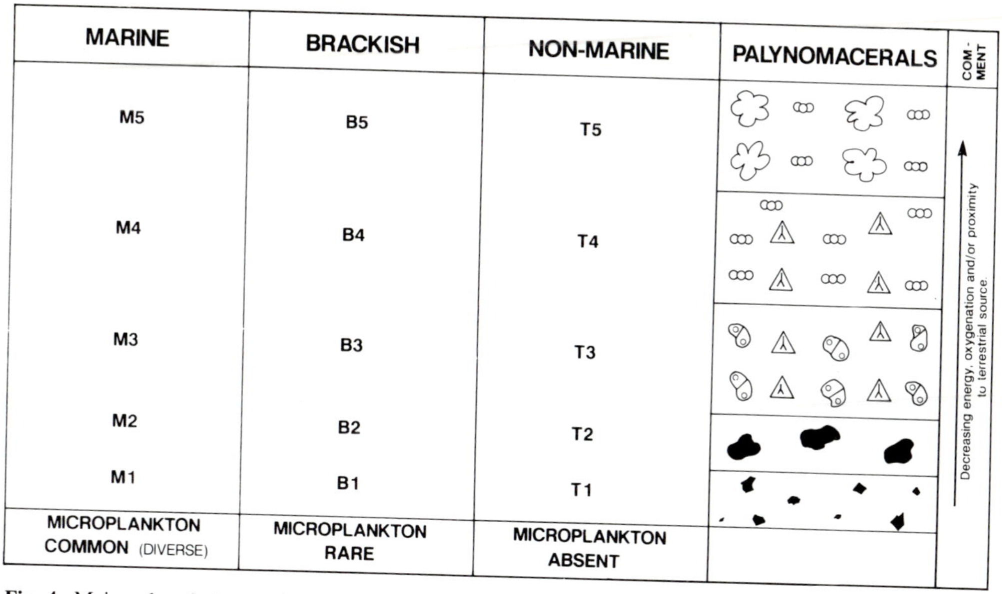

| MARINE | BRACKISH | NON-MARINE | PALYNOMACERALS | COM-MENT |
|---|---|---|---|---|
| M5 | B5 | T5 | | Decreasing energy, oxygenation and/or proximity to terrestrial source. |
| M4 | B4 | T4 | | |
| M3 | B3 | T3 | | |
| M2 | B2 | T2 | | |
| M1 | B1 | T1 | | |
| MICROPLANKTON COMMON (DIVERSE) | MICROPLANKTON RARE | MICROPLANKTON ABSENT | | |

**Fig. 4.** Main palynofacies types.

renders it susceptible to waterlogging. Palynomaceral 1 is not considered to be resistant to physical abrasion and is readily destroyed in high energy palaeoenvironments.

*Palynomaceral 2.* Brown-orange structured or structureless material of irregular shape. It may include structured plant material (some leaf, stem or small rootlet debris), algal detritus and, to a lesser extent, humic gel and resinous substances. Buoyancy is considered to be higher relative to palynomaceral 1 because of its thinner and often lath shaped character.

*Palynomaceral 3.* Pale, relatively thin, irregularly shaped, usually structured material, occasionally bearing stomata. It is considered the most buoyant of palynomacerals 1–3. It may include structured plant material (mainly of leaf origin, which may or may not bear a waxy surface coating), and degraded aqueous plant material.

**Table 2.** *Preparation method*

| |
|---|
| Rock material cleaned and then crushed to 2–5 mm fragments |
| HCl 10% conc., 20 minutes |
| HF 40% conc., 60 minutes |
| HCl ½% conc., washing |
| Heavy liquid separation in Zinc Bromide solution, S.G. 2.2 |
| 15 micron sieve — slide prepared |
| $HNO_3$ 10 minutes |
| Swirled |
| 15 $\mu$m sieve — slide prepared |

*Palynomaceral 4.* Black, or almost black, equidimensional, blade- or needle-shaped material, which is usually uniformly opaque and structureless, but may occasionally show cellular structure. The constituents included in palynomaceral 4 are of many different origins, and include compressed humic gels, charcoal (resulting from forest fires), reworked charcoal and geothermally fusinized (occasionally semi-opaque) material. Opaqueness, or darkening of organic material can also be the result of secondary effects such as staining by migrating fluids within the sediments.

It is possible with the aid of a reflected light system (the light source being placed at an angle to the palynological slide) to recognize several types of surface character associated with palynomaceral 4, some of which correspond closely to shape, as indicated in Table 3.

Blade-shaped palynomaceral 4 has a distinctive appearance, and reflects the break up of larger, oxidized (mainly charcoal) woody debris parallel to the long axis of elongate cellular structure typical in stem material. It can be extremely buoyant, and often transported over long distances. It is particularly resistant to degradation being especially concentrated in high energy palaeoenvironments.

**Table 3.** *Types of opaque palynomacerals*

| Shape | Surface | Origin |
|---|---|---|
| Needle | Dull (smooth surface) | Fungal (in part, personal observation) |
| Blade | Glossy (smooth shiny surface) | Charcoal and reworked geothermally fusinized material |
| Equidimensional | Semi-opaque (shows some cellular structure, or brown colour at margins) | Fusinized or iron stained or thick palynomacerals 1, 2 & 3 |
| Equidimensional | Bright (crystalline appearance) | Mineral |

Needle shaped material is also considered buoyant, but is less likely to be preserved in higher energy settings. Equidimensional forms of palynomaceral 4 are frequently intermediate in character to palynomacerals 1 or 2 and may thus have a relatively lower buoyancy.

*Sporomorphs.* (i) *Non-saccate sporomorphs.* Included in this group are all non-saccate spores and pollen. Less buoyant, large and thick-walled sporomorph types behave sedimentologically as sand or silt-size particles and be most frequent close to their source (Cross *et al.* 1966; Hughes & Moody-Stuart 1967; Batten 1974; Parry *et al.* 1981; Haabib 1982), whereas flimsy, thin walled forms may be carried out and concentrated in distal offshore, finer grained sediments. Such sorting can be seen by comparing the coastal sediments of the Ness and Rannoch Formations with the distal offshore sediments of the Heather and Drake Formations.

(ii) *Saccate sporomorphs.* Included in this group are bisaccate and monosaccate forms. Bisaccates, especially, are known to be extremely buoyant, presumably because of their air bladder and waxy outer layer, rendering them less susceptible to waterlogging (Chaloner 1968). Samples taken by the first author from the upper flotation load of the Rhine were composed almost entirely of gymnospermous bisaccate sporomorphs. This feature was also reported by Faegri & Iverson (1964). Rich concentrates of these forms have been recorded from the remote centres of the major oceans, far from any fluvial influence, thus demonstrating their ability to be transported by air and water currents (Faegri & Iverson 1964; Muller 1959).

(iii) *Ecological relationships.* Diversity can be related to the palaeoenvironment of deposition. Highest diversity, for example, is usually associated with lagoonal settings, but has also been observed in Carboniferous deltaic sediments (Neves 1961). Observations by the author suggest that low diversity, or predominance of one or two types, indicates either an extreme form of sorting, or an association of sporomorphs indigenous to the palaeoenvironment. Some types, irrespective of their morphology, may be locally limited due to the poorly drained nature of the coastal plain area, or the low frequency of their parent plant. Sporomorphs found in high numbers and dominating assemblages from delta top sediments include *Ischyosporites variegatus*, *Corollina* spp., *Densoisporites* spp., *Lycopodiumsporites* spp., *Duplexisporites problematicus*, *Verrucosisporites* spp., and smooth walled forms such as *Cyathidites* spp. Marine and fluvial sediments are generally characterized by moderate or low diversities. Marine facies concentrate the smaller equivalents of the delta top forms and the more buoyant coastal and hinterland forms such as *Chasmatosporites magnolioides*, *Quadraeculina anellaeformis*, together with the saccate forms such as *Cerebropollenites mesozoicus* and bisaccate pollen.

*Structureless (sapropelic) organic matter (SOM).* This group is characterized by essentially structureless material bearing a laminar or granular surface appearance. Composition is variable, but consists generally of a bacterially reworked biomass composed of dispersed residual fragments of the original constituents (marine algae, dinocysts, acritarchs, sporomorphs and woody debris) surrounded by the remains of bacterial bodies (Tyson *et al.* 1979). SOM is preserved only where conditions are anoxic or dysaerobic.

*Amorphous organic matter (AOM).* Characterized by amorphous material lacking any evidence of structure. It is uniformly granular or laminar in surface appearance. Amorphous organic matter bears a superficial resemblance to SOM but is distinguished by its homogeneous composition and the lack of fluorescence in ultra-violet light. In practice, because of their constrasting origins, AOM and SOM do not occur together.

## Palynofacies types

The character and distribution of palynological constituents were studied in detail from work on core material of Jurassic deltaic and marine sediments from the East Shetland Basin, in which a variety of sub-palaeoenvironments could be identified from sedimentological evidence. Similar but less detailed (unpublished) studies have also been made by the author of outcrop material from northeast Yorkshire, and, to a lesser extent, of Recent sea bottom material from Brunei, Rhine, Dutch offshore and Wadden Zee areas in order to provide further background palaeoenvironmental information. From the varying character and distribution of the palynological constituents in sediments of different depositional environments, many palynofacies types have been identified. They are subdivided according to their palynomorph content into marine, brackish and terrestrial associations which are indicated by the appropriate prefix M, B and T respectively (Fig. 4). To each prefix a number is subsequently added which indicates the general proportion and size of constituents. Five distinct constituent associations (1–5) can be attached to each major palaeoenvironmental division (e.g. M1–M5). These associations represent, for each major palaeoenvironmental type, a gradual decrease in energy, oxygenation and proximity to a terrestrial source, as summarized below.

*Palynofacies association 1*. Typically very lean in organic content and virtually barren of palynomorphs. Contains only very small sized palynodebris, usually palynomaceral 4. Where associated with finer grained lithologies, it frequently represents oxidized settings. For example, within terrestrial and brackish palaeoenvironments it may characterize subaerially exposed areas of floodplain, fluvial systems and lake/lagoon margins. When occurring within marine mudstones it can characterize periods of low deposition rates. Higher energy settings are typically represented by coarser lithologies and thus most of the organic content is winnowed out leaving only random small fragments. Since palynomorphs are absent, the marine, brackish or terrestrial nature of the palaeoenvironment can only be suggested by its sequential association with adjacent, more palyniferous facies. Uncertainty in assigning a palynofacies type can be expressed by adding the appropriate prefixes and so suggesting a range of palaeoenvironments e.g. MB1.

*Palynofacies association 2*. Characterized by mainly large, usually equidimensional and irregularly shaped palynomaceral 4. Palynomorphs are absent or very rare. Occurs in all lithologies and represents a range of depositional types similar, but lower in energy or oxidation, to that described for palynofacies type 1. Suggests proximity to a terrestrial source because of large size of palynomaceral constituents and relatively high energy associated with depositional setting.

*Palynofacies association 3*. Characterized by a mixture of large and small palynomacerals 1–4 together with palynomorphs. Typically associated with lower energy depositional settings situated relatively close to a terrestrial source.

*Palynofacies association 4*. Characterized by a predominance of palynomorphs, whether they be miospores or algae. Associated palynomacerals 1–4 are small in size and usually less than 30% of the residue. It represents a further stage of sorting and is believed to typify the more distal parts of ventilated subaqueous lake–marine systems.

*Palynofacies association 5*. Dominated by SOM. It represents the lowest energy setting where sediment–water interface conditions are anoxic or dysaerobic. Such palaeoenvironments may occur in a variety of situations but are usually located near the depocentre which can be part of a marine basin or merely the central part of a relatively small swamp area.

Each constituent association may be subsequently qualified by a suffix (e.g. b, frayed bacterially degraded edges to constituents; d, darkened; e, notably equidimensional; l, notably large; r, rounded; s, exceptionally small; p, common blade-shaped palynomaceral 2–3; t, common blade shaped palynomaceral 4; w, one spore type dominant; y, one dinocyst type dominant). The five constituent associations represent distinct end members within a continuum of variation. Intermediate stages can be represented by indicating the predominant association it may be tending towards in brackets e.g. M2(M3) will have predominantly large sized palynomaceral 4 but will also contain a minor proportion of palynomacerals 1–3 and palynomorphs. More extreme associations can be similarly expressed. For example M4(M2) can describe an even association of palynomorphs and large sized palynomaceral 4. In this way the general palaeoenvironmental setting is indicated by the prefix e.g. M, marine; the depositional energy (and proximity of a fluvial source) by adding the number e.g. M3; and then any special features associated such as oxygenation e.g. $M3^t$.

The approach to palynofacies analysis follows closely that outlined in Whitaker (1984). Below,

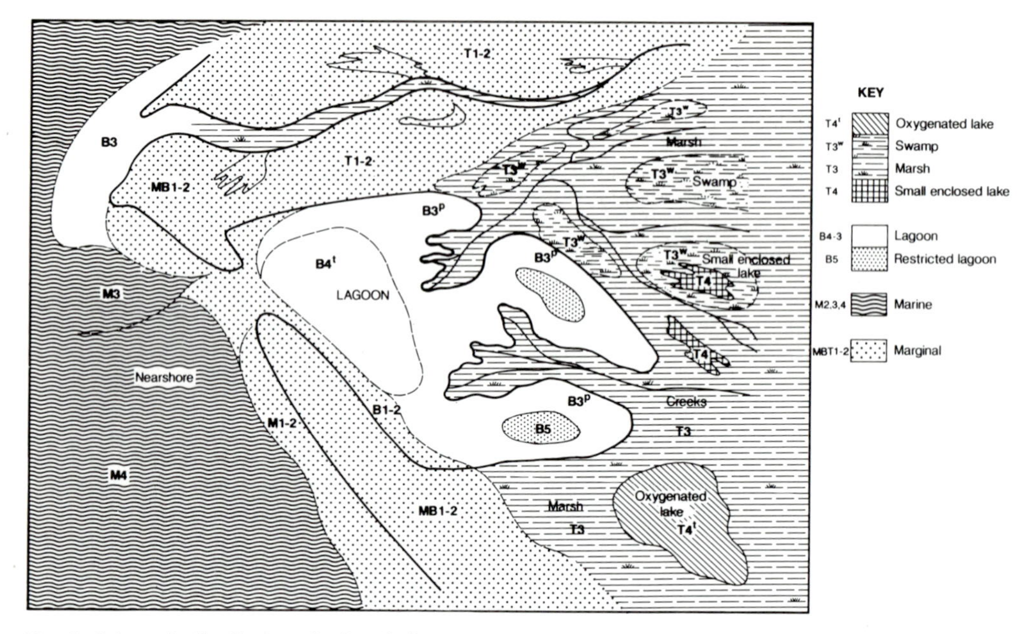


**Fig. 5.** Schematic distribution of palynofacies types.

the principal palynofacies types associated with the most important depositional environments are discussed and illustrated schematically in Fig. 5.

## *Non-marine palynofacies types T1–T5*

Onshore deltaic palaeoenvironments are generally distinguished by the absence of dinocysts, predominance of thick-walled sporomorph types and the often excellent preservation of palynomacerals 2 and 3.

*Mature marsh.* A mature marsh is a low-lying setting characterized by small lacustrine palaeoenvironments surrounded by vegetation and dissected by mainly small fluvial distributaries. Within the basic structure of a delta the marsh area is narrowest at the apex and, in this area, it can be envisaged that fluvial channels are most closely spaced and may include many 'marginal settings' influenced by this fluvial activity. This situation is well developed in the south of the area around Block 3/2. Towards the coast, the marsh areas between fluvial systems become wider and larger lake areas may develop. Palynofacies type T4 is believed to characterize the central parts of these lake areas. They contain a high proportion of buoyant palynomorphs, which include small *Cyathidites* spp. and bisaccate pollen. Local fluctuations in vegetation also frequently allow abundant *Corollina* spp. and *Densoisporites microrugulatus* to concentrate. The high proportion of bisaccate pollen are almost entirely small and appear to be wind derived since they originate predominantly from the hinterland. The palynomorphs may represent up to 75% of the total organic content observed in the palynological preparations. The remaining constituents usually include blade-shaped and small equidimensional palynomaceral 4. Such a 'clean' association appears to reflect not only the distance from direct terrestrial influence but also a high degree of oxygenation. To maintain this degree of oxygenation a large open lake area and perhaps shallow water is suggested. Such areas appear to be represented within the Oich Member of the Ness Formation of well 3/2–4.

The margins of these lake areas do not appear to develop coal swamp palaeoenvironments. This is indicated by the absence of *Callialasporites* and *Araucariacites* spp. They are characterized by palynofacies type T3 containing a significant proportion of palynomacerals 2 and 3. These constituents are frequently bleached T3$^{l}$, and are pale and pitted which is believed to indicate the effects of subaerial degradation and thus very shallow, periodically exposed conditions. Palynomaceral 3 is often characterized by very thin, delicate and structureless material (membranes) suggesting the local vegetation is mainly 'grassy' rather than 'arboraceous'. Such palynofacies are frequently encountered in the southern part of the study area within the Foyers Member of the Ness Formation.

Palynofacies type $T2^t$, containing mainly blade-shaped palynomaceral 4, may also characterize marginal areas. In its pure form it can be indicative of crevasse splay deposition.

Palynofacies type $T2^e$ contains mainly larger sized equidimensional palynomaceral 4. When associated with finer grained organic rich sediments it is believed to represent secondary alteration due to prolonged subaerial degradation. A combination of these types may represent intermediate stages e.g. $T4(T2^t)$ and $T3^t(T2^t)$ which here indicate degrees of influence from 'marginal' processes.

Along the seaward margins of the marsh which border the lagoonal areas transitional facies develop. Palynofacies suggest these areas are frequently swampy. They include palynofacies types $T3^w$, $T3^p$ and $T4^p$. Palynomacerals are often well preserved and sometimes blade-shaped. The excellent preservation and the absence of bacterial degradation suggests an anoxic subaqeous setting associated with swampy conditions. Sporomorphs include common *Callialasporites* and *Araucariacites* spp,. which may represent a local swamp vegetation. Proximity to coastal conditions also allows a proportion of sporomorphs more commonly associated with lagoonal and marine palaeoenvironments to be incorporated. The first of these types to appear is usually *Perinopollenites elatoides*. These assemblages may also include the rare occurrence of tasmanitids. Areas closer to the lagoon are characterized by an increasing number of *Chasmatosporites magnolioides* and *Quadraeculina anellaeformis*. Examples of the above transitional facies are frequently observed within the Enrick Member of the Ness Formation especially in the southern part of the study area.

## *Brackish water and coastal palynofacies types B1–B5*

Distinguished by rare marine palynomorphs or the occurrence of one species in high numbers.

*Lagoonal areas*. Palynofacies are characterized by types B1–5 which represent the likely sequential development from shallow water and exposed marginal areas to the lower energy and anoxic depocentre of a lagoonal setting. Lagoonal sediments within the Brent Group are typically characterized by the rare occurrence of the dinocyst *Nannoceratopsis* spp. together with occasionally common acritarchs, and the freshwater–brackish alga *Botryococcus*. Sporomorph assemblages are usually rich and diverse, dominated by *Perinopollenites elatoides*, *Chasmatosporites magnolioides* and *Quadraeculina anellaeformis*. Lagoonal palynofacies are best represented within the Oich Member, although can occur within the Enrick Member of the Ness Formation.

*Immature marsh–back barrier areas*. Deposition probably occurred in a subaqueous area marginal to a lagoon or coastal barrier. Palynofacies types are variable and include B2, B3, B4, B5, $T3^{(b)}$ and $T4^{(b)}$. Microplankton are frequently represented by 4–5 specimens of *Nannoceratopsis senex* (per slide), acritarchs and tasmanitids which confirm a direct marine influence. The palaeoenvironment can be further distinguished by the common occurrence of *Quadraeculina anellaeformis* and *Chasmatosporites magnolioides* together with minor influxes of *Ischyosporites variegatus* and *Concavissimisporites* spp. The latter two types may represent vegetation that can grow along these partially exposed marginal coastal areas as well as within a mature marsh. This setting is also notable due to the absence of mature marsh indications, for example, of commonly occurring *Corollina* spp., *Cycadopites* sp., *Densoisporites microrugulatus*, *Duplexisporites problematicus*, *Leptolepidites* spp., Spore type A, *Neoraistrickia* spp. and *Uvaesporites argentaeformis*. The absence of these types suggests that the mature marsh vegetation associated with these forms was not established in these areas. Occasional influxes of *Callialasporites* and *Araucariacites* spp. suggest periodic swamp development. The above palynofacies are most frequently represented within the basal part of the Enrick Member to the south of the study area.

## *Marine and coastal palynofacies types M1–M5*

Presence of microplankton, high bisaccate content and the 'frayed' appearance of palynomacerals 2 and 3 are features characteristic of a marine palaeoenvironment. In an offshore direction, palynological constituents derived from a terrestrial source appear to be systematically sorted according to their hydrodynamic properties, degradation characteristics and susceptibility to waterlogging. Microplankton also changes in character in an offshore direction.

In the highest energy settings, such as littoral areas, most of the constituents are winnowed, leaving only low levels of organic material, referred to as palynofacies type M1. Along the shoreface the less buoyant woody material is concentrated either as dispersed organic matter or in clast form. Palynofacies type M2 is typical, characterized by large, frequently darkened, material comparable to palynomaceral 1 and less commonly palynomaceral 2. Sporomorphs and microplankton are usually rare. Such palynofacies usually characterize the middle–upper parts of the Rannoch Formation.

As energy conditions at the sea bottom decrease with depth, the more buoyant constituents begin to settle out, as represented by palynofacies type M3. This type is frequently associated with delta front sedimentation. Large and medium sized, often well preserved, palynomacerals 1, 2 and 3 may predominate with moderate proportions of sporomorphs and microplankton. This palynofacies characterizes the lower part of the Rannoch Formation.

Areas distal from the source are characterized by palynofacies type M4. Plant debris is much reduced in proportion and size, comparable mainly to palynomacerals 2 and 3. Bisaccates may predominate and palynofacies of this type are typical of the upper Drake Formation. The microplankton content is variable. Where sea bottom conditions become relatively stagnant and anoxic, such as in the deeper waters of central basinal areas, structureless sapropelic organic matter may be preserved. Such palynofacies are referred to type M5 and frequently characterize the Heather Formation.

The marine Jurassic sequences from the northern North Sea show several major changes in the general characteristics of the palynological constituents, represented by an alternation of two types of preservation. Certain intervals are characterized by the relatively well preserved appearance of their constituents and include the common occurrence of material comparable to palynomacerals 2 and 3, or in some cases structureless (sapropelic) organic matter, but with infrequent blade-shaped palynomaceral 4. Such intervals are believed to represent relatively enclosed areas where energy conditions at the sediment–waters inferface may be relatively low because current activity from tides or open oceanic conditions is limited. Conversely, other intervals subject to higher oxygenation and current activity are characterized by poorly preserved constituents which are frequently darkened or bleached and dominated by high proportions of blade-shaped palynomaceral 4. These intervals can be correlated, using dinocysts, across the northern North Sea and appear to be relatively isochronous, and thus most likely reflect regional changes of depositional setting.

## Environmental relationships

Using the palaeoenvironmental evidence of the above palynofacies criteria, a sequential development of depositional environments are described for in the southern area of the study (Figs 6 and 7).

The uppermost part of the Dunlin Group and Broom Formation are characterized mainly by palynofacies type M4 and $M4^t$ which suggests deposition in an offshore marine setting with oxygenated bottom conditions. The microplankton content is relatively diverse at this time, and may include *Nannoceratopsis* spp., *Scriniocassis weberi*, *Mancodinium semitabulatum*, *Fromea elongata*, *Parvocysta nasuta*, *Parvocysts barbata*, Dinocyst types 1–4 and acritarchs. Sporomorphs are dominated by *Perinopollenites elatoides* and small bisaccate pollen together with small forms of *Stereisporites*, *Osmundacidites* and *Cyathidites* spp.

In some wells, the uppermost Dunlin Group is characterized by palynofacies type M3 containing approximately 25% of poorly preserved palynomacerals 2 and 3. This suggests a terrestrial influence in this area. The lower part of the overlying Rannoch Formation is characterized mainly by palynofacies types $MB4^t$, MB3 and MB3(MB2) suggesting a brackish nearshore setting strongly influenced by a terrestrial source. The Rannoch and Etive Formations are frequently characterized by the sequence of palynofacies MB4t-MB3-MB2 (MB3)-MB2-MB1 suggesting an increasingly higher energy shore-like character. Palynofacies $MB4^{(t)}$ (MB2) is characterized by abundant *Nannoceratopsis gracilis*, *Nannoceratopsis* aff. *spiculata* and *Nannoceratopsis* cf. *spiculata*. It may also include a pale undifferentiated palynomorph comparable to small sized leiospheres. These forms are difficult, however, to distinguish from degraded *Perinopollenites elatoides*. Other dinocysts can include rare *Mancodinium semitabulatum* and *Parvocysta nasuta*. The distinctive acritarch *Micrhystridium stellatum* can occur together with tasmanitids. The predominance of *Nannoceratopsis* spp. may suggest salinities were abnormal supporting only specific dinocysts. The occasional abundance of *Botryococcus* observed in these facies may also suggest low salinities.

Sporomorphs are dominated by *Perinopollenites elatoides*, bisaccate pollen together with occasional influxes of *Chasmatosporites magnolioides*, *Quadraeculina anellaeformis*, *Cerebropollenites mesozoicus* and *Osmundacidites* spp.

The palynomorph content can be dominated by *Nannoceratopsis* spp. and *P. elatoides* form-

| LITHO UNIT | PALAEO-ENVIRON. | PALYNO-ZONE | PALYNO-FACIES | MAIN CRITERIA |
|---|---|---|---|---|
| HEATHER | Offshore Marine | 11 | M5 | Common *C. chytroeides* Rare *Hystrichogonyaulax sp.* |
| | | 10 | | Common *P. evittii* *Rare Hystrichogonyaulax sp.* |
| | | 9 | M4$^{bt}$ | *Korystocysta ? sp.* Common *Lycopodiumsporites spp.* |
| TARBERT | Marine/ Brackish | 8 | B5$^{bd}$/B2$^{t}$ | Dinocyst type 1, Acritarchs. Common *P. elatoides* and *C. mesozoicus* |
| FOYERS MBR. | Lacustrine | 7 | T4$^{t}$ | Rich sporomorph assemblages. Common *Lycopodiumsporites*, *Corollina* and *Staplinisporites spp.* Rare *Verrucatosporites sp.* |
| | Mature Marsh | 6d | T3 | Common *U. argentaeformis* and *Cycadopites spp.* rare C. minor. |
| | | 6c | | Common *I. variegatus, D. problematicus* *Leptolepidites spp.* |
| | | 6b | | Common *Concavissimsporites spp.* Downhole increase in *Callialasporites spp.* and *P. elatoides.* |
| | | 6a | T3$^{pw}$ | Common *Todisporites spp.* together with *A. grandis.* |
| OICH MBR. | Lagoonal | 5b | B4$^{t}$ | Frequent *N. gracilis* common *C. mesozoicus, Osmundacidites spp.* |
| | | 5a | B5 | Super abundance of *P. elatoides* and *Cythidites spp.* |
| ENRICK MBR. | Swamps | 4 | T3$^{w}$/T2 | Common *Q. anellaeformis, C. magnolioides.* |
| | Lagoonal | | BT3/T3 B4/5 | Common *P. elatoides, Q. anellaeformis, C. magnolioides.* Microplankton. |
| ETIVE RANNOCH | Shoreface | 3 | MB2/1 | Palynomorphs rare or absent. |
| BROOM | Brackish/ Marine | 2 | MB4$^{t}$ | Leiosphere, *Nannoceratopsis spp.* and acritarchs. |
| DRAKE | Offshore Marine | 1 | M4$^{t}$ | *S. weberi, Parvocysta spp. Fromea spp.* |

**Fig. 6.** Main criteria of zonation scheme.

ing up to 80% of the sieved organic content. Blade-shaped and small equidimensional palynomaceral 4 usually make up the remaining constituents. Such associations are believed to represent depositional environments situated along the lower shoreface lateral to any direct

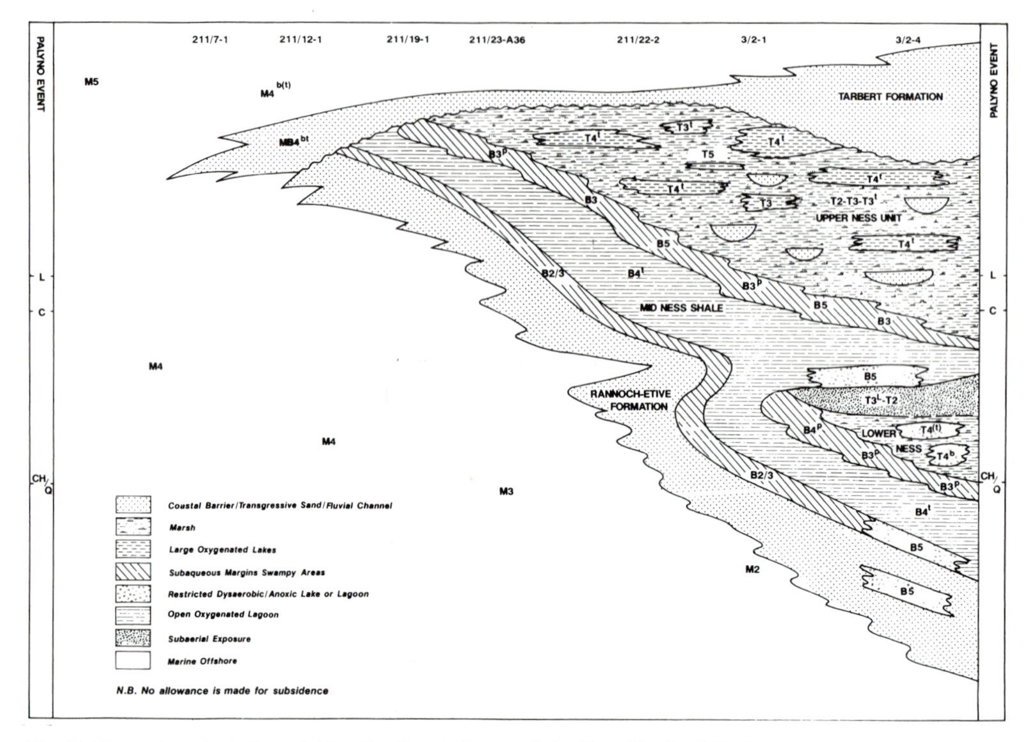


**Fig. 7.** General geological model for the Brent Group of the East Shetland Basin.

fluvial source. Increasing terrestrial influence is indicated by palynofacies type MB3 and the co-occurrence of *Cerebropollenites mesozoicus* and *Osmundacidites* spp. The latter forms would suggest that the terrestrial influence is the result of direct fluvial input and in this way, certain parts of the study of area can be identified as being closer to a distributary mouth.

The uppermost part of the palynofacies sequence is characterized by type MB2 containing mainly large sized equidimensional palynomaceral 4 and darkened palynomaceral 1. This is envisaged as indicating a high-energy shoreface coastal setting.

In most wells a lagoonal area was established at the base of the Enrick Member. The lagoon at this stage appears to have been relatively small and was characterized by type B5, indicating anoxic bottom conditions. The overlying occurrence of palynofacies type B4 and B4$^{t}$ indicates an increase in ventilation probably due to an increasing size of lagoonal area. In the southern part of the study area, a regressive palynofacies sequence was recognized and overlies the lagoonal development represented by palynofacies types B5-B4-B4t-BT3-BT2. This regression probably represents an advance and diachronous progradation of adjacent marginal palaeoenvironments, perhaps immature, weakly developed marsh areas, northward into this area. This regression in some areas developed shallow and subaerial conditions in its uppermost part (i.e. BT3$^{l}$ and BT2). The absence of abundant mature marsh sporomorph types and the predominantly low energy, dysaerobic character of the palynofacies in the lower and middle parts of this interval indicates a mainly protected, periodically swampy setting. This interval comprising the basal lagoonal development and the overlying regressive sequence corresponds approximately to the Enrick Member.

The overlying Oich Member is generally characterized by palynofacies types B5 and B4$^{t}$. The subsequent occurrence of palynofacies types B4$^{t}$ indicate a significant period of transgression and a second development of lagoonal conditions suggesting a period of delta retreat. At this time, therefore, a relatively large area of the Brent delta may have been occupied by this lagoon. In these areas, therefore, the Oich Member could have been virtually isochronous.

The Oich Member in many wells is characterized by a particular sequence of palynofacies types suggesting, after the initial formation of the lagoon, the gradual progradation of transitional and marsh palaeoenvironments.

The base of the sequence is frequently represented by palynofacies types B5 containing almost entirely SOM-P2 constituents. This suggests that as a brackish palaeoenvironment was initiated, anoxic bottom conditions frequently appear to have been established. Bacteria seem to have thrived, and have converted any available woody palynomacerals to SOM-like constituents. Sporomorphs are dominated by abundant *Perinopollenites elatoides* and, in some cases, small sized *Cyathidites* spp. The absence of other sporomorph types is notable and may be a result of adjacent vegetation being temporarily engulfed by the expanding lagoonal waters.

Palynofacies type $B4^{(t)}$ characterizes the middle part of these lagoonal sections. It is dominated by sporomorphs which include bisaccate pollen, *Perinopollenites elatoides*, *Cerebropollenites mesozoicus*, *Chasmatosporites magnolioides*, *Quadraeculina anellaeformis*, *Osmundacidites wellmanii*, sphaeromorph clusters and *Botryococcus*. The forms *Cerebropollenites mesozoicus*, *Osmundacidites wellmanii* and the larger sized bisaccate pollen are believed to represent fluvial input from local distributaries. The other types listed above are believed to represent coastal vegetation. The approaching advance of the mature marsh area is indicated by the frequent occurrence of *Corollina* spp. and *Densoisporites* spp. *Botryococcus* occurrences are thought to be in situ suggesting this form preferred brackish rather than freshwater conditions. Microplankton are usually represented by four to five specimens (per slide) of *Nannoceratopsis senex* together with tasmanitids and acritarchs, indicating direct marine influence. Palynomorphs dominate the palynofacies, the remaining constituents being represented by mainly delicate blade-shaped and needle-shaped palynomaceral 4. This represents a distal association with ventilated bottom conditions. Thus as the lagoon developed in size, conditions at the bottom appear to have become oxygenated.

A period of delta retreat can be identified in well 211/19–6 within the overall progradation of the Etive and Rannoch Formations (see Fig. 7). In this well two developments of the Etive and Rannoch Formations have been identified.

The Foyers Member is generally characterized by palynofacies types T3 and T4 and represents a second major and diachronous period of regression when a more mature marsh setting

**Table 4.** *Sporomorphs related to marsh maturity*

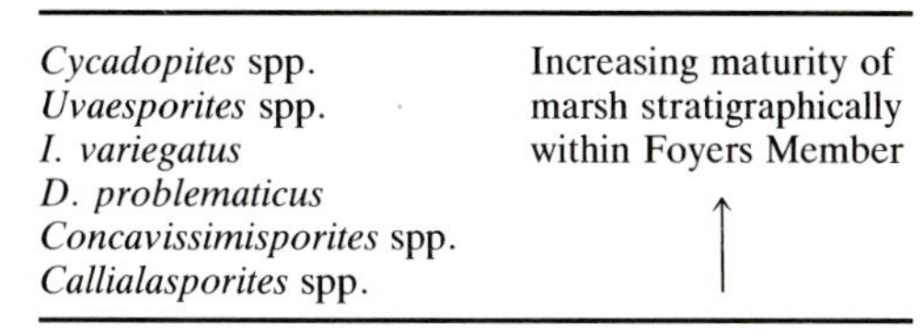

| | |
|---|---|
| *Cycadopites* spp. | Increasing maturity of marsh stratigraphically within Foyers Member |
| *Uvaesporites* spp. | |
| *I. variegatus* | |
| *D. problematicus* | ↑ |
| *Concavissimisporites* spp. | |
| *Callialasporites* spp. | |

prograded across the area. Palynomorphs and palynofacies which characterize the lower part of this interval indicate a transitional period when floras associated with brackish conditions were influential. This is reduced in the middle part of the Foyers Member, where palynomorphs are more dominated by local and exotic types suggesting a mature marsh palaeoenvironment.

Sporomorph assemblages are typically dominated by a single ('exotic' or pteridophyte) type. These types are usually thick walled, well ornamented forms and include *Concavissimisporites–Converrucosisporites* spp., *Uvaesporites argentaeformis*, *Duplexisporites problematicus* and *Ischyosporites variegatus*. The sequential appearance of these forms within the Foyers Member (Table 4) is the same for most wells which suggests they reflect vegetation zones associated with increasing maturity of the marsh (see Fig. 8).

Some assemblages with abundant *Ischyosporites variegatus* and *Duplexisporites problematicus* are bleached suggesting perhaps exposure to subaerial conditions.

Delicate, thin and pale tissues (palynomaceral 3) occur frequently. Such delicate material would not survive transport over large distances and confirms the development of quiet, marsh-like palaeoenvironments.

A change in the nature of the palynomorph assemblage identified in the upper part of the Foyers Member, an interval often characterized by palynofacies type $T4^t$, suggests a change in palaeoenvironment possibly involving larger lacustrine settings with more oxygenated bottom conditions. It seems likely this change indicates the marine incursion associated with the palynofacies types $MB4^{bt}$, $MB5^{bd}$ and $MB2^t$ of the overlying Tarbert Formation.

The overlying Heather Formation is characterized by palynofacies type M5 and $M5(M4^t)$ containing a high proportion of structureless organic matter (SOM) generated at the sea floor where conditions were anoxic. The restriction of circulation at this time appears to have been due to a combination of circumstances involving the depth of water and the

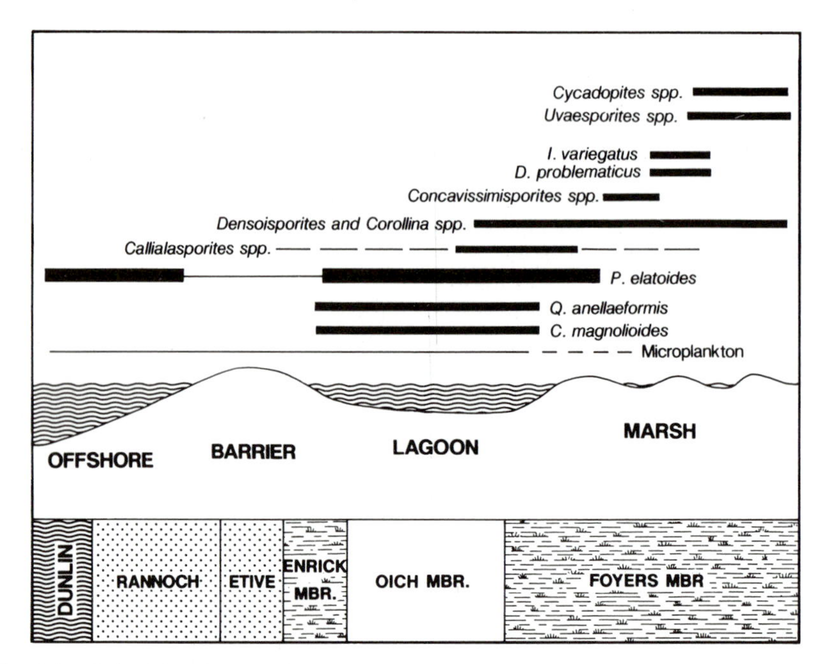


**Fig. 8.** Possible relationships of selected palynomorphs to environment.

enclosed nature of the marine basin during Late Bathonian times (Whitaker 1984). Where sea bottom conditions were periodically ventilated, perhaps as a result of shallowing or periodic storm activity, the SOM is degraded, usually darkened and may be associated with increase in the content of palynomaceral 4. In this latter case the palynofacies is referred to type $M5(M4^{bt})$.

Microplankton diversity and frequency is moderate or poor, characterized by occasional influxes of *Chytroeisphaeridia chytroeides* together with *Hystrichogonyaulax* sp., *Ambonosphaera* sp. and *Pareodinia evitti*, other dinocyst types are less fequent. The relatively sparse dinocyst content may reflect poor circulation of nutrient rich bottom water and, because of the enclosed nature of the basin, low salinities. Fresh water may have been supplied to the basin of the deteriorating Brent-style delta system. This is indicated by the high content of *Cerebropollenites mesozoicus* and *Callialasporites* spp. in the assemblages.

The sequential development of palynofacies described above for the Brent Group is also observed in wells to the north of the study area. Towards the north, however, an increase in microplankton indicates a greater influence of brackish conditions, particularly in the lower part of the Ness Formation. The influence of the regressive phase associated with the Enrick Member is less conspicuous. Finally, in the most northern part of the study area the Ness Formation is reduced and finally pinches out. In this area the Tarbert Formation may immediately overlie the Etive Formation. The gradual disappearance of the Ness Formation northwards is not indicative of a time gap or unconformity but merely an absence of their associated palaeoenvironments (Fig. 9a–b).

## Palynological zonation scheme

Most lithostratigraphical correlations within the Ness Formation employing wireline log criteria are probably made on the basis of facies controlled boundaries, and are, therefore, unlikely to be chronostratigraphic. An important requirement of this study, therefore, was to attempt a correlation of rock units with respect to time. Sporomorph assemblages are also to the same degree facies related and thus in the present study, it was necessary to critically appraise any apparent correlation of sporomorph associations. Dinocysts, tasmanitids and acritarchs also occur sporadically within the sequence and so it was also necessary to consider their use for correlation. The high proportions of core material has allowed the most subtle of changes in the palynomorph distributions to be considered. In this way both tops and bases together with acmes and influxes can be employed.

Wells in the southern part of the study area displayed a regular sequential pattern of assemblage and palynofacies changes allowing a subdivision of the uppermost Dunlin Group, Brent Group and Lower Heather Formation into 11 zones (Fig. 6). These zones however are probably valid only in this part of the East Shetland Basin and, because of their more local rather than regional extent, should be more strictly referred to as palyno-events. These data formed the basis of the zonal scheme described below.

## Palynozone definitions

### *Zone 11; Type Well 3/2–3*

This interval normally comprises a part of the lower Heather Formation. Its principal characteristic is the common occurrence of *Chytroeisphaeridia chytroeides*. Associated dinocysts may include the infrequent occurrences of *Hystrichogonyaulax* sp., *Diacanthum filipicatum* and *Ambonosphaera* sp.

Sporomorphs do not appear to be diagnostic for this zone since they share many features in common with the underlying zone 10. Assemblages are generally characterized by *Cerebropollenites mesozoicus* and *Perinopollenites elatoides* together with *Callialasporites* spp. and bisaccate pollen. Where an upper limit to the zone was observed, the highest common occurrence of *Cerebropollenites mesozoicus* appeared to coincide with *Chytroeisphaeridia chytroeides*.

Palynofacies type M5 characterizes the interval together with occasional types $M5^{dt}$, $M5(M4^{b})$ which indicates a period of offshore marine conditions with a predominantly anoxic sea floor.

### *Zone 10; Type Well 3/2–3*

This interval normally comprises the lower part of the Heather Formation. The highest regular occurrence of *Pareodinia evittii* and *P.* cf. *evittii* define the upper limit of this interval. *Ambonosphaera* sp., *Hystrichogonyaulax* sp., *Pareodinia ceratophora*, *Ctenidodinium gochtii*, and *Chytroeisphaeridia chytroeides* may occur infrequently.

Sporomorphs are characterized by common *Cerebropollenites mesozoicus*, *Perinopollenites elatoides* and *Callialasporites* spp. The highest occurrence of *Chasmatosporites magnolioides* was observed in this zone. The persistent occurrence of *Hystrichogonyaulax* sp. and *Ctenidodinium gochtii* indicates a mid-Bathonian age. This latter age interpretation is based on detailed unpublished work carried out on ammonite dated core material from the well 211/21–1 (Callomon 1975).

Palynofacies type M5 generally characterizes this interval together with the occurrence of types $M5^{dt}(M4^{bt})$ indicating an offshore marine setting with anoxic and sometimes dysaerobic sea bottom conditions.

### *Zone 9; Type Well 211/22–2*

This interval may comprise the lowermost Heather Formation. There is no evidence that this zone occurs within the Tarbert Formation. The zone is distinguished by the highest occurrence of *Korystocysta*? sp. and coarsely ornamented forms of *Batiacasphaera* sp. This zone may also include the highest occurrence of *Parvocysta barbata*. These dinocysts occur in strata no younger than early Bathonian.

Sporomorphs usually increase in diversity (downhole) within this zone. Single specimens of *Lycopodiumsporites austroclavatidites*, *Lycopodiumsporites senimuris*, small *Concavissimisporites* spp., *Leptolepidites* sp. and *Duplexisporites problematicus* together with frequent *Cyathidites* spp. were observed.

The proportion of blade-shaped palynomaceral 4 and levels of deterioration of the SOM content increases down section. SOM present is usually darkened or degraded in some way. This suggests a more dysaerobic or ventilated offshore depositional environment.

### *Zone 8; Type Well 3/2–2*

This zone is limited entirely to the Tarbert Formation. It is distinguished by its high frequency of *Cerebropollenites mesozoicus* and *Perinopollenites elatoides*. Other sporomorphs are usually rare. It is also distinguished by the occurrence of Dinocyst type 1. *Micrhystridium stellatum* may be present.

Whilst not independently dated by ammonites the stratigraphical position of this zone above the highest occurrence of *Nannoceratopsis* spp. and below *Korystocysta*? sp. suggests a Late Bajocian–earliest Bathonian age (Prauss 1989).

Palynofacies $MB4^{bt}$, $MB5^{bd}$ and $MB2^{t}$ characterize this interval. Palynofacies type $MB5^{bd}$ contains constituents intermediate in character to physically degraded palynomaceral 1/2 and bacterially derived SOM. Since the Tarbert Formation comprises in part reworked sediments of the underlying Ness Formation, it seems more likely that physical degradation may be the principal controlling mechanism.

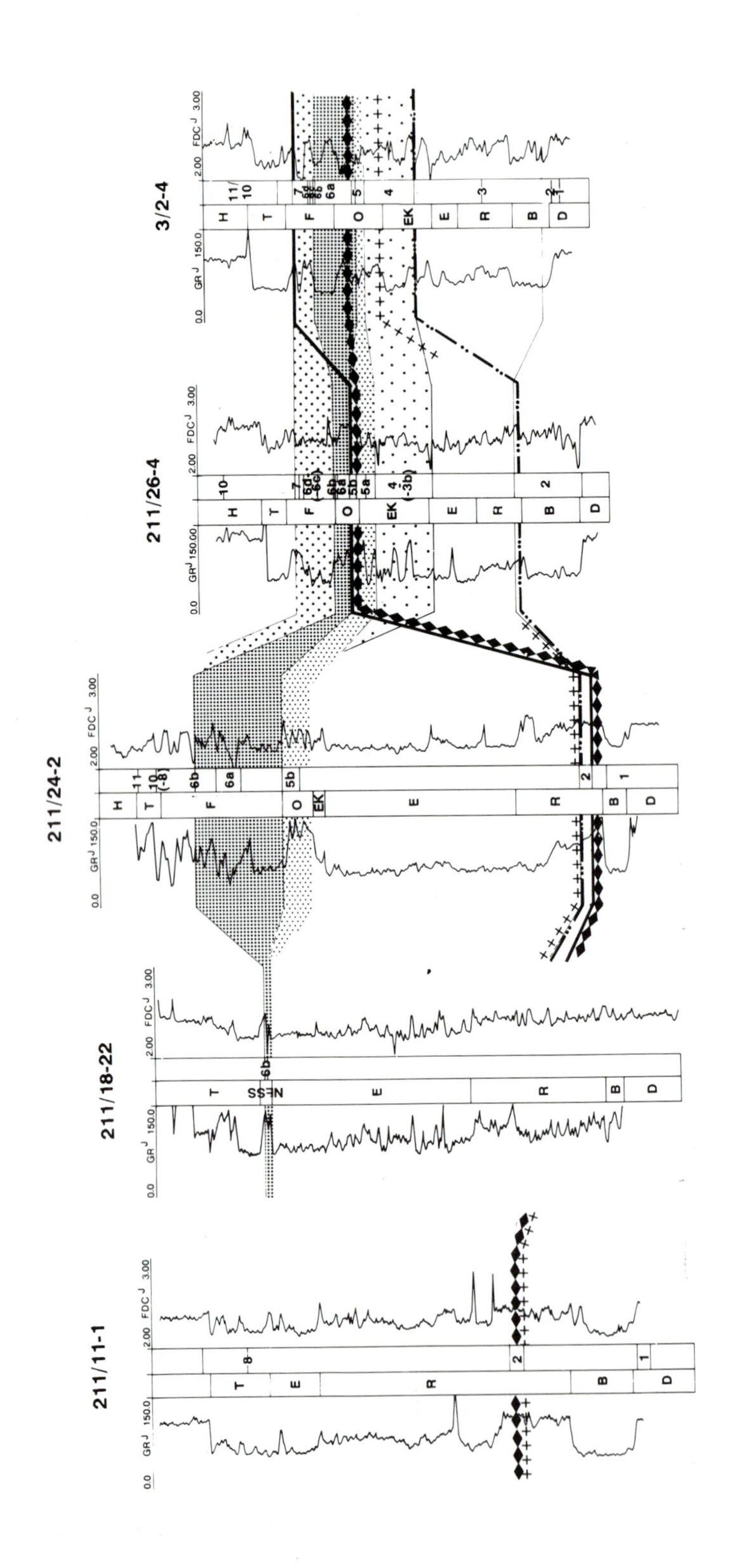
(a)
211/11-1
211/18-22
211/24-2
211/26-4
3/2-4

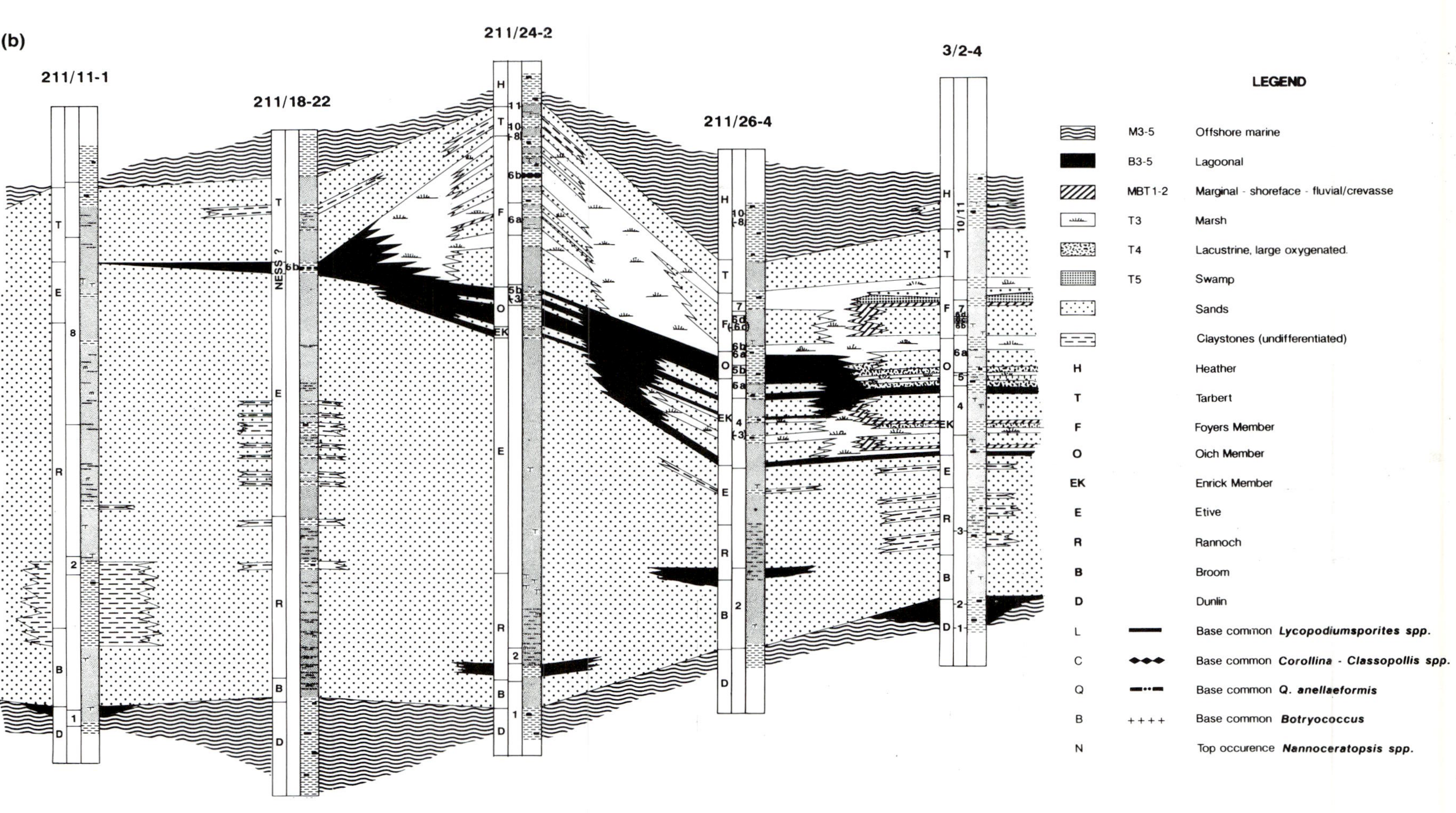


**Fig. 9.** (**a**) N–S section showing relationships of local palynozonation and pseudo-timelines. (**b**) N–S section showing environmental relationships.

Palynofacies type MB2[t] within the depositional circumstances, described above, suggests a winnowed, well-oxygenated setting.

### *Zone 7; Type Well*

This zone is associated always with the Foyers Member. It represents a downhole increase in the frequency of *Cyathidites* spp., *Corollina* spp. and frequently *Densoisporites* spp. In the southern part of the study area, the basal occurrence of commonly occurring (>25 specimens per slide), *Lycopodiumsporites austroclavatidites*, *Duplexisporites problematicus* and *Densoisporites microrugulatus* may be limited to this zone. The rare occurrence of *Verrucatosporites* sp. appears to be limited to this zone.

Microplankton are virtually absent. Single specimens of *Nannoceratopsis senex* were noted in wells 3/2–2 and 3/8–12. If in situ, these occurrences indicate an age no younger than Bajocian (Prauss 1989).

Palynofacies types T4[t] and T2[t] characterize this interval, suggesting a shallow but possibly large, well oxygenated lacustrine setting.

### *Zone 6; Type Well*

This zone may be associated with the Foyers Member but may include part of the Oich Member. It is subdivided into four subzones, but its principal characteristic is the occurrence of palynofacies type T3 and the association of exotic sporomorph types.

*Subzone 6d*. This subzone is limited to the Foyers Member. It is distinguished by the common occurrence of *Uvaesporites argentaeformis*. The common occurrence of *Cycadopites* spp. is also noted at similar horizons and is interpreted as indicative of subzone 6d. *Chomotriletes minor* occurs rarely but appears to be restricted to this zone and the overlying zone 7. *Densoisporites microrugulatus* and *Corollina* spp. may be important elements but, in many cases, *U. argentaeformis* or *Cycadopites* sp. dominate. Microplankton are virtually absent.

Palynofacies include types representing T3 and T4[t]. Palynomacerals 2 and 3 are often very pale and degraded. These palynofacies together with the above sporomorphs indicate a period of predominantly small lacustrine or marginal lacustrine conditions.

*Subzone 6c*. This subzone is also limited to the Foyers Member. It is weakly distinguished by an influx of *Ischyosporites variegatus* which seems distinctive in certain wells from the southern part of the study area. Where this subzone can be delinated, associated sporomorph types include *Duplexisporites problematicus*, *Densoisporites microrugulatus*, *Leptolepidites* sp. and *Corollina* spp. Microplankton were not observed.

Palynofacies types T3 and T4[t] characterize this interval, together with the occasional type T2[t]. Palynomacerals 2–3 are usually degraded and pale. These palynofacies together with the above sporomorphs, indicate a period of predominantly small lacustrine or marginal (fluvial channel) settings.

*Subzone 6b*. This subzone is associated with the Foyers and Oich Members. It is clearly distinguished by the highest common occurrence of *Perinopollenites elatoides*, *Araucariacites australis*, *Callialasporites* spp. and the acme of a *Concavissimisporites* spp. In some wells the top occurrence of *Perinopollenites elatoides* has not been diagnostic. *Leptolepidites* spp. is frequently most common in this interval. A spore type comparable to a darkened degraded form of *Corollina* spp. can be a feature of this zone. Tasmanitids occur more regularly within this zone.

Palynofacies type T3, T3[w] and T4[p] characterize this interval. Palynomacerals 2–3 are often very well preserved and blade-shaped. The high frequency of *P. elatoides* indicates influence from a lagoonal or coastal setting. The common occurrence of *A. australis* and *Callialasporites* spp. indicates a swampy setting. The influx of *Concavissimisporites* spp. suggests the strong influence of a local marginal vegetation. The excellent preservation of the palynomacerals suggests very low energy, subaqueous conditions. Both the palynomorph and palynodebris content indicate a small restricted swampy area close to the edge of a lagoonal area.

*Subzone 6a*. This subzone is associated with the Foyers Member and the uppermost Oich Member. It is only weakly distinguished by the common occurrence of *Todisporites* spp. and the highest occurrence of *Alisporites grandis* and *Eucommiidites*? sp. *Botryococcus* may also be present. Associated forms include *Corollina* spp. and *Densoisporites microrugulatus*, together with commonly occurring *Perinopollenites elatoides*, *Araucariacites australis* and *Callialasporites* spp. The rare occurrence of *Nannoceratopsis senex* and tasmanitids is characteristic of this zone.

Palynofacies T3, T3[w] and T4[p] characterizes this interval. The depositional setting for this interval appears similar to that for subzone 6b, except that the absence of a local exotic flora suggests a predominantly swampy setting situated at the edge of a lagoonal area. Marginal areas close to fluvial channels seem less conspicuous.

The above zone was difficult to distinguish routinely and may only be locally sustainable.

## *Zone 5; Type Well 3/2–2*

This zone is primarily associated with the Oich Member. It is primarily distinguished by the regular occurrence of microplankton associated with palynofacies type $B4^t$.

*Subzone 5b.* This subzone is distinguished in most areas by a distinctive influx of *Nannoceratopsis* spp., *Cerebropollenites mesozoicus*, and in some cases, *Osmundacidites wellmanii*. Alisporites grandis also displays a downhole increase in frequency. The highest regular occurrence of *Quadraeculina anellaeformis* and *Chasmatosporites magnolioides* are frequently associated with this zone. In a few wells these latter types can be common. The frequent occurrence of *Marattisporites* sp., sphaeromorph clusters, and *Neoraistrickia* spp. also characterizes the subzone. *Botryococcus* is often common. The base of this subzone is frequently marked by the basal occurrences of *Corollina* spp. and *Densoisporites microrugulatus*.

Palynofacies $B4^{(t)}$, $T4^{(t)}$ and T3 generally characterize this zone. They usually include a high proportion of palynomorphs together with delicate palynomaceral 4. This association indicates a large open subaqueous palaeoenvironment. Several elements of the palynomorph content suggest a brackish or lagoonal setting. These include the regular occurrence of *Nannoceratopsis* spp., tasmanitids, *Quadraeculina anellaeformis*, *Chasmatosporites magnolioides* and *Botryococcus* together with the abundance of *Perinopollenites elatoides*. The influence of the adjacent marsh areas is indicated by the mature marsh forms such as *Corollina* spp. and *Densoisporites microrugulatus* together with the marginal swamp types *Callialasporites* spp. and *Araucariacites australis*. The sudden influx of *Cerebropollenites mesozoicus* is interpreted as being indicative of fluvial input from the larger rivers.

*Subzone 5a.* This subzone is only weakly distinguished by the absence of subzone 5b and zone 4 indicators. It is usually characterized by the co-occurrence of abundant *Perinopollenites elatoides* and *Cyathidites* spp. and the absence of *Corollina* spp., *Densoisporites microrugulatus*, *Chasmatosporites magnolioides* and *Quadraeculina anellaeformis*.

Palynofacies type B5 characterizes the interval. The abundance of *P. elatoides* clearly suggests proximity to a brackish water setting. The abundance of *Cyathidites* spp. suggests influence from marginal vegetation. The absence of mature marsh types suggests this marginal vegetation is situated near the seaward barrier.

In such situations it is envisaged that adjacent vegetation is limited in development thus explaining the predominance of very buoyant, possibly wind blown forms such as *Perinopollenites elatoides* and small *Cyathidites* spp. The high proportion of SOM associated with the palynofacies indicates stagnant bottom conditions.

## *Zone 4; Type Well 3/2–2*

This zone is usually associated with the Enrick Member and Etive Formation. The zone is weakly distinguished by a downhole change in palynofacies types below zone 5. It is frequently marked in its uppermost part by suggestions of oxidation and possibly subaerial exposure indicated by the occurrence of palynofacies types $T3^w$ or $T3^l$ together with the reappearance of regularly occurring *Chasmatosporites magnolioides* and *Quadraeculina anellaeformis*. Towards the lower part of the zone, the latter two sporomorph types can become abundant, in association with an influx of *Concavissimisporites* spp. and *Ischyosporites variegatus*. *Lycopodiacidites rugulatus* may also occur rarely. *Nannoceratopsis* spp. and tasmantids may also characterize the lower part of this zone.

In the southern part of the study area, *Corollina* spp. and *Densoisporites microrugulatus* are usually absent or infrequent.

Palynofacies type B5 may characterize the lowermost part of this zone. Since this palynofacies type is frequently associated with abundant *Perinopollenites elatoides* it is believed to represent a period of flooding as lagoonal areas are established.

The palynofacies types occurring between the basal B5 event of zone 5a and the lower B5 event of zone 4 represent a regressive-style sequence which include types B4 followed by BT3 and T2. Associated with this possible regression are minor influxes of *Concavissimisporites* spp. and *Ischyosporites variegatus* together with rare *L. rugulatus* which suggest a local marginal vegetation development. A fully mature marsh, however, does not appear to develop since *Corollina*, *Densoisporites* and *Duplexisporites* spp. are not usually conspicuous.

Zone 4 is weakly developed in the northern part of the study area. The Enrick Member is also poorly developed and the regressive palynofacies sequence absent. Abundant *Chasmatosporites magnolioides* and *Quadraeculina anellaeformis* do occur but usually below the base of the Ness Formation.

### *Zone 3; Type Well 3/2–2*

This zone is associated almost exclusively with the Rannoch and Lower Etive Formation. Palynofacies type MB2 is predominant containing only large blade and equidimensional palynomaceral 4 and darkened palynomaceral 1. Palynomorphs are virtually absent and, if present, are usually limited to single occurrences which include *Perinopollenites elatoides*, *Cyathidites* spp. and bisaccate pollen. The occurrence of this palynofacies type B2 in association with the Rannoch Formation lends support to a high energy, well oxygenated shoreface setting.

### *Zone 2; Type Well 3/2–2*

This zone is associated with the lowermost Rannoch and Broom formations together with, in some cases, the uppermost Dunlin Group. It is defined by the highest occurrence of *Nannoceratopsis* aff. *spiculata* and *Nannoceratopsis* cf. *spiculata*, together with commonly occurring *Nannoceratopsis gracilis*. Other diagnostic features can include the abundant occurrence of *Botryococcus* and poorly differentiated palynomorphs with possible affinity to small ?leiospheres. These types are difficult to distinguish from degraded specimens of *Perinopollenites elatoides*. Sporomorphs are not usually diagnostic for the zone and in the north of the study area may include commonly occurring *Chasmatosporites magnolioides*, *Perinopollenites elatoides*, bisaccate pollen, *Cerebropollenites mesozicus*, *Corollina* spp. and *Lycopodiumsporites austroclavatidites*. The occurrence of these two latter two forms suggest an affinity with zones 4 and 5b. If these distributions are substantiated, the relationship of the lower boundary of zones 4 and 5 with zones 2 and 3 will need to be reviewed.

Palynofacies types B3 and B4 are predominant, which together with the above palynomorphs indicate a brackish water nearshore setting. The areas in which high frequencies of *Cerebropollenites mesozoicus* are recorded suggest more direct terrestrial influence from a fluvial source.

### *Zone 1; Type Well 3/2–2*

This zone is associated with the upper part of the Drake Formation in the Dunlin Group. It is distinguished by the highest occurrence of *Scriniocassis weberi*, *Fromea elongata* and Dinocyst type 3. Other dinocyst types may include *Nannoceratopsis* spp., *Mancodinium semitabulatum*, *Parvocysta nasuta* and *P. barbata*. This association indicates a late Toarcian–earliest Bajocian age (Shaw 1983).

Sporomorphs are dominated by small bisaccate pollen and *Perinopollenites elatoides* together with small varieties of *Osmundacidites* spp., *Duplexisporites problematicus*, *Stereisporites* spp., and *Cyathidites* spp. The occurrence of reworked Permo–Triassic striate bisaccates is also a feature of this zone.

Palynofacies type M4$^{t}$ predominates, which together with the above palynomorphs, clearly indicates an offshore ventilated marine shelf palaeoenvironment.

In the wells 211/24–2 and 211/28–1A a lower subzone is suggested by the occurrence of palynofacies type M5 together with common *N. senex* and sphaeromorph clusters. This is indicative of the more anoxic sea bottom conditions associated with the early Toarcian period in the lower part of the Drake Formation.

## Pseudo-time events

The zonal scheme described was established in wells 3/2–2 and 3/2–4. It has been subsequently applied to the 44 wells of this study.

A major problem attached to the zonation scheme, however, is that zones 2–8 are probably facies controlled and have only local chronostratigraphical significance. Many of the criteria for this zonation rely primarily on palynomorph influxes, supported by associated palynofacies characteristics, which are summarized in Fig. 6. However since these zones parallel certain lithological units it is strongly suspected that they are to some degree facies controlled. They are thus diachronous and to some extent they may purely reflect the changing depositional facies of an advancing delta top. As a consequence, the Foyers, Oich and Enrick members will tend to co-occur with particular parts of of palynozone 7, 6, 5 and 4. Similarly the Etive and Rannoch formations will also co-occur with palynozones 2 and 3. Thus in order to establish a regional chronostratigraphic framework for this study, a zonation scheme is required that has correlative elements which demonstrably cut across litho- and palyno-facies boundaries. Several palynomorphs display events within their distribution which appear to fulfil these requirements. The most significant of these 'events' are listed in Fig. 10 and are discussed below.

## *Event CH-Q*

The highest occurrences of *Chasmatosporites magnolioides* and *Quadraeculina anellaeformis* are generally recorded in the Foyers Member of the Ness Formation and occasionally within the Heather Formation. The uppermost part of their range is limited to single occurrences and, because of their close association with coastal facies, this event is likely to be controlled by facies. The basal part of the acme of both *Quadraeculina anellaeformis* and *Chasmatosporites magnolioides*, referred to as events Q and CH respectively, tends to generally occur within the zones 4 or 2. There is however a shift of emphasis (see Fig. 10) from the middle–lower part of the Ness Formation (i.e. zone 4) in the south (e.g. wells 3/3–2, 3/2–3) to a stratigraphic horizon below the Ness Formation (i.e. zone 2) in the north (well 211/22–2, 211/19–1 and 211/19–6). This feature is considered to be chronostratigraphically significant and is referred to to Event CH-Q.

## *Event L*

The base occurrence of commonly occurring *Lycopodiumsporites* spp. (Event L), also shows the same trend occurring at stratigraphically lower horizons from south to north. In certain wells to the south their base common occurrence is limited to the higher part of the Ness Formation and in the north within the Rannoch Formation. The base common occurrence in some wells, however, is difficult to specify. Within the Ness formation, for example, *Lycopodiumsporites* spp. is frequently abundant in zone 7 in the uppermost part and then frequent again for a short interval within the lower part of the Ness Formation. The lower influx is always likely since the Enrick Member is, in part, a

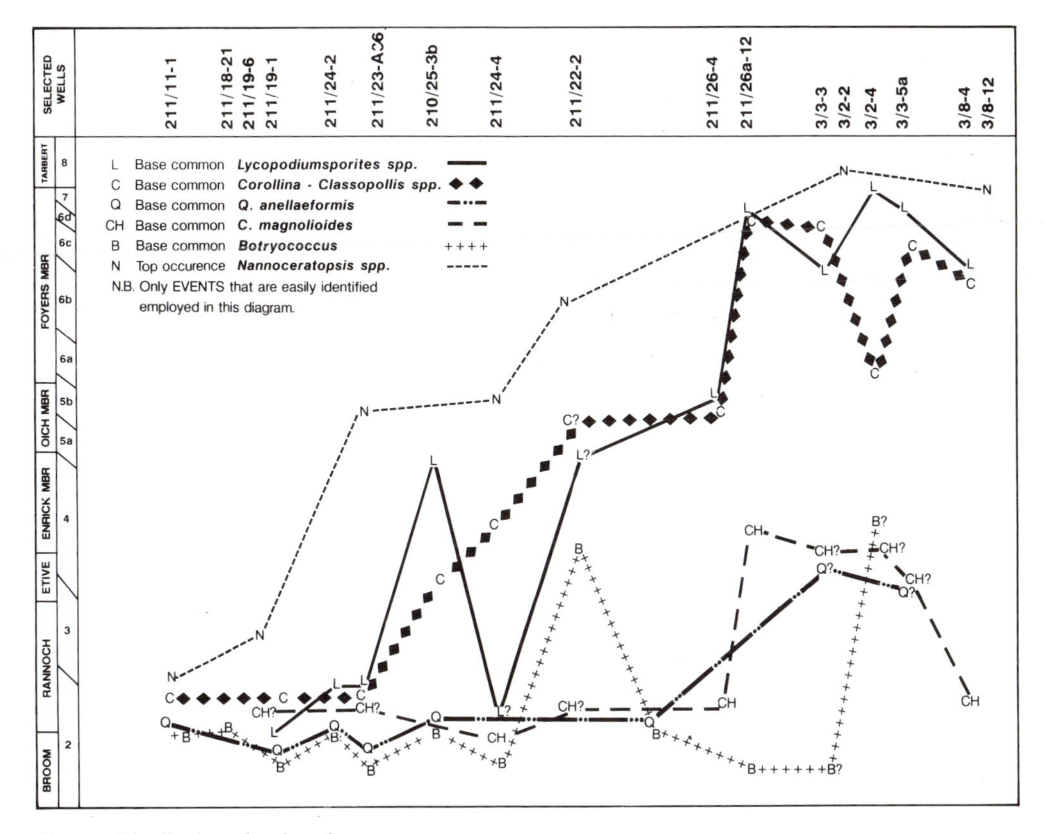


**Fig. 10.** Distribution of main palyno-events.

regressive sequence and thus the base common occurrence has to be judged around this potentially intruding influx.

Despite the above limitation there are several wells in which Event L is reasonably defined. Within the Foyers Member these include wells 3/2–4 (zone 7), 3/3–5 (zone 6d) and 211/26a–12 (zone 6d), the Oich Member of 210/25–3b and within the Rannoch Formation 211/23–A36, 211/24–2 and 211/28–1A.

### *Event C*

Treating *Corollina* and *Classopollis* spp. as a single group, the base of their common occurrence is referred to as Event C. Precisely defining the base of this group in all wells however can be subject to the same difficulties as experienced for Event L. Event C is most clearly identified within the Foyers Member to the south in wells 211/26a–12 and 3/3–5a (zone 6d), within the Oich Member further north in well 211/22–2, the Enrick Member in wells 210/25–3b, 211/26–4 and 211/28–1 and even further north in the Rannoch Formation of wells 211/11–1, 211/24–2, 211/19–1 and 211/23–A36.

### *Event B*

*Botryococcus* shows a distinct tendency to be limited in their common occurrence to the south in the Oich Member in wells 3/4–1, 211/29–3, 211/24–4, 211/29–A16, 210/25–3b, 211/28–1 and 211/23–A36 and to the Enrick Member in wells 3/2–4, 3/3–5a, 211/22–1 and 210/25–5. To the north, however, they are more abundant in the Rannoch Formation as displayed in wells 211/29–1, 211/24–2, 211/11–1, 211/23–A32, 211/18–12 and 211/19–6.

### *Event N*

The highest occurrence of *Nannoceratopsis* spp. Event N) is an obvious example of facies control. Its top common occurrence frequently coincides with the lagoonal character of the Oich Member. This coincidence casts doubt on this boundary having regional chronostratigraphical significance since its upper limit is likely to occur at progressively younger horizons as the delta lagoon diachronously migrated northward with time. Single occurrences are recorded within the upper part of the Foyers Member to the south in wells 3/2–2, 3/8–12 and 211/22–2. If not reworked, these may tentatively approximate to its regional extinction level of latest Bajocian age (Prauss 1989) and thus be time correlative with its uppermost occurrences within the Oich and Enrick Members in wells 211/28–1A, 211/23–A36, 211/24–4, 211/27–11, 211/27–6 and 211/26–12 further north, and within and below the Etive and Rannoch Formations in wells 211/19–6 and 211/11–7. This correlation is attractive since it suggests parts of the Rannoch, Etive and Ness Formation are coeval representing a possible stage of delta progradation.

Within the Brent Group, another dinocyst species, referred to as Dinocyst type 1, is recorded in stratigraphically younger intervals and does not co-occur with *Nannoceratopsis* spp. To the south it is recorded only within the Tarbert Formation, whereas to the north it is found sporadically within the Oich Member (e.g. wells 211/22–1 and 211/12–1). The base occurrence of Dinocyst type 1 may represent an inception event occurring after the extinction of *Nannoceratopsis* spp. This event, however, is likely to be facies controlled and to use these and other sporadic dinocyst occurrences for detailed correlation within the Brent Group seems problematical, although the general trend may be helpful. The likely stratigraphical relationships of *Nannoceratopsis* spp. and Dinocyst type 1 are schematically represented in Fig. 11.

The erratic distribution of common occurrences used to define events, particularly C and L, could in some wells be the result of several factors including local facies control or preparation of sample material. Such potential for discrepancy casts doubt on these events ever being sufficiently accurate for detailed correlation work. The general trend illustrated by particular wells in which the events are easily identified is probably of use. It should be noted that events within the Dunlin Group are not considered relevant since these sediments were not coeval with the Brent Group.

A mechanism that may explain their chronostratigraphical significance includes the way in which a delta progrades with the progressive development and increasing area of mature marsh to the rear. For example, in the early stages of the Brent delta development, marsh areas would be limited and the influence of the associated sporomorph assemblages would be correspondingly small. At a later stage, as the delta advanced, the area of marsh may have increased significantly. The most common types would be flushed out into the lagoon and delta front areas in significant numbers. This could explain the earlier events CH, Q and the distribution of common *Botryococcus* spp. The later event L and to some extent C may represent progressive expansion of more mature parts of

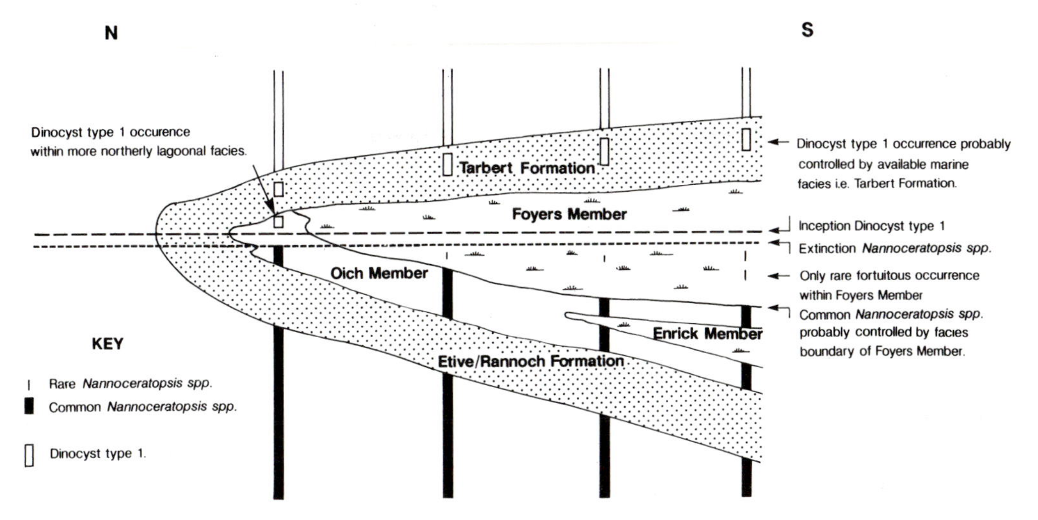


**Fig. 11.** Possible facies controls affecting dinocyst distribution.

the delta as the delta prograded to its maximum extent. The relatively sharp nature of these events may indicate the delta prograded rapidly in two or three pulses.

Assuming these events may correspond closely to time lines, their relationship with the established zones 2 to 7 must be examined. Figures 9a and 10 shows events L and C climbing with respect to other zonal criteria. The events cut across the *Concavissimisporites* spp. influx characterizing zone 6b and the highest common to abundant occurrences of *Alisporites grandis*, *Cerebropollenites mesozoicus*, *Nannoceratopsis* spp. *Araucariacites australis* and *Perinopollenites elatoides* characterizing top zones 5 and 6a. The reason for this divergence to the south of the area, particularly with respect to events CH and Q, is the nature of depositional facies. *Chasmatosporites magnolioides* and *Quadraeculina anellaeformis* are believed to be associated mainly with a brackish setting and thus may not occur frequently at the time of maximum fluvio-lacustrine development during zone 6b–6d times. For reasons explained earlier, the progressive appearance of more mature marsh forms is probably more time significant. If *Concavissimisporites* spp., for example, represents a 'subaqueous' marsh setting with the parent plant poorly represented in the slightly drier mature marsh areas, its occurrence will migrate northwards with the frontal area of the marsh. Consequently the upper limit of zone 6b will tend to be diachronous.

The upper limits of zones 2 and 3 are almost certainly facies controlled and strongly diachronous. The occurrence of events CH, C and L below the Etive Formation in the basal Rannoch Formation is considered time significant, as discussed earlier. It appears therefore, that the bases of zones 5 and 4 are, to the north, within the lower part of the Rannoch Formation.

The above events, where confidently identified, are plotted for each well in Fig. 9a. Figures 10 and 12a illustrate for selected wells their distribution in a typical N–S and W–E direction.

Notwithstanding the problems outlined above, there is still the possibility that events L, C, CH and Q may approximate, in some way, to time lines. The geological significance of these 'pseudo' time lines is discussed below.

## Stratigraphical relationships

Applying regional pseudo-time events to the main palynofacies and lithofacies shows their likely temporal relationships and allows palaeogeographical reconstruction of the delta. Patterns in the deposition of the Brent Group have been previously described by Budding & Inglin (1981). Johnson & Stewart (1985), Brown *et al.* (1987) and Graue *et al.* (1987). The latter publication provides palaeogeographical reconstructions of the Brent delta system as they occur in the Norwegian North Sea for the late Toarcian–Aalenian, Aalenian–Early Bajocian and early Bathonian intervals. The following discussion attempts similar reconstructions for the East Shetland Basin area. A schematic N–S cross section of the Brent Delta showing the likely stratigraphical relationships is illustrated in Figs 9 and 7. Figures 13–16 shows the configuration

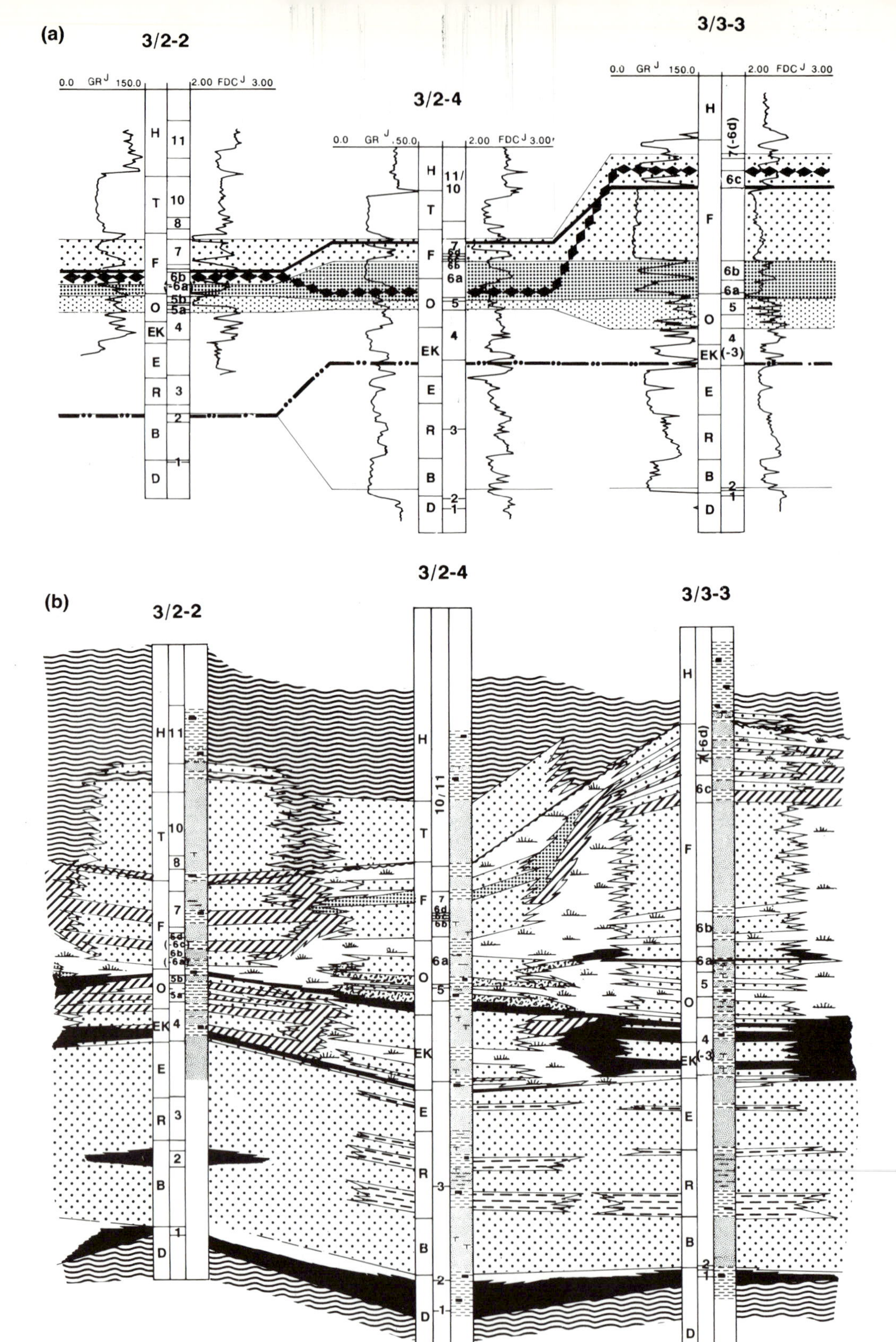


**Fig. 12.** (**a**) N–E section showing relationships of local palynozonation and pseudo-timelines. (**b**) W–E section showing environmental relationships. Key as for Fig. 9.

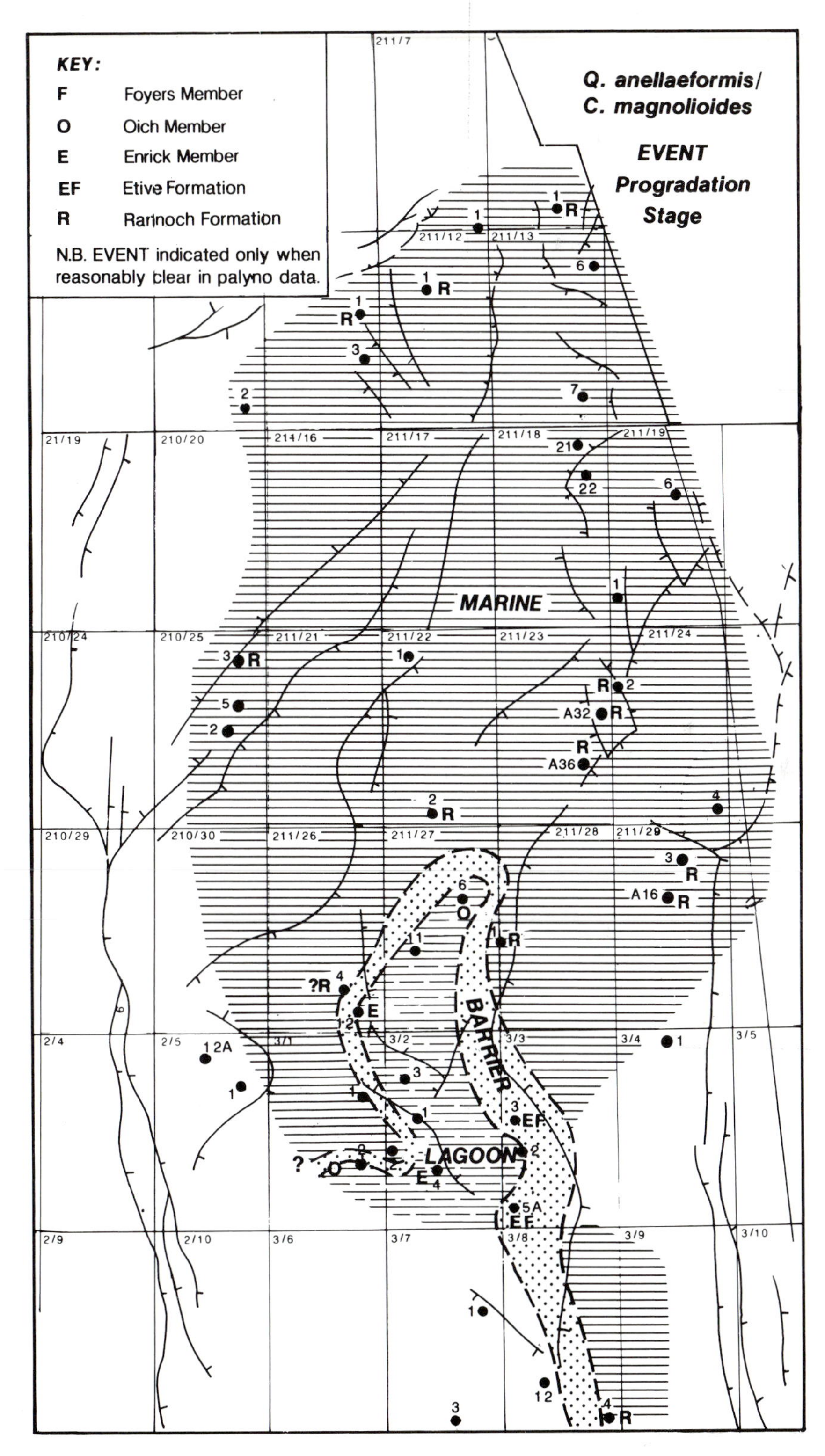


**Fig. 13.** Progradation of Brent Group during event CH times.

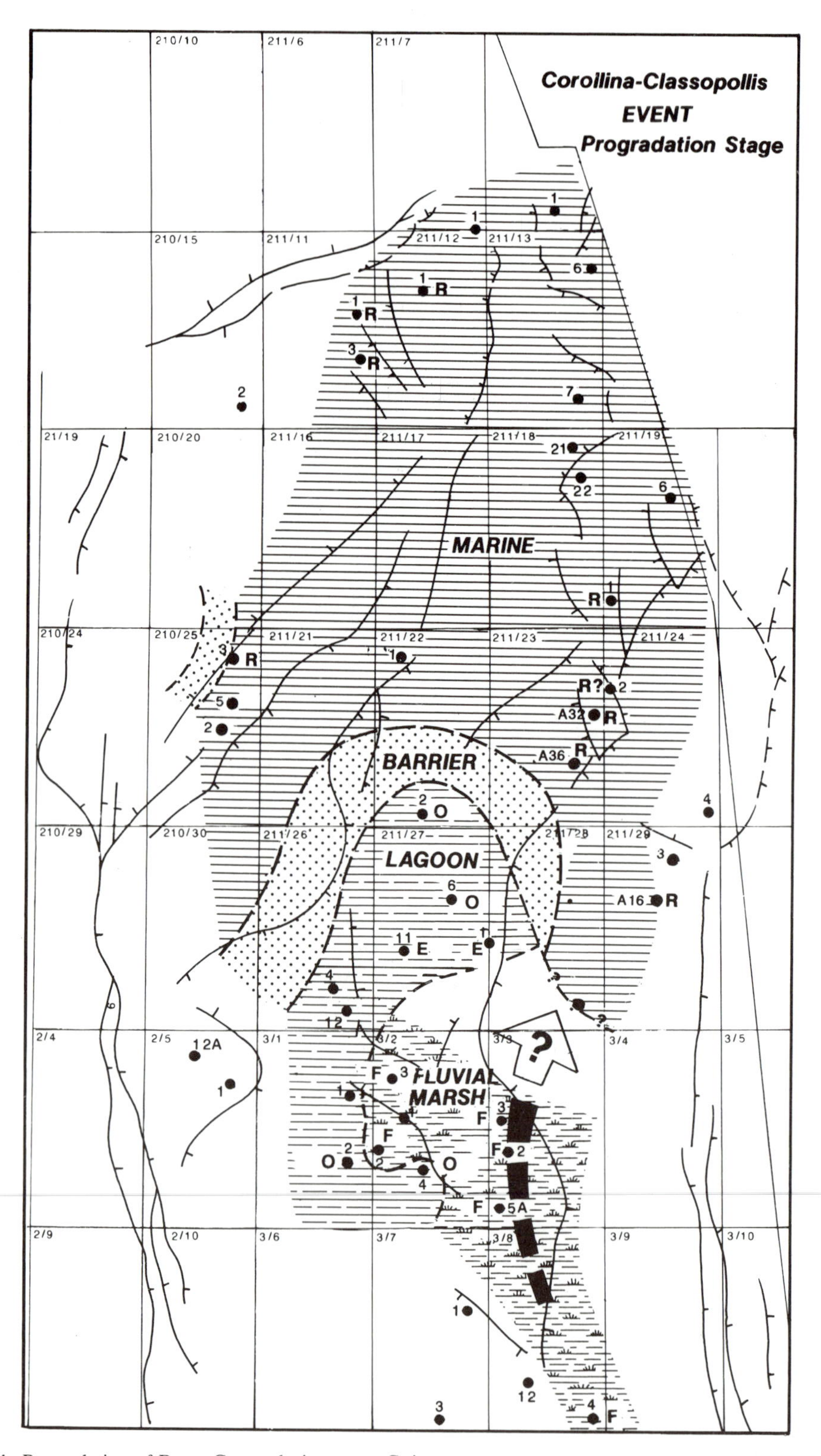


Fig. 14. Progradation of Brent Group during event C times.

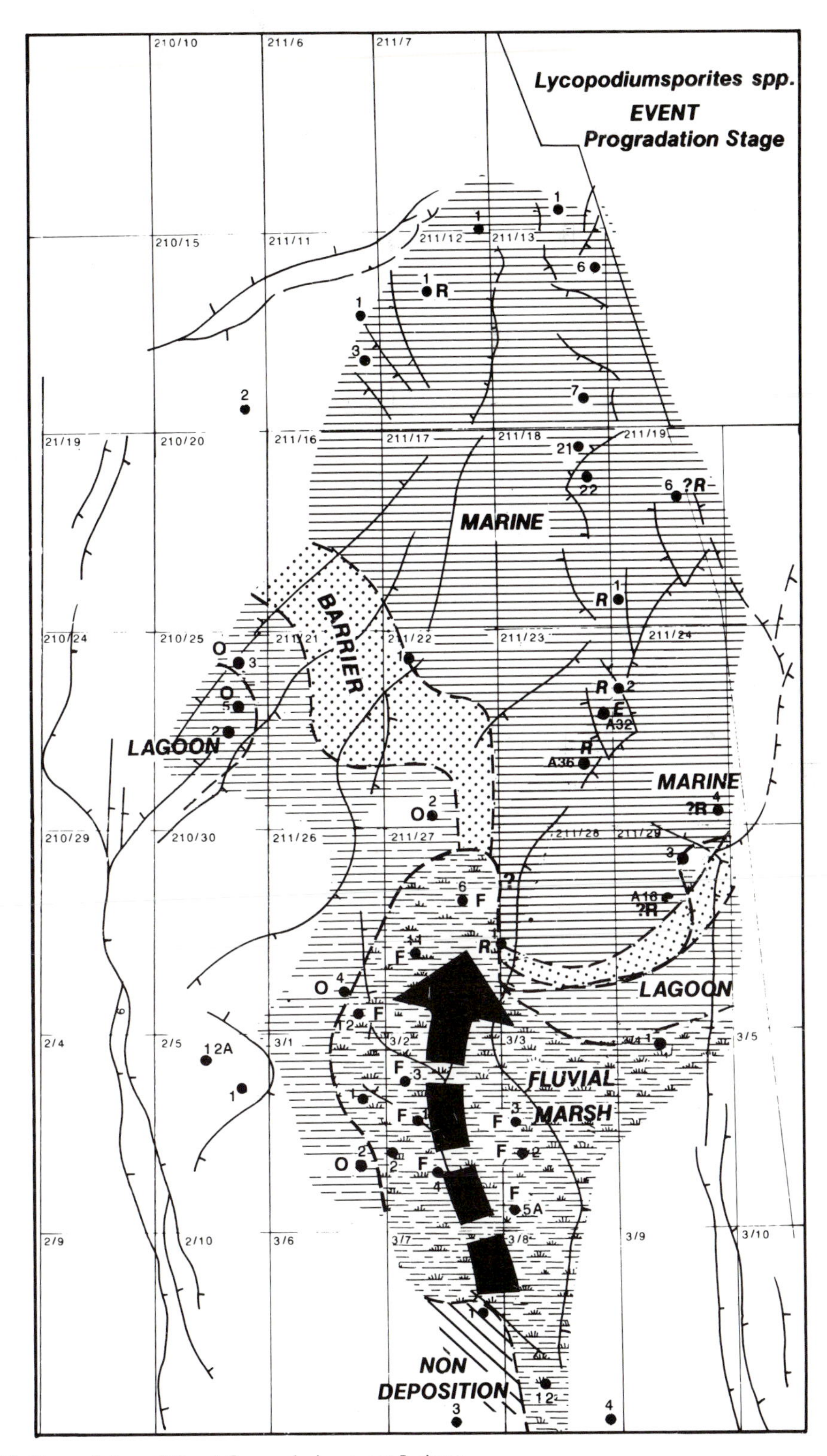


**Fig. 15.** Progradation of Brent Group during event L times.

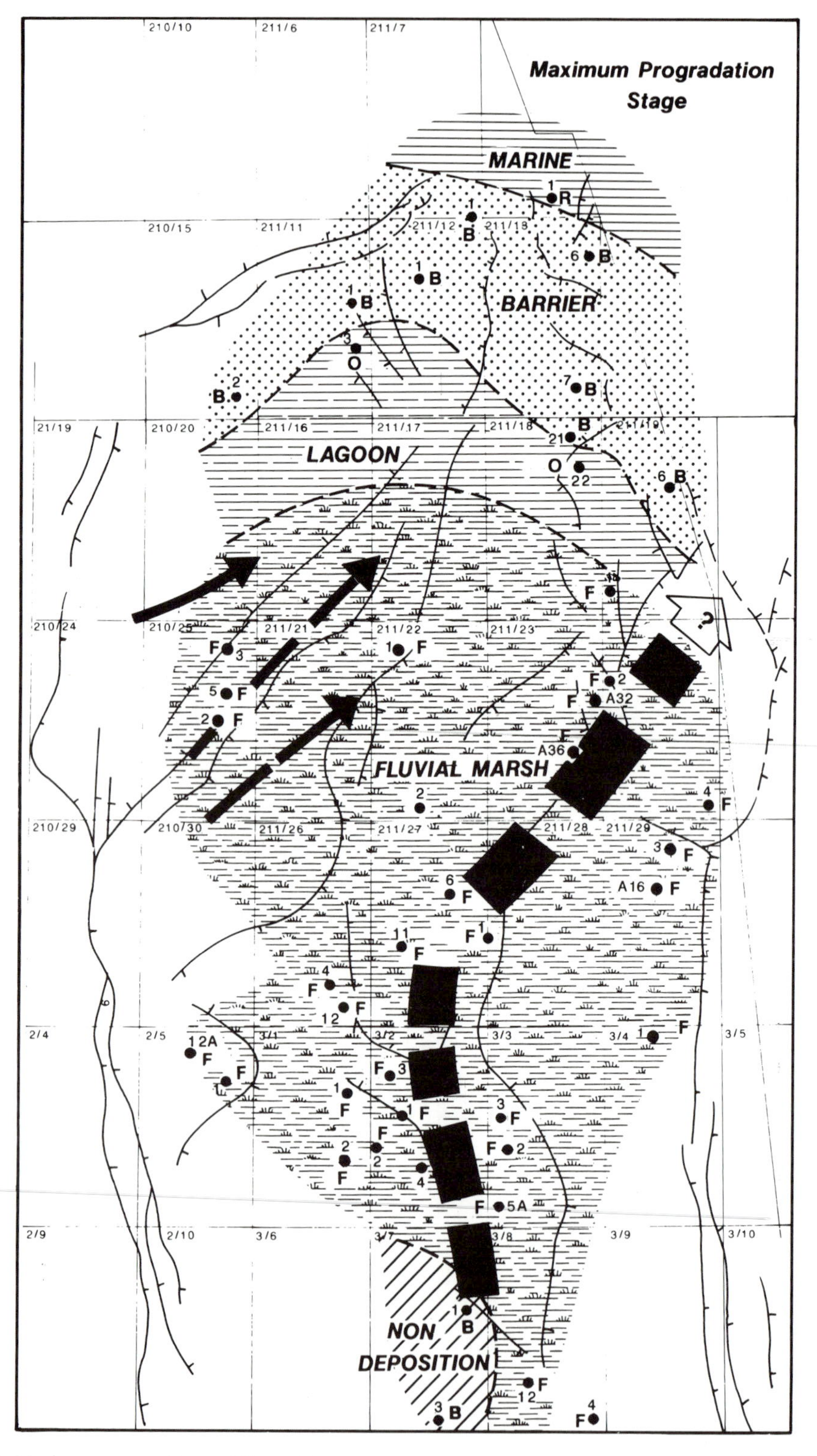


Fig. 16. Maximum progradation of Brent Group.

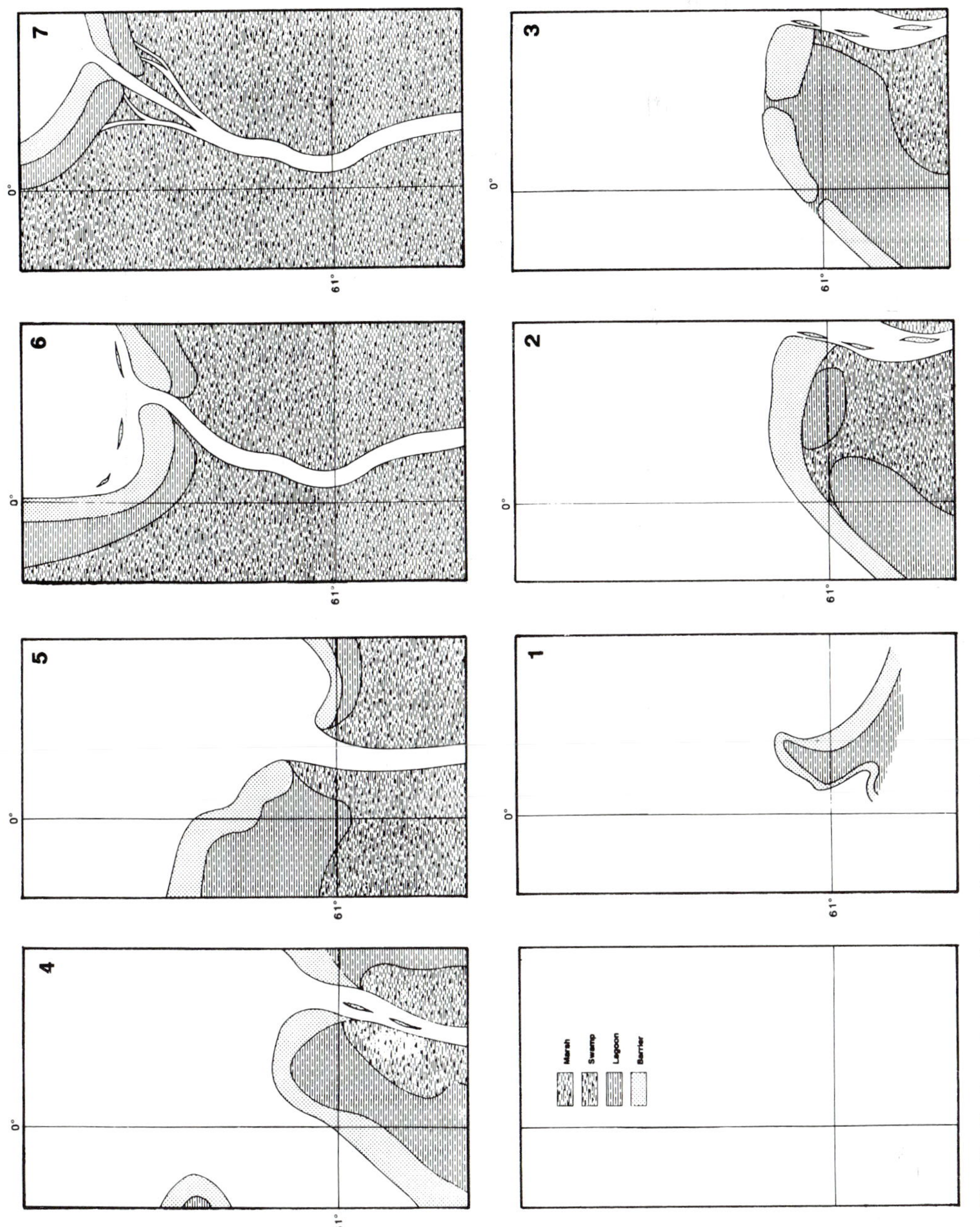


**Fig. 17.** Envisaged stages of delta development.

of the various deltaic facies at the time of events CH-Q, C, L and the stage of maximum progradation. A summary of the likely palaeogeographical evolution of the delta is presented in Fig. 17. These diagrams show several features of geological significance which are likely to affect appreciation of reservoir distribution.

Figures 13 and 17.1 illustrates an early progradation stage for the delta showing the probable coastal and lagoonal configuration at this time. Those reconstructions compare favourably with the early phase (Bajocian–Aalenian) outlined by Graue *et al*. (1987) which suggests the main thrust of the delta was to the east, a fluvial fairway following approximately the same direction of the UK/Norwegian boundary line. As can be seen from Figs 7, 9 and 17.2, in the East Shetland Basin this phase is subsequently followed by further progradation northwards during which immature marsh and swamps encroach into the area. This initial progradation is halted, possibly by a delta lobe switching, or a high marine stand, during which a significant part of the delta was engulfed and a large lagoonal area was formed (Fig. 17.3). This high stand represents a period during which the Oich Member was deposited in the southern part of the study area.

A second and more significant progradation of the delta northwards followed with the development of the Foyers Member. Two of the early stages in this progradation are represented by events C and L. Figures 14 and 17.4 shows the inferred distribution of barrier, lagoonal and mature marsh during event C time weakly indicating a possible fluvial fairway through blocks 3/8 and 3/3. In addition, in the area of the 211/28 block, the close proximity of shore and mature marsh may suggest that a mouth area to this fluvial development was situated close to this site.

During event L times the fluvial fairway shifted to the west and appears to be most conspicuous in the area of block 3/2 and 211/27 (Figs 15 and 17.5). The fluvial mouth lay close to the intersection of blocks 211/22, −27 and −28.

Figure 17.6 represents a minor period of delta retreat during the latter stages of progradation. This is suggested from sedimentological evidence from well 211/19−6 (see Cannon *et al*., this volume) based on the development of an intra-Ness Formation barrier sand.

The maximum stage of progradation is shown in Figs 16 and 17.7. This map shows the maximum occurrence of major sub-palaeoenvironments (i.e. delta front, barrier (Rannoch and Etive Formations), lagoon (Oich Member) and marsh (Foyers Member)). The development of the delta and final location of the coastline at this stage are compatible with Graue *et al*. (1987). The final northward, and then northeastwards, progradation of the delta appears to have merged with more easterly deltaic systems across the Horda Platform, to form a large deltaic plain at this time across this part of the North Sea area. The geographical distribution in the East Shetland Basin of selected palynofacies types at this time are illustrated in Figs 17 and 18. The palynofacies data for these maps were derived by totalling the number of samples characterized by specific palynofacies types (e.g. lagoonal palynofacies types include B3−5) and presenting them as a ratio of the full total of samples from within the Ness Formation in that well. The darker areas of shading indicate the highest concentrations and thus highlight boundaries of maximum areal distribution. Based on this information the northern limit of lagoonal settings appears to be between wells 211/11−3 and 211/18−22. Brackish influence appears to surround a central core area without this influence, represented by blocks 3/2, 3/3 and 211/27 as shown in Fig. 18. This area is characterized by 'marginal' palynofacies (i.e. higher energy or subaerially exposed areas) dominated by palynomaceral 4. These distributions appear to support the existence of the 'central core' of fluvio-marsh development characterizing blocks 211/27, 3/2−2, 3/3−3 and 3/8. This is further substantiated by the concentration of marginal palynofacies to the south of this area in block 3/2 and block 3/8 (Fig. 17).

Placed together these summary maps show support for a progradation emanating from the area of blocks 3/2 and 3/8 and a central core area of probable fluvial activity pushing northward through blocks 211/23 and 211/19 during event C times and then deflected westward toward block 211/27 in event L times. These trends correspond to sandstone prone areas as identified by Cannon *et al*. (this volume) and believed to represent the location of fluvial fairways. To reach the maximum stage of progradation, the delta front then moved rapidly to the northeast, with the Ness Formation reaching as far as block 211/19.

## Reservoir quality and distribution

The nature of the above distribution of palynofacies may have implications with regard to reservoir distribution and performance. The two lagoonal developments (types B4 and B5) of zone 4 and 5 are relatively persistent throughout the area which suggests that the associated finer

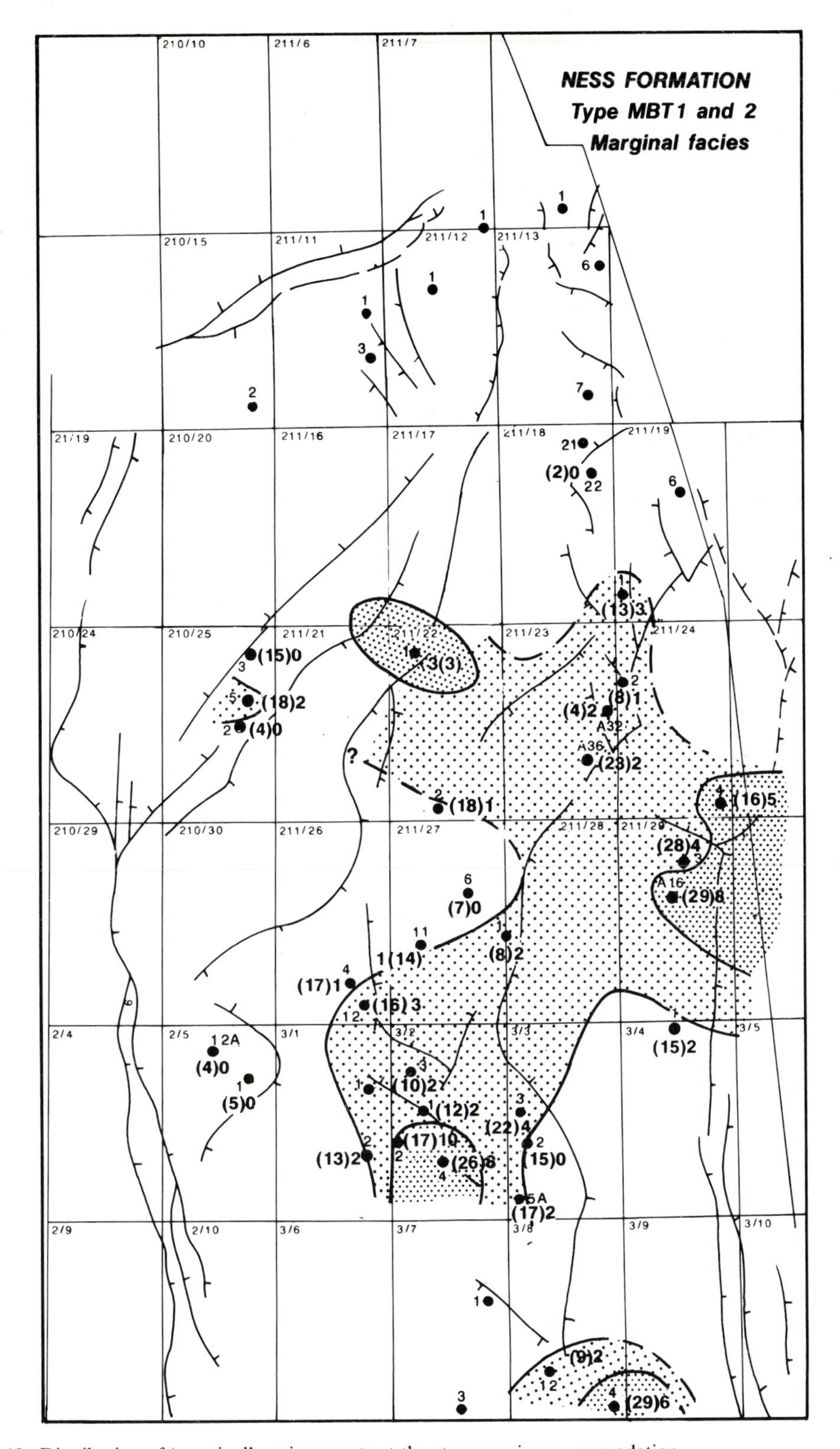


**Fig. 18.** Distribution of 'marginal' environments at the stage maximum progradation.

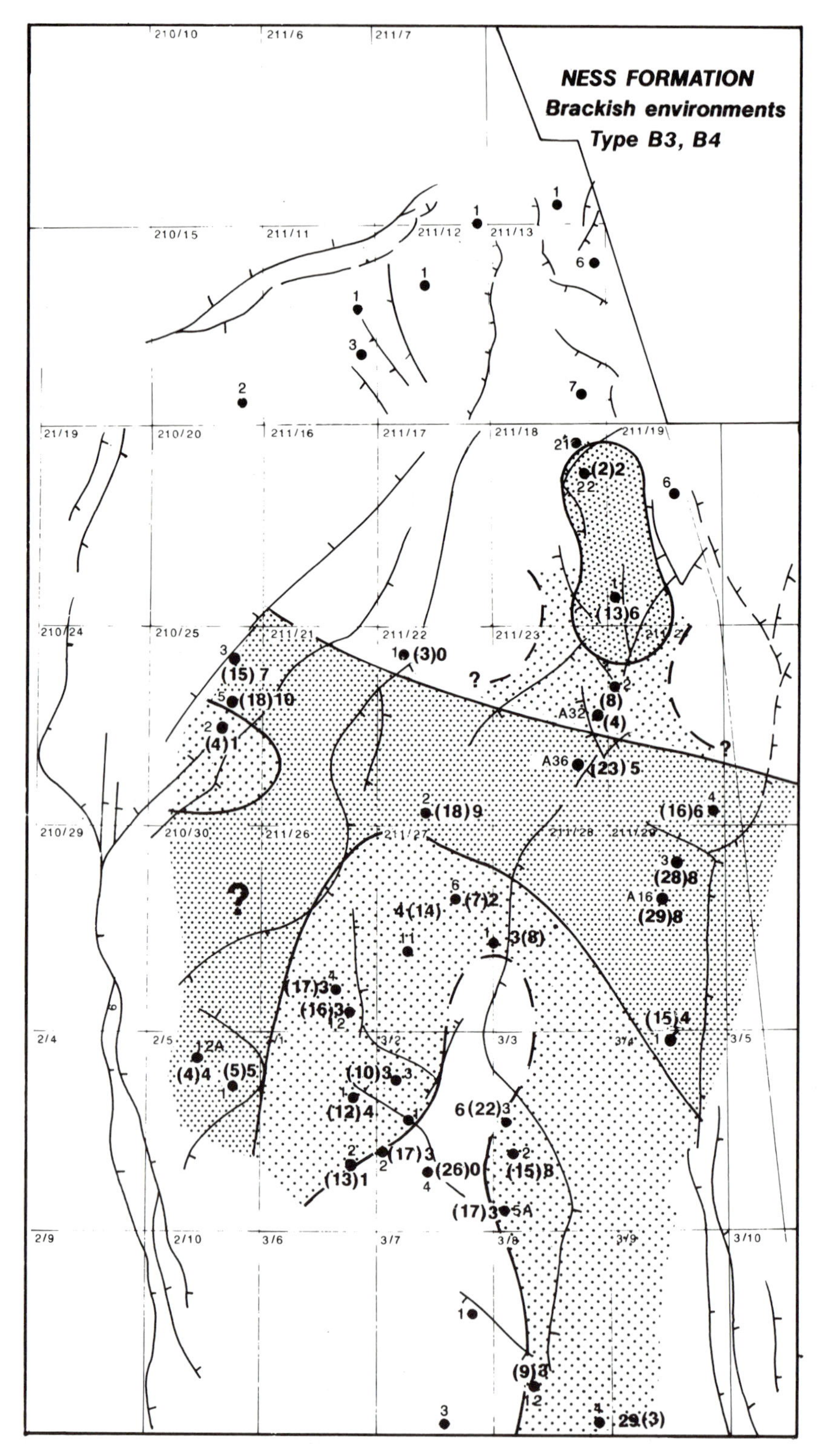


**Fig. 19.** Distribution of brackish environments at the stage of maximum progradation.

grained sediments may act as a barrier to liquid hydrocarbon migration (Figs 9b and 12b). The finer grained sediments of the more oxygenated lacustrine palaeoenvironments characterized by palynofacies type T4[t] may also be relatively persistent. They are likely to be less persistent than the lagoonal sediments but will, to some degree, act as a barrier to liquid hydrocarbon migration. In contrast, the finer grained sediments associated with palynofacies type T3 and T4 are likely to be only locally developed.

The distribution of palynofacies also has implication with regard to the distribution of reservoir sandstones. Fluvial influence, for example appears most notable in blocks 211/27, 3/2–2, 3/3–3 and 3/8 represented by the central core area where 'marginal' facies are most conspicuous. It thus seems logical that fluvial channel sand development is concentrated in these areas. In order to recognize this possibility the sandstones can be more prominently extended towards adjacent wells. Based on the distribution of 'marginal' facies in Fig. 16 and brackish environments in Fig. 19, it appears that the fluvial fairway bears a NNE–SSW trend and that changes in its location, for different stages of delta development, can also be tentatively suggested. This type of information may also be useful in modelling the likely sand concentrations in these and adjacent areas.

Mapping of the various facies in Figs 13–17 also highlights the possible configuration of the reservoir sands associated with the delta-front barrier for specific areas of the East Shetland Basin. For example, during event CH times, the barrier sands occur in the southernmost part of the area, and display a general N–S trend (Fig. 13). As they prograde northward their distribution becomes complex. In the final stages of progradation the barrier sands have a simple WNW and ESE trend.

In conclusion, it appears the combination of the pseudo-time events and palynofacies can provide a basis for the mapping of relevant reservoir facies and can provide valuable clues for reservoir modelling.

## Conclusions

A review of palynomorph and palynofacies distributions from 42 wells in the East Shetland Basin has provided a basis for a detailed biostratigraphical zonation of the Brent Group applicable within local field areas. In addition it has been possible to identify several pseudo-time events within selected sporomorph occurrences which appear to have chronostratigraphical significance over a wider regional area.

Palynofacies characteristics has allowed a better understanding of palaeoenvironmental relationships within the Brent Group. Of special note is the suggestion that the Enrick Member represents a separate but minor regressive phase of the delta top marsh area. This regressive phase is subsequently drowned by a major lagoonal development represented by the Oich Member. The Oich Member appears to be virtually subaqueous over the southern part of the study area. The ratio of higher energy: brackish palynofacies types provided maps showing linear areas of maximum terrestrial influence which may represent fairways of fluvial channel activity.

Application of the pseudo-time events has allowed a series of palaeogeographic reconstructions to be tentatively suggested. They indicate the initial direction of progradation was northwards centred around a fluvial fairway passing through block 3/3. This fairway later appears to have shifted westwards into the 3/2 and 211/27 areas. An additional source from the west influencing block 210/25 later combined with the original southerly source to provide an overall north easterly direction to delta progradation.

The authors wish to express their appreciation to Shell Exploration and Production and Esso Exploration and Production UK Limited for their permission to publish this paper. They also thank Carmel Coyne of the Geochem Group for her valuable work and contribution to this study.

## References

BATTEN, D. J. 1974 Wealden palaeoecology from the distribution of plant fossils. *Proceedings of the Geologists' Association*, **85**, 433–458.

BROWN, S., RICHARDS, P. C. & THOMSON, A. R. J. 1987. Patterns in the deposition of the Brent Group (Middle Jurassic) UK North Sea. *In*: BROOKS, J. & GLENNIE, K. (eds) *Petroleum Geology of North-West Europe*. Graham & Trotman, London, 899–913.

BRYANT, I. D., KANTOROWICZ, J. D. & LOVE, C. F. 1988. The origin and recognition of laterally continuous carbonate-cement horizons in the Upper Lias Sands of southern England. *Marine and Petroleum Geology*, **5**, 108–133.

BUDDING, M. C. & INGLIN, H. F. 1981. A reservoir geological model of the Brent sands in southern Cormorant. *In*: ILLING, L. V. & HOBSON, G. D. (eds). *Petroleum Geology of the Continental Shelf of North-West Europe*. Heyden, London, 326–334.

CALLOMON, J. H. 1975. Jurassic ammonites from the northern North Sea. *Norsk Geologisk Tidsskrift*, **55**, 373–386.

CHALONER, W. G. & MUIR, X. X. 1968. Spores and floras. *In*: MURCHISON, D. & WESTOLL, T. S.

(eds) *Coal and Coal-Bearing Strata*. Oliver & Boyd, Edinburgh, 127–146.

CROSS, A. T., THOMPSON, G. G., & ZAITZEFF, J. B. 1966. Source and distribution of palynomorphs in bottom sediments, southern part of Gulf of California. *Marine Geology*, **4**, 467–524.

DEEGAN, C. E. & SCULL, B. J. 1977. *A proposed standard lithostratigraphic nomenclature for the Central and Northern North Sea*. Report of the Institute Geological Sciences No. **77/25**.

DENISON, C. & FOWLER, R. M. 1980. Palynological identification of facies in a deltaic environment. *Proceedings of the meeting on the Sedimentation of North Sea Reservoir rocks*. Norwegian Petroleum Society, Oslo, Paper XII, 1–22.

CANNON, S. J. C., GILES, M. R., WHITAKER, M. F., PLEASE, P. A. & MARTIN, S. V. 1992. A regional reassessment of the Brent Group, UK Sector, North Sea. *This Volume*.

FAEGRI, K. & IVERSEN, J. 1964. *Textbook of pollen analysis*. Copenhagen, Getijtafels voor Nederland. Yearly. Staatsuitgeverij, 's-Gravenhage.

FISHER, M. J. 1980. Kerogen distribution and depositional environments in the Middle Jurassic of Yorkshire U.K. *In*: *Proceedings of the International Palynological Conference, Lucknow 1976–1977*, **2**, 574–580.

FISHER, M. J. & HANCOCK, N. J. 1985. The Scalby Formation (Middle Jurassic, Ravenscar Group) of Yorkshire: reassessment of age and depositional environment. *Proceedings of the Yorkshire Geological Society*, **45**, 293–298.

GRAUE, E., HELLAND-HANSEN, W., JOHNSEN, J., LØMO, L., NØTTVEDT, A., RØNNING, K., RYSETH, A. & STEEL, R. 1987. Advance and retreat of Brent Delta System, Norwegian North Sea. *In*: BROOKS, J. & GLENNIE, K. (eds) *Petroleum Geology of North-West Europe*. Graham & Trotman, 915–937.

HAABIB, D. 1982. Sedimentary supply origin of Cretaceous black shales. *In*: SCHLANGER, S.O. & CITA, M. B. (eds) *Nature and Origin of Cretaceous Carbon-Rich Facies*. Academic Press, London, 113–127.

HANCOCK, N. J. & FISHER, M. J. 1981. Middle Jurassic North Sea deltas with particular reference to Yorkshire. *In*: ILLING, L. V. & HOBSON, G. D. (eds) *Petroleum Geology of the Continental Shelf of North-West Europe*. Heyden & Son Limited, London, 196–195.

HUGHES, N. F. & MOODY-STUART, J. C. 1967. Palynological facies and correlation in the English Wealden. *Review of Palaeobotany and Palynology*, **1**, 259–268.

JOHNSON, H. D. & STEWART, D. J. 1985. Role of clastic sedimentology in the exploration and production of oil and gas in the North Sea. *In*: BRENCHLEY, P. J. & WILLIAMS, B. P. J. (eds) *Sedimentology: Recent Developments and Applied Aspects*. Geological Society, London, Special Publication, **18**, 249–310.

MULLER, J. 1959. Palynology of recent Orinoco delta and shelf sediments: reports of the Orinoco Shelf expedition; volume 5. *Micropalaeontology*, **5**, 1–32.

NEVES, R. 1961. Namurian plant spores from the southern Pennines, England. *Palaeontology*, **4**, 246–279.

PARRY, C. C., WHITLEY, P. K. J. & SIMPSON, R. D. H. 1981. Integration of palynological and sedimentological methods of facies analysis of the Brent Formation. *In*: ILLING, L. V. & HOBSON, G. D. (eds) *Petroleum Geology of the Continental Shelf of Northwest Europe*. Heyden, London, 205–215.

PRAUSS, M. 1989. Dinocyst–stratigraphy and palynofacies in the Upper Liassic and Dogger of NW Germany. *Palaeontographical Abt. B*, **214**, Lfg. 1–4, 1–124.

SHAW, D. 1983. *A palynological investigation of some Toarcian and lowest Aalenian sediments of the Isles of Skye and Raasay*. MSc manuscript, University of Sheffield.

TYSON, R. V. 1987. The genesis and palynofacies characteristics of marine petroleum rocks. *In*: BROOKS, J. & FLEET, A. J. (eds) *Marine Petroleum Source Rocks*. Geological Society, London, Special Publication, **26**, 47–67.

—— 1989. Late Jurassic palynofacies trends, Piper and Kimmeridge Clay Formations, UK onshore and offshore. *In*: BATTEN, D. J. & KEEN (eds) *Northwest European Micropalaeontology and Palynology*. British Micropalaeontological Society Series, Ellis Horwood, Chichester, 135–172.

—— 1990. Palynofacies analysis. *In*: JENKINS, D. G. (ed.) *Applied Micropalaeontology*, Graham & Trotman 1990.

——, WILSON, R. C. L. & DOWNIE, C. 1979. A stratified water column environmental model for the type Kimmeridge Clay. *Nature*, **277**, 377–380.

WHITAKER, M. F. 1984. The usage of palynology in definition of Troll Field geology. *6th Offshore Northern Seas Conference and Exhibition, Stavanger 1984*, Paper G6.

VAN DER ZWAN, C. J. 1989. Palynostratigraphical principles as applied in the Jurassic of the Troll and Draugen fields, offshore Norway. *In*: COLLINSON, J. D. (ed.) *Correlation in Hydrocarbon Explation*. Graham & Trotman, London, 357–365.

# Palynology as a palaeoenvironmental indicator in the Brent Group, northern North Sea

GWYDION WILLIAMS

*GeoStrat Ltd, Motherwell Business Centre, Dalziel Street, Motherwell ML1 1PJ, UK*

**Abstract:** The Brent Group of the northern North Sea can be subdivided into a series of 12 assemblage units on the basis of palynomorph and kerogen abundances. These assemblages, which can be identified on a regional basis, have been recognized extensively throughout wells of the Brent Group depositional area. They represent changes in the environment, kerogen source and depositional setting and can be linked to lithostratigraphical units. Changes in assemblages through the Brent Group reflect an initial restricted marine environment followed by progradation in a regressive regime. Progradation ceased and the maximum point of regression occurred within the 'Mid' Ness Formation, after which deposition occurred within an abandonment and transgressive regime. These features are defined by palynological data, thus providing input into the regional understanding of Brent Group deposition. The palynological assemblages of the Brent Group are illustrated with reference to two wells: 211/18a-A31 from the Thistle Field and 3/3-5A from the Ninian Field.

Palynomorph assemblages within the Brent Group are largely facies controlled and have an intimate relationship with lithofacies. The identification of these assemblages, therefore, allows for the positive identification of individual lithostratigraphical and reservoir units, particularly in problematic sections where stratigraphical interpretations based on other disciplines are hampered by faulting and truncation or lack of conventional core material and wireline log data, particularly where sidewall core samples are available.

Two cored sections from wells 211/18a-A31 (Thistle Field) and 3/3-5A (Ninian Field) (Fig. 1) which combine to form a complete sequence of Brent Group sediments, have been analysed for palynofacies and palynology. Data from these wells are presented and a series of palynological and kerogen assemblages are described. The aim of subdividing the Brent Group into a series of palynological assemblages is to allow palaeoenvironmental determinations and the 'fingerprinting' and identification of lithostratigraphical and reservoir units on a regional basis. Such a framework of assemblages can then be used to clarify interpretations of problematic sections and as a basis for the comparison of sequences from different parts of the Brent Province. On a more local basis, within a block or field, further potential exists for subdivision of the sequence and the identification of specific environments within the regionally applicable scheme.

## Previous work

Numerous publications exist on the distribution of kerogen and palynomorphs in fluviodeltaic and shallow marine environments, for example, Muller (1959), Traverse & Ginsberg (1966), Fisher (1980), Hancock & Fisher (1981), Dennison & Fowler (1980), Parry *et al.* (1981) and Whitaker (1984). These have dealt with various aspects of the sourcing, transport and deposition of kerogen and palynomorphs in a number of modern and ancient examples. Muller (1959) provided one of the earliest and most thorough investigations of the palynofacies of sediments in a deltaic environment by analysis of recent and modern sediments of the Orinoco. This work provided significant data on the distribution of kerogen and palynomorphs in a deltaic and associated shallow marine environments, illustrating differences in sourcing and variations in transport mechanisms. Traverse & Ginsberg (1966), in a study of the Bahama Banks area, further elucidated the likely hydrodynamic properties of various palynomorph and kerogen types.

A number of other authors gave data on kerogen assemblages in deltaic environments within or associated with the North Sea area. Palynofacies interpretations for the Middle Jurassic sediments of Yorkshire were produced by Fisher (1980) and Hancock & Fisher (1981). In the latter paper, the authors made comparison between the Yorkshire deltaic sediments and other similarly aged deltaic sequences

*From* Morton, A. C., Haszeldine, R. S., Giles, M. R. & Brown, S. (eds), 1992, *Geology of the Brent Group*, Geological Society Special Publication No. 61, pp. 203–212.

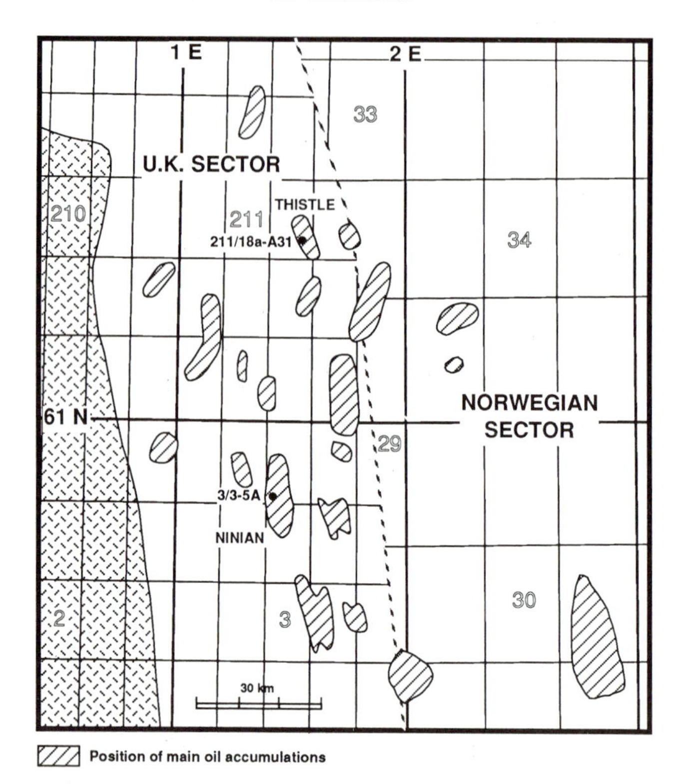


**Fig. 1.** Location map of studied wells indicating oil and gas fields within the Brent Province.

specifically the Brent Group. Both these works identified a number of different fluviodeltaic environments including; channels, sheet sands, interdistributary bays, brackish lagoons and restricted swamps with coals based on a combination of sedimentological and palynofacies analyses. These environments and palynofacies associations can also be identified within the Brent Group sequence. However, the authors suggested caution against too close a comparison between the two areas, as the overall sequences of deposition differ quite markedly.

Whitaker (1984) discussed the usage of palynofacies in Late and Middle Jurassic sediments of the Troll Field, Norwegian North Sea which are in part age equivalent to the sediments discussed in this paper. Parry *et al.* (1981) presented palynofacies data and interpretations from the Brent Group of wells 211/19-4 (Murchison Field) and 211/24-4 (Statfjord Field). Palynological data was used in conjunction with sedimentological interpretations to identify a sequence of environments through the two well sections.

The authors used the relative abundances of kerogen types and palynomorphs, state of preservation and degree of sorting to identify various environments of deposition. The methods of Parry *et al.* (1981) do not differ greatly from those of this paper. However, the approach of this paper is to constrain the interpretation of Brent Group palynofacies and palynomorph assemblages into a sequence of palaeoenvironmental units which is applicable throughout the majority of the Brent Group depositional area.

## Kerogen subdivision

The kerogen subdivision used in this study is as follows. Kerogen, which can be defined as the total organic residue recovered from a sample following hydrofluoric acid maceration, is subdivided into two major elements: humic and sapropelic kerogen (then into the constituent elements of these as indicated in Fig. 2). Humic kerogen and plant cuticle can be characterized by internal structure, preservation, size, shape and sorting to allow further subdivision. Similarly palynomorph groups can be subdivided on broad morphological lines such as heavy ornamented spores and unornamented smooth spores which may differ in either palaeoecological or transportation characteristics.

Kerogen assemblages are therefore described using a combination of categories including:

| KEROGEN | | |
|---|---|---|
| ENVIRONMENT | HUMIC KEROGEN | SAPROPELIC KEROGEN |
| TERRESTRIAL | BLACKWOOD<br>BROWNWOOD | PLANT CUTICLE<br>MEGASPORES<br>SPORES<br>POLLEN |
| FRESHWATER | | FRESH TO BRACKISH WATER ALGAE<br>e.g. BOTRYOCOCCUS sp. |
| BRACKISH - MARINE | | FRESH TO BRACKISH WATER ALGAE<br>e.g. BOTRYOCOCCUS sp.<br>LEIOSPHERES<br>TASMANITES spp. |
| MARINE | | ACRITARCHS<br>DINOCYSTS<br>MICROFORAMINIFERAL TEST LININGS<br>SCOLECODONTS |
| AMORPHOUS ORGANIC MATTER : DEGRADED HUMIC AND SAPROPELIC MATTER | | |

**Fig. 2.** Kerogen subdivisions as used in this paper.

size, structure, preservation, degree of sorting and relative abundance of the various palynomorph groups contained within them.

Amorphous organic matter is derived by the degradation of both humic and sapropelic material and the differentiation of the two types relies on the fluorescence of sapropelic matter.

## Sequence of palynological assemblages and palaeoenvironmental interpretations

The sequence of palynomorph and kerogen assemblages seen through the Brent Group and the palaeoenvironmental interpretations of these environments are illustrated in Fig. 3 and discussed below. In identifying and describing the following assemblages, a number of characteristics of the kerogen and palynomorph assemblages have been used. The broad kerogen subdivisions described above have been used to suggest major changes in depositional energy levels and therefore likely settings within the overall marginal marine to fluviodeltaic regime. Variations in abundances of marine, brackish water and terrestrially derived taxa allow the identification of relative salinity variations, thus defining the major phases of progradation and abandonment. Finally, specific palynomorph abundances probably resulting from major palaeoecological changes, and therefore potentially widespread, have been used to further subdivide the sequence particularly within the Ness Formation. Reference is made to particular samples and intervals in wells 3/3-5A and 211/8a-A31 (Figs 4 and 5) from which these assemblages have been described.

Prior to Brent Group deposition, the fully marine sediments of the Drake Formation, Dunlin Group were deposited. The Drake Formation yields an assemblage of diverse dinocysts and pollen, indicating fully marine conditions in a position distal to sediment source.

### *Assemblage 1: Broom Formation*

Type Sample: 3/3-5A, 10 550 ft (core) (Fig. 6a).

Palynomorphs of Assemblage 1 are dominated by the largely monospecific occurrence of the dinocyst *Nannoceratopsis gracilis* (angular varieties). The monospecific nature of this assemblage suggests a highly restricted marine environment. Low diversities of marine and brackish taxa have been ascribed variously to low salinity, hypersaline and physically restricted environments (Williams & Sarjeant 1967). The lack of significant woody kerogen apart from fine blackwood indicates deposition in a position away from sediment source and suggests a distal, high energy and highly oxidative environment.

The continuation of elements of Assemblage 1 into the basal argillaceous unit of the Rannoch Formation suggests that the Broom Formation is closely related, at least in palynofacies, to the progradational sequence of the Rannoch and Etive formations. The major assemblage change over the Drake Formation–Broom Formation boundary confirms that the latter for-

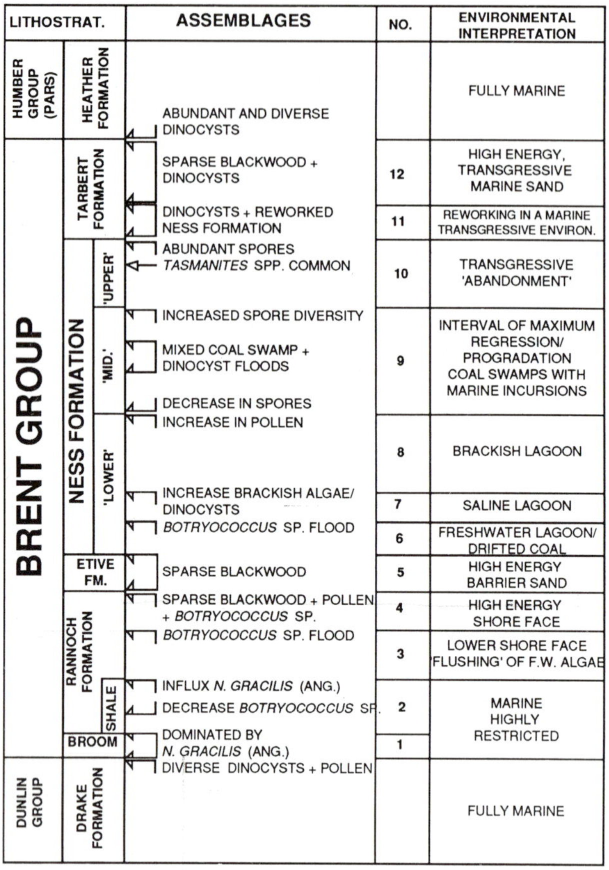

| LITHOSTRAT. | | ASSEMBLAGES | NO. | ENVIRONMENTAL INTERPRETATION |
|---|---|---|---|---|
| HUMBER GROUP (PARS) | HEATHER FORMATION | ABUNDANT AND DIVERSE DINOCYSTS | | FULLY MARINE |
| BRENT GROUP | TARBERT FORMATION | SPARSE BLACKWOOD + DINOCYSTS | 12 | HIGH ENERGY, TRANSGRESSIVE MARINE SAND |
| BRENT GROUP | TARBERT FORMATION | DINOCYSTS + REWORKED NESS FORMATION | 11 | REWORKING IN A MARINE TRANSGRESSIVE ENVIRON. |
| BRENT GROUP | NESS FORMATION 'UPPER' | ABUNDANT SPORES<br>*TASMANITES* SPP. COMMON | 10 | TRANSGRESSIVE 'ABANDONMENT' |
| BRENT GROUP | NESS FORMATION 'MID.' | INCREASED SPORE DIVERSITY<br>MIXED COAL SWAMP + DINOCYST FLOODS<br>DECREASE IN SPORES | 9 | INTERVAL OF MAXIMUM REGRESSION/ PROGRADATION COAL SWAMPS WITH MARINE INCURSIONS |
| BRENT GROUP | NESS FORMATION 'LOWER' | INCREASE IN POLLEN | 8 | BRACKISH LAGOON |
| BRENT GROUP | NESS FORMATION 'LOWER' | INCREASE BRACKISH ALGAE/ DINOCYSTS | 7 | SALINE LAGOON |
| BRENT GROUP | NESS FORMATION 'LOWER' | *BOTRYOCOCCUS* SP. FLOOD | 6 | FRESHWATER LAGOON/ DRIFTED COAL |
| BRENT GROUP | ETIVE FM. | SPARSE BLACKWOOD | 5 | HIGH ENERGY BARRIER SAND |
| BRENT GROUP | RANNOCH FORMATION | SPARSE BLACKWOOD + POLLEN + *BOTRYOCOCCUS* SP. | 4 | HIGH ENERGY SHORE FACE |
| BRENT GROUP | RANNOCH FORMATION | *BOTRYOCOCCUS* SP. FLOOD | 3 | LOWER SHORE FACE 'FLUSHING' OF F.W. ALGAE |
| BRENT GROUP | RANNOCH FORMATION SHALE | INFLUX *N. GRACILIS* (ANG.)<br>DECREASE *BOTRYOCOCCUS* SP. | 2 | MARINE HIGHLY RESTRICTED |
| BRENT GROUP | BROOM | DOMINATED BY *N. GRACILIS* (ANG.) | 1 | MARINE HIGHLY RESTRICTED |
| DUNLIN GROUP | DRAKE FORMATION | DIVERSE DINOCYSTS + POLLEN | | FULLY MARINE |

**Fig. 3.** Sequence of Brent Group palynological and kerogen assemblages.

mation is more closely associated with the Brent Group than the Dunlin Group. This formational boundary can also be seen as a major regressive sequence boundary between fully marine Dunlin Group which yields relatively diverse dinocyst assemblages and restricted Broom Formation sediments.

## *Assemblage 2: 'Rannoch Formation Shale'*

The 'Rannoch Formation Shale' is characterized by an assemblage comprising elements of both the underlying and overlying units. As discussed above, the common occurrence of *Nannoceratopsis gracilis* (angular varieties) extends up from the Broom Formation into the base of the Rannoch Formation.

The common to dominant occurrence (greater than 10% of the palynomorph assemblage) of the fresh to brackish water alga *Botryococcus* sp. is seen within the 'Rannoch Formation Shale' up into the upper part of the Rannoch Formation (Figs 4 and 5). The abundant occurrence of this fresh to brackish water alga within the lower Brent Group is interpreted as an allochthonous flooding event, resulting from the long distance transport of the low density *Botryococcus* sp. colonies from a fresh to brackish water environment. It is possible that this distribution reflects influxes of fresh to brackish water into a marine environment which also facilitated the long distance transport of those forms prior to settling out of suspension into lower shore face and pro-delta sediments. It is probable that this event is more facies related than isochronous and is therefore used to 'fingerprint' this lithostratigraphic unit rather than to provide an approximate time plane.

The co-occurrence of common *N. gracilis* (angular varieties) and common *Botryococcus* sp. which is seen within the 'Rannoch Formation Shale' is interpreted as a transition between 'pro-delta' claystone deposition and progradational sequence of the Rannoch Formation lower shore face sands.

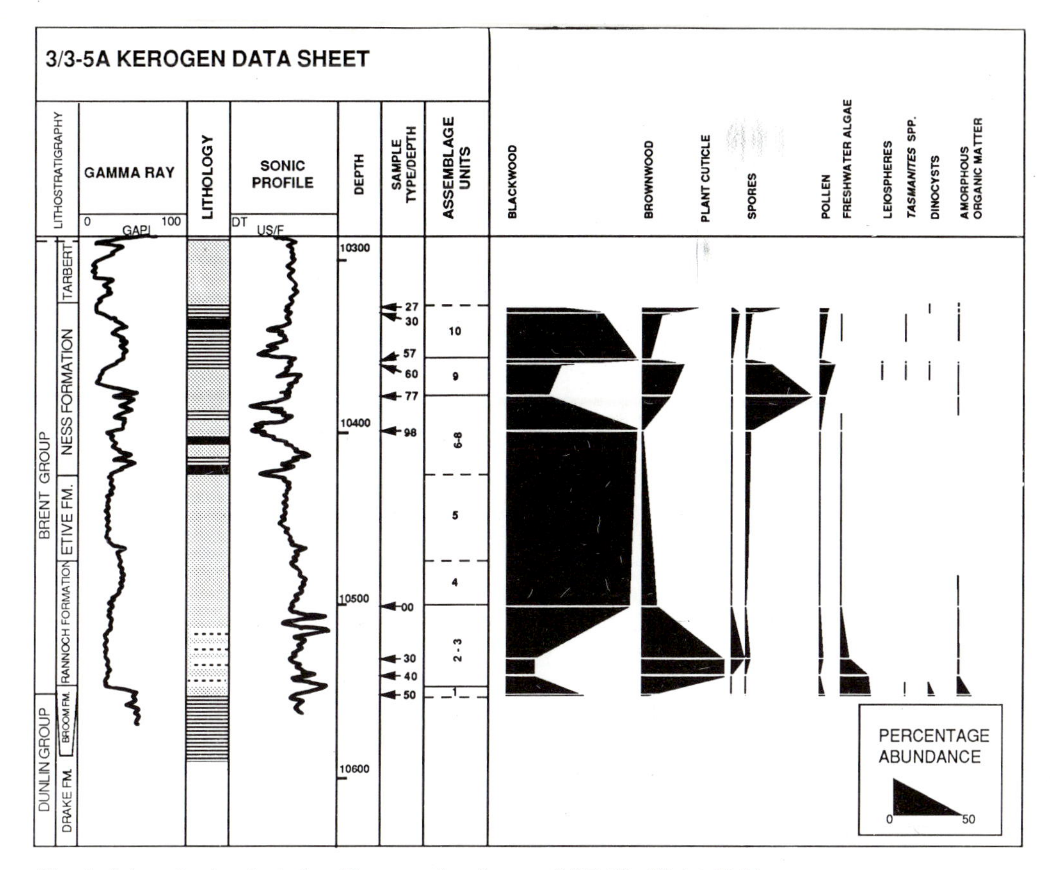


**Fig. 4.** Selected palynological and kerogen data from well 3/3-5A, Ninian Field.

### *Assemblage 3: Rannoch Formation (pars.)*

Type Samples: 211/18a-A31, 10103-10171.5 ft (core) (Figs 5 & 6b).

Assemblage 3 spans the interval of Rannoch Formation above the 'Rannoch Formation Shale' which yields a rich assemblage of the fresh to brackish water alga *Botryococcus* sp. with mixed humic kerogen. The top of this influx occurs within the upper part of the Rannoch Formation and this 'top' probably reflects progressively higher energy levels at this point in the sequence which precluded the settling out of the low density *Botryococcus* sp.. The origin of this fresh to brackish water alga is discussed under Assemblage 2.

### *Assemblage 4: Rannoch Formation (pars.)*

Type Samples: 211/18a-A31, 10057-10064.5 ft (core) (Fig. 5).

The uppermost part of the Rannoch Formation is characterised by a sparse assemblage of black wood, *Botryococcus* sp. and pollen which constitutes Assemblage 4. This assemblage represents deposition in a shore face, possibly upper shore face, environment as indicated by the high energy and highly oxidative conditions suggested by this assemblage.

### *Assemblage 5: Etive Formation*

Type Sample: 211/18a-A31, 10033 ft (core) (Fig. 5).

Regionally the Etive Formation yields a highly impoverished assemblage of blackwood with rare pollen. This assemblage suggests a high energy, possibly barrier sand, environment of deposition.

The presence of local silts and coals within the Etive Formation can result in richer assemblages of pollen within an overall relatively sparse interval.

### *Assemblage 6: 'Lower' Ness Formation (pars.)*

Type Samples: 211/18a-A31, 9956-9962 ft (core) (Figs 5 & 6c).

The basal Ness Formation sediments yield a

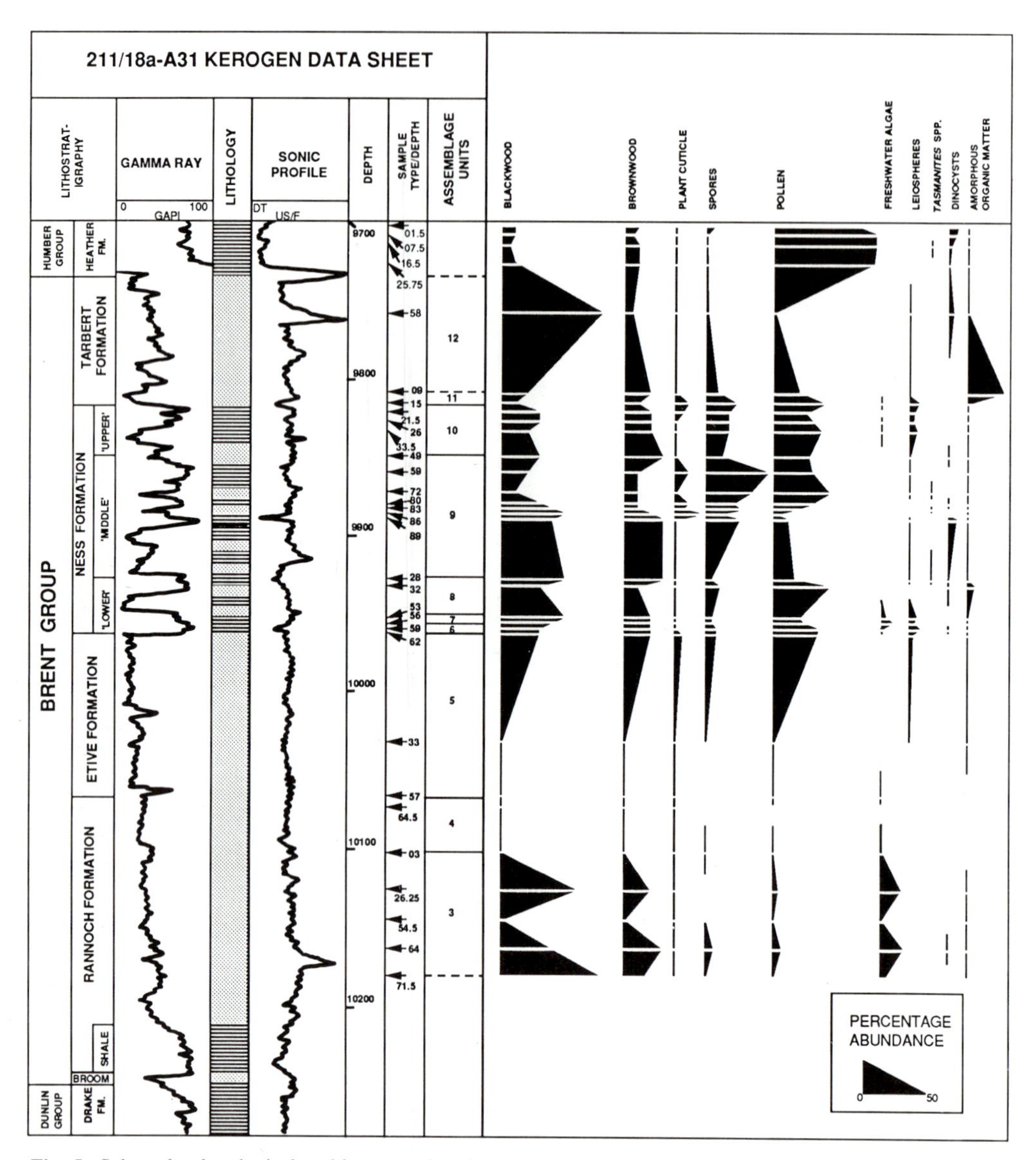


**Fig. 5.** Selected palynological and kerogen data from well 211/18a-A31, Thistle Field.

palynomorph assemblage dominated by the freshwater alga *Botryococcus* sp. and pollen. In this situation the occurrence of fresh to brackish water algae are considered to be in situ due to the association with an overall lagoonal sequence within the 'Lower' Ness Formation. The assemblage is interpreted as having been deposited in a freshwater lagoon with drifted coals.

### *Assemblage 7: 'Lower' Ness Formation (pars.)*

Type Sample: 211/18a-A31, 9953 ft (core) (Figs 5 & 6d).

Following the freshwater assemblages at the base of the Ness Formation, renewed saline influence occurs in Assemblage 7. This palynomorph assemblage is dominated by leiospheres (brackish water algae), pollen and dinocysts and is interpreted as having been deposited in a saline lagoon or interdistributary bay. The occurrence of leiospheres and dinocysts are interpreted as in situ taxa whilst the pollen represent allochthonous taxa borne by air and water. Subassemblages occurring within Assemblage 7 include crevasse splay sand assemblages which consist of highly mixed humic kerogen, spores and pollen and encroachment by coal swamp floras as indicated by spore dominated assemblages.

## *Assemblage 8: 'Lower' Ness Formation (pars.)*

Type Samples: 211/18a-A31, 9926-9953 ft (core) (Figs 5 & 6e).

Assemblage 8 represents a brackish lagoonal environment and is dominated by an abundant pollen and humic kerogen with subordinate dinocyst and brackish water algal elements including leiospheres and tasmanitids. Minor channel and crevasse splay sands introduced humic kerogen and spores into this overall lagoonal assemblage, whilst coal swamp microfloras may also develop.

The pollen grain *Perinopollenites elatoides* commonly dominates palynomorph assemblages within this interval.

## *Assemblage 9: 'Mid' Ness Formation*

Type Samples: 211/18a-A31, 9849-9889 ft (core) (Figs 5, 6f & g).

The base of the 'Mid' Ness Formation as defined on palynological grounds is marked by the up-sequence influx of coal swamp type microfloras dominated by spores. This diverse and abundant spore assemblage spans the entire 'Mid' Ness Formation. There are however, many sub-assemblages identifiable within this interval. The major variation in assemblages is caused by discrete marine dinocyst floods which tend to be restricted to individual shale horizons interdigitated with coals and shales dominated by coal swamp microfloras. The identification of individual dinocyst floods offers potential for local field and block wide correlation within the 'Mid' Ness Formation. Dinocysts present in these events include *Nannoceratopsis gracilis* and taxa comparable to the genera *Microdinium* and *Hystrichodinium*. The predominance of diverse coal swamp type microfloras, typically dominated by floods of *Cyathidites* sp. and *Ischyosporites crateris* indicate that this assemblage represents the maximum point of regression of the Brent Group depositional sequence.

## *Assemblage 10: 'Upper' Ness Formation*

Type Samples: 211/18a-A31, 9815-9833.5 ft (core) (Figs 5 & 6h).

Assemblages recovered from the 'Upper' Ness Formation indicate a progressive decrease in the spore diversity along with an increase in marine and brackish water taxa. This gradational change can be interpreted as an abandonment phase of the depositional system of the Brent Group. As such, Assemblage 10 represents the initial transgression which subsequently led to deposition of the overlying marine Tarbert Formation sands.

A series of events involving the top occurrences of individual spore abundances allows subdivision of this assemblage, providing correlation on a local and subregional level. The overall assemblage is typified by abundant *Cyathidites* spp., and *Ischyosporites crateris* with common *Densoisporites* spp. and the brackish water alga *Tasmanites* sp..

## *Assemblage 11: Tarbert Formation (pars.)*

Type Sample: 211/18a-A31, 9809 ft (core) (Figs 5 & 6i).

This assemblage, at the base of the Tarbert Formation, represents a mix of rare in situ marine dinocysts which typify the Tarbert Formation and reworked 'Upper' Ness Formation taxa mainly consisting of humic kerogen and low diversity spore assemblages. The mixing of these assemblages suggests an erosive contact between the 'Upper' Ness Formation and the transgressive marine Tarbert Formation sands.

## *Assemblage 12: Tarbert Formation (pars.)*

Type Samples: 211/18a-A31, 9758 ft (core) (Figs 5 & 6j).

Assemblage 12 is characterized by sparse black wood, low diversity dinocysts and pollen and is typical of high energy, highly oxygenated deposition in a marginal marine environment. Taxa present in the well studied include the dinocyst *Escharisphaeridia pocockii* and the pollen *Perinopollenites elatoides, Callialasporites dampieri* and bisaccate species. Dinocyst taxa typically present within this assemblage also include *Gongylodinium* sp., *Sentusidinium 'granulosum'* and *S. verrucosum*. These taxa are generally present within low diversity or monospecific dinocyst assemblages, reflecting the marginal marine environment of deposition (Williams & Sarjeant 1967).

Following Brent Group deposition, a return to fully marine argillaceous sediments is identified by a major increase in dinocyst numbers and diversity at the base of the Heather Formation. Specifically, the taxa present include *Gonyaulacysta jurassica, Sirmiodinium grossii* and *Nannoceratopsis pellucida*. At the base of the Heather Formation dinocysts form only a minor proportion of the total assemblage which is dominated by pollen, specifically bisaccate species, *Cerebropollenites mesozoicus* and *Callialasporites dampieri*. However, the dinocyst to

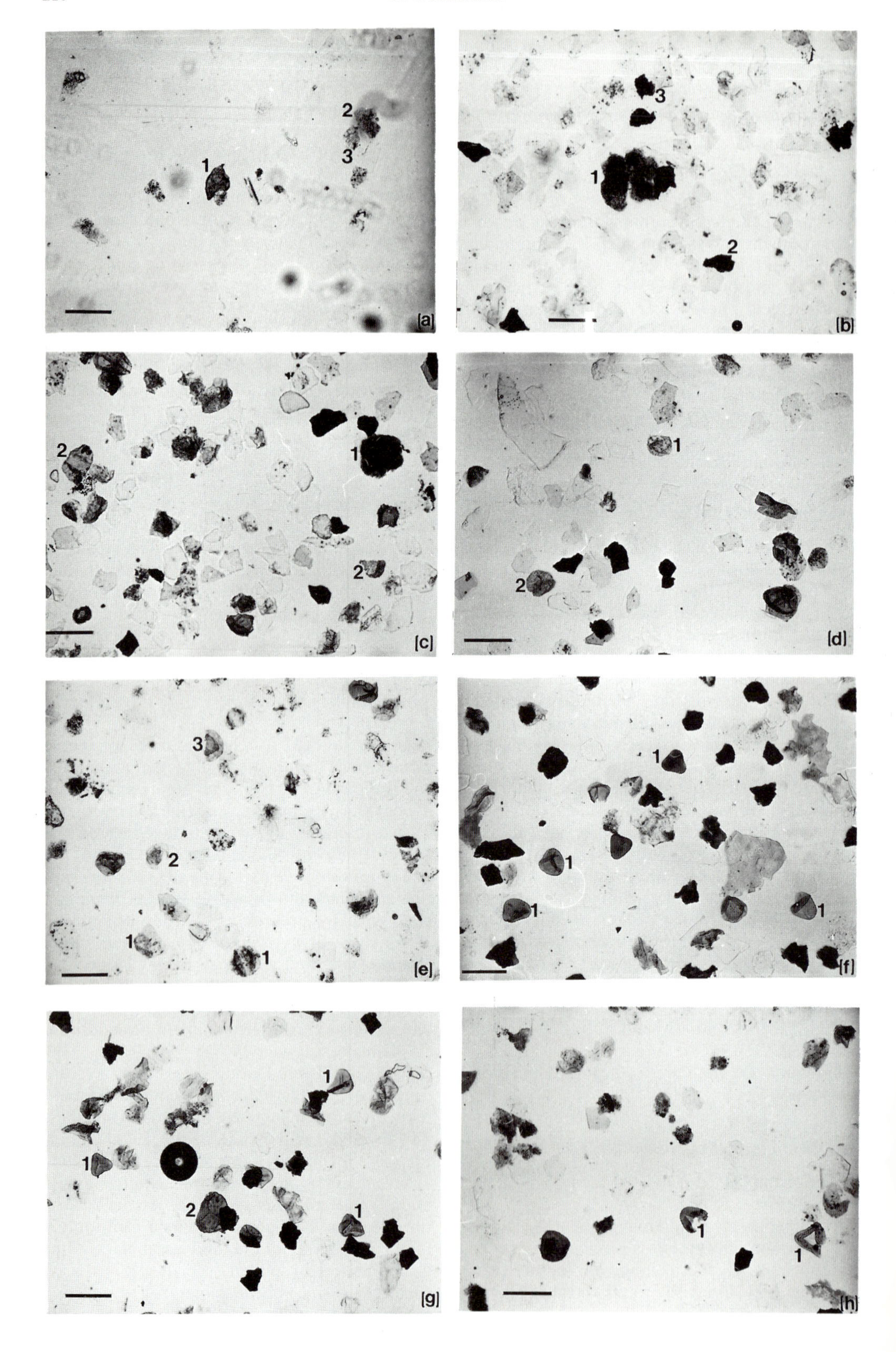
2
3
1
[a]
3
1
2
[b]
2
1
2
[c]
1
2
[d]
3
2
1
1
[e]
1
1
1
1
[f]
1
1
2
1
[g]
1
1
[h]

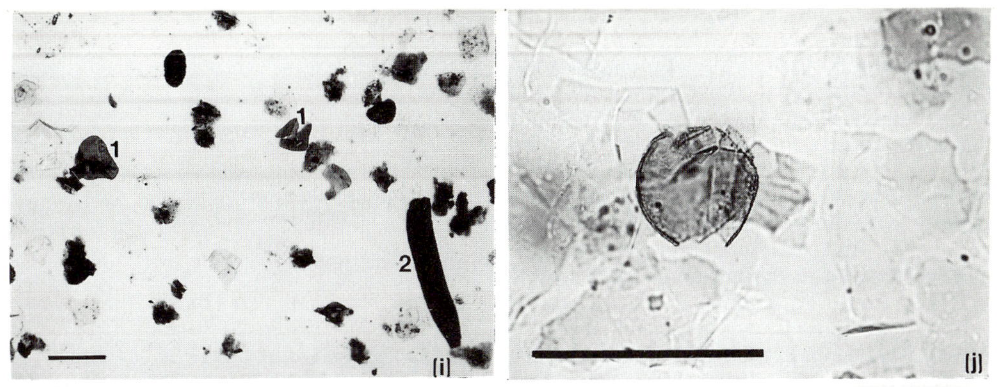


**Fig. 6.** Kerogen photographs (**a**) Assemblage 1, well 3/3-5A 10 550 ft (core). *Nannoceratopsis gracilis* (angular var.) (1) with amorphous organic matter (2) and fine blackwood (humic kerogen). (**b**) Assemblage 3, well 211/18a-A31, 10 164 ft (core). *Botryococcus* sp. (1) (allochthonous freshwater alga) with mixed humic kerogen, blackwood (2) and brownwood (3). (**c**) Assemblage 6, well 211/18a-A31, 9962 ft (core). *Botryococcus* sp. (1) (in situ freshwater alga) with Bisaccate pollen (2). (**d**) Assemblage 7, well 211/18a-A31, 9953 ft (core). Leiospheres (1) (brackish water algae) and pollen (2). (**e**) Assemblage 8, well 211/18a-A31, 9932 ft (core). Abundant pollen; Bisaccate pollen (1), *Perinopollenites elatoides* (2) and *Callialasporites dampieri* (3). (**f**) and (**g**) Assemblage 9, well 211/18a-A31, 9886 ft (core). Abundant and diverse spores. *Cyathidites* spp. (1), *Ischyosporites crateris* (2). (**h**) Assemblage 10, well 211/18a-A31, 9815 ft (core). Low diversity spore assemblage. *Cyathidites* spp. (1). (**i**) Assemblage 11, well 211/18a-A31, (core). Reworking of Ness Formation spores (1) and humic kerogen (2) into Tarbert Formation marine sands. (**j**) Assemblage 12, well 211/18a-A31, 9758 ft (core). *Escharisphaeridia pocockii* (dinocyst).
Scale bar equals 100 μm in all photographs.

pollen ratio increases up through the sequence.

The assemblages described above can be used to 'fingerprint' particular lithostratigraphic units. This intimate relationship between biofacies and lithofacies results from common palaeoenvironmental controls. It is therefore possible to use this relationship to predict the presence of individual lithostratigraphic units on the basis of biofacies alone. However, this indicator should, where possible, always be used in conjunction with electric log and sedimentological interpretations.

## Conclusions

(1) The Brent Group can be subdivided on a regional basis into a series of 12 palynological assemblages.

(2) The assemblages can be used to 'fingerprint' and identify specific lithostratigraphic and reservoir units. This allows the identification of these units in Brent Group sequences where faulting and truncation or lack of conventional core coverage and electric logs hamper the use of log correlation or sedimentological analysis to subdivide the Brent Group. In situations where no conventional core is available, sidewall core samples are the preferred alternative as they do not suffer from the inherent downhole caving problems of ditch cuttings. However, the scheme has also on occasion been applied successfully to data derived from ditch cuttings.

(3) The sequence of assemblages also provides a tool to allow comparison between environments in various parts of the Brent Province.

(4) Progradation of the Brent Group can be indicated palynologically between the base of the Rannoch Formation and the 'Mid' Ness Formation. The point of maximum regression and progradation is associated with the abundant and diverse coal swamp microfloras and coals of the 'Middle' Ness Formation. Following progradation, a phase of abandonment and transgression occurred resulting in the deposition of 'Upper' Ness Formation and Tarbert Formation sediments.

(5) A number of specific environmental interpretations have been made.

(i) The occurrence of a monospecific dinocyst assemblage of *Nannoceratopsis gracilis* (angular varieties) within the Broom and basal Rannoch Formation is interpreted as indicating a highly restricted environment with possibly reduced salinities.

(ii) The abundant occurrence of the freshwater alga *Botryococcus* sp. throughout the lower part of the Rannoch Formation is interpreted as allochthonous and probably represents flushing of a freshwater environment higher on the depositional surface. This interpretation has a bearing on the interpretation of reduced salinities in the basal Rannoch and Broom Formations which may possibly relate to this flushing event.

(iii) The 'Lower' Ness Formation was developed in a predominantly lagoonal environment, initially freshwater, becoming brackish.

(iv) The 'Mid' Ness Formation is interpreted as deposition up to the maximum point of regression. However, continued marine influx is shown by sporadic and restricted dinocyst floods. The 'Upper' Ness Formation is interpreted as representing sedimentation during the initial phase of abandonment and transgression which eventually led to the Tarbert Formation transgression.

(v) The occurrence of relatively rich spore and pollen assemblages at the base of the Tarbert Formation represents reworking of 'Upper' Ness Formation sediments and therefore erosional contact between these formations is suggested where this assemblage occurs.

(6) The sequence of assemblages described in this paper are applicable through most of the Brent Group depositional area. However, due to lithofacies changes some variations in assemblages are likely in peripheral areas such as the northern and eastern limits of deposition.

The author wishes to acknowledge P. Watson and R. Dyer for their comments on this paper, the British Geological Survey for allowing the sampling of the two wells studied and M. J. Fisher for my initial introduction to, and early encouragement in, Brent Group palynological studies.

## Appendix: taxonomic references

### Dinocysts

*Escharisphaeridia pocockii* (Sarjeant 1968) Erkmen & Sarjeant 1980
*Gongylodinium* spp. Fenton *et al.* 1980
*Gonyaulacysta jurassica* (Deflandre 1938*b*) Norris & Sarjeant 1965
*Hystrichodinium* spp. Deflandre 1935 emend. Clarke & Verdier 1967
*Microdinium* spp. Cookson and Eisenack 1960 emend. Stover & Evitt 1978
*Nannoceratopsis gracilis* Alberti 1961 emend. Evitt 1962
*Nannoceratopsis pellucida* Deflandre 1938
*Sentusidinium 'granulosum'* Informal taxon
*Sentusidinium verrucosum* Sarjeant (1968) Sarjeant and Stover 1978
*Sirmiodinium grossii* Alberti 1961 emend. Warren 1973

### Brackish and freshwater algae

*Botryococcus* sp. Kützing 1849
*Tasmanites* sp. Newton 1875

### Spores and pollen

*Callialasporites dampieri* (Balme 1957) Norris 1969
*Cyathidites* spp. Couper 1953
*Densoisporites* spp. Weyland & Krieger 1953
*Ischyosporites crateris* Balme 1957
*Perinopollenites elatoides* Couper 1958

## References

Denison, C. & Fowler, R. M. 1980. Palynological identification of facies in a deltaic environment. *Proceedings of Sedimentation of North Sea reservoir rocks, Geilo*. Norwegian Petroleum Society, Oslo, Paper XII, 1–22.

Fisher, M. J. 1980. Kerogen distribution and depositional environments in the Middle Jurassic of Yorkshire. U.K. *Fourth International Palynological Conference, Lucknow (1976–1977)*, **2**, 574–580.

Hancock, N. J. & Fisher, M. J. 1981. Middle Jurassic North Sea deltas with particular reference to Yorkshire. *In*: Illing, L. V. & Hobson, G. D. (eds) *Petroleum Geology of Continental Shelf of North West Europe*, Heyden & Son, London, 186–195.

Muller, J. 1959. Palynology of Recent Orinoco delta and shelf sediments: Reports of the Orinoco Shelf Expedition, Vol. 5. *Micropaleontology*, **5**(1), 1–32.

Parry, C. C., Whitley, P. K. J. & Simpson, R. D. H. 1981. Integration of palynological and sedimentological methods in facies analysis of Brent Formation. *In*: Illing, L. V. & Hobson, G. D. (eds), *Petroleum Geology of Continental Shelf of North West Europe*, Heyden & Son, London, 206–215.

Traverse, A. & Ginsberg, R. M. 1966. Palynology of the surface sediments of Great Bahama Bank, as related to water movement and sedimentation. *Marine Geology*, **4**, 417–459.

Whitaker, M. F. 1984. The usage of palynostratigraphy and palynofacies in definition of Troll Field geology. In: *Offshore Northern Seas – Reduction of Uncertainties by Innovative Reservoir Geomodelling*. Norwegian Petroleum Society, Article G6.

Williams, D. B. & Sarjeant, W. A. S. 1967. Organic-walled microfossils as depth and shoreline indicators. *Marine Geology*, **5**, 389–412.

# Samarium–neodymium isotopic constraints on the provenance of the Brent Group

EUAN W. MEARNS

*Instituttet for energiteknikk, PO box 40, N-2007 Kjeller, Norway*
*Present address: Isotopic Analytical Services Ltd, PO Box 219, Aberdeen AB9 8LL, UK*

**Abstract:** Samarium–neodymium data for the Brent Group of Gullfaks oilfield provide a detailed record of provenance. The Broom Formation, yielding relatively high Sm-Nd ages in the range 1700–1800 Ma, is interpreted as being derived from a proximal southwesterly source on the Shetland Platform. The Rannoch and lower Etive Formations, having lower Sm-Nd provenance ages (1550–1650 Ma) than the Broom, are interpreted in terms of easterly provenance involving transport westwards to Gullfaks by marine currents. The upper Etive Formation returns to the proximal southwesterly source. This source area grew southwards during initial deposition of the Ness Formation to encompass the Forties volcanic centre resulting in young Sm-Nd provenance ages in some Ness strata. Provenance ages in the upper Etive–Ness section thus have a large range from 1300 Ma to 1800 Ma. It is therefore suggested that the Brent Group was deposited by at least two major river systems: one draining an easterly source in southern Norway and the other draining a proximal southwesterly and southerly source in the Shetland Platform and the central North Sea.

Palaeogeographic reconstructions for Toarcian to Callovian time suggest that the Brent Group formed as the result of a wave dominated barrier shore line prograding first northwards and then retreating southwards through the Viking Graben depression of the northern North Sea (Brown *et al.* 1987; Graue *et al.* 1987; Johnson & Stewart 1985). Brown *et al.* (1987) have suggested that the Broom Formation is not genetically linked to the Brent megasequence but rather represents the submarine portion of westerly derived lateral fan deltas. Along the eastern margin of the Viking Graben, the Oseberg Formation is considered to be the equivalent of the Broom Formation and is interpreted as the submarine portion of easterly derived fan deltas (Graue *et al.* 1987). It is therefore apparent that the Brent megasequence began with deposition of the Rannoch Formation and ended with the deposition of the Tarbert Formation.

References in the literature to 'the Brent delta' (e.g. Budding & Inglin 1981) give the impression that the Brent Group was deposited by a single, large fluvial–deltaic system. Comparisons between 'the Brent delta' and proposed modern analogues such as the Niger and Nile deltas (Budding & Inglin 1981; Johnson & Stewart 1985) perpetuate this idea. While it has not necessarily been the intention of these authors, this has led indirectly to the hypothesis that the Brent unit was deposited by a single large axial river–delta system draining northwards through the Viking Graben.

The alternative hypothesis, that the Brent megasequence was deposited by several smaller fluvial–deltaic systems, is implied from the palaeographic reconstructions presented by Skarpnes *et al.* (1980) and Graue *et al.* (1987). If this latter hypothesis is correct then the distribution of reservoir quality sandstones will be controlled by the positions of individual river systems and their provenance. It is therefore important to develop methods which can discriminate between the above hypotheses and if the latter case proves to be valid, to develop methods for mapping out the spheres of influence of individual fluvial–deltaic systems. This paper focuses on the question of the number of penecontemporaneous fluvial–delta systems which were responsible for the deposition of the Brent Group as recalled by Sm-Nd provenance data (mainly Mearns 1989, additional data from Hamilton *et al.* 1987) and other published provenance information (Morton 1985, 1987; Morton & Humphreys 1983; Morton *et al.* 1989; Hamilton *et al.* 1987).

## The spatial dimensions of the Brent Group and Brent source terrains

A useful starting point in assessing Brent palaeogeography is to consider the dimensions of the

*From* Morton, A. C., Haszeldine, R. S., Giles, M. R. & Brown, S. (eds), 1992, *Geology of the Brent Group*. Geological Society Special Publication No. 61, pp. 213–225.

Brent megasequence and of the source terrains available for its derivation. The presence of Middle Jurassic depocentres to the west of the Shetlands, in the Inner Moray Firth and Norwegian–Danish basins set fairly rigid westerly and southerly limits to the extent of the Brent source area. The easterly limit is less well constrained. However, assuming a watershed in the east close to the present day watershed, yields an area of around 100 000 km$^2$ for the Brent source terrain. This compares with an area of about 25 000 km$^2$ covered by the shallow marine and lower delta plain facies of the Brent megasequence. The delta area:basin area ratios for the Brent Group and a representative selection of modern deltas are plotted together in Fig. 1. This diagram shows that the areal extent of the Brent is extremely large in relation to its potential source terrain when compared with modern delta systems. Only the Klang system has a delta:basin area ratio higher than that of the Brent, while the Mekong, Grijalva, Po and Chao Phraya systems' ratios are similar to that of the Brent. The Nile, in contrast, has a delta of similar area but a basin area which is over an order of magnitude larger.

The majority of river systems with high delta: basin area ratios have wet tropical climates, although there is no consistent relationship between climate, delta type and delta:basin area ratio. What this exercise shows is that the areal extent of the Brent megasequence is extremely large in relation to the area available for its derivation. Furthermore, the calculated delta: basin area ratio is probably a minimum value as

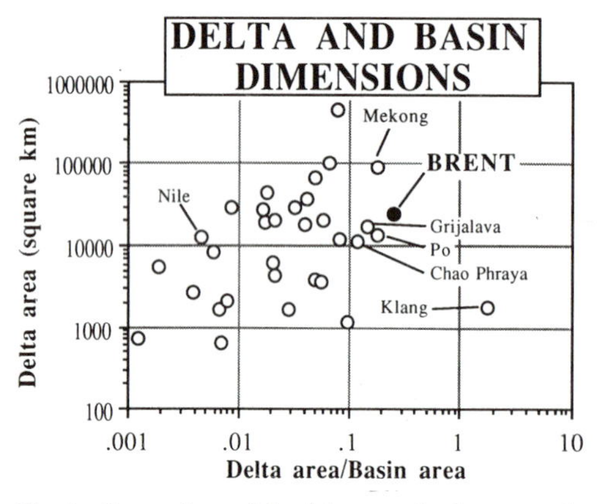


**Fig. 1.** Comparison of the delta area:basin area ratio for the Brent Group with modern fluvial–deltaic systems. Note that only the Klang delta has a higher ratio than the Brent. Data from Coleman & Wright (1981).

it includes source areas to the west, south and east of the Brent province which is an unlikely source configuration for a single river system. If only the area to the south is considered, the basin:delta area ratio for the Brent unit becomes even larger and less plausible. On the other hand, the only way the source area could be extended in the present calculation is to the east. Thus, introducing a major easterly river system helps solve the problem of finding a sufficiently large source area for the Brent sediments.

## Heavy mineral and isotopic constraints on Brent provenance

Published information on the provenance of Brent sandstones is sparse with the exception of a number of detailed garnet geochemistry studies (Morton 1985, 1987; Morton & Humphreys 1983; Morton *et al.* 1989; Hurst & Morton 1988). These studies have demonstrated that garnet provenance varies between the various Brent lithostratigraphic units. For some stratigraphic intervals, garnet populations are similar over restricted geographical areas, e.g. the Thistle and Murchison area (Morton 1987), while in other stratigraphical intervals geographical differences are present. The chemistry of the various garnet populations suggests ultimate provenance predominantly within medium- to high-grade metamorphic rocks. Source terrains could thus be found in either Scotland or Norway or among sedimentary sequences derived from these areas. Morton *et al.* (1989) present garnet data for the Oseberg Formation from Oseberg Field for which an easterly source and westerly transport direction is inferred from dipmeter data and facies distributions (Graue *et al.* 1987). These authors see similarities between the garnet populations of the Oseberg Formation and those of the Etive and Rannoch Formations of Gullfaks, Statfjord and Murchison Fields, which lie to the northeast of Oseberg. On this basis, Morton *et al.* (1989) postulate an easterly source for the Rannoch and Etive Formations in these fields and suggest that longshore transport in the marine environment is responsible for the westerly transport process.

Hamilton *et al.* (1987) discuss the provenance of Rannoch Formation sandstones from non-specified fields on the basis of radiometric dating of detrital minerals and granite clasts. Rb–Sr dates for muscovite gave 440 Ma, which is a common age in the Scottish Caledonides but predates the Scandian uplift and cooling of the south Norwegian Caledonides (410–380 Ma (Lux 1985). However, the mean initial $^{87}Sr/^{86}Sr$ ratio of the muscovites is somewhat lower than that recorded in the Dalradian of the Scottish Caledonides (Dempster 1985) which lead Hamilton *et al.* to suggest a source in the vicinity of the east Shetland Platform where it was believed source rocks of suitable character could exist. Enigmatically, K-Ar dates for the same muscovites ranged between 300 Ma and 400 Ma. This type of date postdates the main Caledonian events of Scotland and is more compatible with the cooling history of the southern Norwegian Caledonides. The Sr isotopic systematics of granite clasts from the Rannoch Formation were also incompatible with a Scottish source.

The problems with finding a suitable Scottish provenance area (Hamilton *et al.* 1987) seen together with Morton's (1987) evidence for easterly provenance could lead to the conclusion that a south Norwegian provenance is more likely for the Rannoch–Etive interval. However, there are major problems with such a conclusion. First, Caledonian granites are rare in southern Norway and secondly much of southern Norway is comprised of Precambrian granites that should yield Precambrian Rb–Sr muscovite dates.

## Samarium–neodymium systematics in sediments

Samarium–neodymium (Sm-Nd) isotope systematics have been described in detail elsewhere (McCulloch & Wasserburg 1978; O'Nions *et al.* 1983; Hamilton *et al.* 1987; Mearns 1989) and thus the intention of this section is to provide a brief overview.

### *Sedimentary geochemistry of Sm and Nd*

Sm and Nd are geochemically very similar and both have extremely low solubilities in water. For example, river water contains about 8 ppb Sm and 40 ppb Nd while seawater contains about 0.8 and 4 ppb respectively. This compares with mean values of about 6 ppm Sm and 30 ppm Nd in the upper continental crust (all data from Taylor & McLennan 1985). Due to their low solubilities, Sm and Nd tend to be transported in the detrital sediment mass while their geochemical similarity has the consequence that Sm and Nd undergo negligible fractionation in the sedimentary cycle. Thus, clastic sediments inherit the Sm-Nd compositions of their source rocks.

### *Sm-Nd provenance ages for sediments*

The long-lived radioactive isotope $^{147}Sm$ decays to the stable isotope $^{143}Nd$ which forms the basis of the Sm-Nd method of radiometric dating. The Sm-Nd isotopic composition of a sediment may be conveniently expressed as a provenance age expressed in millions of years (Mearns 1988). Technically, the provenance age is directly equivalent to the frequently quoted model age parameter $t^{Nd}$ used by isotope geologists (McCulloch & Wasserburg 1978). The term 'crustal residence age', proposed and defined by O'Nions *et al.* (1983), is also widely used to convey the model age of a sediment. Each of these parameters are determined relative to isotopic compositions assumed for the Earth's depleted mantle and as there is no convention as to what value the latter should have, different authors use different values. When comparing provenance ages from different studies it is therefore extremely important to make sure that they are calculated relative to the same model mantle values. In this paper all provenance ages are calculated relative to a mantle with the following model composition:
$^{147}Sm/^{144}Nd = 0.22$ $^{143}Nd/^{144}Nd = 0.51303$ (Mearns 1988).

The provenance age of a sediment or sedimentary rock is a measure of the average time at which the ultimate sediment source first became differentiated from the mantle. Thus, source terrains which may have relatively young historical ages, (e.g. the Caledonian metasediments of Scotland), may yield sediment which has considerably older provenance ages; in Scotland, for example, these may range up to *c.* 1800 Ma (Mearns 1988). This is because the metasediments have inherited their provenance signatures from their ultimate precursors. The only way that the mean Sm-Nd provenance age of a crustal block may be reduced is by the addition of more juvenile mantle-derived material. In the Scottish Caledonides, the intrusion of Caledonian granites, which are partly derived from the mantle (Halliday 1984), has had the effect of giving source areas which have granites younger mean Sm-Nd provenance signatures than those which are comprised only of metasediment (Mearns 1988). These types of considerations are summarized in Fig. 2.

### *Systematics of sandstones and mudstones*

In the Statfjord and Lunde Formations from Snorre Field of the North Sea, the majority of adjacent mudstones and sandstones give similar provenance ages (Mearns *et al.* 1989). Where

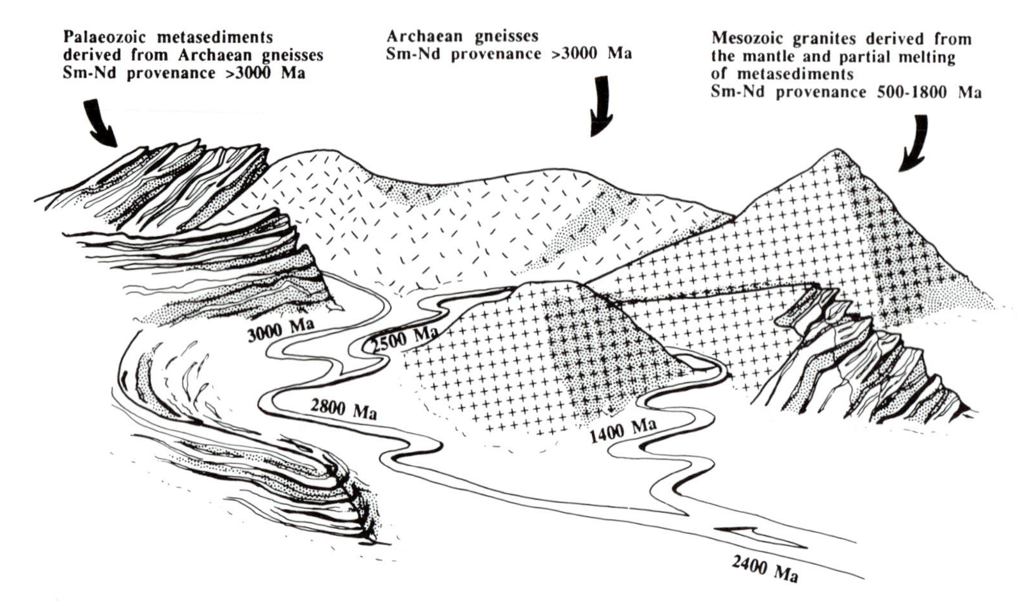


**Fig. 2.** Cartoon sketch showing the principles which determine the Sm-Nd provenance signature of sediment derived from a hypothetical source terrain. Mixing of sediment downstream from a confluence gives the weighted mean provenance age of sediment from the two tributaries. Weighting terms need to include the mass of sediment provided by each tributary and the concentrations of Sm and Nd in the two sources of sediment. This latter variable is likely to be quite uniform between large river systems draining geochemically representative portions of the continental crust.

Note that, in this example, large portions of the source terrain have comparatively young historical ages but that the Sm-Nd provenance signature of sediment in the axial river is comparatively high. This is because, in this hypothetical example, the younger rocks have formed largely through the reworking of precursors which were derived ultimately from Archaean gneisses. In other situations, it could be the case that the source terrain is dominated by rocks which were recently derived from the mantle in which case the sediment eroded would have a lower Sm-Nd provenance age.

differences are recorded, these may be satisfactorily attributed to real differences in provenance. This shows that the Sm-Nd isotopic system is insensitive to factors such as grain size, sorting and diagenetic history. It is therefore justified to interpret variations in Sm-Nd provenance ages in terms of real changes in source area rather than spurious sedimentological effects.

### *Uncertainty*

Methodological uncertainty, arising from arbitrary selection of mantle Sm-Nd composition, exceeds analytical uncertainty (Mearns 1988). The former is estimated to be $\pm 50$ Ma. Thus, differences in provenance ages are only considered to be significant if they exceed 100 Ma.

## Sm-Nd signatures of potential Brent source terrains

It is not possible to estimate the Sm-Nd provenance signatures of source terrains solely on the basis of what is known of their historical development. Rather, prospective sources must be sampled to map out their Sm-Nd compositions. This may be done in one of three ways in order of decreasing certainty:

(1) to analyse samples of modern river sediments from prospective source terrains;
(2) to analyse ancient sediments which have known provenance; in this way the composition of the source may be inferred from that of the ancient sediment;
(3) to analyse a range of samples of the bed-rocks in the prospective source area.

Only methods 1 and 2 are likely to provide representative coverage of the prospective source terrain. Sampling bed-rock preserves small scale heterogeneities in the Sm-Nd record that would normally be homogenized in the sedimentary cycle. Thus, even if a representative number of bed-rock samples are collected, the range in model ages recorded will be much greater than may be seen in river sediments draining the same source. Looking at sediments of known provenance is thus by far the best way of mapping the Sm-Nd compositions of source terrains.

The prospective Brent source terrains around the northern North Sea may be arbitrarily defined as follows (Fig. 3):

(1) the Shetland Platform, referred to as the proximal southwesterly source;
(2) northern Britain east of the Moine thrust, referred to as the distal southwesterly source;
(3) the uplifted mid North Sea, referred to as the southerly source;
(4) the Horda Platform and Utsira High, referred to as the southeasterly source;
(5) southern Norway, referred to as the easterly source;
(6) the submerged continental shelf to the west of the Moine thrust and its northerly extension, referred to as the distal westerly source.

The Sm-Nd isotopic compositions of northern Britain and southern Norway have been mapped in detail by looking at the Sm-Nd compositions of modern river sediments draining these areas (Mearns 1988, and unpublished data). Most major river systems have been examined and these data sets thus provide comprehensive and representative coverage of these areas with the exception of the southwestern fjord coastline of Norway which has not yet been sampled. This provides a direct means of comparison between the Sm-Nd provenance of Brent samples and prospective distal southwesterly and easterly source terrains. The river sediments data are plotted at the top of Fig. 4. The majority of river sediments in both northern Britain and southern Norway have provenance ages between 1400 Ma and 1850 Ma. In both northern Britain and southern Norway a few rivers have lower provenance ages, which reflects local prov-

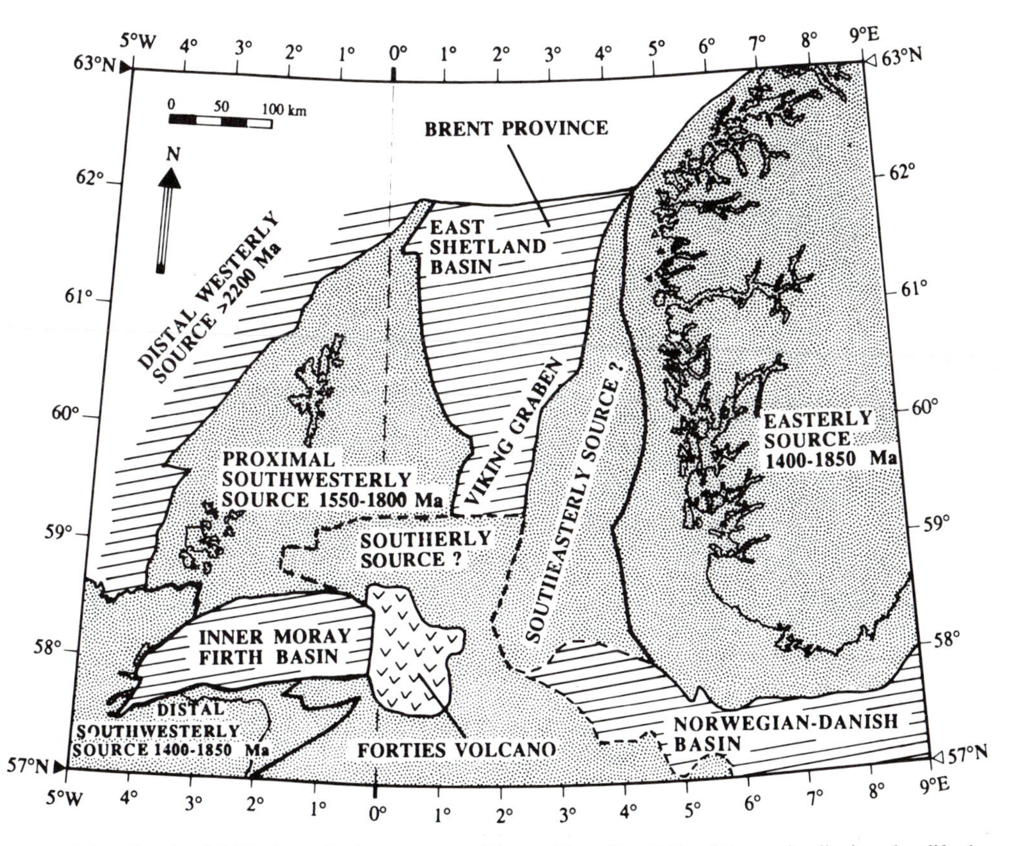


**Fig. 3.** Map showing Middle Jurassic depocentres of the northern North Sea (diagonal ruling) and uplifted highlands (stippled). The source areas are divided into structurally defined blocks and the Sm-Nd provenance signatures are shown where these are known. Note that in this area the majority of the source areas have quite similar provenance signatures with the exception of the distal westerly source where the presence of Archaean gneisses gives a high Sm-Nd provenance age. The presence of volcanic rocks in the Forties region of the southerly source makes it likely that this area will have had a younger provenance age than the surrounding source areas during the Middle Jurassic.

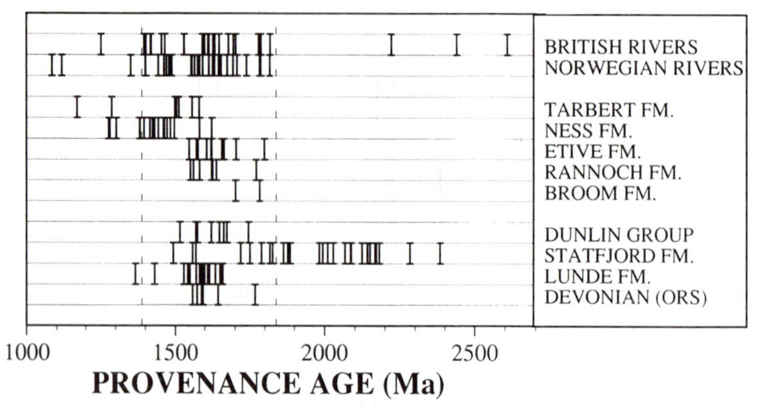


**Fig. 4.** Diagram showing the distributions and ranges of Sm-Nd provenance ages in the various formations of the Brent Group, in the stratigraphically older formations of the East Shetland Basin (Dunlin, Statfjord and Lunde), the Shetland Platform (Old Red Sandstone, ORS), the distal southwesterly source area (British river sediments) and the easterly source (Norwegian rivers). The three British rivers with high provenance ages represent the distal westerly source.

In this diagram each line represents one sample. In the case of the river sediments each line represents one river and these data sets provide good geographical coverage of the source terrains they represent.

The dashed lines span the interval 1400–1850 Ma which characterizes the distal southwesterly and easterly source terrains. Samples which fall outside this interval have provenance outside these areas. The most striking example of the latter is the Statfjord Formation which has provenance dominantly in the distal westerly source area. Some samples of the Ness and Tarbert Formations also fall outside the 1400–1850 Ma interval and inferrably have a source component of young, mantle-derived rocks e.g. Forties volcanic rocks.

Data sources: British rivers, Mearns (1988); Norwegian rivers, Mearns (unpublished results); Brent Group, Dunlin Group and Statfjord Formation, Mearns (1989); Lunde Formation, Mearns *et al.* (1989); Devonian, Mearns & N. Trewin (unpublished results).

enance effects in small drainage basins caused by the presence of Lower Palaeozoic igneous rocks. In terms of the gross regional provenance signature these lower provenance ages can be ignored. In northern Britain, three rivers have provenance ages greater than 2200 Ma. Each of these rivers drains the Archaean foreland to the west of the Moine Thrust and these samples thus provide an indication of the Sm-Nd provenance signature of the distal westerly source.

Unfortunately the Sm-Nd signatures of the distal southwesterly and easterly source terrains are indistinguishable. In contrast, however, the distal westerly source has a distinct signature which would be easily recognised in the Sm-Nd record of sediment derived from it.

During the Middle Jurassic the Inner Moray Firth area was a depocentre and thus the distal southwesterly source can more or less be excluded as a direct source of sediment for the Brent Group. However, as this area probably acted as a source terrain for much of the Upper Palaeozoic sediments of the Shetland Platform and the central North Sea, it is likely that the Sm-Nd provenance signature of the distal southwesterly source will have been partly transmitted to these areas. For example, samples of Old Red Sandstone (ORS) from the East and West Shetland Platform vary between 1550 Ma and 1800 Ma (Fig. 4). This is compatible with the idea that the ORS deposits were derived from the Scottish Caledonian terrain. As much of the Shetland Platform is covered by ORS, 1550–1800 Ma is taken to represent the Sm-Nd provenance signature for the proximal southwesterly source (Fig. 3).

Direct evidence for the Sm-Nd signature of the southerly source is lacking. However, from the foregoing, it could be expected that this should also be in the range 1400–1850 Ma if it is assumed that the sedimentary sequences that comprise this area were derived from northern Britain or southern Norway. Provenance ages for the Late Triassic Upper Lunde Formation in Snorre Field, which is believed to have had a southerly source (Mearns *et al.* 1989; Nystuen *et al.* 1989), fall mainly in the range 1500–1650 Ma which largely supports this contention. However, it is well established that the source characteristics of the central North Sea will have been drastically modified during the Middle Jurassic due to the eruption of the mantle-derived basalts of the Rattray Formation in the vicinity of Forties Field (Fig. 3; Dixon *et al.* 1981). Eynon (1981) has suggested that the processes responsible for this volcanic activity also led to uplift of the central North Sea and thus contributed to establishing the depositional regime of the Brent Group. It can therefore be reasonably expected that the southerly source should have had a somewhat lower range of provenance ages during the Middle Jurassic compared with the Late Triassic deposition of the Lunde Formation.

Direct evidence for the provenance signature of the southeasterly source is also lacking although it is unlikely that this differs significantly from the easterly and proximal southwesterly sources. Finally, the provenance signature of the distal westerly source is based on data for three small rivers draining the northwest Highlands of Scotland. Here, the provenance ages vary between 2200 Ma and 2600 Ma. The high ages reflect the presence of Archaean gneisses in this area. Such rocks also subcrop the Lower Jurassic unconformity of the Rona Ridge to the west of Shetland (Ritchie & Darbyshire 1984; Ridd 1981; Morton *et al.* 1987). Provenance ages >2200 Ma are therefore to be expected from the distal westerly source (Fig. 3). The distal westerly source therefore has a provenance signature quite distinct from the other source areas. However, while this area was available as a source of sediment during the Late Triassic–Early Jurassic, during the Mid-Jurassic it was a depocentre (Morton *et al.* 1987) and can therefore be excluded as a source for the Brent Group on those grounds.

## Sm-Nd data for the Brent Group in Gullfaks oilfield

### *The database*

Ninety three samples from four Gullfaks wells were analysed for Sm-Nd composition by Mearns (1989). These samples spanned the lower Statfjord Formation–Tarbert Formation interval and provide a provenance record for most of the Lower–Middle Jurassic stratigraphy with the exception of parts of the Dunlin Group which were not sampled. The Brent Group is represented by 31 samples spanning the Broom to lower Ness interval in well 34/10–1 and by 26 samples spanning the lower Ness–Tarbert interval in well 34/10–8. The lower Ness interval is therefore represented in two wells and one of the main points made by Mearns (1989) is that the pattern of Sm-Nd provenance ages is similar in the duplicated section. The provenance ages for each formation may be compared with each other and with prospective source terrains in Fig. 4. The stratigraphic evolution of provenance ages throughout the section is shown in Fig. 5.

### *General considerations*

The vast majority of Brent Group samples have provenance ages which fall in the range 1400–1850 Ma (Fig. 4). This shows conclusively that the Brent Group was not derived from the much older distal westerly terrain. In contrast, the provenance signature of the Statfjord Formation leaves little doubt that it was derived from the distal westerly terrain. This indicates major differences in provenance and perhaps drainage directions for the Statfjord Formation and the Brent Group sediments. The Sm-Nd data for the Brent Group are thus largely consistent with the existing ideas of proximal westerly, southerly or easterly provenance for the Brent sediments. The other noteworthy general feature is the marked trend towards lower provenance ages upwards through the Brent section (Figs 4 and 5).

### *Broom Formation*

In Fig. 5 it can be seen that the Broom Formation and adjacent strata, form a minor excursion in the Sm-Nd record towards slightly higher provenance ages. Provenance ages for the Broom Formation (1700–1800 Ma) are at the upper end of the range for southwesterly and easterly sources but could be derived from either and are therefore not diagnostic. However, it has been postulated that the Broom Formation is of westerly provenance as opposed to the Oseberg Formation which has easterly provenance (Graue *et al.* 1987). The 1700–1800 Ma signature is incompatible with derivation from the distal westerly source and is thus taken to indicate the provenance signature of the proximal southwesterly terrain at this time. This conclusion relies heavily on the assumption that the strata at Gullfaks are in fact Broom and not Oseberg Formation. The lowermost sample of the Rannoch Formation has a similar provenance signature to that of the Broom Formation which is consistent with the suggestion of Brown *et al.* (1987) that the lowermost Rannoch locally reworked Broom strata.

### *Rannoch and lower Etive Formations*

Above the lowermost Rannoch sample there is a small but distinct change in the provenance

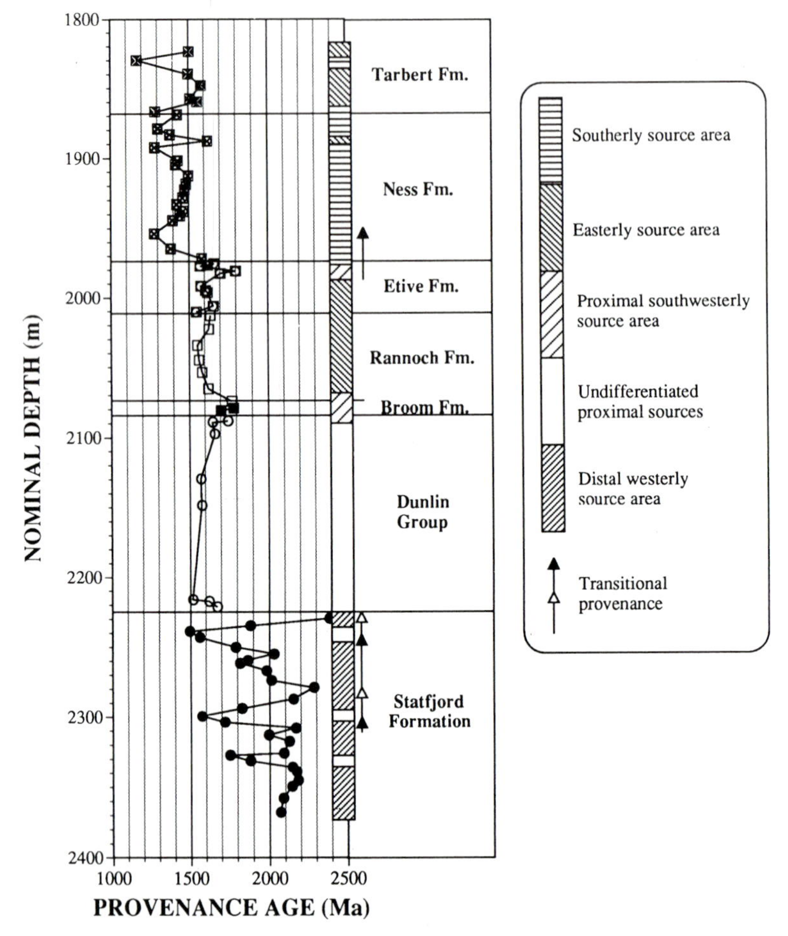


**Fig. 5.** Stratigraphic variations of provenance ages through the Lower–Middle Jurassic section in the Gullfaks oilfield. Note that the Dunlin Group section is incomplete. The section shown comprises sandstones, siltstones and mudstones which give similar provenance ages so long as the provenance is the same. A general feature is an upwards decrease in provenance ages through the section. However, in Snorre Field, which is only 30 km to the north of Gullfaks, the Late Triassic, Upper Lunde Formation has provenance ages in the interval 1350–1650 Ma (Mearns *et al.* 1989) which shows that provenance ages do not necessarily decrease in the direction of younger strata. Note the similarity between Broom and upper Etive provenance ages, uniform provenance ages in the Rannoch and lower Etive Formations and variable provenance ages in the Ness and Tarbert Formations.

signature towards lower provenance ages (1550–1650 Ma, Fig. 5). This lower value persists throughout the Rannoch Formation and the lower half of the Etive Formation. Compared with the rest of the Brent section, this unit is most notable for the uniformity of its provenance ages. The shift in Sm-Nd signature between Broom and Rannoch deposition marks a change in provenance. If the suggestion of Morton *et al.* (1989) is correct then this could indicate a shift from the proximal southwesterly to an easterly source with sediment transport occurring by longshore processes. The uniformity of the provenance ages in the Rannoch and lower Etive is compatible with thorough mixing of detritus in a shallow marine environment. The Sm-Nd data are consistent with these interpretations but at the moment, by no means prove them. A comparison of the Sm-Nd signatures of the Rannoch and lower Etive sections at Gullfaks with the easterly derived Oseberg Formation in Oseberg Field is required to confirm that these two units have a common source area.

### *The upper Etive and lower Ness Formations*

A small but distinct increase in provenance ages in the middle of the Etive Formation signifies a change in source of sediment supply with a return to the provenance ages seen in the Broom Formation (Fig. 5). One possible interpretation is that there was a shift in sediment supply from the postulated easterly long-shore system back to the proximal southwesterly source postulated for the Broom. The shift in provenance coincides with the Gullfaks operators' subdivision of the Etive lithostratigraphic unit into upper and lower units and it therefore seems possible that the change in provenance coincided with a facies change.

One of the most prominent features in the Brent Sm-Nd record is a trend from higher to lower provenance ages which spans the Etive–Ness boundary (Fig. 5). This trend has been recognized in two Gullfaks wells (Mearns 1989) and thus its existence is unequivocal. The trend starts with provenance ages of *c.* 1700–1800 Ma and ends with ages of *c.* 1300 Ma. The fact that this trend is smooth indicates that there has not been any sudden change in provenance but rather a gradual change in source area character which must involve increasing participation of relatively young, mantle-derived rocks.

Numerous samples from the Ness Formation have provenance ages lower than 1400 Ma which makes them unlikely candidates to have been derived from either proximal southwesterly, or easterly source areas. This young provenance signature therefore points towards a southerly source assuming that this source was affected by the extrusion of basalts at this time. It would appear that the return to the proximal southwesterly source area, recognised in the upper Etive Formation, was of a transient nature, as during deposition of the lower Ness Formation the source area extended gradually southwards to eventually incorporate material derived from the Forties area of the mid North Sea. One possible way of viewing this sequence of events is that the maximum extension of the Brent system to the north was simultaneous with the maximum extension of the source terrain to the south. Assuming that the supply of sediment to the Brent system is linked to the areal extent of the source terrain the northwards migration of the Brent shoreline may in this way be linked directly to southwards growth of the southerly source area.

The unit defined by upwards decreasing provenance ages is followed by highly regular but slowly increasing provenance ages through the lower three fifths of the Ness Formation suggesting continuity of sediment supply from the southerly source. However, towards the top of the Ness Formation there is greater variance in the provenance ages, a feature which continues into the Tarbert Formation.

### *Upper Ness and Tarbert Formations*

Graue *et al.* (1987) make the important point that the lower part of the Ness Formation is regressive, having been deposited behind the advancing shoreline, while the upper part is transgressive, having been deposited behind the retreating shoreline. The increased variance in the provenance ages at the top of the Ness Formation could reflect this fundamental change in process. For example, marine incursions across the lower delta plain, which herald the change in shoreline migration, could transport long-shore sediments from the marine environment inland where they would be interdigitated with fluvial sediments derived from the south. A modern example of this type of process has been documented in the inner Moray Firth of Scotland. Here, estuarine muds have lower Sm-Nd provenance ages than the fluvial sands in rivers draining into the firth. This is consistent with the muds being transported landwards by marine currents from the outer Moray Firth (Mearns 1988).

The Tarbert Formation is characterised by provenance ages of *c.* 1500–1600 Ma interdigitated with occasional horizons with significantly lower provenance ages. The 1500–1600 Ma grouping is similar in magnitude to that seen in the Rannoch and lower Etive unit which could indicate a return to the longshore easterly derived sediment source in the marine strata of the Tarbert. The young provenance age spikes could relate to fluvially derived horizons originating from the south that interdigitate with the longshore derived material.

## Synthesis

The preferred interpretations of the present data set are summarized in Fig. 6. Figure 6a showing Broom deposition, shows easterly derivation of the Oseberg Formation and westerly derivation of the Broom Formation and draws heavily on the interpretations of other authors (e.g. Graue *et al.* 1987). The assumed Broom equivalent strata at Gullfaks are shown to have westerly derivation. This, however, is speculative, and Sm-Nd data for the Broom Formation to the west of Gullfaks and from the Oseberg Formation would be necessary to confirm or

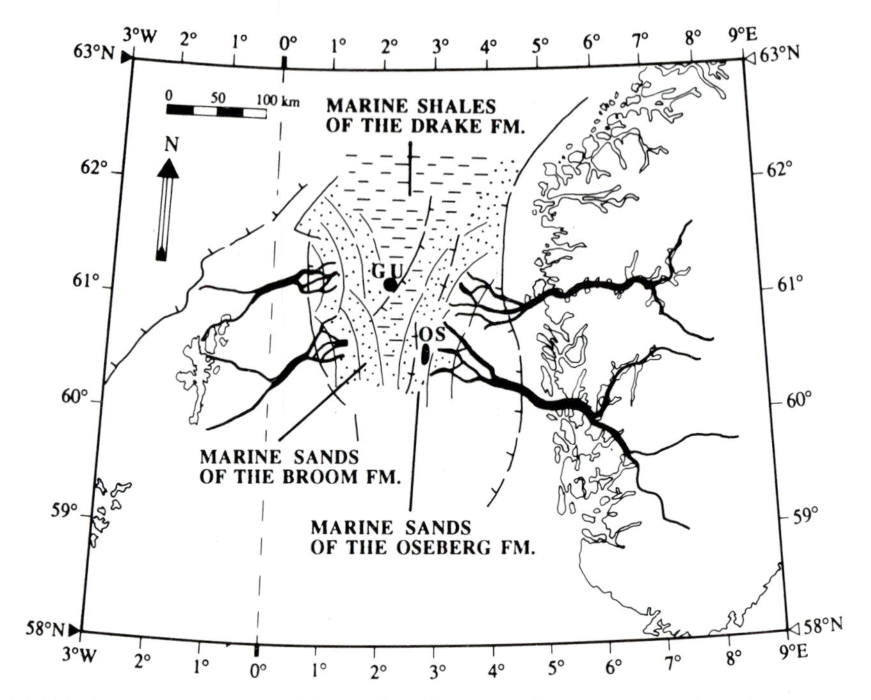


**Fig. 6.(a)** Bajocian palaeogeography of the northern North Sea showing westerly derivation of Broom and Easterly derivation of Oseberg Formation strata. The 'Broom Formation' at Gullfaks is assumed to belong to the westerly provenance. If this turns out to not be the case then many of the interpretations given here will have to be revised. Gu, Gullfaks; OS, Oseberg.

refute this interpretation. In short, Sm-Nd data could be used to map out the spheres of influence of easterly and westerly derived Broom equivalent strata so long as there is a difference in the provenance signatures between the Broom and Oseberg Formations.

In Fig. 6b, illustrating Rannoch–Ness deposition, there are two major fluvial–deltaic systems shown: one draining the proximal southwesterly and southerly sources and the other draining the easterly source. Additional fluvial systems may also be present to the NE and NW of those shown. Detritus deposited by the easterly system is shown to be transported westwards by marine currents. The most significant aspect of this interpretation is that the lower delta plain facies of the axial southerly system migrate over marginal marine strata derived initially from the easterly source. On this basis it may be inferred that the Rannoch–lower Etive sequences to the NW of Gullfaks should contain a component of detritus derived from the axial southerly system. The Sm-Nd provenance signature would depend upon that being provided by the southerly system at the particular time, and as this has varied significantly from *c.* 1800 Ma to *c.* 1300 Ma, quite varied provenance ages could be expected in the Rannoch to the NW of Gullfaks. The Sm-Nd data presented for the Rannoch Formation by Hamilton *et al.* (1987) probably come from a field to the west of Gullfaks and the 5 samples vary between 1440 Ma and 1753 Ma. These data are therefore consistent with what would be expected. The axial southerly system probably derived its detritus initially from the proximal southwesterly (Broom) source terrain before growing southwards along the Viking Graben to eventually draw sediment off the Forties volcano.

A final point which awaits a satisfactory explanation is the consistent Caledonian Rb-Sr dates for detrital muscovites reported by Hamilton *et al.* (1987). From the preceding it should be apparent that it is unlikely that the muscovites were derived as first cycle sediments from the Caledonides of mainland Scotland, and as Caledonian basement is of limited areal extent on the Shetland Platform, first cycle derivation from that area also seems implausible. That leaves two alternative explanations: either the muscovites are derived from the south

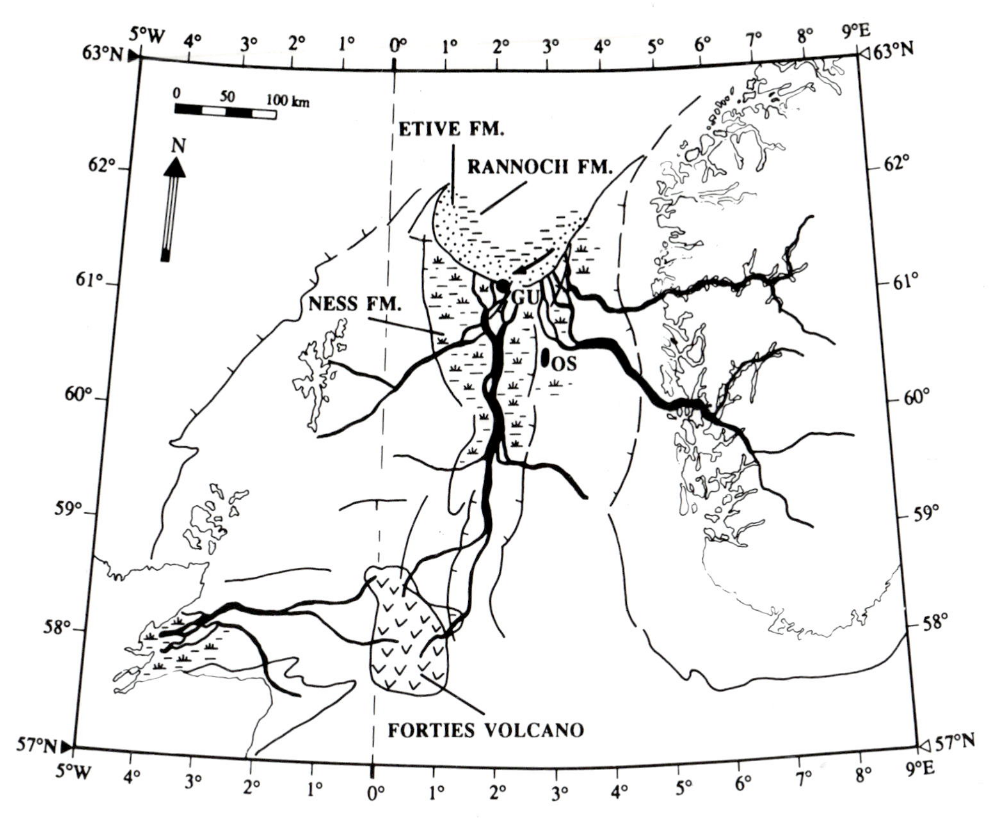


(**b**) Bathonian palaeogeography of the northern North Sea. This model borrows heavily on the interpretation of Morton *et al.* (1989) who recognize similarities between the garnet geochemistry of the easterly derived Oseberg Formation and the Rannoch and Etive Formations in the Gullfaks Field. This led these authors to suggest westward transport of easterly derived sediment in the shallow marine environment denoted here by the arrow. The Sm-Nd data from Gullfaks fit well with this interpretation although further work is required to confirm or refute this hypothesis. One consequence of this model is that the lower delta plain facies of the axial southerly drainage system migrate over shallow marine strata of easterly provenance.

Norwegian Caledonides or they are recycled from sedimentary rocks derived from the Scottish Caledonides, e.g. the ORS deposits. If it is assumed that the locality of these muscovites is to the west of Gullfaks, then in light of the interpretations presented here, a combination of these provenances seems most likely. If this is the case, then the muscovite mineral separates will represent mixtures of diverse populations, some of which may have been recycled several times in the sedimentary environment. The processes of mixing populations combined with possible alteration of the muscovites due to sedimentary recycling would corrupt the isotopic systematics. It is therefore suggested that these dates and initial ratios should be treated as rough estimates. Nevertheless, it would appear that the Rannoch Formation has provenance predominantly within the Caledonian domain which sets important constraints on the areal extent of the source area and this therefore stands as a valuable conclusion.

## Conclusions

Sm-Nd data have provided a detailed picture of how provenance has evolved during deposition of the Brent Group in the Gullfaks Oilfield. At present, regional interpretations are speculative and are constrained mainly by a paucity of such data from other areas. The participation of at least three distinct source terrains is identified. It is suggested that the Broom Formation was derived from the proximal southwesterly source comprising the Shetland Platform. The Rannoch and lower Etive Formations were derived from an easterly source stretching into southern Norway. The upper Etive Formation sees a return to the proximal southwesterly source. During deposition of the Ness Formation this source area was extended southwards to encompass a southerly source that included material derived from the Forties volcanic province. The marine deposition of the Tarbert Formation perhaps sees a return to the easterly source that provided the Rannoch and lower Etive strata. The main consequence of this interpretation is that at least two major river–delta systems were involved in depositing the main Brent cycle. Establishing the spheres of influence of these systems and the probable existence of further drainage systems to the NE and NW of those already proposed awaits further detailed provenance work.

The Sm-Nd work on Gullfaks oil field was supported by the Statoil VISTA program. P. Lowry and A. MacDonald are thanked for useful discussions and for reviewing the first version of the manuscript. C. Smalley and P. Taylor are thanked for thorough and constructive reviews. I. Minciel-Waligora drafted some of the diagrams.

## References

Brown, S., Richards, P. C. & Thomson, A. R. 1987. Patterns in the deposition of the Brent Group (Middle Jurassic) UK North Sea. *In*: Brooks, J. & Glennie, K. (eds) *Petroleum Geology of North West Europe*. Graham & Trotman, London, 899–913.

Budding, M. C. & Inglin, H. F. 1981. A reservoir geological model of the Brent Sands in southern Cormorant. *In*: Illing, L. V. & Hobson, G. D. (eds) *Petroleum Geology of the Continental shelf of North-West Europe*. Heyden and Son, London, 326–334.

Coleman, J. M. & Wright, L. D. 1981. Modern river deltas: Variability of processes and sand bodies. *In*: Broussard, M. L. (ed.) *Deltas*. Houston Geological Society, Houston, 99–149.

Dempster, T. J. 1985. Uplift patterns and orogenic evolution in the Scottish Dalradian. *Journal of the Geological Society, London*, **142**, 111–128.

Dixon, J. E., Fitton, J. G. & Frost, R. T. 1981. The tectonic significance of Post-Carboniferous igneous activity in the North Sea Basin. *In*: Illing, L. V. & Hobson, G. D. (eds) *Petroleum Geology of the Continental shelf of North-West Europe*. Heyden and Son, London, 121–137.

Eynon, G. 1981. Basin development and sedimentation in the Middle Jurassic of the northern North Sea. *In*: Illing, L. V. & Hobson, G. D. (eds) *Petroleum Geology of the Continental Shelf of North-West Europe*. Heyden and Son, London, 196–204.

Graue, E. Helland-Hansen, W., Johnsen, J., Lømo, L., Nøttvedt, A., Rønning, K., Ryseth, A. & Steel, R. 1987. Advance and retreat of Brent Delta System, Norwegian North Sea. *In*: Brooks, J. & Glennie, K. (eds) *Petroleum Geology of North West Europe*. Graham & Trotman, London, 915–937.

Halliday, A. N. 1984. Coupled Sm-Nd and U-Pb systematics in late Caledonian granites and basement under northern Britain. *Nature*, **307**, 229–233.

Hamilton, P. J., Fallick, A. E. & Macintyre, R. M. 1987. Isotopic tracing of the provenance and diagenesis of Lower Brent Group sands, North Sea. *In*: Brooks, J. & Glennie, K. (eds) *Petroleum Geology of North West Europe*. Graham & Trotman, London, 939–949.

Hurst, A. & Morton, A. C. 1988. An application of heavy-mineral analysis to lithostratigraphy and reservoir modelling in the Oseberg Field, Northern North Sea. *Marine and Petroleum Geology*, **5**, 157–169.

Johnson, H. D. & Stewart, D. J. 1985. Role of clastic sedimentology in the exploration and production of oil and gas in the North Sea. *In*: Brenchley, P. J. & Williams, B. P. J. (eds) *Sedimentology: Recent Developments and Applied Aspects*. Geological Society, London, Special Publication, **18**, 249–310.

Lux, D. R. 1985. K/Ar ages from the Basal Gneiss Region, Stadlandet area, Western Norway. *Norsk Geologisk Tidsskrift*, **65**, 277–286.

McCulloch, M. T. & Wasserburg, G. J. 1978. Sm-Nd and Rb-Sr chronology of continental crust formation. *Science*, **200**, 1003–1011.

Mearns, E. W. 1988. A Samarium-Neodymium isotopic survey of modern river sediments from northern Britain. *Chemical Geology (Isotope Geoscience Section)*, **73**, 1–13.

—— 1989. Neodymium isotope stratigraphy of Gullfaks Oil Field. *In*: Collinson, J. D. (ed.) *Correlation in Hydrocarbon Exploration*. Graham & Trotman, London, 201–215.

——, Knarud, R., Ræstad, N. Stanley, K. O. & Stockbridge, C. P. 1989. Samarium-Neodymium isotope stratigraphy of the Lunde and Statfjord Formations of Snorre Oil Field, northern North Sea. *Journal of the Geological Society, London*, **146**, 217–228.

Morton, A. C. 1985. A new approach to provenance studies: electron microprobe analysis of detrital garnets from Middle Jurassic sandstones of the northern North Sea. *Sedimentology*, **32**, 553–566.

——, A. C. 1987. Detrital garnets as provenance and correlation indicators in North Sea reservoir sandstones. *In*: Brooks, J. & Glennie, K. (eds) *Petroleum Geology of North West Europe*. Graham & Trotman, London, 991–995.

—— & Humphreys, B. 1983. The petrology of the Middle Jurassic sandstones from the Murchison Field, North Sea. *Journal of Petroleum Geology*, **5**, 245–260.

——, Stiberg, J. P., Hurst, A. & Qvale, H. 1989. Use of heavy minerals in lithostratigraphic correlation, with examples from Brent sandstones of the northern North Sea. *In*: Collinson, J. D.

(ed.) *Correlation in Hydrocarbon Exploration*. Graham & Trotman, London, 217–230.

MORTON, N., SMITH, R. M., GOLDEN, M. & JAMES, A. V. 1987. Comparative stratigraphic study of Triassic-Jurassic sedimentation and basin evolution in the northern North Sea and north-west of the British Isles. *In*: BROOKS, J. & GLENNIE, K. (eds) *Petroleum Geology of North West Europe*. Graham & Trotman, London, 697–709.

NYSTUEN, J. P., KNARUD, R., JORDE, K. & STANLEY, K. 1989. Correlation of Triassic to Lower Jurassic sequences, Snorre Field and adjacent areas, northern North Sea. *In*: COLLINSON, J. D. (ed.) *Correlation in Hydrocarbon Exploration*. Graham & Trotman, London, 273–289.

O'NIONS, R. K., HAMILTON, P. J. & HOOKER, P. J. 1983. A Nd isotope investigation of sediments related to crustal development in the British Isles. *Earth and Planetary Science Letters*, **63**, 229–240.

RIDD, M. F. 1981. Petroleum geology west of the Shetlands. *In*: ILLING, L. V. & HOBSON, G. D. (eds) *Petroleum Geology of the Continental Shelf of North-West Europe*, Heyden and Son, London, 414–425.

RITCHIE, J. D. & DARBYSHIRE, D. P. F. 1984. Rb-Sr dates on Precambrian rocks from marine exploration wells in and around the West Shetland Basin. *Scottish Journal of Geology*, **20**, 31–36.

SKARPNES, O., HAMAR, G. P. JACOBSSON, K. H. & ORMAASEN, D. E. 1980. Regional Jurassic setting of the North Sea north of the central highs: *In: The Sedimentation of the North Sea Reservoir rocks*. Norwegian Petroleum Society, 13/1–8.

TAYLOR, S. R. & MCLENNAN, S. M. 1985. *The Continental Crust: Its Composition and Evolution*. Blackwell, Oxford.

# Provenance of Brent Group sandstones: heavy mineral constraints

A. C. MORTON

*British Geological Survey, Keyworth, Nottingham NG12 5GG, UK*

**Abstract:** Heavy mineral suites in Brent Group sandstones have been heavily influenced by hydraulic and diagenetic processes in addition to provenance. Geochemical studies of detrital garnet assemblages have shown a great complexity in provenance that argues against a dominant sediment supply northward from an uplifted central North Sea dome. Much of the sediment was fed laterally, particularly in the Broom and Oseberg Formations. The Rannoch, Etive and Tarbert sequences are dominated by similar suites to those in the Oseberg Formation, implying that most of the shoreface and barrier sands were transported longshore from the east. Ness mineralogy is extremely complex. It has a broadly southerly source (possibilities including the southern Shetland Platform and Horda Platform). Close to the western margin of the basin there are strong northern Shetland Platform influences. The lateral and stratigraphic variation in Ness assemblages argues for deposition by several small-scale river systems rather than one single major river. The strong mineralogical variations provide a basis for reservoir subdivision in most of the studied fields, enable the identification of Ness channels that occur at the Etive level, and identify deposits within the Ness sequence that were derived from the barrier sequence into the Ness lagoonal sequence.

The sandstones of the Middle Jurassic Brent Group form the most important hydrocarbon reservoir in northwest Europe. The sequence comprises six formations, Oseberg, Broom, Rannoch, Etive, Ness and Tarbert (Deegan & Scull 1977; Graue *et al.* 1987), which, in broad terms, can be grouped into four main genetic units. At the base of the sequence lie the Broom and Oseberg formations, comprising laterally-fed fan delta systems, Broom from the west and Oseberg from the east. These are succeeded by the Rannoch and Etive formations, comprising a northward-prograding shoreface and barrier complex. This is overlain in turn by the Ness Formation, deposited in a complex of delta-top subenvironments behind the barrier system. The uppermost unit, the Tarbert Formation, comprises stacked barrier bar sequences deposited during the marine transgression marking the end of the Brent depositional cycle. The sequence occupies an embayment in the northern part of the northern North Sea.

In view of the importance of the Brent Group as a hydrocarbon reservoir, there have been many sedimentological evaluations of the sequence both on a local field-wide basis and on a more broad regional basis. However, there has been no serious evaluation of sediment provenance, despite the important constraints this places on evaluation of sedimentological models. The only regional interpretations of sediment provenance have emerged from studies of facies trends within the Brent Group, which have led to the identification of three sources. The fan deltas of the Broom and Oseberg formations are believed to have been laterally sourced from the west (northern Shetland Platform) and east (Norwegian landmass/Horda Platform) respectively (Graue *et al.* 1987), whereas the rest of the Brent deltaic sequence (Rannoch–Tarbert) is believed to have a southerly source on the basis of its overall northward progradation. The location of this general southerly source is the subject of debate. Some interpretations, such as those of Eynon (1981) and Johnson & Stewart (1985) favour a source in the region of the proposed central North Sea domal uplift believed to be associated with the emplacement of the Forties volcanic centre (Ziegler 1981), with sediment transported by a single major river system northward along the line of the subsiding Viking Graben. In contrast, Skarpnes *et al.* (1981, figs 8 and 9) suggest that sediments were carried into the area by a large number of smaller rivers each draining separate source areas to the east, southeast, south, southwest and west. Finally, Richards *et al.* (1988) consider that all sediment was derived transversely, with the sources lying to the southeast (Horda Platform) and southwest (Shetland Platform).

Conventional petrographical constraints on provenance are poor. Although there is much published data on framework composition of Brent sandstones, it is difficult to evaluate such information in provenance terms because there has been extensive modification by diagenetic processes, particularly involving feldspar de-

*From* MORTON, A. C., HASZELDINE, R. S., GILES, M. R. & BROWN, S. (eds), 1992, *Geology of the Brent Group*, Geological Society Special Publication No. 61, pp. 227–244.

pletion, and because provenance-diagnostic lithic fragments are scarce. The small amount of Rb-Sr, Sm-Nd and K-Ar isotopic data available from Rannoch sandstones is consistent with a source within the Scottish–Norwegian Caledonian orogen (Hamilton *et al.* 1987), although the evidence appears to rule out the Scottish mainland. It is impossible to tell on the isotopic data alone whether Brent sands were derived directly from Caledonian basement or indirectly from the same source, by recycling of pre-existing sandstones. Morton & Humphreys (1983) inferred a Shetland Platform source for Brent sandstones of the Murchison Field on the basis of similarities between their heavy mineral suites and those from Palaeocene submarine fan sandstones of the northern North Sea. However, subsequent studies of detrital garnet from the same area provided evidence for major variations in provenance (Morton 1985*a*), requiring sediment input from more than one source. Subsequent studies of detrital garnet from the Thistle, Oseberg, Statfjord and Gullfaks areas (Morton 1987*a*; Hurst & Morton 1988; Morton *et al.* 1989*a*) have also shown that the provenance of the Brent sequence was complex. This paper integrates the existing data with new evidence from wells 211/29–2 and 211/29–3 (Brent Field), 211/26–1 (Cormorant Field), 211/21–3 (North Cormorant Field), 211/26–4 (Eider Field), 211/23–2 (Dunlin Field) and 210/20–1 (north of Tern Field) to provide regional constraints on the provenance of the Brent sandstones. The wells used in the study are shown in Fig. 1.

## Garnet geochemistry

Heavy mineral analysis is one of the most sensitive techniques available for the evaluation of sand provenance. However, the composition of heavy mineral suites in sandstones is known to be strongly dependant not only on provenance but also on the hydraulic conditions at the time of deposition and on the effects of diagenetic processes (Morton 1985*b*). The most common heavy minerals found in the Brent sand are

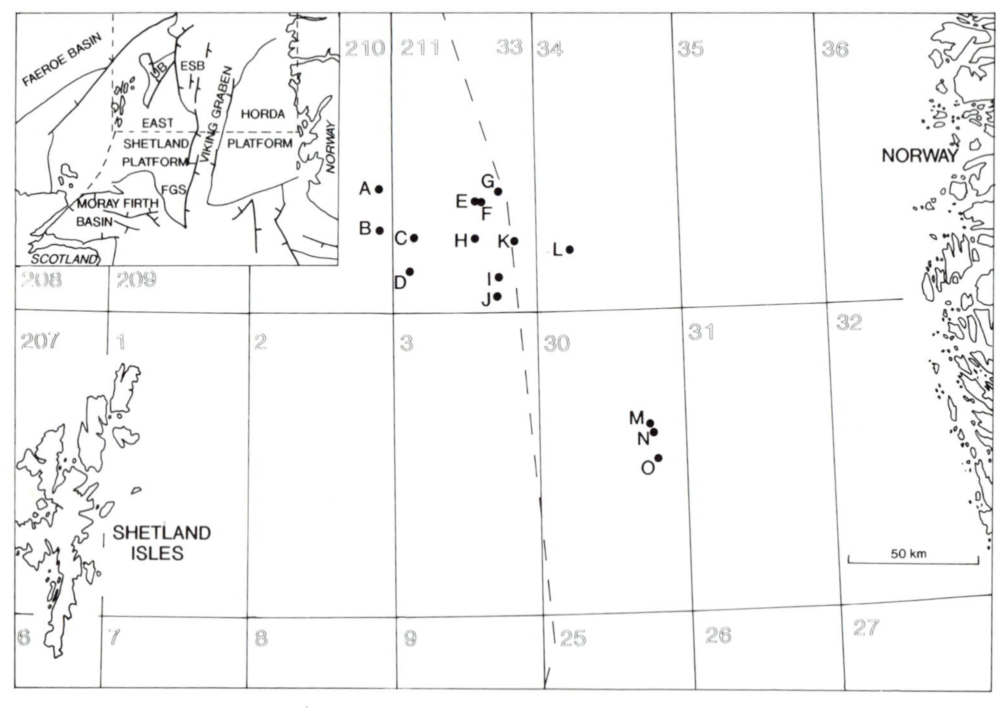


**Fig. 1.** Location of the wells from the Brent province used in this study, with (inset) the structural setting of the northern North Sea area. ESB, East Shetland Basin; UB, Unst Basin; FGS, Fladen Ground Spur. A, 210/20–1; B, 210/25–2 (Tern Field); C, 211/21–3 (North Cormorant Field); D, 211/26–1 (Cormorant Field); E, 211/18– A33 (Thistle Field); F, 211/19–1 (Thistle Field); G, 211/19–4 (Murchison Field); H, 211/23–2 (Dunlin Field); I, 211/29–3 (Brent Field); J, 211/29–2 (Brent Field); K, Statfjord Field; L, Gullfaks Field; M, 30/6–7 (Oseberg Field); N, 30/6–10A (Oseberg Field); O, 30/6–9 (Oseberg Field).

apatite, garnet, rutile, tourmaline and zircon, but their relative abundances fluctuate greatly. Locally, kyanite and staurolite are abundant. Minor phases include chloritoid, spinel (both zinc- and chrome-rich varieties), monazite, titanite, epidote and amphibole. Both grain-size variations, which fractionate the ratio of 'light' to 'heavy' heavy minerals (e.g. apatite/zircon ratios) and diagenesis are major controls on these variations, overprinting and masking variations in the provenance signal. Diagenesis itself is a complex control: apatite dissolution is a meteoric water effect (Morton 1986), whereas removal of amphibole, epidote, titanite, kyanite, staurolite and garnet takes place through high temperature porefluid circulation at depth. Consequently, it is difficult to use conventional heavy mineral data (that is, data acquired by optical analysis of heavy mineral residues) to recognize changes in provenance in sequences such as Brent, in which there are strong variations in grain size and major diagenetic effects. To minimize the effects of such overprinting processes and thus maximise the provenance signal, it is necessary to concentrate on the varieties shown by a single mineral phase (Morton 1985*b*). Studies of the compositional variations shown by detrital garnet have been shown to be particularly effective in evaluating provenance of North Sea reservoir sandstones (Morton 1987*a*), because garnet is widespread in the potential source terrains, is a common component of heavy mineral residues, is relatively stable in deep burial, and has a wide compositional range.

Analysis of detrital garnet suites has been undertaken by electron microprobe analysis, using a Link Systems energy-dispersive X-ray analyser attached to a Microscan V electron microprobe. A total of 50 grains have been analysed from each sample. Details of the analytical method are given by Morton (1985*a*). The observed variations in composition mainly result from changes in proportion of four end-members, almandine ($Fe^{2+}$), pyrope (Mg), grossular (Ca) and spessartine (Mn). Thus, garnet assemblages can be displayed visually on triangular plots using almandine + spessartine (AS), pyrope (P) and grossular (G) as poles.

Different Brent sandstone samples have strikingly different garnet assemblages (Fig. 2), implying considerable complexity in provenance. The nature of this complexity is discussed below. Although AS–P–G triangular plots enable visual comparison between garnet assemblages, the large amount of data now available and the complexity of Brent mineralogy has made it necessary to devise a more suitable way of comparing them. This has been achieved by subdividing AS–P–G space into five fields (A–E, Fig. 3), as described by Morton *et al.* (1989*a*). The boundaries of the individual fields were chosen to surround the most extreme assemblages. The number of garnets that fall into each area was then determined, so that the samples can then be compared visually on a triangular plot using the relative abundances of *A*-, *B*- and *C* + *D* + E-type garnets as poles. Because assemblages dominated by *C*-type garnets also contain minor amounts of *D*- and *E*-type garnets (Fig. 2), *C*- *D*- and *E*-type garnets may normally be considered as part of the same assemblage, referred to here as *Cde* suites. However, as shown in Fig. 2, some assemblages contain *D*- or *E*-type garnets but lack common *C* types (Fig. 4). These are considered separately as they represent a different provenance, and are not plotted on the *A* – *B* *C* + *D* + *E* comparative diagrams. This approach enabled discrimination of lithostratigraphic units in the Oseberg Field (Morton *et al.* 1989*a*), as shown in Fig. 5. Subsequently, the same discrimination has been achieved by more conventional statistical handling techniques such as factor, cluster and discriminant analysis (Stattegger & Morton this volume). This verifies that the simpler method is a valid approach and thus has been applied to the regional data set.

## *Effects of varying hydraulic and diagenetic conditions on garnet suites*

The principal reason for analysing garnet suites by microprobe is to counteract the problems caused by hydraulic and diagenetic processes that beset interpretation of conventional heavy mineral data. However, it is important that the possible effects of these processes on detrital garnets are assessed. Morton (1987*b*) showed that during diagenesis, garnet stability is governed by Ca content, so that as dissolution proceeds the garnets occupy a diminishing area in AS–P–G space, shrinking further from the G pole. A similar pattern is apparent in deeply buried sandstones from offshore New Zealand (Smale & van der Lingen in press). Evidence of garnet dissolution is readily found on optical examination, with the occurrence of etch facets diagnosing ongoing corrosion (Morton *et al.* 1989*b*). Thus, in cases where garnets have large-scale etch facets and occupy only the Ca-poor area of the AS–P–G diagram, diagenetic modification of the suite must be strongly suspected. This situation applies to most samples from the

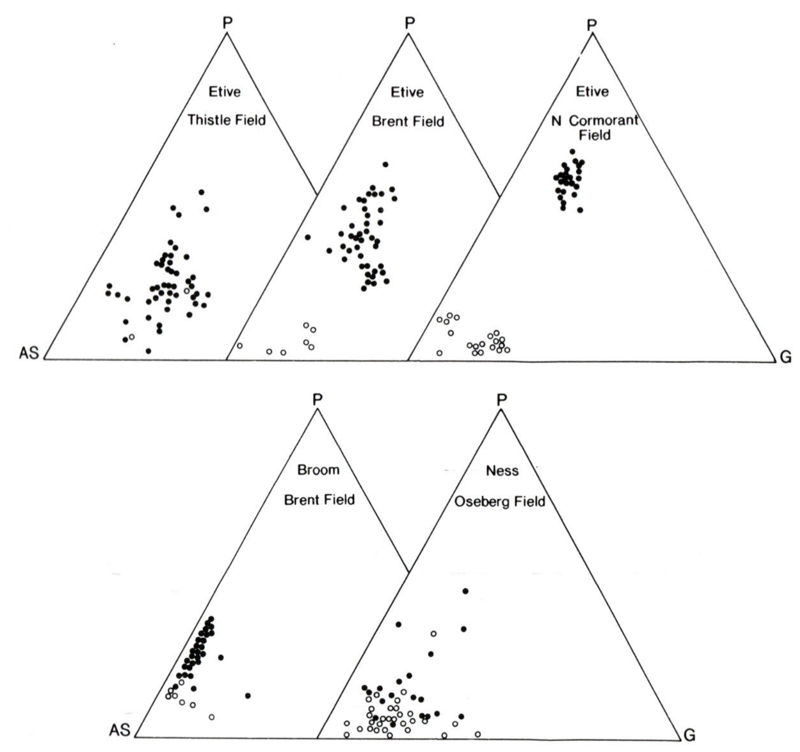


**Fig. 2.** Examples of the variety of garnet geochemical suites found in Brent sandstones from the northern North Sea. AS, almandine + spessartine, P, pyrope; G, grossular. Open circles: spessartine >5%. Filled circles: spessartine <5%.

Eider Field and also to the lowermost part of the Dunlin Field section, and thus these samples were excluded from the provenance analysis.

Consideration of grain size data eliminates the possibility of hydraulic control on garnet suites. As Table 1 shows, samples with *A*-type, *B*-type and *Cde*-type garnets have similar and overlapping ranges in granulometric parameters. Thus, changes in hydraulic conditions could neither cause the generation of a number of different assemblages from a single parent assemblage nor modify one assemblage so that it became another: variations in the proportion of *A*-, *B*- and *C* + *D* + *E*-type garnets are primary and related to variations in provenance.

## Garnet assemblages

The garnet assemblages in the Brent Group are considered here by examining the garnet suites associated with the four main genetic units in turn, thereby building a picture of evolving sources and varying palaeogeography. However, it should be noted that definition of the Ness–Tarbert boundary is problematic in many instances. Indeed, Rønning & Steel (1987) consider that Tarbert and Upper Ness should be considered as a single genetic unit. However, this raises equal, if not greater, problems over the definition of the base of the Upper Ness. Thus, for the purposes of this paper, the Ness and Tarbert are viewed separately, bearing the problems over the boundary definition in mind. There is also some doubt over the Ness–Etive boundary in wells where Ness-related fluvial channels have cut down into the Etive, as in the Oseberg Field (Hurst & Morton 1988; Morton *et al.* 1989*a*). Such occurrences are generally well-marked mineralogically, and for the purposes of this paper, sands at top Etive level that have Ness affinities are discussed in conjunction with undoubted Ness assemblages.

### *Broom and Oseberg formations*

The Broom and Oseberg formations are considered to have been deposited under essentially similar conditions by fan deltas derived respectively from the west (Shetland Platform) and east (Norwegian landmass/Horda Platform), at

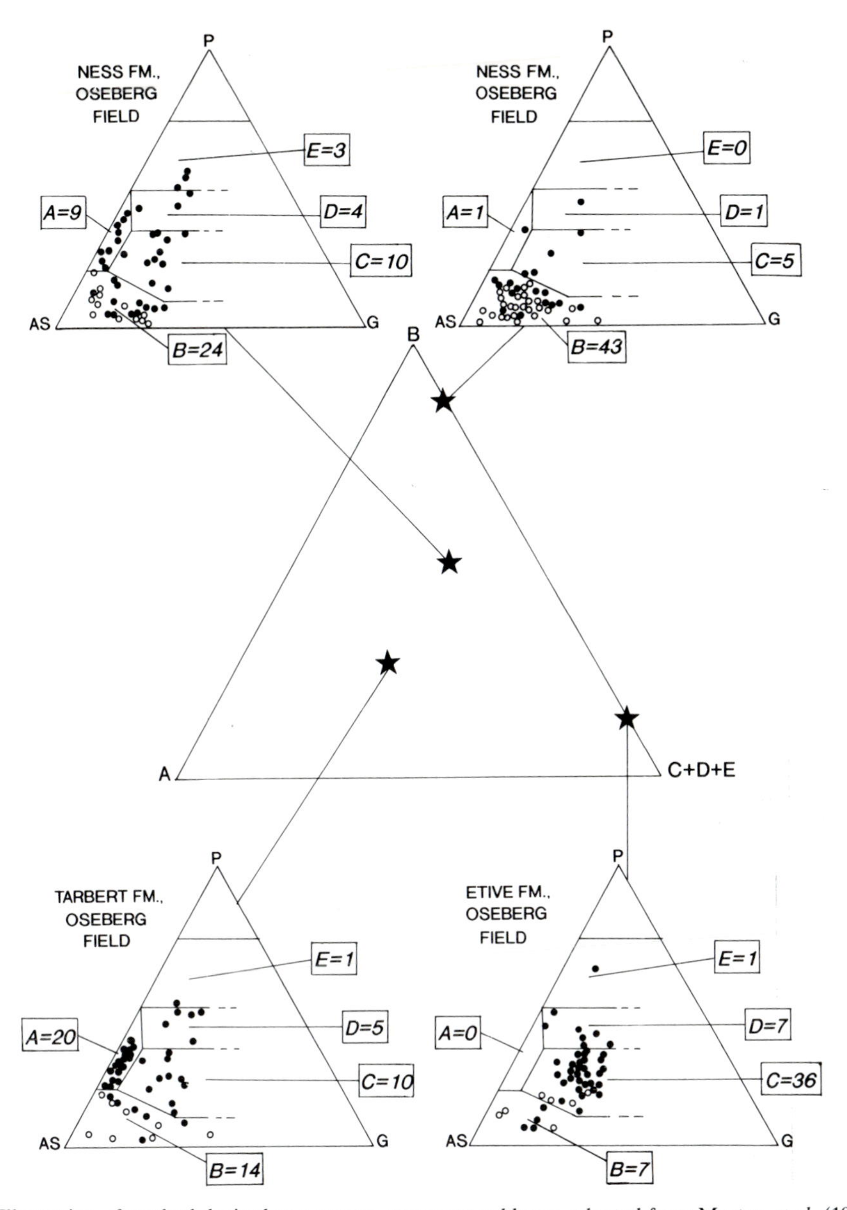


**Fig. 3.** Illustration of method devised to compare garnet assemblages, adapted from Morton *et al.* (1989*a*).

broadly similar times early in the development of the Brent system (Graue *et al.* 1987). Garnet data are now available for the Broom Formation from the Tern, Murchison, Brent and Cormorant fields. All these samples are from the coarse-grained proximal Facies A of Eynon (1981). One probable Broom sample has been studied from North Cormorant, but this is of fine-grained Facies B type. As it bears a closer resemblance to the Rannoch Formation both mineralogically and sedimentologically, it has been considered in conjunction with the Rannoch–Etive interval. Data are also available from the Oseberg Formation of the Oseberg Field (Hurst & Morton 1988; Morton *et al.* 1989*a*).

The garnet suites from these wells are compared in Fig. 6, illustrating a first-order difference between the two fan systems (Fig. 7). This is to be expected considering the clear difference

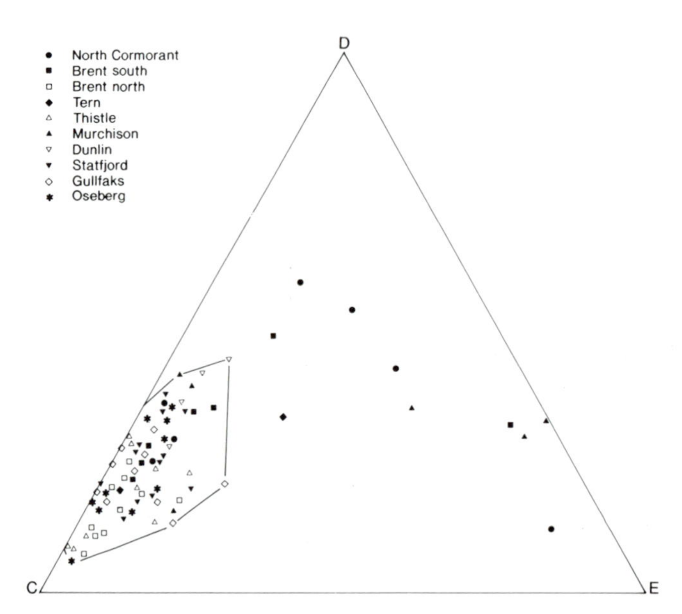


**Fig. 4.** Relative abundances of $C$, $D$ and $E$-type garnets in assemblages with the $C + D + E$ component >50%. See key for explanation of symbols. Note that some samples from the Murchison, Tern, North Cormorant and Brent fields fall outside the main cluster of suites (termed *Cde* suites). These are the *DE* suites discussed in the text.

in source inferred from sedimentological data, but has great significance for the understanding of the overlying shoreface and fluvial sequences. The Broom Formation is dominated by $A$-type garnets (88% in Murchison, 86% in Tern, 81% in Cormorant and 69% in Brent), whereas $A$-type garnets are scarce in Oseberg. In contrast, *Cde* garnets, which predominate in the Oseberg Formation (74–75%) are scarce in the Broom Formation (4–6%). Thus, the material shed from the Shetland Platform is dominated by $A$-type garnets, whereas that from the east (Norway or Horda Platform) is of *Cde* type. A slight regional variation in the nature of the Broom source is suggested by the lower ratio of $A$- to $B$-type garnets in the Brent well compared to the other Broom samples.

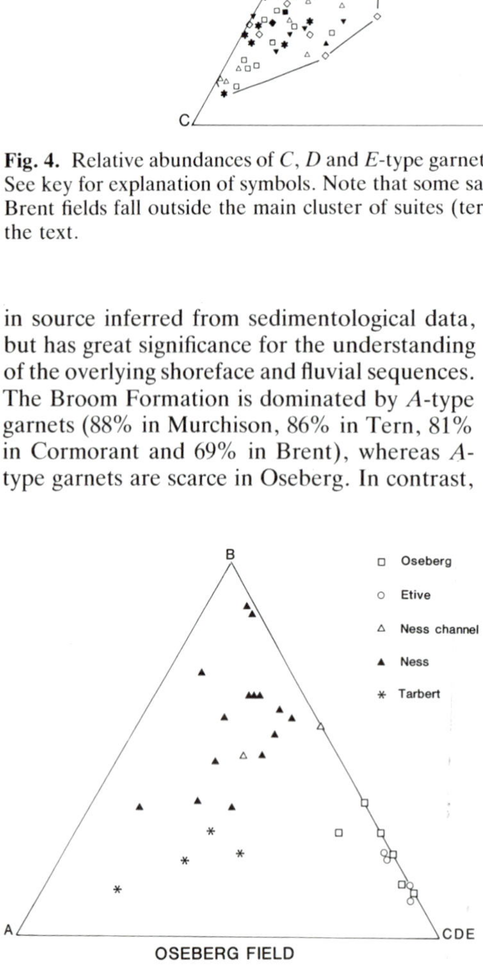

B
Oseberg
Etive
Ness channel
Ness
Tarbert
A
CDE
OSEBERG FIELD

**Fig. 5.** Summary plot showing stratigraphic breakdown of the Brent sequence from the Oseberg Field on the basis of garnet geochemistry. Note the mineralogical identification of a Ness-related sand at top Etive level. (Adapted from Morton *et al.* 1989*a*).

## *Rannoch and Etive formations*

The data set affords a greater geographical coverage of the Rannoch and Etive than of the Broom, with data available from Oseberg, Statfjord, Gullfaks, Cormorant, North Cormorant, Tern, Brent, Thistle and Murchison fields. However, in some cases the stratigraphic coverage is not complete, with a lack of data from Rannoch. This applies to Statfjord, Gullfaks, Tern, Murchison and Thistle, as well as Oseberg where Rannoch is very poorly developed. Despite this, the data provide important insights into the development of the Rannoch and Etive shoreface and barrier system. As Fig. 8 shows, the assemblages mainly cluster close to the $C + D + E$ pole, in a field that compares favourably with that shown by the Oseberg Formation garnets (*Cde* suite). In the northern

**Table 1.** *Comparison of granulometric parameters from selected Brent sandstones showing the similarity between sands with differing garnet assemblages*

| Assemblage | Well | Field | Formation | Mz | IGSD |
|---|---|---|---|---|---|
| A | 211/29–3 | Brent north | Broom | 0.71 | 0.74 |
| A | 211/26–2 | Cormorant | Broom | 1.09 | 1.17 |
| A | 211/23–2 | Dunlin | Tarbert | 1.76 | 0.41 |
| A | 211/29–2 | Brent south | Ness | 2.19 | 0.49 |
| A | 211/21–3 | North Cormorant | Tarbert | 2.61 | 0.52 |
| B | 211/29–3 | Brent north | Ness | 1.40 | 0.92 |
| B | 211/29–2 | Brent south | Ness | 1.90 | 0.62 |
| B | 211/21–3 | North Cormorant | Ness | 2.37 | 0.56 |
| B | 211/29–3 | Brent north | Ness | 3.07 | 0.51 |
| Cde | 30/6–9 | Oseberg | Oseberg | 1.03 | 0.79 |
| Cde | 211/29–3 | Brent north | Etive | 1.63 | 0.73 |
| Cde | 211/29–3 | Brent north | Tarbert | 2.07 | 0.60 |
| Cde | 211/29–3 | Brent north | Rannoch | 2.78 | 0.21 |
| Cde | 211/29–3 | Brent north | Ness | 3.33 | 0.39 |

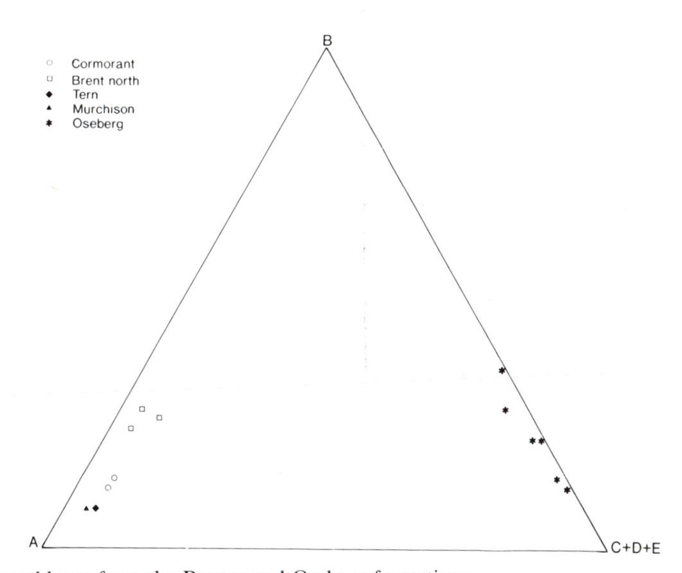


**Fig. 6.** Garnet assemblages from the Broom and Oseberg formations.

Brent well (211/29–3, Fig. 9) and North Cormorant, there is a clear stratigraphic trend from Broom into Rannoch and then into Etive, with an upward progression from *A* domination toward *Cde* domination, with an initial trend toward the *B* pole. In two cases, Cormorant and the southern Brent well (211/29–2), *Cde* domination does not occur. The geographical variation for the entire Rannoch-Etive interval is shown in Fig. 10, which demonstrates the overall *Cde* dominance with a regional trend toward *A*- and/or *B*-rich suites to the southwest.

Thus, the data clearly indicate that with northeastwards progradation of the shoreface system, the early *A*/*B* domination (comparable to Broom) seen in the southwest gives way to *Cde* domination. This either indicates a change in source area or a change in parent rock lithology due to deeper erosion within the same source region. As *A* and *B* suites are prevalent higher in the Brent sequence, the change to *Cde* domination indicates a shift in source area. As these Rannoch–Etive suites are virtually identical to those seen in the Oseberg Formation, it appears that for the most part the shoreface sequence

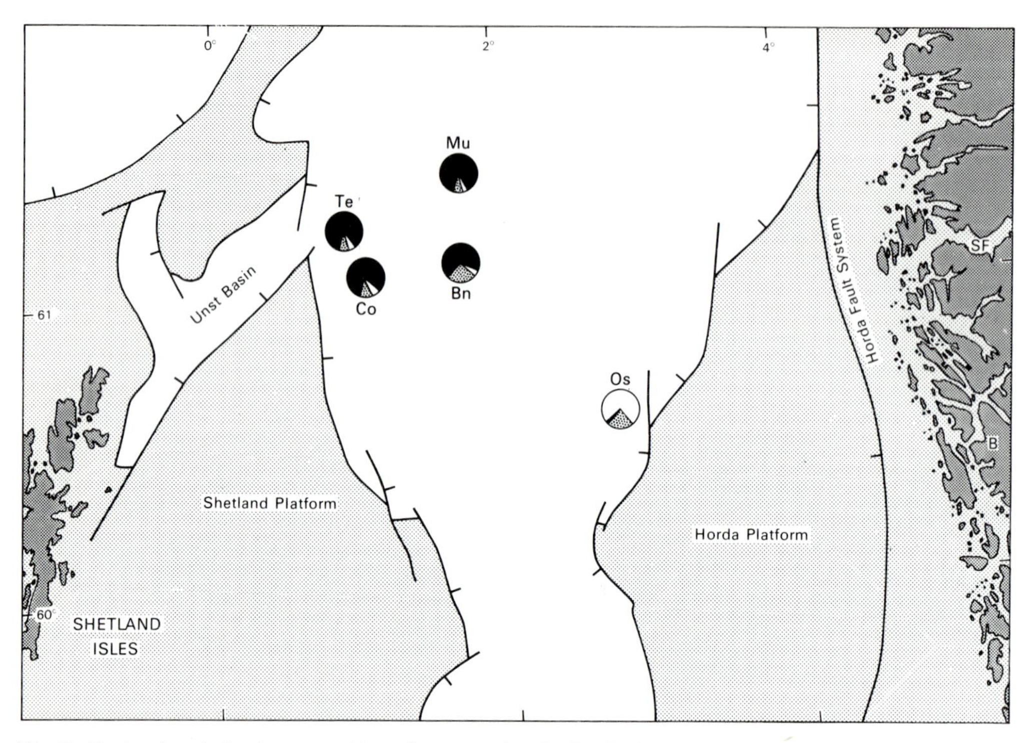


**Fig. 7.** Regional variation in composition of garnet suites in the Oseberg and Broom formations. Black, *A* component; stipple, *B* component; white, *Cde* component. Bn, Brent Field (north); Co, Cormorant Field; Mu, Murchison Field; Os, Oseberg Field, Te, Tern Field.

was sourced longshore, with material carried westward from the Norwegian source. *A*- and *B*-type suites appear to have influenced only the early part of shoreface progradation, with little evidence for involvement in the shoreface at the distal end of the system. This is probably the result of a combination of two factors. Firstly, assuming most sediment was transported longshore, *A*- and *B*-type garnets would be dramatically diluted when incorporated into the shoreface. Secondly, heavy mineral suites dominated by, or rich in, *B*-type garnets are less rich in total garnet than those associated with *Cde* suites, causing a bias in the detection of mixing in assemblages.

## *Ness Formation*

The Ness Formation is the most mineralogically complex interval within the Brent Group (Fig. 11); assemblages include those dominated by *B*- *C*-, *D*- and *E*-type garnets. Such assemblages not only include *Cde* suites as seen in the Etive, Rannoch and Oseberg formations, but also those dominated by *D*- or *E*-types without abundant *C*-types. In addition, *A*-type garnets are also present but rarely dominate assemblages. However, by far the most common suite seen in Ness is dominated by *B*-type garnets, mixed to variable degrees with *A*-type garnets: *Cde*, *D* and *E* suites are found only as special cases. *Cde* suites within the Ness are considered to be related to the shoreface system rather than the fluvial system, as discussed later. They have therefore been excluded from the Ness summary map (Fig. 12). The suites dominated by *D*- and *E*-types appear to represent a single event in Ness development and to not be a part of the normal fluvial system. For this reason, they have also been excluded from the Ness summary map. Thus, the Ness suites are considered in three parts, those strongly influenced by *B*- and *A*-type garnets, those with *Cde* suites, and those dominated by *D*- or *E*-types.

*(i) B- and A-type Ness assemblages.* Most Ness assemblages are rich in the *B* component, as shown in Fig. 11, in which Ness assemblages show a distinct shift toward the *B* pole compared with all other formations. The *A* component is generally subsidiary to *B* except in the western

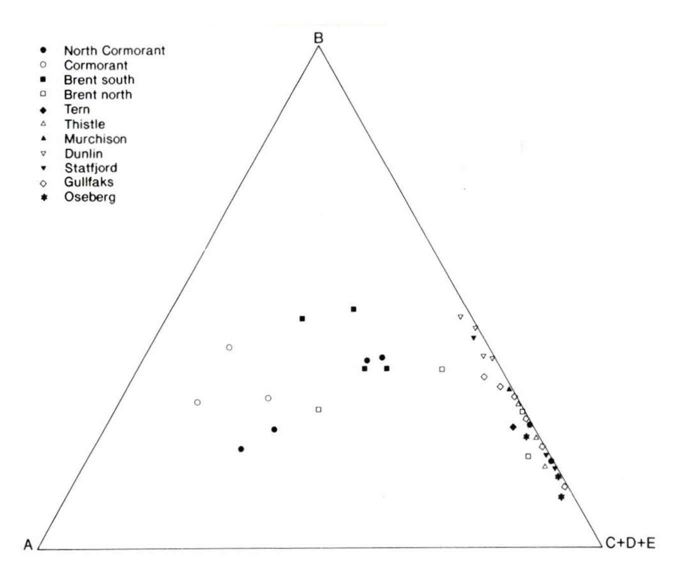


**Fig. 8.** Garnet assemblages from the Rannoch and Etive formations (excluding Ness-related suites at top Etive level).

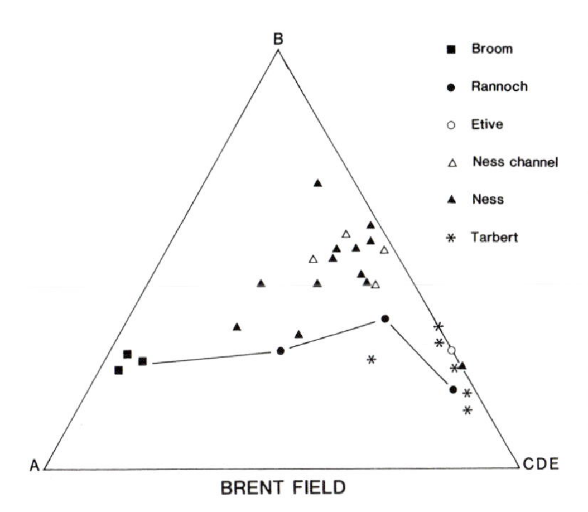


**Fig. 9.** Summary plot showing stratigraphic changes in garnet assemblages from the Brent Field well 211/29–3. Note particularly the progressive change in mineralogy from the Broom into the Etive.

part of the basin; in Tern, Cormorant and the southern Brent well 211/29–2, $A/(A + B)$ exceeds 0.42 (0.427–0.646) whereas in the other areas (Murchison, Dunlin, Thistle, Brent north, Statfjord, Gullfaks and Oseberg) $A/(A + B)$ falls between 0.111 and 0.333. This is consistent with the indications from the Broom Formation, which contains *A*-dominated assemblages that can be traced to the Shetland Platform area to the west. Thus, during Ness deposition, the western area was strongly influenced by sediment derived from the Shetland Platform. This confirms suggestions by Eynon (1981) and Johnson & Stewart (1985) that the western part of the Brent Province was influenced by a fluvial system prograding from the west during Ness sedimentation. The assemblages dominated by *B*-type garnets, in contrast, must be regarded as being of southerly derivation, reflecting the increasing influence of the main fluvial system as the shoreline prograded northward. It is noteworthy that although $C + D + E$ proportions are relatively high in all *A*- and *B*-dominated Ness sediments (minimum of 14% in Oseberg well 30/6–7), they are generally higher in the north (50%) in Thistle, 46% in Gullfaks, 44% in Statfjord; Fig. 12). Thus, although *C*-, *D*- and *E*-type garnets are an integral, if relatively minor component, of the southerly-sourced material, their overall northward increase reflects either increasing amounts of reworking of earlier Rannoch–Etive material (e.g. by channelling of the shoreface sequence by Ness fluvial channels) or increasing degrees of mixing with sediment derived directly from the shoreface system. Although both processes are likely to have been involved, the latter is probably more important, as Ness channelling into Etive appears to become less important toward the north. The local influence of this process can be seen in the Oseberg Field, however, where the two northern wells 30/6–7 and 30/6–10A have low $C + D + E$ values and possess complete, unchannelled Oseberg-Etive sequences,

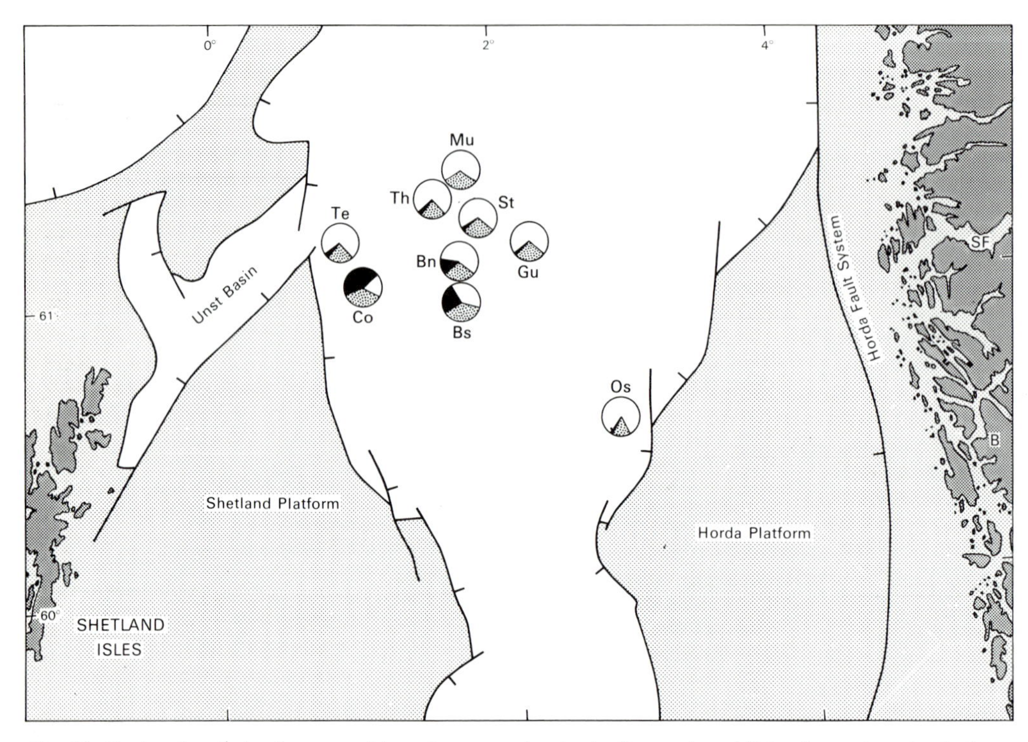


**Fig. 10.** Regional variation in composition of garnet suites in the Rannoch and Etive formations (excluding Ness-related suites at top Etive level). Black, *A* component; stipple, *B* component; white; *Cde* component. Bn, Brent Field (north); Bs, Brent Field (south); Co, Cormorant Field; Gu, Gullfaks Field; Mu, Murchison Field; Os, Oseberg Field; St, Statfjord Field; Te, Tern Field; Th, Thistle Field.

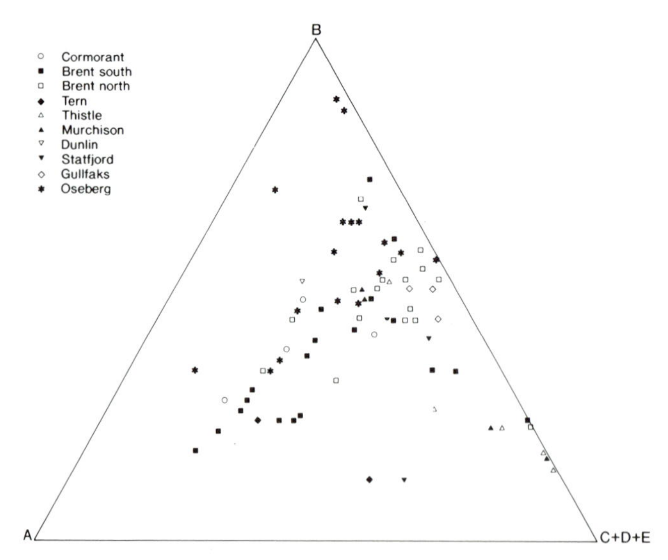


**Fig. 11.** Garnet assemblages from the Ness Formation (including Ness-related suites at top Etive level, but excluding samples with a high *D* + *E* component).

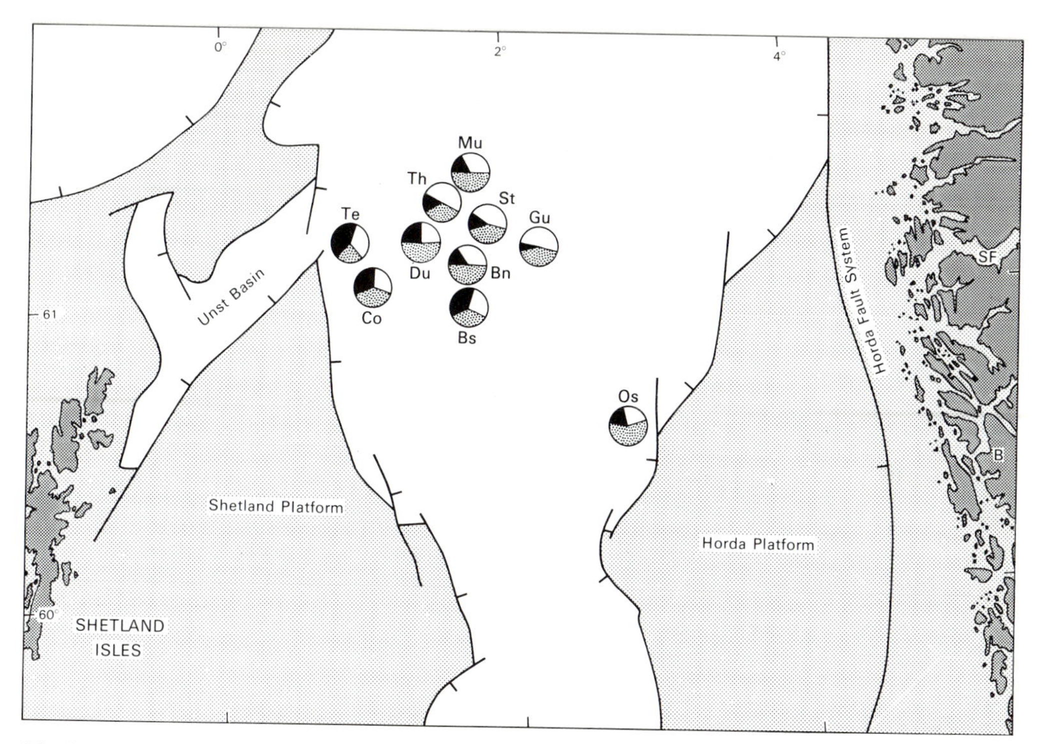


**Fig. 12.** Regional variation in composition of garnet suites in the Ness Formation (including Ness-related suites at top Etive level, but excluding samples with *Cde* and *DE* suites. Black, A component; stipple, *B* component: white, *Cde* component. Bn, Brent Field (north); Bs, Brent Field (south); Co, Cormorant Field; Du, Dunlin Field; Gu, Gullfaks Field; Mu, Murchison Field; Os, Oseberg Field; St, Statfjord Field; Te, Tern Field; Th, Thistle Field.

whereas the southern well 30/6–9, which has higher $C+D+E$ values in Ness, a Ness channel sequence has replaced the Etive and top Oseberg levels (Hurst & Morton 1988).

*(ii) Cde-type Ness assemblages. Cde* assemblages are relatively infrequent in the Ness Formation. They tend to be more common in the north of the area near to the maximum progradational extent of the shoreface system and at the greatest distance from the inferred southerly source of the fluvial system. They are well-developed, for example, in the Ness of the Murchison and Thistle fields, although another example is seen further south, in the Brent Field. Their tendency of occurring in the more distal part of the deltaic complex, together with their mineralogical similarity to the shoreface sands, indicates that Ness sands with *Cde* assemblages were derived by reworking of the barrier sequence. Budding & Inglin (1981) have suggested a number of possible settings for such a phenomenon: flood tidal delta, tidal inlet channel and washover fan. Two examples are shown in Fig. 13, one from the Thistle Field and one from the Brent Field. In Thistle, *Cde* assemblages occur in the lower part of the Ness Formation, both in channels cut down into Etive (in this case preventing a mineralogical distinction between Etive and Ness), and again above a sandstone with a significant *A* component (Fig. 13). This latter situation is also seen in the adjacent Murchison well 211/19–4 (Morton 1987*a*). In the Brent Field, a thin, very fine-grained sand unit near the top of the thick lagoonal Ness sequence is characterized by a *Cde* suite (Fig. 13). Again, this thin but probably laterally widespread sand horizon deposited in the lagoon must have been derived by reworking of barrier sands. The position of this sand near the top of the Ness precludes the possibility that it was reworked from the Etive. In this case, the unit in question was apparently derived from Tarbert shoreface sands, Tarbert in the Brent Field being also characterized by *Cde* suites (see below). This indicates that at this time the

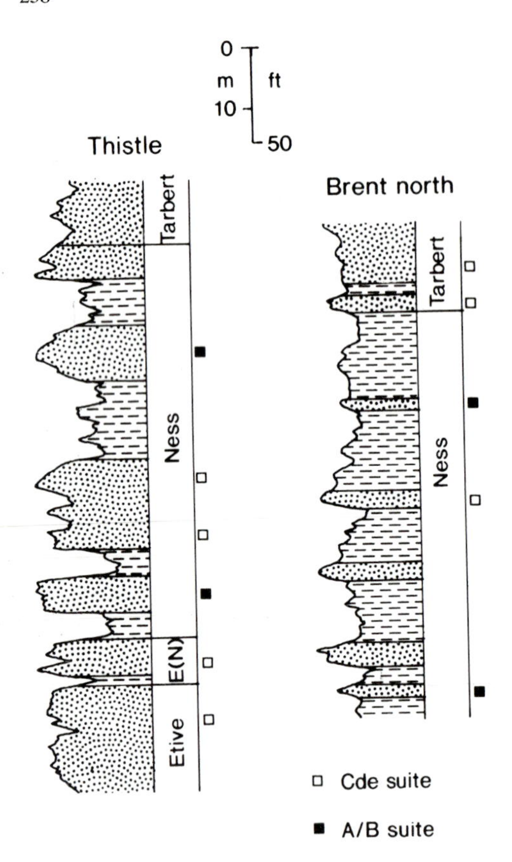


**Fig. 13.** Examples of the occurrence of *Cde* suites within the Ness Formation. E(N), Ness channel at top Etive level.

Tarbert marine transgression was already under way and that the Tarbert shoreface sequence was not far distant to the north.

*(iii) D- and E-type Ness assemblages.* In three, possibly four, wells, distinctive bimodal garnet suites are found, containing abundant *D*- or *E*-type garnets in conjunction with spessartine-rich varieties that fall in field *B*. In all cases, the *DE* suites occur at lowermost Ness or top Etive levels. In Brent south and North Cormorant, both *D*- and *E*-dominated suites occur, in both cases with *E*-type suites succeeding those of *D*-type. Virtually the entire 'Etive' interval in Murchison Field well 211/19–4 comprises *E*-type material, although again the trend toward *E*-domination with time is apparent, with $D/(D+E)$ changing from 0.46 at the base to 0.31 at the top. The only other example seen to date is the occurrence of a *D*-type suite at the base of Ness in the Tern Field, which is not associated with *E*-dominated material. The similarity of the stratigraphic progression and its apparent lack of repetition strongly suggests that the *DE* suites represent a single correlatable sedimentary event (Fig. 14). The direction of transport is very much conjectural at this stage, although the overall change from a general fluvial environment in 211/29–2 and 211/21–3 to a distributary channel setting in 211/19–4 (Brown & Richards 1989) is consistent with a broadly south to north direction, in keeping with the overall pattern of transport previously proposed for the Ness system.

## *Tarbert Formation*

As discussed earlier, the identification of the Tarbert Formation and definition of a Ness–Tarbert boundary is possibly the most difficult lithostratigraphic problem associated with the Brent sequence. Clearly, the effect of an incorrect definition of Tarbert will introduce significant error into the interpretation of Tarbert provenance. However, in most cases the Ness–Tarbert transition is clearly marked mineralogically by a change from the heterogeneous and largely *B*-dominated garnet suites characteristic of Ness, to either *A*- or *Cde*-dominated suites in Tarbert, in some cases with *A* and *Cde* suites coexisting. In Brent, Statfjord and Gullfaks the fluvial–lagoonal Ness with *B*-type garnets gives way to shallow marine shoreface sands with *Cde* assemblages comparable to those in the Rannoch and Etive shoreface sequences (Fig. 15). In other cases (Oseberg, Tern, Dunlin and North Cormorant) the transition is marked by a change from *B*- to *A*-type. In three cases (Brent, Dunlin and Statfjord), *Cde* suites are interbedded with those of *A*-type. The distribution of *A*-dominated and *Cde*-dominated Tarbert is palaeogeographically controlled (Fig. 16); *Cde* suites are dominant in the northeast, but in the west and south the suites are *A*-dominated. Tarbert therefore has a complex provenance, with sediment essentially derived from east and west. Assuming that the oldest Tarbert is that in the northeast, with that in the west and south deposited as the transgression proceeded, the change to *A*-domination presumably reflects the cutting off of the Norwegian *Cde* source as the transgression continued.

## Provenance

The garnet data allow both an assessment of transport direction, by relating the distribution of different assemblages to the facies in which they occur, and an assessment of the nature of the source material from which they were

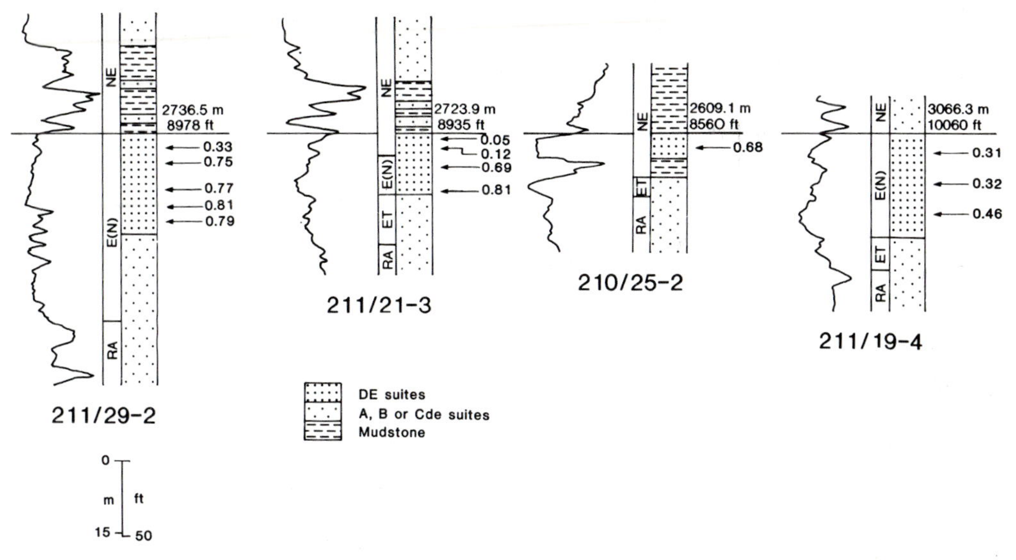


**Fig.14.** *DE* suites in Brent sandstones, showing their similar occurrence at top Etive-base Ness level in the four wells in which they have been documented, and the similarity in stratigraphic trend of the $D/(D + E)$ ratio. Datum is top '*DE* event' (log depth).

derived. By combining both types of information, a detailed picture of provenance can be established.

In broad terms, as described above, four distinct garnet suites are present: *A*-dominated, *B*-dominated, *Cde*-type and *DE*-type (the latter strictly consisting of two different assemblages, but as they are temporally and spatially associated they are here considered together).

## *A-dominated assemblages*

These occur in the coarse grained facies of the Broom Formation, in Tarbert sandstones in the west, south and southeast of the Brent province,

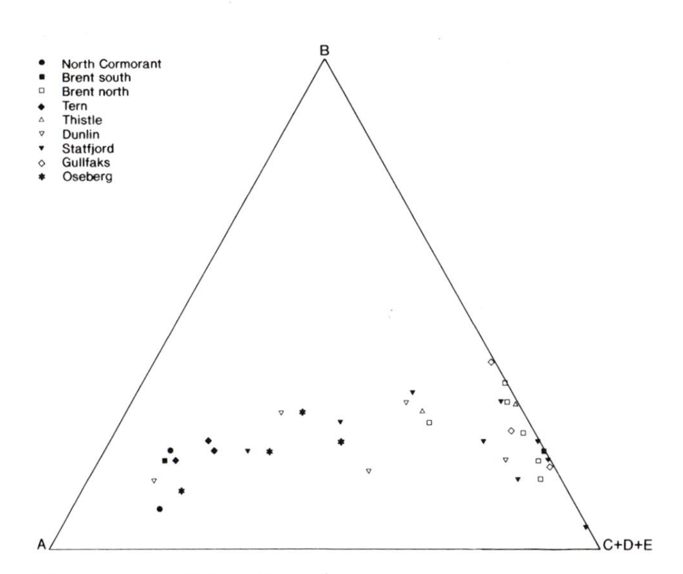


**Fig. 15.** Garnet assemblages from the Tarbert Formation.

**Fig. 16.** Regional variation in composition of garnet suites in the Tarbert Formation. Black, A component; stipple, *B* component; white, *Cde* component. Bn, Brent Field (north); Bs, Brent Field (south); Du, Dunlin Field; Gu, Gullfaks Field; Nc, North Cormorant Field; Os, Oseberg Field; St, Statfjord Field; Te, Tern Field; Th, Thistle Field.

and in Ness sandstones in the west. Their occurrence in the Broom Formation indicates a source on the Shetland Platform to the west, and this is consistent with the occurrence of similar suites adjacent to the Platform during succeeding phases of Brent sedimentation. The garnets that fall into field A are comparatively unusual in metamorphic rocks. Of the 78 almandine–pyrope analyses provided by Deer *et al.* (1982) only six fall into this field, representing garnets from granulites, gneisses and migmatites. However, low-Ca garnets are the most stable in sediments, and are likely to be concentrated during weathering and diagenesis. It is probable, therefore, that *A*-dominated garnet suites indicate a reworked sandstone provenance. Possible candidates include sandstones of Permo-Triassic, Carboniferous and Devonian age, as well as low-grade metasediments, which may also contain etched and rounded detrital garnets (Allen & Mange-Rajetzky 1992). However, as data on garnet mineralogy of potential source rocks are not presently available, it is not possible to identify which is the more likely candidate.

## B-*dominated assemblages*

The *B*-dominated assemblages that characterize the Ness are distinctly heterogeneous, far more than the *A*, *Cde* and *DE* suites (Fig. 11), with markedly variable *A*, *B* and *Cde* components. Although to some extent this reflects intrabasinal mixing, it appears to be an intrinsic feature of Ness sandstones, in both regional and stratigraphic terms. It is generally accepted that the Ness fluvial sequence has a broadly southerly source; for example, Richards *et al.* (1988) infer both southwesterly and southeasterly sources, whereas Johnson & Stewart (1985) indicate derivation from the south. Therefore, the source of the *B*-type suites lay in a broadly southerly direction, although the precise location cannot be determined without data on possible source rocks. The lateral and temporal variability of the Ness suites indicates that either the rivers sourcing the Ness were of limited size, each draining small areas with mineralogies that were broadly comparable but different in detail, or that the mineralogy of the source area changed with time. The lack of any obvious inter-field correlations tends to favour the former, but the degree of variability within individual fields, such as Oseberg (Hurst & Morton 1988; Morton *et al.* 1989*a*) also suggests a degree of the latter. Thus, it appears that the source of the *B* suites was laterally and vertically heterogeneous. Sedimentological studies indicate this source to lie, broadly speaking, to the south of the Brent area, implying that they are most likely to be reworked from pre-existing sediments. The lack of data on potential source lithologies prevents a definite assessment of the age of this source. However, it is noteworthy that Permo-Triassic sandstones of the central North Sea have broadly comparable garnet suites (Morton 1987**a**).

### Cde-*dominated assemblages*

Because *Cde* suites characterise the Oseberg Formation, deposited by a fan-delta system prograding from the east or southeast into the basin, the source of the *Cde* suite can be constrained to lie on the Norwegian margin of the basin. This is further suggested by the presence of identical suites in shallow marine Sognefjord Formation sandstones of the Troll Field (Morton, unpublished data), which lie in blocks 31/2, 31/3 and 31/6 very close to the Norwegian coast between Bergen and the Sognefjord. The range of garnet compositions shown by the *Cde* suites is closely comparable to that shown by Wright (1938) for garnets from amphibolite-facies rocks, and it is therefore suggested that the *Cde* suites were derived from amphibolite-facies metamorphic basement on the Norwegian landmass, such as that located in the area to the north of Bergen (Bryhni & Sturt 1985).

### DE-*dominated assemblages*

The provenance of the *DE* association is the most difficult to define. As discussed earlier, its occurrence within the overall southerly- or westerly-derived Ness fluvial system would be most compatable with a source to the south or west. However, the sedimentological relationships between these sands and typical *B*-type Ness sands are not known, and thus this assumption is likely to be invalid. The overall progression from *D*- to *E*-dominated suites with time is suggestive of unroofing, and the similarity of the assemblages to Wright's (1938) eclogitic and peridotitic assemblages suggests a high-grade ultramafic source. The distinctive association of *D*- and *E*-type garnets with chrome spinel is consistent with this hypothesis. It is unlikely, therefore, that the source lay within the existing basinal area, unless the *DE* suites are themselves reworked, which is unlikely on account of their relative purity. On mineralogical grounds, it appears most likely that the source lay in Norway or the Shetland Platform, where ophiolite and eclogite complexes are found within the Caledonides (Bryhni & Sturt 1985).

## Implications for the regional sedimentological model

The garnet geochemical study has established that the provenance of the Brent sands was considerably more complex than that suggested by earlier regional sedimentological models, which indicated that the source was an uplifted central North Sea dome, with transport northward along the line of the Viking Graben (Eynon 1981; Ziegler 1981; Johnson & Stewart 1985). The present study shows that longshore sediment transport from east to west was of considerable importance. Most of the Rannoch, Etive and Tarbert was sourced in this way, and several Ness sand units were also derived from the same source by reworking of barrier sands. Only in the Ness is southerly-derived detritus dominant.

The importance of longshore processes is well known in many coastal sequences, and major mineralogical differences can readily be observed in such cases. For example, Darby (1985) showed that along the estuary-dominated US Atlantic coast, the barrier sands have different mineralogies to the bay and river sands, implying that the barrier sands were mostly sourced longshore. Beach and adjacent river sands of the Rio de Janiero area (Brazil) also have different sources (Morton, unpublished data). Such situations are unlikely in more fluvially-dominated deltas such as the Mississippi: thus, the mineralogical findings tend to uphold the view that the Brent sands were the product of a wave-dominated delta (Johnson & Stewart 1985) rather than of a fluvially-dominated delta. This, coupled with the evidence for considerable lateral and vertical heterogeneity within Ness, indicates that the sediments were deposited by several small-scale rivers and that the sequence might best be viewed as a prograding coastal complex rather than a true deltaic system. In this respect, the mineralogical data favour the palaeogeographic reconstructions suggested by Skarpnes *et al.* (1980) and Richards *et al.* (1988), with a number of small-scale rivers draining relatively small areas, rather than those of Eynon (1981), Ziegler (1981) and Johnson & Stewart (1985), which essentially involve a single major river draining a larger source.

## Conclusions

Consideration of the garnet geochemistry of sandstones from the four genetic depositional units has provided insight into the patterns of sediment transport and dispersal within the Brent sequence. In particular the following points are evident.

(1) The fan delta deposits of the Broom and Oseberg formations were fed laterally from west and east respectively, the Broom characterized by *A* suites and the Oseberg by *Cde* suites.

(2) The northward-prograding Rannoch–Etive shoreface system was largely fed longshore from east to west, and is characterized largely by *Cde* suites. The earliest Rannoch–Etive sequences, in the south of the basin, are locally fed and have *A* and *B* suites.

(3) The 'delta-top' Ness Formation contains mostly southerly- and westerly-derived detritus, with *B* suites predominant. *A* suites are found in the western part of the basin, in close proximity to the Shetland Platform. Several sandstone units are characterized by *Cde* suites derived from the barrier sequence to the north as washover, tidal channel or tidal delta deposits. The Ness also contains *D* and *E* suites locally, believed to represent a single, correlatable depositional event, from sources unknown at this stage.

(4) The Tarbert transgressive sequence can be identified mineralogically, which may assist with its definition in uncertain cases. In the northeast it contains *Cde* suites, whereas in the west and south it contains *A* suites. Tarbert therefore has a complex provenance with material derived from both east and west. Assuming that the northeastern Tarbert is older than that to the west and south, the change to *A*-domination presumably reflects the cutting-off of the Norwegian *Cde* source as the transgression continued.

The sediment provenance has been ascertained by combining the intrabasinal patterns of garnet distribution, the nature of the source rock lithologies as inferred by the garnet compositions and published sedimentological interpretations of the Brent sequence. *Cde* suites were probably derived from amphibolite-facies metamorphic basement rocks on the Norwegian landmass, probably in the area immediately north of Bergen. *A* suites were derived from the northern part of the Shetland Platform. Their restricted composition is likely to reflect reworking of diagenetically-modified sandstones of possibly Permo-Triassic, Carboniferous or Devonian age, or even from low-grade metasediments. *B* suites had a broadly southerly source. Their general heterogeneity suggests derivation from pre-existing sediments, possibly of Permo-Triassic age by comparison with the central North Sea. Finally, *D* and *E* suites have uncertain provenance. Their association with Ness indicates a possible southerly derivation. However, their geochemistry and their association with chrome spinel indicates that they have an ultramafic source, and the area where this lay is not readily identifiable. It is unlikely that it lay withi: the existing basinal area unless the assemblages were reworked from earlier sediments. The purity of the assemblages makes the latter an unlikely possibility, and thus a source from Caledonian ophiolites on the Norwegian landmass or Shetland Platform is preferred.

It is clear that the observed garnet geochemical variations are entirely independent of hydraulic processes, as each assemblage type is found in coarse, medium- and fine-grained facies. This confirms the value of varietal studies in producing data which, because they are independant of hydraulic processes, are particularly provenance-sensitive. Diagenetic modifications of garnet suites have been identified in the Eider Field and in lower part of the Dunlin section, where garnet etching is extensive and the compositional range of the garnet suite is restricted. These results are consistent with the earlier observations of the effects of burial diagenesis on garnet suites in the central North Sea Palaeocene.

Finally, although the location of source areas can be inferred by combining intrabasinal distribution of garnet data, published sedimentological interpretations and garnet compositional data from basement lithologies, direct identification of sources can only be made by reference to mineralogical data from possible source rocks. It is crucial to the continued updating of Brent depositional models that regional data on potential source rocks are acquired. The lack of data on source material is not only a problem in this context, but is also a feature of many other provenance studies, because it is now frequently the case that considerably more detailed information exists for the sediments than for the source terrains.

The new data presented in this paper was acquired as part of a project sponsored by Shell UK Exploration and Production, and their financial support is gratefully acknowledged. The manuscript has benefitted greatly from the detailed reviews and constructive comments of M. Mange-Rajetzky, K. Stattegger, J. Marshall, R. Maskall, M. Giles, S. Brown and R. Knox. I am also indebted to C. Hallsworth and K. Moore for their excellent technical and laboratory support. The paper is published with the approval of the Director of the British Geological Survey (NERC).

## References

ALLEN, P. A. & MANGE-RAJETZKY, M. A. 1992. Sedimentary evolution of a Devono-Carboniferous rift basin, Clair Field, offshore northwestern UK: impact of changing provenance. *Marine and Petroleum Geology*, in press.

BROWN, S. & RICHARDS, P. C. 1989. Facies and development of the Middle Jurassic Brent Delta near the northern limit of its progradation, UK North Sea. *In*: WHATELEY, M. K. G. & PICKERING K. T. (eds) *Deltas: sites and traps for fossil fuels*. Geological Society, London, Special Publication, **41**, 253–267.

BRYHNI, I. & STURT, B. A. 1985. Caledonides of southwestern Norway. *In*: GEE, D. G. & STURT, B. A. (eds) *The Caledonide Orogen – Scandinavia and related areas*. Wiley, New York, 89–107.

BUDDING, M. C. & INGLIN, H. F. 1981. A reservoir geological model of the Brent Sands in southern Cormorant. *In*: ILLING, L. V. & HOBSON, G. D. (eds) *Petroleum geology of the continental shelf of northwest Europe*. Heyden & Son, London, 326–334.

DARBY, D. A. 1985. Trace elements in ilmenite: a way to discriminate provenance or age in coastal sands. *Bulletin of the Geological Society of America*, **95**, 1208–1218.

DEEGAN, C. E. & SCULL, B. J. 1977. *A standard lithostratigraphical nomenclature for the central and northern North Sea*. Report of the Institute of Geological Sciences **77/25**.

DEER, W. A., HOWIE, R. A. & ZUSSMAN, J. 1982. *Rock-forming minerals, volume 1A: orthosilicates*. Longman, London.

EYNON, G. 1981. Basin development and sedimentation in the Middle Jurassic of the North Sea. *In*: ILLING, L. V. & HOBSON, G. D. (eds) *Petroleum geology of the continental shelf of northwest Europe*. Heyden & Son, London, 196–204.

GRAUE, E., HELLAND-HANSEN, W., JOHNSEN, J., LØMO, L., NØTTVEDT, A., RØNNING, K., RYSETH, A. & STEEL, R. J. 1987. Advance and retreat of the Brent delta system, Norwegian North Sea. *In*: BROOKS, J. & GLENNIE, K. (eds) *Petroleum geology of North West Europe*. Graham & Trotman, London, 915–937.

HAMILTON, P. J., FALLICK, A. E., MACINTYRE, R. M. & ELLIOTT, S. 1987. Isotopic tracing of the provenance and diagenesis of Lower Brent Group sands, North Sea. *In*: BROOKS, J. & GLENNIE, K. (eds) *Petroleum geology of North West Europe*. Graham & Trotman, London, 939–949.

HURST, A. R. & MORTON, A. C. 1988. An application of heavy-mineral analysis to lithostratigraphy and reservoir modelling in the Oseberg Field, northern North Sea. *Marine and Petroleum Geology*, **5**, 157–169.

JOHNSON, H. D. & STEWART, D. J. 1985. Role of clastic sedimentology in the exploration and production of oil and gas in the North Sea. *In*: BRENCHLEY, P. J., & WILLIAMS, B. P. J. (eds) *Sedimentology: recent developments and applied aspects*. Geological Society, London, Special Publication, **18**, 249–310.

MORTON, A. C. 1985*a*. A new approach to provenance studies: electron microprobe analysis of detrital garnets from Middle Jurassic sandstones of the northern North Sea. *Sedimentology*, **32**, 553–566.

—— 1985*b*. Heavy minerals in provenance studies. *In*: ZUFFA, G. G. (ed.) *Provenance of arenites*. Reidel, Dordrecht, 249–277.

—— 1986. Dissolution of apatite in North Sea reservoir sandstones: implications for the generation of secondary porosity. *Clay Minerals*, **21**, 711–733.

—— 1987*a*. Detrital garnets as provenance and correlation indicators in North Sea Jurassic sandstones. *In*: BROOKS, J. & GLENNIE, K. (eds) *Petroleum geology of North West Europe*. Graham & Trotman, London, 991–995.

—— 1987*b*. Influences of provenance and diagenesis on detrital garnet suites in the Paleocene Forties sandstone, central North Sea. *Journal of Sedimentary Petrology*, **57**, 1027–1032.

—— & HUMPHREYS, B. 1983. The petrology of the Middle Jurassic sandstones from the Murchison Field, North Sea. *Journal of Petroleum Geology*, **5**, 245–260.

——, STIBERG, J. P., HURST, A. & QVALE, H. 1989*a*. Use of heavy minerals in lithostratigraphic correlation, with examples from Brent sandstones of the northern North Sea. *In*: COLLINSON, J. (ed.) *Correlation in hydrocarbon exploration*. Graham & Trotman, London, 217–230.

——, BORG, G., HANSLEY, P. L., HAUGHTON, P. D. W., KRINSLEY, D. H. & TRUSTY, P. 1989*b*. The origin of faceted garnets in sandstones: dissolution or overgrowth? *Sedimentology*, **36**, 927–942.

RICHARDS, P. C., BROWN, S., DEAN, J. M. & ANDERTON, R. 1988. A new palaeogeographic reconstruction for the Middle Jurassic of the northern North Sea. *Journal of the Geological Society, London*, **145**, 883–886.

RØNNING, K. & STEEL, R. J. 1987. Depositional sequences within a 'transgressive' reservoir sandstone unit: the Middle Jurassic Tarbert Formation, Hild area, northern North Sea. *In*: KLEPPE, J. *et al.* (eds) *North Sea Oil and Gas Reservoirs*. Graham & Trotman, London, 169–176.

SKARPNES, O., HAMAR, G. P., JAKOBSSON, K. H. & ORMAASEN, D. E. 1981. Regional Jurassic setting of the North Sea north of the central highs. *In*: *The sedimentation of the North Sea reservoir rocks, XIII*. Norsk Petroleumsforening, Oslo,

1–8.
SMALE, D. & VAN DER LINGEN, G. J. in press. Differential leaching of garnet grains at a depth of 3.5 km in Tane-1, western platform, New Zealand. *New Zealand Geological Survey Research Notes*.
STATTEGGER, K. & MORTON, A. C. 1992. Statistical analysis of garnet compositions and lithostratigraphic correlation: Brent Group sandstones of the Oseberg Field, northern North Sea. *This volume*.
WRIGHT, W. I. 1938. The composition and occurrence of garnets. *American Mineralogist*, **23**, 436–449.
ZIEGLER, P. A. 1981. Evolution of sedimentary basins in Northwest Europe. *In*: ILLING, L. V. & HOBSON, G. D. (eds) *Petroleum geology of the continental shelf of northwest Europe*. Heyden & Son, London, 3–39.

# Statistical analysis of garnet compositions and lithostratigraphic correlation: Brent Group sandstones of the Oseberg Field, northern North Sea

K. STATTEGGER[1] & A. C. MORTON[2]

[1] *Geologisch-Palaontologisches Institut und Museum der Universitat Kiel, Olshausenstrasse 40, D-2300 Kiel, Germany*

[2] *British Geological Survey, Keyworth, Nottingham NG12 5GG, UK*

**Abstract:** Garnet compositions from sandstones of the Middle Jurassic Brent Group of the Oseberg Field, northern North Sea, have been treated statistically in order to assess objectively their value in stratigraphic classification and grouping and for provenance analysis. The data set consists of the compositions of 1550 garnet grains from three wells, as determined by electron microprobe analysis. The statistical techniques used include summary statistics, exploratory data analysis, analysis of correlation structure and of subcompositions, discriminant analysis and fuzzy clustering. General univariate statistics provide the characteristic properties of the data distribution pattern. Almandine is the prevailing component, followed by pyrope, grossular and spessartine. Variation in the bulk data is generally low except for many high spessartine outliers. Stepwise linear discriminant analysis is the first step in stratigraphic grouping of the samples and in end-member modelling. Using all 1550 garnet grains, group classification and separation is efficient for the Tarbert Formation and the Etive–Oseberg interval, but not for the Ness Formation. End-member analysis by fuzzy c-means clustering generates similar stratigraphic garnet logs to those generated by discriminant analysis, the advantage of the method being that fuzzy clustering allows a dynamic interpretation by assuming that the calculated cluster centres represent the garnet compositions of source rocks or source areas. There is an overall increase in almandine with time: the Etive–Oseberg interval is characterized by high pyrope contents: the Ness Formation has high grossular and spessartine contents and depletion of pyrope: and the Tarbert Formation is depleted in grossular. Therefore multivariate end-member modelling of garnet compositions has provided objective criteria for classifying and differentiating samples. The resulting grain and sample distribution patterns provide a clearer picture of the stratigraphic variations than those obtained by using the geochemically-defined garnet end members.

Varietal studies of garnet compositions by means of electron microprobe is a valuable tool in correlating clastic rock sequences. This technique has some advantages over conventional heavy-mineral analysis and has been successfully applied in Mesozoic and Tertiary sandstones of the North Sea area (Morton 1985, 1987; Hurst & Morton 1988; Morton *et al.* 1989*a*). Stratigraphic variations in garnet compositions during the filling history of a sedimentary basin reflect changes in the source which provided the clastic detritus. Garnet distributions in sediments are comparable to conventionally-determined heavy mineral distributions, in that they represent mixing of garnet distributions from different source rocks. Statistically, this raises the question of mixing, unmixing and end members. The data set comes from cored sequences of Middle Jurassic Brent Group sandstones from the Oseberg Field, northern North Sea (Fig. 1). The Brent sequence is a complex unit comprising sediments deposited in a broadly paralic setting during a regressive–transgressive couplet, and forms the most important hydrocarbon reservoir unit in NW Europe. Graue *et al.* (1987) have described the Brent sequence in the Oseberg Field. It consists of a basal thick, very coarse grained sandstone of fan-delta origin (Oseberg Formation), overlain by a coarsening upward sequence (Rannoch and Etive formations) deposited in a shoreface setting during the progradational phase. The succeeding Ness Formation is a complex of sandstones, mudstones and coals deposited in a broadly delta-top setting, and the uppermost unit (Tarbert Formation) marks a return to shallow marine conditions as a result of the marine transgression that terminated delta progradation. Hurst & Morton (1988) and Morton *et al.* (1989*a*) considered that there were major changes in provenance during Brent Group deposition in the Oseberg area on the basis of garnet geochemis-

*From* Morton, A. C., Haszeldine, R. S., Giles, M. R. & Brown, S. (eds), 1992, *Geology of the Brent Group*. Geological Society Special Publication No. 61, pp. 245–262.

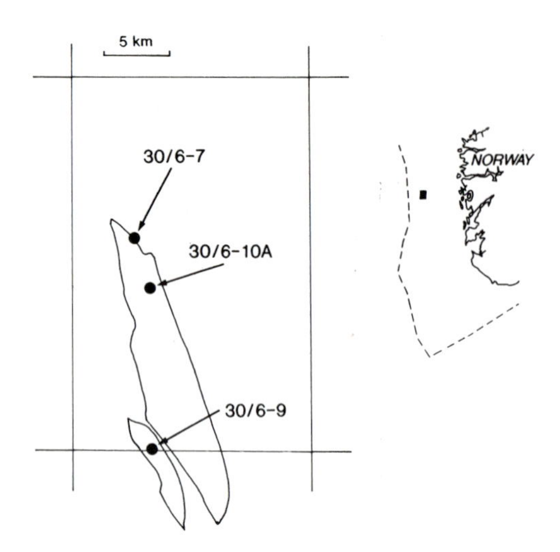


**Fig. 1.** Location of the Oseberg Field, northern North Sea, showing the position of the three studied wells.

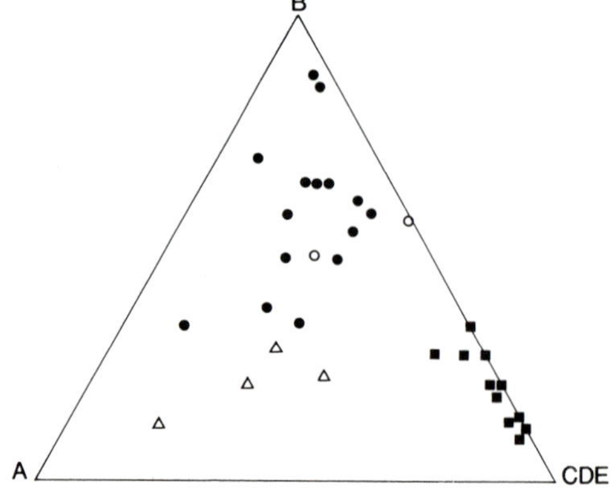


**Fig. 2.** Distinction of Tarbert, Ness and Etive–Oseberg samples, using the subfield method outlined by Morton *et al.* (1989*a*).

try supported by evidence from conventional heavy mineral data and a varietal study of tourmaline. The data set concerned in this paper comprises that used by Morton *et al.* (1989*a*), the results of analyses of 31 samples from 3 wells (9 from 30/6–7, 11 from 30/6–9 and 11 from 30/6–10A) over the depth range 2462–2775 m (Fig. 1). Fifty garnet grains were hand picked from the heavy mineral residues, which were derived by gravity-settling of the 63–125 $\mu$m fraction through bromoform. Studies of polished thin sections of garnets of this grain size have shown that compositional zoning is either absent or of minor importance (Morton 1985; Morton *et al.* 1989*b*). The garnets were analysed using a Link Systems energy-dispersive X-ray analyser attached to a Geoscan electron microprobe, giving rise to a total of 1550 individual analyses. Garnet compositions are expressed in terms of the variations in the four end-members almandine, spessartine, pyrope and grossular, which sum to 100%. Hurst & Morton (1988) and Morton *et al.* (1989*a*) displayed the compositional range of each sample on triangular plots that use almandine + spessartine (AS), pyrope (P) and grossular (G) as poles (Fig. 2). Hurst & Morton (1988) compared and differentiated samples on a purely visual basis, but Morton *et al.* (1989*a*) devised a method for graphical comparison of the data. This involved dividing the AS-P-G triangle into subfields and scoring the number of garnets that fell into each area. Every sample was then plotted on a summary diagram (Fig. 2), allowing them to be directly compared. This indicated that there are stratigraphically significant variations in garnet geochemistry, with three groups (the Etive–Oseberg interval, the Ness Formation and the Tarbert Formation) distinguished (Fig. 2). The aim of this paper is to check whether these groupings can be sustained statistically and to evaluate the advantages of multivariate data grouping and end-member modelling in stratigraphic correlation and provenance studies. This needs first a careful evaluation of the general structure of the data set.

## The statistics

The purpose of the statistical treatment of the garnet data was to devolve the garnet compositions into multivariate end members by unmixing models. This involved a number of procedures.

### *Elementary univariate statistics and exploratory data analysis*

Summary statistics, histograms and box-plots are useful for the recognition of data structures (cf. Henley 1981; Velleman & Hoaglin 1981; Rock 1988). Summary statistics include the most important statistical parameters such as measures of central tendency and variation in a data set. Other parameters describe the shape of the frequency distribution relative to the normal distribution. The frequency distribution can be graphically displayed by histograms.

Exploratory data analysis treats the variables of a data set in an ordinary general manner and requires no theoretical data distribution. Therefore one can easily compare several variables in a data set. The box-plot (described in Fig. 4) is an effective way of displaying univariate statistics graphically and provides a good description of the variables, particularly for the detection of outliers and asymmetric behaviour.

### *Analysis of correlation structures and analysis of subcompositions from the garnet data*

Geochemical compositions obey the constant-sum constraint that is of special importance in cases such as this, where the compositions are expressed in terms of a small number of variables. Spurious correlations may occur and outliers strongly influence the correlation structure. The same problems arise in the modelling of end members to evaluate the mixing proportions of a geochemical data set. These problems can be overcome in part by appropriate data transformations, which preserve the initial relationships between variables while avoiding the effects of closure (Chayes 1971; Aitchison 1984, 1986; Pawlowsky & Stattegger 1988). In order to determine the relative magnitudes of subsets of the variables under investigation, it is necessary to extract subcompositions.

### *Discriminant analysis*

The method of stepwise linear discriminant analysis is a valuable tool for discrimination of several predefined sample groups on the basis of their variables (Jennrich 1977). It maximizes the differences among the groups by linear classification functions selecting the most discriminating variable among the groups at each step; likewise a variable will be deleted if its discrimination power in comparison to other variables becomes too low. In addition, canonical variables are computed: these are composed of a linear combination of the original variables and a constant. The canonical variables are the coordinate axes of the multivariate sample space. They allow optimal separation among the sample groups (Reyment *et al.* 1984).

The efficiency of the method depends heavily on the data structure of the predefined sample groups (Krzanowski 1977) because it requires normally distributed variables. If this condition does not hold, one should use kernel functions for discriminant analysis (Remme *et al.* 1980).

### *End-member modelling*

*Factor analysis.* Multivariate compositions are the mixing proportions of multivariate end members. Q-mode factor analysis uses a similarity matrix among samples to extract factors from the data matrix. These factors are modified to end members and each sample can be expressed in terms of the end members (Klovan & Imbrie 1971; Klovan & Miesch 1976; Full *et al.* 1981). The method was successfully applied in the determination of source rocks from heavy mineral data by Stattegger (1987).

However, there may be problems in determining both definite positive end members and positive sample compositions to represent these end members. This is because end members models from factor analysis can be seriously affected by extreme sample compositions which may become end members. As a result, sample points may be placed outside the calculated mixing polytope, which consequently provides a poor description of the mixing proportions.

*Fuzzy clustering.* A new approach to end-member modelling is the fuzzy c-means clustering method (Full *et al.* 1982; Bezdek *et al.* 1984; Granath 1984; Stattegger 1989). Fuzzy c-means clustering provides a robust method for grouping geochemical data into c clusters and evaluating their mixing proportions. It relies on the collective properties of the data and estimates the similarity of each observation to each group instead of the exclusive classification performed by hierarchical clustering methods. Group or cluster centres are formed according to the density of data and can be used as end members. Outliers have only minor influence in the location and composition of group centres, because the method does not require normally-distributed data.

The membership for an observation depends on the distance (Euclidean, Diagonal or Mahalanobis norm) from the observation to the cluster centre. The system is hard if the membership takes exclusively the values 0 or 1, but is fuzzy if it takes any value between 0 and 1, so that one observation can share all c groups. The weighting exponent (from 1 to infinity) serves to guide the fuzziness of the system; a hard system has a weighting exponent of 1. Measures of fuzziness are partition coefficient and partition entropy. The values for partition coefficient range from 1/c to 1; hard clustering has a value of 1. The partition entropy lies between 0 and ln(c), and hard clustering has a value of 0. Fuzzy partitions expressed by membership functions can be used to detect multivariate outliers: extreme samples have low memberships of all groups and are positioned at great distances to the cluster centres.

## Results

### Summary statistics and exploratory data analysis

Summary statistics of the garnet varieties from the 1550 analysed grains are given in Table 1. Almandine (al) is the most abundant variety, with an average value of 56.6%: pyrope (py) contributes 21.6%, grossular (grs) 15.8% and spessartine (sp) 6.0%. The histograms in Fig. 3 show the frequency distributions. Almandine has a near-symmetric distribution close to the normal distribution. This is consistent with the statistical parameters (Table 1). Pyrope and grossular show several moderate peaks, the most prominent at low percentages in a moderately positively skewed distribution. Spessartine has a strongly positively skewed distribution. The box-plots (Fig. 4) display the variability and outliers shown by each garnet variety. Almandine, pyrope and grossular have almost no outliers, whereas spessartine has 235 high value outliers. Table 1 and Fig. 3 show that extreme values do not greatly influence the data structures of the three most abundant garnet varieties, whereas spessartine, the smallest component, is severely affected by its many high value outliers. Thus, in the subsequent statistical treatment, almandine, pyrope and grossular can be treated by multivariate procedures which assume normally distributed variables, but difficulties may arise applying such procedures to spessartine. The combined variations of almandine, pyrope and grossular are of the same order of magnitude as the variations in spessartine because of the closed four-variable system involved. Elementary statistics of the garnet varieties of individual samples calculated from the 50 grains analysed per sample are summarized in Table 2. Each sample corresponds closely to the general features of the garnet compositions described above. Therefore, variability among the individual samples is only moderate and there is considerable overlap between samples.

### Correlation structure and subcompositions of the system almandine–pyrope–grossular–spessartine

The correlation structure of the four garnet varieties consists of highly significant negative correlations between the four variables, apart from a weak negative correlation between grossular and spessartine. The negative correlations are the effect of strong closure of the data: the compositions are expressed in terms of only four variables, so that variations in one variable strongly influence the variations in the other variables.

The interdependence of variations among variables is clearly shown by considering subcompositions of the four-component system (Fig. 5). Figure 5a shows the relative proportions of the high percentage variables almandine, pyrope and grossular in relation to the predictive area containing 95% of all 1550 grains (Aitchison 1986). The corresponding graph of the low percentage variables pyrope, grossular

**Table 1.** *Summary statistics of the bulk garnet compositions in percentages. Data from all 1550 analysed grains included*

| | almandine | pyrope | grossular | spessartine |
|---|---|---|---|---|
| Average | 56.58 | 21.56 | 15.82 | 6.04 |
| Median | 57.4 | 20.8 | 16.4 | 1.9 |
| Mode | 60.8 | 3.3 | 3.3 | 0.9 |
| Geometric mean | 55.23 | 0 | 12.76 | 0 |
| Variance | 137.19 | 180.04 | 75.60 | 94.03 |
| Standard deviation | 11.71 | 13.42 | 8.69 | 9.70 |
| Standard error | 0.298 | 0.349 | 0.221 | 0.246 |
| Minimum | 21.4 | 0 | 1.0 | 0 |
| Maximum | 83.7 | 64.4 | 44.9 | 70.9 |
| Range | 62.3 | 64.4 | 43.9 | 70.9 |
| Lower quartile | 48.6 | 9.3 | 7.4 | 1.0 |
| Upper quartile | 65.4 | 31.5 | 22.3 | 5.5 |
| Interquartile range | 16.8 | 22.2 | 14.9 | 4.5 |
| Skewness | −0.355 | 0.436 | 0.139 | 2.767 |
| Kurtosis | −0.377 | −0.601 | −0.769 | 8.698 |

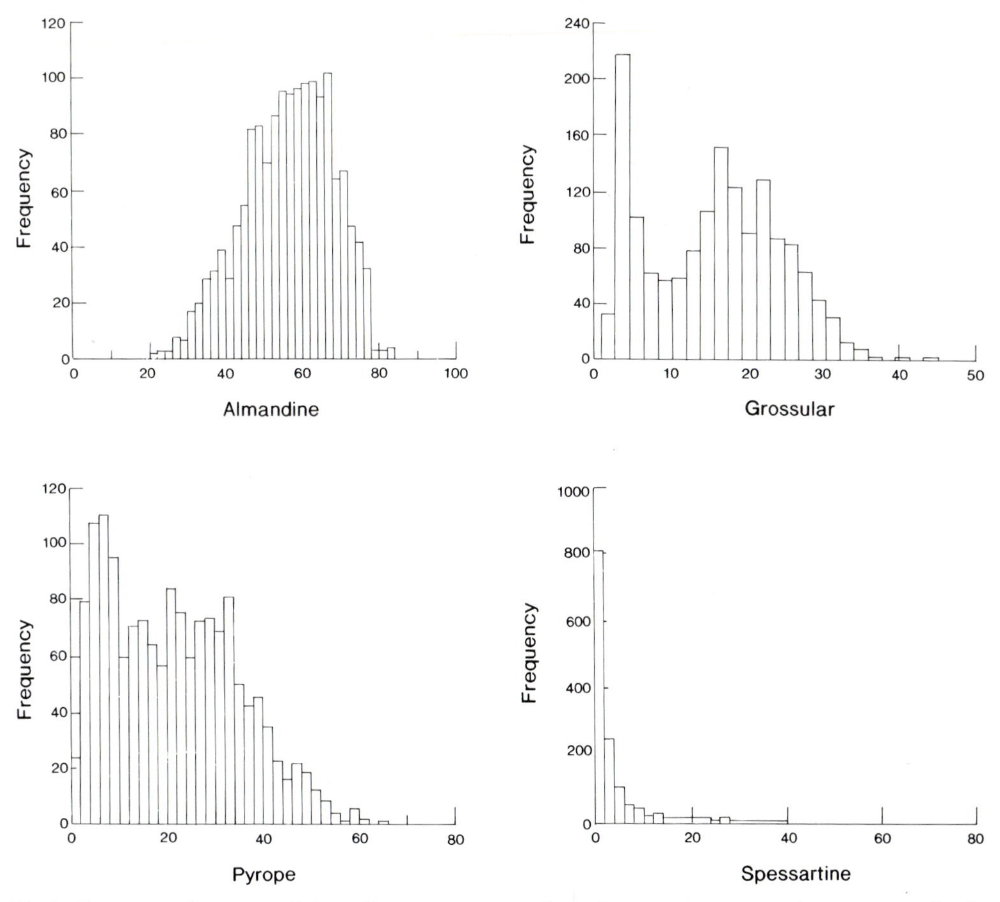


**Fig. 3.** Frequency histograms of almandine, pyrope, grossular and spessartine contents in percentages for the 1550 garnet grains.

and spessartine is shown in Fig. 5b. The data points in the triangle with almandine are concentrated towards the almandine end-point and cover the almandine-rich part. In contrast, the data points in the triangle with spessartine are spread almost over the whole area. The other two possible ternary subcompositions have triangular distributions between the two extreme ones shown in Fig. 5. The ternary subsystems therefore show that spessartine significantly enlarges the variability of the data set.

## *Data grouping and end-member modelling*

*Grouping by discriminant analysis*. Each sample was classified by its lithostratigraphic position into 3 groups (Etive–Oseberg, Ness and Tarbert). As described by Hurst & Morton (1988) and Morton *et al.* (1989*a*), two samples originally assigned to the Etive Formation in Well 30/6–9 were found to have garnet assemblages typical of the Ness Formation and were therefore regarded as sampling a Ness-type fluvial sandstone that has eroded into the Etive interval. This conclusion is compatible with sedimentological observations on the core (Hurst & Morton 1988; Morton *et al.* 1989*a*). These samples have therefore been assigned to the Ness group in this analysis. The data set comprises four samples from the Tarbert interval, 15 from the Ness interval and 12 from the Etive–Oseberg interval. Elementary statistics of the stratigraphic groups based on the individual garnet grains (Table 3) show that there is little compositional difference between the stratigraphic groups. The general garnet stratigraphy of the three wells (Fig. 6) relies on the mean garnet compositions of the raw data from each sample (Table 2). Variations are small within and between the wells because the cal-

**Table 2.** *Basic statistics of the individual samples in percentages based on 50 grains per sample*

| | | Well N30/6–7 | | | | | | | | |
|---|---|---|---|---|---|---|---|---|---|---|
| Sample | | 1T 2648.6 m | 2N 2680.3 m | 3N 2690.4 m | 4N 2718.8 m | 5N 2731.5 m | 6E 2736.0 m | 7E 2753.95 m | 8E 2764.3 m | 9E 2775.5 m |
| Almandine | $\bar{x}$ | 62.7 | 62.0 | 62.3 | 58.8 | 63.3 | 53.1 | 49.2 | 52.6 | 56.0 |
| | $s$ | 9.8 | 10.8 | 10.3 | 12.4 | 8.9 | 12.7 | 10.0 | 9.9 | 11.2 |
| | $C$ | 0.16 | 0.17 | 0.16 | 0.21 | 0.14 | 0.24 | 0.20 | 0.19 | 0.20 |
| Pyrope | $\bar{x}$ | 27.3 | 14.4 | 10.7 | 16.9 | 26.3 | 27.3 | 29.1 | 27.6 | 24.7 |
| | $s$ | 9.4 | 12.0 | 7.9 | 11.8 | 12.0 | 12.9 | 12.6 | 12.5 | 14.3 |
| | $C$ | 0.34 | 0.83 | 0.73 | 0.70 | 0.45 | 0.47 | 0.43 | 0.45 | 0.58 |
| Grossular | $\bar{x}$ | 6.9 | 12.8 | 15.1 | 12.8 | 7.0 | 16.4 | 19.3 | 17.0 | 14.0 |
| | $s$ | 5.2 | 8.1 | 6.2 | 6.1 | 5.4 | 6.5 | 4.6 | 5.2 | 5.5 |
| | $C$ | 0.75 | 0.63 | 0.41 | 0.48 | 0.77 | 0.40 | 0.24 | 0.31 | 0.39 |
| Spessartine | $\bar{x}$ | 3.1 | 10.7 | 11.9 | 11.5 | 3.4 | 3.2 | 2.4 | 2.8 | 5.3 |
| | $s$ | 7.1 | 12.7 | 11.7 | 14.6 | 5.3 | 6.2 | 5.2 | 4.2 | 9.72 |
| | $C$ | 2.30 | 1.19 | 0.99 | 1.28 | 1.58 | 1.92 | 2.21 | 1.53 | 1.82 |

**Table 2.** *Cont.*

| | | Well N30/6–10A | | | | | | | | | | |
|---|---|---|---|---|---|---|---|---|---|---|---|---|
| Sample | | 1T 2481.75 m | 2T 2486.55 m | 3N 2498.5 m | 4N 2510.75 m | 5N 2519.25 m | 6N 2546.9 m | 7N 2548.86 m | 8N 2552.55 m | 9N 2557.1 m | 10E 2574.95 m | 11E 2579.0 m |
| Almandine | $\bar{x}$ | 57.2 | 59.5 | 57.9 | 57.0 | 58.4 | 58.9 | 54.0 | 55.6 | 58.2 | 51.7 | 51.3 |
| | $s$ | 10.8 | 11.6 | 9.7 | 12.9 | 13.0 | 10.6 | 13.3 | 13.0 | 11.1 | 8.2 | 9.9 |
| | $C$ | 0.19 | 0.19 | 0.17 | 0.23 | 0.22 | 0.18 | 0.25 | 0.23 | 0.19 | 0.16 | 0.19 |
| Pyrope | $\bar{x}$ | 25.7 | 24.5 | 20.5 | 11.0 | 21.6 | 15.5 | 16.3 | 18.6 | 17.9 | 24.8 | 24.5 |
| | $s$ | 11.5 | 12.7 | 12.7 | 9.8 | 15.3 | 11.9 | 14.5 | 15.1 | 14.5 | 10.9 | 11.3 |
| | $C$ | 0.45 | 0.52 | 0.62 | 0.90 | 0.71 | 0.77 | 0.89 | 0.81 | 0.81 | 0.44 | 0.46 |
| Grossular | $\bar{x}$ | 14.5 | 12.9 | 13.4 | 17.6 | 15.3 | 18.6 | 18.5 | 15.2 | 15.3 | 21.4 | 22.0 |
| | $s$ | 10.8 | 10.4 | 8.3 | 8.6 | 8.7 | 8.8 | 9.2 | 9.2 | 9.2 | 6.7 | 7.3 |
| | $C$ | 0.74 | 0.80 | 0.62 | 0.49 | 0.57 | 0.47 | 0.50 | 0.60 | 0.60 | 0.31 | 0.33 |
| Spessartine | $\bar{x}$ | 2.6 | 3.1 | 8.2 | 14.4 | 4.7 | 7.0 | 11.2 | 10.6 | 8.6 | 2.1 | 2.2 |
| | $s$ | 3.3 | 6.1 | 10.4 | 14.7 | 7.6 | 9.7 | 14.1 | 13.8 | 12.9 | 2.3 | 2.3 |
| | $C$ | 1.28 | 1.95 | 1.27 | 1.02 | 1.61 | 1.39 | 1.26 | 1.30 | 1.50 | 1.08 | 1.04 |

**Table 2.** *Cont.*

| Sample | | Well N30/6–9 | | | | | | | | | | |
|---|---|---|---|---|---|---|---|---|---|---|---|---|
| | | 1T 2462.0 m | 2N 2489.0 m | 3N 2495.95 m | 4N 2509.0 m | 5N 2542.65 m | 6E 2558.0 m | 7N 2569.85 m | 8E 2574.75 m | 9E 2595.0 m | 10E 2610.0 m | 11E 2617.5 m |
| Almandine | $\bar{x}$ | 60.1 | 53.7 | 59.9 | 60.0 | 56.7 | 61.8 | 51.2 | 50.1 | 55.1 | 52.9 | 52.1 |
| | $s$ | 9.1 | 12.6 | 10.0 | 11.4 | 13.2 | 12.0 | 12.6 | 8.4 | 11.1 | 9.4 | 11.8 |
| | $C$ | 0.15 | 0.23 | 0.17 | 0.19 | 0.23 | 0.19 | 0.25 | 0.17 | 0.20 | 0.18 | 0.23 |
| Pyrope | $\bar{x}$ | 24.3 | 19.5 | 25.2 | 24.3 | 21.0 | 21.0 | 19.6 | 25.1 | 17.8 | 22.6 | 23.2 |
| | $s$ | 9.8 | 15.4 | 13.5 | 13.3 | 15.5 | 13.2 | 15.5 | 10.3 | 9.8 | 12.6 | 13.6 |
| | $C$ | 0.40 | 0.79 | 0.54 | 0.55 | 0.74 | 0.64 | 0.79 | 0.41 | 0.55 | 0.56 | 0.58 |
| Grossular | $\bar{x}$ | 13.1 | 16.6 | 10.8 | 13.0 | 15.6 | 13.0 | 19.2 | 20.7 | 21.2 | 20.9 | 19.6 |
| | $s$ | 11.9 | 8.1 | 7.9 | 9.1 | 7.3 | 8.8 | 6.4 | 6.9 | 8.2 | 6.6 | 7.9 |
| | $C$ | 0.91 | 0.49 | 0.73 | 0.70 | 0.47 | 0.68 | 0.33 | 0.33 | 0.39 | 0.31 | 0.40 |
| Spessartine | $\bar{x}$ | 2.5 | 10.2 | 4.1 | 2.7 | 6.7 | 4.2 | 10.0 | 3.6 | 5.9 | 3.6 | 5.1 |
| | $s$ | 3.0 | 12.7 | 6.6 | 2.7 | 10.1 | 4.6 | 11.7 | 6.7 | 8.6 | 4.1 | 9.8 |
| | $C$ | 1.21 | 1.25 | 1.59 | 1.00 | 1.51 | 1.12 | 1.17 | 1.85 | 1.44 | 1.15 | 1.92 |

Samples are referred to by core depth, number and lithostratigraphic affinity (T, Tarbert; N, Ness; E, Etive–Oseberg). $\bar{x}$ = arithmetic mean, $s$ = standard deviation, $C$ = coefficient of variation.

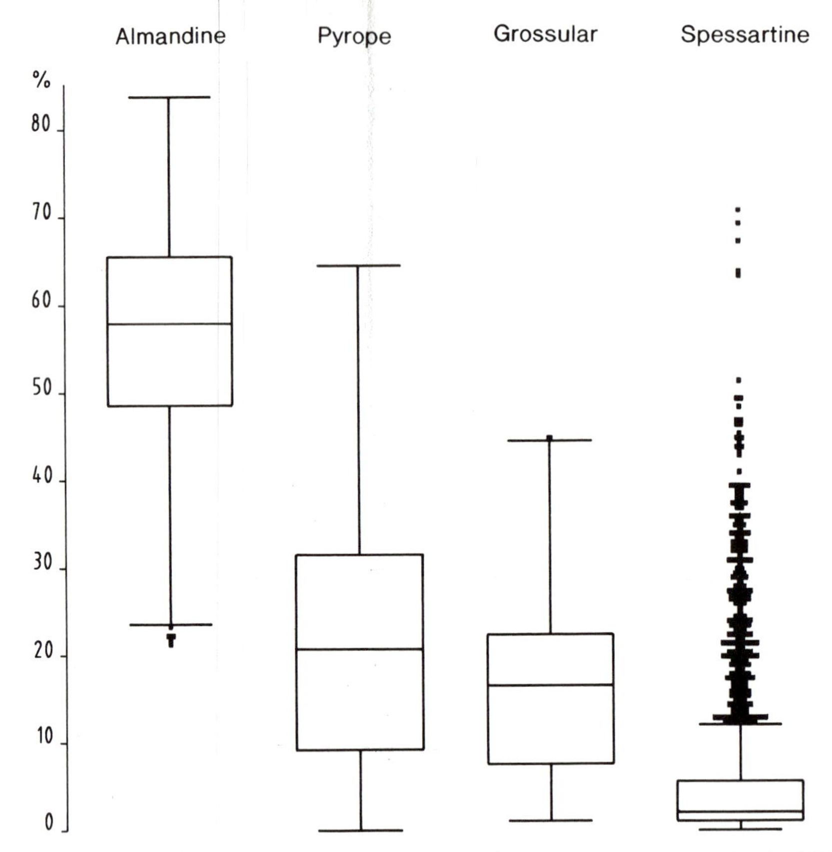


**Fig. 4.** Box plots of almandine, pyrope, grossular and spessartine on a percentage scale for the 1550 garnet grains. The plot divides the data into four areas, the quartiles. The second and the third quartile are in the box, which also displays the position of the median line. The first and the fourth quartiles are displayed as linear extensions of the box until specified boundary values or the highest and lowest values are reached. Outliers that fall outside these boundary values are marked by dots. Note the many high-value outliers of spessartine outside the high-boundary value.

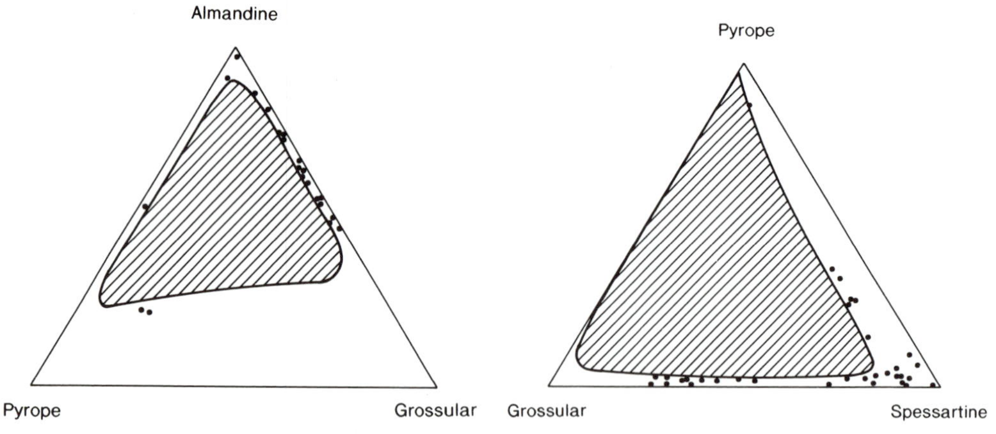


**Fig. 5.** Variability of garnet composition within the 95% predictive area for the 1550 garnet grains. Outliers are marked by dots. (**a**) Subcomposition almandine–pyrope–grossular; (**b**) Subcomposition pyrope–grossular–spessartine.

**Table 3.** *Basic statistics of the garnet compositions for the three lithostratigraphic sample groups in percentages*

| | | Almandine | Pyrope | Grossular | Spessartine |
|---|---|---|---|---|---|
| Tarbert | $\bar{x}$ | 59.9 | 25.4 | 11.9 | 2.8 |
| $n$ = 200 | $s$ | 10.5 | 10.9 | 10.2 | 5.1 |
| | 95% conf. band | 39.2–80.5 | 4.0–46.8 | 0–31.9 | 0–12.8 |
| Ness | $\bar{x}$ | 58.5 | 18.6 | 14.5 | 8.4 |
| $n$ = 750 | $s$ | 11.8 | 13.8 | 8.5 | 11.6 |
| | 95% conf. band | 35.4–81.6 | 0–45.6 | 0–31.2 | 0–31.1 |
| Etive–Oseberg | $\bar{x}$ | 53.1 | 24.0 | 18.7 | 4.2 |
| $n$ = 600 | $s$ | 11.1 | 12.9 | 7.3 | 7.2 |
| | 95% conf. band | 31.3–74.9 | 0–49.3 | 4.4–33.0 | 0–18.3 |

$n$ = number of grains, $\bar{x}$ = arithmetic mean, $s$ = standard deviation, 95% confidence band is the range that includes 95% of the grains.

culation of mean values causes a strong smoothing effect compared with the original data, and there are no significant univariate differences among the stratigraphic formations, as tested by analysis of variance. Therefore linear stepwise linear discriminant analysis was used in order to optimize the separation between the predefined stratigraphic sample groups based

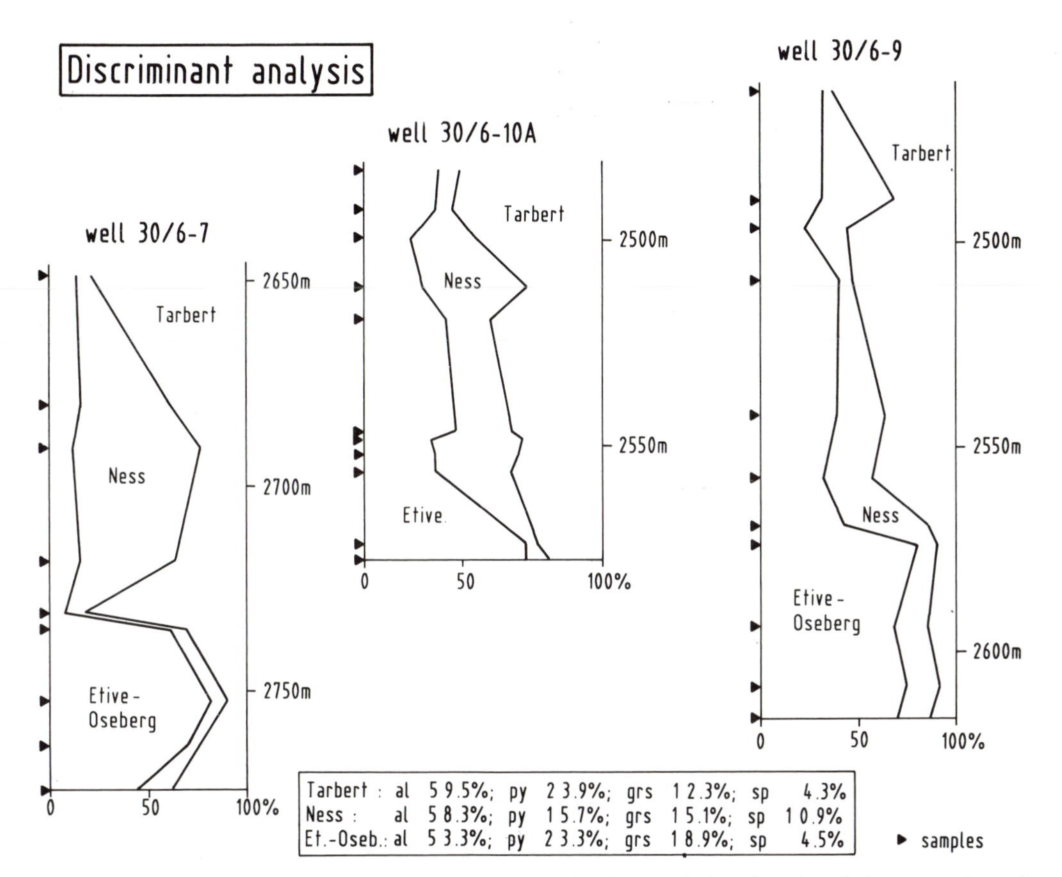


**Fig. 6.** Garnet stratigraphy of the Brent Group interval in the three wells based on the relative proportions of almandine, pyrope, grossular and spessartine derived from the sample mean compositions (compare Table 2).

on all garnet varieties. No transformation was made on the data since the largest three of the four variables can be assumed to have normal distributions.

The first analysis was performed on the 31 mean compositions. The low variability of the mean values mentioned above causes little dispersion and overlap among samples and sample groups, causing problems in group separation. However, two variables, almandine and spessartine, provide significant discrimination of the three groups (Table 4, Fig. 7). All Tarbert samples are classified correctly, but 4 of the 15 Ness samples and 2 of the 13 Etive–Oseberg samples also share the Tarbert group, as seen from the sample distribution on the first and second canonical variable using almandine and spessartine (Fig. 7). All three sample groups are well defined and separated, with only the Tarbert group containing additional samples from the two other groups.

A discriminant analysis using all 1550 garnet grains and each of the 31 samples as an individual group yields a similar but less clear grouping pattern. In this case, grossular, pyrope and almandine (in that order) are discriminant variables which contribute to the canonical variables (Table 5). Seven samples dominate the grouping process, attracting the majority of all grains (74%). The other samples are not effective in the grouping process, containing less than their 50 original grains; indeed, ten samples have less than 5 grains each after the grouping process. The sample distribution on the first and second canonical variable (Fig. 8) contains the centre of each sample. Major groups are formed according to the stratigraphic setting of the samples. Samples are added to a group only if they are classified correctly by at least one half of its grains (Table 5). The Tarbert group is again formed by the 4 Tarbert samples, with 3 additional Ness samples. The Ness group contains only 5 of the 15 Ness samples, with 7 Ness samples positioned between the Ness and Tar-

**Table 4.** *Main results of the discriminant analysis of the 31 sample means compositions related to the three predefined stratigraphic groups*

| | *Group means in percentages with standard deviations in brackets* | | | |
|---|---|---|---|---|
| | Almandine | Pyrope | Grossular | Spessartine |
| Tarbert | 59.9 | 25.4 | 11.9 | 2.8 |
| $n = 4$ | (2.2) | (1.4) | (3.4) | (0.3) |
| Ness | 58.5 | 18.6 | 14.5 | 8.4 |
| $n = 15$ | (2.8) | (4.7) | (3.0) | (3.5) |
| Etive–Oseberg | 53.1 | 23.9 | 18.7 | 4.2 |
| $n = 12$ | (3.3) | (3.3) | (3.0) | (2.2) |

*Discriminant variables with* F-*values*

| | | |
|---|---|---|
| Almandine | 13.52 | $F = 9.12$ |
| Spessartine | 9.94 | Degrees of freedom: 2,26 |

F-*matrix of group separation*

| | Tarbert | Ness | |
|---|---|---|---|
| Ness | 5.89 | | $F = 5.53$ |
| Etive–Oseberg | 7.51 | 20.54 | Degrees of freedom: 2,26 |

*Classification-matrix*
number of cases classified into group

| | Tarbert | Ness | Etive–Oseberg |
|---|---|---|---|
| Tarbert | 4 | 0 | 0 |
| Ness | 4 | 11 | 0 |
| Etive–Oseberg | 2 | 0 | 10 |
| % of correctly classified samples | 100 | 73 | 83 |

*Canonical variables*

$v_1 = 0.312\text{almandine} + 0.375\text{spessartine} - 22.472$

$v_2 = -0.198\text{almandine} + 0.116\text{spessartine} + 13.284$

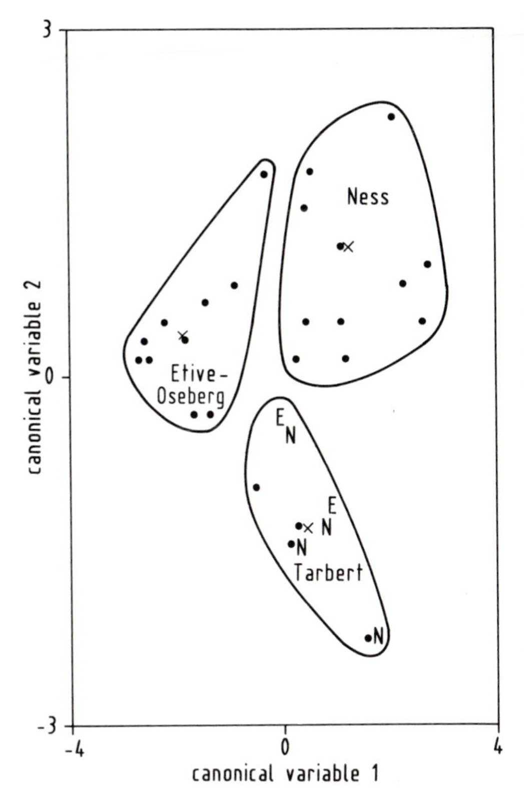


**Fig. 7.** Discriminant model of the mean garnet compositions from the 31 samples in relation to the first and second canonical variables. Group centres are marked by ×, sample mean compositions by dots. N and E refer to samples from the Ness and Etive–Oseberg intervals respectively, that plot outside their stratigraphic group.

**Table 5.** *Main results of the discriminant analysis using each sample as a separate group*

*Discriminant variables with* F-*values*

| | | |
|---|---|---|
| Grossular | 12.45 | $F$ = 2.04 |
| Pyrope | 9.05 | Degrees of freedom: 29,1518 |
| Almandine | 3.90 | |

*Samples most frequently joined by individual grains*

| Sample | Number of grains | % of the total sum of grains |
|---|---|---|
| Well 30/6–7/7E | 233 | 15.0 |
| Well 30/6–7/1T | 207 | 13.4 |
| Well 30/6–7/5N | 169 | 10.9 |
| Well 30/6–10A/4N | 164 | 10.6 |
| Well 30/6–10A/6N | 143 | 9.2 |
| Well 30/6–9/9E | 116 | 7.5 |
| Well 30/6–10A/11E | 108 | 7.0 |

*Sample distribution and stratigraphic grouping*

| | Tarbert | Ness | Etive–Oseberg |
|---|---|---|---|
| % of correctly classified samples | 100 | 33 | 75 |

*Canonical variables*

$v_1 = -0.006$ almandine $+ 0.039$ pyrope $+ 0.128$ grossular $- 2.555$

$v_2 = 0.071$ almandine $+ 0.104$ pyrope $+ 0.039$ grossular $- 6.861$

bert groups. The Etive–Oseberg group contains 9 of the 12 samples, with the remaining 3 dispersed between the three groups. Thus, grouping and classification is more difficult when all 1550 garnet grains are used, particularly in the case of the Ness group.

A third discriminant analysis was performed using all 1550 garnet grains and the three predefined stratigraphic groups. In this analysis, samples were grouped on the basis of the individual grain classifications. Several runs led to some rearrangement of samples: sample 5N of well 30/6–7 and all Ness samples of well 30/6–9 were put into the Tarbert group instead of the Ness group, whereas samples 5N and 6N of well 30/6–10A were placed into the Etive–Oseberg group instead of the Ness group. The main results of the final run after this rearrangement are shown in Table 6 and Fig. 9, the latter in relation to the well stratigraphy. The group means are similar to those calculated from the sample means but have much higher standard deviations due to the variations between the individual grains. Again, there are changes in the discriminatory variables. Almandine, although the most dominant and most regularly distributed variable in relation to the normal distribution, is not sensitive in separation of the stratigraphic groups. Instead, separation relies on the minor components grossular, pyrope and spessartine (in that order), as these show greater differences (Fig. 5b). Group separation is significant although the number of correctly classified garnet grains is not very high. Classification is best for the Etive–Oseberg group. Of particular interest is the well stratigraphy generated from these results (Fig. 8). Each sample is represented by the percentages of individual grains classified into the three stratigraphic groups. The Tarbert and Etive–Oseberg groups have distinct maxima near the top and the base of the wells respectively. The Ness group has a clear maximum only in well 30/6–7. Thus, the grossular-rich Etive–Oseberg interval is separated from the grossular-poor Tarbert group by the Ness garnets, which are more pyrope-poor and spessartine-rich compared with the other

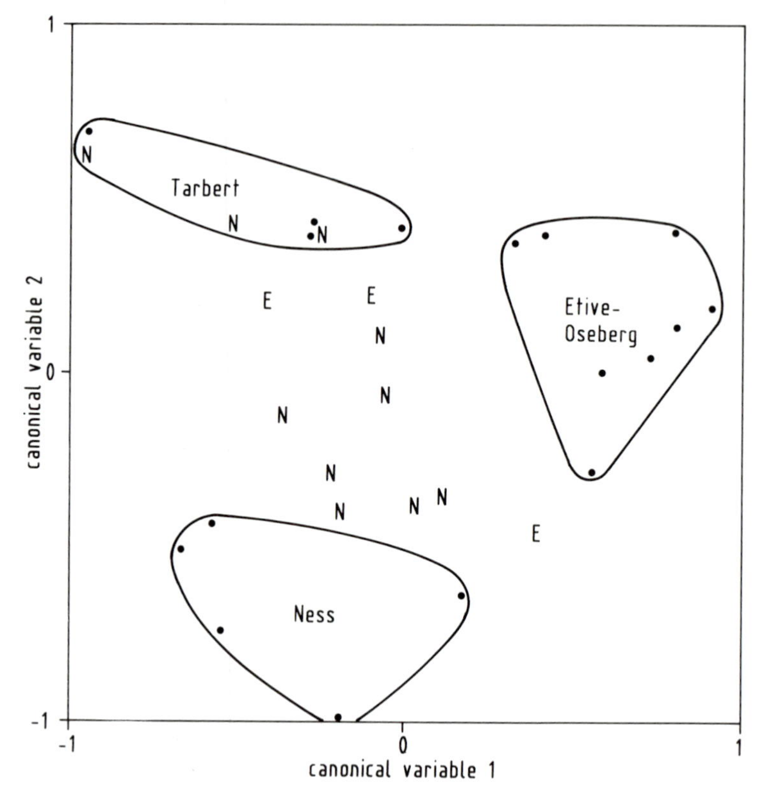


**Fig. 8.** Discriminant model taking each sample consisting of 50 grains as a separate group. Major stratigraphic groups are formed by those samples that are classified by at least 50% of their grains that belong to this group. Sample centres are marked by dots. N and E refer to samples from the Ness and Etive–Oseberg intervals respectively, that plot outside their stratigraphic group.

groups, but are in general more variable and less distinctly defined.

*End-member modelling*. The grouping and classification of the garnet compositions by means of discriminant analysis raises the question of the source-related garnet end-members responsible for the stratigraphic variations. Garnet compositions in clastic sediments are mixing proportions which should relate to meaningful end-members. End-member analysis by Q-mode factor analysis did not work successfully because even after several runs it was not possible to ascribe positive values to all end members. Therefore, some negative composition loadings with negative percentages resulted, probably due to the strong closure effect of the data in connection with extreme value compositions.

Fuzzy clustering is a more robust method of end-member analysis compared with factor analysis methods, because end members are seen as positive definite cluster centres and hard and soft partitions can be calculated from the data. The raw data comprising the 1550 garnet compositions were used. The clustering algorithm is not sensitive to the many outliers of spessartine, and so the non-normality of spessartine, which is the smallest component, does not significantly influence the location and composition of the cluster centres. The data were partitioned into three groups according to their lithostratigraphy. As discriminant analysis and Table 3 show, the three groups overlap each other considerably. Therefore, it is a major problem to obtain an optimal final partition of the variable space for the data set. Such a partition should be sufficiently hard for one group or cluster to contain, as far as possible, only one stratigraphic group in its core. The weighting exponent can be changed to achieve more fuzziness in order to extend the membership of one item to several groups, especially if it is located at great distances from the cluster

**Table 6.** *Main results of the discriminant analysis using all 1550 garnet grains related to the three predefined stratigraphic groups*

*Group means in percentages with standard deviations in brackets*

| | Almandine | Pyrope | Grossular | Spessartine |
|---|---|---|---|---|
| Tarbert | 59.5 | 23.9 | 12.3 | 4.3 |
| $n = 500$ | (11.3) | (12.9) | (9.2) | (7.2) |
| Ness | 58.3 | 15.7 | 15.1 | 10.9 |
| $n = 400$ | (12.0) | (12.8) | (8.4) | (13.2) |
| Etive–Oseberg | 53.3 | 23.3 | 18.9 | 4.5 |
| $n = 650$ | (11.1) | (13.1) | (7.3) | (7.7) |

*Discriminant variables with* F-*values*

| | | |
|---|---|---|
| Grossular | 91.61 | $F = 6.94$ |
| Pyrope | 76.25 | Degrees of freedom: 2,1545 |
| Spessartine | 25.62 | |

F-matrix of group separation

| | Tarbert | Ness | |
|---|---|---|---|
| Ness | 58.52 | | $F = 5.46$ |
| Etive–Oseberg | 67.86 | 77.64 | Degrees of freedom: 3, 1545 |

*Classification-matrix*
Number of cases classified into group

| | Tarbert | Ness | Etive–Oseberg |
|---|---|---|---|
| Tarbert | 275 | 84 | 141 |
| Ness | 100 | 192 | 108 |
| Etive–Oseberg | 120 | 92 | 438 |
| % of correctly classified samples | 55 | 48 | 67 |

*Canonical variables*

$v_1 = 0.048$ pyrope $+ 0.113$ grossular $- 0.020$ spessartine $- 2.698$
$v_2 = 0.012$ pyrope $- 0.068$ grossular $- 0.075$ spessartine $+ 1.267$

centres. The resulting fuzzy partitions are very useful for end-member modelling.

Heuristically, it was found after several runs that a weighting exponent of 1.5 provided optimal classification of the 1550 garnet grains into three groups. The partition coefficient of 0.7691 and the partition entropy of 0.4135 indicate moderate fuzziness (Bezdek *et al.* 1984). A clear three group structure is recognizable, with three end-members or cluster centres defined (Fig. 10). The first cluster centre (EM1) has high almandine (66%), intermediate pyrope (23%), low grossular (8%) and spessartine (3%) proportions. The second (EM2) has high grossular (23%) and spessartine (12%) proportions, with intermediate almandine (57%) and low pyrope (9%). The third (EM3) is in high pyrope (36%), intermediate in grossular (18%) and low in almandine (45%) and spessartine (2%). Each sample was classified according to the percentage of predominant memberships to the clusters. The grain density of the three predefined stratigraphic intervals are shown on triangular plots in relation to the three cluster centres (Fig. 10). The Etive–Oseberg interval is dominated by EM3, has a minor EM2 influence, and shows mixing between EM2 and EM3 and to a lesser extent between EM2 and EM1. The grain density distribution of the Ness Formation is governed by EM2 with minor contributions from EM3 and EM1 and shows a significant degree of mixing between EM2 and EM1. The Tarbert Formation is dominated by EM1, with additional contributions from EM3 and to a lesser extent from EM2. Mixing occurs mainly between EM1 and EM3 with a marked mixing gap.

Stratigraphic trends are shown in Fig. 11. Each sample is represented by grain percentages expressing the dominant affinities of the indi-

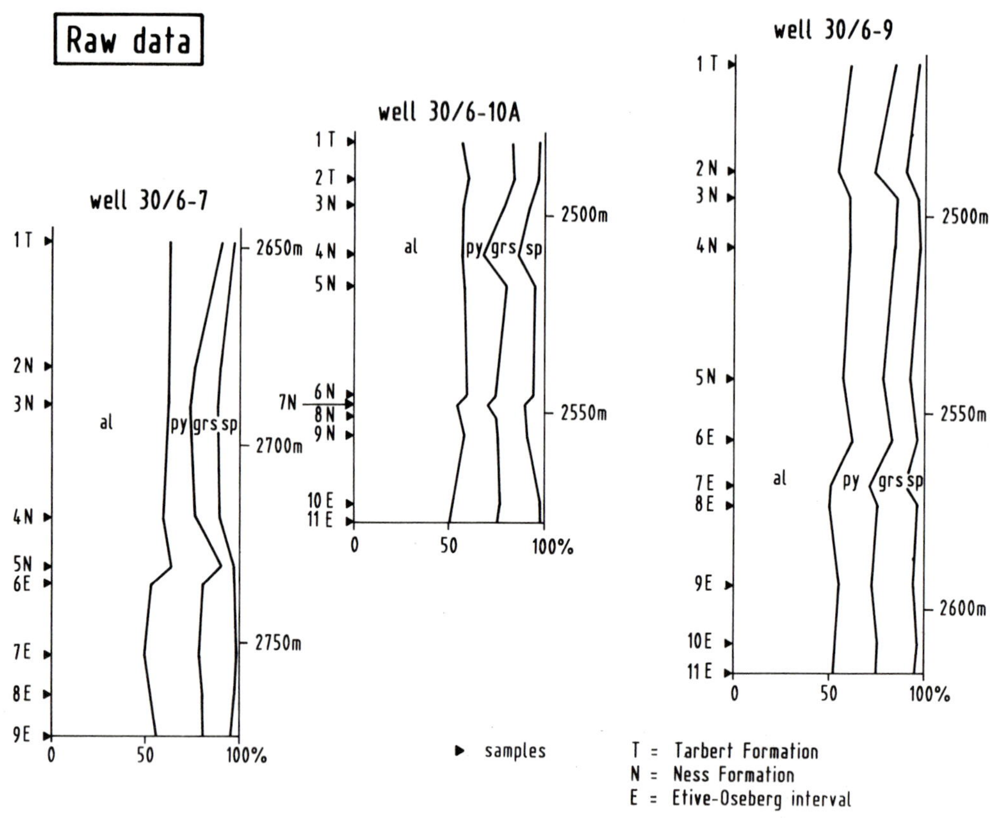


**Fig. 9.** Stratigraphic model of the three investigated wells from stepwise linear discriminant analysis using all 1550 garnet grains and stratigraphic grouping into Tarbert, Ness and Etive–Oseberg intervals. Each sample is represented by its grain percentages for the three stratigraphic groups. Group mean compositions are framed.

vidual grains to the end members. Well 30/6–7 shows the best relationship between the results of the clustering and the lithostratigraphy: EM3 corresponds to the Etive–Oseberg interval, followed by the maximum development of EM2 corresponding to the Ness Formation. There is a gradation towards EM1 dominance in the upper part of the Brent sequence (Tarbert Formation), although a Tarbert-like input occurs between Etive and Ness. The general trends of upward-increasing EM1 and upward-decreasing EM3 are also seen in 30/6–9 and 30/6–10A, but the EM2 maxima in the Ness interval are less marked and in part obscured by early Tarbert-like EM1 inputs. In general, the Etive-Oseberg interval is characterized by high pyrope contents, the Ness by high grossular and spessartine and low pyrope, and the Tarbert by low grossular. Superimposed on this is an overall increase in almandine. If the end members are interpreted in terms of garnet compositions of different source rocks or source areas, the distinct lithostratigraphic changes seen particularly in well 30/6–7 indicate the influence of three distinct source areas, rather than unroofing of a single source area with lateral and vertical variations in metamorphic lithology. Consideration of the regional distribution of the garnet assemblages, as described by Morton (this volume), shows that unroofing cannot adequately explain the stratigraphic pattern, particularly as the EM3 component that characterises the Etive–Oseberg interval reappears at the top Brent level in the NE of the Brent province and the EM1 component found in the Tarbert Formation in Oseberg is also seen at the base of the Brent (Broom Formation) in the western part of the Brent province.

The stratigraphic pattern found by fuzzy clustering is remarkably similar to that found by discriminant analysis, even though the two methods are quite different. Although the compositions of the cluster centres are slightly more extreme in the discriminant analysis than in

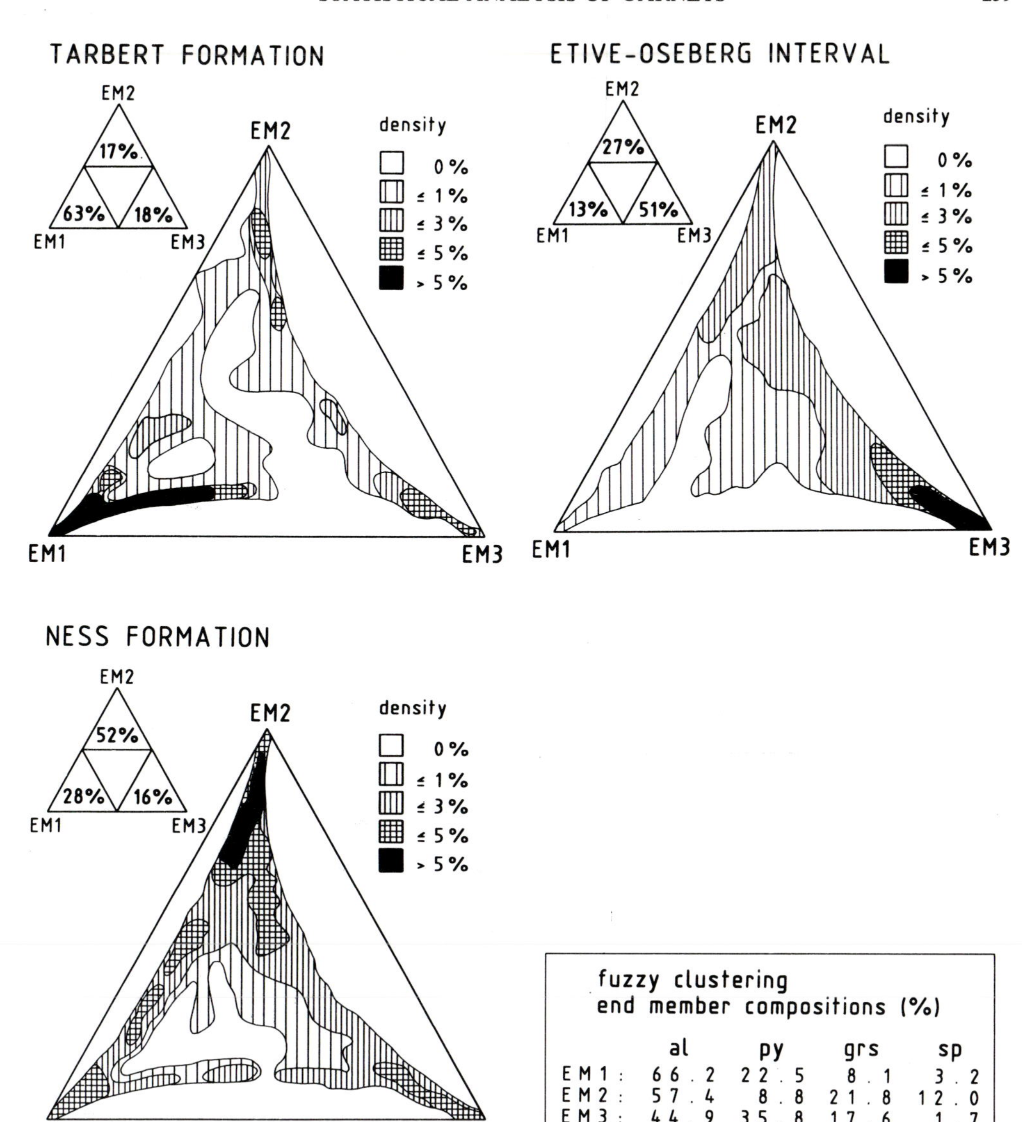


fuzzy clustering
end member compositions (%)

| | al | py | grs | sp |
|---|---|---|---|---|
| EM1: | 66.2 | 22.5 | 8.1 | 3.2 |
| EM2: | 57.4 | 8.8 | 21.8 | 12.0 |
| EM3: | 44.9 | 35.8 | 17.6 | 1.7 |

**Fig. 10.** Stratigraphic garnet grain density patterns from fuzzy c-means clustering in relation to three clusters, the fuzzy end-members EM1, EM2 and EM3. End member compositions are framed. Density percentages are based on the number of grains in one 1% portion of the total area in relation to all grains in that triangle, using the sample with predominant membership to one specific cluster. The small triangles contain the grain percentages with over 50% affinity to one particular end member. **(a)** Tarbert Formation; **(b)** Ness Formation; **(c)** Etive–Oseberg interval.

fuzzy clustering, only grossular has a different ranking. This results in a better definition of the Ness group using fuzzy clustering because the group identification is stronger. Both methods confirm that the stratigraphic subdivision of the sequence on the basis of garnet geochemistry is valid, including the assignation of two 'Etive' samples to the Ness group, confirming their fluvial affinities. However, fuzzy clustering is more efficient than discriminant analysis because of its objective grouping which does not require any predefined sample groups and because of its more definite grouping pattern. Finally, the results of both the discriminant analysis and the fuzzy clustering confirm that the grouping of the samples on the basis of the

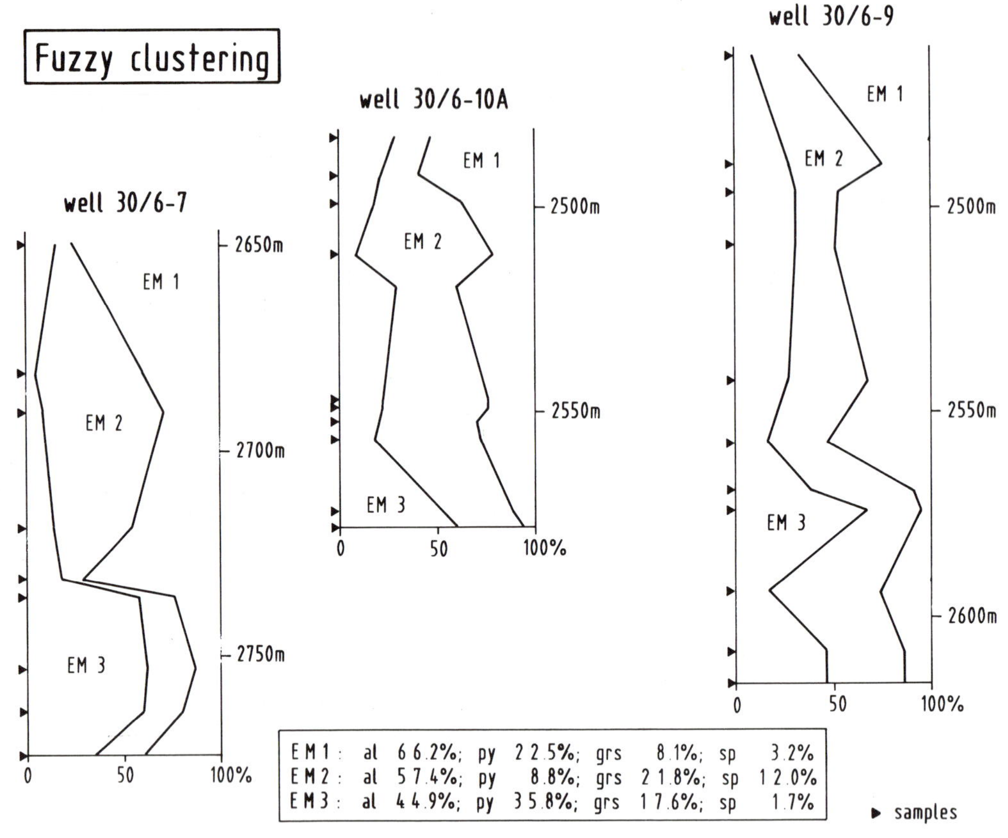


**Fig. 11.** Stratigraphic model of the three investigated wells from fuzzy c-means clustering. Fuzzy end members EM1, EM2 and EM3 are framed. Each sample is described in terms of the percentage of grains that belong to each end member.

subdivision of AS-P-G space into subfields, as proposed by Morton *et al.* (1989*a*), can be sustained statistically. The subfield method distinguished the same three groups (Etive–Oseberg, Ness and Tarbert), with the Etive–Oseberg interval characterized by high pyrope and grossular contents, the Ness by high total almandine plus spessartine, and the Tarbert by low grossular. Furthermore, the greater scatter of the Ness grouping was also recognized by this method, with Ness samples occupying a wider area of the summary plot (Fig. 2) than either the Etive–Oseberg or the Tarbert samples. It is therefore clear that for preliminary grouping of the data and broad distinction of sources, the subfield method is a valid approach.

## Conclusions

Statistical analysis of garnet compositions is a valuable tool in classifying and differentiating samples. In this example, a variety of methods were used to evaluate differences in garnet geochemistry in samples from the Middle Jurassic Brent Group of the Oseberg Field, northern North Sea. Three lithostratigraphic units (the Etive–Oseberg interval, the Ness Formation and the Tarbert Formation) have statistically significant differences in garnet geochemistry related to variations in source lithology during deposition. Elementary statistics, exploratory data analysis, correlation structure and subcompositions determine the characteristic properties of the garnet data in preparing multivariate strategies. Almandine is the prevailing component followed by pyrope, grossular and spessartine. Variation in the bulk data is generally low apart from many high value spessartine outliers.

Multivariate grouping by stepwise linear discriminant analysis can be based either on predefined sample classification or by taking each sample as separate group. Stratigraphic classification of samples is easily made when calculated from the mean sample compositions. When the full data set is used, without the strong smoothing effect provided by the mean values, the classification is less clear: it is efficient for the Etive–Oseberg interval and the Tarbert Formation but the Ness Formation is not clearly defined. End-member analysis by fuzzy c-means clustering is more efficient. It generates similar stratigraphic garnet logs with clear stratigraphic variations, but in addition allows a dynamic interpretation in assuming that end members represent garnet compositions of specific source rocks or source areas. Thus stratigraphic trends in the three wells exhibit the source relations of the individual garnet grains. These trends suggest input from three distinct source areas, one for the Etive–Oseberg interval, one for the Ness Formation and one for the Tarbert.

In general, multivariate end member modelling of garnet compositions provides objective criteria for classifying and differentiating samples. Discriminating between predefined sample groups and evaluating multivariate end members from individual grain compositions result in grain and sample distribution patterns which clearly reflect stratigraphic variations. This has considerable advantages over the distribution patterns obtained from the pure, univariate geochemical garnet end members, especially in cases where the data set does not contain strong variations and individual data points are close together in the variable space.

We are grateful to M. Giles and L. Gaarenstroom for their thorough and constructive reviews of the manuscript. This paper is published with the approval of the Director, British Geological Survey (NERC)

## References

AITCHISON, J. 1984. Reducing the dimensionality of compositional data sets. *Mathematical Geology*, **16**, 637–650.

—— 1986. *The Statistical Analysis of Compositional Data*. Chapman and Hall, London.

BEZDEK, J. C., EHRLICH, R. & FULL, W. 1984. FCM: the fuzzy c-means clustering algorithm. *Computers & Geosciences*, **10**, 191–204.

CHAYES, F. 1971. *Ratio Correlation*. University of Chicago Press, Chicago.

FULL, W. E., EHRLICH, R. & KLOVAN, J. E. 1981. EXTENDED QMODEL – objective definition of extended end members in the analysis of mixtures. *Mathematical Geology*, **13**, 331–344.

——, —— & BEZDEK, J. C. 1982. FUZZY QMODEL – a new approach for linear unmixing. *Mathematical Geology*, **14**, 259–270.

GRANATH, G. 1984. Application of fuzzy clustering and fuzzy classification to evaluate the provenance of glacial till. *Mathematical Geology*, **16**, 283–302.

GRAUE, E., HELLAND-HANSEN, W., JOHNSEN, J., LOMO, L., NOTTVEDT, A., RONNING, K., RYSETH, A. & STEEL, R. J. 1987. Advance and retreat of Brent delta system, Norwegian North Sea. *In*: BROOKS, J. & GLENNIE, K. (eds) *Petroleum Geology of North West Europe*. Graham & Trotman, London, 915–937.

HENLEY, S. 1981. *Nonparametric Geostatistics*. Applied Science Publishers, London.

HURST, A. & MORTON, A. C. 1988. An application of heavy-mineral analysis to lithostratigraphy and reservoir modelling in the Oseberg Field, northern North Sea. *Marine and Petroleum Geology*, **5**, 157–169.

JENNRICH, R. I. 1977. Stepwise discriminant analysis. *In*: ENSLEIN, K., RALSTON, A. & WILF, H. S. (eds) *Statistical Methods for Digital Computers*. Wiley, New York, 328–342.

KLOVAN, J. E. & IMBRIE, J. 1971. An algorithm and FORTRAN-IV program for large scale Q-mode factor analysis of compositional data. *Mathematical Geology*, **3**, 61–67.

—— & MIESCH, A. T. 1976. Extended CABFAC and QMODEL computer programs for Q-mode factor analysis of compositional data. *Computers & Geosciences*, **1**, 161–178.

KRZANOWSKI, W. J. 1977. The performance of Fisher's linear discriminant function under non-optimal conditions. *Technometrics*, **19**, 191–208.

MORTON, A. C. 1985. A new approach to provenance studies: electron microprobe analysis of detrital garnets from Middle Jurassic sandstones of the northern North Sea. *Sedimentology*, **32**, 553–566.

—— 1987. Detrital garnets as provenance and correlation indicators in North Sea reservoir sandstones. *In*: BROOKS, J. & GLENNIE, K. (eds) *Petroleum Geology of North West Europe*. Graham & Trotman, London, 991–995.

—— 1992. Provenance of the Brent Group: heavy mineral constraints. *This volume*.

——, BORG, G., HANSLEY, P. L., HAUGHTON, P. D. W., KRINSLEY, D. H. & TRUSTY, P. 1989*b*. The origin of faceted garnets in sandstones: dissolution or overgrowth. *Sedimentology*, **36**, 927–942.

——, STIBERG, J. P., HURST, A. & QVALE, H. 1989*a*. Use of heavy minerals in lithostratigraphic correlation, with examples from Brent sandstones of the northern North Sea. *In*: COLLINSON, J. (ed.) *Correlation in hydrocarbon exploration*. Graham & Trotman, London, 217–230.

PAWLOWSKY, V. & STATTEGGER, K. 1988. Cokriging of compositional data – a case study from modern stream sands. *Science de la Terre, Serie Informatique*, **28**, 267–280.

REMME, J., HABBEMA, J. D. F. & HERMANS, J. A. 1980. A simulative comparison of linear, quadratic and kernel discrimination. *Journal of Statistics and Computer Simulation*, **11**, 87–105.

REYMENT, R. A., BLACKITH, R. E. & CAMPBELL, N. A. 1984. *Multivariate Morphometrics*. Aca-

demic Press, London.
ROCK, N. M. S. 1988. Summary statistics in geochemistry: a study of the performance of robust estimates. *Mathematical Geology*, **20**, 243–276.
STATTEGGER, K. 1987. Heavy minerals and provenance of sands: modeling of lithological end members from river sands of Northern Austria and from sandstones of the Austroalpine Gosau Formation (Late Cretaceous). *Journal of Sedimentary Petrology*, **57**, 301–310.
—— 1989. Mixing, outliers and end members: geochemical compositions of detrital garnets from Jurassic sandstones of the North Sea. *28th Geological Congress, Abstracts*, **3**, 172.
VELLEMAN, P. F. & HOAGLIN, D. C. 1981. *Applications, Basics and Computing of Exploratory Data Analysis*. Duxbury Press, Boston.

# Diagenetic processes in the Brent Group (Middle Jurassic) reservoirs of the North Sea: an overview

KNUT BJØRLYKKE, TOR NEDKVITNE, MOGENS RAMM & GIRISH C. SAIGAL

*Department of Geology, University of Oslo, PO Box 1047, Blindern, 0316 Oslo 3, Norway*

**Abstract:** The Brent Group is extensively cored in the northern Viking Graben and East Shetland Basin in the range between 1.8 and 4.5 km. It therefore offers the opportunity to study diagenetic processes arrested at different burial depths and temperatures. The sandstones of the Brent Group remain poorly indurated and contain only small amounts of quartz cement down to burial depths of about 2.5–3.5 km. The quartz cementation increases with depth reflecting progressive silica cementation sourced by pressure solution. Dissolved feldspar and authigenic kaolinite occur in the shallowest of the Brent Group reservoirs and secondary porosity shows no significant increase with depth. Petrographic analyses show a strong depletion of K-feldspar with depth. This can be explained by albitization of K-feldspar at intermediate burial depths (2.5–3.5 km) and dissolution of K-feldspar during illitization of kaolinite on deeper burial (below 3.7–4.0 km). Extensive quartz cementation and formation of illite are the main causes of an observed rapid permeability reduction below 4 km burial depth. K-Ar dates of authigenic illite indicate that illite formed mostly in Early Tertiary at rather shallow burial depth. Illite formed by alteration of kaolinite and K-feldspar seems, however, to increase very significantly at present burial depths of 3.5–4.0 km, suggesting a more recent formation. Emplacement of oil probably does not stop all diagenetic reactions in the remaining residual pore water, and precipitation of illite and quartz may continue in a closed system.

The Brent Group is an important reservoir rock in the North Sea Basin and has been extensively drilled in the major fields of the East Shetland Basin and the northern Viking Graben. This allows us to study the diagenesis of these sandstones at burial depths from less than 2 km to more than 4 km, at widely different degrees of overpressure, and from both above and below the oil–water contact. A substantial literature on the diagenesis of the Brent Group has already accumulated. It is unfortunate that many of these papers do not indicate the wells or even the fields on which the studies were conducted. However, this literature can not be ignored since it contains important data and conclusions. Although there may be significant regional variations in the initial sediment composition and burial history, there are sufficient common factors to justify some generalizations about the diagenetic processes. The Brent Group was deposited in a shallow marine and deltaic environment (Fig. 1) (e.g. Brown *et al.* 1987; Graue *et al.* 1987; Fält *et al.* 1989; Helland-Hansen *et al.* 1989) and the sedimentary facies are important as a background for understanding the diagenetic processes.

The following diagenetic processes influence on reservoir properties of the Brent Group sandstones and will be discussed in this paper: (1) feldspar dissolution and precipitation of kaolinite; (2) carbonate cementation; (3) albitization of K-feldspar; (4) quartz cementation; (5) precipitation of authigenic illite.

Published literature on the mineralogical composition and the diagenesis of the Brent Group will be reviewed with emphasis on the processes listed above and on the discussion on formation of secondary porosity at deep burial. Furthermore, we will present new petrographic data for the Brent Group sandstones from several fields in the Norwegian sector (i.e. the Gullfaks, Veslefrikk, Hild, Oseberg and Huldra Fields; Fig. 2). These data will form the basis for a general discussion on the diagenesis of the Brent Group. Particular emphasis will be on the establishment of a sequence of mineral dissolution and precipitation, and reservoir quality versus depth trends.

## Mineralogical composition of the Brent Group sandstones

Brent Group sandstones are dominated by arkoses and subarkoses with rare occurrences of sublitharenites and litharenites. XRD analy-

*From* Morton, A. C., Haszeldine, R. S., Giles, M. R. & Brown, S. (eds), 1992, *Geology of the Brent Group*. Geological Society Special Publication No. 61, pp. 263–287.

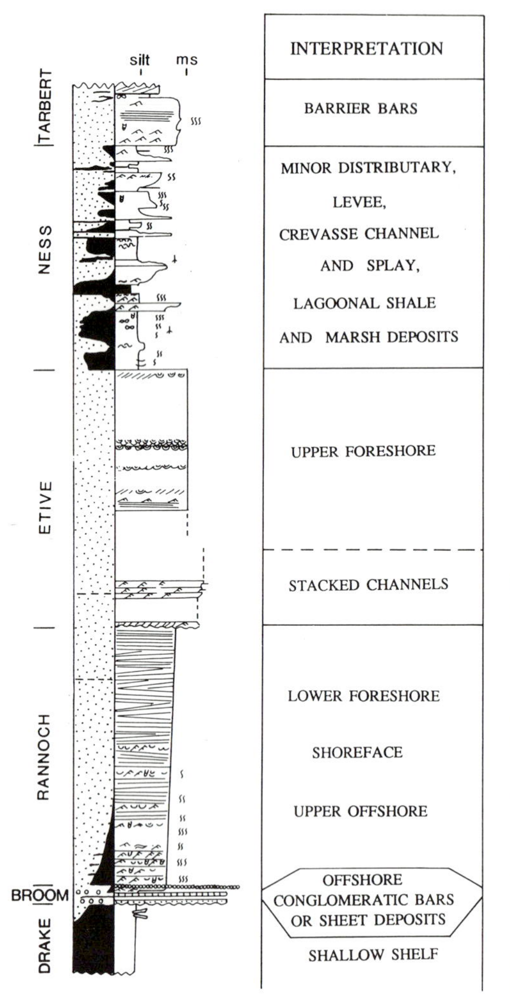


**Fig. 1.** Stratigraphy and sedimentology of the Brent Group based on cores from well 33/9–A23, Statfjord Field (from Bjørlykke & Brendsdal 1986).

ses of bulk samples from several fields (Gullfaks, Veslefrikk, Oseberg, Hild and Huldra) are given in Table 1. Average quartz (dominantly monocrystalline) content varies between 40 and 90%. The feldspar content shows both regional and depth related variations. The understanding of the regional variation is not fully understood and requires more data. Comparing average K-feldspar and albite concentrations between cores from variable depths, from the Oseberg–Veslefrikk area in the southeast and the Gullfaks area in the northwest (Table 1, Fig. 3), shows no significant regional variations in the K-feldspar content. Both in the Oseberg–Veslefrikk area and the Gullfaks area, the K-feldspar content decreases from average values near 10% at 2.5–3 km to values below the detection limit of XRD below about 3.5–4 km. The albite concentration shows no distinct trend versus depth but seems to be higher in the Gullfaks area than in the Oseberg–Veslefrikk area. In some of the wells from block 30/6, the albite concentration is very low, particularly in samples from the Etive Formation. The samples from the Tarbert Formation from the Hild Field deviate both from samples from the Gullfaks and the Oseberg–Veslefrikk areas by being more quartzitic with very low concentrations of albite and K-feldspar. According to Harris (1989), the feldspar content increases northwards from an orthoquartzitic composition in the Lyell Field (3500 m) through the Hutton Field (3000 m) to the Statfjord Field (2580 m). The Hutton Field sandstones contain only about 3% feldspar (Scotchman *et al.* 1989) while in the Statfjord Field, the feldspar (dominantly K-feldspar) content varies between 10 and 25% (Bjørlykke & Brendsdal 1986; Kittilsen 1987). The K-feldspar distribution within single wells shows systematically higher concentrations in the marine and marginal marine sandstones compared to the fluvial Ness sandstones. Furthermore, the open marine Rannoch, Tarbert and Oseberg Formations contain more K-feldspar than the Etive and Ness Formations. The analogous relationship with respect to albite is somewhat different, as the Etive Formation seems to have less albite than all the other formations.

Lithic clasts include mostly quartzite and partly altered volcanic rocks. Clasts of siderite, probably representing winnowed concretions, are common particularly in the Rannoch and the Ness Formations. Siderite is often partly replaced by ankerite at greater depth (Fig. 4).

The dominant clay minerals are illite, kaolinite and small amounts of chlorite. These minerals occur both as detrital clasts and authigenic minerals, but standard XRD analyses can not distinguish between these two. This is, however, possible using petrographic microscope and SEM. An exact quantification of the authigenic phases are difficult, but in the clean, good reservoir sandstones there are very little clastic clay minerals. The Etive Formation in the deepest well studied (*c.* 4.2 km) comprises little kaolinite relative to illite while an opposite trend is seen in the shallower reservoirs.

## Provenance

The immature composition of the sandstones, which often contain some altered biotite in ad-

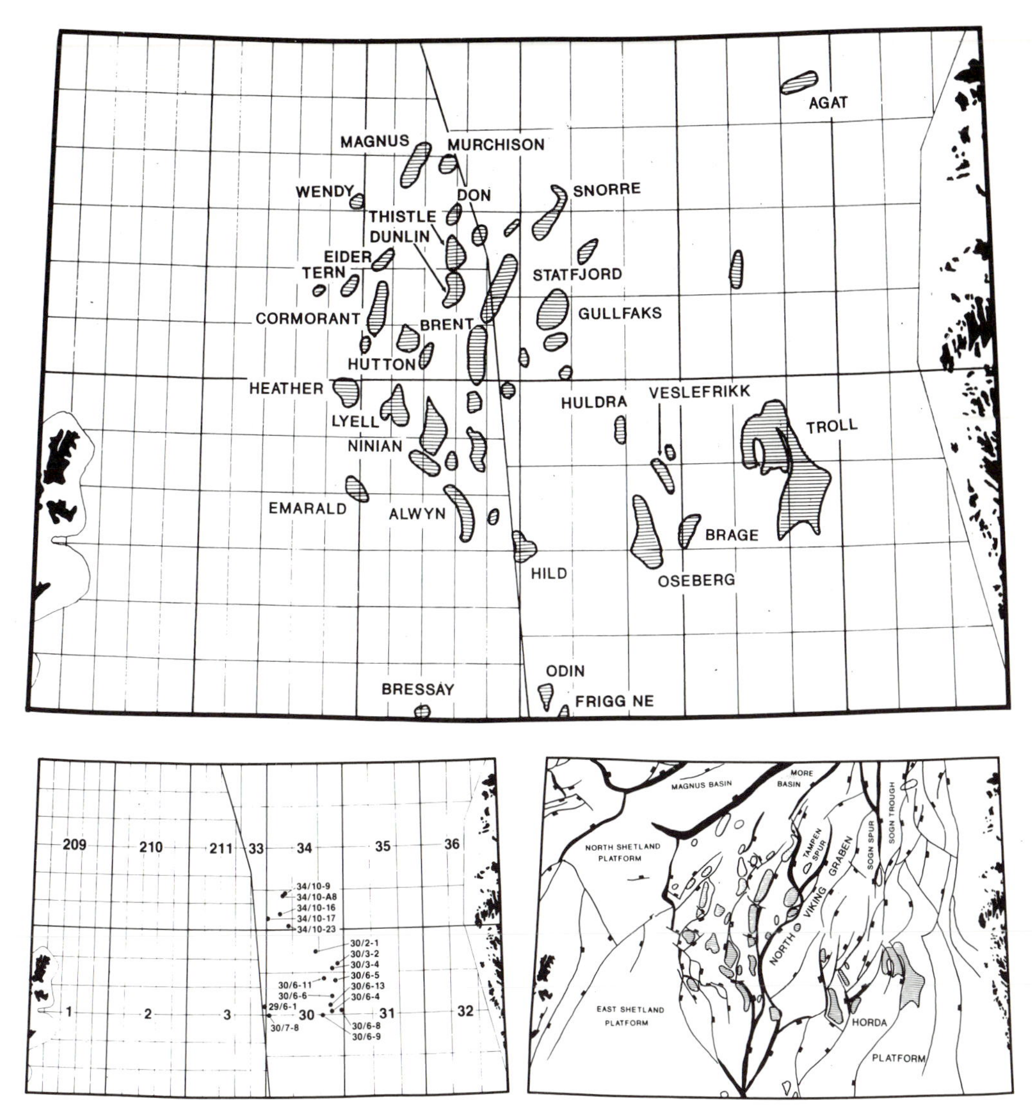


**Fig. 2.** Location map for the major oil and gas fields in the North Viking Graben and wells discussed in this paper.

dition to muscovite, suggests a proximal basement provenance. This is probably related to tectonic uplift and erosion into the Caledonian or Precambrian basement along the margins of the basin.

Rb-Sr datings of muscovites from the Brent Group (wells unspecified) gave Caledonian ages (400–470 Ma) which suggest derivation from Caledonian metamorphic rocks or Precambrian rocks subjected to Caledonian metamorphism (Hamilton *et al.* 1987*a*). They also found that Rb-Sr, Sm-Nd and K-Ar analyses of lithic clasts agreed well with data from basement rocks in the Shetland region. Analyses of detrital garnets (Morton 1985) suggest that the sediments are derived mainly from basement rocks of the Orkney–Shetland platform and of the Norwegian mainland and that the uplifted Moray Firth area could not have been the main source.

Sm-Nd isotope analyses on bulk samples from the Gullfaks Field indicate different ages for the sources of the Etive, Rannoch and Ness Formations (Mearns 1989). A systematic primary difference in the mineral composition of the Ness Formation compared to the Etive and Rannoch

**Table 1.** *Mineralogical composition of the Brent Group sandstones based on XRD analyses carried out by M. Ramm*

| Field (well) Formation | Interval *N* | 14 Å Chlorite | 10 Å Illite/Mica | 7 Å Kaolinite | 4.26 Å Quartz | 3.24 Å K-Feldspar | 3.18 Å Albite | 3.04 Å Calcite | 2.89 Å Dolom./Anker. | 2.79 Å Siderite | 2.71 Å Pyrite |
|---|---|---|---|---|---|---|---|---|---|---|---|
| Gullfaks (34/10–9) | | | | | | | | | | | |
| Etive | 1851 – 1863 | 0 ± 1 | 4 ± 2 | 21 ± 5 | 42 ± 4 | 18 ± 4 | 12 ± 4 | 1 ± 0 | 1 ± 0 | 1 ± 1 | 1 ± 1 |
| | *N* = 6 | 0 – 2 | 2 – 8 | 12 – 28 | 37 – 48 | 14 – 23 | 7 – 18 | 0 – 1 | 0 – 1 | 0 – 3 | 0 – 3 |
| Gullfaks-South (34/10–16) | | | | | | | | | | | |
| Ness | 3274 – 3381 | 0 ± 0 | 2 ± 1 | 10 ± 5 | 67 ± 8 | 5 ± 4 | 14 ± 5 | 1 ± 1 | 0 ± 0 | 2 ± 2 | 0 ± 0 |
| | *N* = 15 | 0 – 0 | 0 – 4 | 0 – 23 | 56 – 85 | 0 – 16 | 4 – 24 | 0 – 4 | 0 – 1 | 0 – 8 | 0 – 1 |
| Gullfaks-$\beta$ (34/10–17) | | | | | | | | | | | |
| Tarbert | 2685 – 2716 | 0 ± 0 | 3 ± 2 | 21 ± 9 | 51 ± 11 | 15 ± 4 | 7 ± 2 | 1 ± 0 | 1 ± 0 | 1 ± 1 | 1 ± 0 |
| | *N* = 6 | 0 – 0 | 0 – 7 | 7 – 32 | 37 – 64 | 5 – 21 | 4 – 11 | 0 – 1 | 0 – 1 | 0 – 3 | 0 – 2 |
| Ness | 2721 – 2933 | 0 ± 0 | 2 ± 2 | 13 ± 7 | 67 ± 12 | 8 ± 3 | 8 ± 4 | 0 ± 0 | 1 ± 2 | 1 ± 1 | 0 ± 0 |
| | *N* = 15 | 0 – 1 | 0 – 8 | 4 – 35 | 45 – 85 | 0 – 12 | 2 – 17 | 0 – 1 | 0 – 7 | 0 – 3 | 0 – 0 |
| Etive | 2935 – 2937 | 0 ± 0 | 1 ± 0 | 6 ± 2 | 76 ± 6 | 10 ± 2 | 7 ± 2 | 0 ± 0 | 1 ± 0 | 1 ± 0 | 0 ± 0 |
| | *N* = 5 | 0 – 0 | 0 – 1 | 5 – 9 | 69 – 85 | 5 – 12 | 4 – 10 | 0 – 1 | 0 – 1 | 0 – 1 | 0 – 0 |
| Rannoch | 2937 – 2964 | 0 ± 0 | 3 ± 1 | 18 ± 5 | 51 ± 8 | 12 ± 2 | 9 ± 1 | 1 ± 1 | 4 ± 4 | 2 ± 1 | 0 ± 0 |
| | *N* = 7 | 0 – 1 | 2 – 4 | 10 – 24 | 39 – 86 | 10 – 17 | 8 – 11 | 0 – 1 | 0 – 12 | 1 – 3 | 0 – 0 |
| Gullfaks-$\gamma$ (34/10–23) | | | | | | | | | | | |
| Tarbert | 4083 – 4118 | 0 ± 0 | 8 ± 5 | 8 ± 5 | 62 ± 12 | 0 ± 0 | 16 ± 3 | 0 ± 0 | 2 ± 2 | 3 ± 2 | 1 ± 1 |
| | *N* = 7 | 0 – 0 | 1 – 18 | 2 – 17 | 48 – 83 | 0 ± 0 | 9 – 20 | 0 – 0 | 0 – 6 | 0 – 6 | 0 – 3 |
| Ness | 4124 – 4233 | 0 ± 0 | 3 ± 2 | 14 ± 7 | 67 ± 12 | 0 ± 0 | 12 ± 7 | 0 ± 0 | 0 ± 0 | 3 ± 3 | 0 ± 0 |
| | *N* = 8 | 0 – 0 | 1 – 7 | 8 – 29 | 43 – 85 | 0 ± 0 | 3 – 21 | 0 – 0 | 0 – 0 | 0 – 10 | 0 – 0 |
| Etive | 4238 – 4281 | 0 ± 0 | 4 ± 4 | 1 ± 1 | 81 ± 9 | 0 ± 0 | 9 ± 3 | 1 ± 1 | 1 ± 1 | 3 ± 3 | 0 ± 0 |
| | *N* = 9 | 0 – 0 | 1 – 14 | 0 – 4 | 65 – 90 | 0 ± 0 | 4 – 14 | 0 – 4 | 0 – 4 | 0 – 9 | 0 – 0 |
| Rannoch | 4284 – 4307 | 0 ± 0 | 4 ± 2 | 3 ± 2 | 77 ± 7 | 0 ± 0 | 14 ± 3 | 0 ± 0 | 0 ± 0 | 2 ± 2 | 0 ± 0 |
| | *N* = 7 | 0 – 1 | 1 – 7 | 1 – 8 | 66 – 89 | 0 ± 0 | 9 – 20 | 0 – 0 | 0 – 0 | 0 – 6 | 0 – 0 |
| Hild (29/6–1) | | | | | | | | | | | |
| Tarbert | 4220 – 4335 | 0 ± 0 | 3 ± 2 | 4 ± 3 | 92 ± 6 | 0 ± 0 | 1 ± 1 | 0 ± 1 | 0 ± 0 | 0 ± 0 | 0 ± 0 |
| | *N* = 13 | 0 – 0 | 0 – 7 | 0 – 10 | 73 – 100 | 0 – 0 | 0 – 3 | 0 – 3 | 0 – 1 | 0 – 2 | 0 – 1 |
| (30/7–8) | | | | | | | | | | | |
| Tarbert | 4062 – 4152 | 0 ± 0 | 2 ± 2 | 6 ± 7 | 91 ± 8 | 0 ± 0 | 0 ± 0 | 0 ± 1 | 0 ± 0 | 0 ± 0 | 0 ± 0 |
| | *N* = 12 | 0 – 0 | 0 – 5 | 0 – 19 | 76 – 100 | 0 – 0 | 0 – 2 | 0 – 4 | 0 – 1 | 0 – 0 | 0 – 1 |

**Table 1.** *Continued*

| Field (well) Formation | Interval $N$ | 14 Å Chlorite | 10 Å Illite/Mica | 7 Å Kaolinite | 4.26 Å Quartz | 3.24 Å K-Feldspar | 3.18 Å Albite | 3.04 Å Calcite | 2.89 Å Dolom./Anker. | 2.79 Å Siderite | 2.71 Å Pyrite |
|---|---|---|---|---|---|---|---|---|---|---|---|
| Huldra (30/2–1) | | | | | | | | | | | |
| Ness | 3698 – 3756 | 0 ± 0 | 2 ± 1 | 12 ± 5 | 71 ± 7 | 0 ± 0 | 14 ± 7 | 0 ± 1 | 0 ± 0 | 1 ± 2 | 0 ± 0 |
| | $N$ = 12 | 0 – 0 | 1 – 4 | 6 – 21 | 59 – 82 | 0 – 0 | 7 – 30 | 0 – 3 | 0 – 0 | 0 – 7 | 0 – 1 |
| Etive | 3759 – 3770 | 0 ± 0 | 5 ± 4 | 7 ± 5 | 76 ± 10 | 0 ± 0 | 9 ± 2 | 0 ± 1 | 0 ± 0 | 2 ± 3 | 0 ± 0 |
| | $N$ = 3 | 0 – 0 | 1 – 10 | 2 – 15 | 67 – 90 | 0 – 0 | 7 – 11 | 0 – 1 | 0 – 0 | 0 – 7 | 0 – 0 |
| Rannoch | 3778 – 3792 | 1 ± 2 | 4 ± 3 | 21 ± 8 | 58 ± 14 | 1 ± 1 | 11 ± 4 | 0 ± 0 | 0 ± 0 | 4 ± 2 | 0 ± 0 |
| | $N$ = 4 | 0 – 4 | 1 – 8 | 10 – 29 | 42 – 75 | 0 – 3 | 6 – 16 | 0 – 0 | 0 – 0 | 4 –2 | 0 – 1 |
| Oseberg-$\alpha$ (30/6–6) | | | | | | | | | | | |
| Ness | 2921 – 2952 | | | | | | | | | | |
| | $N$ = 2 | 0 – 0 | 1 – 1 | 20 – 33 | 43 – 48 | 7 – 15 | 0 – 21 | 0 – 2 | 0 – 1 | 2 – 3 | 0 – 4 |
| Etive | 2932 – 2952 | 0 ± 0 | 3 ± 2 | 19 ± 8 | 78 ± 5 | 14 ± 4 | 2 ± 2 | 1 ± 0 | 1 ± 0 | 1 ± 0 | 1 ± 0 |
| | $N$ = 7 | 0 – 0 | 1 – 5 | 11 – 32 | 45 – 75 | 7 – 19 | 0 – 6 | 0 – 1 | 0 – 1 | 0 – 1 | 0 – 1 |
| (30/6–8) | | | | | | | | | | | |
| Tarbert | 3033 – 3060 | 1 ± 2 | 3 ± 2 | 19 ± 6 | 55 ± 14 | 9 ± 2 | 11 ± 4 | 1 ± 1 | 0 ± 0 | 1 ± 1 | 1 ± 1 |
| | $N$ = 4 | 0 – 4 | 0 – 5 | 12 – 28 | 37 – 71 | 6 – 10 | 6 – 17 | 0 – 2 | 0 – 1 | 0 – 1 | 0 – 2 |
| Etive | 3152 – 3163 | | | | | | 5 – 6 | | | | |
| | $N$ = 2 | 0 – 0 | 3 – 6 | 10 – 17 | 62 – 63 | 9 – 11 | | 0 – 1 | 2 – 4 | 1 – 1 | 0 – 0 |
| Oseberg-$\gamma$ (30/6–9) | | | | | | | | | | | |
| Ness | 2463 – 2543 | 0 ± 0 | 1 ± 1 | 14 ± 7 | 61 ± 15 | 8 ± 3 | 13 ± 10 | 0 ± 0 | 0 ± 0 | 2 ± 2 | 0 ± 1 |
| | $N$ = 4 | 0 – 0 | 0 – 3 | 2 – 23 | 48 – 88 | 5 – 13 | 0 – 29 | 0 – 0 | 0 – 1 | 0 – 5 | 0 – 1 |
| Etive | 2556 – 2605 | 0 ± 0 | 1 ± 0 | 9 ± 3 | 72 ± 6 | 10 ± 3 | 5 ± 1 | 0 ± 0 | 1 ± 0 | 1 ± 0 | 1 ± 0 |
| | $N$ = 5 | 0 – 0 | 1 – 2 | 5 ± 14 | 62 – 80 | 5 – 15 | 5 – 7 | 0 – 1 | 0 – 1 | 0 – 1 | 0 – 1 |
| (30/6–11) | | | | | | | | | | | |
| Ness | 3453 – 3458 | | | | | | | | | | |
| | $N$ = 2 | 0 – 0 | 1 – 8 | 2 – 11 | 82 – 87 | 0 – 5 | 0 – 0 | 0 – 0 | 0 – 0 | 0 – 2 | 1 – 1 |
| Etive | 3459 – 3508 | 0 ± 0 | 5 ± 4 | 2 ± 2 | 83 ± 9 | 8 ± 3 | 0 ± 1 | 0 ± 0 | 0 ± 0 | 0 ± 1 | 1 ± 1 |
| | $N$ = 10 | 0 – 0 | 1 – 12 | 0 – 7 | 63 – 94 | 4 – 14 | 0 – 2 | 0 – 1 | 0 – 1 | 0 – 2 | 0 – 3 |
| Oseberg-$\alpha$ (30/6–13) | | | | | | | | | | | |
| Ness | 2586 – 2628 | 0 – 0 | 1 ± 1 | 11 ± 4 | 66 ± 6 | 10 ± 2 | 11 ± 2 | 0 ± 0 | 0 ± 0 | 1 ± 1 | 0 ± 0 |
| | $N$ = 5 | 0 – 0 | 0 – 2 | 6 – 18 | 56 – 75 | 8 – 12 | 8 – 15 | 0 – 1 | 0 – 1 | 0 – 3 | 0 – 1 |
| Etive | 2643 – 2659 | 0 ± 0 | 2 ± 1 | 14 ± 2 | 67 ± 3 | 14 ± 2 | 0 ± 0 | 0 ± 0 | 1 ± 0 | 0 ± 0 | 1 ± 1 |
| | $N$ = 9 | 0 – 1 | 1 – 4 | 11 – 17 | 62 – 73 | 11 – 18 | 0 – 0 | 0 – 1 | 0 – 1 | 0 – 1 | 0 – 3 |

**Table 1.** *Continued*

| Field (well) Formation | Interval $N$ | 14 Å Chlorite | 10 Å Illite/Mica | 7 Å Kaolinite | 4.26 Å Quartz | 3.24 Å K-Feldspar | 3.18 Å Albite | 3.04 Å Calcite | 2.89 Å Dolom./Anker. | 2.79 Å Siderite | 2.71 Å Pyrite |
|---|---|---|---|---|---|---|---|---|---|---|---|
| Oseberg-$\beta$ (30/6–5) | | | | | | | | | | | |
| Ness | 2844 – 2851 | 0 ± 0 | 3 ± 1 | 11 ± 7 | 71 ± 7 | 3 ± 3 | 10 ± 4 | 0 ± 0 | 0 ± 0 | 1 ± 0 | 0 ± 0 |
| | $N$ = 4 | 0 – 0 | 2 – 4 | 5 – 22 | 62 – 80 | 0 – 8 | 5 – 15 | 0 – 0 | 0 – 0 | 0 – 1 | 0 – 1 |
| Etive | 2872 – 2893 | 0 ± 0 | 1 ± 1 | 7 ± 2 | 83 ± 7 | 3 ± 3 | 5 ± 3 | 0 ± 0 | 0 ± 0 | 0 ± 0 | 0 ± 0 |
| | $N$ = 8 | 0 – 0 | 0 – 5 | 5 – 12 | 71 – 90 | 0 – 10 | 2 – 10 | 0 – 1 | 0 – 1 | 0 – 1 | 0 – 0 |
| Oseberg | 2896 – 2930 | 0 ± 0 | 1 ± 2 | 11 ± 2 | 71 ± 7 | 8 ± 3 | 4 ± 2 | 0 ± 0 | 3 ± 3 | 1 ± 2 | 0 ± 0 |
| | $N$ = 10 | 0 – 0 | 0 – 5 | 8 – 13 | 56 – 81 | 6 – 14 | 0 – 6 | 0 – 1 | 0 – 9 | 0 – 7 | 0 – 1 |
| Veslefrikk (30/3–2) | | | | | | | | | | | |
| Ness | 2829 – 2840 | 0 ± 0 | 2 ± 1 | 9 ± 2 | 76 ± 3 | 2 ± 4 | 9 ± 2 | 0 ± 0 | 0 ± 0 | 2 ± 1 | 0 ± 0 |
| | $N$ = 5 | 0 – 0 | 1 – 3 | 6 – 12 | 73 – 79 | 0 – 10 | 6 – 13 | 0 – 0 | 0 – 0 | 0 – 4 | 0 – 2 |
| Etive | 2897 – 2943 | 0 ± 0 | 1 ± 0 | 7 ± 3 | 69 ± 7 | 12 ± 2 | 6 ± 3 | 0 ± 0 | 1 ± 1 | 2 ± 2 | 1 ± 1 |
| | $N$ = 9 | 0 – 0 | 0 – 1 | 3 – 13 | 52 – 82 | 9 – 15 | 3 – 10 | 0 – 0 | 0 – 4 | 0 – 7 | 0 – 4 |

Quantitative estimates are based on correlations to a series of external standards and use of an iterative procedure to correct for bulk mass attenuation coefficients. (The complete procedure is described by Ramm, 1990). Upper numbers are the average of the N analyses while lower numbers show the total range.

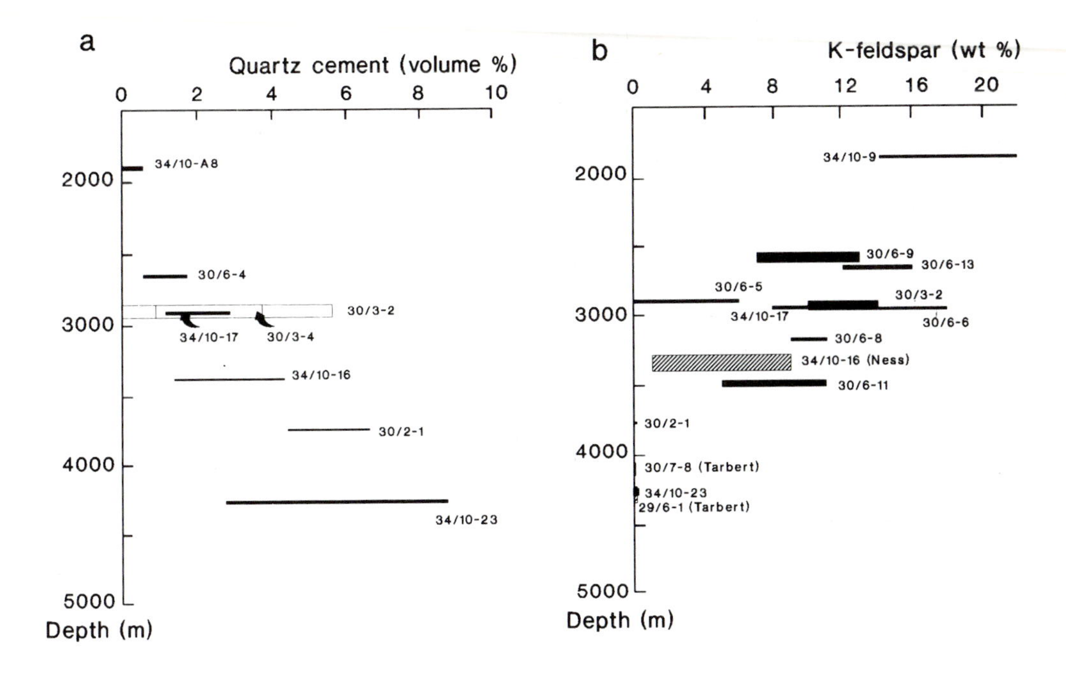


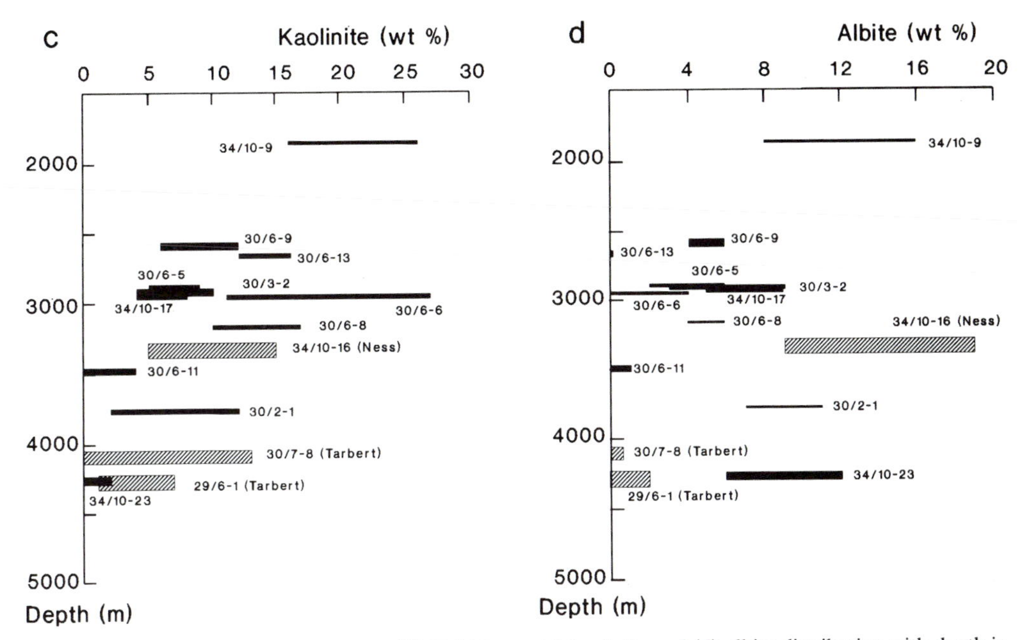


**Fig. 3.** Plots showing (**a**) quartz cement, (**b**) K-feldspar, (**c**) kaolinite and (**d**) albite distribution with depth in the Brent Group sandstones examined in this study. Note that the unspecified bars represent samples from the Etive Formation. Quartz cement content is based on petrographic point count analyses (Table 2) while K-feldspar, kaolinite, and albite data represent XRD analyses (Table 1).

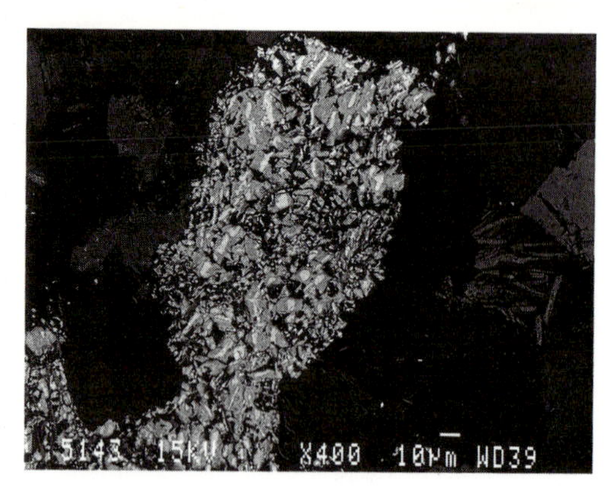


**Fig. 4.** BSE photomicrograph showing replacement of siderite (in a carbonate clast) by ankerite. Siderite occurs as light grey irregular cores that are surrounded by dark grey ankerite rhombs (Well 34/7–14, 2230 m).

Formations is contradictory to a simple model where these units are time equivalents in a prograding clastic wedge and suggests that strike oriented transport along the shoreline was important. If that was not the case, the sediments supplied through the channels of the Ness Formation would have the same composition as those deposited in the delta front and slope environment represented by the Etive and the Rannoch Formations.

Volcanic clasts are observed as a minor component but may often be difficult to recognize due to later diagenetic alteration. Inside a carbonate cemented layer in the Ness Formation of the Statfjord Field, basaltic (tholeiitic) fragments are preserved indicating an eruption at only a few km distance (Malm *et al.* 1979). In this case early carbonate cementation sealed off the Jurassic volcanic glass and prevented it from dissolving.

## Diagenesis

The changes in texture and quantity of authigenic minerals with increasing burial depth provide important information to aid the reconstruction of the diagenetic processes. This is particularly important as neither textural relationships between authigenic minerals nor isotopic data consistently give unique answers about the sequence of diagenetic processes. Most of the diagenetic processes studied are probably not diagenetic events occurring over a short period of time, but represents slow processes towards mechanical stability and thermodynamic equilibrium. Quartz cement is a good example of this as its abundance shows a steady increase with depth. Since Brent Group sandstones buried to shallower depth than 1.8 km have not been studied, early diagenetic processes will always be overprinted by later ones and are therefore more difficult to study. Sulphides and siderite are early diagenetic minerals and K-feldspar overgrowths also appear to have formed at shallow depths. Early diagenetic carbonate minerals, which might have been sourced by dissolution of biogenic carbonate, probably dissolved at greater burial depths and reprecipitate as poikilotopic cement. Kaolinite cement and dissolved feldspar occur in the shallowest reservoirs and seem not to be formed at deep burial.

Although the various sandstones in the Brent Group may differ somewhat in their primary composition and facies, the main diagenetic processes are considered to be rather similar for all formations. In the following, each of the processes considered essential for the development of reservoir properties will be discussed.

## Feldspar dissolution and precipitation of kaolinite

Formation of secondary porosity at shallow burial depth due to dissolution of feldspars and precipitation of authigenic kaolinite are important features of the Brent Group. This was first noted by Blanche & Whitaker (1978), Hancock (1978*a*), and Hancock & Taylor (1978) (fields not specified). Their conclusion that kaolinite was formed early at shallow burial depth, before quartz and illite cementation, was supported by later studies from the Statfjord Field (Bjørlykke *et al.* 1979) and Hild Field (Lønøy *et al.* 1986). $\delta^{18}O$ (SMOW) of kaolinite from the Heather and Huldra Fields (Glasmann *et al.* 1989*a*, *b*), are consistent with the formation temperatures less than 60°C in meteoric water with a composition of $\delta^{18}O$ between −6 and −8‰ (SMOW).

Secondary porosity caused by feldspar dissolution is easier to quantify petrographically than dissolved carbonate cement, particularly where relict grains or clay rims are well developed, as is often the case in the Brent Group sandstones. Petrographic point counting data from the Etive Formation in the Veslefrikk, Oseberg, and Gullfaks Fields indicate that the secondary porosity comprises 2–7% of the total rock volume (Table 2, Fig. 5). Similar results have been obtained from the Statfjord, Hutton and Lyell Fields (Harris 1989), the Huldra Field (Nedkvitne & Bjørlykke 1992) and the Snorre-B structure (Saigal 1990).

**Table 2.** *Petrographic point counting analyses from the Etive Formation carried out by T. Nedkvitne and M. Ramm*

| Field (Well) | Interval $N$ | A Framework Grains | B Quartz Cement | C Calcite Cement | D Siderite Cement | E Kaolinite Cement | F Inter-Macro Porosity | G Inter-Micro Porosity | H Intra-Macro Porosity | I Intra-Micro Porosity | J Matrix | K Inter Granular Volume | L Total Porosity | M Secondary Porosity |
|---|---|---|---|---|---|---|---|---|---|---|---|---|---|---|
| Gullfaks (34/10–A8) | 1890 – 1920 | 52.2 ± 7.2 | 0.3 ± 0.3 | 4.5 ± 8.4 | 0.2 ± 0.3 | 5.7 ± 3.8 | 30.7 ± 12.7 | 2.4 ± 1.3 | 1.7 ± 0.9 | 1.2 ± 1.9 | 1.2 ± 1.0 | 45.0 ± 7.1 | 34.3 ± 12.0 | 2.3 ± 1.2 |
| | $N$ = 16 | 39.0 – 65.0 | 0.0 – 0.7 | 0.0 – 30.7 | 0.0 – 0.7 | 2.0 – 17.3 | 2.0 – 51.0 | 0.3 – 4.3 | 0.3 – 3.3 | 0.0 – 5.7 | 0.0 – 4.0 | 33.7 – 56.0 | 6.9 – 53.5 | 0.7 – 5.2 |
| Gullfaks-South (34/10–16) | 3399 – 3411 | 67.4 ± 3.6 | 2.9 ± 1.5 | 1.6 ± 3.0 | 1.5 ± 2.0 | 6.7 ± 2.2 | 5.6 ± 4.2 | 5.2 ± 1.2 | 1.1 ± 0.8 | 1.1 ± 0.7 | 6.9 ± 4.4 | 30.4 ± 4.1 | 10.5 ± 4.7 | 1.7 ± 1.1 |
| | $N$ = 13 | 61.7 – 73.7 | 0.7 – 5.3 | 0.0 – 10.3 | 0.0 – 6.3 | 3.0 – 10.0 | 1.0 – 13.7 | 2.7 – 7.0 | 0.0 – 3.0 | 0.0 – 2.3 | 0.0 – 15.0 | 23.0 – 37.0 | 5.3 – 18.5 | 0.0 – 4.2 |
| Gullfaks-$\beta$ (34/10–17) | 2934 – 2937 | 68.3 ± 2.1 | 2.1 ± 0.9 | 0.3 ± 0.6 | 0.1 ± 0.1 | 3.2 ± 2.0 | 17.8 ± 4.6 | 3.5 ± 0.6 | 1.7 ± 0.6 | 0.9 ± 0.5 | 2.3 ± 0.7 | 29.2 ± 2.4 | 21.9 ± 4.1 | 2.1 ± 0.8 |
| | $N$ = 4 | 66.3 – 71.7 | 0.7 – 3.0 | 0.0 – 1.3 | 0.0 – 0.3 | 1.0 – 6.3 | 11.3 – 22.3 | 2.7 – 4.3 | 1.0 – 2.7 | 0.3 – 1.7 | 1.3 – 3.3 | 26.3 – 32.0 | 15.7 – 25.6 | 1.3 – 3.5 |
| Gullfaks-$\gamma$ (34/10–23) | 4243 – 4281 | 61.3 ± 5.8 | 5.8 ± 3.0 | 4.4 ± 7.5 | 1.7 ± 1.9 | 5.7 ± 3.5* | 8.5 ± 3.5 | 6.2 ± 2.2 | 3.0 ± 1.5 | 1.2 ± 0.5 | 2.4 ± 3.9 | 34.6 ± 6.5 | 15.2 ± 3.6 | 3.5 ± 1.6 |
| | $N$ = 16 | 50.3 – 70.0 | 0.3 – 12.3 | 0.0 – 29.7 | 0.0 – 6.0 | 0.0 – 11.7 | 1.7 – 13.7 | 1.3 – 9.3 | 1.3 – 6.0 | 0.3 – 2.3 | 0.0 – 13.3 | 25.0 – 47.7 | 6.3 – 20.0 | 1.7 – 7.2 |
| Veslefrikk (30/3–2) | 2879 – 2947 | 65.2 ± 5.9 | 2.6 ± 3.1 | 7.4 ± 13.3 | 0.4 ± 1.0 | 2.8 ± 1.9 | 7.8 ± 4.5 | 3.4 ± 1.9 | 2.2 ± 1.7 | 1.3 ± 1.2 | 6.9 ± 7.2 | 31.3 ± 6.7 | 13.1 ± 6.4 | 2.9 ± 2.2 |
| | $N$ = 15 | 56.3 – 74.0 | 0.0 – 10.0 | 0.0 – 40.7 | 0.0 – 4.0 | 0.0 – 7.0 | 0.0 – 14.3 | 0.0 – 6.0 | 0.0 – 6.3 | 0.0 – 4.7 | 0.0 – 24.7 | 22.7 – 43.7 | 0.0 – 20.0 | 0.0 – 8.7 |
| (30/3–4R) | 2878 – 2949 | 70.2 ± 7.1 | 2.4 ± 1.4 | 2.3 ± 5.4 | 0.6 ± 1.4 | 4.4 ± 1.9 | 9.3 ± 3.4 | 2.9 ± 1.6 | 1.8 ± 0.8 | 1.4 ± 0.7 | 4.9 ± 6.4 | 26.6 ± 7.1 | 13.7 ± 2.9 | 2.5 ± 0.9 |
| | $N$ = 16 | 57.0 – 84.3 | 1.0 – 7.0 | 0.0 – 22.0 | 0.0 – 5.0 | 1.7 – 8.0 | 1.3 – 14.0 | 0.3 – 6.7 | 0.3 – 3.0 | 0.0 – 2.7 | 0.0 – 24.0 | 12.3 – 40.0 | 7.1 – 17.8 | 0.3 – 3.7 |
| Oseberg (30/6–4) | 2660 – 2680 | 70.1 ± 3.3 | 1.2 ± 0.6 | 0.0 ± 0.0 | 0.0 ± 0.0 | 2.7 ± 1.0 | 15.3 ± 3.5 | 4.8 ± 1.1 | 3.0 ± 0.9 | 1.3 ± 1.0 | 1.7 ± 1.7 | 25.6 ± 3.6 | 21.5 ± 2.7 | 3.6 ± 1.1 |
| | $N$ = 15 | 64.3 – 76.3 | 0.3 – 2.7 | 0.0 – 0.0 | 0.0 – 0.0 | 1.0 – 4.7 | 7.7 – 21.3 | 2.7 – 6.3 | 1.7 – 4.7 | 0.0 – 3.3 | 0.0 – 5.7 | 17.3 – 30.0 | 16.5 – 26.0 | 2.0 – 6.3 |

300 points have been identified in each thin section as: (A) Framework grains, (B–E) intergranular cement, (F–I) inter- and intra granular micro and macro porosity, and (J) matrix. Inter Granular Volume (K) is calculated as $K = 100 - (A + H + I)$, Total Porosity (L) is calculated as $L = F + H + 0.5 \times (E + G + I) + 0.1 \times J$ and Secondary Porosity (M) as $M = H + 0.5 \times I$. Upper numbers are the average of the N analyses while lower numbers show the total range.

FRAMEWORK GRAIN DISSOLUTION POROSITY

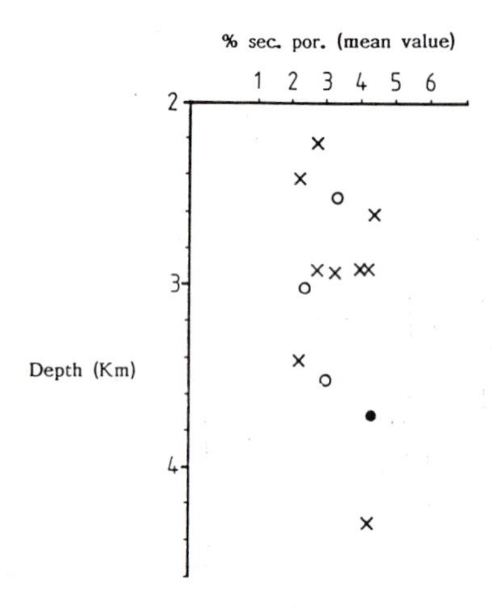


Data base:

o Hutton, Statfjord & Lyell Field; Harris (1989)

• Huldra Field; Nedkvitne & Bjørlykke (1992)

× Veslefrikk, Oseberg, Brage & Gullfaks Field; this study

**Fig. 5.** Plot of secondary porosity versus depth for the Brent Group sandstones.

Hancock & Taylor (1978) and Sommer (1978) related the dissolution of feldspar and the precipitation of kaolinite to the Cimmerian uplift while Bjørlykke (1983) and Bjørlykke & Brendsdal (1986) in addition pointed out the importance of feldspar leaching and precipitation of kaolinite both shortly after deposition of the Brent Group and after the tectonic uplift and erosion during Late Jurassic and Early Cretaceous. Shanmugam (1984, 1988) also stressed the association of secondary porosity with unconformities. Bjørkum *et al.* (1990) found, however, no significant increase in the percentage of kaolinite in the Rannoch Formation towards the unconformity in the Gullfaks Field and interpreted most of the kaolinite in the Rannoch Formation to be of clastic origin. This may be true for the lower, more fine-grained part of the Rannoch Formation, but clear evidence of authigenic kaolinite replacing mica and expanding into the available pore space is common (Bjørlykke & Brendsdal 1986). It is, however, difficult to assess the importance of the leaching associated with the Cimmerian unconformity compared to meteoric water flushing shortly after deposition. The flux of meteoric water through sandstones below the Cimmerian unconformity would depend very much on the duration and relief of the subarial exposure of the different rotated fault blocks. Curtis (1983) argued that meteoric water should be expected to be neutralized near the soil surface and therefore not be able to leach along the meteoric flow pathways into the basin. Meteoric water will become saturated relatively rapidly with respect to carbonate, but may remain undersaturated with respect to feldspar for a longer time due to slower kinetic reaction rates (Bjørlykke *et al.* 1989). The hydrolysis of feldspar does not require acidic pore water but, as stressed by Bjørkum *et al.* (1990), protons have to be supplied for feldspar dissolution and kaolinite precipitation to proceed. It is possible that carbon dioxide and other acids are supplied from the sediments along the meteoric water flow pathways. In this way the mica and feldspar leaching and kaolinite precipitation will continue over longer distances.

The rate of feldspar hydrolysis is strongly temperature dependent and is $10^2$ to $10^3$ times more rapid at deeper burial (100°C) compared to shallow burial (20°C) (Giles 1987; Helgeson *et al.* 1984; Knaus & Woley 1986). At higher temperature, equilibrium between minerals and pore water is therefore reached much more quickly, limiting the distance over which leaching is likely to occur due to pore water flow.

In the Upper Jurassic shelf sandstones of the Ula Field in the norwegian Central Graben the amounts of feldspar dissolution and authigenic kaolinite are much more variable and generally lower than in Middle Jurassic sandstones (Larese *et al.* 1984). In the Upper Jurassic sandstones of the Fulmar Field, these features are totally absent (Johnson & Stewart 1985; Johnson *et al.* 1986). This may be explained by less effective meteoric water flushing of shelf sandstones than deltaic and shallow marine deposits. Triassic and Permian sandstones from the North Sea, representing continental arid environments, may contain significant amounts of authigenic kaolinite, but illite is normally much more abundant (Glennie *et al.* 1978; Hancock 1978*b*; Waugh 1978). The characteristic development of secondary porosity and abundance of authigenic kaolinite in the Brent Group are therefore probably related to the depositional environment. Studies of heavy minerals such as garnets in the Brent Group have provided important information about the provenance as well as diagenesis (Morton 1985, 1987; Morton *et al.* 1989). The distribution of apatite is facies dependent within the Brent Group and Morton (1986) interpreted apatite to have been selectively dissolved at shallow burial by meteoric water flushing in the fluvial facies.

## Carbonate cements

Carbonate cements in reservoir sandstones have received particular interest because carbonate cemented intervals may act as barriers to fluid flow in the reservoir. In the Brent Group the carbonate cemented intervals are generally thin, usually less than 1 m and usually more abundant in the marine part of the Brent Group (i.e. Rannoch, Etive) than in the delta top environment of the Ness Formation. This is well demonstrated in studies of the Heather Field (Glasmann *et al.* 1989*a*). Point countings and XRD-data from the NW Hutton Field also show that the 'sub-littoral sheet sand' and the 'wave-dominated delta front sand' contain more carbonate cement than 'distributary mouth bar' and 'crevasse splay lobe' facies (Scotchman *et al.* 1989).

The dependence of the amount of carbonate cement on facies is even more apparent when the sandstones of the Brent Group are compared with other reservoir sandstones of different ages and facies. Sandstones representing marine shelf facies, e.g. in the Troll Field (Kantorowicz 1987), the Agat Field (Saigal & Bjørlykke 1987) and the Draugen Field (Bjørlykke *et al.* 1986), generally have higher concentrations of carbonate cement than the Brent Group. The fact that the distribution of carbonate cement seems to be facies related suggests that the cement is derived dominantly from biogenic sources or early carbonate cement formed on the sea floor. Such a pattern would not be expected if carbonate cement was transported into the reservoir from an outside source.

The interpretation of the oxygen isotopic data of the carbonate cement depends on the assumptions made about the isotopic compositions of the pore water. Data from the Brent Group of the Heather Field give an average $\delta^{18}O$ of $-11‰$ (PDB) which translates into 40–50°C assuming pore water of meteoric water origin ($\delta^{18}O$ $-4$ to $-6‰$ SMOW) (Table 3). The carbonate cement in the Gullfaks Field is characterized by positive $\delta^{13}$ C values (+5.8 to +12.5‰ PDB) (Saigal & Bjørlykke 1987), indicating that the cement to be formed at shallow depth in the fermentation zone (Irwin *et al.* 1977). Since the temperature then must have been rather low (<30–50°C), the corresponding $\delta^{18}O$ values ($-5$ to $-6‰$ SMOW) should be close to the composition of the meteoric water flowing into the Gullfaks Field.

Very negative $\delta^{13}C$ values ($-20$ to $-28‰$ PDB) like those measured from the Heather Field (Glasmann *et al.* 1989*a*) are probably too negative for thermogenetically derived carbon (Galimov 1980). This suggests that the carbonate cement may have been at least partly derived from sea floor cement precipitated by bacterial oxidation (Irwin *et al.* 1977). If the carbonate cemented layers or concretions are derived from dissolution of locally derived biogenic carbonate and early cement, the observed carbonate cement may represent the last stage of possibly several dissolution and precipitation processes more or less in situ. A critical question is how the more stable carbonate phases would precipitate from the dissolving less stable carbonate minerals. If transport is controlled by diffusion, one would expect the distance to be rather short since the degree of supersaturation with respect to the precipitating minerals would decrease away from the areas of dissolution. The carbon isotopic values vary greatly (Table 3) suggesting that in most cases multiple sources of carbon including thermogenic carbon may have contributed. The wide scatter of the $\delta^{13}C$ values indicates that the pore water was not well homogenized on a larger scale during calcite cementation and that the carbon isotopic composition of calcite is dominated by the local supply of organic carbon. If calcite was precipitated using $Ca^{2+}$ supplied externally by leached plagioclase and thermogenic carbon, one would have expected the $\delta^{13}C$ values to reflect this more closely. This mechanism would also not explain observed relationships between facies and abundance of carbonate cements. Walderhaug *et al.* (1989), in their study on the Upper Jurassic Fensfjord Formation from the Brage Field, observed a large scatter in the isotopic composition of the carbonate cement and concluded that dissolution of plagioclase could not have supplied sufficient $Ca^{2+}$ to account for the carbonate cement. In the Lower Cretaceous sandstones of the Agat Field, which represent a deep water facies, carbonate cement is abundant and probably derived from pelagic carbonate fossils and has $\delta^{13}C$ values close to zero. This indicates that the marine carbon signature has been preserved even if the $\delta^{18}O$ values ($-9$ to $-12‰$, PDB) show that the carbonate has recrystallized at greater depth.

The strontium and oxygen isotope values are often rather similar for different types of carbonate cements in the Brent Group, as reported by Hamilton *et al.* (1987*b*), but the carbon values are very different suggesting that they have not precipitated from the same type of pore water (Table 3). The $^{87}Sr/^{86}Sr$ ratio varies from 0.711 to 0.7114, which is more radiogenic then the

**Table 3.** *Published data on the stable and strontium isotopic composition of calcite cement from the Brent Group*

| Field | Well | Depth (m) | Type | $\delta^{13}C$ (PDB) | $\delta^{18}O$ (PDB) | $^{87}Sr/^{86}Sr$ | Source |
|---|---|---|---|---|---|---|---|
| Heather | – | 3321.95 to 3482.50 | Calc 1 + 2 | −4 to −27 | −8.1 to 15.0 | – | Glasmann *et al*. (1989*a*) |
| | 2/5–2 | 3321.95 | Calc 2 | – | – | 0.7119 | " |
| | 2/5–3 | 3433.23 | Calc 2 | – | – | 0.7115 | " |
| | | 3448.96 | Calc 2 | – | – | 0.7115 | " |
| | 2/5–H18 | 3482.50 | Calc 1 + 2 | – | – | 0.7112 | " |
| Veslefrikk | 30/3–4 | 2903.50 | Calc 1 + 2 | −16.5 | −7.5 | – | Glasmann *et al*. (1989*b*) |
| | | 2925.50 | Calc 1 + 2 | −24.1 | −7.6 | – | " |
| | | 2926.40 | Calc 1 + 2 | −17.0 | −10.0 | – | " |
| | | 2940.20 | Calc 1 + 2 | −13.7 | −10.3 | – | " |
| | | 2941.00 | Calc 1 + 2 | −17.9 | −8.5 | – | " |
| – | – | – | Streaks | −16.3 to −14.5 | −13.5 to −13.3 | 0.7116–0.7140 | Hamilton *et al*. (1987) |
| – | – | – | Nodules | −19.5 to 16.5 | −11.1 to −10.7 | 0.7120 | " |
| – | – | – | Doggers | −9.1 to −0.5 | −12.8 to −9.3 | 0.7110 | " |
| – | – | – | Veins | −11.0 | −9.6 | 0.7123 | " |

Calc 1, microsparry – sparry calcite
Calc 2, poikilotopic calcite

Jurassic sea water, and Hamilton *et al.* (1987*b*) concluded that the excess radiogenic strontium is derived by leaching of silicate minerals during meteoric water flushing. Differences in the isotopic composition of strontium, which are sometimes found in closely spaced samples of carbonate cement (Table 3 & 4), is also evidence of poor mixing of pore water if precipitation occurred at the same time.

The rather narrow range in the $\delta^{18}O$ values (−9 to −13‰ PDB) observed by Hamilton *et al.* (1987*b*) suggests that precipitation of the carbonate cements occurred over a limited temperature range. The cement could then have been sourced by less stable carbonate cement, precipitated at lower temperatures, or biogenic carbonate. The isotopic signature of carbon reflects varying degree of exchange with $CO_2$ in the pore water.

Siderite occurs in sideritic benches within shale layers in the fluvial or brackish Ness Formation and on a microscopic scale in association with altered biotite. The latter association was observed during the point counting of thin sections from the Etive Formation for this study and has been reported in the Rannoch Formation of the Statfjord Field (Bjørlykke & Brendsdal 1986) and the Ness Formation of the Snorre Field (Saigal 1990). Ankerite is also quite common, frequently as a poikilotopic cement, and is interpreted as a late diagenetic phase that may have replaced earlier calcite and siderite cements (Bjørlykke & Brendsdal 1986).

In summary, the distribution of carbonate cements in the Brent Group is mostly controlled by the amount of initial carbonate, mainly biogenic, buried in the sequence. The depositional environment and lithofacies are therefore the main factors controlling the carbonate distribution.

## Diagenetic albitization of K-feldspar

Albitization of feldspars in sandstone reservoirs from different parts of the world has been recognized by several workers (e.g. Boles 1982; Walker 1984; Gold 1987; Saigal *et al.* 1988; Morad *et al.* 1990) and is regarded as an important diagenetic process, which can significantly modify the original grain framework composition. In the North Sea reservoirs, diagenetic albitization of feldspars has recently been reported by Saigal *et al.* (1988), Saigal (1990) and Morad *et al.* (1990) from Jurassic, Lower Cretaceous, Tertiary and Triassic sandstones. The present study of the Brent Group reservoir sandstones suggests that albitization of K-feldspar is also important in these sandstones.

Integrated optical microscopy, cathodoluminescence (CL) and SEM study show that detrital K-feldspar grains have been subjected to varying degree of replacement by albite in different wells. Generally, samples shallower than *c.* 2 km do not show albitization, while partly albitized K-feldspar grains are found in samples from greater depths (Fig. 6). For detailed textural description of the albitized grains, see Saigal *et al.* (1988). SEM studies reveal that authigenic albite occurs as tiny to blocky euhedral crystals (1–60 μm) that have grown parallel to the cleavage planes of the parent grain and as overgrowths or pore filling cement. The tiny to

**Table 4.** *Oxygen, carbon and strontium isotopic composition of calcite cements from the Brent Group (Gullfaks field) and its equivalents in other fields on the Norwegian continental shelf*

| Field | Well | Depth (m) | Type* | $\delta^{13}C$ (PDB) | $\delta^{18}O$ (PDB) | $^{87}Sr/^{86}Sr$ |
|---|---|---|---|---|---|---|
| Gullfaks | 34/10–3 | 1930.9 | Calc 1 | +12.56 | −6.24 | 0.71221 |
| | | 1942.1 | Calc 1 | +11.33 | −6.38 | 0.71210 |
| | 34/10–4 | 1852.0 | Calc 1 | +5.81 | −5.08 | 0.71174 |
| | | 1892.0 | Calc 2 | +9.84 | −9.22 | 0.71182 |
| Midgard | 6407/2–2 | 2498.9 | Calc 1 | −17.6 | −7.8 | 0·70977 |
| | | 2509.7 | Calc 2 | −26.73 | −10.34 | 0·70977 |
| Tyrihans | 6407/1–3 | 3685.3 | Calc 2 | −25.93 | −10.60 | 0·71343 |
| Brage (U. Jurassic) | 31/4–2 | 2222.7 | Calc 1 | −8.55 | −4.52 | 0·70955 |
| Gudrun (U. Jurassic) | 15/9–3 | 3523.4 | Calc 2 | −8.20 | −9.91 | 0.70886 |
| | 15/3–3 | 4262.2 | Calc 2 | −2.69 | −9.61 | 0.70993 |
| | | 4268.2 | Calc 2 | −3.66 | −10.45 | 0.71112 |

* Calc 1, microsparry – sparry calcite
Calc 2, Poikilotopic calcite

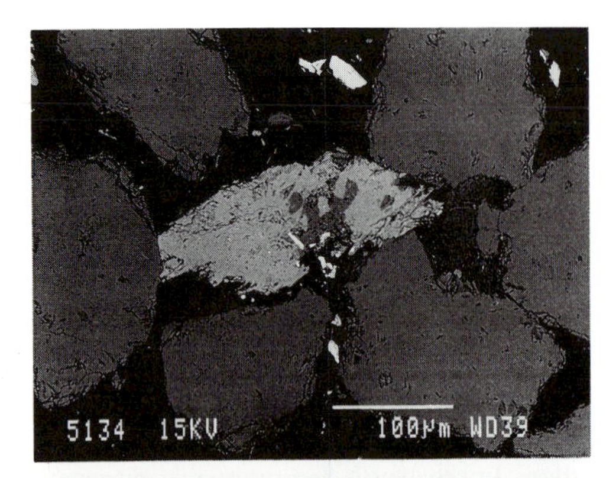

**Fig. 6.** BSE photomicrograph showing partial (**a**) and significantly (**b**) albitized K-feldspar grains. Note dark grey patches are albite, while light grey represent relict K-feldspar. Surrounding dark grains are quartz, while white crystals in intergranular pores are of Fe-oxides (Well 34/7–14, 2229 m).

blocky albite crystals are probably dominant in the Brent Group sandstones e.g. in wells 34/7–12 and 34/7–14 (Saigal 1990). Under CL, diagenetic albite is non-luminescent, while detrital feldspars show blue luminescence. Microprobe analyses carried out on a few albite crystals show that they are virtually pure (≥99.7% Ab).

Saigal *et al.* (1988) and Aagaard *et al.* (1990) concluded that albitization of K-feldspar starts at about 65°C and is almost complete by 105°C. The process involves dissolution of K-feldspar and precipitation of albite and requires low potassium activities. K-feldspar distribution trends established for several wells on the basis of XRD analyses carried out during this study show depletion of K-feldspar in samples from depths between 2.5 and 3.7 km (Fig. 3). It is suggested, therefore, that diagenetic albitization of K-feldspar occurred in the Brent Group at temperatures greater than 65°C and is partially responsible for the loss of K-feldspar at intermediate burial. In the Huldra Field, diagenetic albite occurs only as pore-filling cement and overgrowths on plagioclase grains (Nedkvitne & Bjørlykke 1992).

## Quartz cementation

Quartz cementation and pressure solution are responsible for most of the porosity loss at deeper burial (3–5 km) (Table 2, Fig. 3a). Where the Brent sandstones are buried to less than 3 km, as in the Statfjord and the Gullfaks Fields, the sandstones are generally loosely cemented, frequently to the extent that the sandstone cores easily disintegrate. Furthermore, studies of the Brent Group sandstones from the Statfjord Field (at 2.5 km burial depth) using conventional optical microscopy, SEM, and cathodoluminescence show that the amount of quartz overgrowth is only about 1–3% of the total rock volume. Kittilsen (1987) found that some intervals in the Brent Group of the Statfjord Field are almost devoid of cement and are so poorly cemented and compacted that sand production during oil recovery is a problem. Kittilsen (1987) also showed that the shear strength of the Brent sandstones at shallow burial depth is not only a function of the amount of quartz cement but also of other factors such as the mechanical compaction (number of contacts per grain) and grain size. The fine-grained sandstones tend to have higher shear strength than the coarser sandstones. In the Gullfaks area the percentage of quartz cement increases from less than 1% at 1.8 km burial depth to more than 10% in individual samples at 4.2 km depth (Table 2). In the Hutton Field a progressive increase in the degree of quartz overgrowth is observed from an average of about 10% quartz overgrowth at 3430 m to about 18% at 3886 m (Scotchman *et al.* 1989). They also observed that sandstones of the Ness Formation are more quartz cemented than those of the Rannoch Formation and that the fine grained sandstones are more severely cemented than the coarse grained ones. In other fields where the Brent Group has been buried to less than about 3 km, the amount of quartz cement is also rather moderate (Table 2). In the Veslefrikk Field, however, the sandstones have somewhat more quartz cement than in other reservoirs at about the same depth. This may be explained by the fact that this field only has a very slight overpressure compared to the highly overpressured Statfjord and Gullfaks Fields. Minor amounts of early quartz cement may have been derived from the silica released from the dissolution of other types of unstable minerals or rock fragments, or from incipient stylolitization and pressure solution against mica. It is, however, clear that in the case of the Brent Group the precipitation of quartz cement at shallow burial depth (<2.5 km) is too small to significantly influence the reservoir properties.

Blanche & Whitaker (1978) and Bjørlykke *et al.* (1979) pointed out that the reduced pressure solution and high porosities of the sandstones from the Statfjord Field were due to the high overpressures reducing the net stress to a level equivalent with only 1–1.5 km of overburden under hydrostatic conditions. However, pressure solution is influenced not only by net stress but also by temperature, which increases the rate of quartz dissolution and precipitation (e.g. Angevine & Turcotte 1983). Recent fluid inclusion studies of the Brent sandstone from the Huldra Field show that quartz cementation occurs mainly in the temperature range of 115–155°C (Glasmann *et al.* 1989*b*). In the Greater Alwyn area, fluid inclusion measurements indicate temperatures of quartz cementation between 120–140°C (Jourdan *et al.* 1987), similar to those in the Huldra Field. Walderhaug (1990) reported homogenization temperatures for inclusions in quartz overgrowth from Jurassic sandstones from the Haltenbanken area ranging from 92–118°C. Since hydrocarbon inclusions homogenized within a similar range of temperatures and below an upper limit corresponding closely to the present day bottom hole temperature, Walderhaug (1990) concluded that quartz cementation may have continued after oil emplacement in the reservoir. It is difficult to estimate the oil saturation at the time when the inclusions were formed but, if quartz cement postdates the main filling phase, we have to assume that quartz overgrowths can continue to form even at only 10–20% water saturation. This would require the silica to be derived internally from pressure solution and one would expect the presence of a hydrocarbon phase to retard the diffusion from areas of dissolution to the quartz overgrowth surfaces.

Convective pore water flow (Haszeldine *et al.* 1984; Wood & Hewett 1984) requires a rather high permeability (Wood & Hewett 1984, assumed 10 D). Even in highly porous and permeable sandstones the conditions for Rayleigh convection are probably rarely met, due to the presence of thin shales or carbonate cemented intervals (Bjørlykke *et al.* 1988). Recent calculations by Palm (1990) show that the critical Rayleigh number for convection may be exceeded in highly permeable sandstones of about 50 m thickness. The Etive Formation might approach these requirements as the average permeability of the 156 m thick sequence of the Etive Formation in the Statfjord Field is 2.5 D (Roberts *et al.* 1987). Palm's (1990) calculations, however, also show that if the critical Rayleigh number was exceeded by only 10%, the convective flow would transport enough silica to add 10% quartz cement in the upper part of the formation and remove about the same amount by dissolution from the lower part in about 10 Ma. The fact that this pattern of quartz cementation has not been observed in the Brent Group suggests that Rayleigh convection has not been important. Any convective flow which would supply silica to precipitate quartz cement would also necessarily remove carbonate cement since it has the reverse solubility/temperature trend to that of quartz. In conclusion, Rayleigh convection has not been predominant in the Brent Group.

At the depth intervals where intensive quartz cementation is observed (3–5 km), the permeability is generally too low for Rayleigh convection to occur. It is important to note that other mechanisms for importing silica cement into sandstones, such as focused compactional pore water flow, would also be most effective at shallow depth where the water fluxes would be highest. At depths below 3 km most of the dewatering of shales has already occurred and the pore water flux generated by compaction is limited.

Quartz cementation increase significantly at depths exceeding 3 km in the Brent Group (Table 2, Fig. 3a). Similar results have been found in the Jurassic sandstones from the Haltenbanken area (Bjørlykke *et al.* 1986). Larese *et al.* (1984), in their study of the Jurassic sandstones from the Norwegian Shelf, presented petrographic evidence suggesting that the quartz cement was mainly sourced by pressure solution from both grain contacts and stylolites. Cathodoluminescence images provide a good basis for quantifying grain to grain pressure solution (Houseknecht 1988), but dissolution along microstylolites and clay laminae is more difficult to measure. Since quartz precipitation and dissolution from internal sources such as pressure solution will be inhomogeneously distributed, it is difficult to prove on a petrographic basis that there is no import or export of silica. We think that there is no need to call on external sources of silica to explain the cementation of the Brent Group sandstones and that transport of large volumes of silica in the deep subsurface is theoretically difficult to explain. We accept, however, that this is a point which is difficult to prove. It is nevertheless clear that quartz cementation is essentially a late diagenetic process occurring below 3 km burial. This is confirmed by the increasing amount of quartz cement with depth (Fig. 3a), the scarce amount of quartz cement in the shallow reservoirs, and by the microthermometric studies on fluid inclusions from other North Sea sandstones. The presence of clay and mica may enhance pressure solution but clay linings around most grains will discourage both dissolution and quartz precipitation.

## Precipitation of authigenic illite

In addition to the formation of quartz overgrowths, the precipitation of illite is the most important diagenetic reaction controlling reservoir properties at deeper burial. Authigenic illite, in particular, severely reduces the permeability. It is therefore important to understand the factors controlling the distribution of illite in sandstones. Illite may form diagenetically by alteration of smectite, through a gradual transition from highly expandable smectite to increasingly ordered mixed layer minerals and towards pure illite, over a temperature range from 80–110°C (Eslinger & Pevear 1988). Very little diagenetic illite or smectite are found in the shallowest (1.8 km), best sorted Brent sandstone reservoirs where the temperatures would not have exceeded 80°C. We must therefore assume that rather little smectite formed during early diagenesis of the Brent sandstones. The reason for this is probably that the Brent Group contained little amorphous silica or volcanic fragments, so that the silica content in the pore water was too low for smectite to form. High concentrations of illite which significantly affect the reservoir properties are only found in deeper reservoirs (below 3.7–4 km) and this illite is observed to replace kaolinite. Authigenic illite at shallower depths is probably formed by alteration of detrital mica or by transformation of small amounts of smectite.

Determination of the formation age of authigenic clay minerals is a well established technique (Eslinger & Pevear 1988). One of the first to point out the significance of dating illite in the Brent Group was Sommer (1975, 1978), who argued that authigenic illite would be formed during oil and water migration but not after oil emplacement into the reservoirs. The age of the fine fibrous illites would therefore date the oil emplacement and Sommer (1975, 1978) reported that the youngest K-Ar dates were 45–55 Ma, which are similar to the values obtained by subsequent studies. Thomas (1986) found that illite was much more abundant below the oil–water contact than above in the Statfjord Formation and Brent Group of the Heimdal Field in the Viking Graben (wells 25/4–1 and 25/4–5). In the shallowest, oil saturated, well at the crest of the structure (25/4–1, 3200 m), very little authigenic illite was found. Two samples gave an Eocene age (40–44 Ma) which was taken to represent the time of the oil emplacement. In contrast, illitization of kaolinite has taken place in the deeper well (25/4–5) where the Brent reservoir is at 3700 m. One date from this well indicated an Oligocene age (28 Ma). The author interpreted this difference to be mainly due to the continued growth of illite in the deeper well below the oil–water contact. In the Hutton Field the amount of illite also increases dramatically at burial depths below 12000 ft (3658 m) and is not related to a palaeo-oil–water contact (Scotchman *et al.* 1989). Although the abundance of authigenic illite is clearly much greater at the same present day burial depth in nearby areas, Scotchman *et al.* (1989) concluded that illitization occurred much earlier, prior to oil migration and at a burial depth of about 12000 ft (2450 m). This conclusion was based on illite dates of about 40 Ma in the oil saturated zone. Illites from the water zone gave, however, only slightly younger age (average 2 Ma) than those from the oil saturated zone. In their petrographic description, Scotchman *et al.* (1989) state that illite precipitation is the last diagenetic phase after quartz cementation. The implication would then be that the diagenetic processes stopped 40 Ma ago when the reservoir was at least 1 km shallower than at the present day. It seems, however, difficult to understand why authigenic illite and quartz would not continue to form during the last one km of burial.

Early Tertiary ages on authigenic illites were also obtained from Brent sandstones of the Greater Alwyn area (Jourdan *et al.* 1987; Hogg *et al.* 1987). Illite dates around 40–50 Ma pose many problems because they imply that illitization took place in Early Tertiary when the reservoirs were buried to only 2–3 km. Since authigenic illite only replaces kaolinite in well sorted sandstones at depths greater than 3.5–4 km at the present time, higher geothermal gradients must have prevailed in this area in Early Tertiary. Jourdan *et al.* (1987) therefore concluded that the illite had precipitated from circulating hot fluid from the deeper parts of the basin.

In Haltenbanken, where reservoirs are also found over a wide range of burial depths, illitization of kaolinite and dissolution of K-feldspar occur at about 3.7 km and 130–

140°C (Bjørlykke *et al.* 1986; Ehrenberg & Nadeau 1989). Glasmann *et al.* (1989*b*) found a wide range of illite dates from the Brent Group sandstones of the Huldra and the Veslefrikk Fields. The youngest ages are, however, around 30 Ma and the authors suggest that this represents the time of oil emplacement for the Veslefrikk Field, and that the Huldra Field was filled from the Late Palaeocene (58 Ma) to the Late Eocene (38 Ma). Also Ehrenberg & Nadeau (1989) found that the K-Ar ages of the illites in the Middle Jurassic Garn Formation from the Haltenbanken area ranged from 55 to 31 Ma, which would correspond to a burial depth of about 2–2.5 km. They did not accept these dates as true ages for the illites and concluded that they may be too old as a result of contamination by older clastic silicate minerals. No significant change in the degree of illitization across the oil–water contact was found and illite was not consistently older in the hydrocarbon filled parts of the reservoir. Illites from the Upper Jurassic sandstones of the Piper and the Tartan Fields represented a rather wide range of K-Ar ages, ranging from 140–30 Ma, with younger ages in the water saturated zones (Burley & Flisch 1989).

The available K-Ar dates of diagenetic illite from the Brent Group, and equivalent formations in other areas, range from Late Cretaceous to Middle Tertiary with the highest frequency around 40–60 Ma (Hamilton *et al.* 1989). The finer fractions of illite had younger ages and relatively constant strontium isotopic ratios, consistent with a relatively closed system diagenesis. If the illite had formed by transformation of smectite or clastic mica, the dates would correspond more closely to the burial history. If, however, the illites had formed from kaolinite, high geothermal gradients would be required. Were all the fields analyzed influenced by a similar degree of hydrothermal activity in Early Tertiary i.e. in connection with the Eocene volcanism? Or perhaps a regional heating episode affected not only the Viking Graben and the East Shetland Basin, but the whole of the North Sea and the Haltenbanken area? Large volumes of rock require a long cooling time and regional heating could therefore not have been a short event. If temperatures up to 120–140°C were reached in the Early Tertiary then that would have altered the maturity of the kerogen accordingly. A significant isostatic uplift should also be expected following such a dramatic 'thermal event'. Once the high geothermal gradients were established on a regional scale, the rate of cooling of the crust would be slow and could be calculated using standard basin modelling techniques. Given the heat capacity of sediments, a short heating event necessarily has to involve rather small volumes of sediments and would therefore be on a local scale.

The high temperatures implied in the illitization have been attributed to the introduction of hot compactional fluids (Glasmann *et al.* 1989*a*, *b*). This seems to be unlikely because the faults around the reservoirs were active mainly in Late Jurassic and in Early Cretaceous whereas the Cenozoic is largely unfaulted (Lovell 1990). In Eocene there was no rapid subsidence or very high sedimentation which could cause such highly focused compactional flow. The North Sea has in the Late Neogene experienced the most rapid subsidence since the Jurassic and the geothermal gradients are remarkably stable (Hermanrud 1986). The K-Ar dates of 40–50 Ma, however, coincide with the Late Palaeocene, Early Eocene vulcanism, and deposition of volcanic tuffs (Balder Formation). Heating from hydrothermal waters associated with this volcanism is therefore a possibility that should be explored further. If the dates reflect hydrothermal events, they can, however, not be related to the timing of oil emplacements but only taken as maximum ages.

In the Brent Group authigenic illite forms primarily by dissolution of kaolinite and K-feldspar (Bjørlykke 1983) and the process is probably kinetically controlled and requires a certain temperature (130–140°C) (Ehrenberg & Nadeau 1989). In the absence of K-feldspar, however, kaolinite is stable to much higher temperatures. In the Hild Field, where the K-feldspar concentration is very low, kaolinite is still stable and illitization has not taken place even at temperatures of about 150–160°C (Lønøy *et al.* 1986). In the Garn Formation from the Haltenbanken area, the abundance of illite is much reduced locally where K-feldspar is absent. Also from the Stø Formation in the Barents Sea, similar dependence on K-feldspar for diagenetic illite growth is observed (Ramm in prep.). Illitization seems thus to depend on the locally available sources of $K^+$ and images a rather closed diagenetic system where $K^+$ is not supplied from sources more distant than a few meters. This is evidence of limited pore water circulation during deeper diagenesis.

## Formation of secondary porosity at deep burial

Secondary porosity is well developed in the shallowest wells (1.8 km) and there is no significant increase in the total volume of secondary pores with increasing burial depth (Fig. 5). There is, however, an increase in the relative percentage of secondary porosity relative to total porosity with burial depth, as the primary pores between quartz grains tend to be occluded first (Bjørlykke *et al.* 1986). The development of secondary porosity often does not cause an increase in the overall porosity because of the associated precipitation of other authigenic minerals. At deep burial, illitization of kaolinite is associated with dissolution of K-feldspar, but with little gain in net porosity.

Schmidt & McDonald (1979) included sandstones from the Brent Group in their examples of sandstones with secondary porosity. They claimed that most of the observed porosity in the Brent Group formed as a result of leaching by acids released from maturing source rocks at great (*c.* 3 km) depth. The importance of organic acids from source rocks during late diagenesis has been stressed by Curtis (1983), Surdam (1984) and Burley *et al.* (1985). Liberation of acids from maturing kerogen is, however, not likely to have had an important impact on the diagenesis of the Brent Group. In contrary, the pH of pore waters in clastic reservoirs from the Gulf Coast and the Norwegian Continental shelf, including the Brent Group, seems to have been controlled by equilibria between aluminosilicate minerals (Smith & Ehrenberg 1989). Liberation of $CO_2$ from maturing kerogen is thus likely to cause precipitation of carbonates and alteration of feldspars to clay minerals, which does not result in any improvements in reservoir quality. Furthermore, the fact that secondary porosity is well developed in the shallow reservoirs, are at considerable distances to mature source rock, argues against the organic acid induced secondary porosity model (Bjørlykke *et al.* 1988).

## Porosity and permeability trends related to burial depth

Porosity–depth trends in Jurassic reservoirs from the North Sea have been published previously by Selley (1978), Gluyas (1985) and Bjørlykke *et al.* (1989). He-porosities and horizontal permeabilities of the Etive Formation from 10 wells are plotted against depth in Fig. 7. The porosity plot also include the linear best fit lines from the Norwegian sector of the North Sea and the Haltenbanken area, off Mid Norway, reported by Bjørlykke *et al.* (1989). The Etive Formation porosity data plot at or close to the general trend previously reported (Fig. 7a). Substantial deviations are found mainly for data from the highly overpressured sandstones from the Gullfaks and Huldra Fields which plot to the right of the best fit lines, whereas data for the Etive Formation in the Veslefrikk Field plot slightly to the left of these lines.

The permeability distribution decrease systematically with increasing depth. All the Etive Formation sandstones have good reservoir properties with permeabilities higher than 100 mD at depth shallower than 3.7 km. However, an abrupt shift towards lower permeabilities is observed between 3.7 and 4.2 km. This shift, which is about 2 to 3 orders of magnitude, can not easily be accounted for by considering the more gentle decrease in porosities by compactional processes over this interval. The most probable reason is the formation of authigenic illite replacing pore filling kaolinite which occurs at about this depth and temperature (*c.* 130°C) (Fig. 7c).

Using the results of the point counting reported in Table 2 and following the concepts of Houseknecht (1987), Fig. 8 illustrates the texture and porosity evolution of the Etive Formation with increasing depth. The Etive Formation of well 34/10-A8 has intergranular volumes varying between 30 and 45%, indicating textures little affected by compaction despite burial to almost 2 km. In wells 30/6–4 and 34/10–17 where the Etive Formation is buried to approximately 2660 and 2940 m respectively, the intergranular volume is reduced to between 20 and 30% without very much associated cementation. The Etive Formation in the two wells from the Veslefrikk Field and in 34/10–16 has about the same amount of intergranular volume, but overall somewhat higher amounts of cement. In many of these samples, quartz cement makes up about five percent of the volume, partly explaining the displacement towards the right relative to samples from wells 30/6–5 and 34/10–17. Samples plotting outside the main clusters are mainly those having a high concentration of carbonate cement, and these plot towards the upper right corner in Fig. 8. Burial from 3 to 4 km results in a further displacement to the right as displayed by samples from well 34/10–23. This displacement is caused by more extensive quartz cementation.

It is worth noticing that in these examples the intergranular volume does not decrease very much by compactional processes after burial to about 2.5 km. An additional decrease in intergranular volume would have been expected if intergranular pressure solution was the main mechanism for porosity destruction during deep

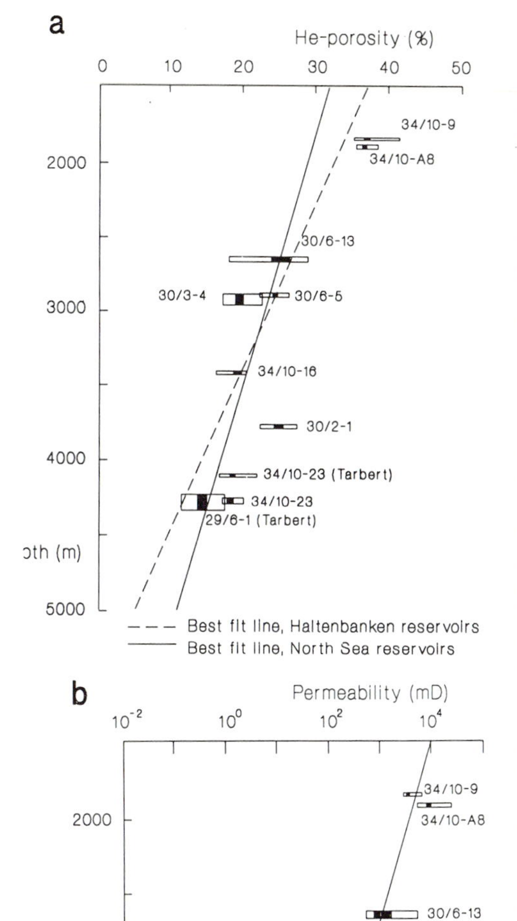


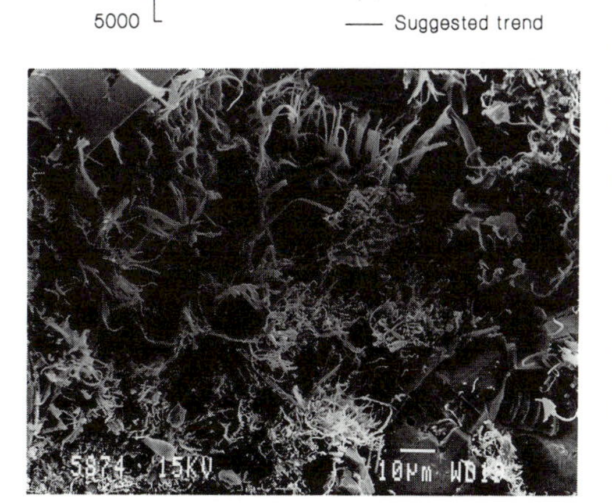


burial. It is more probable therefore, that the porosity reduction after burial to 3 km is mainly due to infilling of quartz cements within the pore space. The source of this cement is most likely to be from dissolution of silica along stylolites, clay lamina and against micas followed by short distance diffusion before precipitation.

Many of the large petroleum reservoirs in the north Viking Graben are situated within highly overpressured domains. Overpressure retards mechanical and chemical compaction and preserves porosity by reducing effective stresses. The quantitative effect on the Brent Group sandstones reservoir quality is, however, not fully understood. The reservoirs of the Gullfaks structure (at about 1.8–2 km) are overpressured today, with pore pressures near 300–330 bars, (i.e. effective stress on grain contacts corresponds to burial depths less than 1 km in a hydrostatic pressured sandstone). The low amount of textural changes in these sandstones due to compaction can therefore be explained in terms of the overpressure. However, the reservoir of the Gullfaks-Gamma structure (at about 4 km, e.g. well 34/10–23) is also highly overpressured. The pore pressure is here about 770 bars, so the effective stress on grain contacts corresponds to a burial depths of less than 1 km in hydrostatic settings. Clearly the porosity of these sandstones is lower than normally found at less than 1 km burial so there has been more compaction than expected from the effective stress alone. At shallow burial the porosity reduction occurs by mechanical compaction, which is a function of grain strength and net stress, whereas during deeper burial the porosity destruction is mainly due to chemical compaction. At pore pressures approaching the lithostatic pressure, the rate of porosity destruction is thus retarded relatively to that in

**Fig. 7.** Porosities (**a**) and horizontal permeabilities (**b**) in the Etive Formation as a function of burial depth. (**c**) Illustrates completely illitized kaolinite in the Etive Formation well 34/10–23, 4261 m. Overall porosity trends for the North Sea reservoir sandstones ($\phi = 41.0 - 0.0060 \times \text{depth}$) and the Haltenbanken reservoir sandstones ($\phi = 51.1 - 0.0092 \times \text{depth}$) are from Bjørlykke *et al.* (1989) and represented by continuous and broken lines, respectively. Note that the permeabilities drop significantly at a depth between 3.8 and 4.2 km. The full range of bars indicate the variation between the median (50 percentile) and the 95 percentile He-porosities/horizontal permeability. The central filled bars indicate the variation between the 70 and 80 percentile porosities/permeabilities.

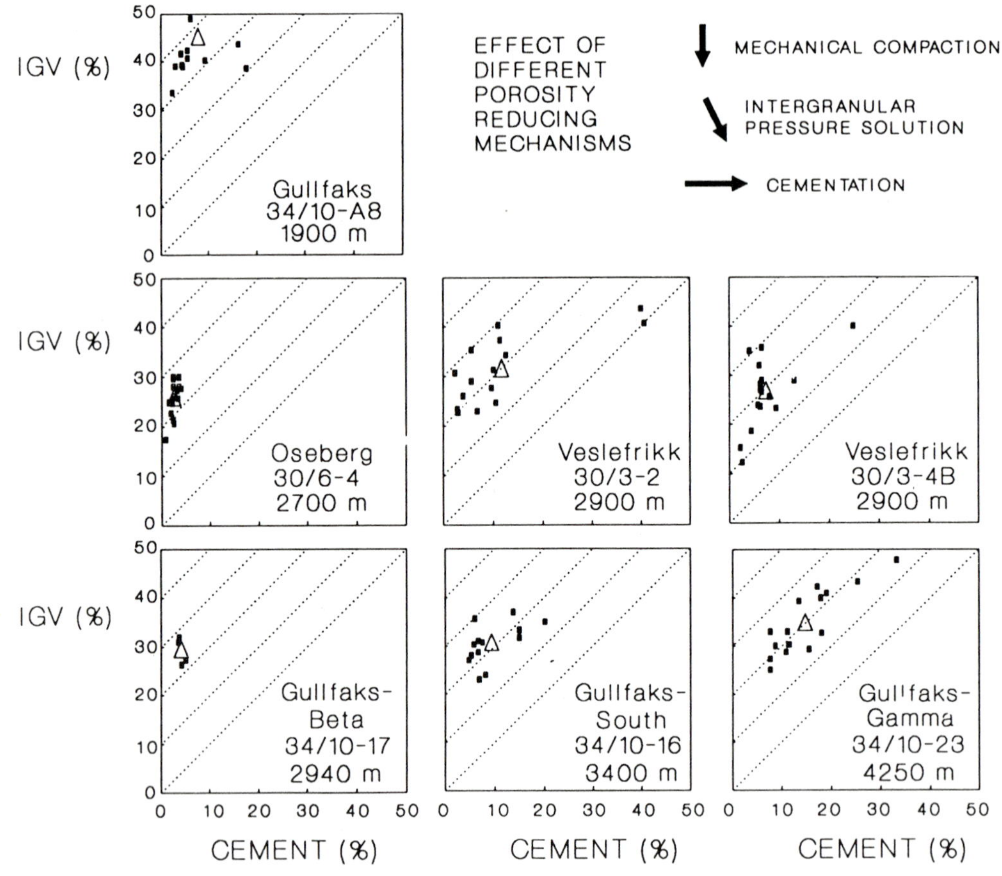


**Fig. 8.** Plots of intergranular volume versus cement in the Etive Formation sandstones from seven wells illustrating the effects of mechanical compaction, chemical compaction, and cementation on porosity reduction. Note that diagonal dotted lines indicate equal amounts of intergranular porosity and matrix.

hydrostatic settings, but the rate of porosity destruction by chemical compaction is enhanced by increasing temperature.

## Conclusions

The sandstones of the Brent Group are of rather heterogeneous composition, particularly with respect to feldspar content. An important parameter controlling diagenetic reactions at deeper burial is differences in the primary composition of the sandstones. The distribution of carbonate is facies dependent. The amount of biogenic carbonate, carbonate lithoclasts or early marine cements determines the volume of carbonate cements precipitated. In addition, limited amount of carbonate may be generated from dissolution of calcium bearing silicate minerals like plagioclase. The fact that the distribution of carbonate cements is facies controlled within the Brent Group and also in Upper Jurassic sandstones, suggests that the cement is derived dominantly from biogenic carbonate and/or early marine cement. A local dissolution/reprecipitation model is supported by the heterogeneous isotopic composition of carbonate cements which does not imply a high degree of mobility and mixing of the pore waters precipitating carbonate cements.

The main diagenetic changes affecting reservoir quality of the Brent Group sandstones are correlated to approximate depth of occurrence in Fig. 9. Secondary porosity due to leaching of feldspars in the Brent Group is well developed in the shallowest of the reservoirs and does not show an increase with depth. Some secondary porosity is however generated at deep burial

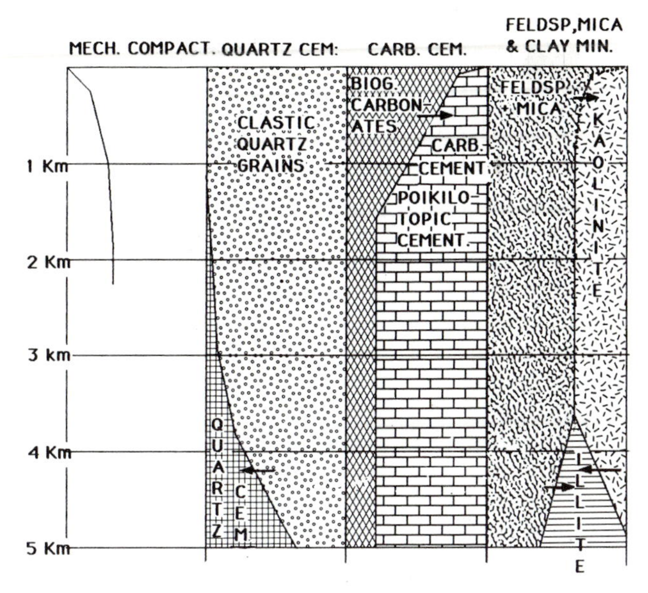


**Fig. 9.** Schematic summary of the most important diagenetic reactions in the Brent Group. During meteoric water flushing, feldspar and mica are dissolved and kaolinite precipitate in an open diagenetic system. During burial diagenesis, the reactions are essentially isochemical. Clastic quartz are dissolved to make quartz cement and biogenic carbonate are dissolved and precipitated as carbonate cement. K-feldspar and kaolinite are thermogenically unstable when present together and illite forms at temperatures sufficiently high to overcome the activation energy of illitization (120–140°C). In the absence of K-feldspar or kaolinite this reaction will not proceed and the reservoir properties will be less affected.

when K-feldspar is dissolved during illitization of kaolinite, but this causes very little enhancement of the total porosity. The relative percentage of secondary porosity to the total porosity, however, increases with depth as the primary porosity is selectively cemented first, mainly by quartz cement. The fact that feldspar leaching and formation of authigenic kaolinite is pervasive in the fluvial to marginal marine Brent Group, and is much more restricted in many offshore sandstones representing different age and facies, suggests that this process is related to meteoric water flushing at relatively shallow depth. Dissolution processes at greater depth would not be expected to be facies dependent.

K-feldspars show a strong depletion with increasing burial which is due to albitization of K-feldspar at intermediate depth (2.5–3.5 km) and dissolution during illitization of kaolinite at deep burial (3.7–4 km).

The Brent Group sandstones contain only small amounts of quartz cement down to a burial depth of about 3 km. Below 4 km the quartz cement has occluded a high percent of the primary porosity and stylolites are normally developed. A selective preservation of the secondary pores formed by leaching of feldspar will produce relatively large pores, with pore throats becoming increasingly narrower, due to progressive quartz cementation, resulting in low permeabilities.

The growth of diagenetic illite may cause very significant permeability reductions. K-Ar dating of diagenetic illite indicate that illite formed in Early Tertiary times at shallow depth (2–2.5 km). However, very little illite is found at these depths in good reservoir sandstones at the present day. High concentration of authigenic illite are only found at depths below 3.5 km, and the distribution seems to be controlled by the present day burial depth. The significance of the K-Ar dates of diagenetic illite and their relationships to the emplacement of oil is poorly understood. Fluid inclusion data suggest that quartz cement may have continued to form after oil emplacement in the irreducible water. This may be the case for the formation of illite from kaolinite and K-feldspar since this is an isochemical reaction.

More studies of porosity depth trends are needed to quantify the processes causing changes in the reservoir properties of the Brent Group sandstones with increasing burial so that reliable porosity prediction can be made.

This research has been supported by VISTA, a research cooperation between the Norwegian Academy of Science and Letters and Den Norske Stats Oljeselskap (Statoil). Den Norske Stats Oljeselskap (Statoil) is sincerely thanked for providing thin sections and porosity/permeability data from the Brent Group. Support from the Norwegian Research Council (NAVF) to T. Nedkvitne is acknowledged. Saga Petroleum a.s. is thanked for permission to publish data from the Snorre-B structure and R. Knarud for his comments on the diagenetic studies of the Brent Group sandstones from the Snorre-B structure. G. C. Saigal wishes to thank Nordic Ministerråd for financial support. We also thank B. L. Berg for technical assistance and preparation of figures.

## References

AAGAARD, P., EGEBERG, P. K., SAIGAL, G. C., MORAD, S. & BJØRLYKKE, K. 1990. Diagenetic albitization of detrital K-feldspar in Jurassic, Lower Cretaceous and Tertiary clastic reservoir rocks from offshore Norway. II. Formation water chemistry and kinetic considerations. *Journal of Sedimentary Petrology*, **60**, 575–581.

ANGEVINE, C. L. & TURCOTTE, D. L. 1983. Porosity reduction by pressure solution: A theoretical model for quartz arenites. *Geological Society of America, Bulletin*, **94**, 1129–1134.

BJØRKUM, P. A., MJØS, R., WALDERHAUG, O. & HURST, A. 1990. The role of the Late Cimmerian unconformity for the distribution of kaolinite in the Gullfaks Field, northern North Sea. *Sedimentology*, **37**, 396–406.

BJØRLYKKE, K. 1983. Diagenetic reactions in sandstones. *In*: PARKER, A. & SELLWOOD, B. W. (eds) *Sediment Diagenesis*, NATO. ASI Series, Reidel Publishing Company, 169–213.

——, ELVERHØI, A. & MALM, O. A. 1979. Diagenesis in Mesozoic sandstones from Spitsbergen and the North Sea – A Comparison. *Geologische Rundschau*, **68**, 1152–1171.

—— & BRENDSDAL, A. 1986. Diagenesis of the Brent Sandstone in the Statfjord Field, North Sea. *In*: GAUTIER, D. L. (ed.) *Roles of organic matter in sediment diagenesis*. Society of Economic Paleontologists and Mineralogists, Special Publication, **38**, 157–168.

——, AAGAARD, P., DYPVIK, H., HASTINGS, D. S. & HARPER, A. S. 1986. Diagenesis and reservoir properties of Jurassic sandstones from the Haltenbanken area, Offshore Mid Norway. *In*: SPENCER, A. M. *et al.* (eds) *Habitat of Hydrocarbons on the Norwegian Continental Shelf*. Norwegian Petroleum Society, Graham & Trotman, 275–386.

——, MO, A. & PALM, E. 1988. Modelling of thermal convection in sedimentary basins and its relevance to diagenetic reactions. *Marine and Petroleum Geology*, **5**, 388–351.

——, RAMM, M. & SAIGAL, G. C. 1989. Sandstone diagenesis and porosity modification during basins evolution. *Geologische Rundschau*, **78**, 243–268.

BLANCHE, J. B. & WHITAKER, J. H. Mc. D. 1978. Diagenesis of part of the Brent Sand Formation (Middle Jurassic) of the northern North Sea Basin. *Journal of the Geological Society, London*, **135**, 73–82.

BOLES, J. R. 1982. Active albitization of plagioclase, Gulf Cost. *American Journal of Science*, **282**, 165–180.

BROWN, S., RICHARDS, P. C. & THOMSEN, A. R. 1987. Patterns in the deposition of Brent Group (Middle Jurassic) U.K. North Sea. *In*: BROOKS, J. & GLENNIE, K. W. (eds) *Petroleum Geology of North West Europe*, Graham & Trotman, 899–913.

BURLEY, S. D. & FLISCH, M. 1989. K-Ar Geochronology and timing of detrital I/S clay illitization and authigenic illite precipitation in the Piper and Tartan Fields, outer Moray Firth, UK North Sea. *Clay Minerals*, **24**, 285–315.

——, KANTOROWICZ, J. D. & WAUGH, B. 1985. Clastic diagenesis. *In*: BRENCHLEY, P. J. & WILLIAMS, B. P. J. (eds) *Sedimentology: Recent Developments and Applied Aspects*. Geological Society, London, Special Publication, **18**, 189–225.

CURTIS, C. D. 1983. Link between aluminum mobility and destruction of secondary porosity. *American Association of Petroleum Geologists Bulletin*, **63**, 380–384.

EHRENBERG, S. N. & NADEAU, P. H. 1989. Formation of diagenetic illite in sandstones of the Garn Formation, Haltenbanken area, Mid-Norwegian Continental shelf. *Clay minerals*, **24**, 233–253.

ESLINGER, E. & PEVEAR, D. 1988. Clay Minerals for Petroleum Geologists and Engineers. *Society of Economic Palaeontologists and Mineralogists; Short course Notes*, 22.

FÄLT, L. M., HELLAND, R., JACOBSEN, V. W. & RENSHAW, D. 1989. Correlation of transgressive-regressive depositional sequences in the Middle Jurassic Brent/Vestland Group megacycle, Viking Graben, Norwegian North Sea. *In*: COLLINSON, J. D. (ed.) *Correlation in Hydrocarbon Exploration. Norwegian Petroleum Society*, Graham & Trotman, 191–200.

GALIMOV, E. M. 1980. $^{13}C/^{12}C$ in Kerogen. *In*: DURAND, B. (ed.) *Kerogen*. Editions Technip, Paris, 270–299.

GILES, M. R. 1987. Mass transfer and problems of secondary porosity creation in deeply buried hydrocarbon reservoirs. *Marine and Petroleum Geology*, **4**, 188–201.

GLASMANN, J. R., LUNDEGARD, P. D., CLARK, R. A., PENNY, B. K. & COLLINS, I. D. 1989*a*. Geochemical evidence for the history of diagenesis and fluid migration: Brent Sandstones, Heather Field, North Sea. *Clay Minerals*, **24**, 255–264.

——, CLARK, R. A., LARTER, S., BRIEDIS, N. A. & LUNDEGARD, P. D. 1989*b*. Diagenesis and hydrocarbon accumulation, Brent Sandstones (Jurassic), Bergen High, North Sea. *American Association of Petroleum Geologists Bulletin*, **73**, 1341–1360.

GLENNIE, K. W., MUDD, G. C. & NAGTEGAAL, P. C.

1978. Depositional environment and diagenesis of Permian Rotliegendes sandstones in Leman Bank and Sole Pit areas of the UK southern North Sea. *Journal of Geological Society, London*, **135**, 25–34.

GLUYAS, J. G. 1985. Reduction and prediction of sandstone reservoir potential, Jurassic, North Sea. *Royal Society of London, Philosophical Transactions*, **A315**, 187–202.

GOLD, P. B. 1987. Textures and geochemistry of authigenic albite from Miocene sandstones, Louisiana Gulf Coast. *Journal of Sedimentary Petrology*, **57**, 353–362.

GRAUE, E., HELLAND-HANSEN, W., JOHNSEN, J., LIMO, L., NØTTVEDT, A., RØNNING, K., RYSETH, A. & STEEL, R. 1987. Advance and retreat of Brent delta system of Norwegian North Sea. *In*: BROOKS, J. & GLENNIE, K. W. (eds) *Petroleum geology of North West Europe*, Graham & Trotman, 915–938.

HAMILTON, P. J., FALLICK, A. E., MACINTYRE, R. M. & ELLIOT, S. 1987*a*. Isotopic tracing of the provenance and diagenesis of Lower Brent Group sands North Sea. *In*: BROOKS, J. & GLENNIE, K. W. (eds) *Petroleum Geology of North West Europe*, Graham & Trotman, 939–949.

——, BRINT, J., HASZELDINE, R. S., FALLICK, A. E. & BROWN, S. 1987*b*. Formation of calcite cemented zones, Brent Group, North Sea. *Terra Cognita*, **7**, 342.

——, KELLEY, S. & FALLICK, A. E. 1989. K-Ar dating of illite in hydrocarbon reservoirs. *Clay Minerals*, **24**, 215–231.

HANCOCK, N. J. 1978*a*. Diagenetic modelling in the middle Jurassic Brent Sand of the Northern North Sea. *European Offshore Petroleum Conference and Exhibition, London, Oct. 1978*, 275–280.

—— 1978*b*. Possible cause of Rotliegend sandstone diagenesis in northern West Germany. *Journal of Geological Society, London*, **135**, 35–40.

—— & TAYLOR, A. M. 1978. Clay mineral diagenesis and oil migration in the Middle Jurassic Brent Sant Formation. *Journal of Geological Society, London*, **135**, 69–72.

HARRIS, N. B. 1989. Diagenetic quartzarenite and destruction of secondary porosity: An example from the Middle Jurassic Brent sandstone of Northwest Europe. *Geology*, **17**, 361–364.

HASZELDINE, R. S., SAMSON, I. M. & CORNFORD, C. 1984. Quartz diagenesis and convective fluid movements: Beatrice Oilfield, UK Northern North Sea. *Clay Minerals*, **19**, 391–402.

HOUSEKNECHT, D. W. 1987. Assessing the relative importance of compaction processes and cementation to reduction of porosity in sandstones. *American Association of Petroleum Geologists Bulletin*, **71**, 633–642.

—— 1988. Intergranular pressure solution in four quartzose sandstones. *Journal of Sedimentary Petrology*, **58**, 228–246.

HELGESON, H. C., MURPHY, W. M. & AAGAARD, P. 1984. Thermodynamic and kinetic constraints on reaction rates among minerals and aqueous solutions. II. Rate constants, effective surface area, and the hydrolysis of feldspar. *Geochimica et Cosmochimica Acta*, **48**, 2405–2432.

HELLAND-HANSEN, W., STEEL, R., NAKAYAMA, K. & KENDALL, C. G. St. C. 1989. Review and computer modelling of the Brent Group stratigraphy. *In*: WHATELEY, M. K. G. & PICKERING, K. T. (eds) *Deltas, Sites and Traps for Fossil Fuels*. Geological Society, London, Special Publications, **41**, 237–252.

HERMANRUD, C. 1986. On the importance of petroleum generation of heating effects from compaction driven water: an example from the northern North Sea. *In*: BURRUS, J. (ed.) *Thermal modelling of sedimentary basins*. Edition Technip, Paris, 247–269.

HOGG, A. J. C., PEARSON, M. J., FALLICK, A. E., HAMILTON, P. J. & MACINTYRE, R. M. 1987. Clay mineral and isotopic evidence for control on reservoir properties of Brent Group Sandstones, British North Sea. *Terra Cognita*, **7**, 342.

IRWIN, H., CURTIS, C. & COLEMAN, M. 1977. Isotopic evidence for source of diagenetic carbonates formed during burial of organic rich sediments. *Nature*, **269**, 209–213.

JOHNSON, H. D. & STEWART, D. J. 1985. Role of clastic sedimentology in the exploration and production of oil and gas in the North Sea. *In*: BRENCHLEY, P. J. & WILLIAMS, B. P. J. (eds) *Sedimentology: Recent developments and Applied Aspects*. Geological Society, London, Special Publications, **18**, 249–310.

JOHNSON, H. D., MACKAY, T. A. & STEWARD, D. J. 1986. The Fulmar oil field (Central North Sea): geological aspects of its discovery, appraisal and development. *Marine and Petroleum Geology*, **3**, 99–125.

KANTOROWICZ J. D., BRYANT, I. D. & DAWINS, J. M. 1987. Controls on Geometry and distribution of carbonate cements in Jurassic Sandstones: Bridport Sands, Southern England and Viking Group, Troll Field, Norway. *In*: MARSHALL, J. D. (ed.) *Diagenesis of Sedimentary Sequences*. Geological Society, London, Special Publications, **36**, 103–118.

KITTILSEN, J. E. 1987. *Sandproduksjonen fra oljereservoarer belyst en sedimentologisk og diagenetic undersøkelse av materiale fra Statfjordfeltet*, Cand. Scient Thesis, University of Oslo.

KNAUS, K. G. & WOLEY, T. J. 1986. Dependence of albite dissolution kinetics on pH and time at 25°C and 70°C. *Geochimica et Cosmochimica Acta*, **50**, 3481–2497.

LARESE, R. E., HASKELL, N. L., PREZBINDOWSKI, D. R. & BEJU, D. 1984. Porosity development in selected Jurassic sandstones from the Norwegian and North Seas, Norway – an overview. *In*: SPENCER, A. M. *et al.* (eds) *Petroleum Geology of the North European Margin*. Norwegian Petroleum Society, Graham & Trotman, 81–95.

LØNØY, A., AKSELSEN, J. & RØNNING, K. 1986. Diagenesis of a deeply buried sandstone reservoir: Hild Field, Northern North Sea. *Clay Minerals*, **21**, 497–511.

LOVELL, J. P. B. 1990. Cenozoic. *In*: GLENNIE, K. W.

(ed.) *Introduction to the petroleum geology of the North Sea*. Blackwell. 273–293.

MALM, O. A., FURNES, H. & BJØRLYKKE, K. 1979. Volcanoclastics of Middle Jurassic age in the Statfjord oil-field of the North Sea. *Neues Jahrbuch für Geologie und Paläontologie, Monadshefte*, **10**, 607–618.

MEARNS, E. W. 1989. Neodymium isotope stratigraphy of Gullfaks oilfield. *In*: COLLINSON, J. D. (ed.) *Correlation in Hydrocarbon Exploration*. Norwegian Petroleum Society, Graham & Trotman. 201–215.

MORAD, S., BERGAN, M., KNARUD, R. & NYSTUEN, J. P. 1990. Albitization of detrital plagioclase in Triassic Reservoir sandstones from the Snorre Field, Norwegian North Sea. *Journal of Sedimentary Petrology*, **60**, 411–425.

MORTON, A. C. 1985. A new approach to provenance studies: electron microprobe analysis of detrital garnets from Middle Jurassic sandstones of the Northern North Sea. *Sedimentology*, **32**, 553–566.

—— 1986. Dissolution of Apatite in North Sea Jurassic sandstones: Implications for the generation of secondary porosity. *Clay Minerals*, **21**, 711–733.

—— 1987. Detrital garnets as provenance and correlation indicators in the North Sea. *In*: BROOKS, J. & GLENNIE, K. W. (eds) *Petroleum Geology of North West Europe*. Graham & Trotman, 991–995.

——, STIBERG, J. P., HURST, A. & QVALE, H. 1989. Use of heavy minerals in lithostratigraphic correlation, with examples from Brent Sandstones of the Northern North Sea. *In*: COLLINSON, J. D. (ed.) *Correlation in Hydrocarbon Exploration*. Norwegian Petroleum Society,, Graham & Trotman, 217–230.

NEDKVITNE, T. & BJØRLYKKE, K. 1992. Secondary porosity in the Brent Group (Middle Jurassic), Hildra field, North sea: implication for predicting lateral continuity of sandstones? *Journal of Sedimentary Petrology*, **62**, in press.

PALM, E. 1990. Rayleigh convection, mass transport and change in porosity in layers of sandstone. *Journal of Geophysical Research*. **95B**, 8675–8679.

RAMM, M. 1990. *Quantitative mineral analysis of sandstone samples by XRD*. Department of Geology, Oslo, Norway, Internal Publication **63**.

ROBERTS, J. D., MATHISEN, A. S. & HAMPSON, J. M. 1987. Statfjord. *In*: SPENCER, A. M. *et al.* (eds) *Geology of the Norwegian Oil and Gas Fields*. Graham & Trotman, 319–340.

SAIGAL, G. C. 1990. *Petrology, diagenesis and reservoir properties of the Brent Group sandstones from well 34/7–14*. Internal report, Saga Petroleum a.s. April 1990.

——, MORAD, S., BJØRLYKKE, K., EGEBERG, P. K. & AAGAARD, P. 1988. Diagenetic albitization of detrital K-feldspar in Jurassic, Lower Cretaceous, and Tertiary Clastic reservoir rocks from offshore Norway, I. Texture and origin. *Journal of Sedimentary Petrology*, **58**, 1003–1013.

—— & BJØRLYKKE, K. 1987. Carbonate cements in clastic reservoir rocks from offshore Norway – Relationship between isotopic composition, textural development and burial depth. *In*: MARSHALL, J. D. (ed.) *Diagenesis of Sedimentary Sequences*. Geological Society of London, Special Publications, **36**, 313–324.

SCHMIDT, V. & MCDONALD, D. A. 1979. The role of secondary porosity in the course of sandstone diagenesis. *In*: SCHOLLE, P. A. & SCHLUGER, P. R. (eds) *Aspects of diagenesis*. Society of Economic Paleontologists and Mineralogists, Special Publications, **26**, 209–225.

SCOTCHMAN, I. C., JOHNES, L. H. & MILLER, R. S. 1989. Clay diagenesis and oil migration in Brent Group sandstones of NW Hutton field, UK North Sea. *Clay Minerals*, **24**, 339–374.

SELLEY, R. C. 1978. Porosity gradients in the North Sea oil-bearing sandstones. *Journal of the Geological Society, London*, **135**, 119–132.

SHANMUGAM, G. 1984. Secondary porosity in sandstones. Basic contribution of Chepikov and Savkevich. *American Association of Petroleum Geologists Bulletin*, **68**, 106–107.

—— 1988. Origin, recognition, and importance of erosional unconformities in sedimentary Basins. *In*: KLEINSPEHN, K. L. & PAOLA, C. (eds) *New Perspectives in Basin Analysis*. Springer Verlag, New York, 83–108.

SMITH, J. T. & EHRENBERG, S. N. 1989. Correlation of carbon dioxide abundance with temperature in clastic hydrocarbon reservoirs: Relationship to inorganic chemical equilibrium. *Marine and Petroleum Geology*, **6**, 129–135.

SOMMER, F. 1975. Histoire diagenétique d'une serie grésuse de Mer Du Nord. Dation de L'introduction des hydrocarbons. *Revue de L'institut Francais du Pétrol*, **30**, 729–741.

—— 1978. Diagenesis of Jurassic sandstones in the Viking Graben. *Journal of the Geological Society, London*, **135**, 63–67.

SURDAM, R. C. 1984. The chemistry of secondary porosity. *In*: MCDONALD, D. A. & SURDAM, R. C. (eds) *Clastic diagenesis*. American Association of Petroleum Geologists, Memoir, **37**, 127–150.

THOMAS, M. 1986. Diagenetic sequences and K/Ar dating in Jurassic sandstones, Central Viking Graben: Effects on reservoir properties. *Clay Minerals*, **21**, 695–710.

WALDERHAUG, O. 1990. A fluid inclusion study of quartz cemented sandstones from offshore Mid-Norway – possible evidence for continued quartz cementation during oil emplacement. *Journal of Sedimentary Petrology*, **60**, 203–210.

——, BJØRKUM, P. A. & NORDGÅRD BOLÅS, H. M. 1989. Correlation of calcite-cemented layers in shallow-marine sandstones of the Fensfjord Formation in the Brage Field. *In*: COLLINSON, J. D. (ed.) *Correlation in Hydrocarbon Exploration*. Norwegian Petroleum Society, Graham & Trotman, 367–375.

WALKER, T. R. 1984. Diagenetic albitization of potassium feldspar in arkosic sandstones. *Journal of Sedimentary Petrology*, **54**, 3–16.

WAUGH, B. 1978. Authigenic K-feldspar in British Permo-Triassic sandstones. *Journal of the Geological Society, London*, **135**, 51–56.

WOOD, J. R. & HEWETT, T. A. 1984. Reservoir diagenesis and convective fluid flow. *In*: SCHOLLE, P. A. & SCHLUGER, P. R. (eds) *Aspects of diagenesis*. Society of Economic Paleontologists and Mineralogists, Special Publications, **26**, 209–225.

# The reservoir properties and diagenesis of the Brent Group: a regional perspective

M. R. GILES[1], S. STEVENSON[2], S. V. MARTIN, S. J. C. CANNON[2], P. J. HAMILTON[3], J. D. MARSHALL[4], & G. M. SAMWAYS[5]

*Shell UK Exploration & Production, Shell-Mex House, Strand, London WC2R ODX, UK*

*Present addresses: [1] RR/21, KSEPL, Shell Research, Voolmerlaan 6, 2281GD Rijswijk, Netherlands*

*[2] Geochem Group, Chester Street, Chester, UK*

*[3] CSIRO, 51 Delhi Road, North Ryde, NSW, Australia*

*[4] Department of Earth Sciences, University of Liverpool, P.O. Box 147, Liverpool L69 3BX, UK*

*[5] Badley, Ashton & Associates, Winceby House, Winceby, Horncastle, Lincs LN96BP, UK*

**Abstract:** A regional study of Brent Group diagenesis and reservoir properties has been undertaken in order to determine the main controls on porosity and permeability in the sandstone reservoirs. Data from 44 wells from block 211/7 in the north to block 3/8 in the south and spanning current depths from 6700 to 13 400 ft include 9000 porosity, permeability and grain density determinations, quantitative petrographic information from 850 thin sections, and stable isotopic and K/Ar analyses of authigenic phases.

The diagenesis of the sediments is similar across the study area and most of the diagenetic phases occur in all formations. The sequence of precipitates and dissolution events reflects early porewater evolution in shallow burial environments and later reactions which were essentially isochemical and controlled largely by increasing burial depth and therefore temperature. Early diagenetic products include siderite, calcite, chlorite and vermicular kaolinite. Only where the calcite cements form concretions or cemented horizons have they a significant effect on reservoir properties. Local dissolution of feldspars and carbonate cements took place on the crest of some fault blocks inferred to have been emergent during the Jurassic and this has caused local enhancement of porosity in some crestal wells.

In general, the porosity of each of the reservoir facies decreases systematically with depth but permeability only starts to decrease significantly at depths greater than 10 200 ft (3109.0 m). The general decrease in porosity can be attributed to compaction, together with burial cementation by quartz and iron-rich carbonates. Secondary porosity resulting from feldspar dissolution is increasingly common at depth but there is no net increase in porosity as much of the dissolution was evidently accompanied by the precipitation of authigenic quartz initially with kaolinite but at greater depths with illite. The systematic changes in porosity and the decrease in permeability are compatible with thermally driven dissolution of feldspar and the local reprecipitation of the authigenic silicates; the decrease in permeability corresponds to the presence of increased quantities of illite at depth.

In November 1987, prior to the 11th UK Offshore Licensing Round, Shell Expro decided to embark on a major review of the Brent Group north of latitude 60°40′. The aim of this study was to establish a predictive model for reservoir properties and to systematically examine the remaining exploration potential of the area. This paper together with its sister papers (Cannon *et al.*, Hamilton *et al.*, Whitaker *et al.* all this volume) represent some of the results of this multidisciplinary study.

Cores from producing fields and exploration wells were studied from both Shell and competitors' blocks across the study area shown in Fig. 1. Cores from 22 blocks were examined, stretching from block 3/8 in the south to 211/7 in the north. Much of the data used in the study has not been released and consequently well names will not be identified.

The study was based on a complete re-

*From* Morton, A. C., Haszeldine, R. S., Giles, M. R. & Brown, S. (eds), 1992, *Geology of the Brent Group*. Geological Society Special Publication No. 61, pp. 289–327.

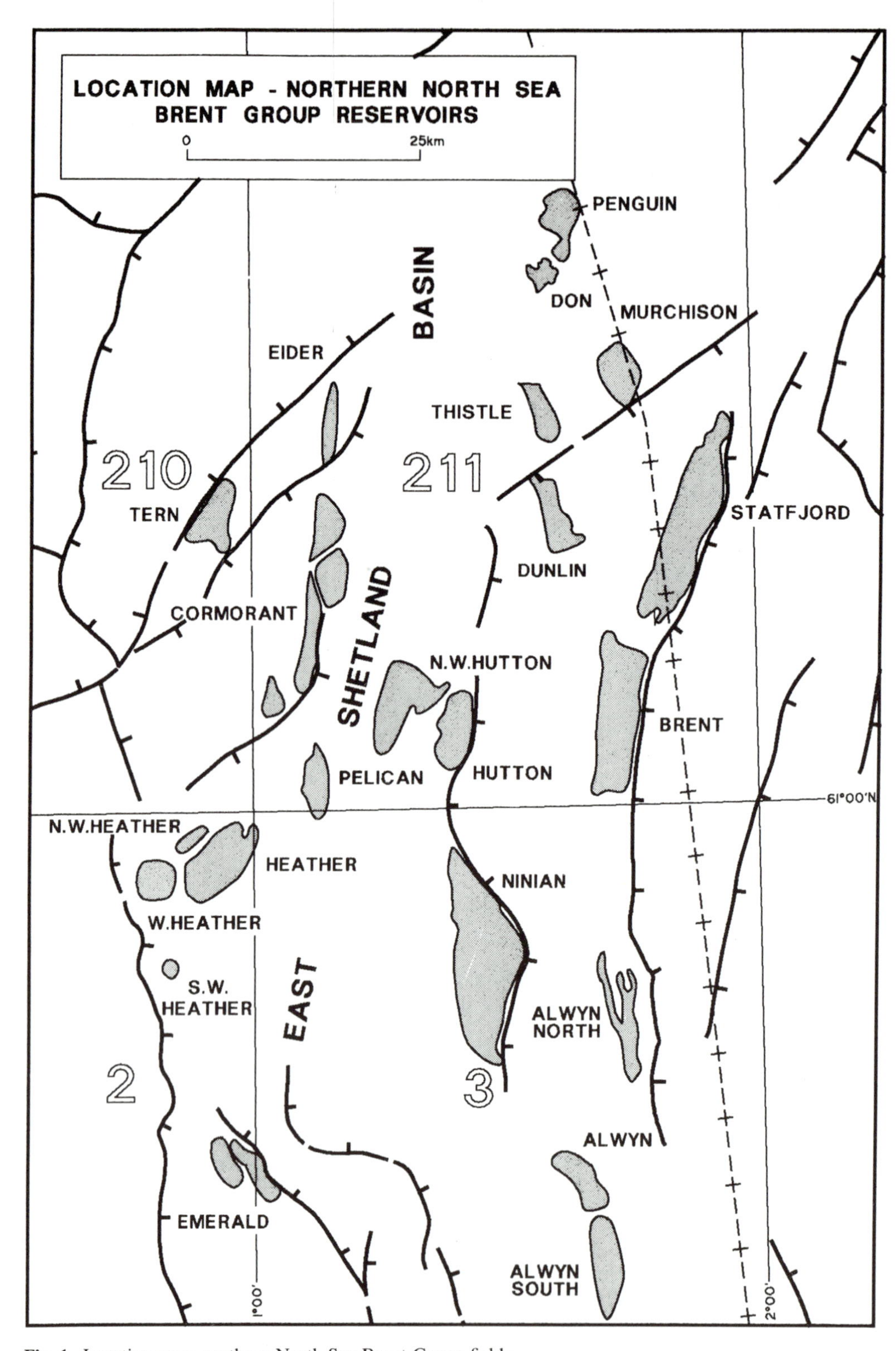


**Fig. 1.** Location map: northern North Sea Brent Group fields.

appraisal of existing Shell reports and data from both Production and Exploration Departments. In addition a large body of additional data has been collected as outlined in the methods section. The project has attempted to integrate reservoir property data, petrographic data and isotopic data into a coherent whole. A strong emphasis has been placed on identifying statistically valid trends in both the reservoir property and diagenetic data. These trends have been used to identify the significant diagenetic processes modifying the reservoir characteristics of the rocks.

For nearly two decades the Brent Group has been one of the most prolific North Sea reservoir sequences, and a considerable body of diagenetic data has been published (e.g. Blanche & Whitaker 1978; Bjørlykke *et al.* 1979; Kantorowicz 1984*b*; Hamilton *et al.* 1987; Jourdan *et al.* 1987; Glasmann *et al.* 1989*a*, *b*; Scotchman *et al.* 1989). Most of these previous studies are based on single wells or individual fields. In contrast the data set amassed for this publication is, for the first time, both large and regional in extent.

## Methods

### *Data collection*

Samples collected for petrographic study were taken, where possible, from core plug trims so that direct comparisons could be made with measured reservoir properties. Where this was not possible, samples were taken from a point adjacent to the points at which core plugs had been taken. All cores selected for study were also described in detail so that formation, facies and depositional environment could be assigned to each sample.

A total of 633 thin sections were prepared from samples with representative reservoir properties. All thin sections were impregnated with blue-dyed epoxy resin so that pore space could be easily identified. Furthermore, each thin section was stained to aid identification of carbonate phases and K-feldspars using alizarin red S/potassium ferricyanide and HF/ sodium cobaltinitrite respectively. Each thin section of the primary data set (633 thin sections) was subsequently point-counted by a single operator (Stevenson) in order to maintain consistency. Representative samples were described in detail. In addition to the primary data set, additional point-count results from previous studies were also included where appropriate, giving a total data set of some 850 samples. All abundances point-counted from thin section are reported as percentages of bulk volume.

The thin section, XRD, SEM and cathodoluminescence work was carried out at the Geochem Laboratories at Chester.

XRD data was also compiled using methods similar to those outlined by Brown & Brindley (1980) using a Phillips PW 1710 automated powder diffraction system. Bulk and less than 2 $\mu$m fractions were studied. Samples subject to XRD analysis were generally also investigated using SEM methods employing a Phillips 515 scanning electron microscope equipped with an EDAX PB9100 energy dispersive spectrometer utilising a ECON IV light element detector.

Selected carbonate and quartz cemented samples were examined under cathodoluminescence using a Technosyn 8200 Mk II Cold Cathode Luminescence system.

150 C & O isotopic measurements on selected carbonate cements were made at University of Liverpool as part of the project. The methods used are reported in Samways (1990).

K/Ar ages on diagenetic illites were measured on samples from 11 wells from across the Brent province and a review made of all published dates. The methods employed and the results are discussed in a separate paper in this volume (Hamilton *et al.*).

A total of 9000 porosity, horizontal air permeability and grain density measurements were used in the study for statistical analysis. Each measurement was categorized according to its well name, formation, facies and pore fill. These measurements were then used to assess regional variations in reservoir quality.

### *Approach adopted*

The analysis of the data in the following sections relies heavily on the assumption that samples taken from selected wells are sufficient in number to yield statistically significant trends. Wells were chosen covering as wide a geographic area as possible. The reservoir property study used data from 44 wells, with one or more wells being chosen from each of the following blocks: 2/5, 3/1, 3/2, 3/3, 3/8, 210/15, 210/30, 211/7, 211/11, 211/13, 211/14, 211/16, 211/18, 211/19, 211/21, 211/22, 211/24, 211/26, 211/27, 211/28 and 211/29. Within an individual block, wells with the greatest amount of core data were chosen. Reservoir property data from 836 ft (254.82 m) of Tarbert core, 2834 ft (863.81 m) of Ness, 2180 ft (664.47 m) of Etive and 2621 ft (798.89 m) of Rannoch Formation core have been incorporated in the study.

Samples for the diagenetic study were taken from most of the cores used for the reservoir property study. The study has involved 35 primary wells (from which the 633 new thin sections were prepared) and 25 secondary wells for which data was already available. Thin sections from one or more wells from each of the following blocks were examined: 2/5, 3/1, 3/2, 3/3, 3/7, 3/8, 210/15, 211/11, 211/12, 211/13, 211/16, 211/18, 211/19, 211/21, 211/22, 211/23, 211/24, 211/26, 211/27, 211/28 and 211/29.

Data from both the reservoir property and the diagenetic study were examined in order to discover whether geographical provinces of reservoir properties or diagenetic cements existed within the study area. However, as will be discussed below, the main variation appears to be with depth.

## Regional trends in reservoir properties

### *Regional trends in porosity*

Porosity/depth diagrams were constructed for separate facies of the Tarbert, Ness, Etive and Rannoch Formations. The porosity and permeability data were taken from conventional core analysis results and represent both oil and water bearing zones. A formation average trend was also prepared for each of the above formations. These trends have subsequently been used as a method of identifying zones of anomalously high or low porosities. The method for their construction is outlined in the flow chart given in Fig. 2. Outliers, as defined in this figure, refer to data points lying beyond the 95% confidence limits of the data.

*Etive Formation.* The main reservoir facies within the Etive Formation generally consist of massive, parallel-bedded, rippled or cross-bedded sandstones. Massive, cross-bedded and parallel-bedded sandstones exhibit similar porosity/depth characteristics (Fig. 3). Rippled sandstones generally have a slightly different porosity/depth trend from the others, showing lower porosity for a given depth. Initial porosities in sandstones are of the order of 35–50% (Beard & Weyl 1973) depending on their sorting. Few sandstones encountered in reservoirs have primary porosities of this order, hence porosity must fall fairly rapidly to values of 30–35% which are commonly encountered in the shallower Brent reservoirs. Porosity/depth trends are generally assumed to be exponential (e.g. Baldwin & Butler 1985); however, for the depth range and data set available for this study a simple linear trend is a good approximation to the data. The porosity/depth trends result from the combined effects of initial compaction and later cementation. Once

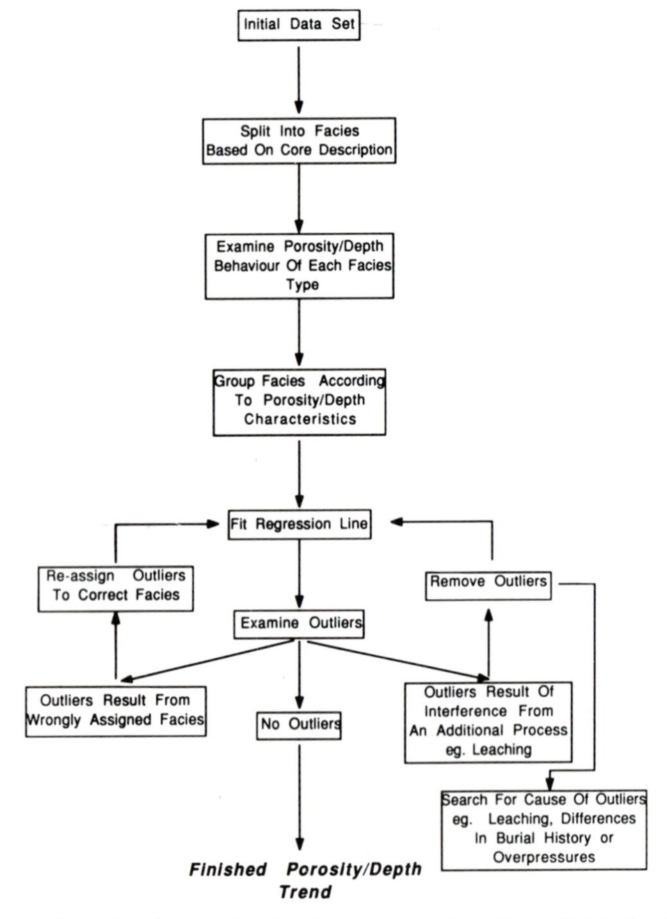


**Fig. 2.** Flow diagram illustrating the steps involved in the construction of porosity/depth relationships.

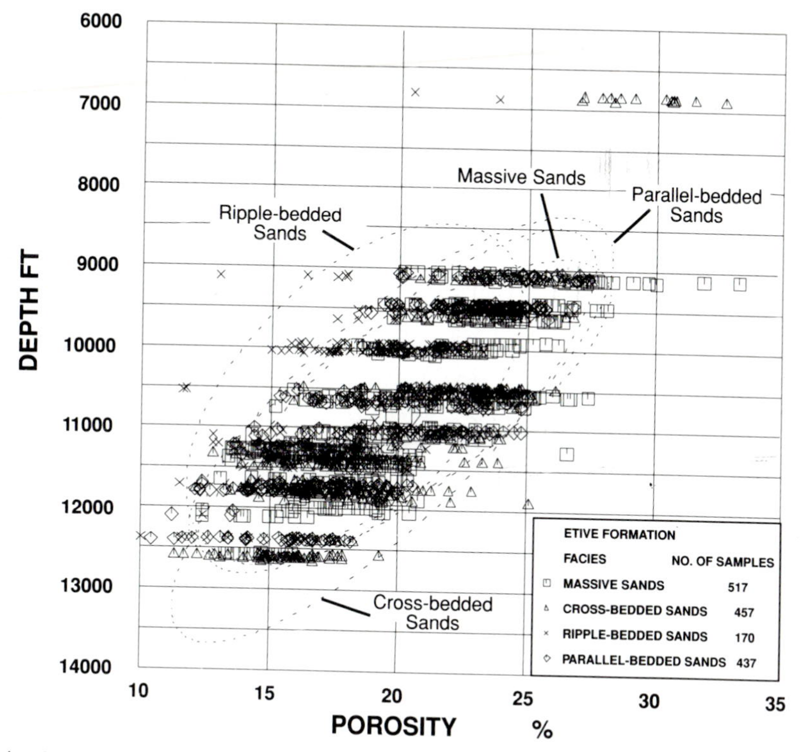


**Fig. 3.** Facies dependence of porosity loss with depth for Etive Formation reservoir quality facies. The ellipses give the 90% confidence interval for each facies.

cementation effects overtake compaction, the driving force behind the linear fall in porosity is more likely to be temperature rather than depth, as most inorganic chemical reactions which do not involve a gaseous phase are more sensitive to temperature than they are to pressure. However, as no unique temperature can be associated with a given sample, present day burial depth has been used as a surrogate variable.

Regression analysis of the massive, cross-bedded and parallel-bedded sandstone data gives a porosity gradient of about 2.8% per thousand feet (Fig. 4a). The solid hyperbolic curves define the 90% confidence region to the position of the regression line and the outer dotted hyperbolic curves give the 90% confidence region to porosity estimated from depth. This trend may be considered as the 'Etive best reservoir sand' trend.

In Fig. 5a the average porosity is plotted against depth for the whole formation. The trend shows that the average rate of fall of porosity for the whole formation (2.5% per thousand feet) is only slightly less than that given by the 'best reservoir sands'. This result is not surprising considering that the Etive Formation is comparatively uniform sand.

*Ness Formation.* The Ness Formation is much more heterogeneous than the Etive, with more diverse sandstone types, such as rootletted and heterolithic sands. The best reservoirs are found in cross-bedded fluvial sands. Massive and parallel-bedded sands show a similar porosity/depth trend to the cross-bedded sands, although the data are more scattered. The porosity/depth trend for the Ness cross-bedded sands is given in Fig. 4b and reveals that porosity falls at about 3.1% per thousand feet. If all Ness sandstones are considered, a porosity gradient of 2.5% per thousand feet is arrived at (Fig. 5b). Cross-bedded sands are well sorted and have a higher initial porosity than more heterogeneous clay rich sands. Hence they will tend to show a steeper rate of porosity decline because the total porosity destroyed will be greater than that for the more poorly sorted sands.

*Tarbert Formation.* Interpretation of porosity/depth trends within the Tarbert Formation is limited by the lack of uniformly distributed data throughout the depth range investigated. Cross-bedded and massive sandstones of the Tarbert Formation exhibit similar porosity/depth relations, whereas the more clay rich bio-

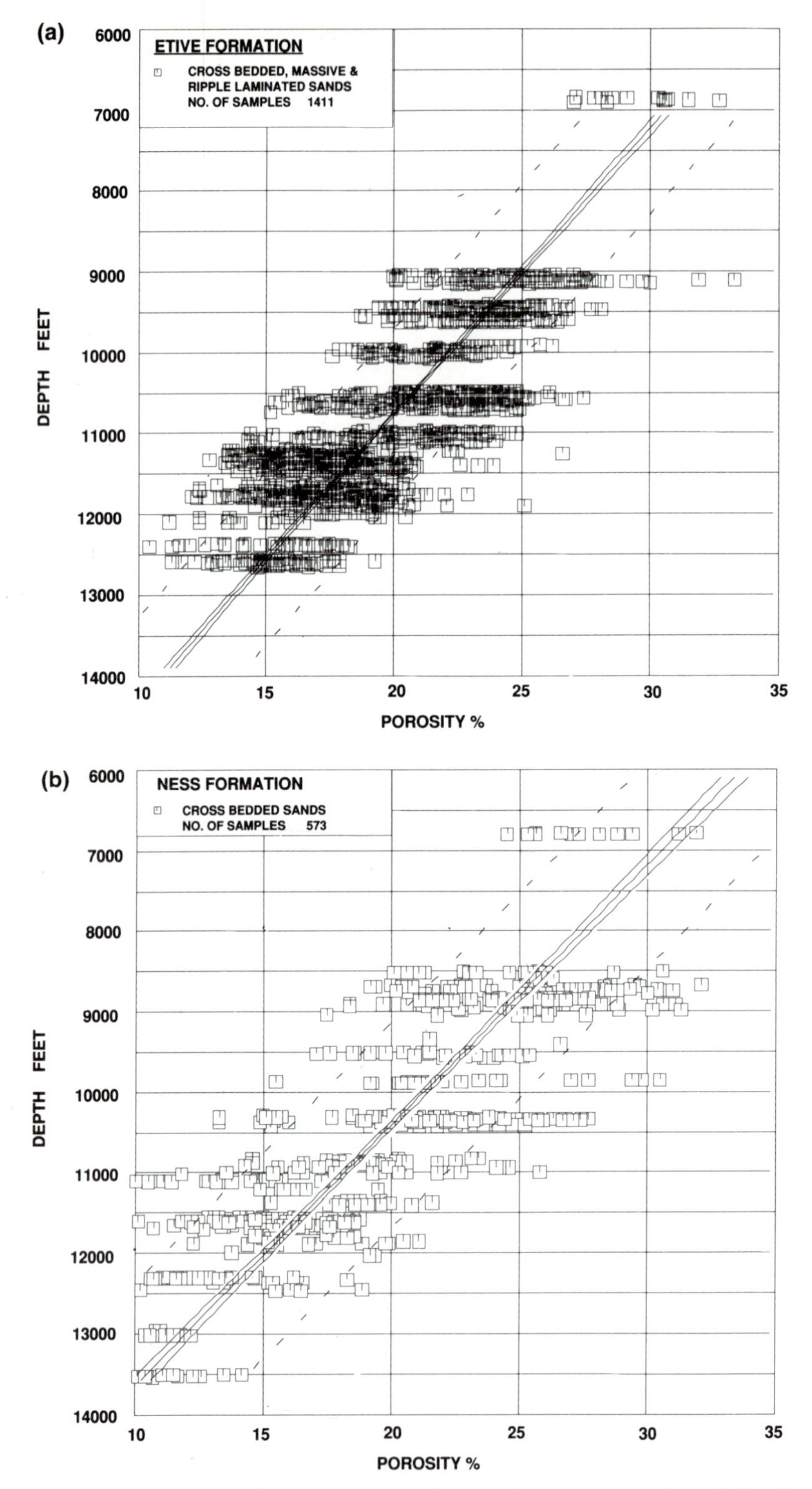


**Fig. 4.** Porosity/depth relations for (**a**) Etive Formation (cross-bedded, massive and parallel-bedded) sands. Note the approximately linear fall of porosity with burial depth, and (**b**) Ness Formation (cross-bedded) sands. Data points are porosity measurements made on one inch diameter core plugs as part of routine conventional core analysis. The inner solid line is the regression line for porosity as a function of depth. The two solid hyperbolic curves give the 90% confidence region to the position of the regression line, and the outer two dashed hyperbolic curves give the 90% confidence interval to porosity estimated from burial depth.

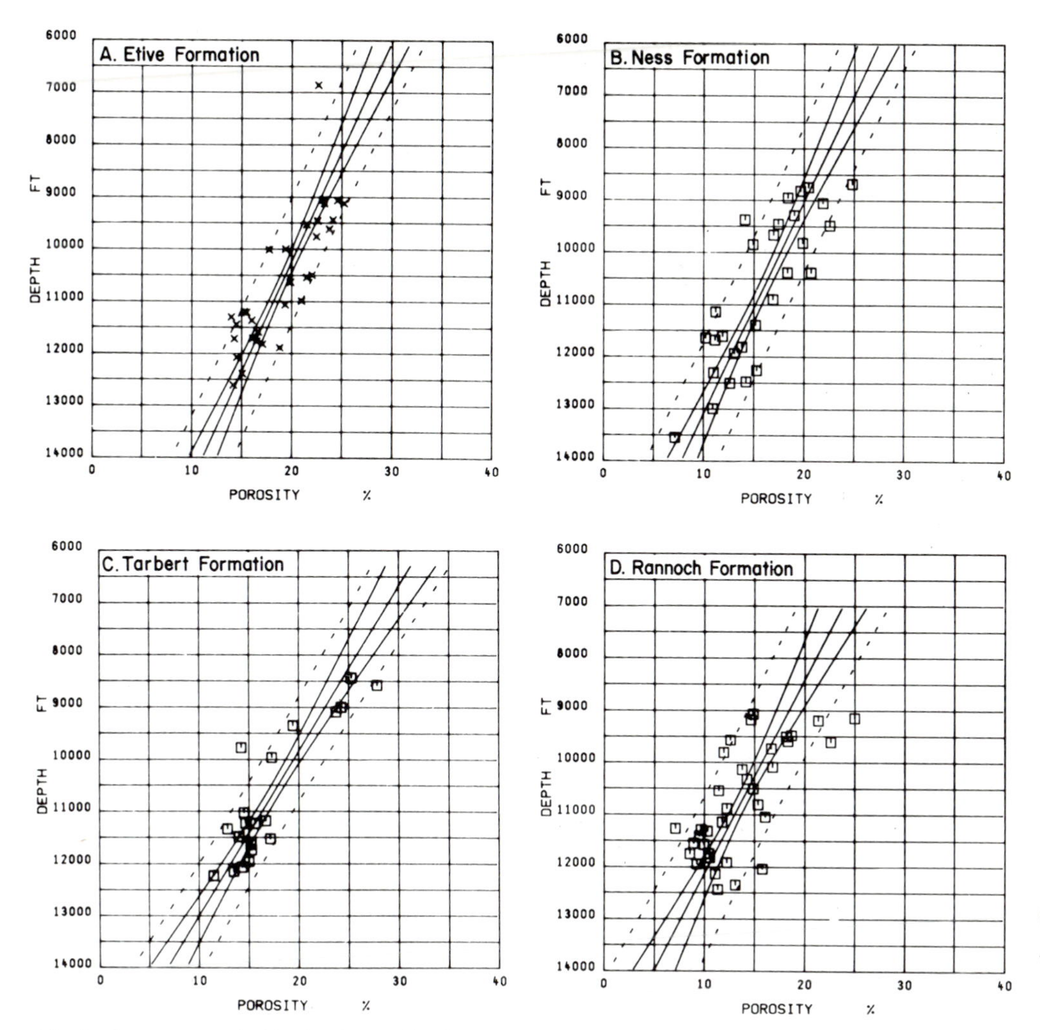


**Fig. 5.** Average porosity/depth relationships for whole formations as calculated from core data. Note the approximately linear fall of average porosity with increasing burial depth.
The inner solid line is the regression line for porosity as a function of depth. The two solid hyperbolic curves give the 90% confidence region to the position of the regression line, and the outer two dashed hyperbolic curves give the 90% confidence interval to porosity estimated from burial depth.

turbated and rippled sandstones lose porosity with depth somewhat faster. Unfortunately, there are few data on which to base a trend for the massive and cross-bedded Tarbert sandstones. Available data suggests that a good fit to the data from the massive and cross-bedded sands is obtained using the Etive 'best reservoir sand' trend. The trend for the average sandstone porosity for this formation is shown in Fig. 5c. The average porosity decreases by about 3.2% per thousand feet.

*Rannoch Formation*. The Rannoch Formation was not studied in any great detail. The average fall in porosity with depth is about 2.7% per thousand feet (Fig. 5d), but starts from a lower initial porosity than any of the other formations. Thus, the porosity of the Rannoch is always below that of the other formations; however, the difference between the formations narrows with increasing burial depth.

*Porosity anomalies*. During the course of porosity/depth analysis a number of wells were identified which displayed anomalous behaviour. During the construction of the Etive Formation 'best reservoir sand' trend, two wells 'A' & 'B' (Fig. 6a) were found to show higher porosities than would have been expected for their burial depth. Analysis of the Ness Formation porosity/depth data also identified the above two wells as anomalous, plus a further well 'C' (Fig. 6b).

The common factor is that all these wells lie in crestal or near-crestal structural locations. All three wells have a thick sand directly below an unconformity. Not all crestal wells have unusually high porosities, and no wells which are significantly downdip show any comparable effect. Such high porosities could result from a number of processes such as, unusual facies development (i.e. extremely well sorted sands), leaching by meteoric waters, early hydrocarbon emplacement, or late leaching perhaps related to the fluid migration up crestal faults. These points will be discussed further in the section on diagenesis.

The enhancement in porosity does not automatically result in an increase in permeability. The permeabilities encountered in the shallow well 'A' are probably higher than expected for the burial depth whereas those from the deeper well 'B' are not. The difference is almost certainly the result of illite growth in the deeper reservoirs (see section on diagenesis).

*Effects of hydrocarbon emplacement.* The notion that early hydrocarbon emplacement will suspend diagenesis and consequently result in the preservation of higher porosities or permeabilities in the hydrocarbon as opposed to the water leg is widely held (see for example Füchtbauer 1967). In single well studies this is often difficult to prove as a result of facies changes which may occur over the same interval as the hydrocarbon/water contact. There is also the problem of obtaining statistically meaningful data from any one facies type that spans both the hydrocarbon and water legs.

Assuming that the differential diagenesis hypothesis is correct, then samples from the oil legs should show, on average, higher porosities than those from the water legs. Thus, if a comparison is made between the expected porosity calculated from the regression lines (e.g. Fig. 4) with the actual porosities measured on the core plugs, then the water leg samples should show a distribution biased toward lower than expected porosities. Should such a result be obtained, then the porosity/depth regression analysis carried out would be invalid as the data would consist of two populations. Figures 7a & b show the distribution of the porosity residuals (difference between actual porosity and expected porosity calculated from the 'best sand' regression lines), for oil and water leg samples of the Ness and Etive Formations respectively. Both figures are similar, and suggest that differential diagenesis has had little effect on porosity. In turn, it weakens arguments for the high porosities seen in the crestal wells discussed above as being the result of early hydrocarbon emplacement.

## *Regional permeability/depth relationships*

The permeability/depth relationships shown by the cross-bedded sands of the Etive Formation (Fig. 8a) demonstrate a feature of regional importance for Etive, Ness and Tarbert reservoirs, which will later be shown to be the result of illite growth. At depths shallower than 10200 ft (3109.0 m) permeability is independent of depth, but below this depth permeability falls off very rapidly. This feature is common to all facies of the Etive Formation (Fig. 8b) and can also be seen in data from the Ness (Fig. 9a) and Tarbert (Fig. 9b) Formations. Data for any given depth show considerable scatter which reflects differences in porosity, grain size or diagenesis. This scatter is greatest for the Ness Formation and is due to the larger range of grain sizes present and the presence of a number of permeability anomalies which will be discussed in more detail later. The sudden discontinuity in the permeability/depth dependance is in marked contrast to the smooth decrease in porosity with depth (Fig. 4).

Permeability values for intervals can be calculated from the results of production tests. These results are generally more dependent on the average permeability of an interval than on individual plug measurements, thus for commercial purposes it is useful to study averaged trends rather than the per facies trends. The averaging techniques adopted for a given formation are those that are believed to give the closest approximation to permeability values, calculated from production test results for a given formation, and no physical significance is assigned to the type of averaging used. Furthermore the form of the trends does not depend on the averaging technique used, only the absolute values of permeability. Etive Formation sandstones (Fig. 10a) (only cores with more than 20 permeability values have been used) display a trend similar to that seen in earlier plots, which is exemplified with a distinct break at about 10200 ft (3109.0 m) and below that depth a linear trend of rapidly falling average permeabilities with increasing burial depth. The trend for the Etive is not exceptional, and trends for the Ness (Fig. 10b) and Tarbert (Fig. 10c) are similar. The Rannoch,

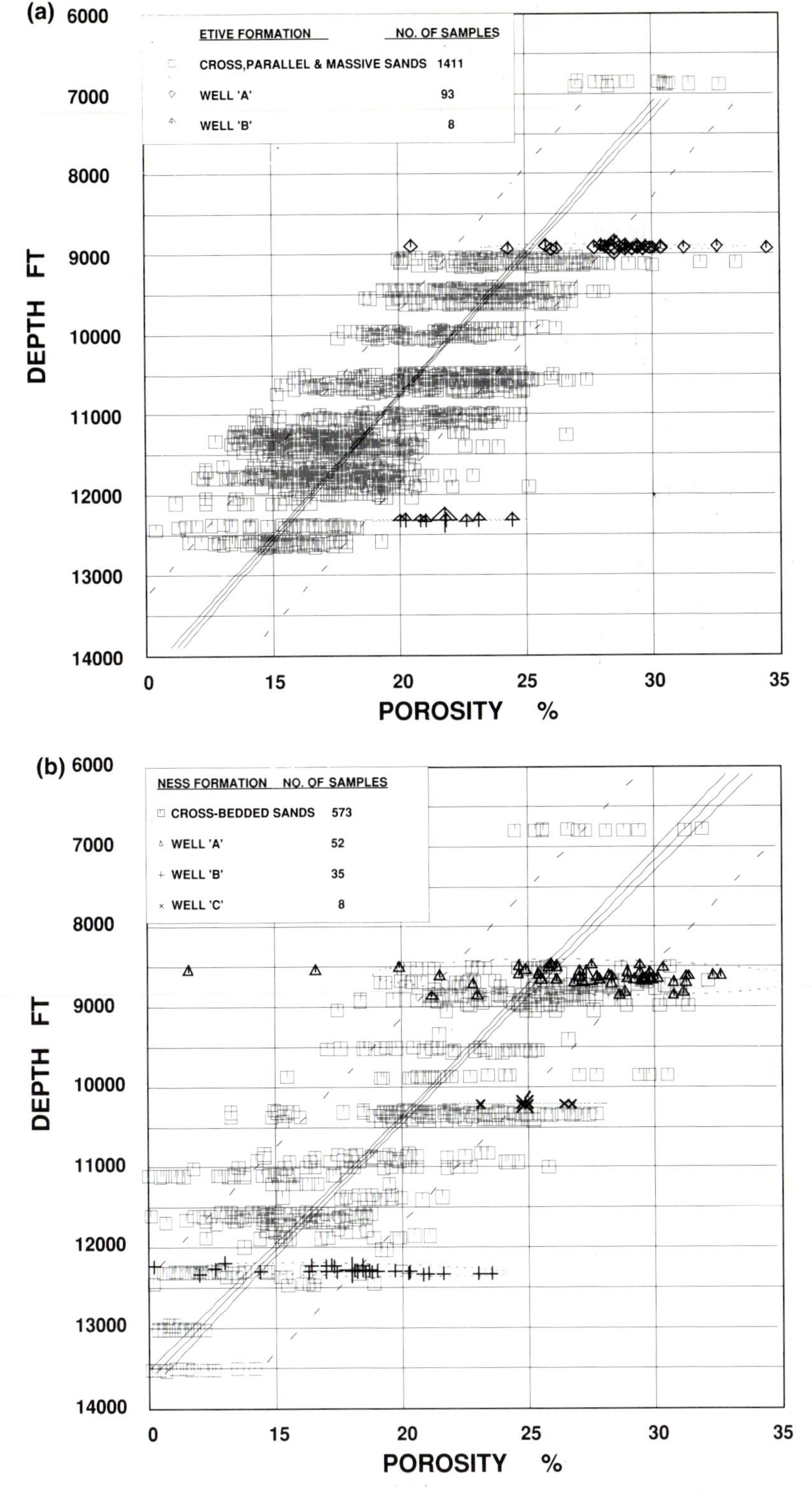


**Fig. 6.** (**a**) Comparison of porosity data from the Etive Formation of two crestal wells with the normal trend of decreasing porosity with increasing burial depth. Wells A and B both show higher porosities than would be expected from their burial depth.
The inner solid line is the regression line for porosity as a function of depth. The two solid hyperbolic curves give the 90% confidence region to the position of the regression line, and the outer two dashed hyperbolic curves give the 90% confidence interval to porosity estimated from burial depth.
(**b**) Comparison of porosity data from the Ness Formation of three crestal wells with the normal trend of decreasing porosity with increasing burial depth. Wells A, B and C show higher porosities than would be expected from their burial depth. Solid and dashed lines as for (a).

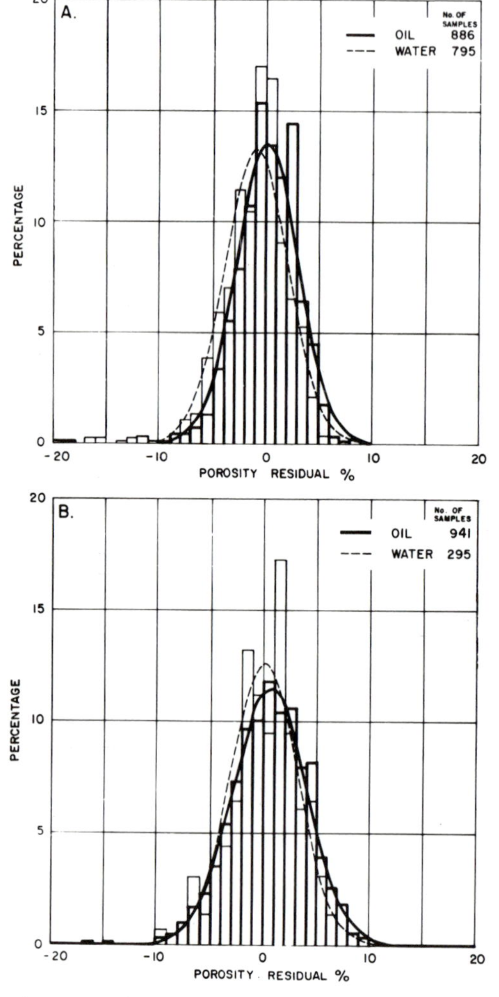


**Fig. 7.** Porosity differences in relation to pore fill. The porosity residual (difference between actual and calculated porosity) is plotted for oil and water leg sandstones of the Etive Formation (**a**) and Ness Formation (**b**). Both of these figures show that there is no systematic difference between the porosity that can be expected in hydrocarbon zones as opposed to water bearing zones.

however, shows a steady decrease in permeability with depth (Fig. 10d).

*Permeability anomalies*. The Etive interval from three wells, 'D', 'G' & 'K' (Fig. 11a) and one Ness interval (Fig. 11b), well 'D' display unusually low permeabilities for their depths. The common feature of these intervals is that they are all water bearing. So far no hydrocarbon-bearing well has been found which exhibits this effect. However, not all water bearing sequences have lower than expected permeabilities. The majority of water bearing intervals fall on the normal trend defined from the data set as a whole. The cause of these anomalously low permeabilities are believed to be continued illite growth within the water bearing intervals as will be discussed later.

The second type of anomaly, showing higher than expected permeabilities, occurs in the Ness Formation of two wells 'E' & 'F' (see Fig. 11b). However, the significance of these permeability anomalies should not be over estimated, because out of 2834 Ness Formation core plugs studied, there are only 11 anomalous values! The core from the Ness Formation of well 'F' displays a 5 ft (1.52 m) thick channel sand which has 4 permeability values in the range 40–265 mD. This is some 2–3 orders of magnitude higher than would be expected for its depth of 13 403 ft (4085.28 m). In contrast a second 13 ft (3.96 m) thick channel follows a normal permeability/depth trend with permeabilities generally below 10 mD and in only one case reaching 40 mD. Cores from well 'E' show an even more remarkable situation. In this well between 12 461 ft (3798.2 m) and 12 480 ft (3803.95 m) are two active channel fill sequences stacked one upon the other. The upper sand body is approximately 10 ft (3.05 m) thick and displays permeabilities from 16 to 485 mD, the maximum permeability being some 2–3 orders of magnitude higher than expected. The lower sand body, which is 9 ft (2.74 m) thick has permeabilities between 1 and 29 mD and generally less than 10 mD, which is approximately what would be expected for the depth.

These permeability anomalies are not the result of the averaging process. A geometric mean permeability/depth trend constructed for the Ness cross-bedded and massive sands is not significantly different from the arithmetic average of the whole formation, because the arithmetic mean is biased toward high values which characterise these facies.

The expected porosities and permeabilities of the two intervals can be compared with their actual values in order to determine if the high permeabilities are associated with unusually high porosities. This has been done by subtracting the expected porosity or permeability, calculated on the basis of the regional trends (using a geometric average permeability trend, and the Ness cross-bedded facies porosity/depth trend) from their actual values. The results of these calculations show that the high permeabilities are not associated with higher than expected porosities. The cause will be discussed further in the section on diagenesis.

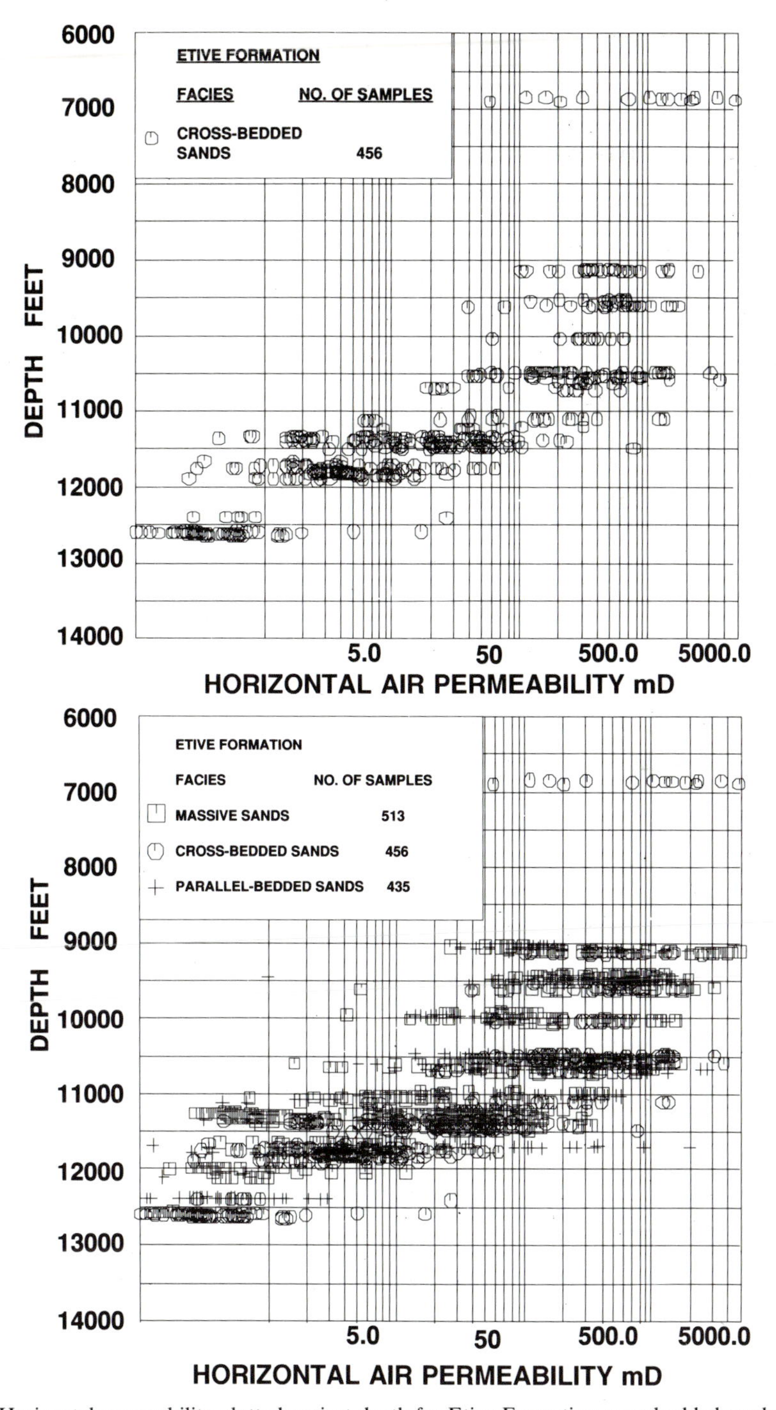


**Fig. 8.** (**a**) Horizontal permeability plotted against depth for Etive Formation cross-bedded sands. The figure shows that at depths in excess of 10 000 ft (3048.04 m) permeability becomes a function of burial depth. (**b**) Horizontal permeability plotted against depth for Etive Formation cross-bedded, massive and parallel bedded sands. The dependence of permeability on burial depth below 10 000 ft (3048.04 m) shown by the cross-bedded sands of the Etive Formation is also displayed by other sand facies of the Etive Formation.

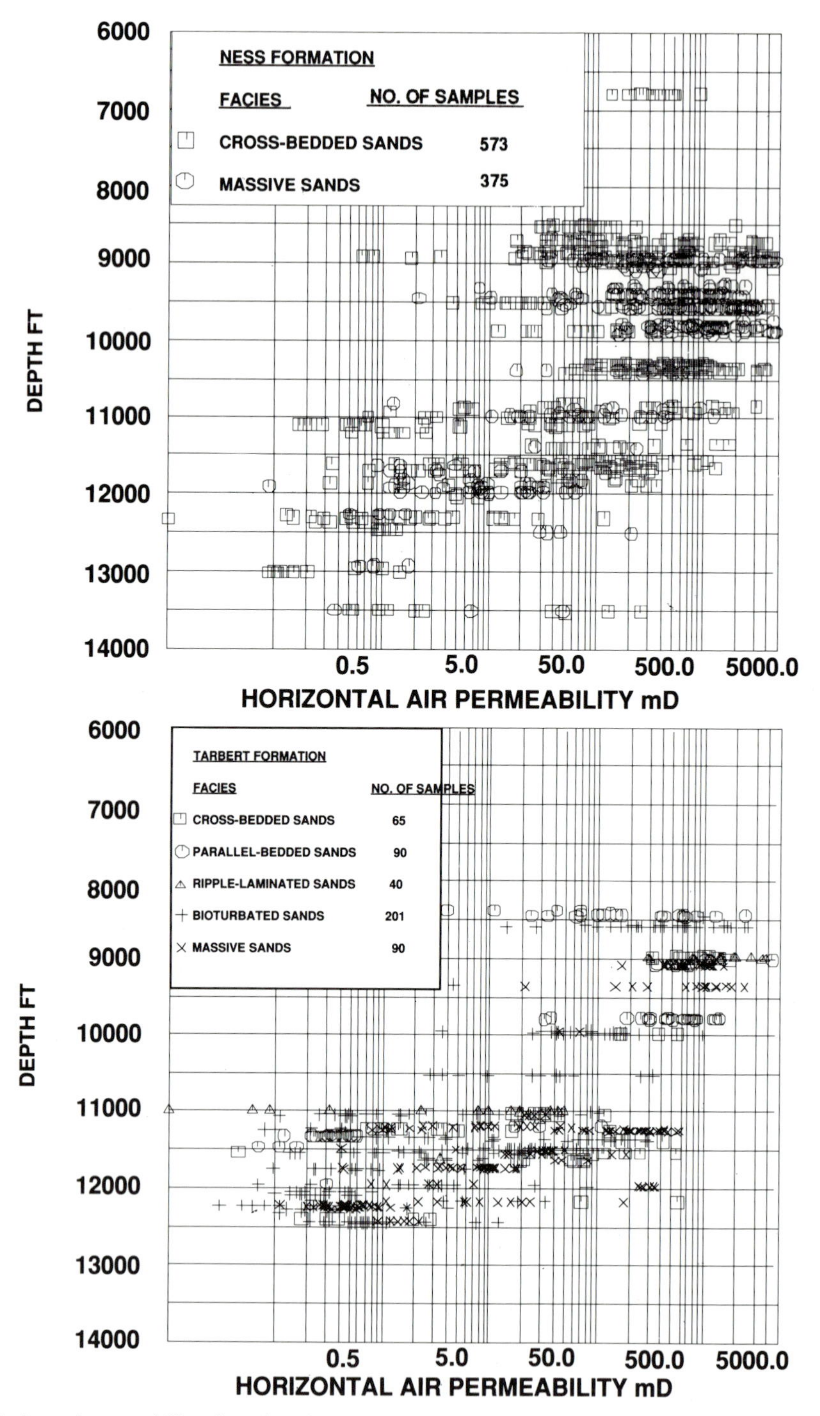


**Fig. 9.** Horizontal permeability plotted against depth for Ness Formation reservoir facies (**a**) and Tarbert Formation reservoir facies (**b**). Note the change in permeability behaviour below 10 000 ft (3048.04 m) as already shown for the Etive Formation in Fig. 8a & b.

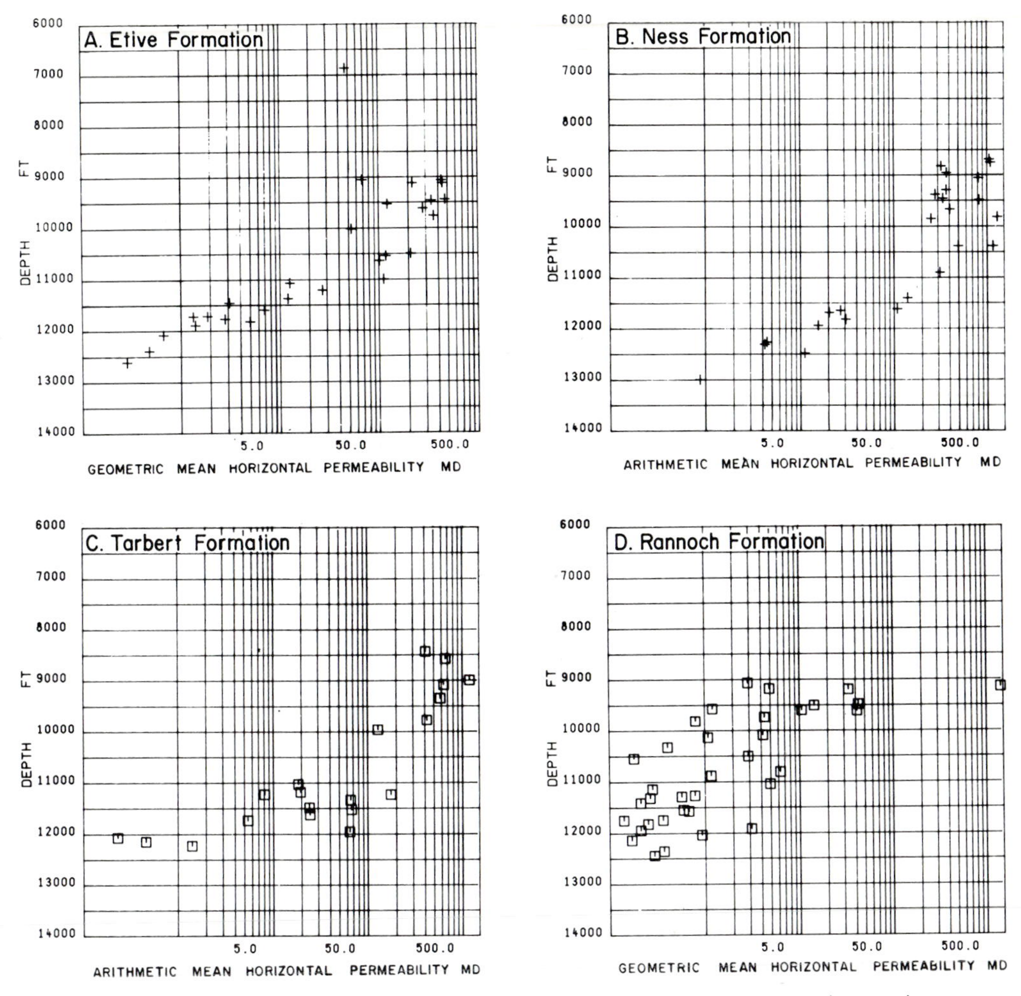


**Fig. 10.** Average horizontal permeability plotted against depth for the Etive Formation (geometric means plotted) (**a**). Ness Formation (arithmetic means plotted) (**b**), Tarbert Formation (arithmetic means plotted) (**c**) and the Rannoch Formation (geometric means plotted) (**d**).

*Effects of hydrocarbon emplacement on permeability*. Hydrocarbon emplacement has a significant impact on the permeability in deeply buried Brent Group reservoirs as shown by a comparison of the difference between the actual and expected permeabilities for hydrocarbon and water bearing sequences of Etive reservoirs (Fig. 12a & b). Water bearing samples from depths in excess of 10 200 ft (3109.0 m) are shifted to the lower than expected permeabilities (Fig. 12b) relative to the hydrocarbon bearing samples. However, at depths shallower than 10 200 ft (3109.0 m) (Fig. 12a) there is no significant difference between the hydrocarbon and water bearing samples. Similar effects can also be observed for Ness samples. As will be shown in the section on diagenesis, the difference in permeabilities between the hydrocarbon and water bearing samples is the result of preferential growth of illite in the water leg. Previous authors (e.g. Hancock & Taylor 1978; Thomas 1986; Kantorowicz 1990) have also noted differences in the degree of illitization between the oil and water legs of Brent Group wells.

*Formation pressures*. Formation pressure data show that the pressures are generally compartmentalized such that the lowest overpressures are found in fields on the western fringe of the

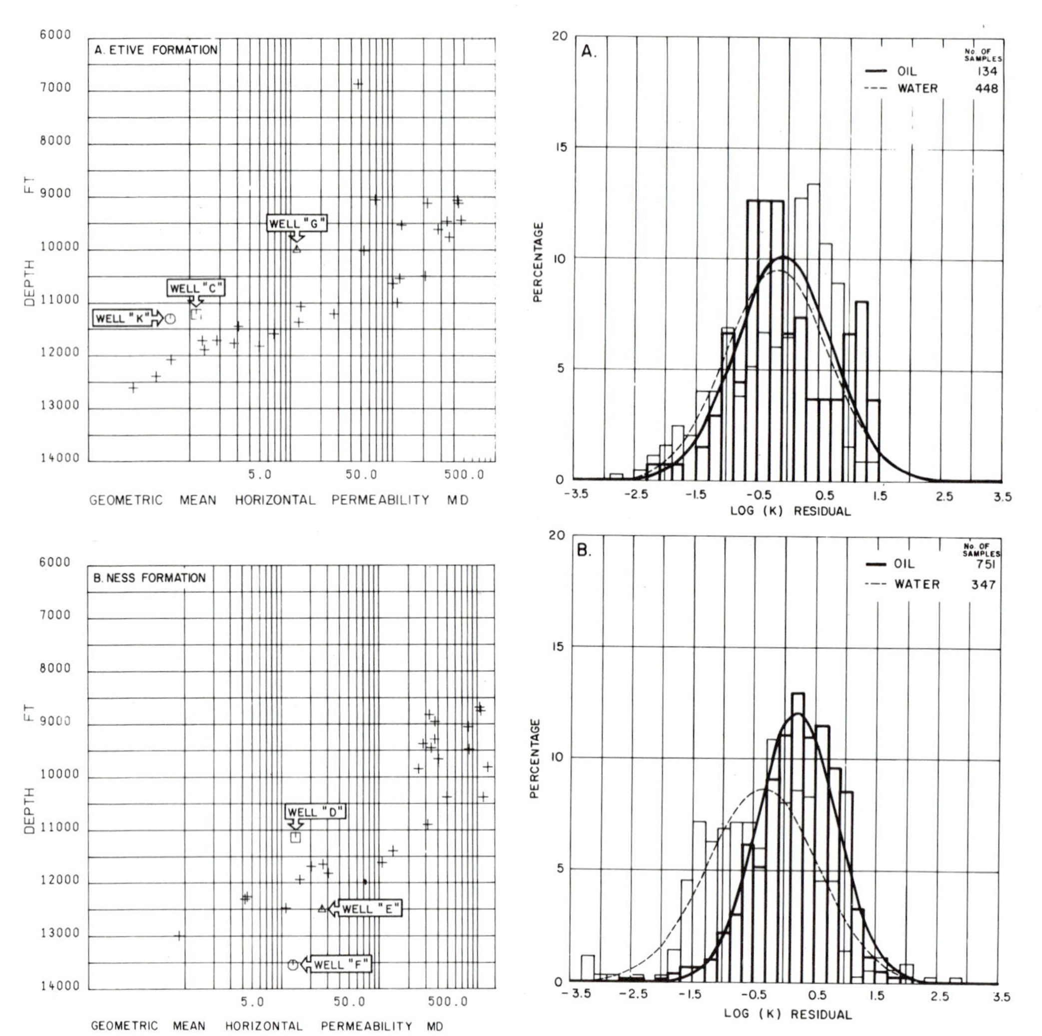


**Fig. 11.** Intervals of unusually low or high permeability within the Etive Formation (**a**) and the Ness Formation (**b**).

**Fig. 12.** Histogram of horizontal permeability residuals (difference between actual permeability of core plugs and calculated permeability) for oil and water bearing Etive Formation samples from shallower than 10 200 ft (3109.0 m) (**a**) and deeper than 10 200 ft (3109.0 m) (**b**).
Data from less than 10 200 ft (3109.0 m) (**a**) shows little difference between the oil and water zone samples, whereas, at depths greater than 10 200 ft (3048.04 m) (**b**) the distribution of data from water zone samples is strongly skewed toward lower than expected permeabilities.

East Shetland Basin, whereas the highest overpressures are found in the Viking Graben. Thus fields such as Heather are near hydrostatically pressured, while reservoirs near to the graben centre are overpressured by over 3000 psi in blocks such as 3/14. Within each pressure cell there is a trend of increasing pressure with burial depth that approximates to the hydrostatic gradient. The wells looked at in this study come from a range of overpressures from near hydrostatic to nearly 3000 PSI. The scatter in the reservoir property plots (Figs 4, 5, 8–10) and the diagenetic trends which will be presented later cannot be correlated with overpressures.

## Diagenesis of Brent reservoirs

### *Significant diagenetic phases*

Major diagenetic events which have had a serious impact on reservoir properties include cementation by carbonate cements (siderite, non ferroan calcite, ferroan calcite, ferroan dolomite and ankerite), kaolinite, illite, authigenic quartz and secondary porosity creation dominantly from feldspars but with a minor contribution from the dissolution of carbonate cements. Minor cementing phases include:- anatase, pyrite, chlorite and albitization. The overall sequence of cements is the same irrespective of formation or area (Fig. 30), although some of the phases are more abundant in one formation than another.

## *Authigenic quartz*

Syntaxial quartz overgrowths are the dominant pore filling cement in most sandstones and abundances sometimes exceed 28% of the rock volume. Highest abundances of authigenic quartz are to be found in the Ness Formation, followed by the Tarbert and finally the Etive and Rannoch Formations (Fig. 14a–c). Variations in quartz cementation between different sandstone facies is minimal for any one formation, with the exception of the generally lower quantities present in the heterolithic sands. For instance in the Ness Formation, the cross-bedded and massive sands have only a slightly wider distribution of authigenic quartz than shown by the other facies (Fig. 13). This observation appears to be quite general, and the facies relationships of other diagenetic cements between sands of any one formation are subdued.

The abundance of authigenic quartz is strongly related to the current depth of burial (Fig. 14a–d). Above 9000 ft (2743.23 m) only minor authigenic quartz occurs, while below this depth there is a distinct trend of increasing quantities of authigenic quartz cement with increasing burial depth (Fig. 14a–d). Thus, the primary control on the distribution of authigenic quartz appears to be burial depth, and by implication temperature. The form of the authigenic quartz/depth trend clearly differs between formations (Fig. 14a–d), with the Ness Formation showing the strongest trend and the Etive Formation the weakest. The Ness trend (Fig. 14c), if extrapolated back to the zero on the authigenic quartz axis would give a depth of about 7600 ft (2316.51 m). Taking this as the starting depth for the burially related authigenic quartz cementation (assuming a geothermal gradient of 30°C/km, known to be reasonable for the province, and allowing for the water depth) a temperature of about 70°C is arrived at for the onset of quartz cementation. A significant amount of cementation (>5%) occurs from 9000 ft (2743.23 m) onwards, equivalent to a temperature in excess of 84°C. However, it is probable that not all the authigenic quartz has a burial origin and that some of it formed as a result of the early dissolution of feldspars (see below).

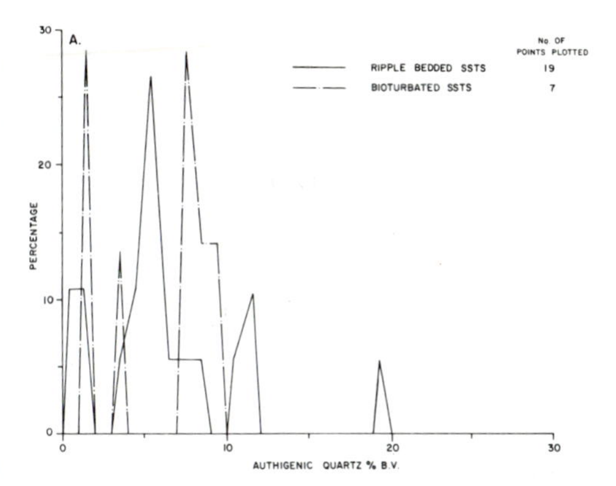


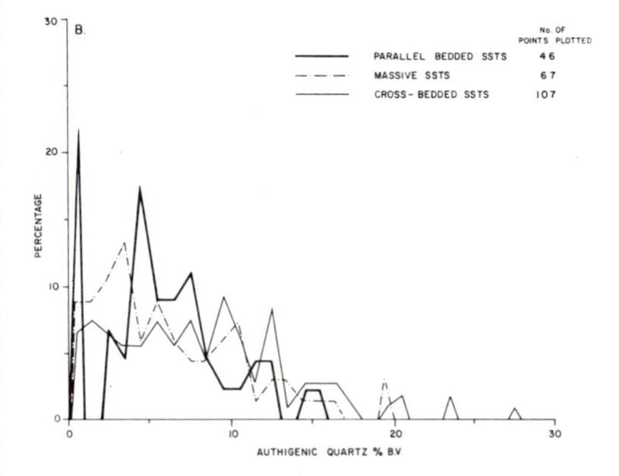


**Fig. 13.** Relationship between authigenic quartz abundance (in % bulk volume) and reservoir facies of the Ness Formation. *y*-axis is percentage of samples.

Cathodoluminescence investigations show that grain to grain pressure solution within the sandstones of all formations is minimal. However, experience from other areas, notably the Permo-Carboniferous sands of the Middle East, leads the authors to suspect that widespread pressure solution may be occurring in the siltstones and shales, or along muddy laminae within the sandstones.

Possible differences in the abundance of authigenic quartz above and below the oil/water contact are more difficult to assess. By far the majority of samples from the Tarbert or the Ness Formations come from oil-bearing intervals, and comparing these to samples from water bearing intervals, there appears to be no significant difference in the degree of authigenic quartz cementation between the two zones. In the case of the Etive Formation there is both a large number of samples and a reasonable

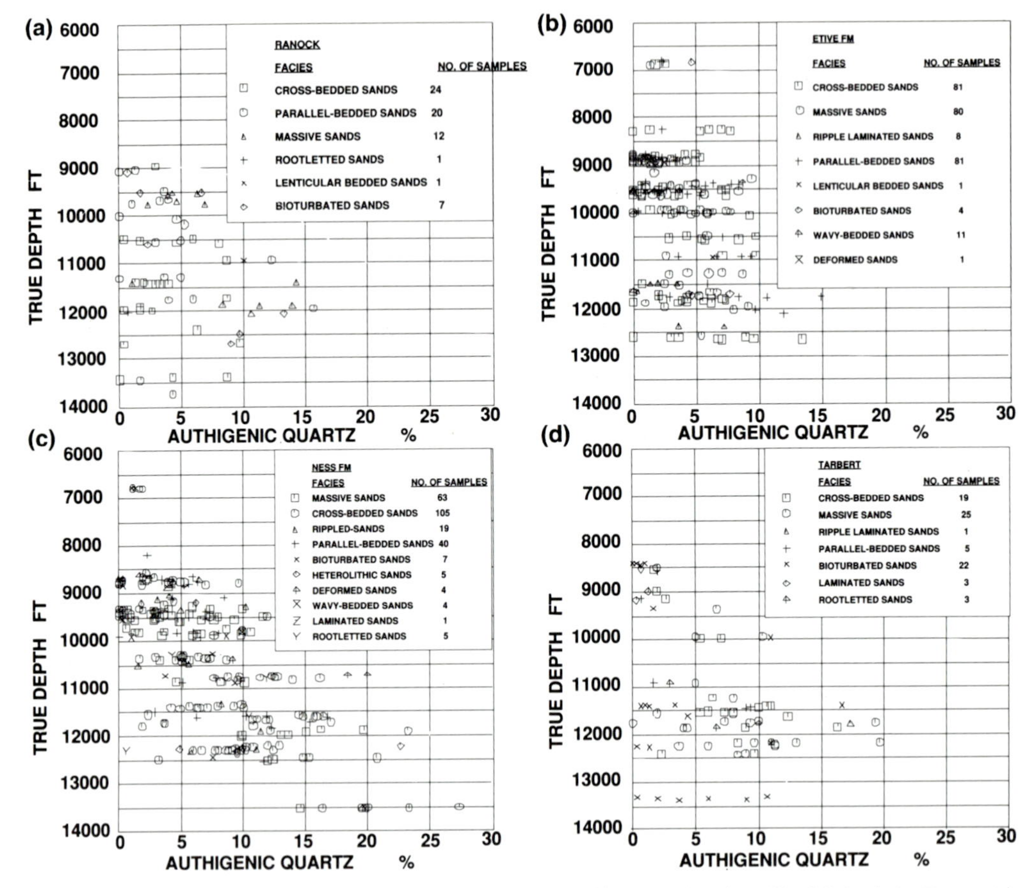


**Fig. 14.** Relationship between authigenic quartz abundance (in % bulk volume) and burial depth for Rannoch, Etive, Ness and Tarbert Formations.

distribution of samples between the oil and water zones, making this data set a better test of any possible differences. The mean amount of authigenic quartz in the oil bearing wells is 3.4% (standard deviation 2.9%), whereas the water bearing wells show a mean of 3.7% (standard deviation 3.1%). Clearly, therefore, there are no dramatic differences in authigenic quartz abundances between the two.

### *Kaolinite*

Kaolinite occurs in two forms, an early vermicular form and a later blocky form. The abundance of kaolinite can exceed 20% in the Etive Formation but is generally less than 15% in all the other formations. Within the Ness Formation there is a tendency for the highest kaolinite abundances to be found within the massive and cross-bedded sands; however, this facies dependence is absent from other formations. The data from all formations show no clear increase in kaolinite content in the water legs of wells as compared to oil legs. A reduction in the abundance of kaolinite with increasing depth is visible in the data (see Fig. 15) and this is believed to be due to the growing stability field of illite as the temperature increases (Sass *et al.* 1987). The point-count results do not distinguish between the two types of kaolinite present.

The exact position of the blocky kaolinite in the diagenetic sequence is difficult to assess. Based on petrographic relationships the vermicular kaolinite appears to be older than the blocky kaolinite, but how much older is a much more difficult question. Glasmann *et al.* (1989*a*), studied the oxygen isotopic composition of this type of kaolinite in the Heather Field and concluded that it formed at temperatures of 45–60°C from a meteoric pore water.

Blocky kaolinite rarely shows signs of alter-

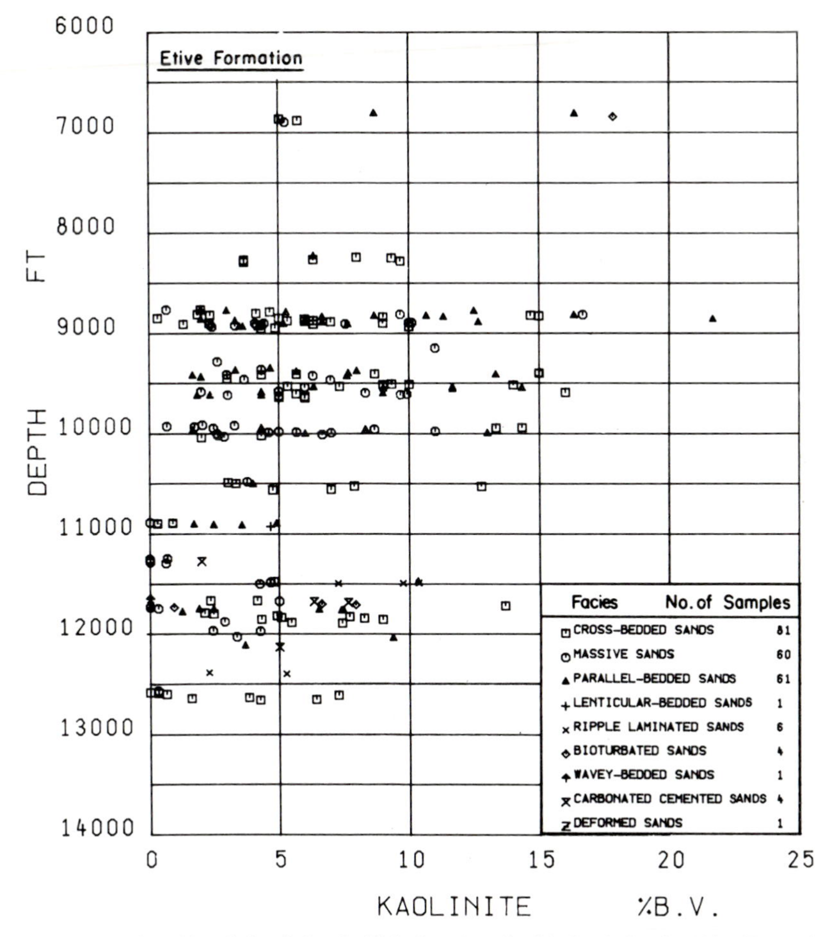


**Fig. 15.** Distribution of authigenic kaolinite (in % bulk volume) with depth for the Etive Formation.

ation to illite, whereas vermicular kaolinite does. This difference is probably related to the higher specific surface of vermicular kaolinite as a result of its expanded nature, which makes it more reactive.

### *Illite*

Fibrous illite is especially common in the deeper Brent reservoirs and is known to have a disastrous effect on permeability as seen in studies of various reservoirs (Stalder 1973; Pallatt *et al.* 1984; Cocker 1986; de Waal *et al.* 1988; Kantorowicz 1990). Illite occurs with a variety of growth habits including grain coating, platy, honeycomb, acicular (needle like) and fibrous. Platy and honeycomb illites are often associated with illitized clay flakes. The distribution of fibrous illite could be more widespread than is evident from this work as none of the samples studied were critically point-dried. Air drying has been shown to disrupt the delicate in situ texture of diagenetic illite (McHardy *et al.* 1982), leaving behind a fabric which is unrepresentative of the original rock, and consequently will have significant effects on core analysis (Pallatt *et al.* 1984; Cocker 1986; de Waal *et al.* 1989; Kantorowicz 1990). Air permeabilities measured on air dried, illite bearing samples of Brent sandstones, are likely to be on average 3.5 times higher (Kantorowicz 1990) than if the same sample had been critically point-dried. It is therefore likely that permeabilities in Etive, Ness and Tarbert sands are likely to be even lower than the trends shown in Figs 8–10 would suggest.

The main evidence on the distribution of illite comes from point-count results. However, what has been point-counted is the total abundance of illite, and not its growth form, which is impossible to identify from thin sections. As Kantorowicz (pers. comm.) has demonstrated, the abundance of fibrous illite in two Cormorant Field wells (wells 'G' & 'H') is less than 1% but its presence significantly reduces the gas permeabilities.

The point-count results suggest that the ranges of illite abundance in the Brent Group formations fall in the following order: Tarbert > Rannoch ≈ Etive > Ness. Illite does not exceed 10% B.V., and with the exception of the Rannoch Formation the modal abundance lies in the range 0–0.5%. In the case of the Rannoch, however, the modal value is in the range 1–1.5% B.V. Within the Etive Formation, the cross-bedded and massive sands have the widest range of illite abundances (Fig. 16) and this pattern of abundances also holds for the Ness and Rannoch Formations. In the Tarbert, however, the bioturbated sands have the widest range of illite contents, which may reflect a high detrital component of mixed layer clays.

The abundance of illite shows a strong relationship with depth (Fig. 17). Illite abundances can exceed 2% in all formations at depths in excess of 10 000 ft (3048.04 m) and show a wider range of abundances at greater depths. The Etive Formation (Fig. 17) appears to show a wider range of illite abundance at a shallower depth than the Ness Formation. The onset of increasing illite abundance (Fig. 17) occurs at approximately the same depth as the change in the permeability/depth behaviour of the main reservoir sands (Figs 8–10).

Illitization of vermicular kaolinite is frequently observed using SEM techniques in the more deeply buried samples. Illitization of the later blocky kaolinite is more rarely observed, although this may be due to differences in reactivity between the two kaolinite growth habits. The relative proportion of illite to kaolinite increases dramatically below 10 500 ft (3200.44 m) in the Ness Formation (Fig. 18). Similar behaviour can also be demonstrated for the Etive and Tarbert Formations.

X-ray diffraction data from the less than 2 μm fraction also lend support to the observation of increasing illite abundance with depth (Fig. 19a). This indicates that illite is becoming stable at depths below 10 500 ft (3200.44 m). In addition, illite/smectite mixed layer clays show a progressive increase in the proportion of illite with depth (Fig. 19b). These mixed layer clays are present as detrital grains, early diagenetic (?) coatings to detrital grains or from detrital clays which infiltrated or were bioturbated into the sediment.

The degree to which kaolinite is being transformed into illite is not clear. SEM studies do show that vermicular kaolinite or finely crystalline kaolinite can become unstable and consequently illitized during burial (Kantorowicz 1984*a*). The main nuclei for fibrous illite growth

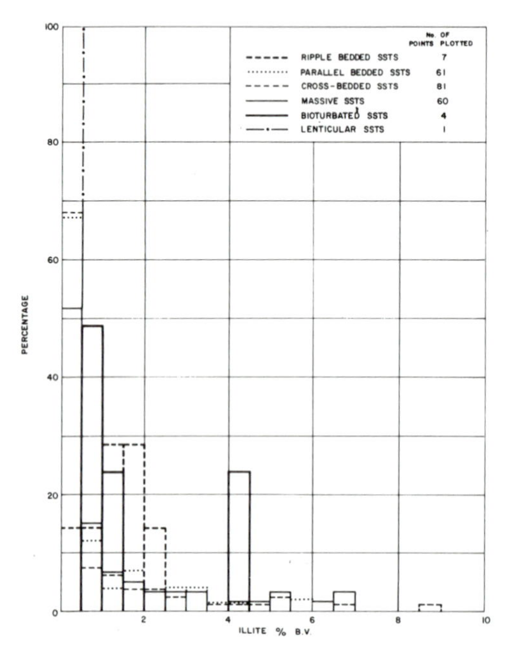


**Fig. 16.** Variation of authigenic illite abundance (in % bulk volume) in relation to facies differences within the sands of the Etive Formation. *y*-axis is percentage of samples.

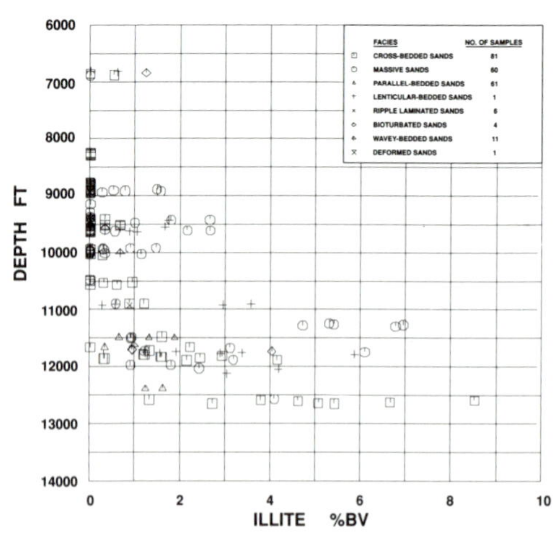


**Fig. 17.** Distribution of illite (in % bulk volume) with burial depth in sands of the Etive Formation. Note the increase in illite abundance at depths in excess of 10 000 ft (3048.04 m).

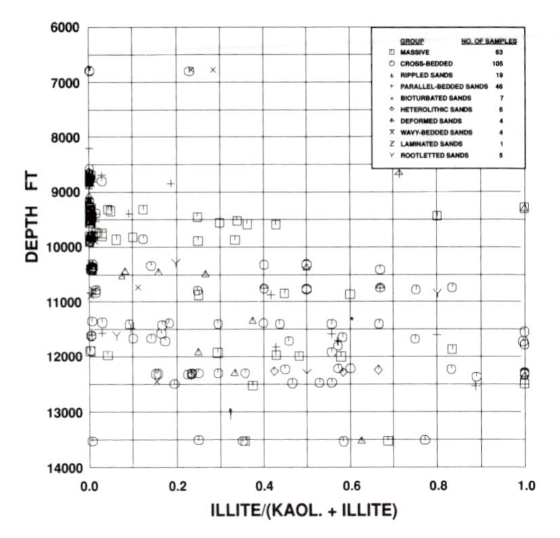


**Fig. 18.** Relative proportions of illite and kaolinite versus depth for Ness Formation sands. At depths down to 10 000 ft (3048.04 m) most samples contain no illite, however, at greater depths illite forms an increasing percentage of the clay mineral assemblage. Data from point counting of thin sections.

appear to have been the already existing illitic grain coatings. Thus the main phase of illite growth was not as a replacement of kaolinite, but from ions supplied by the pore fluid, possibly from the dissolution of kaolinite, and which had then been transported to sites of existing illitic grain coatings.

SEM observations suggest that illite growth habit changes with current burial depth. At about 9000 ft (2743.23 m), original detrital illitic clay coats start to recrystallize. Below 9900 ft (3017.56 m), box-work and honeycomb textures become common. Illitic fibres up to 1–2 μm long may protrude from ragged illitic plates that form the detrital grain coating. This type of illite development can extend downwards from 9900 ft to 13 400 ft (3017.56–4084.37 m). Illite fibres of 5 μm or longer commonly occur at depths of 11 000 ft (3352.84 m) or greater. Some of the longest illite fibres (generally 10–15 μm, but may reach 20 μm) are found growing from detrital illitic clay coatings in the more argillaceous sandstones (e.g. bioturbated, deep Tarbert reservoirs).

In the Etive Formation in some wells, such as well 'G' (211/26-CA30), fibrous illite is found at anomalously shallow depths (≤ 10 000 ft, 3048.04 m). It is possible that in general fibrous illite starts to appear at a much shallower depth but has been destroyed as a result of air drying of the samples. As was outlined in the reservoir properties section, anomalies of this type are most frequent in the Etive. The authigenic illites of well 'G' have been studied in detail by Kantorowicz (1990), who has employed critical point drying techniques and has recognized a

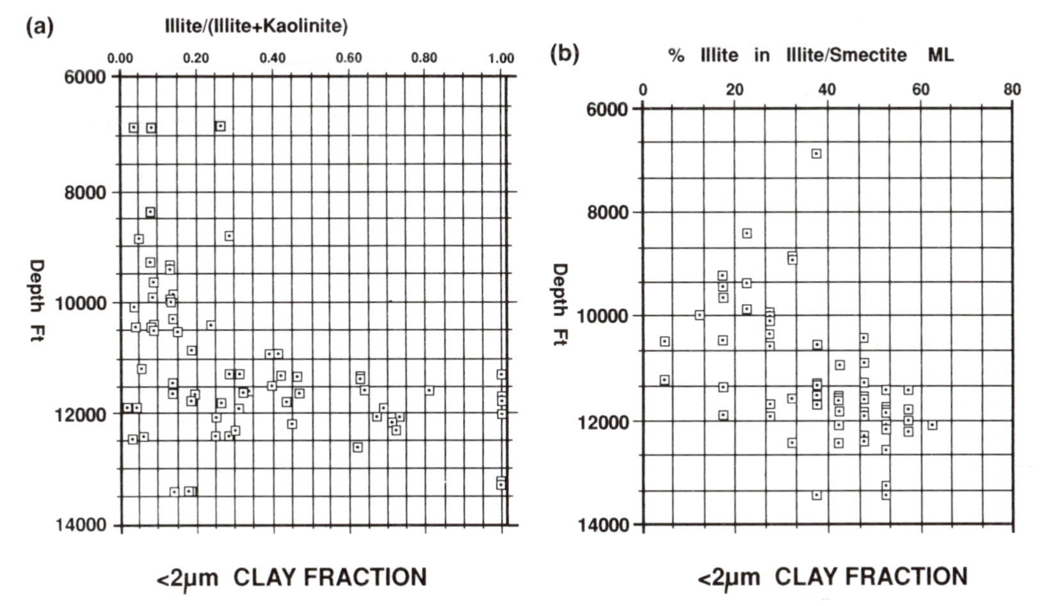


**Fig. 19.** (**a**) Relative proportions of illite and kaolinite versus depth as derived from XRD data. Note the increase in the proportion of illite at depths in excess of 10 000 ft (3048.04 m). (**b**) Relative proportions of illite in illite/smectite mixed layer clays as derived from XRD data.

number of growth habits including pore lining illite which coats detrital grains, blooming illite, pore bridging illite and illite that replaces kaolinite. Kantorowicz interprets the pore lining illites to have formed prior to hydrocarbon charge, whereas the pore filling illites formed during or after hydrocarbon charge and are therefore restricted to the deeper parts of the oil column and the water-bearing sandstones below. Pore lining illites frequently form nuclei from which the pore bridging illites grow. The pore lining illites are probably the result of the progressive transformation of detrital illite/smectite clay coatings. The origin of these detrital illitic clay coatings is uncertain, however, it is posible that some at least are the result of infiltration of clays from the overlying Ness Formation.

Etive intervals at or below 10 000 ft (3048.04 m) show a greater range of illite abundances in the water legs than in the oil legs (Fig. 20); other formations do not appear to show the same effect. The Etive Formation is, however, the only formation from which there is a good spread of samples with approximately equal numbers available from the oil and water legs. If the Etive Formation is more affected than the other formations it is difficult to understand why. One possible explanation is that the high degree of sand continuity in the Etive Formation makes it easier to introduce (via the pore water) the potassium needed for the growth of illite. This is more difficult in the Ness Formation with its discontinuous channel sand bodies. However, should the K ions necessary come from dissolution of feldspars within the Etive Formation itself, then the previous explanation would be invalid.

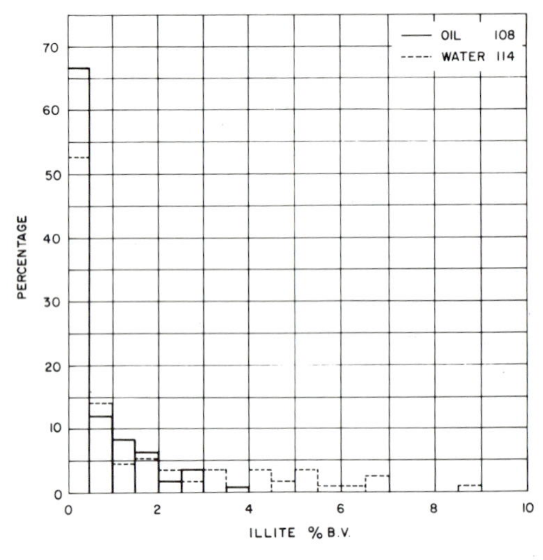


**Fig. 20.** Differences in illite abundance (in % bulk volume) related to oil and water legs within the Etive Formation. The figure shows a wider distribution of illite abundances in water zone samples. *y*-axis is percentage of samples.

K/Ar datings of diagenetic illites have been used by various authors (e.g. Lee *et al.* 1985; Hamilton *et al.* 1987, 1989, this volume; Jourdan *et al.* 1987; Glasmann *et al.* 1989*a*, *b*; Scotchman *et al.* 1989) to attempt to date the emplacement of hydrocarbons, based on the assumption that the age of the finest diagenetic illite fraction approximates to the timing of oil emplacement. The age of the coarsest diagenetic illite fraction is assumed to give an age which will tend toward the timing of the onset of illite formation. Unfortunately this latter age is often difficult to obtain because of contamination of samples with mica, detrital illite or K-feldspar. In this study authigenic illites from 11 wells from all over the UK portion of the Brent Delta have been dated and the results are discussed in detail in Hamilton *et al.* (this volume). However, the main results relevant to the present paper are discussed below.

From the point of view of obtaining an estimate of the lowest temperature associated with the formation of authigenic illite the data from the Tern field well 'I' and Cormorant Field well 'G' (211/26-CA30) are the most interesting as illite occurs in both at depths less than 10 000 ft (3048.04 m). SEM work on well 'I' demonstrates that fibrous illite is the latest diagenetic precipitate. Taking the almost certainly incorrect view that this phase has only recently started to grow in the Etive Formation at a present burial depth of 8350 ft (2545.11 m), which is almost certainly its maximum burial depth, then it must have formed at a temperature close to the present-day in-situ temperature of about 85°C. K/Ar dates from diagenetic illites of this well are unreliable due to high levels of contamination (Hamilton *et al.* this volume). An even lower temperature for illite growth can be arrived at using the illite dates from the Etive of well 'G' (211/26-CA30). It is assumed that these give the time at which oil migrated through the system, halting illitization. The most reliable K/Ar dates from diagenetic illites from this well give ages of 41–36 Ma (Hamilton *et al.* this volume). The geohistory plot for this well (Fig. 21) indicates temperatures of about 75°C at 41–36 Ma. This temperature is the minimum temperature so far recorded for the onset of illite generation within the Brent Province. No other well as shallow as this has been identified with significant illite growth.

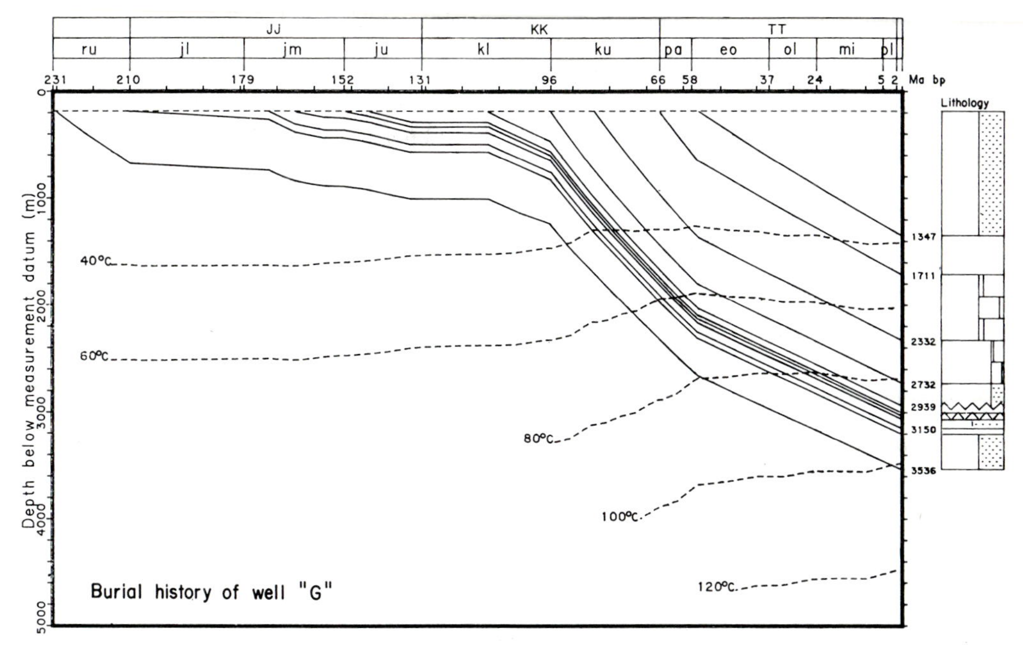


**Fig. 21.** Burial history of well G. Solid lines show the bases of the modelled intervals and dashed lines are palaeo-isotherms. Time scale is used is taken from Harland *et al.* (1982). Abbreviations used: JJ, Jurassic; KK, Cretaceous; TT, Tertiary; ru, Upper Triassic; jl, Lower Jurassic; jm, Middle Jurassic; Kl, Lower Cretaceous; ku, Upper Cretaceous; pa, Palaeocene; eo, Eocene; ol, Oligocene; mi, Miocene; pl, Pliocene.

Depth plots (Figs 17, 18 & 19) indicate that illite is not common until the depth exceeds 10 000 ft (3048.23 m) which is equivalent to a temperature of about 100°C. Illite dominates the clay mineral assemblage of the <2 $\mu$m clay fraction at burial depths greater than 12 000 ft (3657.64 m) (Fig. 19a) and therefore temperatures in excess of 115°C. By about 13 000 ft (3962.45 m) (equivalent to a temperature of about 125°C) K-feldspars, the source of the K for illite growth, have generally been consumed (see Fig. 27).

In conclusion, illite growth begins under favourable conditions at temperatures as low as 75°C, becomes significant by 100°C, and may ultimately be limited by the availability of K from K-feldspars by temperatures of about 125°C. These temperatures are lower than have been observed elsewhere in the North Sea, for instance by Ehrenberg & Nadeau (1989). However, a significant feature here is that the transformation process is a time/temperature dependent process (see Siever 1983; Giles & de Boer 1990), although the exact kinetics are not known in detail. Thus the difference between the two areas may lie in their burial histories.

K/Ar dates from diagenetic illites within the Brent Group have not yielded recent ages, which is surprising as there is no apparent reason why illite growth in the Brent province should have ceased. The youngest date recorded as part of this study was 17 Ma from an Etive sample (Hamilton *et al.* 1991), from a well in block 211/11 (well 'K'). Given that there has been no subsequent uplift of the East Shetland Basin, and overpressures were probably established sometime in the early Tertiary and hence the hydrodynamic regime has remained constant, there is no reason why illite should not be forming today. An explanation of why K/Ar dates from diagenetic illites cannot yield dates of zero may be found in Hamilton *et al.* (1991). In the more deeply buried wells there appears to be a relationship between the reliable illite dates and the estimated onset of hydrocarbon migration in the adjacent source rocks (see Hamilton *et al.* this volume) which is calculated to occur at temperatures of 105–120°C from geohistory analysis.

## *Chlorite*

The presence of minor amounts of chlorite was detected both by XRD and SEM. Chlorite may be seen as small plates coating grains, but rarely as pore-filling clay. In contrast to Kantorowicz

(1984*b*) who claims that chlorite is restricted to the Broom Formation, SEM, XRD and thin section studies show that it can also be found in the Tarbert, Ness, Etive and Rannoch Formations. Its widespread distribution suggests that it is of burial diagenetic origin. It is never in any great abundance and has negligible effects on reservoir properties.

### *Carbonate cements*

Authigenic carbonates in the Brent Group sandstones occur as local concretions and as more dispersed patches of cement (Thomas 1986; Lønøy *et al.* 1986; Hamilton *et al.* 1987; Brint *et al.* 1988; Glasmann *et al.* 1989*a*, *b*; Samways & Marshall, in prep).

The main carbonate cements recognized in this study are: non-ferroan calcite, ferroan calcite, siderite, ferroan dolomite and ankerite. The variety and polygenetic origin of many the carbonate cements within the Brent Group causes difficulties in unravelling the depth relationships of the carbonate cement types from petrographic data alone. Consequently, the discussion presented below will concentrate on using the C and O isotope measurements to constrain the origins of Brent Group carbonate cements.

Within the Tarbert, Ness or Etive Formations there is little difference in carbonate cementation between facies. Both the Etive and Ness Formations contain very little carbonate cement, with over 80% of all samples containing less than 1% B.V. Thus in both of these formations carbonate cementation has a generally negligible impact on reservoir properties. The Tarbert Formation contains a generally higher quantity of carbonate cement, but over 50% of all samples still contain less than 1% B.V. Petrographic data from both the Etive and Rannoch Formations appears to show slightly enhanced levels of carbonate cementation in water-bearing sandstones when compared to their oil-bearing equivalents.

Early diagenetic calcite concretions are particularly common in the shoreface sandstones of the Rannoch Formation where they may coalesce to form effective permeability barriers. Such permeability barriers can give rise to significant problems during the production history of a field. Overall the Rannoch Formation contains a higher percentage of samples which have significant amounts (>5%) of carbonate cement than any other formation. Consequently carbonate cementation is a major factor reducing reservoir properties in the Rannoch Formation. Siderite concretions also occur, and are common in the Tarbert and Broom sandstones.

Early diagenetic calcites are generally non-ferroan although the early concretionary calcites from the Rannoch are weakly ferroan. Non-ferroan calcite occurs in all the 'marine' influenced formations (Broom, Rannoch, Etive, Tarbert) but not in the Ness Formation. Early calcites occur as a blocky to poikilotopic cement which appears highly replacive with respect to the detrital grains and which can occupy as much as 30% of the rock. Contacts between detrital grains are rare and it is apparent that most of this calcite formed before significant burial compaction. Where the early calcite is non-ferroan it is commonly surrounded and therefore succeeded by a ferroan calcite cement zone. Elsewhere the ferroan calcite can be shown to fill dissolution features in the non-ferroan cement: it also fills secondary pores and acts as an independent pore-filling cement. With the exception of the weakly ferroan calcite concretions in the Rannoch Formation and the ferroan zone on the non-ferroan calcite crystals, which are evidently early, most of the ferroan calcite is interpreted to have formed during later sediment burial as it often encloses well-developed quartz overgrowths and blocky kaolinite. Some of the ferroan calcite also evidently post-dates extensive sediment compaction which suggests a significant time lapse occurred between the two phases of calcite cementation.

Siderite occurs mainly as an early (syn-depositional), environmentally-related cement in the fluvial sediments of the Ness Formation and in micaceous Rannoch sandstones. It occurs as small nodules or as discrete clusters of rhomb-shaped crystals which are frequently associated with altered biotite grains, secondary pores resulting from the dissolution of feldspars and with mudstone clasts. The biotite alteration probably supplied the iron for the siderite and may have provided a chemical microenvironment conducive to carbonate precipitation (Bjørlykke & Brendsal 1986). Siderite also occurs as a later diagenetic cement. These later siderites are much rarer than the early precipitates, and they are characterized largely by having a much more negative oxygen isotopic composition than the early siderites. Their occurrence demonstrates later renewal of the source of iron and magnesium.

Ferroan dolomite, non-ferroan dolomite and ankerite occur as rhombic crystals or crystal mosaics (Kantorowicz 1984*b*). Early diagenetic precipitates occur as displacive fringe cements on micas. Later diagenetic cements occur as relatively coarsely crystalline rhomb pore-filling cements. Occurrence of ankerite is reputed to be enhanced in the water-leg of some wells (Kantorowicz 1984*b*) although the present authors have no data to support this claim.

*Concretionary carbonates and carbonate cemented horizons: isotopic data.* Stable isotopic data from fully cemented sandstones forming concretions and discrete concretionary horizons has enabled us to determine changes in pore fluid composition and temperature during burial. Only the general trends from the concretions will be discussed here as a more detailed discussion is in preparation (Samways & Marshall, in prep).

Samways & Marshall (in prep.) have examined a large number of carbonate concretions and carbonate-cemented horizons from the Brent province. Petrographic criteria and stable isotopic compositions have been used to characterize carbonate cements and enable the sources of carbonate and conditions of cementation to be identified. The degree of compaction prior to cementation (minus-cement porosity) and post-compaction deformation of laminae around concretion margins have been used to establish the relative timing of carbonate cements. Trends in isotopic composition within concretionary and cemented horizons have been used to monitor changes in the isotopic composition of the pore fluids during carbonate precipitation. The general pattern of isotopic results is common across the province and will be discussed with reference to the examples from the Dunlin and Cormorant Fields (Figs 22 and 23). Early concretionary carbonates,

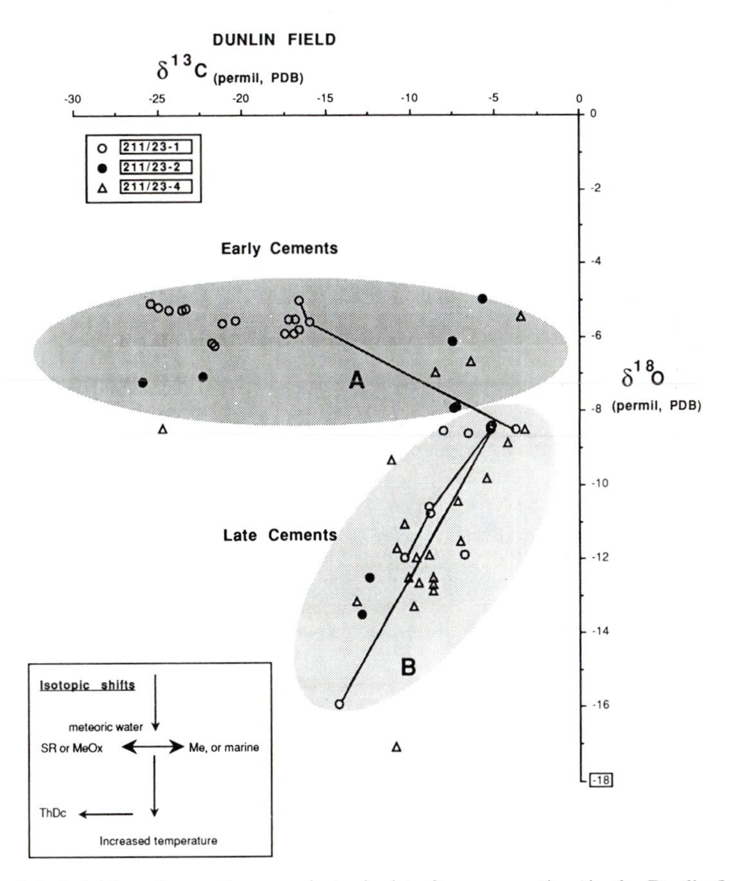


**Fig. 22.** Cross-plot of stable carbon and oxygen isotopic data from concretions in the Dunlin field. Early and late diagenetic carbonates plot in two discrete fields. Trend lines A and B represent trends in isotopic composition in samples from individual concretions.
Insert shows the isotopic shifts associated with different processes. Sr, sulphate reduction; MeOx, oxidation of methane; Me, $CO_2$ produced during methanogenesis; ThDc, $CO_2$ produced by the thermal decarboxylation of organic matter.

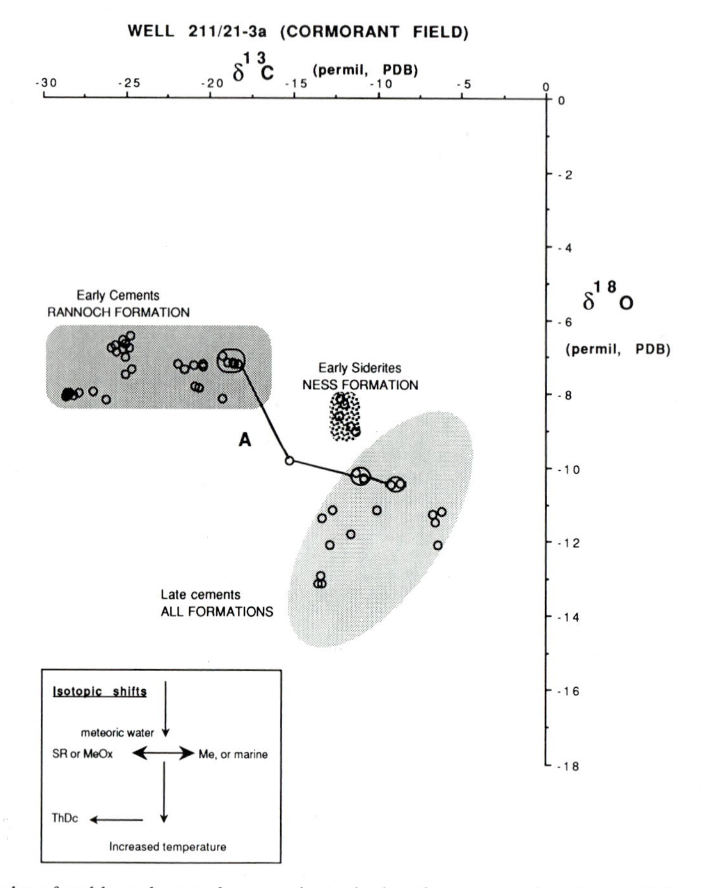


**Fig. 23.** Cross-plot of stable carbon and oxygen isotopic data from concretions in a well from the Cormorant field. Early and late diagenetic carbonates plot in three discrete fields defined by early cements from the Rannoch Formation, Early siderites from the Ness and late cements which are similar in all formations. Trend line A represents trends in isotopic composition in samples from an individual concretion.
Insert shows the isotopic shifts associated with different processes. Sr, sulphate reduction; MeOx, oxidation of methane; Me, $CO_2$ produced during methanogenesis; ThDc, $CO_2$ produced by the thermal decarboxylation of organic matter.

common in the Rannoch Formation, are dominated by microsparry to poikilotopic ferroan calcite. Negative carbon isotope values (Figs 22 & 23) indicate an organic source of carbon, possibly derived from sulphate reduction ($-25‰$ $\delta^{13}C_{PDB}$, Irwin *et al.* 1977) or oxidation of bacteriogenic methane. Oxygen isotope compositions of the concretions range from approximately $-5‰$ to $-8‰$ $\delta^{18}O_{PDB}$. The strong post-cementation compaction fabrics often observed around these concretions indicate near-surface cementation, within 100 m of the sediment water interface. Temperatures at such shallow depths must have been relatively low (<30°C) and precipitation must therefore have occurred from pore fluids with negative oxygen compositions. Such fluids are likely to have been brackish or meteoric in origin (local rainwater is likely to have had $\delta^{18}O_{SMOW}$ of $-5‰$ to $-7‰$, Marshall & Ashton 1980) so it is likely that this cementation took place in shallow aquifers in relatively fresh or perhaps brackish porewaters (see discussion below).

Sideritic concretions common in the Ness Formation and within micaceous sediments of the Rannoch Formation, have similar oxygen isotope compositions to the early calcite concretions (Fig. 22). However, siderite carbon isotope values are consistently more positive than the ferroan calcite dominated concretions (Fig. 23). This is thought to reflect mixing between isotopically negative bicarbonate sources and isotopically positive methanogenic bicarbonate (with $\delta^{13}C_{PDB}$ = +20‰, Irwin *et al.* 1977).

Early concretionary carbonate cements often form the nucleus for later overgrowths with quite different isotopic composition (Trend A, Figs 22 and 23). In the examples illustrated, the later cements have more positive carbon isotope values, possibly due to the inclusion of higher proportions of isotopically positive methanogenic bicarbonate or marine bicarbonate ($\delta^{13}C_{PDB}$ = 0 to +3) sourced from the dissolution of bioclasts. In addition the oxygen isotope composition is more negative, indicative of increasing temperature of precipitation. The late carbonate cement at the end of Trend A (Figs 22 and 23) falls within the isotope compositional field occupied by the majority of late diagenetic nodular and more dispersed cements in other formations (Figs 22 and 23).

The later carbonate concretions, composed of ferroan calcite, siderite, ferroan dolomite and ankerite have isotopic compositions falling within Field B of Figs 22 and 23. The ferroan calcite cemented horizons may be composed of more than one generation of carbonate cement, each with distinctive petrographical and isotope characteristics. Petrographically later phases consistently exhibit more negative carbon isotope compositions and more negative oxygen isotope compositions than the earlier precipitates (Trend B, Fig. 22). The more negative oxygen isotope values are interpreted as indicating an increase in the temperature of precipitation whilst the more negative carbon isotope values are thought to reflect the influx of isotopically negative carbon related to thermal decarboxylation of organic matter (with $\delta^{13}C_{PDB}$ = −25‰$_{PDB}$, Irwin *et al.* 1977).

The petrographical and isotopic trends recognized in these cemented horizons are generally not systematic and centrifugal as has been recognized by some authors (e.g. Coleman & Raiswell 1981; Irwin 1980) and therefore the cements are unlikely to provide a continuous record of the evolution of diagenetic pore fluids.

*Isotopic data from carbonate cements: regional pattern*. The isotopic data accumulated in the course of this study were determined predominantly from non-concretionary samples. Figure 24 shows the isotopic data and dominant carbonate mineralogy. The distribution of data points indicates a similar pattern to that determined from the concretions and that most of the cements occur within the 'late cement' field of Figs 22 and 23.

Early carbonate cements exhibit a very wide variation in carbon isotope composition. The highly positive carbon isotope composition of some early siderite and mixed ferroan calcite/siderite samples (with $\delta^{13}C_{PDB}$ up to +20‰$_{PDB}$) clearly confirms a dominance of methanogenic bicarbonate in these cements. Some non-ferroan calcites exhibit the most negative carbon isotope values in this sample set, possibly confirming the relationship between non-ferroan calcite and sulphate reduction where iron would preferentially be incorporated into pyrite. This is compatible with the observation that non-ferroan calcite is restricted to the 'marine' influenced formations (Broom, Rannoch, Etive, Tarbert) where sulphate reduction and pyrite formation would have been most likely.

The late carbonate cements, characteristically with $\delta^{18}O_{PDB}$ more negative than −8‰, include ferroan calcite, ankerite and siderite, and occur in all formations. The 'patchy' nature of cements sampled in this study precludes the identification of mineralogical and isotopic trends except in the Lyell Field where petrographical relationships between non-ferroan calcite and ferroan calcite indicate that ferroan calcite is the later phase (see above). The transition from non-ferroan to ferroan calcite is accompanied by more positive carbon isotope compositions (from −25 to −10‰ PDB) possibly reflecting a mixing of isotopically negative sulphate reduction and methanogenic bicarbonate. A parallel decrease in oxygen isotope composition from (−9 to −15‰) again probably reflects an increase in temperature of precipitation. The trend is similar to Trend A described above (Fig. 22). A similar mineralogical and isotopic trend has been recognized in cements from well 211/26–4 (Pelican Field); here, though, the ferroan calcites with relatively positive carbon values are succeeded by later ferroan calcites which have more negative carbon and oxygen values. This reversal is similar to that described from the concretion data (trend A to trend B in Figs 22 and 23).

*Isotopic composition of pore fluids responsible for precipitation of carbonate cements*. Where there are independent indicators of likely diagenetic temperatures it is possible to calculate the isotopic composition of diagenetic fluids in equilibrium with cements using standard fractionation equations (Friedman & O'Neil 1977). As has already been discussed, if the early concretionary cements (Fig. 22) precipitated at a maximum of 30°C the isotopic composition of calcite would indicate cementation in a fluid with a negative isotopic composition intermediate between meteoric water and marine water (Fig. 24). The non-ferroan calcites which exhibit a wide range of $\delta^{18}0$ are also likely to have formed during early burial from a similar pore fluid at temperatures ranging from 30°C to 60°C. The later ferroan calcites may also have precipitated from similar meteoric-derived pore fluids at increased temperatures (Fig. 24), and this has been suggested by Hamilton *et al.* (1987) and Glasmann *et al.* (1989*a*, *b*). However, the indication from the carbon isotopic data that some late carbonate cements contain decarboxylation carbon may indicate the input of a pore fluid, perhaps expelled from more deeply buried marine shales which were undergoing compaction and thermal decomposition of organic matter (e.g. Glasmann *et al.* 1989*b*). Such fluids are likely to have had more positive oxygen isotopic compositions. Some authors have assumed that such compactional fluids will have retained a composition identical to Jurassic marine water i.e. $-1.2‰_{SMOW}$ (e.g. Kantorowicz, 1984*b*). Alternatively, marine pore fluids are perhaps more likely to have evolved towards more positive values through water-rock interaction (Yeh & Savin 1977; Wilkinson *et al.* in prep.). Support for an evolution of fluid compositions towards more positive values comes from the few direct measurements of oxygen isotopic compositions reported for North sea oil well formation waters in Brent sandstones: such values range from +1 to $+3.5‰_{SMOW}$ (Saigal & Bjørlykke 1987 and references cited therein).

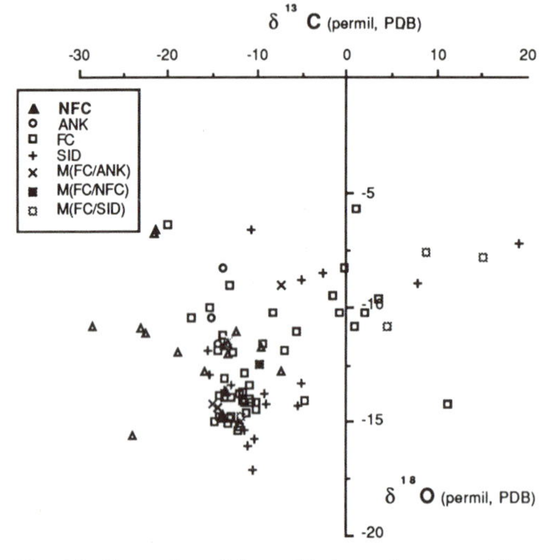


**Fig. 24.** Cross-plot of the stable isotopic composition of cements from the Brent province analyzed for this study. Mineral compositions are: NFC, non-ferroan calcite; ANK, ankerite; FC, ferroan calcite; SID, siderite; M, mixed mineralogy samples (analysed for total $CO_2$).

In well 'K' in the northern part of the Brent province late diagenetic cements in water-bearing sandstones have $\delta^{18}O$ values of around −15‰ (Fig. 25). The present formation temperature is approximately 95°C and geohistory analysis indicates that this is probably the hottest that the sediments have been. At current temperatures the cements would be in isotopic equilibrium with pore-waters with a $\delta^{18}O$ of approximately −2 (v. SMOW). Such water values are reasonable for fluids resulting from a mixture of meteoric-derived and marine-derived water. Without being able to determine the magnitude or timing of any shift in compositions due to water/rock interaction it is not possible to make any inference about the proportions of the two fluids. However, the derived water value is more negative than the published water analyses from deeper North Sea oil wells and accordingly suggests that the cements may have formed during burial whilst pore fluid composition was still evolving.

It is possible to test the compatibility of the isotopic data and the evidence from geohistory analysis and illite dates. For example in a well from the Pelican Field (well 'L') the illite ages suggest that cementation ceased at 40 Ma at which time the temperature (derived from geohistory analysis) would have been around 105°C. Any isotopic temperature estimate should therefore be less than this value. The late cements in well L have minimum $\delta^{18}O$ values of $-13_{PDB}$. The equilibrium relationships for the whole range of likely fluid compositions are shown on Fig. 25. The cements clearly must have formed at less than the 105°C as this is almost the maximum temperature for the most extreme fluid composition. There is of course an almost infinite number of water compositions and temperatures that would be in equilibrium with the cement composition. However, if it is assumed that the latest cements formed immediately prior to hydrocarbon migration, as is likely on petrographic grounds, the measured carbonate values would indicate that the pore fluid composition had indeed evolved towards the positive values encountered at present in Brent reservoirs in the Norwegian sector of the North Sea (Saigal & Bjørlykke 1987).

## *Secondary porosity and feldspar distribution*

Secondary porosity is ubiquitous within the Brent Group, largely as a result of feldspar dissolution and to a far lesser extent from the dissolution of carbonates.

Porosity/depth relationships (Fig. 6) demonstrate that some crestal wells have significantly higher porosities than would be expected, and samples display an open virtually uncompacted grain framework with pitted surfaces to incipient quartz overgrowths seen using the SEM. Such textures may be interpreted as due to the regeneration of some primary porosity by dissolution of a calcite cement which had previously supported the grain framework. As only a few crestal wells show any evidence for this type of porosity enhancement it is tempting to associate this phenomena with leaching by meteoric water during erosion of the crestal areas. Any such dissolution would probably have taken place during the Bathonian times when active rifting resummed in the area. These findings contradict those of Bjørkum *et al.* (1990) who looked at the role of the late Cimmerian unconformity on the distribution of kaolinite in the Gullfaks Field and concluded that 'transformation of significant amounts of feldspar and mica to kaolinite probably did not take place within the sandstones which at present underlie the unconformity'. It is tempting therefore to search for other explanations for the enhanced porosities. These might include unusual facies development (i.e. extremely well-sorted sands), early hydrocarbon emplacement, or late leaching. There is no evidence for any unusual facies development. Early hydrocarbon emplacement could account for the preservation of the high porosities. However, if the K/Ar dates from diagenetic illites are taken to indicate timing of hydrocarbon emplacement into the reservoir, then this occurred between 60–17 Ma (Palaeocene to Miocene, see Hamilton *et al.* this volume for a review of the available dates) which is hardly early in the diagenetic history of a Middle Jurassic sandstone. By this time the sediments in well 'B' were buried to depths of around 2.5 km. These ages are broadly in agreement with those given by Glasmann *et al.* (1989*a*) for the kitchen areas of the Heather Field, who believe that hydrocarbon generation began in latest Cretaceous–early Palaeocene, reaching a peak during the late Palaeocene/early Eocene. Early hydrocarbon emplacement is therefore a very unlikely possibility. Late dissolution of feldspars does occur and is discussed below. However, it does not appear to create any enhancement in porosity.

As all alternative explanations are fraught with difficulties, it is worth examining once more the hypothesis of leaching by meteoric waters. The work of Bjørkum *et al.* (1990) concentrated on feldspar dissolution, whereas the petrographic evidence points to possible carbonate cement dissolution. Studies of modern aquifers demonstrate that undersaturations and thus dissolution of carbonate minerals can extend considerable distances downdip, see for example studies of the Tertiary Floridan carbonate aquifer by Plummer *et al.* (1983) or the Triassic East Midlands aquifer studied by Edmunds *et al.* (1982). It seems reasonable, therefore, to conclude that these anomalies are due to leaching by meteoric waters. One unanswered legitimate question is why all crestal areas do not show porosity enhancement. One possible explanation for this could lie in the interconnectivity of the sands below the unconformity; poorly interconnected sands may impede penetration of meteoric water into the formation. The elevation of the crests above sea level and the length of exposure is also unclear. Modelling of footwall uplift (Yielding *et al.* this volume) has enabled a relationship to be derived between the size of Brent fault blocks and the amount of uplift. However, it is not possible to determine whether they were subaerially exposed or for how long the exposure might have lasted.

The fraction of secondary porosity relative to the total porosity in all formations shows a pronounced increase with increasing depth (Fig. 26). In the Etive Formation (Fig. 26) nearly half of the total porosity is of secondary origin by 12 000 ft (3657.64 m). At depths shallower than 10 500 ft (3200.44 m) the amount of secondary porosity is variable but generally below 40% of the total pore space. This trend is not unique to the Brent Delta but has been seen elsewhere, for instance in the Tertiary of the Gulf Coast (Loucks *et al.* 1979). The secondary porosity is predominantly as a result of feldspar dissolution. Secondary pores are present in all samples, and as demonstrated by the porosity/depth trends most samples belong to the population which show a smooth reduction of porosity with depth, indicating that the formation of secondary porosity had no overall impact on the total porosity of the rock. This suggests that the secondary porosity is largely redistributional in nature as outlined Giles & de Boer (1990).

The trend of an increasing fraction of secondary porosity with depth is paralleled by

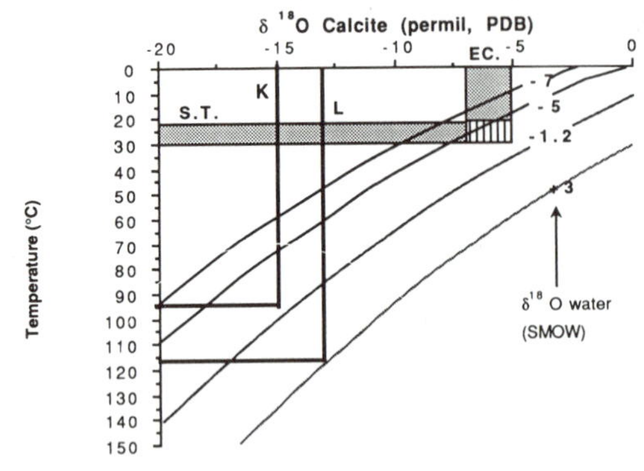


**Fig. 25.** Isotope equilibrium diagram relating measured oxygen isotopic composition of calcites to temperature and isotopic composition of the formation fluid. EC, early diagenetic calcites from the Rannoch Formation. ST, estimated near surface temperatures. The box with heavy vertical ornament indicates the likely range of temperatures and fluid compositions for the cements. K and L, measured isotopic compositions for the latest calcite cements encountered respectively in a well in the northern part of the Brent province and in a well from the Pelican field (see text for further discussion).

decreasing abundance of feldspars and kaolinite (see Figs 27a, b and 15) with increasing depth. These trends are largely independent of formation or facies. Figures 27a & b show the distribution of feldspars in the Ness and Etive Formations. In both cases the range of feldspar

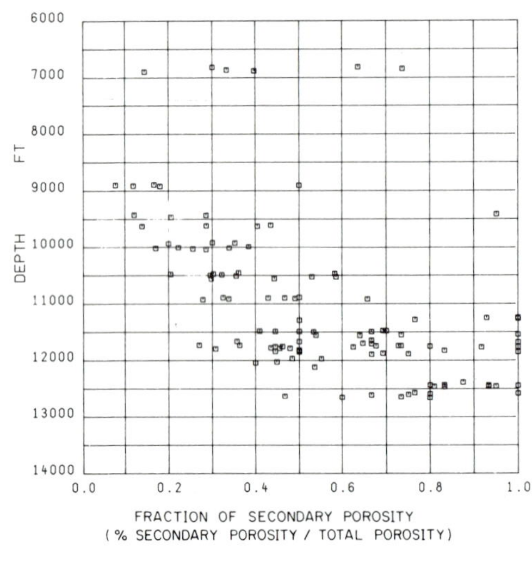


**Fig. 26.** Variation of the fraction of the total porosity which is formed by the dissolution of feldspars and carbonate cements with depth. Data from Etive Formation sands.

abundance falls with depth. In the case of sands of the Ness Formation feldspar contents of 5% or more are rare below 10 500 ft (3200.44 m). This relationship indicates steady decomposition of feldspars with increasing depth and temperature.

The distribution of secondary porosity on a facies basis is similar for the Etive and Ness Formations. The massive and cross-bedded sands have a wide range of secondary porosity contents, but a significant number of samples, particularly the deeper samples, have more than 90% of their porosity of a secondary origin. The parallel bedded sands have fewer samples with such a high proportion of secondary porosity, and there is a distinct peak in the distribution between 40 and 60%. This might be expected as the high permeabilities in the cross-bedded and massive sands would allow greater fluid flow through these intervals and consequently more dissolution.

## *Multivariate statistics*

Multivariate statistics have been used extensively to examine the structure of the data set. Factor analysis has been used in an attempt to reduce the number of variables contributing to the variance of the data set by grouping related variables together into new variables or 'factors', which are linear combinations of the old variables. In the best case, variables contributing to a given factor have a causal relationships to one another. However, this ideal result is not always achieved and many small negative correlations arise between variables as a result of the constraint that a modal analysis must total 100%.

Factor analysis of the data set as a whole brings out the depth dependence of porosity, permeability, authigenic quartz, illite, and secondary porosity. It also emphasizes other relationships such as the negative relationship between porosity and authigenic quartz.

All of these relationships are readily apparent from the cross-plots presented so far. A more interesting problem is to try to go beyond the obvious relationships apparent in the data set and to attempt to identify variables causing deviation from the expected reservoir properties. These associations are not readily visible when the analysis is carried out with a data set dominated by the depth dependence of reservoir properties. Consequently the deviations from the expected (i.e. depth related) reservoir properties were used in the analysis, rather than the reservoir properties themselves, to remove the overwhelming effect of depth. The reservoir

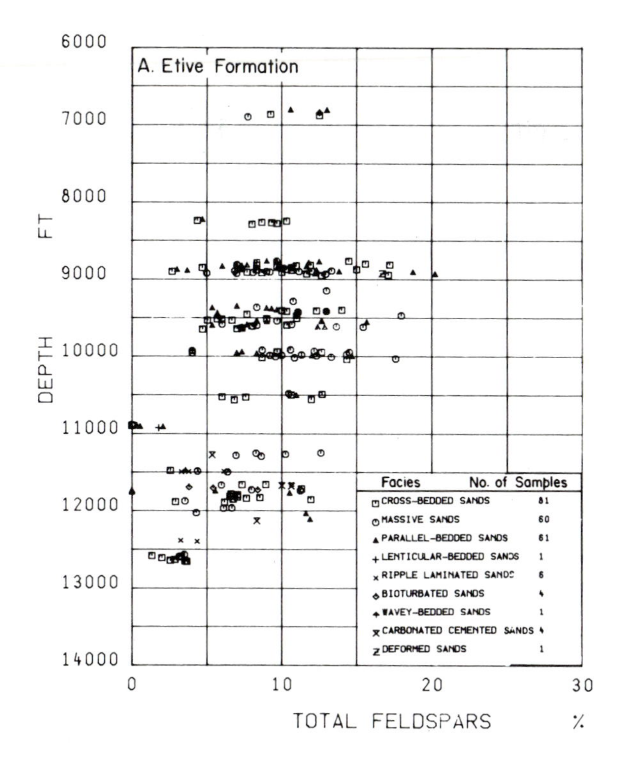


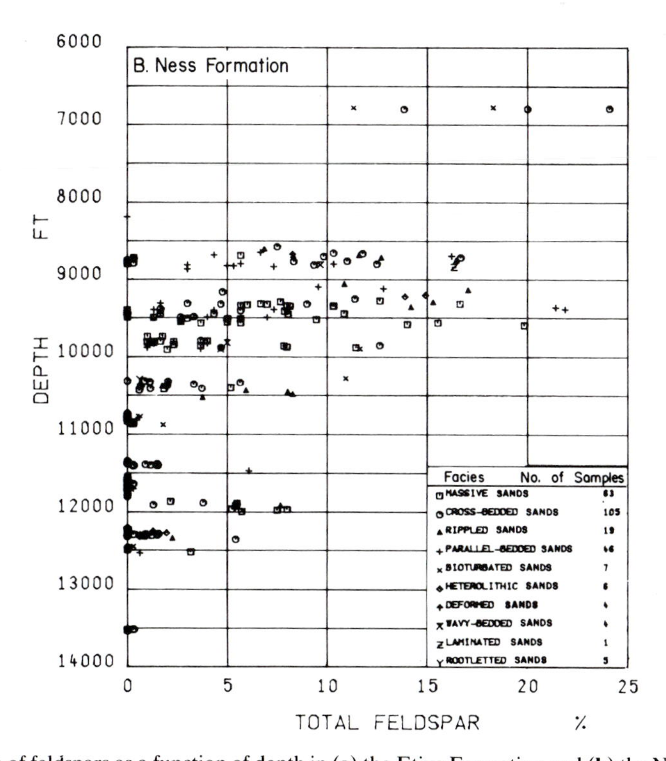


**Fig. 27.** Distribution of feldspars as a function of depth in (**a**) the Etive Formation and (**b**) the Ness Formation.

property variables used were therefore the porosity or $\log_{10}$ permeability residuals (difference between actual measured values and those expected on the basis of depth). This was done to emphasize the anomalies in the data set. Each formation was analyzed separately. Furthermore, given the discontinuity in permeability behaviour occurring at about 10200 ft (3109.0 m), the data sets were divided into two depth ranges; i.e. those shallower than 10200 ft (3109.0 m) and those deeper.

*Evaluation of data from greater than 10200 ft (3109.0 m).* Multivariate statistical analysis of the data set confirms all the previously discussed relationships. For instance, Fig. 28a shows the results of a factor analysis of the Ness Formation sands from greater than 10200 ft (3109.0 m) after exclusion of all heavily (>5%) carbonate cemented samples. The first factor and the one carrying the most variance is composed of three principle variables, the $\log_{10}$ permeability residuals and the quantity of matrix and rock fragments. The relationship of the variables within this factor is such that abnormally high $\log_{10}$ permeability residuals correspond to unusually low quantities of matrix and rock fragments. This is an obvious association and no amount of tinkering with the input data to enhance the relative importance of the permeability anomalies would give a different result. Hence, unusually permeable sandstones within the Ness Formation appear to have the lowest quantities of detrital clays.

The second factor (Fig. 28a) is composed of the variables; authigenic quartz, detrital quartz, kaolinite and depth. This is at least in part a closure effect, as the reduction in the amount of detrital quartz must automatically occur if the abundance of any other component is increased, and the total remains 100%. The relationship between authigenic quartz and burial depth has already been discussed; however, it is interesting that increasing authigenic quartz correlates partly with increasing kaolinite.

The third factor (Fig. 28a) shown is somewhat artificial. If all carbonate cemented samples are left in, then they strongly influence the factor analysis and dominate the factor related to porosity residuals. By removing heavily carbonate cemented samples from the analysis, it was hoped to bring out other variables which might influence porosity. This factor is tentatively interpreted to suggest that small amounts of carbonate cement might help preserve porosity, possibly by limiting the effects of compaction. The fourth factor (Fig. 28a) is another related to porosity residuals and simply shows that the precipitation of opaques, principally pyrite, can reduce porosity.

The fifth factor (Fig. 28a) is composed of the variables depth, illite and secondary porosity. It is interpreted as showing the link between burial depth, secondary porosity from the dissolution of feldspars, and the formation of illite as a by-product of feldspar dissolution as previously discussed.

Grain size, a variable that is usually expected to influence permeability is notable for its absence. Although included in the analysis, no relationship emerged between permeability residual and grain size.

The results from a study of the Etive Formation (Fig. 28b) are slightly different and it should be remembered that no intervals of anomalously high permeability were found in this formation. In all, four factors could be identified. The first factor is composed of the variables illite, grain size, porosity residual and permeability residuals. The variables within this factor are correlated such that higher than expected permeabilities correlate with higher than expected porosities and/or higher than expected grain sizes. The effect of illite is also clearly visible, with increasing amounts of illite correlating with lower than expected permeabilities.

The second factor (Fig. 28b) consists of mica, feldspar and detrital quartz. This factor is an expression of the closure problem (i.e. the total of a modal analysis must always be 100%) with high mica and/or feldspar abundances correlating with low quantities of detrital quartz.

The third factor (Fig. 28b) comprises the variables depth, fraction of secondary porosity and illite. This factor resembles the fifth factor from the Ness factor analysis study (Fig. 26a). The factor simply expresses the relationship that increased burial depth correlates with an increased proportion of secondary porosity, higher illite abundances, and less feldspar.

The fifth factor (Fig. 28b) is composed of the variables kaolinite and matrix. The negative correlation between kaolinite and matrix is the result of the point-counting process itself. When the point-count operator can unambiguously identify a clay mineral, there will be less matrix point-counted.

*Evaluation of data from less than 10200 ft (3109.0 m).* Multivariate analysis of data from the Ness Formation from depths less than 10200 ft (3109.0 m) suggests that 5 factors are important to reservoir quality (Fig. 28c). Correlations amongst the variables of the first factor suggests that clean sands with high detrital

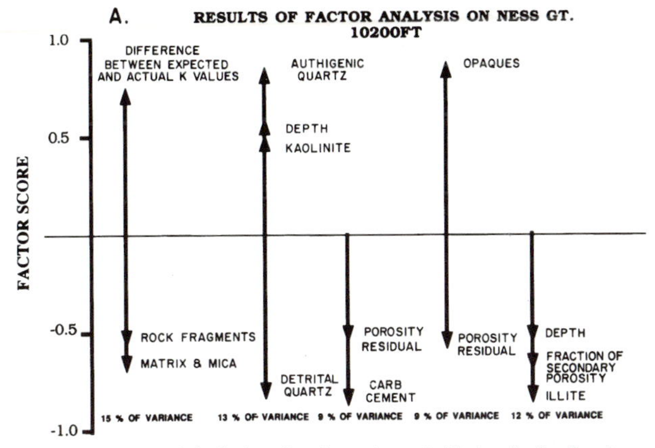


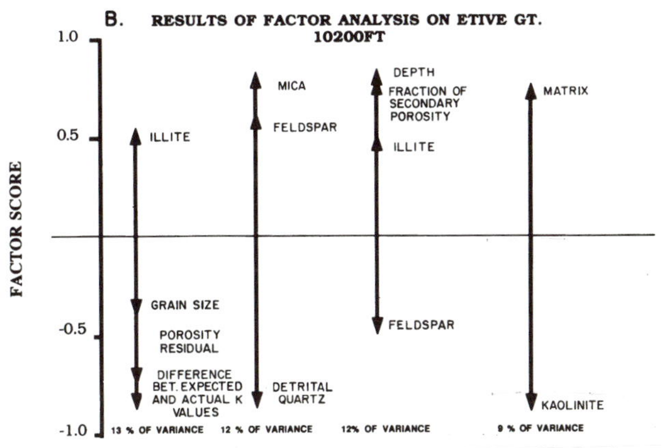


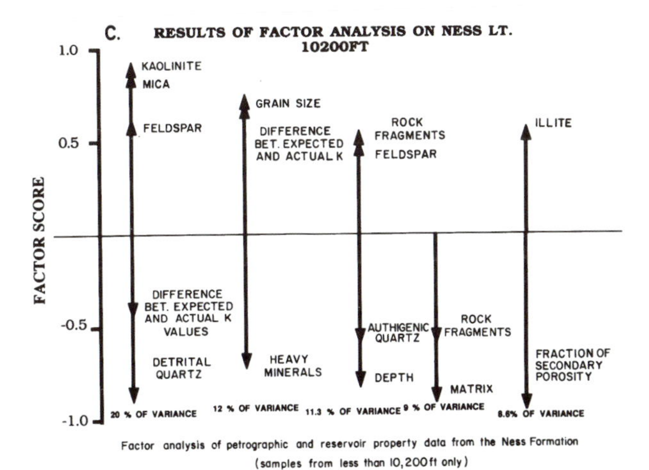


**Fig. 28.** Factor analysis of petrographic and reservoir property data from (**a**) the Ness Formation and (**b**) the Etive Formation; samples from greater than 10200 ft only. (**c**) Factor analysis of petrographic and reservoir property data from the Ness Formation (samples from less than 10200 ft only).

quartz content, low mica and kaolinite have a higher than expected permeability. The second factor is similar, with coarse grain sands showing a higher than expected permeability. Heavy minerals also show a relationship to this factor as the abundance of heavy minerals is strongly grain size controlled. The main controls on permeability are thus inferred to be, the amount of kaolinite, mica and feldspar and the grain size. Kaolinite and mica control permeability by increasing the proportion of micro porosity, which in turn increases the tortuosity and specific surface of the pore network. The third factor reflects the disappearance of rock fragments and feldspars with depth and the steady increase in authigenic quartz cement. The fourth factor shows a positive relationship of matrix and rock fragments. The fifth factor analysis contrasts with the results from the deeper samples. Here illite and secondary porosity do not correlate, probably because the point-counted abundances reflect the presence of detrital illitic clay particles or grain coating illite and not the more damaging varieties of authigenic illite.

*Origins of the permeability anomalies.* In the case of the Etive Formation no wells are known to exhibit abnormality high permeability but at least 3 wells (Fig. 11a) have unusually low permeability. Factor analysis supports the hypothesis that illite is the cause of the low permeabilities. Higher than expected permeabilities are related to higher than expected porosities or, unusually coarse grain sizes.

The high permeability anomalies of the Ness Formation correspond to cleaner than normal sandstones. These sandstones contained less detrital feldspar and consequently had less capability to produce clay minerals particularly illite. It is worth noting that the $<2$ $\mu$m clay mineral fraction from the unusually permeable zones have a low abundance of illite relative to kaolinite. Other intervals which are not anomalous also show low values; however, the presence of less illite can only help improve the permeability.

### *Diagenetic sequence and thermal evolution*

Petrographic, isotopic and burial history data can be combined to place constraints on the temperatures at which the diagenetic phases formed. For instance, the isotopic data from carbonate phases indicates that the phase of burial related ferroan calcite cement began to form about 45°C onwards, and only ceased forming when hydrocarbons were emplaced. K/Ar dates from diagenetic illites indicate that growth occurred from at least 75°C onwards, continued in some areas until fairly recently and may even be growing today.

The authigenic quartz cement/depth trend may be used to constrain the onset of large scale quartz cementation to temperatures in excess of 70°C. The quantity of authigenic quartz precipitated only starts to become significant at temperatures above 84°C. This conclusion is in conflict with the results of Jourdan *et al.* (1987). These authors and others (e.g. Scotchman *et al.* 1989) all agree that there is a down dip increase in the volume of authigenic quartz which is in line with the results of the present study. Jourdan *et al.* (1987), in a paper on the Alwyn Fields, relate authigenic quartz cementation to hot fluids circulating upwards from the deepest part of the basin, possibly using faults as conduits (Thomas 1986). Glasmann *et al.* (1989*a*) in a study of the Heather Field, concluded that the majority of the authigenic quartz was burial related; however, the hot fluids hypothesis was invoked to explain the origin of authigenic quartz seen in two wells (2/5–4 & 2/5–8b) from the field. Both of these studies rely for their evidence on fluid inclusion temperatures measured in authigenic quartz. In the light of the depth relationship shown by quartz cementation presented here, a number of serious problems with the hot fluids hypothesis need to be discussed.

(1) Can hot fluids migrating account for the linear depth trend of authigenic quartz cementation?

(2) Where is so much formation water coming from late in the burial history of the sequence? The main phase of compactional fluids from shales is generated early in the first 2 km of burial (Bjørlykke 1984). Thus, most of the compactionally-derived waters will have been expelled before the main phase of quartz cementation.

(3) If hot fluids moving up faults are responsible, then why is intense authigenic quartz cementation not seen in the crests of many other fields particularly shallow fields such as Brent and Cormorant which abut major faults, and have deep basins on their downthrown sides?

(4) Why is the Ness Formation more cemented than the Etive? The Etive Formation, being a more continuous sandstone body, should have been the major migration pathway for formation waters.

(5) If large volumes of hot fluids had been involved then there should be maturity anomalies in source rocks associated with their passage.

The mechanism proposed by Jourdan *et al.* (1987) assumes hot silica-saturated fluids escape from the adjacent basinal areas possibly partly via fault conduits (see Jourdan *et al.* 1987, plate 3), and that as these fluids moved through the Alwyn reservoir they cool and as a result precipitated silica until reservoir and fluid reached thermal and chemical equilibrium. The amount of silica which can be precipitated by a hot fluid will be dependent on the temperature difference between the source of the hot fluids and the final reservoir temperature. Some idea of the duration and volumes of fluids required for large scale authigenic quartz cementation of this type can be found in the mass balance calculations of Blatt (1979). Blatt's (1979) calculations show that to cause large-scale quartz cementation, very large volumes of water are required, far more than would be commonly available during deep burial diagenesis, and such processes would have to have durations of millions of years. Jourdan *et al.*'s (1987) data show that fluid inclusions from authigenic quartz cements of 3/14a-4 indicate temperatures of formation of up to 140°C occurred prior to hydrocarbon migration at 55–45 Ma. Such temperatures would have required a considerably higher geothermal gradient than is presently suggested for the area and would, given the time scales required for large scale quartz cementation, have been expected to result in a maturity anomaly in the overlying Kimmeridge source rocks. Vitrinite data from well 3/14a-4 may be used to compare modelled maturity with the actual measurements. Figures 29a & b show the computed geohistory graph and a comparison of the modelled and measured vitrinite data. These figures show two important points. First, there is no maturity anomaly and the vitrinite data is explained by a simple burial model. Secondly the present day temperatures in the Middle Jurassic sequence are about the same as the temperatures obtained from the fluid inclusion studies. The fact that present day down-hole temperatures are close to those from the fluid inclusion studies must cast considerable doubt on the significance placed on fluid inclusion measurements. Osborne & Haszeldine (pers. comm.) have studied the data generated from numerous fluid inclusion studies of North Sea samples and have concluded that many of the fluid inclusions from authigenic quartz overgrowths have been reset. The strongest piece of evidence cited by these authors is the very good correlation they obtain between fluid inclusion homogenization temperatures and present day burial depths.

Thus given the authigenic quartz/depth trend, the lack of any known maturity anomalies and the doubt cast on many fluid inclusion studies, it would seem that there is no reliable evidence for heat pulses in the Paleocene or early Eocene, as suggested by Jourdan *et al.* (1987) and others.

### *Burial diagenetic model*

The sequence of diagenesis observed in the Brent Group of the North Viking Graben is largely related to depth of burial and therefore temperature. The diagenetic scheme of the Brent Group sandstones, shown together with the associated temperatures, is presented in Fig. 30a and is put into its burial related perspective in Fig. 30b. Any diagenetic model must explain the following:

(1) the steady decrease in porosity on a macroscopic scale with burial depth (Figs 4a, b & 5) and the discontinuity in the permeability/depth (Figs 8–10) behaviour;
(2) the increase in the abundance of illite (Figs 17–19) and the increasing fraction of the total porosity formed by secondary porosity below 10000 ft (3048.04 m) (Fig. 26);
(3) the decline in the abundance of feldspars with depth (Fig. 27);
(4) the relationship between depth and authigenic quartz cementation (Fig. 14).

The smooth decrease of porosity (on a macroscopic scale) with burial depth is certainly associated primarily with compaction, and then increasing amounts of cementation, in particular authigenic quartz cementation. The discontinuity that is apparent in the permeability/depth relations and not in the porosity/depth relations clearly points to a cause other than large scale cementation for sudden decline in permeability. It seems reasonable, therefore, to divorce quartz cementation from the other diagenetic events which cause the change in permeability behaviour. This is not to say that authigenic quartz cementation would not have reduced permeability. However, it would have done so in proportion to the drop in porosity, as occurs at depths shallower than 10000 ft (3048.04 m). The change in permeability/depth relations correlates with the increasing abundance of illite over the depth range 10000–12000 ft (3048.04–3657.64 m). The relatively small quantities of illite involved, coupled with the decrease in abundance of kaolinite over the same interval results in virtually no impact on the porosity/depth relationships, but a drastic change in the permeability/depth behaviour.

The development of an increased proportion of secondary porosity is not reflected in an

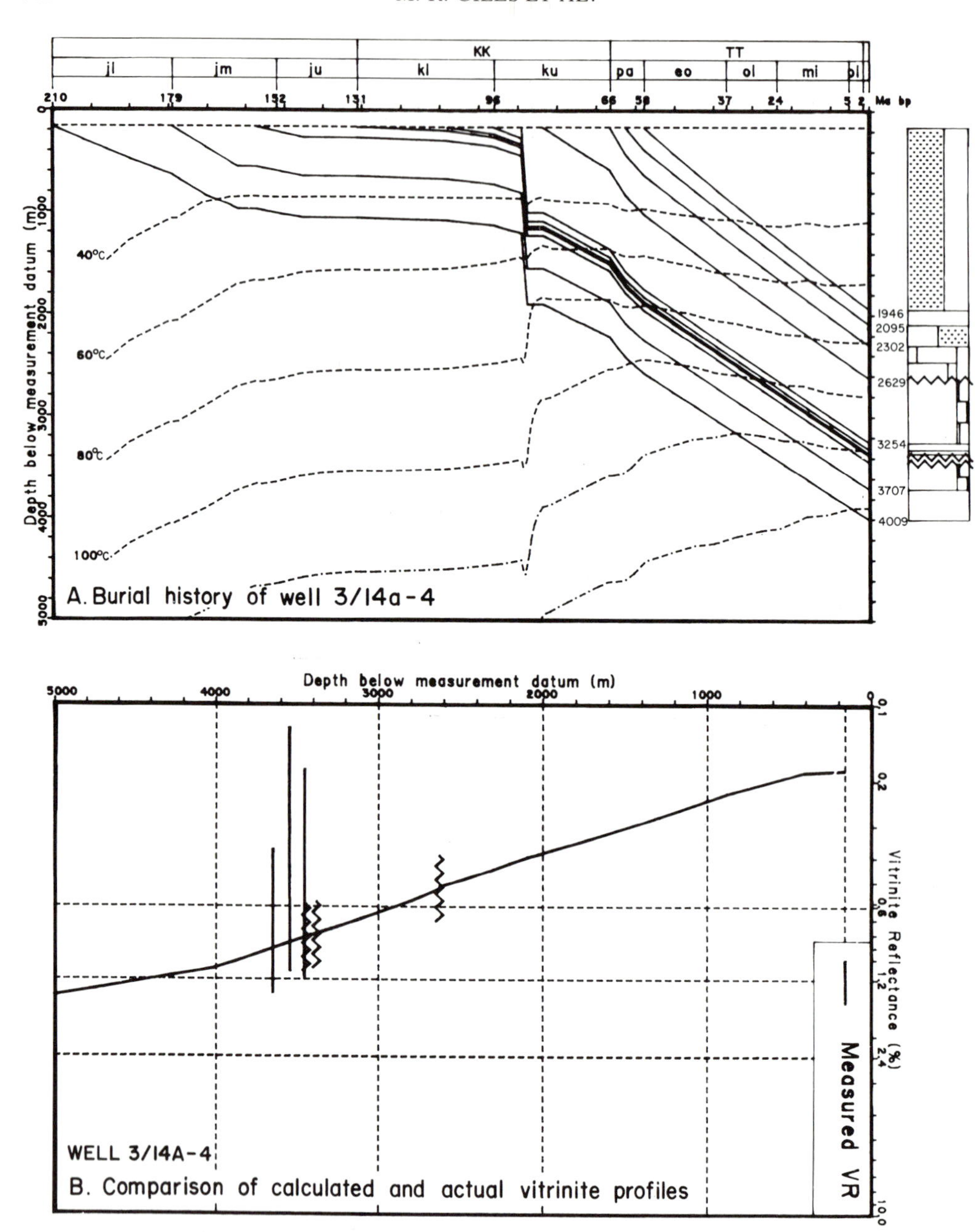


**Fig. 29.** (**a**) Burial history of well 3/14a-4. Solid lines show the bases of the modelled intervals and dashed lines are palaeo-isotherms. Time scale used is taken from Harland *et al.* (1982). Abbreviations used: JJ, Jurassic; KK, Cretaceous; TT, Tertiary; ru, Upper Triassic; jl, Lower Jurassic; jm, Middle Jurassic; Kl, Lower Cretaceous; ku, Upper Cretaceous; pa, Palaeocene; eo, Eocene; ol, Oligocene; mi, Miocene; pl, Pliocene. (**b**) A comparison of calculated (solid line shown increasing with depth) and actual vitrinite profiles (solid horizontal lines). Crenellated lines show the position of unconformities.

increase in the total porosity of the rock. Thus, the secondary porosity must be redistributional in origin (Giles 1987), i.e. primary porosity is being swapped for secondary through the local reprecipitation of dissolution products in primary pore space. Clearly the formation of secondary porosity from the dissolution of feldspars and the formation of illite are linked in

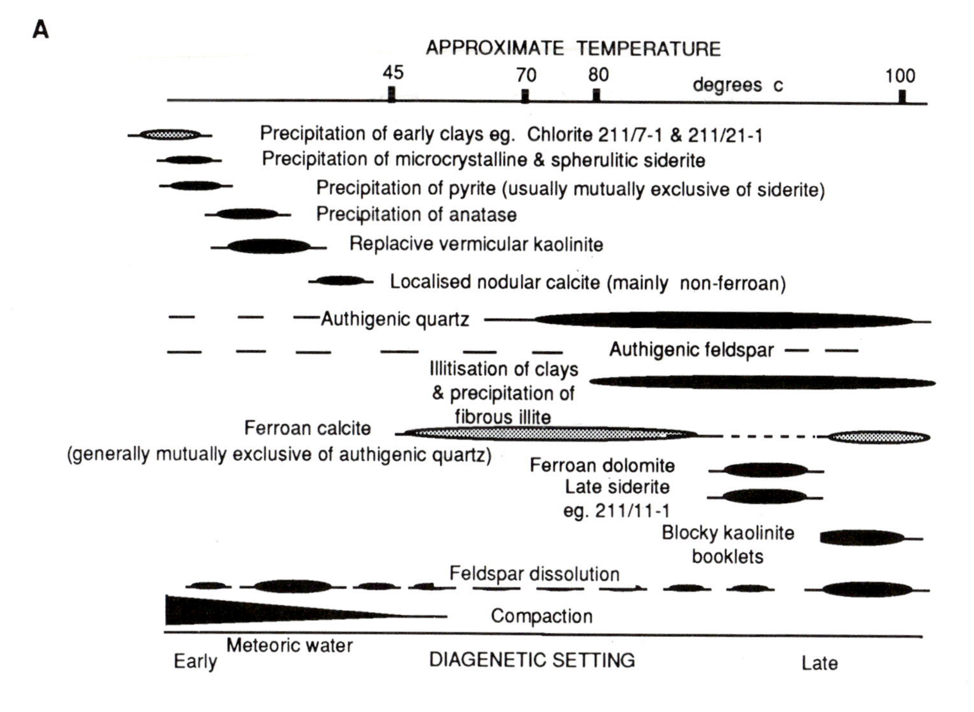


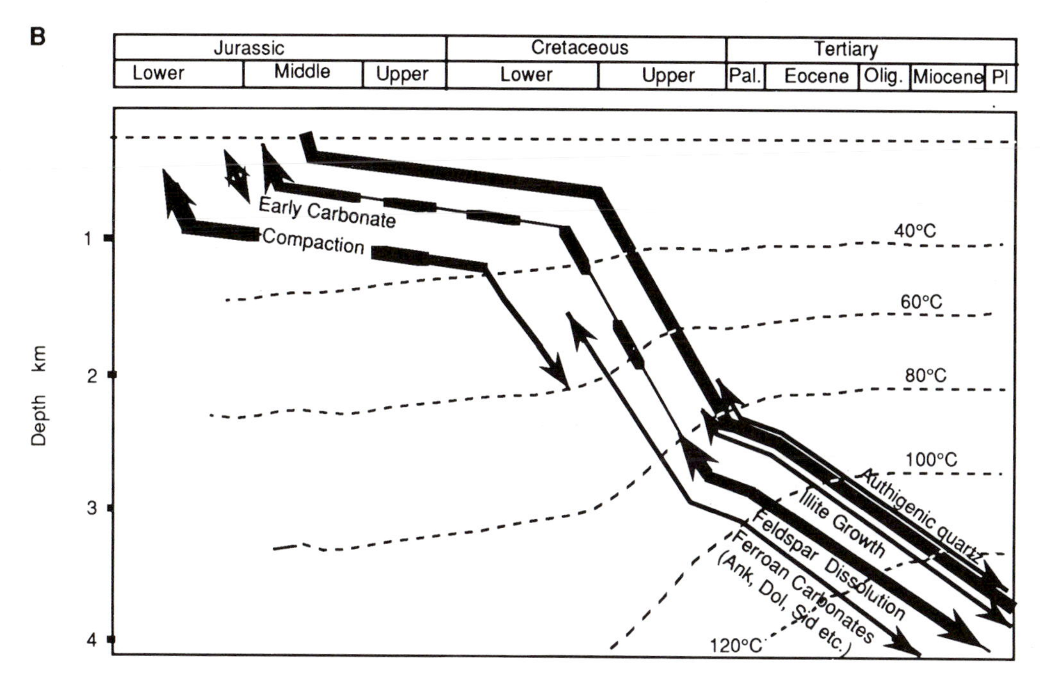


**Fig. 30.** Summary of the diagenetic history of Brent Group sediments. (**A**) Relative order of diagenetic events as deduced from the petrographic study. (**B**) Diagenetic and burial histories.

some way. Huang *et al.* (1986) have suggested a possible link between feldspar dissolution and illite formation based on experiments carried out at 200°C. Such a link is also implicit in the work of Bjørkum & Gjelsvik (1987) who used thermodynamic data to attempt to understand the distribution of kaolinite, K-feldspar and illite in North Sea sediments. The model for the decline of feldspars, increase in the proportion of secondary porosity and growth of illite used here is taken from Giles & de Boer (1990). This model is based on the thermodynamic instability of feldspars and the kinetics of feldspar dissolution. Feldspars, with the exception of the pure end members albite and microcline, would not normally be stable during diagenesis. Sediments, however, contain feldspars of various composition, not just the pure end members. As seen in this study there is a tendency for the abundance of feldspars to fall with increasing depth. This is compatible with the fact that the stability of feldspars is kinetically controlled, and that with increased temperature the rate of dissolution will increase. Experimental data on feldspar decomposition (see Giles & de Boer 1990, for details) indicates that feldspars will decompose not just in acid pore waters, as is popularly believed (see Schmidt & MacDonald 1979, and as suggested for Brent Group sediments by Scotchman *et al.* 1989), but also in neutral or alkaline pore waters. Feldspar decomposition at depths shallower than 10 200 ft (3109.0 m) results in the production of secondary porosity and the precipitation of kaolinite and small amounts of authigenic quartz. However, the relative stability field of kaolinite shrinks with increasing temperature (see Sass *et al.* 1987). Thus, with increased burial, providing sufficient potassium is available, the pore fluid compositions will move into the stability field of illite, or remain around the point of mutual stability of illite and kaolinite. The evidence for the latter possibility may be found in the apparent coexistence of blocky kaolinite and illite in Brent reservoirs. However, it is unclear from petrographic evidence if the blocky kaolinite is in equilibrium with its pore fluid or whether it is a metastable phase. In the Brent Group, illite appears to have begun growing in earnest at about 80°C. Deeper burial led to the dissolution of more feldspar and thus, to an increased proportion of secondary porosity and illite, blocky kaolinite and minor amounts of authigenic quartz. The amount of illite formed was probably related to the availability of potassium which in turn was controlled by the abundance of the potassium feldspars.

The model for feldspar dissolution and illite growth discussed above would appear quite general and therefore applicable to many other basins. However, the pore fluid chemistry has an important role to play, in particular the relative proportion of Na and K. If there is a high Na content to the brines, as occurs in basins where salt domes penetrate the reservoir sequence, as in the Central North Sea or Gulf Coast, clay minerals such as illite may become less stable than authigenic albite and consequently albitization may become widespread. Such a course of events is clearly predicted by the activity diagrams to be found in Helgeson *et al.* (1978).

The relationship of authigenic quartz to burial depth is probably unrelated to the above. The volumes of authigenic quartz that can be produced from the dissolution of feldspars (see Giles & de Boer 1990) are insufficient to account for the volumes of late authigenic quartz cement seen in these sandstones. The cause of this trend of increasing authigenic quartz cementation with increasing depth is not known, but it is likely that many reservoir units show such a trend on a regional scale. However, what is clear from the work of Blatt (1987) and Giles (1987) is that the silica cannot be transported far. In a study of Permo-Carboniferous sediments from the Middle East a similar depth relationship was observed by one of the authors (Giles); in this case a cathodoluminescence study showed intense pressure solution was occurring in siltstones and finer grained sediments, and that the surrounding sandstones were heavily quartz cemented. It is possible therefore that increasing amounts of pressure solution occurring in siltstones and shales of the Ness, Heather and Dunlin may have supplied the authigenic quartz.

## Conclusions

(1) The Etive, Ness and Tarbert reservoirs show a steady loss in porosity with burial depth that reflects a primary control by compaction, with an additional input from depth related authigenic quartz cementation.

(2) Enhancement of porosity occurs only at the crests of a few structures. It is likely that this is caused by leaching of early carbonate cements by meteoric waters.

(3) A discontinuity exists in the permeability/depth relations at about 10 200 ft (3109.0 m) for the Etive, Ness and Tarbert Formations. Above 10 200 ft (3109.0 m) permeability is independent of burial depth, below 10 200 ft log permeability falls linearly with burial depth. This change is permeability/depth behaviour is brought about

by the increased rates of destruction of feldspar with depth, the precipitation of illite as a product of feldspar dissolution and the conversion of the pore structure from dominantly primary to dominantly secondary.

(4) Zones of anomalously low permeability exist in the Ness and in particular the Etive formations of some wells. These zones are commonly to be found at depths in excess of 10 200 ft (3109.0 m) and are due to preferential illite growth in the water leg.

(5) Extremely rare zones of anomalously high permeability exist in Ness channel sands at depths in excess of 10 200 ft (3109.0 m). These channel sands contain lower than expected quantities of illite. This results from them being unusually mineralogically mature thus containing lower quantities of reactive minerals which may convert to illite. Nowhere are these permeability anomalies sufficiently thick to constitute exploration targets.

The authors wish to thank Shell UK Exploration & Production and Esso UK Exploration & Production for permission to publish this paper. We would like to mention the support from colleagues, particularly L. Wakefield, John Marshall, D. den Hartog Jager, J. Millson, M. Lewis and the whole of the past and present members of the northern Exploration team and of the Exploration sedimentology group of Shell Expro. The support and encouragement from Esso is also gratefully acknowledged particularly from B. Vining. This paper would not be possible without the tremendous support of B. Hanncock, A. Morris and K. Francis who undertook the drafting of the figures.

Finally the authors would like to thank the reviewers for their critical and helpful comments.

## References

BALDWIN, B. & BUTLER, C. O. 1985. Compaction curves. *Bulletin of the American Association of Petroleum Geologists*, **69**, 622–626.

BEARD, D. C. & WEYL, J. K. 1973. Influence of texture on porosity and permeability of unconsolidated sands. *Bulletin of the American Association of Petroleum Geologists*, **57**, 349–369.

BJØRKUM, P. A. & GJELSVIK, N. 1987. An isochemical model for the formation of authigenic kaolinite, K-feldspar, and illite in sediments. *Journal of Sedimentary Petrology*, **58**, 506–511.

——, MJØS, R., WALDERHAUG, O & HURST, A. 1990. The role of the late Cimmerian unconformity for the distribution of kaolinite in the Gullfaks Field, northern North Sea. *Sedimentology*, **37**, 395–406.

BJØRLYKKE, K. 1984. Formation of secondary porosity: How important is it? *In*: MCDONALD, D. A. & SURDAM, R. C. (eds) *Clastic Diagenesis*. American Association of Petroleum Geologists Memoir, **37**, 277–286.

—— & BRENDSAL, A. 1986. Diagenesis of the Brent Sandstone in the Statfjord Field, North Sea. *In*: GAUTIER, D. L. (ed.) *Roles of Organic Matter In Sediment Diagenesis*. Society of Economic Paleontologists and Mineralogists, Special Publication, **38**, 157–167.

——, ELVERHOI, A. & MALM, A. O. 1979. Diagenesis in Mesozoic sandstones from Spitsbergen and North Sea. *Geologische Rundschau*, **68**, 1152–1171.

BLANCHE, J. B. & WHITAKER, J. H. MC D. 1978. Diagenesis of part of the Brent Sand Formation (Middle Jurassic) of the northern North Sea. *Journal of the Geological Society, London*, **135**, 73–82.

BLATT, H. V. 1979. Diagenetic processes in sandstones. *In*: SCHOLLE, P. A. & SCHLUGER, P. R. (eds) *Aspects of Diagenesis*. Society of Economic Paleontologists and Mineralogists, Special Publication, **26**, 141–157.

BRINT, J. F., HASZELDINE, R. S., FALLICK, A. E., HAMILTON, P. J. & BROWN, S. 1988. Carbonate cemented zones in the Brent Sandstone Group, Northern North Sea. *British Sedimentological Research Group Annual Meeting, Abstracts*.

BROWN, G. & BRINDLEY, G. W. 1980. X-ray difraction procedures for clay mineral identification. *In*: BRINDLEY, G. W. & BROWN, G. (eds) *Crystal Structures of Clay Minerals and their X-Ray Identification*. Publication of the Mineralogical Society, London, 305–359.

CANNON, S. J. C., GILES, M. R., WHITAKER, M. F., PLEASE, P. M. & MARTIN, S. V. 1992. A regional review of the Brent Group, UK sector, North Sea. *This volume*.

COCKER, J. D. 1986. Authigenic illite morphology: appearances can be deceiving. (Abs.) *Bulletin of the American Association of Petroleum Geologists*, **70**, 575.

COLEMAN, M. L. & RAISWELL, R. 1981. Carbon oxygen and sulphur isotope variations in concretions from the Upper Lias of north-east England. *Geochimica et Cosmochimica Acta*, **45**, 329–340.

EDMUNDS, W. M., BATH, A. H. & MILES, D. L. 1982. Hydrochemical evolution of the East Midlands Triassic sandstone aquifer, England. *Geochimica et Cosmochimica Acta*, **41**, 1097–1103.

EHRENBERG, S. N. & NADEAU, P. H. 1989. Formation of diagenetic illite in sandstones of the Garn Formation, Haltenbank Area, Mid-Norwegian Continental shelf. *Clay Minerals*, **24**, 233–253.

FRIEDMAN, I. & ONIEL, J. R. 1977. Compilation of stable isotope fractionation factors of geochemical interest. *In*: FLEISCHER, M. (ed.) *Data of geochemistry*. Geological Survey Professional Paper **440-KK**, 1–12.

FÜCHTBAUER, H. 1967. Influence of different types of diagenesis on sandstone porosity. *Proceedings of the 7th World Petroleum Congress, Mexico*, **2**, 353–369.

GILES, M. R. 1987. Mass transfer of problems of secondary porosity creation in deeply buried hydrocarbon reservoirs. *Marine and Petroleum Geology*, **4**, 188–204.

—— & DE BOER, R. B. 1990. Origin and significance of redistributional secondary porosity. *Marine and Petroleum Geology*, **7**, 378–397.

—— & MARSHALL, J. D. 1986. Constraints on the development of secondary porosity in the subsurface: Re-evaluation of processes. *Marine and Petroleum Geology*, **3**, 243–255.

GLASMANN, J. R., LUNDEGARD, P. D., CLARK, R. A., PENNY, B. K. & COLLINS, I. D. 1989*a*. Isotopic evidence for the history of fluid migration and diagenesis: Brent Sandstone Heather Field, North Sea. *Clay Minerals*, **24**, 255–284.

——, CLARK, R. A., LARTER, S., BRIEDIS, N. A. & LUNDEGARD, P. D. 1989*b*. Diagenesis and Hydrocarbon accumulation, Brent sandstone (Jurassic), Bergen High area, North Sea. *Bulletin of the American Association of Petroleum Geologists*, **73**, 1341–1359.

HAMILTON, P. J., FALLICK, A. E., MACINTYRE, R. M. & ELLIOT, S. 1987. Isotopic tracing of the provenance and diagenesis of Lower Brent Group sands, North Sea. *In*: BROOKS, J. & GLENNIE, K. W. (eds) *Petroleum Geology of North West Europe*. Graham & Trotman, London, 939–949.

——, GILES, M. R. & AINSWORTH, P. 1992. K-Ar Dating of illites in Brent Group reservoirs – A regional perspective. *This volume*.

——, KELLEY, S. & FALLICK, A. E. 1989. K-Ar dating of illite in hydrocarbon reservoirs. *Clay Minerals*, **24**, 215–231.

HANCOCK, N. J. & TAYLOR, A. M. 1978. Clay mineral diagenesis and oil migration in the Middle Jurassic Brent Sandstone Formation. *Journal of the Geological Society, London*, **135**, 69–72.

HARLAND, W. B., COX, A. V., LLEWELLYN, P. G., PICKTON, C. A. G., SMITH, A. G. & SMITH, D. G. 1982. *A geologic time scale*. Cambridge University Press.

HELGESON, H. C., DELANY, J. M., NESBITT, H. W. & BIRD, D. K. 1978. Summary and critique of the thermodynamic properties of rock forming minerals. *American Journal of Science*, **278A**, 1–229.

HUANG, W. L., BISHOP, A. M. & BROWN, R. W. 1986. The effect of fluid/rock ratio on feldspar dissolution and illite formation under reservoir conditions. *Clay Minerals*, **21**, 585–601.

IRWIN, H. 1980. Early diagenetic carbonate precipitation and pore-fluid migration in the Kimmeridge Clay of Dorset, England. *Sedimentary*, **27**, 577–591.

IRWIN, H., CURTIS, C. & COLEMAN, M. 1977. Isotopic evidence for sources of diagenetic carbonates formed during burial of organic-rich sediments. *Nature*, **269**, 209–213.

JOURDAN, A., THOMAS, M., BREVART, O., ROBSON, P., SOMMER, F. & SULLIVAN, M. 1987. Diagenesis as the control of the Brent sandstone reservoir properties in the Greater Alwyn area (East Shetland Basin). *In*: BROOKS, J. & GLENNIE, K. W. (eds) *Petroleum Geology of North West Europe*. Graham & Trotman, London, 951–961.

KANTOROWICZ, J. D. 1984*a*. The nature, origin and distribution of authigenic clay minerals from Middle Jurassic Ravenscar and Brent Group Sandstones. *Clay Minerals*, **19**, 359–375.

—— 1984*b*. The origin of authigenic ankerite from the Ninian Field, UK North Sea. *Nature*, **315**, 214–216.

—— 1990. The influence of variations in illite morphology on permeability of Middle Jurassic Brent Group sandstones, Cormorant Field, UK North Sea. *Marine and Petroleum Geology*, **7**, 66–74.

——, EIGNES, M. R. P., LIVERA, S. E., VAN SCHIJNDEL-GOSTNER, F. S. & HAMILTON, P. J. 1992. Integration of Petroleum engineering studies of producing Brent Group Fields to predict reservoir properties the Pelican Field, North Sea. *This volume*.

LEE, M., ARONSON, J. L. & SAVIN, S. M. 1985. K/Ar dating of Rotliegendes Sandstone, Netherlands. *Bulletin of the American Association of Petroleum Geologists*, **68**, 1381–1385.

LØNØY, A., AKSELSEN, J. & RØNNING, K. 1986. Diagenesis of a deeply buried sandstone reservoir: Hild Field, Northern North Sea. *Clay Minerals*, **21**, 497–511.

LOUCKS, R. G., DODGE, M. M. & GALLOWAY, W. E. 1979. Importance of leached porosity in Lower Tertiary sandstone reservoirs along the Texas Gulf Coast. *Transactions of the Gulf Coast Association of Geological Societies*, **29**, 164–171.

MARSHALL, J. D. & ASHTON, M. 1980. Isotopic and trace element evidence of submarine lithification of hardgrounds in the Jurassic of eastern England. *Sedimentology*, **27**, 271–289.

MCHARDY, W. J., WILSON, M. J. & TAIT, J. M. 1982. Electron microscope and x-ray diffraction studies of filamentous illite clay from sandstones in the Magnus Field. *Clay Minerals*, **17**, 23–29.

PALLAT, N., WILSON, J. & MCHARDY, B. 1984. The relationship between permeability and morphology of diagenetic illite in reservoir rocks. *Journal of Petroleum Technology*, (Dec.), 2225–2227.

PLUMMER, L. N., WIGLEY, T. M. L., PARKHURST, D. L. & THORSTENSON, D. C. 1983. Development of reaction models for ground-water systems. *Geochimica et Cosmochimica Acta*, **47**, 665–686.

SAIGAL, G. C. & BJØRLYKKE, K. 1987. Carbonate cements in clastic reservoir rocks from offshore Norway-relationships between isotopic composition, textural development and burial depth. *In*: MARSHALL, J. D. (ed.) *Diagenesis of Sedimentary Sequences*. Geological Society, London, Special Publication, **36**, 313–324.

SAMWAYS, G. M. 1990. *Carbonate cementation in the Brent Group, Middle Jurassic, UK Northern North Sea*. PhD thesis, University of Liverpool.

SASS, B. M., ROSENBERG, P. E. & KITTRICK, J. A. 1987. The stability of illite/smectite during diagenesis: An experimental study. *Geochimica et Cosmochimica Acta*, **51**, 2103–2116.

SCHMIDT, V. & MCDONALD, D. A. 1979. The role of secondary porosity in the course of sandstone diagenesis. *In*: SCHOLLE, P. A. & SCHLUGER, P. R. (eds) *Aspects of Diagenesis*. Society of Economic Paleontologists and Mineralogists, Special Publication, **26**, 175–207.

SCOTCHMAN, I. C., JOHNES, L. H. & MILLER, R. S. 1989. Clay mineral diagenesis and oil migration in the Brent Group, N. W. Hutton Field, U.K. North Sea. *Clay Minerals*, **24**, 339–374.

SIEVER, R. 1983. Burial history and diagenetic kinetics. *Bulletin of the American Association of Petroleum Geologists*, **67**, 684–691.

STALDER, P. J. 1973. Influence of crystallographic habit and aggregate structure of authigenic clay minerals on sandstone permeability. *Geologie en Mijnbouw*, **52**, 217–222.

THOMAS, M. 1986. Diagenetic sequences and K/Ar dating in Jurassic sandstones, Central Viking Graben: Effects on reservoir properties. *Clay Minerals*, **21**, 695–710.

WAAL, J. A. DE, DICKER, A. I. M., KANTOROWICZ, J. D. & BIL, K. J. 1989. Petrophysical analysis of core analysis of sandstones containing delicate illite. *Log Analyst*, **29**, 317–331.

WHITAKER, M. F., GILES, M. R. & CANNON, S. J. C. 1992. Palynological review of the Brent Group, UK Sector, North Sea. *This volume*.

YEH, H. W. & SAVIN, S. M. 1977. Mechanism of burial metamorphism of argillaceous sediments: 3. O-Isotope evidence. *Geological Society of America Bulletin*, **88**, 1321–1330.

YIELDING, G., BADLEY, M. & ROBERTS, A. 1992. The structural evolution of the Brent Province. *This volume*.

# The fate of feldspar in Brent Group reservoirs, North Sea: a regional synthesis of diagenesis in shallow, intermediate, and deep burial environments

J. REED GLASMANN

*Unocal Science & Technology, 376 South Valencia Avenue, Brea, California 92621, USA*

**Abstract:** Feldspar dissolution is an important process affecting the reservoir quality of Brent Group sandstones in the northern North Sea and shows a strong relationship to reservoir temperature and pore water chemistry. In low temperature, brackish, shallow to moderately buried reservoirs (2400–3700 m, reservoir temperature less than 100–120°C) feldspar is generally a major detrital component whose abundance is affected by local variations in facies, extent of early meteoric leaching, and kaolinitization. In deeply buried Brent reservoirs (3700–4700 m) where present burial temperatures exceed 130–140°C detrital K-feldspar is absent and sandstones are characterized by major illitization of early formed kaolinite, precipitation of fibrous illite, extensive quartz cementation, and variable albitization. K-feldspar dissolution apparently supplied elements necessary for illitization, which, according to K-Ar evidence, occurred during early Tertiary burial, associated with initial stages of hydrocarbon accumulation and brine migration. Illite K-Ar ages are unaffected by feldspar contamination in deep reservoirs and the narrow range of measured ages for widely ranging illite size-separates indicates that the major dissolution/illitization reaction was a geologically rapid process. In contrast, illite K-Ar ages from shallow and intermediate-depth reservoirs range from Palaeocene to Early Miocene and often have considerable size dependency due to small amounts of feldspar contamination in coarser fractions. Variable contamination and its effect on measured K-Ar ages often precludes establishing the history of illitization and a well-defined relationship to hydrocarbon accumulation in intermediate-depth reservoirs. Although some feldspar is dissolved during early burial history by low temperature circulation of meteoric water, only deeply buried Brent reservoirs are characterized by the complete diagenetic removal of feldspar; however, the secondary porosity generated by late dissolution is of little benefit to reservoir quality because accompanying cementation results in major loss of permeability.

Detrital feldspar (dominantly K-feldspar) typically comprises from 10 to 20% of Brent Group sandstones, yet in the deep Viking Graben the complete absence of feldspar has been noted in several fields (Glasmann *et al.* 1989*a*; Harris 1989; Lonoy *et al.* 1986). In some cases feldspar abundance has been linked to facies, with transgressive sands generally having lower feldspar content due to reworking and physical destruction prior to burial (Glasmann *et al.* 1989*c*). Grain dissolution is also an important mechanism affecting feldspar abundance and is strongly dependent upon facies (permeability) and pore fluid chemistry (Hogg 1989). Early flushing of the Brent by meteoric water has been invoked as the most important agent of feldspar dissolution (Bjorlykke & Brendsdal 1986; Thomas 1986; Morton & Humphreys 1983; Hancock & Taylor 1978; Sommer 1978), although late-stage dissolution is also important (Glasmann *et al.* 1989*a*; Harris 1989; Liewig *et al.* 1987; Curtis 1978, 1983). In the deeply buried Hild field, the absence of K-feldspar was attributed to provenance factors (Lonoy *et al.* 1989); however, late diagenetic dissolution in the deep Brent, such as documented at nearby Huldra (Glasmann *et al.* 1989*a*) and Lyell fields (Harris 1989), suggests that the provenance hypothesis is probably incorrect.

The diagenetic removal of K-feldspar from shales and sandstones in a number of sedimentary basins shows a relationship to burial temperature (Lundegard & Trevena 1990; Milliken *et al.* 1989; Pearson & Small 1988; Land 1984; Aronson & Hower 1976). The temperature range for complete dissolution generally lies between 130 and 160°C and is usually accompanied by clay precipitation and albitization of K-feldspar and plagioclase (Lundegard & Trevena 1990; Morad *et al.* 1990; Glasmann *et al.* 1989*a*; Milliken *et al.* 1989; Saigal *et al.* 1988; Land 1984). In the northern North Sea, feldspathic sandstones of the Middle Jurassic Brent Group occur in a wide range of burial, thermal, and pore fluid environments and provide a natural laboratory to study the fate of feldspar during diagenesis. This report examines the diagenesis of K-feldspar in shallow

*From* Morton, A. C., Haszeldine, R. S., Giles, M. R. & Brown, S. (eds), 1992, *Geology of the Brent Group*. Geological Society Special Publication No. 61, pp. 329–350.

(<2800 m), intermediate (2800–3700 m), and deep (<3700 m) burial environments and discusses the relative importance of early and late diagenetic processes to feldspar abundance. The goal of this study was to understand the timing of feldspar dissolution and clay precipitation events in the Brent sandstone and their relationship to reservoir temperature and hydrocarbon accumulation.

## Methods

Conventional core analyses of samples from 10 fields were combined with published data from several other fields to provide the data base for this study (Table 1). 120 thin sections for petrographic analysis were made after impregnating rock samples with blue-dyed epoxy and a two-stage feldspar staining technique was employed to aid identification of K-feldspar and plagioclase (Barker & Reynolds 1984; Friedman 1971; Dickson 1965). Scanning electron microscopy (SEM) was used to identify diagenetic components. Clay minerals separated from crushed core samples by dispersion in distilled water were characterized by X-ray powder diffraction analysis (XRD), transmission electron microscopy (TEM), oxygen isotope, and K-Ar analysis (Glasmann *et al.* 1989*c*). Silt and clay were separated into the following size-ranges for analysis (based on equivalent spherical diameter settling time): 20–15, 15–10, 10–5, 5–2, 2–0.5, 0.5–0.2, 0.2–0.1, and <0.1 $\mu$m. Oxygen was liberated from silicates by fluorination (Clayton & Mayeda 1963), following the drying procedures of O'Neil & Kharaka (1976). Kaolinite oxygen isotopic values were corrected for the presence of quartz and feldspar contaminants (Eslinger 1971; Jackson 1979; Savin & Lee 1988). The illite oxygen isotope fractionation equation of Yeh & Savin (1977) was used as modified by Glasmann *et al.* (1989*c*). Illite K-Ar analyses were made by P. Damon, University of Arizona, Tuscon, and J. Aronson, Case Western Reserve University, Cleveland, Ohio. Ages were calculated using the $^{40}$K abundance and decay constants of Steiger & Jager (1977). Carbon dioxide was liberated from calcite for isotopic analyses by the method of McCrea (1950).

Subsidence, thermal, and hydrocarbon maturation histories were modeled using one-dimensional geohistory analysis as described by Glasmann *et al.* (1989*c*). Because calculated temperature histories are sensitive to selected input variables (surface temperature, subsidence rate, heat flow history, lithology) a variety of modeling conditions were used, constrained by the requirement that calculated bottomhole temperatures and predicted vitrinite reflectance curves closely match available measured data. Once a close match was achieved, modeled thermal histories were used to help interpret diagenetic history and the relationship of diagenetic processes to hydrocarbon maturation and accumulation.

## Results and discussion

### *Shallow burial environments (<2800 m)*

Brent Group sandstones are buried less than 2800 m in several northern North Sea fields (Gullfaks, Snorre, Statfjord, Hugin, Brent, Oseberg, Munin; Fig. 1). In addition to published data sources (Table 1), 104 samples from three fields (Gullfaks, 14 samples; Statfjord, 62 samples: Oseberg, 28 samples) were included in the shallow Brent data base. Subsidence history analysis of shallow Brent reservoirs (Fig. 2) indicates a prolonged period of minimal burial and/or erosion from mid-Jurassic deposition until the late Cretaceous, followed by more rapid late Cretaceous/Tertiary subsidence and burial. During the first 100 Ma following Brent deposition, maximum burial depths of shallow Brent reservoirs were generally less than 1 km and maximum burial temperatures were probably less than 60°C (Fig. 3). Because several of the shallow Brent reservoirs are situated adjacent to major hydrocarbon kitchens of the deep northern Viking Graben, late Cretaceous/early Tertiary hydrocarbon generation and migration in the graben (Thomas *et al.* 1985; Goff 1983), followed by early Tertiary hydrocarbon accumulation in peripheral structures (e.g., Alwyn, Brent, Statfjord, Oseberg; Jourdan *et al.* 1987; Thomas *et al.* 1985; Field 1985) had an important effect on limiting the duration of reservoir diagenesis in hydrocarbon-charged sandstones. Thus, excluding compaction, the diagenesis of shallow, hydrocarbon-saturated Brent reservoirs is generally unaffected by cementation associated with Tertiary burial and study of these reservoirs provides the best opportunity to understand early diagenetic controls influencing feldspar distribution.

Shallow, hydrocarbon-saturated Brent reservoirs are generally poorly cemented (except where cemented by early carbonate), friable, kaolinitic (both blocky and vermiform varieties), and are characterized by widely ranging feldspar abundance (Fig. 4). Tarbert sandstones often contain less feldspar than underlying Ness, Etive, or Oseberg sandstones (Glasmann *et al.* 1989*a, c*), probably due to reworking and mech-

**Table 1.** *Data sources used for regional evaluation of Brent Sandstone diagenesis, northern North Sea*

| Investigator | Field Studied | Wells | Depth | Stratigraphic Unit | Isotopic Data | Petrographic Data |
|---|---|---|---|---|---|---|
| Bjorkum *et al.* 1990 | Gullfaks | 34/10−1, −4, −14 | 1800−2200 m | Rannoch Fm. | NA | A |
| Bjorlykke & Brendsdal 1986 | Statfjord | 33/9−A29 | 2450−2650 m | Brent Grp | NA | NA |
| Blackbourn 1984 | NA | NA | NA | Ness, Etive Fms | NA | NA |
| Blanche & Whitaker 1978 | NA | NA | NA | Brent Grp | NA | NA |
| Brint *et al.* 1988 | Thistle/Murchison | A | 2650−2900 m | Brent Grp | O,C,K-Ar,D/H | A |
| Glasmann *et al.* 1989*a* | Heather | A | 2960−3670 m | Brent Grp | O,C,Sr,K-Ar,D/H FW | A |
| Glasmann *et al.* 1989*b* | Huldra, Veslefrikk | A | 2902−4133 m | Brent Grp | O,C,K-Ar FW | A |
| Hallett 1981 | Thistle | NA | NA | Brent Grp | NA | NA |
| Hamilton *et al.* 1987 | Thistle, Murchison Dunlin | A | A | Brent Grp | O,C,Sr,K-Ar,D/H | A |
| Hancock & Taylor 1978 | NA | NA | NA | Brent Grp | Na | NA |
| Harris 1989 | Statfjord, Hutton Lyell | NA | 2500−3500 m | Brent Grp | Na | A |
| Hogg 1989 | Alwyn South | 3/14a−7, −8, −9, −11 | 3552−3830 m | Brent Grp | O,C,K-Ar,D/H | A |
| Jourdan *et al.* 1987 | Greater Alwyn | A | A | Brent Grp | K-Ar | NA |
| Kantorowicz 1984 | Ninian | 3/3−1, −2, −3 | 3138−3245 m owc at 3210.5 m | Brent Grp | NA | NA |
| Liewig *et al.* 1987 | Alwyn | NA | 3238−3470 m | Brent Grp | K-Ar,Rb-Sr | NA |
| Lonoy *et al.* 1986 | Hild | NA | 3800−4400 m | Tarbert Fm. | O,C (Calcite) FW | NA |
| Malley *et al.* 1986 | Alwyn | NA | 3200−3700 m | Brent Grp | NA | NA |
| Morton & Humphreys 1983 | Murchison | 211/19−3, −4 | ave. 3500 m | Brent Grp | NA | NA |
| Scotchman *et al.* 1989 | NW Hutton | A | 3810−3963 m | Brent Grp | K-Ar | A |
| Thomas 1986 | Heimdal | 25/4−1, −5 | 3179−3750 m 3287−4000 m | Brent Grp Statfjord Fm. | K-Ar,O,C | NA |

Key: A Tabular data reported in publication; NA, quantitative data not listed in publication; FW, formation water isotope data reported; owc, oil−water contact. Isotopic data are from clays (kaolinite & illite) and calcite. Additional petrographic, clay mineral, and isotopic data from Statfjord, Gullfaks Gamma, Hild, Huldra, Oseberg, and central Viking Graben fields are included in this study (unpublished Unocal data).

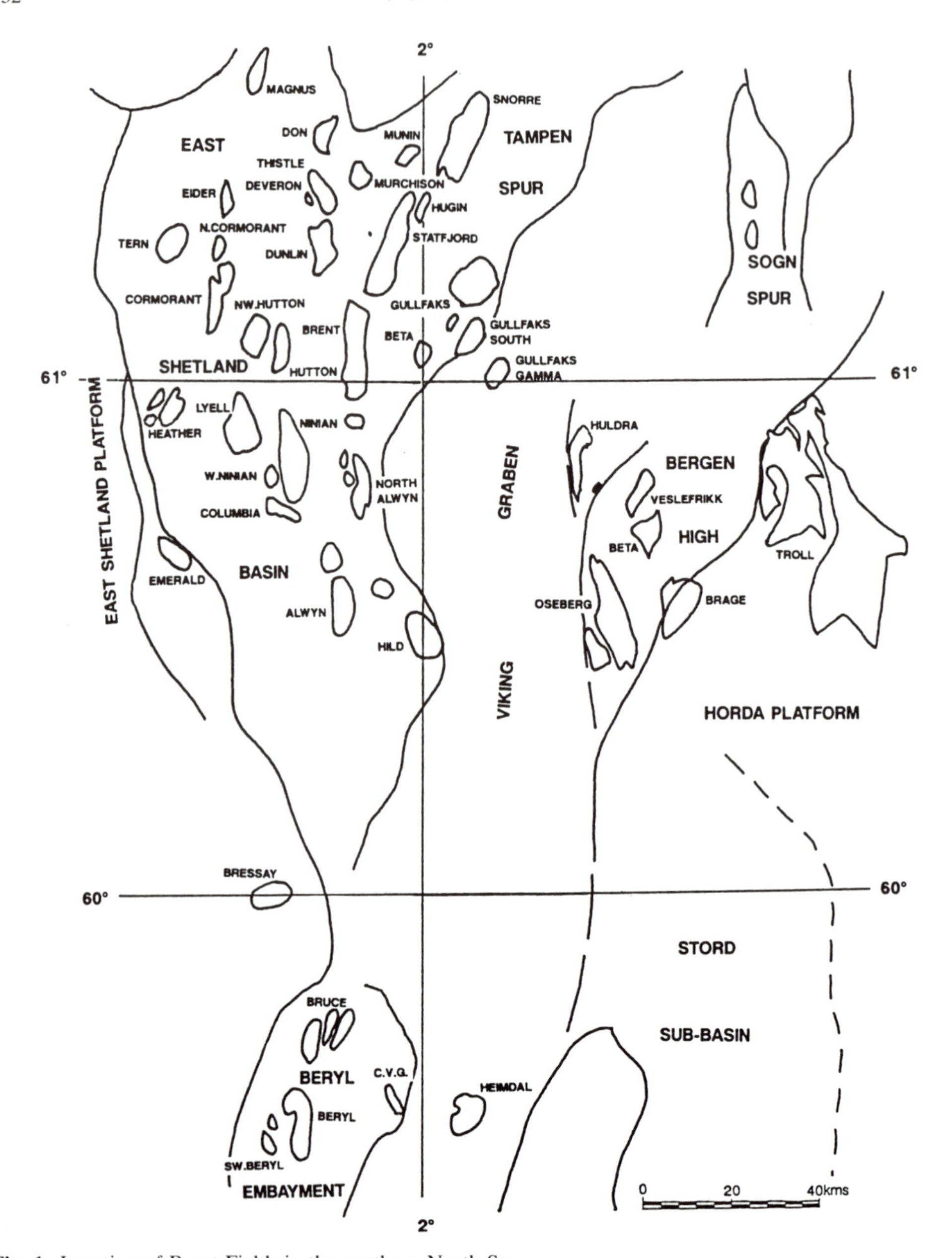


**Fig. 1.** Location of Brent Fields in the northern North Sea.

anical destruction of feldspar during marine transgression. The paragenetic sequence of shallow reservoirs includes a very early (syndepositional?) episode of K-feldspar overgrowth (not present in all fields) and alteration of detrital mica (chiefly biotite) to large vermiform kaolinite, followed by local calcite cementation. A subsequent phase of feldspar dissolution, precipitation of blocky kaolinite, and minor development of quartz overgrowths also occurs (Figs 5, 6 & 7). The petrographic evidence upon which this paragenetic sequence is based includes the presence of K-feldspar overgrowths and coarse kaolinite within calcite cements, the absence of compaction of calcite cemented zones (signifying early cementation), and the presence of significant feldspar dissolution, blocky kaolinite, and quartz cement out-

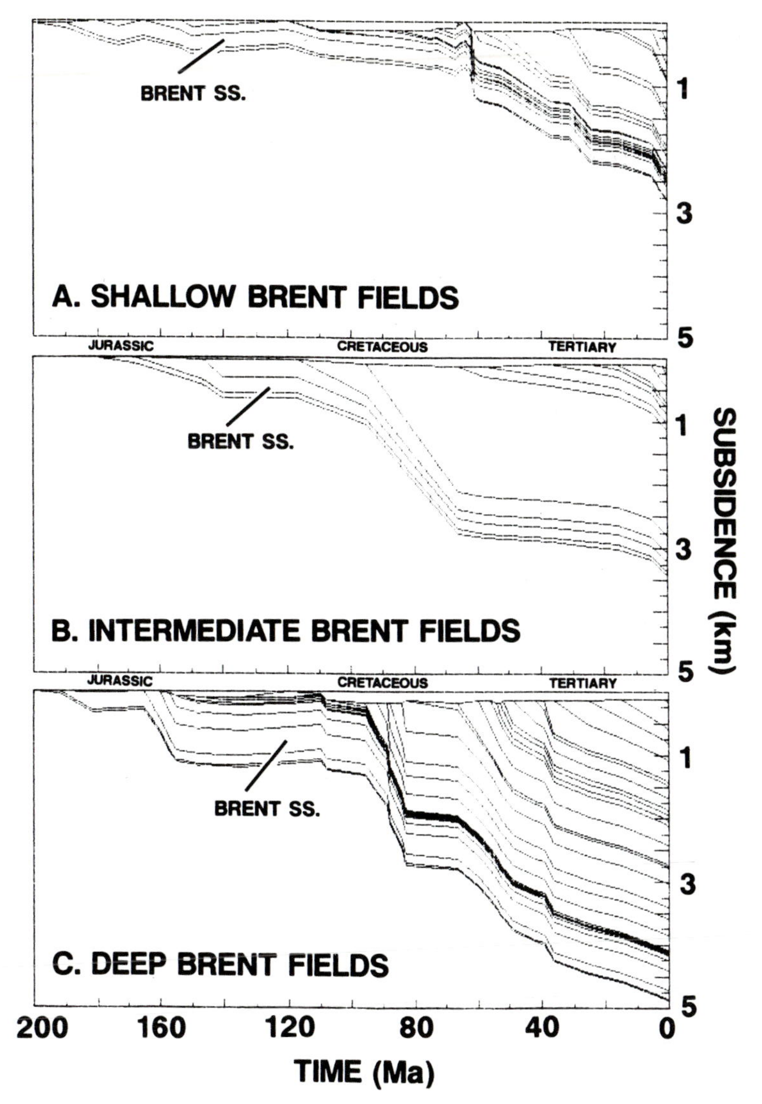


**Fig. 2.** Subsidence history diagrams for the Brent Sandstone in different burial environments: (**A**) shallow (Oseberg Field, 30/9–2); (**B**) intermediate burial (Heather Field, 2/5–2); (**C**) deep burial (central Viking Graben, 24/6–1).

side calcite nodules (Glasmann *et al.* 1989*a, b*; Hamilton *et al.* 1987). In some cases, kaolinite may have a detrital origin (e.g. Rannoch Fm. in Gullfaks Field, Bjorkum *et al.* 1990), although petrographic distinction of detrital and authigenic kaolinite in fine-grained sandstones is often difficult. Fibrous illite, which is a common late-stage diagenetic phase in deeper reservoirs, is generally absent from the shallow Brent, especially in hydrocarbon-saturated reservoir zones (Table 2); however, traces of lath-shaped illite and illite overgrowths on detrital grains suggest that shallow reservoirs have undergone incipient stages of illitization (Fig. 7).

*Clay mineral and isotopic evidence of diagenetic environment.* Isotopic studies of calcite and kaolinite cements provide additional evidence of low temperature diagenesis and indicate the importance of meteoric water during initial Brent cementation (Glasmann *et al.* 1989*a, c*; Hogg 1989; Brint *et al.* 1988; Hamilton *et al.* 1987). Kaolinite is the most abundant diagenetic clay in shallow Brent reservoirs, except in the

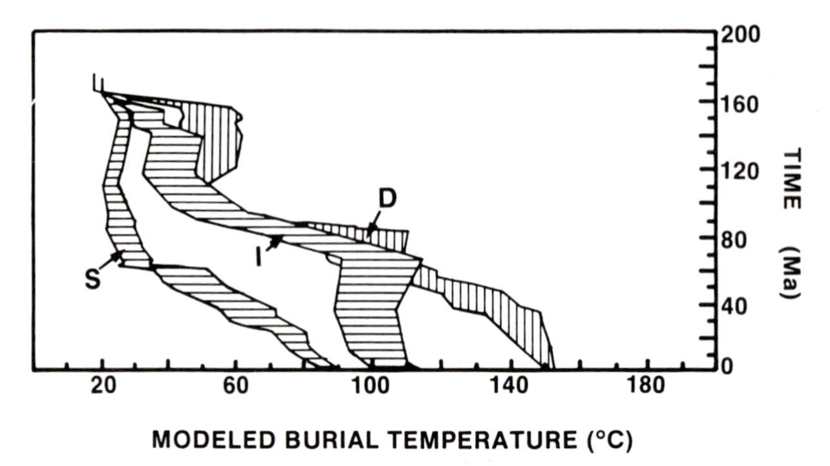


**Fig. 3.** Calculated temperature history curves for the Brent Sandstone based on subsidence histories illustrated in Fig. 2. Present thermal gradients were assumed constant to calculate minimum reservoir temperatures. Early Cretaceous rifting was assumed for rift heat flow models, which resulted in higher palaeotemperatures than constant gradient models. S, shallow (Oseberg); I, intermediate (Heather); D, deep (central Viking Graben).

<0.1 $\mu$m clay fraction (Table 2), and its oxygen isotopic composition (*c.* 15–17‰, Figs 8, 9) is consistent with an interpretation of low temperature precipitation from unmodified Jurassic/early Cretaceous meteoric water (Fig. 8, <40–45°C, *c.* −7‰ (SMOW), Hamilton

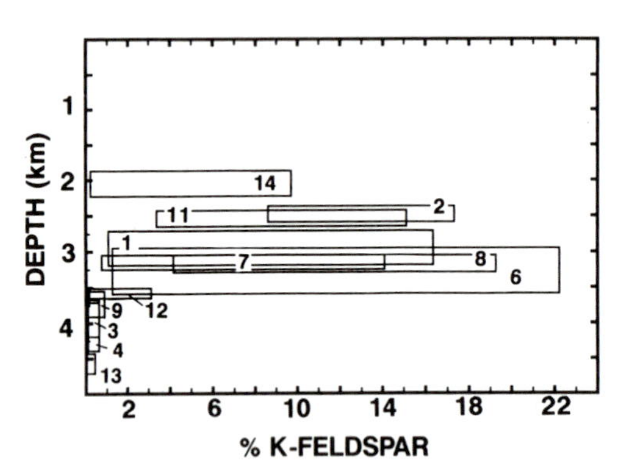


**Fig. 4.** Feldspar abundance in the Brent Sandstone as a function of present depth of burial, northern North Sea. Key to fields and data sources: 1, Veslefrikk; 2, Oseberg; 3, Huldra, 4, Hild; 5, Heimdal (Thomas 1986); 6, Heather; 7, Hutton (Harris 1989; Scotchman *et al.* 1989); 8, Murchison (Hamilton *et al.* 1987; Morton & Humphreys 1983); 9, Alwyn South (Hogg 1989); 10, Alwyn North (Hogg 1989); 11, Statfjord (Harris 1989; Bjorlykke & Brendsdal 1986); 12, Lyell (Harris 1989); 13, central Viking Graben, 14, Gullfaks (Bjorkum *et al.* 1990). Data for fields 1 and 3 are from Glasmann *et al.* 1989*a*. Field 7 data are from Glasmann *et al.* 1989*c*. Data for fields 2, 4, and 13 are unpublished Unocal data.

*et al.* 1987). While kaolinite oxygen isotopic composition varies slightly from field to field, at a given field, isotopic composition may be very homogeneous (e.g. Oseberg Field, $\delta^{18}O$ = 15.0 ± 0.5‰, $n$ = 17, Glasmann unpublished data; Glasmann *et al.* 1989*c*), indicating that kaolinite precipitation occurred over a large area from pore water of similar temperature and isotopic composition. Thus, the abundance of kaolinite, its homogeneous isotopic character (Fig. 9), and the association of kaolinite with major leaching of detrital feldspar indicate that diagenesis occurred during a period of thorough flushing of the Brent by meteoric water (Glasmann *et al.* 1989*c*; Bjorlykke & Brendsdal 1986; Hancock & Taylor 1978; Sommer 1978). Such conditions are easy to envision where the Brent is truncated by major unconformity surfaces (e.g., Gullfaks, Statfjord and Oseberg fields), but less easy to understand where the Brent was not breached by late Cimmerian erosion (e.g., Heather and Hutton fields; Glasmann *et al.* 1989*c*; Scotchman *et al.* 1989). Detrital kaolinite may also occur in low energy facies of the Brent (e.g. Rannoch Fm., Gullfaks Field, Bjorkum *et al.* 1990) and, because it probably formed during surface weathering, its isotopic composition should be characteristically heavy (>16‰, Glasmann unpublished data, Gullfaks Field, Fig. 9). As will be shown, kaolinite isotopic data from deeper Brent reservoirs are not consistent with simple kaolinite precipitation at an unconformity (Glasmann *et al.* 1989*a*, *c*).

Illite is typically a trace or minor component

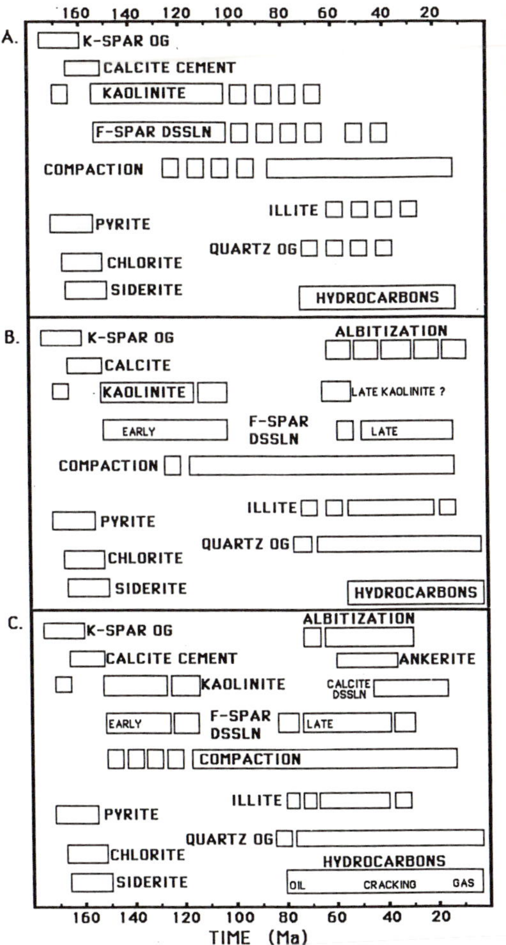


**Fig. 5.** Generalized paragenetic sequence of the Brent Sandstone based on integration of published reservoir studies (Table 1). (**A**) Shallow Brent fields (<2800 m). (**B**) Intermediate depth Brent fields (2800–3700 m). (**C**) Deep Brent fields (>3700 m).

of total diagenetic clay in shallow Brent reservoirs and is most abundant in the finest clay fraction (Table 2). TEM analyses of different clay fractions show the presence of illite as discrete laths associated with hexagonal plates of kaolinite in the <0.1 $\mu$m fraction, and as lath-shaped overgrowths on detrital micas in coarser clay separates (Fig. 7A). Illite separates coarser than 0.1 $\mu$m are generally contaminated with clay-size remnants of partially dissolved feldspar (Fig. 7B) and are thus unsuitable for K-Ar dating. The delicate, embayed appearance of K-feldspar fragments in TEM photos is characteristic of samples regardless of dispersion treatment (mild sonification in distilled water or pH 5 buffer, no sample grinding) and probably results from in situ grain dissolution during meteoric diagenesis (compare particle morphology of dispersed feldspar fragment in Fig. 7B with in situ partially dissolved feldspar, Fig. 9A, B).

The general absence of illite in hydrocarbon-saturated samples from Statfjord and Oseberg fields precluded K-Ar measurement of the timing of illite diagenesis and identification of vertical trends in illite age that may be correlated to hydrocarbon accumulation. Mineralogically pure illite from the water zone of the Brent at Oseberg field yielded a K-Ar age of 53.0 ± 0.9 Ma, implying an Early Eocene minimum age of hydrocarbon accumulation. Dahl *et al.* (1985) suggested that hydrocarbon accumulation at Oseberg may have begun during late Cretaceous time, with sourcing from the adjacent Oseberg 'kitchen' (Thomas *et al.* 1985). While the age of water-zone illite is not usually indicative of hydrocarbon accumulation history, the older age of Oseberg illite relative to water-zone illite from nearby Huldra and Veslefrikk fields (Glasmann *et al.* 1989*a*) suggests that the water-zone illite at Oseberg has not been influenced by late Tertiary diagenesis. The absence of 'young' water-zone illite at Oseberg suggests that illite diagenesis is not an ongoing process. If illitie precipitation was associated with changes in formation water chemistry related to initial hydrocarbon accumulation, a lack of 'nutrient' recharge coupled with low reservoir temperatures may have restricted continued late Tertiary illite growth.

The timing of illite diagenesis in Oseberg field corresponds with calculated reservoir temperatures in the range of about 50–60°C (Fig. 3), considerably cooler than temperatures associated with illitization in deeper fields (Giles *et al.* this volume; Glasmann *et al.* 1989*a*, *c*) and considerably cooler than the 75–80°C temperature for onset of illite precipitation in the Brent Sandstone suggested by Hamilton *et al.* (this volume). The variation in range of initial illitization temperature may reflect inaccuracies in palaeotemperature reconstruction or indicate the influence of local formation water chemistry. The absence of well-developed, long, thick illite particles (compare Fig. 7A and 7C) in shallow reservoirs suggests that elements for crystal growth were limited compared to deeper Brent reservoirs. With an unlimited supply of dissolved 'nutrients', a broad spectrum of illite crystals would precipitate (Fig. 7C). Because of this apparent limitation in the supply of dissolved elements required for crystal growth, an open-system diagenetic model based on large-

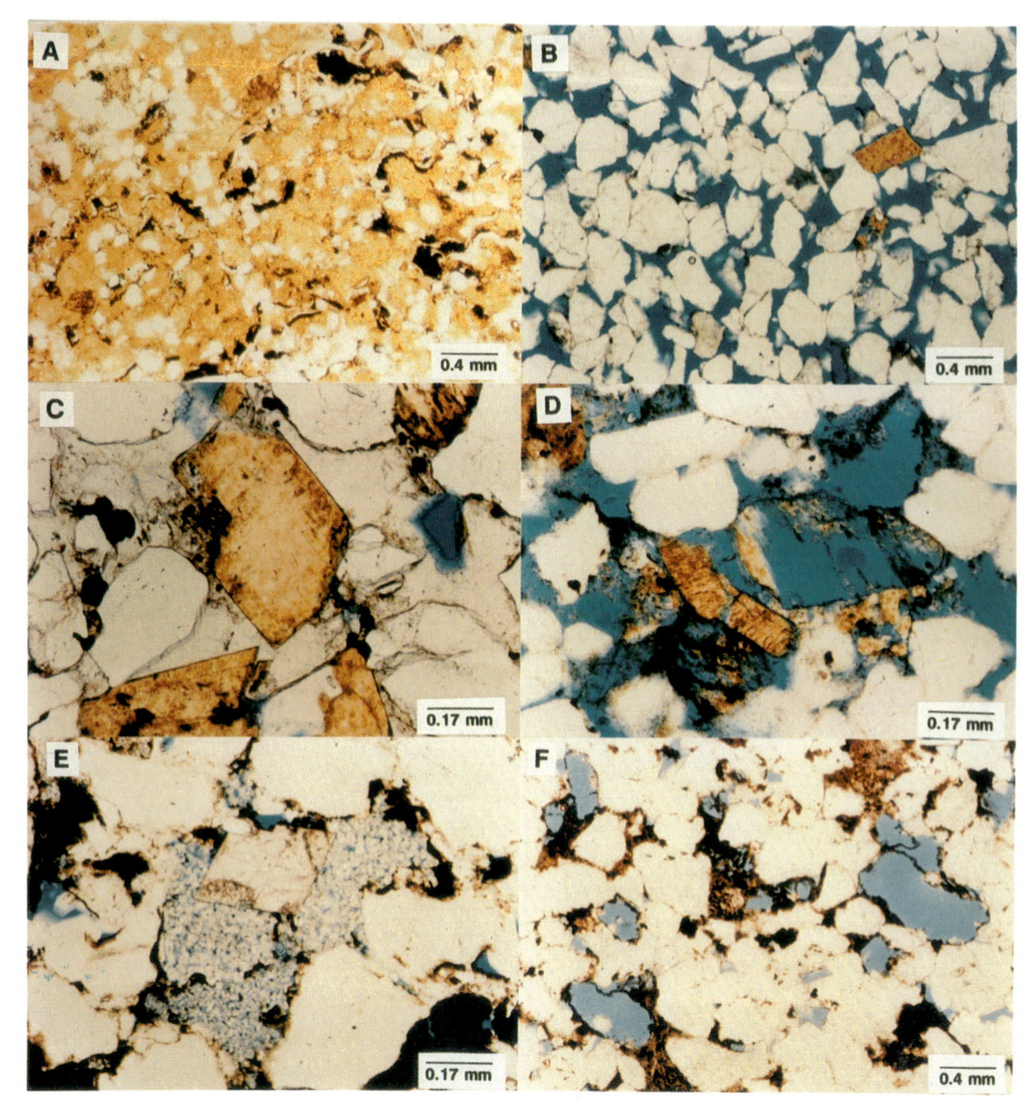


**Fig. 6.** Photomicrographs illustrating various aspects of feldspar diagenesis in Brent Sandstones, northern North Sea. All thin sections stained for both K-feldspar (yellow) and plagioclase (pink). All depths reported are core depths. (**A**) Abundant detrital K-feldspar with early K-feldspar overgrowths, Rannoch Fm, Heather Field, 2/5–9, 3350 m. (**B**) Low feldspar content of transgressive shoreface sandstone, Oseberg Field, 30/9–2, 2600 m. Note low degree of compaction and absence of cements. (**C**) K-feldspar grains with well developed overgrowths surrounded by early pore-filling calcite cement. Post-calcite dissolution has dissolved some K-feldspar and is associated with kaolinite precipitation. Etive Fm, Heather Field, 2/5–2, 3383 m. (**D**) Extensive dissolution of feldspar, Etive Fm, West Heather, 2/5–8b, 2880 m. Quartz overgrowths are moderately abundant and contain fluid inclusions whose mean homogenization temperature indicates late quartz precipitation (Glasmann *et al.* 1989*b*). (**E**) Late ankerite has partially enveloped earlier kaolinite formed from kaolinitization of detrital feldspar. Compaction of kaolinitized grains and inclusion of kaolinite within quartz overgrowths (Fig. 11c) indicates that kaolinitization occurred prior to significant compaction, quartz cementation, and precipitation of late carbonate. Hild Field, 30/4–2, 3907 m. (**F**) Complete dissolution of feldspar in deep burial environments associated with extensive quartz cementation and illitization of kaolinite, central Viking Graben, 24/6–1, 4496 m. Opaque material consists of bitumen residue left after in situ cracking of liquid hydrocarbons to gas.

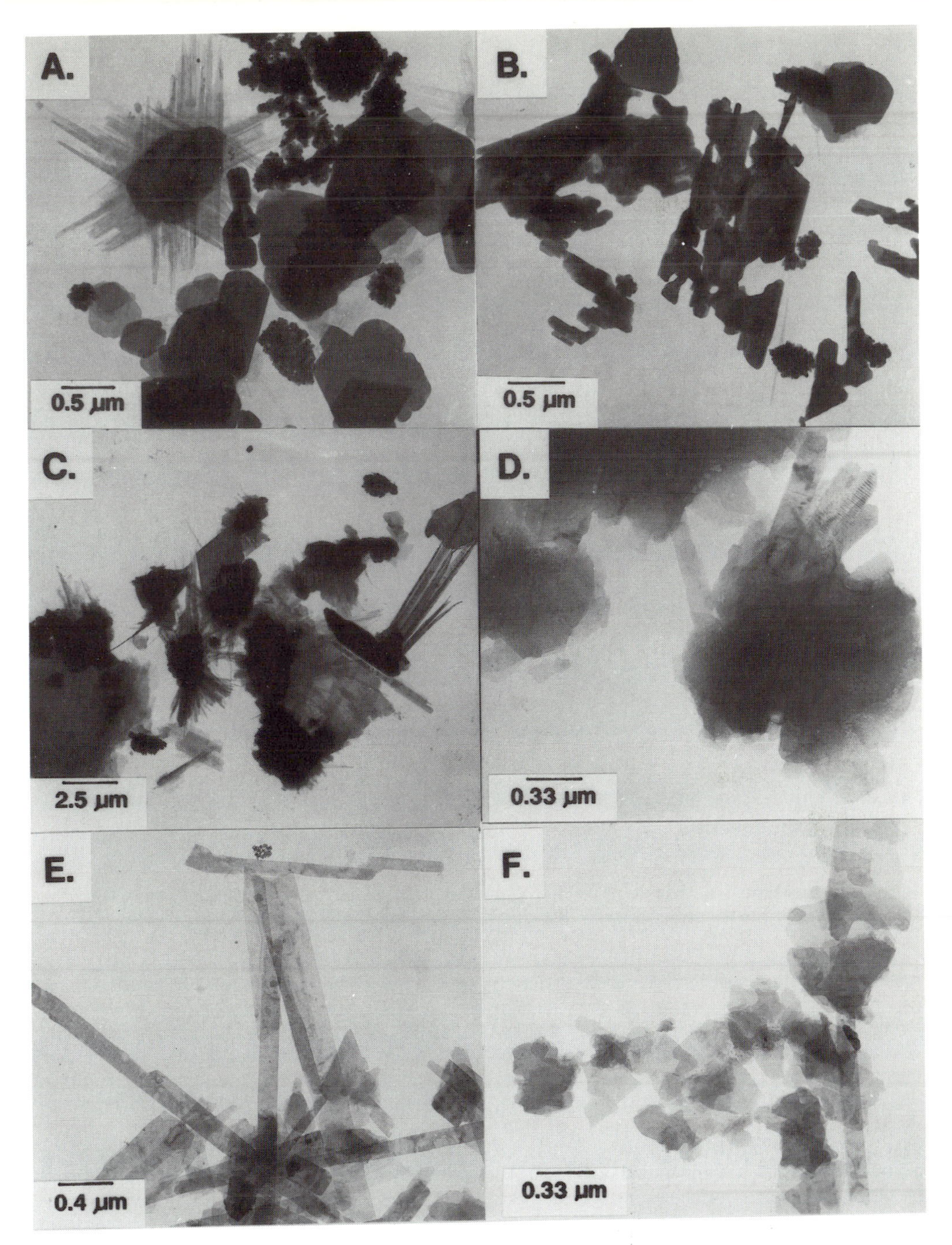


**Fig. 7.** Transmission electron micrographs of dispersed clay, Brent Sandstone, northern North Sea. (**A**) Syntaxial illite overgrowth on detrital mica flake, hexagonal plates of kaolinite and corroded clay-sized feldspar, Oseberg Fm., 2.0–0.5 µm, Oseberg Field. (**B**) Corroded clay-size feldspar resulting from congruent dissolution of feldspar during meteoric diagenesis. Oseberg Fm., 2–0.5 µm, Oseberg Field. (**C**) Aggregates of lath-shaped diagenetic illite and illite laths growing from feldspar dissolution fragment. Bundles of illite laths are not floccules, but represent intergrown crystal masses that resist dispersion, even after repeated ultrasonic treatment. Large platy flakes are partially dissolved kaolinite. Etive Fm., 5–2 µm, Hild Field. (**D**) Illitized kaolinite showing both pseudomorphous, platy illite and euhedral illite laths, central Viking Graben, 2.0–0.5 µm, 4496 m. (**E**) Floccule of illite laths, Ness Fm., 0.2–0.1 µm, central Viking Graben, 4494 m. (**F**) Platy illite resulting from illitization of kaolinite, Ness Fm., central Viking Graben, 0.2–0.1 µm.

**Table 2.** *Clay mineralogy of Brent Sandstones from shallow, intermediate, and deep burial environments, North Sea*

| Sample | Depth (m) | Size (μm) | Smectite (%) | Chlorite (%) | Illite (%) | I/S (%) | Kaolinite (%) | Other min.* |
|---|---|---|---|---|---|---|---|---|
| 33/12–1 | 2483 | 20–10 | – | tr | 5 | – | 80 | Q1, F, B1, P, S |
| | | 10–5 | – | tr | 5 | – | 85 | Q, F, B, P, S |
| | | 5–2 | – | tr | 5 | – | 85 | Q, F, B, P, S |
| | | 2.0–0.5 | tr | tr | 5 | – | 85 | Q, F, B, P, S |
| | | 0.5–0.2 | 5 | – | 5 | 10 | 80 | B |
| | | <0.2 | 10 | – | tr | 50 | 40 | |
| 30/9–2 | 2638.5 | 15–10 | – | – | <5 | tr | 85 | Q1, F1 |
| | | 10–5 | – | – | <5 | tr | 90 | Q1, F1 |
| | | 5–2 | – | <5 | <5 | tr | 90 | Q, F1 |
| | | 2.0–0.5 | – | tr | <5 | tr | 95 | Q, F1 |
| | | 0.5–0.1 | – | – | tr | 10 | 85 | Q, F |
| | | <0.1 | – | – | – | 90 | 10 | |
| 2/5H4 | 3275 | 15–2 | – | – | tr | – | 85 | Q1, F1 |
| | | 2.0–0.5 | – | – | <5 | tr | 90 | Q1, F |
| | | 0.5–0.1 | – | – | 40 | 40 | 15 | Q1 |
| | | <0.1 | – | – | – | 100 | – | |
| 30/4–2 | 3897 | 10–5 | – | – | tr | – | 95 | Q, F |
| | | 5–2 | – | – | 5 | – | 90 | Q, F |
| | | 2.0–0.5 | – | tr | 25 | 5 | 65 | Q |
| | | 0.5–0.2 | – | – | 60 | 20 | 20 | |
| | | <0.2 | – | – | 60 | 35 | 5 | |
| 30/2–2 | 4030.4 | 15–2 | – | 15 | 35 | – | 50 | Q |
| | | 2.0–0.5 | – | 5 | 25 | – | 70 | Q |
| | | 0.5–0.1 | – | 15 | 5 | 30 | 50 | |
| | | <0.1 | – | – | 5 | 95 | – | |
| 24/6–1 | 4496.4 | 15–5 | – | – | 85 | – | – | Q |
| | | 5–2 | – | – | 90 | – | – | Q |
| | | 2.0–0.5 | – | – | 95 | – | – | Q |
| | | 0.5–0.2 | – | – | 100 | – | – | |
| | | 0.2–0.1 | – | – | 100 | – | – | |

* Q, quartz; F, feldspar; B, = barite; P, pyrite; S, siderite. Smectite and barite in 33/12–1 samples are from drill mud contamination. 1 = 5–10%, 2 = 10–25%, no number = <5%.

scale late circulation of K-enriched compaction water is probably inappropriate for shallow Brent illitization. If a closed-system diagenetic model is used, then local supply of 'nutrients' for illite growth would rely upon late dissolution of K-sources (K-feldspar and mica). At low temperature, kinetic control of feldspar dissolution results in a slow supply of elements for illitization and explains the paucity of illite in shallow burial environments. The illite-precipitation reaction appears to have begun as overgrowths on detrital mica particles (Fig. 7A), which offered nucleation sites for initial illite crystallization.

In summary, shallow hydrocarbon-saturated Brent Group reservoirs provide the best opportunity to study processes important to early diagenesis and evaluate the role of provenance, facies, and meteoric diagenesis in feldspar distribution. Significant local variation in feldspar abundance often correlates with facies and grain-size (Glasmann *et al.* 1989*c*; Scotchman *et al.* 1989), and petrographic evidence of early diagenetic feldspar dissolution is abundant (Bjorlykke & Brendsdal 1986; Shanmugam 1988). The oxygen isotopic composition of kaolinite precipitated in association with early diagenetic dissolution of feldspar strongly supports meteoric diagenesis. In spite of major early dissolution, considerable detrital feldspar remains in shallow burial environments (Fig. 4). The continued presence of detrital feldspar, even after prolonged meteoric diagenesis, indicates that meteoric diagenesis is an unlikely mechanism for the complete absence of feldspar in deeper Brent reservoirs, which generally have a briefer history of meteoric flushing (Glasmann *et al.* 1989*a*). Furthermore, the paucity of illite in shallow Brent reservoirs strongly suggests that neither late Tertiary feldspar dissolution nor external supply of nutrients contributed significantly to illitization.

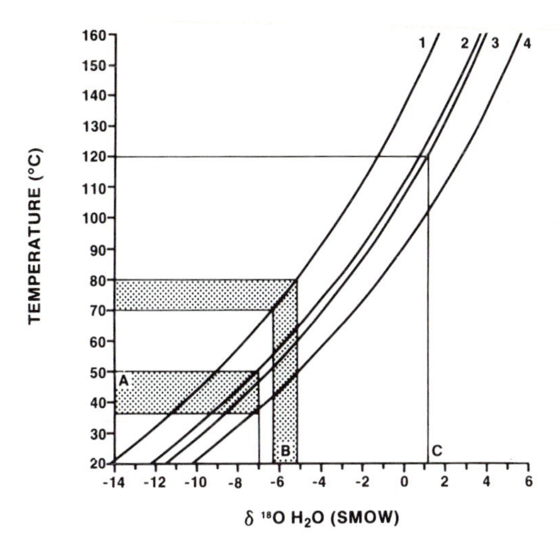


**Fig. 8.** Temperature versus pore water $\delta^{18}$O diagram for diagenetic kaolinite and illite, Brent Sandstone, northern North Sea. Curves 1, 2, & 4 represent 12, 14, & 16‰ kaolinite, respectively. Curve 3 represents 13‰ illite. Low temperature kaolinitization is consistent with unmodified Jurassic meteoric water (A). If a component of late, higher temperature kaolinite is assumed, only slight modification of original meteoric pore water is indicated (B). Fluid from which illite precipitated was generally 4–7‰ heavier than pore fluid during kaolinite precipitation (C).

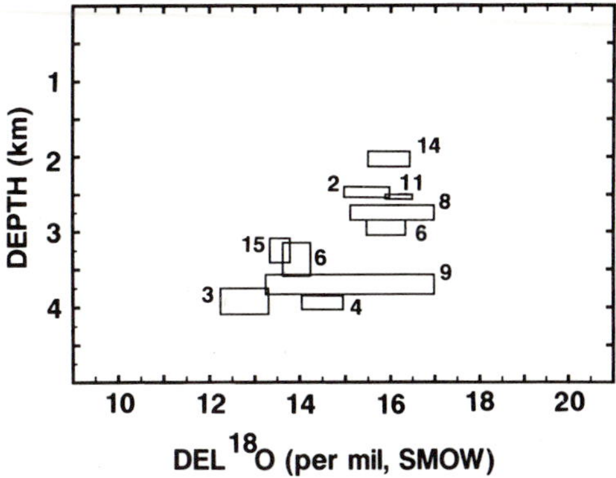


**Fig. 9.** Oxygen isotopic composition of kaolinite from several fields as a function of reservoir depth, Brent Sandstone, northern North Sea. Key to data locations the same as in Fig. 4. Sample set 15, Gullfaks Gamma Field. Thistle/Murchison isotopic data are from Brint *et al.* 1988, Alwyn data are from Hogg 1989. All other data are from Glasmann *et al.* (1989*a, c*) and Glasmann (unpublished data).

This indicates that late Tertiary diagenesis in the shallow Brent is largely a closed-system process that reflects the low temperature stability of K-feldspar (Aagaard *et al.* 1990).

## *Intermediate burial environments (2800–3700 m)*

The subsidence history of the Brent in intermediate burial environments (2800–3700 m) differs from shallow Brent reservoirs by the occurrence of a moderately thick upper Cretaceous section of marls and shales (Fig. 2). The early thermal history of intermediate depth Brent reservoirs is slightly warmer than that of shallow Brent fields (Fig. 3), reflecting burial beneath Oxfordian and Kimmeridgian mudstones that drape an unconformity at the top of the Brent (Glasmann *et al.* 1989*c*; Scotchman *et al.* 1989). Higher temperatures relative to the shallow Brent occurred during the Upper Cretaceous in response to post-rift subsidence and the thermal blanketing effect of added Cretaceous sediments (Fig. 3). Intermediate Brent reservoirs are more quartz and illite cemented than shallow reservoirs and show a trend of increasing cement abundance with increasing depth of burial (Giles *et al.* this volume; Hogg 1989; Scotchman *et al.* 1989). Feldspar abundance is generally similar to that of the shallow Brent, although it decreases in fields buried below 3500 m (Fig. 4; Glasmann *et al.* 1989*a*; Harris 1989; Hogg 1989; Saigal *et al.* 1988). Petrographic evidence of feldspar dissolution is abundant (Fig. 6) and distinct facies-related variations in feldspar abundance have been reported (Glasmann *et al.* 1989*c*; Scotchman *et al.* 1989; Hogg 1989).

Kaolinite is the dominant diagenetic clay mineral in intermediate Brent reservoirs (Table 2). SEM analyses clearly indicate that kaolinite formed prior to the major phase of quartz and illite cementation (Fig. 10). Illite is distributed through a wide range of clay particle sizes and is more abundant than in shallow Brent reservoirs (Table 2). Lath-shaped illite is the dominant crystal morphology (Figs 7 and 10) and lath length and thickness increase with increasing clay particle size, correlating with decreasing expandability of the illitic phase as determined by XRD (Glasmann *et al.* 1989*c*; McHardy *et al.* 1982). Parallel oriented bundles of illite laths, apparently grown from a common nucleation point, are common in coarse clay fractions, as are illite overgrowths on detrital grains (Fig. 7C). Such illite morpho-types are distinctly different in appearance from flocculated aggregates common in finer-grain fractions

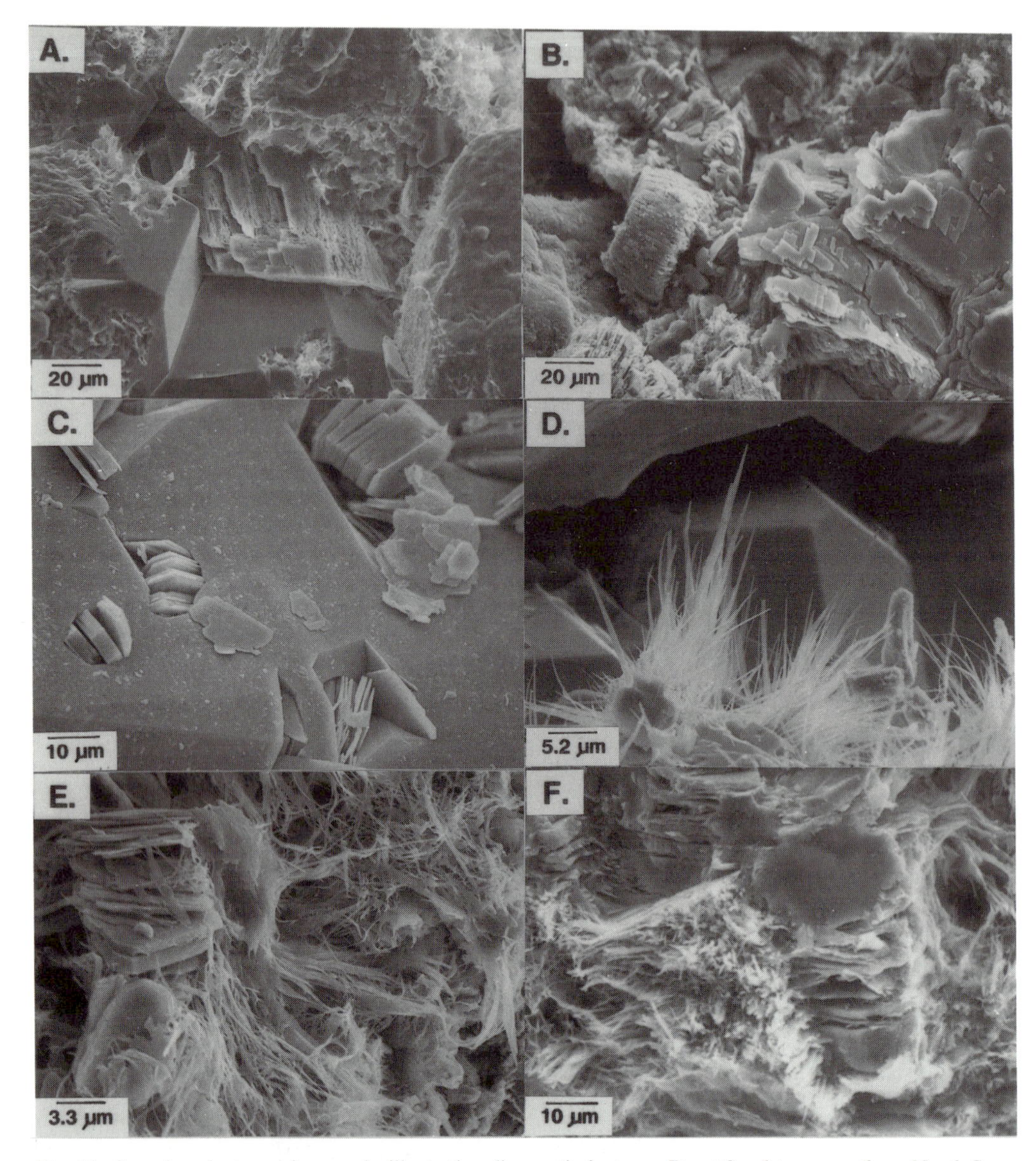


**Fig. 10.** Scanning electron micrographs illustrating diagenetic features, Brent Sandstone, northern North Sea. (**A**) K-feldspar overgrowth (partially dissolved) that has been partially enveloped by a later stage of quartz and illite cementation. Heather Field. (**B**) Partially dissolved K-feldspar and pore-filling kaolinite, Oseberg Field. Note morphologic similarity of feldspar dissolution textures to textures identified in TEM photos (Fig. 7B). (**C**) Early kaolinite that has been enveloped by a later phase of quartz cementation, Hild Field. (**D**) Tufts of fibrous illite growing from a substrate of partially dissolved K-feldspar. Note similarity with fibrous masses in TEM photos (Fig. 7C), Huldra Field. (**E**) Matted fibrous pore-bridging illite that has overgrown early-precipitated kaolinite, Huldra Field. (**F**) Completely illitized kaolinite booklet surrounded by dense mass of fibrous illite, central Viking Graben.

(Fig. 7E). The increased abundance of illite in the <0.5 µm clay correlates with a relative decrease in the abundance of fine-grain kaolinite compared to shallow Brent fields (Table 2). Clay-sized feldspar dissolution fragments are often present in the >0.5 µm size fraction and result in serious contamination of illite K-Ar ages (Glasmann *et al.* 1989*a*; Liewig *et al.* 1987);

however, the <0.1 $\mu$m size fractions typically consist of mineralogically pure illitie (Glasmann *et al.* 1989*c*; Hamilton *et al.* 1989).

*Clay mineral and isotopic evidence of diagenetic environment.* Brent kaolinite in intermediate burial environments is slightly lighter than shallow Brent kaolinite (Fig. 9), having an average $\delta^{18}O$ of about 14‰. While simple, low-temperature (30–40°C) meteoric leaching seems appropriate to explain the isotope chemistry of shallow Brent kaolinite, higher diagenetic temperatures (40–55°C) are inferred from stable isotopic analyses of kaolinite in intermediate Brent burial environments, if similar meteoric pore waters are assumed (Fig. 8). These elevated temperatures are inconsistent with kaolinitization directly beneath an unconformity (Bjorlykke 1984; Blanche & Whittaker 1978) and indicate that a portion of the kaolinite formed after early burial (Glasmann *et al.* 1989*a*, *c*; Hogg 1989; Scotchmann *et al.* 1989; Jourdan *et al.* 1987; Hancock & Taylor 1978; Sommer 1978).

Subsidence and thermal history models suggest that late Jurassic–early Cretaceous syn-rift burial was sufficient to produce mild warming of the Brent in intermediate burial environments relative to shallower diagenetic settings (Fig. 3). Syn-rift tectonism exposed many structures to erosion and provided topographic relief to drive meteoric recharge of the Brent aquifer (Glasmann *et al.* 1989*c*; Hogg 1989). Gravitational flow may result in the deep penetration of meteoric water into sedimentary basins (Bethke 1985, 1989; Lundegard & Land 1986) and greatly affect the temperature and isotopic composition of basin fluids (Longstaffe 1989). Gravitational meteoric recharge of the Brent was possible until Cretaceous marine submergence of regionally exposed highs on the East Shetlands and Horda Platforms and locally exposed islands within the Viking Graben. The duration of meteoric recharge may have been greater than 50 Ma and would have resulted in thorough flushing of the Brent with meteoric water. Extensive flushing of the Brent is consistent with the high water/rock ratio required to produce extensive feldspar leaching and kaolinitization without accompanying illite and quartz cementation. Basin-wide kaolinite oxygen isotopic composition varies about 2‰ (Fig. 9), suggesting that homogeneous fluid chemistry existed on a regional scale, consistent with a model of extensive meteoric flushing. While some have suggested that deep penetration of meteoric water has little effect on diagenesis (Bjorkum *et al.* 1990; Scotchman *et al.* 1989; Curtis 1983), these contentions do not seem supported by the growing body of available clay mineral isotopic data (Glasmann *et al.* 1989*c*; Hogg 1989; Longstaffe 1989).

In lieu of meteoric kaolinitization, others have suggested that kaolinite in the Brent precipitated in response to circulation of acidic pore water expelled from adjacent marine shales prior to hydrocarbon accumulation (Scotchman *et al.* 1989; Curtis 1983). The late dissolution and kaolinitization of 5–10% K-feldspar in the Brent would require a tremendous volume of shale-derived compaction water (Lundegard & Land 1986) and would result in complete displacement of early entrained meteoric pore water; however, the presence of brackish, isotopically light pore water in intermediate burial environments (salinities <1.5–2% NaCl, $\delta^{18}O$ from 0 to −5‰; Egeberg & Aagaard 1989; Glasmann *et al* 1989*c*; Bjorlykke *et al.* 1988) does not support a model of regional late flushing of the Brent in intermediate burial environments by saline, isotopically-evolved compaction water (Jourdan *et al.* 1987). Even if organic acids are invoked to augment feldspar dissolution, the removal of 5% feldspar in the Brent would exceed by many times the available reactive capacity of primary compaction water, without major fluid circulation (Bjorlykke *et al.* 1988; Lundegard & Land 1986). Because of such volumetric problems, focused flow, convection, and recycling of compaction water have been suggested as means of obtaining the required water/rock ratio for deep dissolution/cementation reactions (Bjorlykke 1984; Haszeldine *et al.* 1984; Wood & Hewett 1984), although none of these models adequately explains the regional isotopic homogeneity of kaolinitization in the Brent.

If a major phase of late kaolinitization of feldspar resulted from circulation of acidic marine shale compaction water, the isotopic composition of kaolinite should support this hypothesis in a manner consistent with other paragenetic evidence. Marine shale pore water, whose initial oxygen isotopic composition is close to that of sea water (0‰, SMOW, but probably *c.* −1‰ in the Jurassic; Sellwood *et al.* 1987), evolves with depth towards heavier values in response to shale diagenesis and rock/water interaction (Lundegard 1989; Yeh & Savin 1977). Thus, compaction water from marine shales can be assumed to have an oxygen isotopic composition greater than 0‰ (values of 2–4‰ are common, Yeh & Savin 1977). If all of the kaolinite ($\delta^{18}O$ = 12–14‰) precipitated from a water heavier than 0‰, the temperature of precipitation would exceed 100°C (Fig.

8). Such high formation temperatures were achieved subsequent to oil accumulation in most intermediate burial environments and are inconsistent with paragenetic observations that indicate that kaolinite precipitation precedes illitization, quartz cementation (Fig. 9C) and hydrocarbon accumulation. This paragenetic constraint limits the most likely maximum temperature of kaolinite precipitation to 70–80°C (Hogg 1989; Scotchman *et al.* 1989), and indicates that precipitation occurred from isotopically light pore water (lighter than −4‰ for 14‰ kaolinite, Fig. 8). Such isotopically light pore waters are unlike any known compaction water derived from marine shale (see Longstaffe 1989 for review of pore water isotopic evolution). Thus, major kaolinite precipitation was probably not associated with large-scale injection of acidic, isotopically heavy compaction water into the reservoir prior to hydrocarbon accumulation. Late kaolinitization may accompany albitization of detrital feldspar (Morad *et al.* 1990; Saigal *et al.* 1988), but the limited extent of albitization in the Brent at depths <3500 m suggests that this process has contributed very little to the total amount of kaolinite in intermediate burial environments (Saigal *et al.* 1988). At greater burial depth and higher temperature, the albitization of feldspar yields illite as a reaction product (Morad *et al.* 1990). The significance of albitization of deeply buried Brent sandstones will be discussed in greater detail in a later section.

Another alternative explanation of the moderate shift in kaolinite isotopic composition with depth (Fig. 9) is the possibility of depth/temperature-related isotopic re-equilibration. The extent of isotopic exchange in kaolinite depends on water/rock ratio and temperature and has a greater impact on hydrogen isotopic composition than oxygen (Longstaffe 1989). The impact of isotopic exchange is difficult to evaluate in the Brent, but would be more pronounced if large scale circulation of hot compaction water occurred. Since the likelihood of regional-scale, late-stage circulation of compaction fluid is low in the Brent in shallow and intermediate burial environments, (Bjorlykke *et al.* 1988) and generally not indicated by available formation water chemical data (Glasmann *et al.* 1989*a, c*; Egeberg 1988), isotopic exchange reactions probably had minimal effect on the isotopic composition of kaolinite. Thus, the constraints imposed by temperature, pore water chemistry, and kaolinite paragenesis suggest that kaolinite precipitation was predominantly related to late Jurassic–early Cretaceous meteoric recharge of the Brent aquifer.

Illite in intermediate depth Brent reservoirs occurs in all clay size fractions (Table 2) and K-Ar age determinations generally indicate an early Tertiary onset of precipitation (Fig. 11). The considerable scatter of K-Ar ages of mineralogically pure <0.1 μm illite primarily reflects the influence of local hydrocarbon accumulation history in a given reservoir. For example, hydrocarbon-saturated samples from reservoirs adjacent to major kitchen areas of the Viking Graben yield the oldest measured illite K-Ar ages (Fig. 11, Alwyn North and South, Hild, Huldra, Heimdal, and central Viking Graben; data from Glasmann *et al.* 1989*c*; Hogg 1989; Thomas 1986; unpublished Glasmann data). The absence of 'young' illite in the oil zones of near-graben fields suggests that oil accumulation caused a cessation of illite precipitation (Lee *et al.* 1985). The trend of measured ages is consistent with the predicted early maturation and migration of hydrocarbons out of the deep graben kitchen areas. Depth-related trends in measured illite ages have been related to progressive termination of diagenesis down structure in response to hydrocarbon accumulation (Glasmann *et al.* 1989*a, c*; Lee *et al.* 1989). Fields located farther from the graben, or farther from local kitchen areas, generally have younger measured illite ages (e.g., Heather, Veslefrikk, and Hutton fields, North Alwyn versus Alwyn South, Fig. 11; data of Glasmann *et al.* 1989*a, c*; Hogg 1989; Scotchman *et al.* 1989). In some cases, such as at Hutton and Alwyn South (Scotchman *et al.* 1989; Hogg 1989) illite precipitation appears unrelated to hydrocarbon accumulation, and measured K-Ar ages do not have a discernable depth-trend. The general restriction of illite ages to the Palaeogene indicates that illitization in intermediate burial environments was associated with palaeoburial depths in the range of 2000–3000 m and reservoir temperatures in excess of 75–85°C (Figs 2 and 3; Hamilton *et al.* this volume).

Pore water isotopic composition at the time of illite precipitation can be estimated by modeling formation temperature during the Palaeogene (Figs 3 and 8). For 30–50 Ma illite, reservoir temperatures in intermediate burial environments were generally in the range of 90–110°C (Fig. 3) and indicate pore water isotopic compositions between −2 to +3 permil for illite with $\delta^{18}O$ = 13–17‰ (Glasmann *et al.* 1989*c*; Hogg 1989). Such pore waters are considerably heavier than −8 to −5‰ pore waters involved in kaolinite cementation and suggest that kaolinite and illite could not have formed simultaneously. Modern pore water from inter-

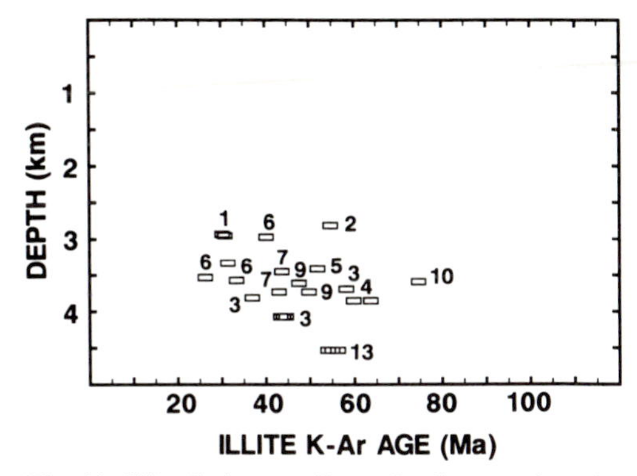


**Fig. 11.** Illite K-Ar ages, Brent Sandstone, plotted against present depth of burial. Data sources listed in Table 1 and key to fields studied is same as Fig. 4.

mediate depth Brent reservoirs typically has $\delta^{18}O$ values in the range of $-1$ to $+1‰$ (Glasmann *et al.* 1989*c*; Egeberg 1988). Thus, present pore water isotopic composition falls within the range predicted from the timing and temperature constraints of late Tertiary illitization.

The isotopic evolution of Brent pore water with increasing burial may result from in-situ rock–water interaction in a closed system, or reflect moderate addition of isotopically evolved compaction water during Tertiary illitization. In a closed system, reaction between detrital K-feldspar, kaolinite, and quartz (Huang *et al.* 1986) yields illite and quartz cement. Pore water would become progressively heavier through isotope exchange reactions as detrital feldspar and low temperature kaolinite dissolved. Since the precipitation of illite removes K from solution, the closed system reaction will continue until K sources are consumed, unless the system is 'opened' or equilibrium conditions are established. If K is provided from an external source, the precipitation of illite may be decoupled from late K-feldspar dissolution. Partial mixing of deep compaction water with older meteoric pore water in intermediate burial environments could result in modern pore water chemical and isotopic trends (Egeberg & Aagaard 1989; Glasmann *et al.* 1989*c*; Hogg 1989). A general increase in Cl with depth is consistent with a deep seated source of Cl and meteoric recharge at shallower depth (Egeberg & Aagaard 1989). Pore water salinity may vary on a local scale due to compartmentalization resulting from faulting, such as seen at Heather Field, where G Block and West Heather pore waters are several times saltier than the main part of the field (Glasmann *et al.* 1989*c*). Addition of saline compaction water has also been invoked to explain the weak albitization of detrital feldspar seen in several intermediate depth Brent fields (Hogg 1989; Saigal *et al.* 1988). Export of K from albitized sandstones to adjacent shales promotes illitization of smectite (Aagaard *et al.* 1990), which has been shown to be synchronous with early Tertiary illite cementation of the Brent in the Bergen High area (Glasmann *et al.* 1989*b*). These observations imply that diagenesis in intermediate burial environments occurs under at least partially open conditions, and that the degree of 'openness' varies from field to field, or even locally within fields in response to faults and reservoir heterogeneity.

Quartz overgrowth fluid inclusion data indicating migration of hot saline water into the intermediate burial environment further support the heterogeneity of pore water chemistry in intermediate Brent burial environments. Though often not representative of the temperature or salinity of modern Brent pore water (Glasmann *et al.* 1989*c*; Hogg 1989; Jourdan *et al.* 1987), inclusion data do indicate the nature of migrating compaction fluids which mixed with trapped palaeometeoric water to produce the present Brent pore water. The localization of fluid inclusions near major faults and high permeability zones (Burley *et al.* 1989; Glasmann *et al.* 1989*c*) indicates that saline compaction fluids probably have a deep, pre-Jurassic source.

Whether closed or open diagenesis prevails, both systems suggest that illite should be forming in the present burial environment, yet the youngest illite K-Ar ages reported generally lie between 17–25 Ma (Hamilton *et al.* this volume; Glasmann *et al.* 1989*c*). The absence of very young illite K-Ar ages reflects the effect of mixing of recently-formed illite with older illite. New illite precipitated as overgrowth on a precursor crystal will result in a reduction of the measured age of the crystal, but will not yield the age of the added material. Thus, the younger illite K-Ar age in water zones relative to the hydrocarbon zone is evidence of a long and ongoing history of diagenesis. More problematic is the occurrence of early Tertiary illite ages in water-saturated Brent, implying an absence of Neogene illitization (e.g. *c.* 50 Ma illite in G Block versus 27 Ma illite in the water zone of northwest Heather, Glasmann *et al.* 1989*c*). Since continued illite precipitation requires continuous replenishment of essential 'nutrients', the occurrence of old illite in water-saturated sandstones indicates that nutrients were cut off. Cutting off of nutrients is not consistent with closed system diagenesis, especially when considerable detrital K-feldspar still exists in the host sandstone (6–14% in G Block, Heather Field, Glasmann *et al.* 1989*c*).

In summary, available petrographic data, combined with clay mineral and formation water isotopic data suggest that an early phase of feldspar dissolution occurred during shallow burial in response to circulation of low temperature meteoric pore water ($\delta^{18}O$ −8 to −5‰, Fig. 8) that resulted in precipitation of kaolinite. The interrelationships of burial history, illite precipitation, hydrocarbon accumulation, formation temperature, and evolution of pore water isotopic composition do not support significant regional precipitation of late-stage kaolinite from hot, acidic basin compaction water. As Cretaceous subsidence progressed, meteoric recharge was cut off, the diagenetic environment changed from open to predominantly closed, and the system gradually warmed with continued burial. During early Tertiary time, rapid compaction resulted in mixing of palaeometeoric pore water with minor amounts of saline, isotopically evolved water. As temperatures increased, chemical reaction of K-feldspar and kaolinite to produce illite accelerated, aided by onset of albitization from import of Na from compacting shales and/or migration of evaporitic brine (Aagaard *et al.* 1990; Hamilton *et al.* this volume). Differential fluid mixing is evident from the variable salinity of modern pore water in intermediate Brent reservoirs, which ranges from slightly brackish to slightly more saline than sea water (Egeberg & Aagaard 1989; Glasmann *et al.* 1989*c*; Hogg 1989). The occurence of long, thick, well-crystallized fibrous illite in intermediate burial environments suggests that 'nutrients' for illite growth were readily available, in contrast to the poorly developed illite of shallow Brent fields. Thus, the major differences between shallow and intermediate Brent diagenesis revolve around the increased salinity and higher temperature of deeper reservoirs and the impact of these variables on K-feldspar stability and clay precipitation.

## *Deep burial environments (>3700 m)*

Detrital K-feldspar is generally absent in Brent reservoirs whose present burial depth exceeds 3700 m (Fig. 4). These deep reservoirs are characterized by relatively high temperatures (130–60°C, Fig. 3) and are generally situated in axial graben settings, where rapid early Cretaceous subsidence followed late Cimmerian rifting. Associated with the absence of K-feldspar is the partial to complete illitization of kaolinite (Table 2, Figs 7 and 10), the formation of abundant fibrous and platy illite, quartz cement (Figs 6, 7 and 10; Glasmann *et al.* 1989*a*; Giles *et al.* this volume; Hogg & Fallick pers. comm.), late carbonate, and albitization of detrital feldspar (Saigal *et al.* 1988). Platy illite morphologies are particularly abundant in fine clay separates of the deep Brent (Fig. 7), and apparently develop from the pseudomorphous replacement of precursor kaolinite (Fig. 10; Glasmann 1989). The secondary porosity generated by the diagenetic removal of detrital feldspar in deep Brent reservoirs is not associated with an increase in reservoir quality, primarily because of the extensive illitization and quartz cementation that accompanies feldspar destruction (Giles *et al.* this volume). The temperature of late feldspar removal in the Brent (present reservoir temperature greater than 130°C) is similar to the temperature range of complete feldspar dissolution reported in other sedimentary basins (Lundegard & Trevena 1990; Milliken *et al*, 1989; Land 1984), suggesting that the complete dissolution of feldspar in reservoir sandstones may have application as a crude geothermometer; however, to correctly interpret the temperature of dissolution, the timing of dissolution and thermal history of the well must be known.

The late dissolution of K-feldspar in the deep Brent was apparently synchronous with early Tertiary hydrocarbon accumulation (Glasmann *et al.* 1989*a*). Maturation modeling and illite K-Ar ages suggest that hydrocarbon accumulation began at least by Palaeocene time (Ungerer *et al.* 1990; Glasmann *et al.* 1989*a*; Thomas *et al.* 1985; Field 1985; Goff 1983). Since the dissolution of feldspar in deep reservoirs is closely linked with late illitization, measured K-Ar ages of diagenetic illite can be used to constrain the timing and temperature of feldspar dissolution. Calculated temperature history models for the deep Brent suggest that formation temperatures during illite precipitation ranged from 120 to 140°C (Fig. 3), slightly cooler than present thermal conditions.

*Isotopic evidence of diagenetic environment.* Isotopic reconstruction of diagenetic environment in the deep Brent is complicated by clay recrystallization and sample contamination. Mineralogically pure samples of kaolinite are difficult to obtain from deeply buried Brent sandstones because of the abundance of coarse crystalline illite (Fig. 7C, E) and destruction of kaolinite during illitization (Hogg 1989; Hancock & Taylor 1978; Sommer 1978). The isotopic composition of contaminated kaolinitic clay samples will be skewed towards lighter values by the addition of a high temperature illite component. Thus, the difference in the average $\delta^{18}O$ of kaolinite from South Alwyn (15.0 ± 2.5‰, Hogg 1989), Hild (14.3 ± 0.5‰, Glasmann unpublished data), and Huldra (12.0 ± 0.5‰, Glasmann *et al.* 1989*a*; Fig. 9) probably reflects the abundance of illite in the Huldra samples (Table 2). The isotopic composition of relatively uncontaminated kaolinite in deep burial environments is generally 1–2‰ lighter than kaolinite from shallower fields (Fig. 9), suggesting that a portion of the kaolinite in deep reservoirs may have formed at higher temperature or from heavier pore water (Fig. 8); however, even with the lightest kaolinite (12‰, Huldra Field, Fig. 9), pore water during kaolinite formation was clearly of meteoric origin. On a regional scale, the isotopic data for kaolinite from all burial environments show an average scatter less than about 5‰ (Fig. 9). This regional homogeneity in isotopic composition is further evidence that major kaolinitization could not have occurred after Cretaceous subsidence and development of thermo/chemical differences between the various Brent fields.

Illite isotopic data from deeply buried Brent reservoirs are generally several permil lighter than associated kaolinite (e.g., 12.8‰ illite v. 15‰ kaolinite at South Alwyn (Hogg 1989)), consistent with precipitation at higher temperature from isotopically heavier pore water (Fig. 8). Whereas the isotopic composition of illite from intermediate burial environments shows considerable scatter (e.g. 13.2–17.3‰ at Heather, Glasmann *et al.* 1989*c*), the range of illite $\delta^{18}O$ values from the deep Brent is more narrow (12.8 ± 0.7‰ at South Alwyn, 13.0‰ at Huldra, 13.4‰ at 24/6–1 (data from Hogg 1989; Glasmann *et al.* 1989*a*, and unpublished Glasmann data, respectively)). The greater isotopic homogeneity of illite in the deep Brent suggests that illite formation has been more strongly influenced by deep diagenetic processes than by processes inherited during progressive burial. If illitization reflects a temperature-related kinetic control, then the maximum rate of illite precipitation would have occurred during rapid early Tertiary burial and associated heating (Figs 2 & 3). Thus, minor amounts of illite formed during earlier burial would be followed by a flush of late illitization and the isotopic variability inherited during lower temperature illite growth would be overprinted by the chemistry of deep precipitation reactions.

The isotopic homogeneity of illite in deep Brent reservoirs could also reflect important depth-related differences in pore water chemistry. Deep Brent pore water is typically more saline and isotopically heavier than sea water and much more saline than pore waters in shallow and intermediate depth reservoirs (Egeberg & Aagaard 1989; Bjorlykke *et al.* 1988). The high salinity of deep Viking Graben pore water has been related to long-distance lateral and vertical migration of Permian or Triassic evaporitic brine (Egeberg & Aagaard 1989; Egeberg 1988). Partial mixing of evaporitic brine with meteoric pore water explains the variable salinity and isotopic composition of intermediate depth Brent pore water (Egeberg & Aagaard 1989); however, palaeometeoric pore water seems to have been completely displaced from the deep Brent. Migration of brine into the deep Brent aquifer has had a major impact on reservoir diagenesis and the distribution and migration history needs to be better documented.

The late dissolution of K-feldspar in the deep Brent may have occurred by one of several mechanisms: closed-system reaction (Huang *et al.* 1986), open-system dissolution involving organic acids (Jourdan *et al.* 1987), or albitization by highly sodic brine migration (Saigal *et al.* 1988). The closed-system model, which appropriately explains illitization of feldspar in shallow and intermediate Brent diagenetic environments, falls short of explaining the observed relationship between K-feldspar dissolution, illitization, albitization, and the saline formation water chemistry that characterizes the deep Brent (Aagaard *et al.* 1990). If illitization followed the closed-system model, the reaction would progress slowly from its initiation at a temperature of about 75°C and accelerate with rising temperature until complete consumption of reactants (primarily K-feldspar and kaolinite) stopped the reaction. Because more extensive illitization results during higher temperature reaction, measured illite K-Ar ages will be biased towards the final period of illite precipitation and K-Ar data will probably have a narrow age distribution, especially if heating was rapid (Fig. 12; Lee *et al.* 1989). However, a similar narrow illite K-Ar age pattern would result from a brine migration event that facilitated a pulse of albitization and consumed available K-feldspar. Since illite diagenesis during albitization requires an Al source (kaolinite or smectite), illite precipitation in the deep Brent would continue until either all the K-feldspar or all the kaolinite was consumed, assuming imported Al is minimal. The absence of Neogene illite K-Ar ages in water-saturated deep Brent reservoirs (Fig. 12) indicates an absence of late Tertiary illite and implies that nutrients for illitization were exhausted during early Tertiary

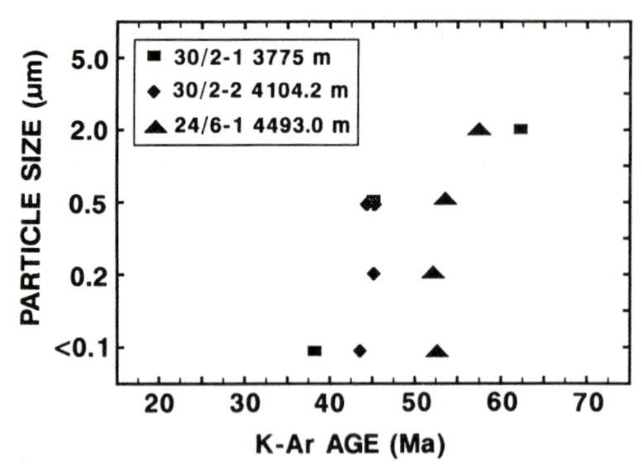


**Fig. 12.** Measured K-Ar age of different illite size separates from deep Brent reservoirs in which feldspar has been completely dissolved. Overlapping ages of fine- and coarse-grained illitic clay reflect the lack of detrital contamination and indicate that illite crystallization occurred at a rapid rate.

burial. Furthermore, explosive crystallization is indicated by the overlapping ages of coarse and fine-grained illite (Fig. 12). Such explosive crystal growth would not be expected of closed-system diagenesis and is further evidence that illitization was probably related to a geologically rapid alteration of pore water chemistry.

The high salinity (2 to 5 times sea water salinity) and evolved isotopic composition (+1.3 to +3.9‰) of deep pore waters in clastic reservoirs of the Viking Graben indicate a replacement of early-trapped meteoric pore water by evaporitic brine (Egeberg & Aagaard 1989) and point to a Tertiary episode of open diagenesis. Permian evaporites may occur at great depth in the Viking Graben (Egeberg & Aagaard 1989) and upward migration through the Triassic via major faults is one possible migration scenario for entry of brine in to the Brent. Brine migration may have been initiated by basin compression associated with Laramide tectonics and rapid early Tertiary sedimentation/compaction. Brine migration does not preclude illitization of feldspar by closed-system reaction. Brine chemistry may be 'open' with respect to a particular element, yet 'closed' with respect to K necessary for illitization. If 'open' and a majority of K was derived from the brine, the requirement for intraformational supply of K through feldspar dissolution would be reduced or eliminated and illite precipitation would be decoupled from feldspar dissolution. If 'closed', K-feldspar in the Brent would continue to provide the major source of elements required for illitization and, once consumed, illitization would cease, regardless of continued brine migration.

Albitization of detrital feldspar in the Brent shows a relationship to hydrocarbon accumulation and late carbonate precipitation, indicating an early Paleogene age for onset of brine migration (Glasmann *et al.* 1989*a, b*; Saigal *et al.* 1988; Nedkvitne 1987). Albitization alone is apparently insufficient to explain the complete destruction of K-feldspar in the deep Brent. The total amount of albitized K-feldspar in the deep Brent (average about 6%, Saigal *et al.* 1988) is only about half of the average amount of late feldspar dissolution (10–12% inferred from Fig. 4). Therefore, it seems likely that a combination of albitization, more 'ideal' closed-system (Huang *et al.* 1986), and open-system reactions (Glasmann *et al.* 1989*b*) were active during diagenesis of the deep Brent.

Glasmann *et al.* (1989*b*) presented evidence that vertical escape of K-bearing fluids from the Brent facilitated illitization of overlying shales in the Bergen High area during latest Cretaceous time. The transport of excess K liberated during albitization of detrital feldspar in the Brent to adjacent shales is predicted by mass balance calculations for deep diagenesis (Aagaard *et al.* 1990). Vertical fluid migration may have been related to hydraulic fracturing associated with release of overpressure (Buhrig 1989). Illite K-Ar data and crystal morphology suggest that illitization of shales occurred during a period of relatively high shale porosity at approximately the same time that the underlying Brent was undergoing initial illitization (Glasmann *et al.* 1989*b*). These observations indicate the complexity of diagenetic reactions in the deep Brent. The net effect of multifaceted attack on detrital feldspar in the deep Brent is complete dissolution (Fig. 4).

In summary, the absence of K-feldspar in deep Brent reservoirs is related to elevated early Tertiary burial temperature (>120°C) and modification of older meteoric pore water by Late Cretaceous/Early Tertiary introduction of isotopically evolved, saline brine. High $a_{Na+}/a_{K+}$ in the brine and high formation temperature promoted albitization of detrital feldspar, which released K ions, promoting illitization of early-formed kaolinite and major precipitation of fibrous, pore-bridging illite. Illite oxygen isotope data support an interpretation of high temperature formation (100–140°C) from isotopically heavy pore water (>0‰). Additional K-feldspar dissolution occurred in the deep Brent that was apparently unrelated to albitization, probably reflecting a continuance of late dissolution processes active in shallower Brent fields. Although significant secondary porosity may be generated during deep burial by feldspar dissolution, the precipitation of abundant illite and quartz cement seriously reduces reservoir quality. Complete dissolution of K-feldspar removes the major source of contamination from illite clay separates and allows K-Ar dating of a wide range of illite crystal sizes without problems related to detrital Ar contamination. The overlapping ages of both coarse- and fine-grained illitic clays indicate that illitization was a geologically rapid process and is indirect evidence of geologically rapid early Tertiary feldspar dissolution in the deep Brent.

## Conclusions

Reactions involving K-feldspar have played an important role in the diagenetic evolution of middle Jurassic Brent Group sandstones in the northern Viking Graben and show a strong relationship to depth of burial, temperature, and formation water chemistry. Early feldspar dissolution during shallow burial ($<1$ km) resulted from aggressive circulation of meteoric water that was probably driven by late Cimmerian syn-rift topography. This phase of feldspar dissolution is characterized by the precipitation of kaolinite ($\delta^{18}O = +15$ to $+17‰$) at temperatures between 20 and 55°C from isotopically light meteoric water ($-8$ to $-5‰$). Shallow leaching by meteoric water was extensive, but did not result in complete removal of detrital feldspar, as evidenced by the abundance of feldspar in shallow and intermediate-depth Brent fields. Cretaceous subsidence and burial of the Brent by marine shales changed the pore water system from one of aggressive meteoric circulation to static pore water whose composition was slowly modified by the gradual escape of compaction water from the basin. The brackish modern pore water chemistry of the Brent in shallow and intermediate burial environments suggests that migration of compaction water was rather limited. During this later phase of subsidence and changing pore water chemistry, K-feldspar became increasingly unstable as temperature increased and imported Na initiated albitization reactions. Initial illitization begins slowly at temperature around 65–75°C, but increases in rate with higher temperature, linked to the increased availability of K ions resulting from reaction of K-feldspar and kaolinite. Onset of albitization of detrital feldspar during latest Cretaceous/early Tertiary burial also provided K ions for illite precipitation. Hydrocarbon accumulation arrested illitization and extent of feldspar dissolution in some instances. In deep burial environments, late-stage dissolution of feldspar results in complete removal of K-feldspar. Dissolution was probably promoted by a combination of high temperature, high $a_{Na+}/a_{K+}$, and introduction of evaporitic brine and was accompanied by albitization, illitization of kaolinite, extensive precipitation of fibrous illite, quartz cementation, and precipitation of late carbonate cement. The complete dissolution of K-feldspar has application as a crude geothermometer and is evidence of palaeotemperatures in excess of 120° during early Tertiary hydrocarbon accumulation in deep Brent fields. Secondary porosity generated by complete dissolution of feldspar in the deep Brent generally does not benefit reservoir quality due to the reduction of permeability caused by extensive illite precipitation.

I thank J. Wood for technical assistance during this study. Early versions of the manuscript benefitted from reviews by P. Lundegard and B. Penny. G. Ytreland and J. Ellice-Flint provided assistance and encouragement during initial phases of the project. Finally, I thank Unocal for support of this study and encouragement to publish these results.

## References

Aagaard, P., Egeberg, P. K., Saigal, G. C., Morad, S., & Bjorlykke, K. 1990. Diagenetic albitization of detrital K-feldspars in Jurassic, Lower Cretaceous, and Tertiary clastic reservoir rocks from offshore Norway, II: Formation water chemistry and kinetic considerations. *Journal of Sedimentary Petrology*, **60**, 575–581.

Aronson, J. L. & Hower, J. 1976. The mechanism of burial metamorphism of argillaceous sediments 2. radiogenic argon evidence. *Geological Society of America Bulletin*, **87**, 738–744.

Barker, C. E. & Reynolds, T. J. 1984. Preparing doubly polished sections of temperatures-sensitive sedimentary rocks. *Journal of Sedimentary Petrology*, **54**, 635–636.

Bethke, C. M. 1985. A numerical model of compaction-driven groundwater flow and heat transfer and its application to the paleohydrology of intracratonic sedimentary basins. *Journal of Geophysical Research*, **90**, 6817–6828.

—— 1989. Modeling subsurface flow in sedimentary basins. *Geologische Rundschau*, **78**, 129–154.

Bjorkum, P. A., Mjos, R., Walderhaug, O. & Hurst, A. 1990. The role of the late Cimmerian unconformity for the distribution of kaolinite in the Gullfaks Field, northern North Sea. *Sedimentology*, **37**, 395–406.

Bjorlykke, K. 1984. Formation of secondary porosity: How important is it? *In*: Mc Donald D. A. & Surdam, R. C. (eds) *Clastic Diagenesis*. American Association of Petroleum Geologists Memoir, **37**, 277–286.

—— & Brendsdal, A. 1986. Diagenesis of the Brent Sandstone in the Statfjord Field, North Sea. *In*: Gautier, D. L. (ed.) *Roles of Organic Matter in*

*Sediment Diagenesis*. Society of Economic Paleontologists Mineralogists Special Publication, **38**, 157–167.

——, MO, A. & PALM, E. 1988. Modelling of thermal convection in sedimentary basins and its relevance to diagenetic reactions. *Marine and Petroleum Geology*, **5**, 338–351.

BLACKBOURN, G. A. 1984. Diagenetic history and reservoir quality of a Brent sand sequence. Clay *Mineralogy*, **19**, 377–389.

BLANCHE, J. B. & WHITAKER, J. H. McD. 1978. Diagenesis of part of the Brent Sand Formation (Middle Jurassic) of the northern North Sea basin. *Journal of the Geological Society, London*, **135**, 73–82.

BRINT, J. F., HASZELDINE, R. S. HAMILTON, P. J. & FALLICK, A. E. 1988. Isotope diagenesis and fluid movement in Brent sandstones, northern North Sea. *In*: *Clay Diagenesis in Hydrocarbon Reservoirs and Shales*. Conference Abstracts. Clay Minerals Group/Mineralogical Society, Cambridge, UK, 2.

BUHRIG, C. 1989. Geopressured Jurassic reservoirs in the Viking Graben: modelling and geological significance. *Marine & Petroleum Geology*, **6**, 31–48.

BURLEY, S. D., MULLIS, J. & MATTER, A. 1989. Timing diagenesis in the Tartan Reservoir (UK North Sea): constraints from combined cathodoluminescence microscopy and fluid inclusion studies. *Marine & Petroleum Geology*, **6**, 98–120.

CLAYTON, R. N. & MAYEDA, T. K. 1963. The use of bromine pentafluoride in the extraction of oxygen from oxides and silicates for isotopic analysis. *Geochimica et Cosmochimica Acta*, **27**, 43–52.

——, O'NEIL, J. R. & MAYEDA, T. K. 1972. Oxygen isotope exchange between quartz and water. *Journal of Geophysical Research*, **77**, 3057–3067.

CURTIS, C. D. 1978. Possible links between sandstone diagenesis and depth-related geochemical reactions occurring in enclosing mudstones. *Journal of Geological Society, London*, **135**, 107–117.

—— 1983. Geochemistry of porosity enhancement and reduction in clastic sediments. *In*: BROOKS, J. (ed.) *Petroleum Geochemistry and Exploration of Europe*. Geological Society, London, Special Publication, **12**, 113–125.

DAHL, B., NYSAETHER, E., SPEERS, G. C. & YUKLER, A. 1987. Oseberg area — integrated basin modeling. *In*: BROOKS, J. & GLENNIE, K. (eds) *Petroleum Geology of North West Europe*. Graham & Trotman, London, 1029–1038.

DICKSON, J. A. D. 1965. A modified staining technique for carbonates in thin section. *Nature*, **219**, 587.

EGEBERG, P. K. 1988. *Water–rock interaction in the diagenetic environment; deep sea sediments and deeply buried chalks and sandstones*. PhD, Dissertation, University of Oslo, Norway.

—— & AAGAARD, P. 1989. Origin and evolution of formation water from oil fields on the Norwegian shelf. *Applied Geochemistry*, **4**, 131–142.

ESLINGER, E. V. 1971. *Mineralogy and oxygen isotope ratios of hydrothermal and low-grade metamorphic argillaceous rocks*. PhD Dissertation, Case Western Reserve University, Cleveland, Ohio.

FIELD, J. D. 1985. Organic geochemistry in exploration of the Northern North Sea. *In*: DORE, A. G. *et al.* (eds) *Petroleum Geochemistry and Exploration of the Norwegian Shelf*. Norwegian Petroleum Society, Graham & Trotman, 39–57.

FRIEDMAN, G. M. 1971. Staining. *In*: CARVER, E. (ed.) *Procedures in Sedimentary Petrology*. John Wiley & Sons, Inc., New York, 511–531.

GILES, M. R., STEVENSON, S., MARTIN, S. V., CANNON, S. J. C., HAMILTON, P. J., MARSHALL, J. D. & SAMWAYS, G. R. 1992. The reservoir properties and diagenesis of the Brent Group: a regional perspective. *This volume*.

GLASMANN, J. R. 1989. Illite crystal morphology as an indicator of diagenetic environment. *American Association of Petroleum Geologists Bulletin*, **73**, 539.

——, CLARK, R. A., LARTER, S., BRIEDIS, N. A. & LUNDEGARD, P. D. 1989*a*. Diagenesis and hydrocarbon accumulation, Brent Sandstone (Jurassic), Bergen High area, North Sea. *American Association of Petroleum Geologists Bulletin*, **73**, 1341–1360.

——, LARTER, S., BRIEDIS, N. A. & LUNDEGARD, P. D. 1989*b*. Shale diagenesis in the Bergen High area, North Sea. *Clays and Clay Minerals*, **37**, 97–112.

——, LUNDEGARD, P. D., CLARK, R. A., PENNY, B. K. & COLLINS, I. D. 1989*c*. Geochemical evidence for the history of diagenesis and fluid migration: Brent Sandstone, Heather field, North Sea. *Clay Minerals*, **24**, 255–284.

GOFF, J. C. 1983. Hydrocarbon generation and migration from Jurassic source rocks in the E. Shetlands Basin and Viking Graben of the northern North Sea. *Journal of the Geological Society, London*, **140**, 445–474.

HALLETT, D. 1981. Refinement of the geological model of the Thistle field. *In*: ILLING, L. V. & HOBSON, G. D. (eds) *Petroleum Geology of the Continental Shelf of NW Europe*. Heyden, London, 315–325.

HAMILTON, P. J., BLACKBOURN, G. A., MCLACHLAN, W. A. & FALLICK, A. E. 1987. Isotopic tracing of the provenance and diagenesis of Lower Brent Group sands: *In*: BROOKS, J. & GLENNIE, K. (eds). *Petroleum Geology of North West Europe*. Graham & Trotman, London, 939–949.

——, GILES, M. R. & AINSWORTH, P. 1992. K-Ar dating of illites in Brent Group reservoirs: a regional perspective. *This volume*.

——, KELLY, S. & FALLICK, A. E. 1989. K-Ar dating of illite in hydrocarbon reservoirs. *Clay Minerals*, **24**, 215–232.

HANCOCK, N. J. & TAYLOR, A. M. 1978. Clay mineral diagenesis and oil migration in the Middle Jurassic Brent Sand Formation. *Journal of the Geological Society, London*, **135**, 69–72.

HARRIS, N. B. 1989. Diagenetic quartzarenite and destruction of secondary porosity: an example from the Middle Jurassic Brent sandstone of northwest Europe. *Geology*, **17**, 361–364.

HASZELDINE, R. S., SAMSON, I. M. & CORNFORD, C. 1984. Quartz diagenesis and convective fluid movements: Beatrice Oilfield, UK, North Sea. *Clay Minerals*, **19**, 391–402.

HOGG, A. J. C. 1989. *Petrographic and isotopic constraints on the diagenesis and reservoir properties of the Brent Group Sandstones, Alwyn South, northern U.K. North Sea*. PhD dissertation, University of Aberdeen, Scotland.

HUANG, W. L., BISHOP, A. M., & BROWN, R. W. 1986. The effect of fluid/rock ratio on feldspar dissolution and illite formation under reservoir conditions. *Clay Minerals*, **21**, 585–602.

JACKSON, M. L. 1979. *Soil chemical analysis–advanced course*. 2nd Edition, 11th printing: Published by the author, Madison, WI, 53705.

JOURDAN, A., THOMAS, M., BREVART, O., ROBSON, P., SOMMER, F. & SULLIVAN, M. 1987. Diagenesis as the control of the Brent Sandstone reservoir properties in the greater Alwyn area (East Shetlands Basin). *In*: BROOKS, J. & GLENNIE, K. (eds) *Petroleum Geology of North West Europe*. Graham & Trotman, London, 951–961.

KANTOROWICZ, J. 1984. The nature, origin and distribution of authigenic clay minerals from Middle Jurassic Ravenscar and Brent Group Sandstones. *Clay Minerals*, **19**, 359–375.

LAND, L. S. 1983. The application of stable isotopes to studies of the origin of dolomite and to problems of diagenesis of clastic sediments. *In*: ARTHUR, M. A. (ed.) *Stable Isotopes in Sedimentary Geology*. Society of Economic Paleontologists and Mineralogists Short Course No. **10**, Chapter 4.

LAND, L. S. 1984. Frio Sandstone diagenesis, Texas Gulf Coast: a regional isotopic study. *In*: MC DONALD, D. A. & SURDAM, R. C. (eds) *Clastic Diagenesis*. American Association of Petroleum Geologists Memoir, **37**, 47–62.

LEE, M. ARONSON, J. L. & SAVIN, S. M. 1985. K-Ar dating of the time of gas emplacement in Rotliegendes Sandstone, Netherlands. *American Association of Petroleum Geologists Bulletin*, **69**, 1381–1385.

——, —— & —— 1989. Timing and conditions of Permian Rotliegendes Sandstone diagenesis, southern North Sea. *American Association of Petroleum Geologists Bulletin*, **73**, 195–215.

LIEWIG, N., CLAUER, N. & SOMMER, F. 1987. Rb-Sr and K-Ar dating of clay diagenesis in Jurassic sandstone oil reservoir, North Sea. *American Association of Petroleum Geologists Bulletin*, **71**, 1467–1474.

LONGSTAFFE, F. J. 1989. Stable isotopes as tracers in clastic diagenesis. *In*: HUTCHEON, I. E. (ed.) *Short Course in Burial Diagenesis*. Mineralogical Association of Canada, **15**, 201–278.

LONOY, A., AKSELSEN, J. & RONNING, K. 1986. Diagenesis of a deeply buried sandstone reservoir: Hild field, northern North Sea. *Clay Minerals*, **21**, 497–511.

LUNDEGARD, P. D. 1989. Temporal reconstruction of sandstone diagenetic histories. *In*: HUTCHEON, I. E. (ed.) *Short Course in Burial Diagenesis*. Mineralogical Association of Canada, **15**, 161–200.

—— & LAND, L. S. 1986. Carbon dioxide and organic acids: their role in porosity enhancement and cementation, Paleogene of the Texas Gulf Coast. *In*: GAUTIER, D. L. (ed.) *Roles of Organic Matter in Sediment Diagenesis*, SEPM Special Publication, **38**, 129–146.

—— & TREVENA, A. S. 1990. Sandstone diagenesis in the Pattani Basin (Gulf of Thailand): history of water-rock interaction and comparison with the Gulf of Mexico. *Applied Geochemistry*, **5**, 669–685.

MALLEY, P., JOURDAN, A. & WEBER, F. 1986. Study of fluid inclusions in the silica overgrowths of the North Sea reservoir sandstones: A possible new diagenetic history of the Brent of the Alwyn area. *Conple Rendus de l'Academie des Sciences, Paris*, **302**, Serie II, 9, 653–658.

MCCREA, J. M. 1950. On the isotopic chemistry of carbonates and a paleotemperature scale. *Journal of Chemistry & Physics*, **18**, 849–857.

MCHARDY, W. J. WILSON, M. J. & TAIT, J. M. 1982. Electron microscope and X-ray diffraction studies of filamentous illitic clay from sandstones of the Magnus Field. *Clay Minerals*, **17**, 23–39.

MILLIKEN, K. L., MCBRIDE, E. F. & LAND, L. S. 1989. Numerical assessment of dissolution versus replacement in the subsurface destruction of detrital feldspars, Oligocene Frio Formation, South Texas. *Journal of Sedimentary Petrology*, **59**, 740–757.

MORAD, S., BERGAN, M., KNARUD, R. & NYSTUEN, J. P. 1990. Albitization of detrital plagioclase in Triassic reservoir sandstones from the Snorre Field, Norwegian North Sea. *Journal of Sedimentary Petrology*, **60**, 411–425.

MORTON, A. C. & HUMPHREYS, B. 1983. The petrology of Middle Jurassic sandstones from the Murchison field, North Sea. *Journal of Petroleum Geology*, **5**, 245–260.

NEDKVITNE, T. 1987. *Klastisk diagenese og dannelse av sekundaer porositet i Brentgruppen pa Huldra Feltet*. PhD Dissertation, University of Oslo, Norway.

O'NEIL, J. R. & KHARAKA, Y. K. 1976. Hydrogen and oxygen isotope exchange reactions between clay minerals and water. *Geochimica et Cosmochimica Acta*, **40**, 241–246.

PEARSON, M. J. & SMALL, J. S. 1988. Illite–smectite diagenesis and paleotemperatures in northern North Sea Quaternary to Mesozoic shale sequences. *Clay Minerals*, **23**, 109–132.

SAIGAL, G. C., MORAD, S., BJORLYKKE, K., EGEBERG, P. K., & AAGAARD, P. 1988. Diagenetic albitization of detrital K-feldspar in Jurassic, Lower Cretaceous, and Tertiary clastic reservoir rocks from offshore Norway, I. Textures and origin. *Journal of Sedimentary Petrology*, **58**, 1003–1013.

SAVIN, S. M. & LEE, M. 1988. Isotopic studies of phyllosilicates. In: BAILEY, S. W. (ed.) *Hydrous Phyllosilicates (exclusive of micas). Reviews in Mineralogy*, **19**, 189–223.

SCOTCHMAN, I. C., JOHNS, L. H. & MILLER, R. S. 1989. Clay diagenesis and oil migration in Brent Group sandstones of NW Hutton field, UK North Sea. *Clay Minerals*, **24**, 339–374.

SELLWOOD, I. C., SCOTT, J. JAMES, B., EVANS, R. & MARSHALL, J. 1987. Regional significance of 'dedolomitization' in Great Oolite reservoir facies of southern England. *In*: BROOKS, J. & GLENNIE, K. (eds) *Petroleum Geology of North West Europe*. Grahm & Trotman, London, 129–137.

SHANMUGAM, G. 1988. Origin, recognition, and importance of erosional unconformities in sedimentary basins. *In*: KLEINSPEHN, K. L. & PAOLA, C. (eds). *New Perspectives in Basin Analysis*. Springer Verlag, New York, 83–108.

SOMMER, F. 1978. Diagenesis of Jurassic sandstones in the Viking Graben. *Journal of the Geological Society, London*, **135**, 63–68.

STEIGER, R. H. & JAGER, E. 1977. Subcommission on geochronology: Convention on the use of decay constants in geo- and cosmochronology. *Earth and Planetary Science Letters*, **36**, 359–362.

THOMAS, B. M., MOLLER-PEDERSEN, P., WHITAKER, M. F. & SHAW, N. D. 1985. Organic facies and hydrocarbon distributions in the Norwegian North Sea. *In*: DORE, A. G., EGGEN, S. S., HOME, P. C., MARNE, R. & THOMAS, B. M. (eds). *Geochemistry in Exploration of the Norwegian Shelf*. Norwegian Petroleum Society, Graham & Trotman, London, 3–26.

THOMAS, M. 1986. Diagenetic sequences and K-Ar dating in Jurassic sandstones, central Viking Graben: Effects on reservoir properties. *Clay Minerals*, **21**, 695–710.

UNGERER, P., BURRUS, J., DOLIGEZ, B., CHENET, P. Y. & BESSIS, F. 1990. Basin evaluation by integrated two-dimensional modeling of heat transfer, fluid flow, hydrocarbon generation, and migration. *American Association of Petroleum Geologists Bulletin*, **74**, 309–335.

WOOD, J. R. & HEWETT, T. A. 1984. Reservoir diagenesis and convective fluid flow. *In*: MC DONALD, D. A. & SURDAM, R. C. (eds) *Clastic Diagenesis*. American Association of Petroleum Geologists Memoir, **37**, 99–111.

YEH, H.-W. & SAVIN, S. M. 1977. Mechanism of burial metamorphism of argillaceous sediments: 3. O-isotope evidence. *Geological Society of American Bulletin*, **88**, 1321–1330.

# Burial diagenesis of Brent sandstones: a study of Statfjord, Hutton and Lyell fields

NICHOLAS B. HARRIS

*Exploration Research and Services Division, Conoco Inc., Ponca City, Oklahoma 74603, USA*

**Abstract:** Brent sandstone compositions vary systematically with increasing depth and temperature in three North Sea fields. At Statfjord field (*c.* 2500 m), sandstones are arkoses, lithic arkoses and subarkoses. At Hutton field (*c.* 3050 m), sandstones are subarkoses and quartzarenites, whereas at Lyell field (*c.* 3500 m), sandstones are mostly quartzarenites. Compositions are transformed by pervasive dissolution of plagioclase at relatively shallow depth, dissolution of K-feldspar at greater depth, and alteration of detrital mica to kaolinite or illite. Quartz cement is a minor constituent at Statfjord, mostly forming early during meteoric water flushing. Quartz cement abundance increases significantly at Hutton and particularly at Lyell, where quartz cement comprises up to 13% of rock volume; the increase in quartz cement coincides with dissolution of detrital quartz along stylolites and microstylolites at temperatures greater than 90–100°C as well as the dissolution of feldspar.

The dominant clay at Statfjord is kaolinite; at Hutton, illite forms as a pore-lining precipitate and, at Lyell, also by alteration of previously formed kaolinite. The transition from a kaolinite-dominated to an illite-dominated sandstone is probably controlled by several factors, among these the dissolution of K-feldspar, which may be due either to an influx of organic acids from the overlying Heather and Kimmeridge Clay Formations or to thermodynamic instability of the assemblage kaolinite-illite-K-feldspar at temperatures greater than 100°C. Elevated silica activity due to quartz dissolution on stylolites and microstylolites may also induce illite formation.

Chemical analysis of sandstone samples shows the Brent undergoes systematic loss of sodium and, to a lesser extent, potassium during burial diagenesis. The sodium loss is consistent with observed plagioclase dissolution. The potassium loss indicates that the potassium in illite is derived from the Brent itself, probably from dissolved K-feldspar, and not from an external source such as a basinal brine.

Sandstones of the Brent Group in the northern North Sea rank among the most important petroleum reservoirs of the world, hosting seven accumulations with in-place reserves in excess of one BBO (Coustau *et al.* 1988) and many more smaller fields. At shallow depths, the Brent exhibits remarkable reservoir properties, with porosities commonly in excess of 30% and permeabilities in excess of 1000 mD (for example, Gullfaks field; Erichsen *et al.* 1987); consequently shallow Brent wells are typically prolific producers. For example, early wells at Statfjord (2500 m) flowed initially at *c.* 30 000 BOPD, with higher rates recorded in some wells (Roberts *et al.* 1987). However, reservoir properties degrade rapidly with depth. Porosities decrease from an average of 27% at Statfjord field (Roberts *et al.* 1987) to approximately 16% at Lyell. Permeabilities also decrease rapidly with depth. The Brent at Statfjord, for example, averages 2500 mD permeability, whereas the Brent at 3500 m has average permeabilities more than a full order of magnitude lower (e.g NW Hutton; Scotchman *et al.* 1989). Diagenesis of Brent sandstones merits study, therefore, simply by virtue of its impact on Brent exploration and production.

Studies of Brent diagenesis also have implications beyond their immediate relevance to the northern North Sea. The Brent province is relatively well constrained in terms of its burial and thermal histories. The Brent can thus be tested for open or closed system diagenesis. Have the sandstones been flushed by dilute meteoric water, or conversely, have they been charged by highly concentrated brines? Are components necessary to the formation of authigenic minerals imported from some external source, or are they derived from the sandstone itself? These questions are crucial both to our understanding of fluid flow in sedimentary basins and to the inherent predictability of sandstone reservoir properties.

This paper describes the diagenesis of the Brent sandstones in three fields, Statfjord, Hutton and Lyell (Fig. 1), which represent

*From* Morton, A. C., Haszeldine, R. S., Giles, M. R. & Brown, S. (eds), 1992, *Geology of the Brent Group*. Geological Society Special Publication No. 61, pp. 351–375.

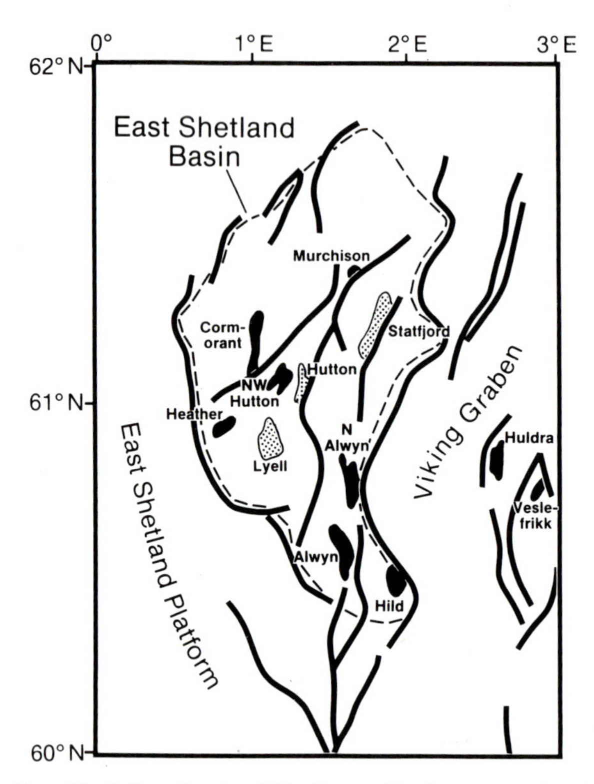


**Fig. 1.** Map of the northern North Sea, showing fields discussed in the paper and major tectonic features. Statfjord, Hutton and Lyell fields are indicated by cross-hatched pattern. Other fields mentioned are indicated by ruled pattern.

relatively shallow, medium and deep levels of burial respectively. While these fields differ by at most 1200 m in burial depth, they exhibit striking variation in style of diagenesis and consequent reservoir properties. Three aspects of Brent diagenesis are of particular interest: dissolution of feldspar; formation of quartz cement and associated stylolitization; and authigenesis/diagenesis of the clay minerals kaolinite and illite. The fates of these minerals are linked by a complex network of chemical reactions. The observed transition from a kaolinite-dominated to an illite-dominated clay assemblage can be related in part to changes to silica activity and $[K^+]/[H^+]$. Problems of mass balance of chemical components in the Brent, particularly with respect to sodium and potassium are addressed with whole rock geochemical data.

## Geology of the Brent Group

The Brent is a clastic unit of late Toarcian to Bathonian age (Graue *et al.* 1987), found in the East Shetland Basin, Viking Graben and Bergen High areas of the northern North Sea. It consists of five formations which, in ascending stratigraphic order, are (Fig. 2): Broom, Rannoch, Etive, Ness, and Tarbert. The Broom generally consists of coarse-grained, poorly sorted sandstone. Its origin is complex and includes fan delta, fluvial, mouth bar, beach, and offshore facies (Brown & Richards 1990; Kirk 1980; Eynon 1981; Nagy *et al.* 1984; Budding & Inglin 1981). Recent studies (Brown & Richards 1990) suggest that the Broom is time-equivalent to the Oseberg Formation in the Bergen High area, and that these formations represent detritus shed from the west and east sides of the Viking Graben, respectively.

The Rannoch, Etive and Ness Formations represent facies of a large, northward-prograding delta system more than 225 km long (Brown *et al.* 1987; Graue *et al.* 1987). The Rannoch consists of fine-grained, silty and micaceous sandstones, deposited in a shallow marine shoreface setting (Brown *et al.* 1987). The Etive consists of clean, fine- to coarse-grained sand-

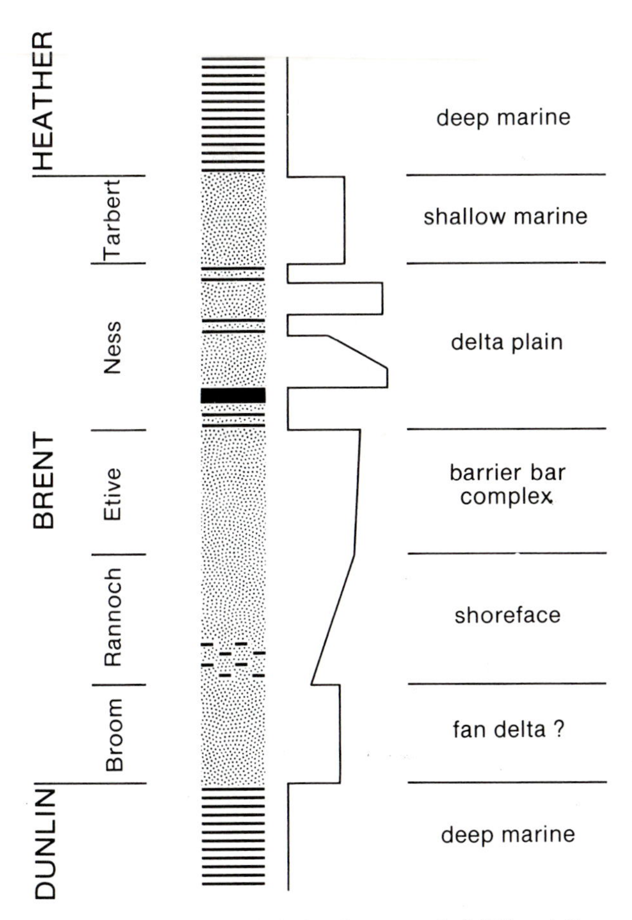


**Fig. 2.** Stratigraphy of the Brent Group, modified after Brown *et al.* (1987) and Graue *et al.* (1987).

stone deposited on a tidally-modified barrier-bar complex (Brown *et al.* 1987). The Ness is a complex unit, consisting of fine- to coarse-grained sandstones, shales and coals, deposited on a delta front and delta plain (Brown *et al.* 1987; Graue *et al.* 1987; Cannon *et al.* this volume). The Tarbert Formation consists of marine sandstones; it marks the start of the marine transgression that culminated in deposition of the Middle to Upper Jurassic marine shales, the Heather and Kimmeridge Clay Formations (Brown *et al.* 1987).

The provenance(s) of the Brent sandstones is not yet well established. While regional sedimentological and biostratigraphic studies clearly show a systematic northward progradation and southward regression of the Brent delta, geochemical studies based on garnet composition (Morton 1985, this volume) suggest considerable contributions from high-grade metamorphic rocks of the East Shetland platform and the Fennoscandian shield.

## Fields in this study

Statfjord, Hutton and Lyell fields represent shallow, moderate and deep levels of burial in the Brent province, averaging approximately 2500 m, 3050 m and 3500 m burial depth subsea, respectively. Average temperatures, pressures and number of samples from each Brent formation are summarized in Table 1. Temperatures range from approximately 90°C at Statfjord to as high as 120°C at Lyell. While this temperature range is relatively narrow, it brackets the 100° isotherm which, in the East Shetland Basin (discussed later), is a critical temperature for stylolite formation and quartz grain dissolution. Pore pressure gradients, calculated from drill

**Table 1.** *Fields, reservoir conditions and number of samples studied*

| | Statfjord | Hutton | Lyell |
|---|---|---|---|
| Depth of oil column | 2360–2586 m ss | 2803–3078 m ss | 3385–3644 m ss |
| (samples from) | (2417–2708) | (2819–3017) | (3385–3578) |
| Temperature | 89–93°C | 106–109°C | 107–121°C |
| Pressure | 383 bar at 2469 m ss | 433 bar at 2940 m ss | 493 bar at 3515 m ss |
| | | No. of samples | |
| Broom | 1 | 5 | 4 |
| Rannoch | 13 | 7 | 5 |
| Etive | 16 | 20 | 9 |
| Ness | 14 | 14 | 13 |
| Tarbert | 0 | 0 | 4 |

m ss, = metres sub-sea

stem tests are virtually identical in all three fields, averaging 0.147 bar $m^{-1}$.

## Methods

The study is largely based on petrographic data from optical microscope, X-ray diffraction (XRD) and scanning electron microscope (SEM) examination. Data on the detrital and diagenetic composition of sandstones were obtained from standard thin sections. Whole rock XRD data are similar, but for the most part, are not presented here because lithic grains and mica cannot be readily distinguished from matrix clay and detrital quartz from quartz cement with this technique. The thin sections were stained for potassium and plagioclase feldspar. A minimum of 400 points per thin section were counted.' Sandstone compositions are classified by the system of McBride (1963). In data tabulations, quartz includes monocrystalline and polycrystalline quartz and chert (present at most in trace amounts). 'Mica and lithics' is dominated by biotite and muscovite, with small amounts of sedimentary and volcanic rock fragments, garnet, hornblende and titanite or leucoxene. Pores within dissolved feldspar were treated as porosity and, therefore, not included in estimates of rock compositions; however, the type of dissolved feldspar was recorded for purposes of reconstructing sandstone compositions. Matrix material is not represented in compositional diagrams; however, the data set includes only matrix-poor sandstones.

Data on the clay mineral component of the sandstones were obtained from SEM and clay XRD analysis of the $<2\mu$ m size fraction. Calculated clay compositions are considered semi-quantitative and are based on I/Ic values obtained from analysis of 50/50 mixtures of corundum and kaolinite, illite and chlorite standards. Samples were run both before and after glycolation to test for expandability; only small amounts of slightly expandable clay were found.

Whole rock chemical compositions of sandstones were obtained by X-ray fluorescence (XRF) analysis of pressed powder samples, following a peroxide wash to remove residual oil in some samples. Several US Geological Survey, National Bureau of Standards and Conoco rock standards were used, including sandstone, limestone, dolomitic limestone, granite, granodiorite, and pure quartz. Matrix corrections were not used because standards of similar composition to the unknowns were used.

## Petrography of Brent Sandstones

### *Statfjord field*

Statfjord samples came from four wells, two of which penetrated the Brent above the oil–water contact and two of which largely penetrated the Brent below the oil–water contact. Well locations are shown in Fig. 3.

Brent sandstones at Statfjord field range in composition from arkose to lithic arkose to subarkose (Fig. 4A and Table 2). The Ness and Etive sandstones are typically the most quartzose, usually subarkosic. Rannoch sandstones are richer in mica and feldspar, usually lithic arkoses. In general, quartz content correlates positively with grain size. The Ness and Etive sandstones are the coarsest (Fig. 4A) whereas the Rannoch sandstones are the finest.

Compositional variation in the Brent at Statfjord is, in part, probably a depositional effect. In modern depositional systems, Odom *et al.* (1976) showed that feldspar content decreases with increasing grain size. This effect may be enhanced by diagenetic dissolution of feldspar, since the higher permeability of coarser sandstones will result in a greater flux of formation water; consequently they could undergo more feldspar dissolution.

*Feldspar dissolution.* Approximately 15% of feldspar grains are partially or completely dissolved (Fig. 5A). Dissolution of other detrital grains contributes relatively little to the total

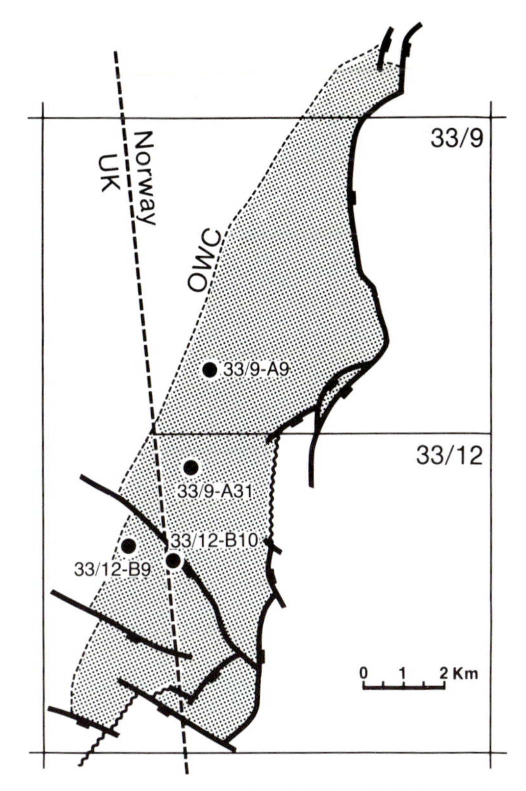


**Fig. 3.** Location of Statfjord field wells in the study Modified after Roberts *et al.* (1987). OWC is oil–water contact.

porosity of the rock. The resulting secondary porosity averages 3.6% of the rock volume. Bjørlykke & Brensdal (1986) also concluded that feldspar dissolution contributes relatively little to the total porosity of the sandstones. The volume of secondary porosity resulting from feldspar dissolution is essentially constant in the wells studied here, showing no apparent increase toward the Brent erosional subcrop (Harris 1990*a*).

Petrographic data indicate that plagioclase is much more susceptible than K-feldspar to dissolution. This is shown in Fig. 6A, in which the tips of the arrows represent the present composition of the sandstones (following feldspar dissolution) and the base of the arrows represents the reconstructed composition of the sandstones, based on identification of feldspar remnants. If plagioclase were the only feldspar dissolved, the arrows would point directly away from the plagioclase apex; if plagioclase and K-feldspar were dissolved in proportion to their abundance in the sandstone, the arrows would point directly toward the quartz apex. The orientation of the arrows indicates that disproportionate amounts of plagioclase are dissolved. Bjørlykke & Brensdal (1986) have also described preferential dissolution of albitic lamellae in perthitic feldspar in the Brent at Statfjord Field. Rare grains of albitized feldspar were tentatively identified by SEM on the basis of chemical composition and morphology; however, these may be relatively pure, partially dissolved detrital albite.

*Quartz cement.* Small amounts of quartz cement are present (Fig. 5B and C), averaging approximately 0.5% of rock volume, also noted by Bjørlykke & Brensdal (1986). Some of the quartz cement clearly formed early, predating calcite in a 'dogger' (Fig. 5C). Elsewhere, the timing of quartz cementation is less clear. Quartz cement at Statfjord probably formed by meteoric flushing during Upper Jurassic and/or lowermost Cretaceous (Bjørlykke & Brensdal 1986). Meteoric water, charged with atmospheric $CO_2$ and slightly acidic; dissolved feldspar, and precipitated kaolinite and quartz cement. Quartz grain pressure-solution is absent except in one relatively deep sample (Fig. 5D) and cannot have significantly contributed to quartz cementation.

*Clays.* Kaolinite is the dominant clay mineral in the Brent at Statfjord field (Fig. 5E), averaging 3.0% of the rock volume (thin section point count data). In mica-poor sandstones, kaolinite content ranges between 0 and 3.5%, occurring as isolated pseudo-hexagonal plates and as long 'worms' (Fig. 5E). Kaolinite content is considerably higher (up to 8.9%) in micaceous sandstones in the Rannoch and Ness, where it is formed by alteration of biotite and muscovite. Of the micas, biotite is more susceptible to such alteration, commonly forming siderite as a by-product.

Small amounts of illite are present in micaceous sandstones in water-leg wells. It forms small wisps draping the ends of mica flakes (Fig. 5F) and less commonly kaolinite.

*Other diagenetic features.* K-feldspar overgrowths are common, though not abundant, amounting to less than 0.5% of rock volume (Fig. 5B). They are generally contemporaneous with or pre-date quartz cement. They are often dissolved in preference to the detrital cores, suggesting that the overgrowths formed early, possibly due to dissolution of plagioclase feldspar and alteration of mica to kaolinite. Formations later may have become undersaturated with respect to K-feldspar.

**Table 2.** *Mineralogical composition of Brent sandstones at Statfjord, Hutton and Lyell fields, showing average composition and ranges for each formation in the three fields*

| | Statfjord field | | | | Hutton field | | | |
|---|---|---|---|---|---|---|---|---|
| | Broom | Rannoch | Etive | Ness | Broom | Rannoch | Etive | Ness |
| *Detrital constituents* | | | | | | | | |
| Quartz | 76.1% | 59.3% (51.5–71.2) | 77.1% (51.6–89.6) | 79.2% (66.5–87.4) | 76.9% (67.0–83.1) | 66.1% (36.3–86.2) | 80.0% (64.8–94.3) | 91.1% (83.7–97.0) |
| Plagioclase | 1.6% | 9.3% (4.7–14.5) | 5.9% (2.0–16.7) | 5.8% (2.1–11.7) | 6.5% (4.7–10.4) | 9.5% (4.5–17.1) | 6.3% (0.6–13.8) | 4.0% (1.3–8.5) |
| K-feldspar | 20.3% | 16.6% (13.4–23.4) | 12.7% (6.3–20.6) | 12.1% (4.7–20.9) | 11.0% (5.4–18.8) | 12.1% (7.5–15.2) | 10.7% (4.9–19.6) | 3.2% (0.4–11.5) |
| Rock fragments | 2.0% | 4.0% (0.7–7.7) | 1.4% (0.0–4.0) | 0.7% (0.0–3.4) | 5.1% (0.0–10.4) | 2.1% (0.3–5.6) | 1.6% (0.0–5.9) | 0.7% (0.0–2.9) |
| Muscovite | 0.0% | 5.2% (1.2–11.3) | 1.5% (0.0–8.5) | 1.7% (0.0–7.6) | 0.4% (0.0–1.1) | 10.6% (0.0–33.8) | 2.3% (0.0–13.2) | 0.6% (0.0–2.1) |
| Biotite | 0.0% | 3.8% (0.0–8.6) | 1.0% (0.0–8.8) | 0.7% (0.0–2.4) | 0.2% (0.0–0.4) | 0.7% (0.3–1.3) | 0.1% (0.0–0.3) | 0.0% (0.0–0.0) |
| *Diagenetic constituents* | | | | | | | | |
| Quartz OGs | 1.1% | 0.4% (0.0–1.7) | 0.7% (0.0–4.1) | 0.4% (0.0–2.2) | 0.8% (0.0–2.0) | 0.6% (0.0–1.7) | 0.9% (0.30–1.9) | 1.5% (0.0–2.8) |
| K-feldspar OGs | 0.0% | 0.0% (0.0–0.0) | 0.06% (0.0–0.40) | 0.09% (0.0–0.40) | 0.0% | 0.0% (0.0–0.0) | 0.0% (0.0–0.0) | 0.0% (0.0–0.0) |
| Calcite | 15.5% | 1.3% (0.0–5.3) | 0.2% (0.0–1.7) | 0.02% (0.0–0.3) | 16.6% (2.3–48.2) | 7.3% (0.3–41.4) | 1.3% (0.0–9.3) | 0.2% (0.0–1.5) |
| Siderite | 0.0% | 0.5% (0.0–2.6) | 0.2% (0.0–0.9) | 0.4% (0.0–2.1) | 0.0% (0.0–0.0) | 0.1% (0.0–0.4) | 0.1% (0.0–0.2) | 0.1% (0.0–0.2) |
| Kaolinite | 0.0% | 5.4% (3.1–9.0) | 2.6% (0.2–7.1) | 2.5% (0.0–7.2) | 5.0% (0.6–8.3) | 2.6% (0.0–4.9) | 2.1% (0.2–5.0) | 1.2% (0.0–3.8) |
| Pyrite | 1.7% | 1.3% (0.0–2.2) | 1.8% (0.0–7.0) | 1.19% (0.2–4.4) | 0.1% (0.0–0.2) | 1.9% (0.5–4.8) | 0.7% (0.0–3.3) | 0.7% (0.0–3.6) |
| *Porosity* | | | | | | | | |
| Primary Porosity | 5.7% | 25.8% (17.0–32.4) | 26.2% (7.6–33.4) | 30.3% (21.7–39.3) | 11.7% (0.0–18.8) | 14.1% (0.0–26.5) | 21.7% (2.5–28.1) | 25.4% (13.4–29.8) |
| Secondary Porosity | 11.5% | 2.9% (1.2–4.6) | 3.1% (1.2–6.3) | 1.7% (0.4–2.8) | 3.6% (0.0–5.7) | 2.5% (0.2–7.8) | 2.4% (0.7–5.5) | 2.1% (0.4–6.0) |

Table 2 (*Continued*)

| | Lyell field | | | | |
|---|---|---|---|---|---|
| | Broom | Rannoch | Etive | Ness | Tarbert |
| *Detrital constituents* | | | | | |
| Quartz | 90.1% (80.9–98.4) | 78.8% (66.6–89.2) | 90.4% (84.4–94.4) | 96.7% (88.1–99.7) | 95.4% (92.4–97.6) |
| Plagioclase | 1.0% (0.3–1.3) | 1.3% (0.7–2.0) | 1.9% (0.6–5.6) | 0.6% (0.0–3.6) | 0.9% (0.3–1.4) |
| K-feldspar | 6.3% (0.7–12.2) | 7.7% (6.9–9.2) | 5.8% (1.2–9.6) | 1.6% (0.0–6.7) | 2.4% (1.0–5.7) |
| Rock Fragments | 1.0% (0.3–2.0) | 1.1% (0.3–1.7) | 0.9% (0.0–2.9) | 0.5% (0.0–2.5) | 1.0% (0.3–2.3) |
| Muscovite | 1.2% (0.0–4.3) | 9.8% (1.2–18.5) | 0.8% (0.3–1.7) | 0.4% (0.0–2.4) | 0.2% (0.0–0.6) |
| Biotite | 0.2% (0.0–0.6) | 1.3% (0.0–4.3) | 0.1% (0.0–0.6) | 0.1% (0.0–1.2) | 0.2% (0.0–0.7) |
| *Diagenetic constituents* | | | | | |
| Quartz OGs | 5.3% (2.1–8.6) | 3.8% (2.3–5.5) | 2.7% (1.6–4.3) | 6.3% (0.5–13.3) | 6.7% (5.0–8.9) |
| K-feldspar OGs | 0.0% (0.0–0.0) | 0.0% (0.0–0.0) | 0.0% (0.0–0.0) | 0.0% (0.0–0.0) | 0.0% (0.0–0.0) |
| Calcite | 3.1% (0.0–12.5) | 1.0% (0.0–1.8) | 0.1% (0.0–0.8) | 2.52% (0.0–38.6) | 0.1% (0.0–0.3) |
| Siderite | 0.0% (0.0–0.0) | 0.5% (0.0–1.1) | 0.1% (0.0–0.6) | 0.1% (0.0–0.4) | 0.0% (0.0–0.0) |
| Kaolinite | 5.3% (2.8–7.3) | 5.2% (2.7–8.0) | 3.0% (1.1–6.9) | 1.0% (0.0–2.6) | 5.1% (4.7–5.5) |
| Pyrite | 1.1% (0.0–2.2) | 2.9% (1.3–5.7) | 1.2% (0.5–3.5) | 0.6% (0.0–2.2) | 0.2% (0.0–0.5) |
| *Porosity* | | | | | |
| Primary Porosity | 14.7% (11.9–17.7) | 12.5% (8.9–15.0) | 14.7% (9.6–20.8) | 12.0% (0.0–20.9) | 15.6% (11.4–18.5) |
| Secondary Porosity | 2.6% (1.8–4.0) | 3.4% (2.5–3.9) | 3.7% (1.8–4.9) | 2.2% (0.0–5.8) | 3.2% (2.6–4.1) |

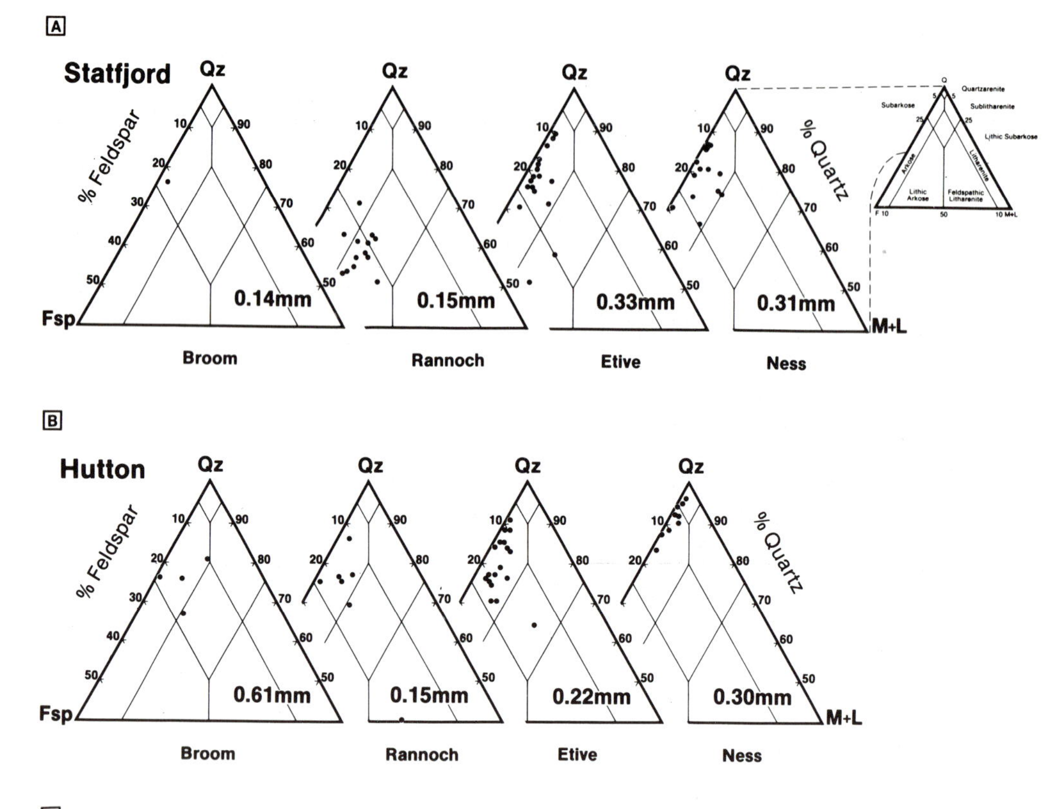


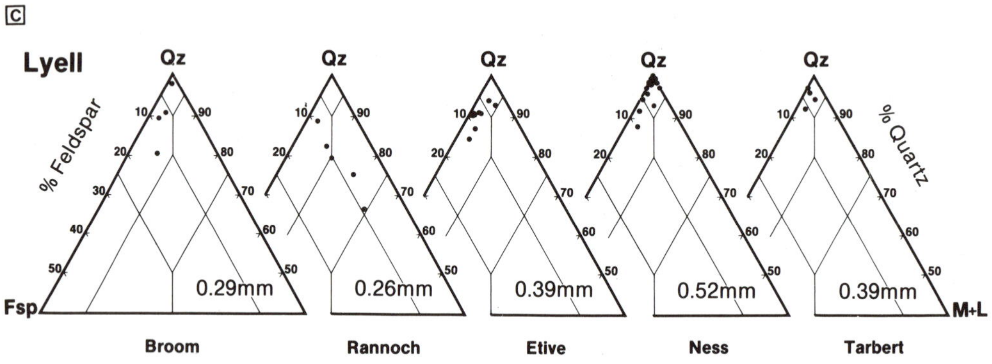


**Fig. 4.** Composition of Statfjord, Hutton, and Statfjord field sandstones, depicted on a quartz–feldspar–lithics + mica diagram. Classification scheme of McBride (1963) is used. Average grain size for each formation is noted. (**A**) Statfjord; (**B**) Hutton; (**C**) Lyell.

Calcite is a minor constituent at Statfjord field. Rare samples in the Broom are tightly cemented by poikilotopic calcite (Fig. 5C), which formed at the surface or at very shallow burial depths. The calcite in these doggers is locally partially dissolved, creating moderate volumes (5%) of secondary porosity. However, there is no evidence for the past presence of large volumes of calcite cement, as suggested by Schmidt & McDonald (1979), such as the widespread occurrence of corroded calcite or volumes of undercompacted sand grains. Small amounts of siderite (0–2%) occur throughout the Brent.

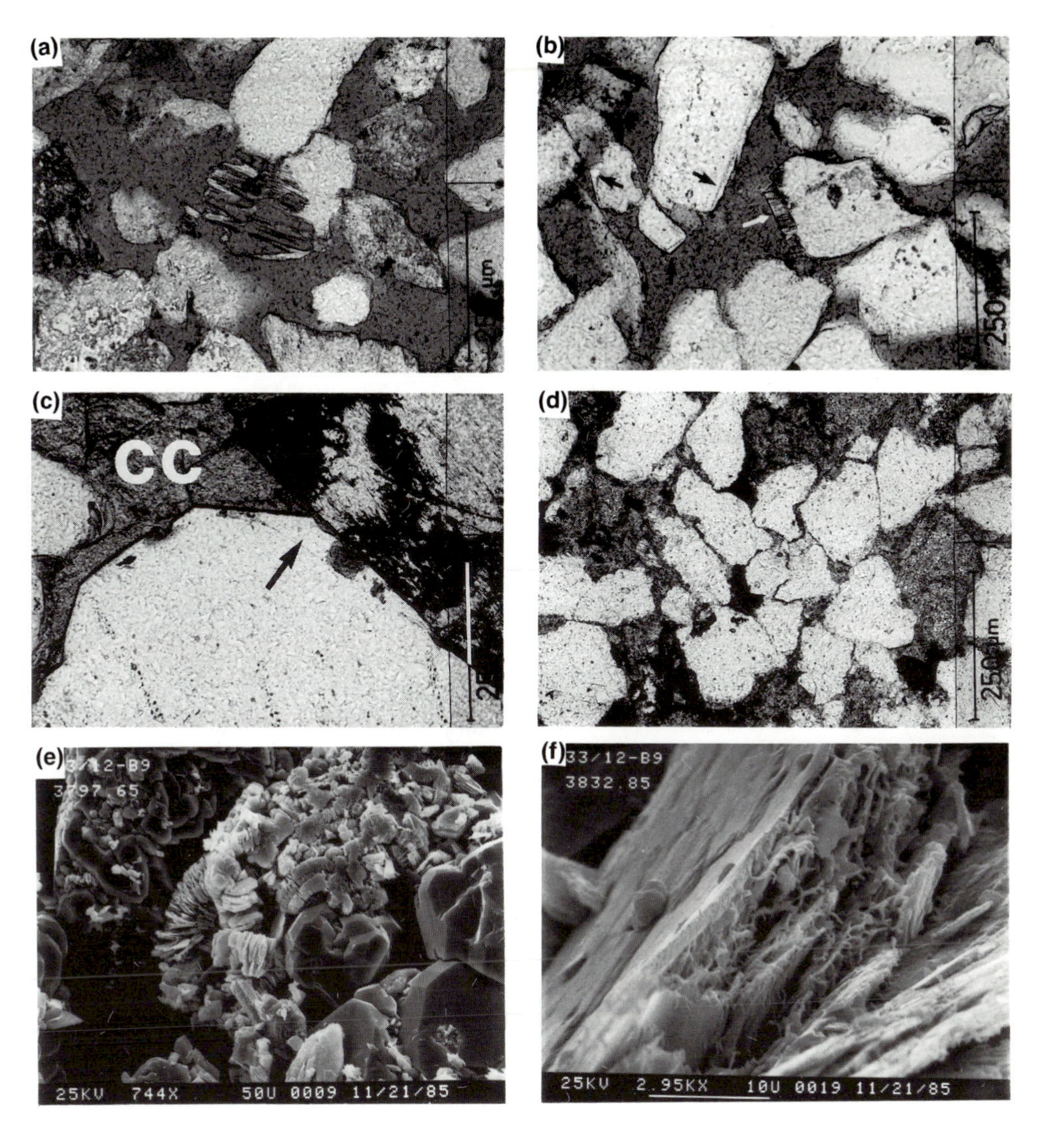


**Fig. 5.** Thin-section and SEM photographs of Statfjord field sandstones. (**A**) Partially dissolved plagioclase grain (centre), from Ness Formation, 2498.0 m ss. (**B**) Porous sandstone from Etive Formation, 2638.9 m ss. Quartz grain in upper left (black arrow) has thin rim of quartz cement. K-feldspar grain just to right of centre (white arrow) has partially dissolved rim of K-feldspar cement. (**C**) Calcite-cemented sandstone from Broom Formation, 2708.6 m ss. Rim of quartz cement on quartz grain (arrow) predates calcite cement (CC). (**D**) Relatively low porosity sandstone (9.7% porosity) from Etive Formation, 2661.9 m ss. Quartz grain contacts show high degree of pressure solution. (**E**) Kaolinite filling pores in Rannoch Formation sandstone, 2692.3 m ss. Thin rims of quartz cement on quartz grains are present in lower right. Scale bar in this and subsequent SEM photos at bottom. (**F**) Wispy illite fibres growing on edge of biotite grain. From Rannoch Formation, 2713.0 m ss.

Pyrite cement is a minor feature at Statfjord, averaging one to two percent of the rock volume. It typically occurs as small, euhedral crystals and less commonly as small, early formed nodules of poikilotopic cement.

## *Hutton field*

Sandstone samples from Hutton field came from nine wells, including 6 above the oil–water contact and 3 from below the oil–water contact (Fig. 7).

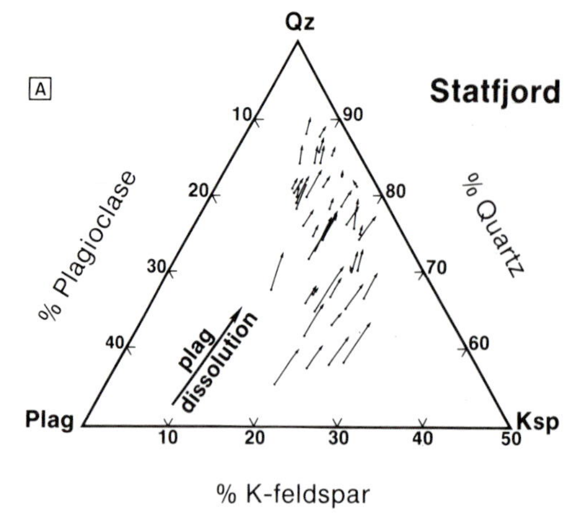


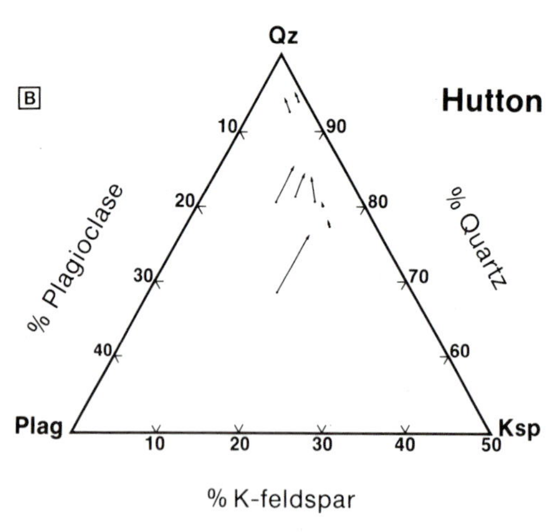


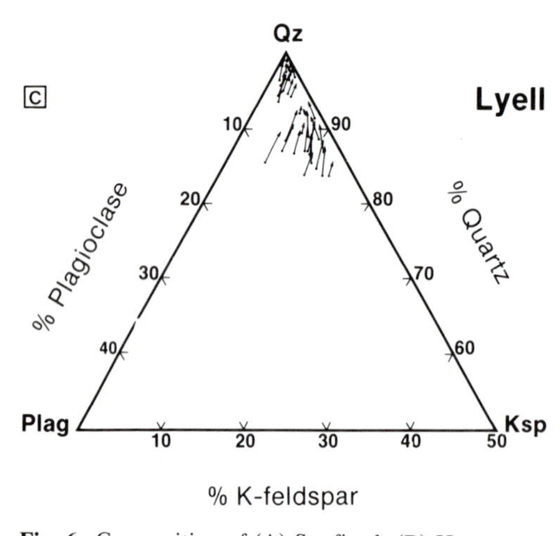


**Fig. 6.** Composition of (**A**) Statfjord, (**B**) Hutton, and (**C**) Lyell field sandstones, depicted on a quartz–K-feldspar–plagioclase triangle, showing the effects of feldspar dissolution. Tips of arrows represent present composition of sandstones. Base of arrows represents reconstructed composition, based on identification of feldspar remnants.

*Sandstone composition.* Sandstone compositions at Hutton field are considerably more quartzose than at Statfjord field (Fig. 4B). As at Statfjord, Ness sandstones at Hutton are the most quartzose, ranging from quartzarenite to subarkose. Etive sandstones are subarkoses, whereas Rannoch sandstones range from subarkose to feldspathic litharenite. In general, the Hutton samples contain 20–50% less feldspar than Statfjord samples in equivalent facies. Mica content is also slightly lower, probably the result of continued alteration of mica to kaolinite.

*Feldspar dissolution.* Much of the feldspar is partially or completely dissolved, resulting in an average of 2.4% secondary porosity (S.D. = 1.74%), very similar to Statfjord field. There is no systematic variation in the abundance of secondary porosity relative to the position of the Brent subcrop (Harris 1990*a*). Plagioclase is preferentially dissolved in relatively plagioclase-rich samples; the resulting feldspar assemblage is extremely K-feldspar-rich (Fig. 6B). No albitized feldspar was identified.

*Quartz cement.* Quartz cement is more abundant at Hutton than at Statfjord field, amounting to an average of 1.0% (S.D. = 0.70%), with a maximum of 2.8%. It forms syntaxial overgrowths on detrital grains (Fig. 8A) and is generally more abundant in the Ness and Etive than in the finer grained, micaceous Rannoch.

The formation of increased volumes of quartz cement at Hutton field coincides with the development of stylolites and microstylolites (Fig. 8B and C), an association noted previously by Blanche & Whitaker (1978). Many quartz grains, particularly in the Rannoch Formation, have highly interdigitated, pressure-solved contacts (microstylolites). These are characteristic of catalysed pressure solution, but differ from normal grain-to-grain pressure solution, which tends to produce flat grain contacts. In addition, but less commonly, quartz grains are highly dissolved along throughgoing dissolution surfaces (stylolites). Both microstylolites and stylolite seams contain mica and clay, which appear to have catalysed the dissolution of quartz. Some also contain opaque and organic matter. Sandstones containing relatively little mica (i.e. the Ness) are less pressure-solved and contain the most abundant quartz overgrowths, while the more clay- and mica-rich sandstones (i.e. the Rannoch) are more pressure solved and contain less quartz cement (Table 2). This relationship suggests that silica is exported from the Rannoch to the Etive and Ness.

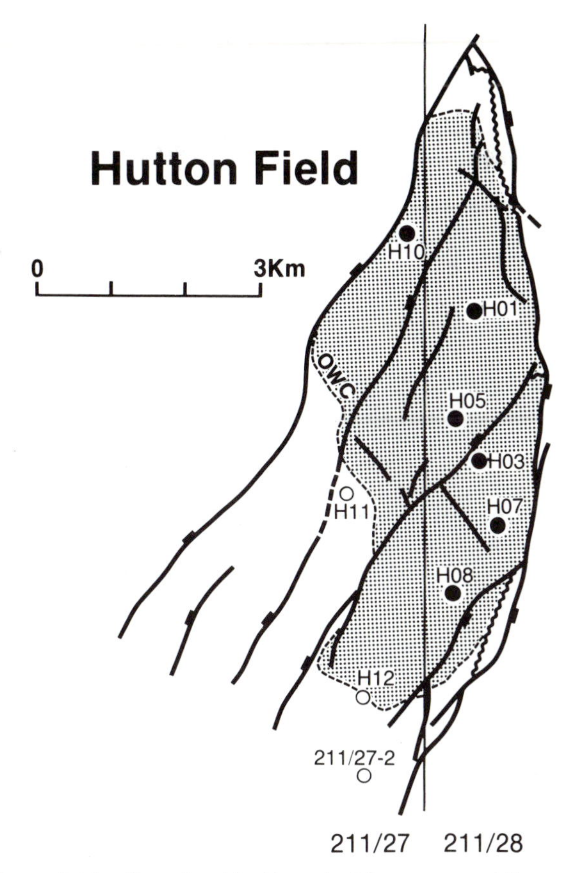


**Fig. 7.** Locations of Hutton field wells analysed in this study. Map courtesy of Conoco (UK) Ltd. OWC is oil–water contact.

*Clays.* Kaolinite is the dominant clay at Hutton field, forming small pseudo-hexagonal plates (Fig. 8D), long 'worms' and matted flakes formed by alteration of mica (Fig. 8E). It averages 2.1% of rock volume (thin section point count data), ranging as high as 8.25% (Table 2). Fibrous illite is more abundant than at Statfjord field, particularly below the oil–water contact. It is also present above the oil–water contact in some sandstones with high detrital clay content. Illite formed contemporaneously with quartz cement (Fig. 8F). It is detrimental to permeability, locally choking off pore throats.

*Other.* As at Statfjord field, calcite is a minor constituent. It occurs in doggers and as overgrowths on clay ooids in the Broom Formation. Small amounts of pyrite (up to 4.8%) and siderite (up to 0.5%) are also present. No K-feldspar overgrowths were noted.

## *Lyell field*

Sandstone samples at Lyell came from four wells, three of which penetrated the Brent above the oil–water contact; the other penetrated the Brent below the oil–water contact (Fig. 9).

*Sandstone composition.* Sandstones at Lyell field are strikingly more quartzose than those at Hutton field (Fig. 4C). As with the other fields, Ness sandstones are both the coarsest grained and most quartz-rich, typically quartzarenite. Etive and Tarbert samples range from subarkose to quartzarenite, while Rannoch samples range from subarkose to sublitharenite.

*Feldspar dissolution.* Secondary porosity, mostly due to feldspar dissolution (Fig. 10A), averages 2.9% (S.D. = 1.45%), virtually identical to abundances at Statfjord and Hutton.

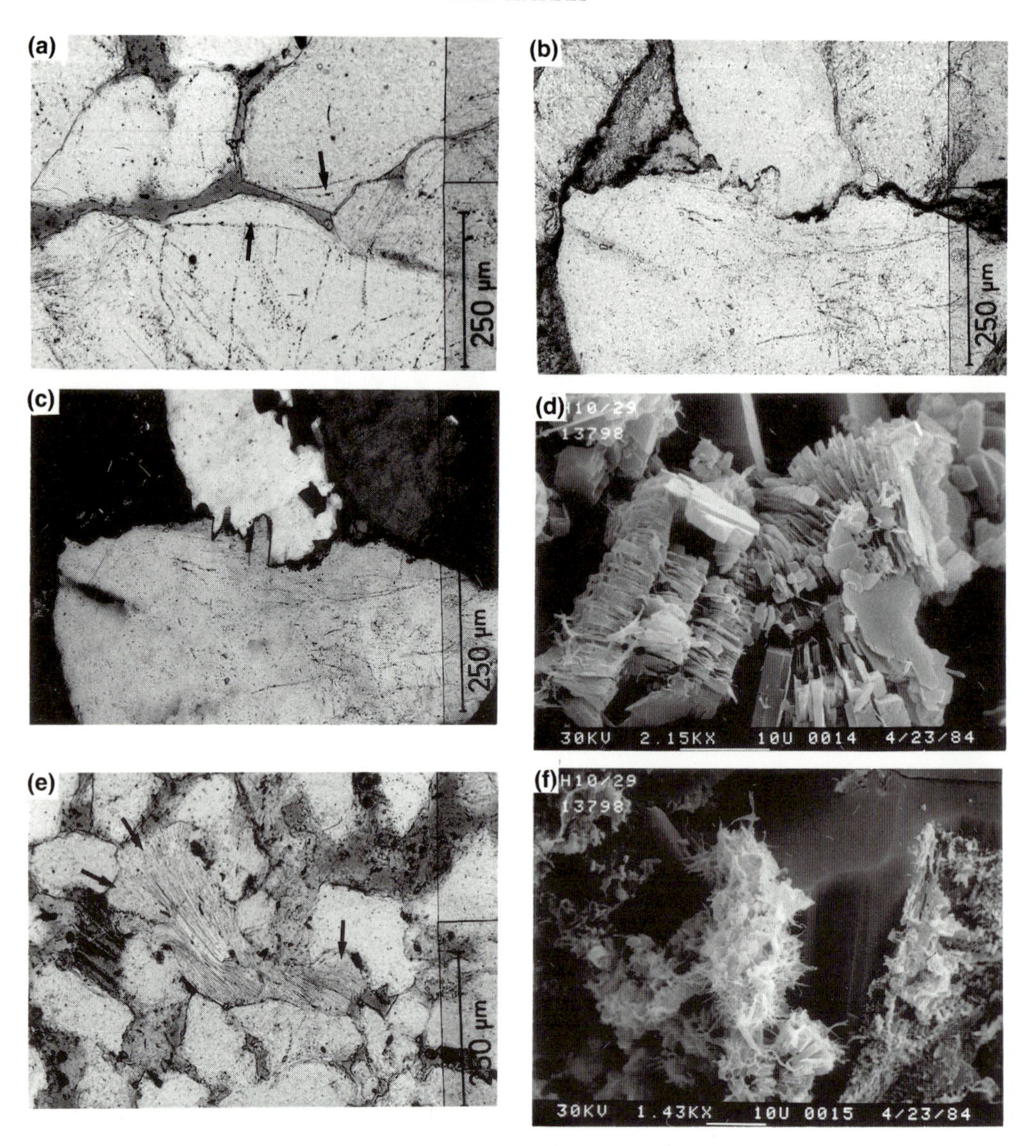


**Fig. 8.** Thin-section and SEM photographs of sandstones from Hutton field. (**A**) Quartz-cemented sandstone from Ness Formation, 2975.5 m ss. (**B**) and (**C**) Plane-polarized and cross-nicols view of a microstylolite contact between quartz grains. Note highly interdigitated contact. From Broom Formation, 3001.3 m ss. (**D**) Kaolinite filling pore in Ness Formation, 3072.8 m ss. A few illite fibres overgrow kaolinite (e.g. in lower left and lower right parts of photo. (**E**) Mica altering to kaolinite (arrow). From Ness Formation, 2987.4 m ss. (**F**) Fibrous illite forming contemporaneously with and post-dating quartz cement. From Ness Formation, 3072.8 m ss.

Most Lyell feldspars show evidence of dissolution.

The feldspar assemblage in the Lyell sandstones is dominated by K-feldspar (Fig. 6C). In most samples, greater than 80% of the feldspar is K-feldspar, and in many samples, virtually no plagioclase is present. The rock thus consists essentially of quartz, K-feldspar and the clays kaolinite and illite (see below). Dissolution affects K-feldspar to a much greater extent at Lyell than at Statfjord and Hutton; however, plagioclase is still preferentially dissolved. No albitized feldspar was identified.

*Quartz cement*. Quartz cement is much more abundant at Lyell (Fig. 10B) than at Hutton, ranging from 2% to 13% of rock volume and averaging 5.2% (S.D. = 3.20%) (Table 2). The

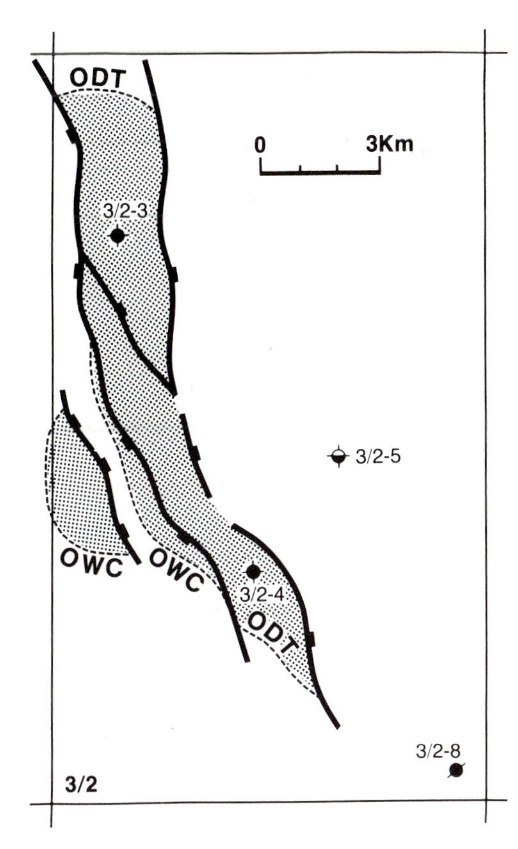


**Fig. 9.** Location of Lyell field wells analyzed in this study. Map courtesy of Conoco (UK) Ltd. OWC is oil–water contact. ODT is oil-down-to.

relatively coarse grained Ness sandstones typically contain large amounts of cement, whereas the finer grained sandstones of the Rannoch contain relatively little. The latter, however, are intensely pressure-solved and probably provide some of the silica responsible for the quartz cement in the Ness and Etive. Both stylolites and microstylolites are present (Fig. 10C and D), the former more abundant than at Hutton field. It may be that given the more extensive pressure solution at Lyell field, many microstylolites linked together to become throughgoing stylolites.

*Clays.* The Lyell sandstones contain subequal amounts of kaolinite and illite, both above and below the oil–water contact. As at Statfjord and Hutton, the kaolinite forms books, long 'worms', and as an alteration product of mica. The average kaolinite abundance is 2.85%, based on thin section point counts (S.D. = 2.30%), essentially identical to that at Statfjord and Hutton field (Table 2).

Illite occurs as fine fibres and ribbons (Fig. 10E). It also appears to be forming by alteration of older kaolinite (Fig. 10F). Pseudohexagonal kaolinite plates are commonly corroded, with illite fibres growing from the plate edges and ribbons peeling off the middle of plates. The plates themselves contain potassium (determined by the energy dispersive method during SEM analysis), which is indicative of illitization since kaolinite contains no potassium.

*Other.* Poikilotopic calcite cement is a minor constituent of the Brent at Lyell field. In rare samples, it is partially dissolved, though there is no evidence that large amounts of calcite have been removed from the rock. Siderite is an uncommon constituent of the rock, amounting to at most 1% of the rock in a few samples.

## Porosity trends

Primary porosities in the Brent decrease systematically with depth, from an average of 27.3% at Statfjord to 20.6% at Hutton to 13.3% at Lyell (Fig. 11). Secondary porosity is essentially constant in the three fields; therefore, total porosity decreases by 14 porosity units per 1000 m from 2500 m to 3500 m burial depth. Assuming that the initial porosity of the sandstones was 40% (Beard & Weyl 1973) and that it consisted entirely of primary porosity, porosity decreased at an average of 5.1 porosity units per 1000 m down to 2500 m (Statfjord level), but then decreased much more rapidly in the depth range of 2500–3500 m. Total porosity at Statfjord (27.3% primary porosity plus 3.6% secondary porosity) is higher than the 22% predicted by Scherer's (1987) method, for that burial depth, sandstone composition, degree of sorting, and overpressure (pressure above hydrostatic). The higher-than-expected porosities are probably a function of the burial history, the Brent having remained at fairly shallow depths until the Late Cretaceous. Porosity at Lyell field is predicted more accurately (19% predicted porosity versus 16.2% actual total porosity).

Porosity is lost through a combination of mechanical compaction (the mechanical rearrangement of sand grain packing), chemical compaction (i.e. stylolites and microstylolites) and cementation. Mechanical compaction alone can reduce porosity to 26% in perfectly rounded and sorted sandstones (Fraser 1935; Graton & Fraser 1935), somewhat less in sandstones with poorer sorting. The degree to which the differ-

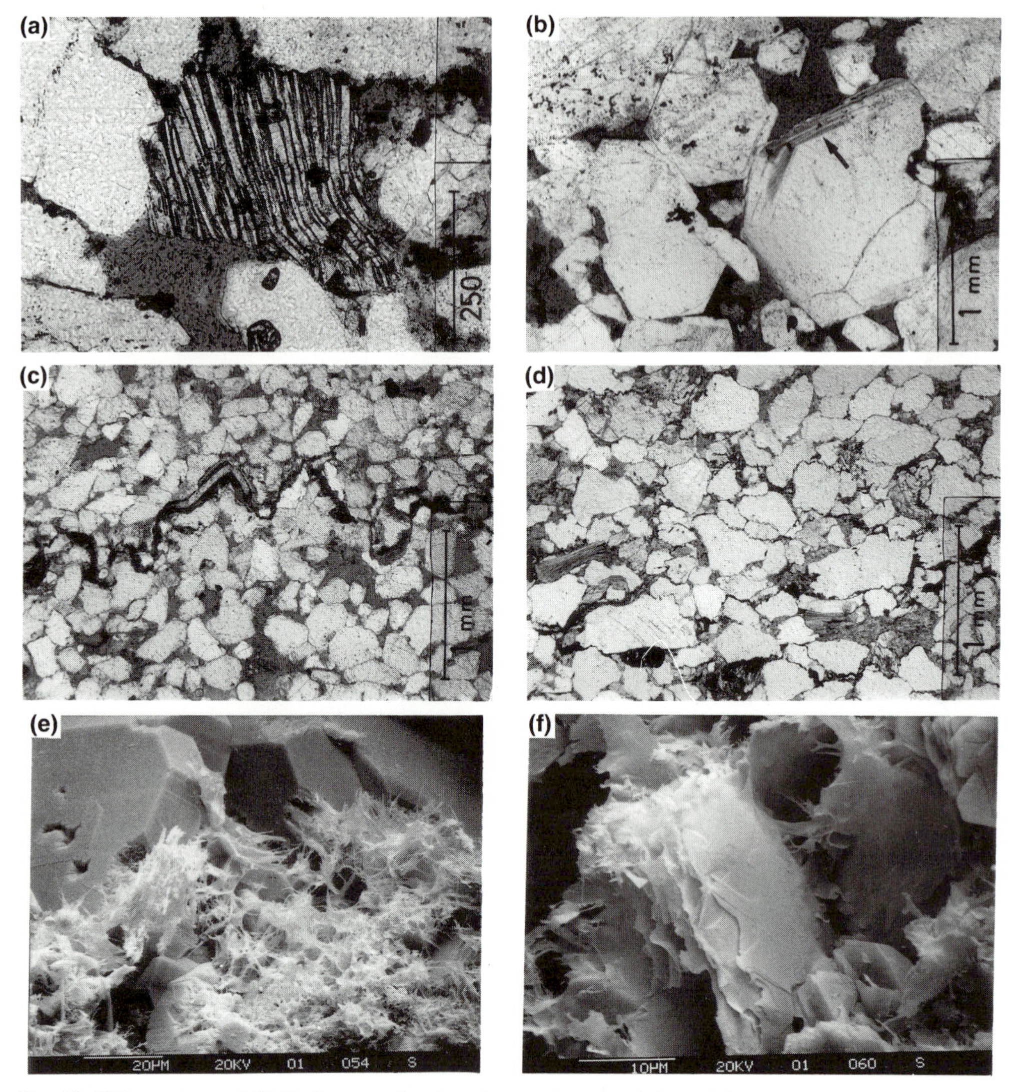


**Fig. 10.** Thin section and SEM photographs of sandstones from Lyell field. (**A**) Partially dissolved, crushed plagioclase grain from Ness Formation, 3413.9 m ss. (**B**) Quartz-cemented sandstone from Ness Formation 3492.1 m ss. Cement on grain just to right of centre contains submicroscopic oil-filled inclusions. (**C**) Stylolite from Ness Formation, 3386.2 m ss. Mica, detrital and authigenic clay and opaque minerals are concentrated along stylolite seam. (**D**) Microstylolites in Ness Formation, 3412.8 m ss. (**E**) Fibrous illite, formed contemporaneously with quartz cement, from Etive Formation, 3419.4 m ss. (**F**) Illite, formed by alteration of pre-existing kaolinite. Note ragged nature of plate edges, fibrous overgrowth and ribbons peeling off centre of plates. Plates contain significant potassium. From Etive Formation, 3419.4 m ss.

ent processes act can be assessed by plotting primary porosities against intergranular volume (the volume of primary porosity plus volume of cement, also called minus-cement volume). This plot (Fig. 12) is similar to that described by Houseknecht (1987). A sandstone which loses porosity entirely by compaction follows a path with a slope of one through the origin. Statfjord field sandstones from the Ness and Etive Formation fall near the 'pure compaction' line, reflecting the small amount of cement in these samples. Samples from the Rannoch contain more cement, reflecting the early alteration of mica to kaolinite. The rare calcite doggers in the Brent (Broom Formation) have very low porosities but high intergranular volumes, plot-

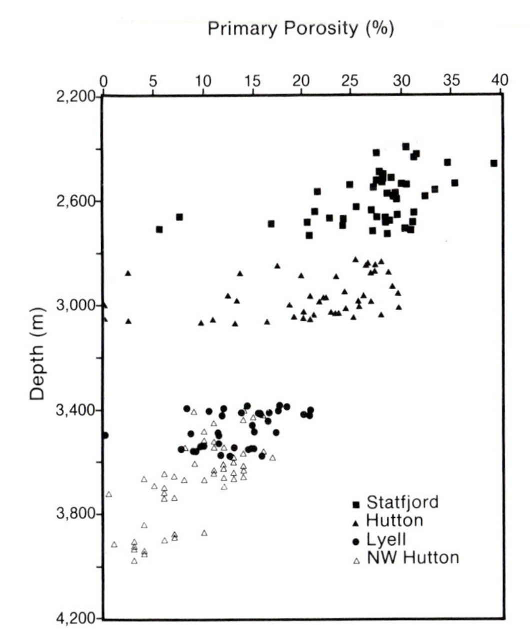


**Fig. 11.** Primary porosity versus depth in Brent Group at Statfjord, Hutton and Lyell fields. Data from NW Hutton (Scotchman *et al.* 1989) are also shown; they are strikingly consistent with data in this study.

ting in the upper left part of the diagram; this reflects early cementation. One unusually micaceous sample from the Etive Formation at Statfjord, noted before (Fig. 5D), is highly compacted, with porosities and intergranular volumes near 10%. This sample, from a relatively deep part of Statfjord field at 2660 m ss, has undergone considerable chemical compaction.

Hutton and Lyell data shift toward lower intergranular volume and primary porosity and show a somewhat greater spread. Many samples from the Ness, Etive, and Rannoch Formations plot near the compaction line, some with primary porosities of 10% or less. These samples are relatively fine grained with extremely low primary porosities and intergranular volumes and have undergone intense chemical compaction (for example, Figs 8B and 10D). Other samples, particularly from the Ness, have more cement (generally quartz) and higher intergranular volumes. The presence of both extremely compacted sandstones and heavily cemented sandstones in the same fields suggests that much of the quartz cement is locally derived and that the Brent itself is the source of the quartz cement. Compaction on stylolites, which would not be evident on Fig. 12, probably also contributed significant amounts of silica. The silica was presumably redistributed within the Brent by diffusion and/or small-scale convection.

## Summary of mineralogical trends

### *Feldspar*

The total feldspar content of the Brent sandstones decreases systematically with increasing burial depth, from an average of 21.1% (S.D. = 6.62%) of detrital constituents at Statfjord to 13.7% (S.D. = 7.01%) at Hutton to 5.0% (S.D. = 4.18%) at Lyell field (Fig. 13A). Data from other well-documented Brent fields are consistent with this trend and are also shown in Fig. 13A, including NW Hutton (Scotchman *et al.* 1989), Heather (Glasmann *et al.* 1989*b*), Huldra and Veslefrikk (Glasmann *et al.* 1989*a*).

The trend of decreasing feldspar with depth may result from diagenetic alteration or variation in provenance and facies. Variation in source terrain and degree of depositional reworking, for example, could affect the feldspar content of the sandstones. Morton (this volume) has argued for significant vertical and lateral variation in the provenance of Brent sediments, based on garnet compositions. These different source terrains might be expected to produce sands of different composition. Reworking in the depositional environment might result in downstream or down-current sands being more feldspar-rich.

While it is clearly important to distinguish the effects of provenance from those of diagenesis, it is virtually impossible to prove such distinctions in an absolute sense, since no pristine rocks are preserved in the deepest fields. However, four different, albeit circumstantial, arguments suggest that the loss of feldspar is probably the result of diagenetic processes. First, the decrease of feldspar with depth is consistent over a wide geographic area, suggesting that the decrease is neither a function of variability of source terrain or depositional reworking. Second, compositional trends are consistent in all Brent formations studied here, despite clear differences in depositional environment and possible differences in source terrain. Third, grain size data (Fig. 14) show that feldspar content decreases from Statfjord to Lyell despite similar grain sizes in similar facies. Lastly, aluminium/silica ratios are similar among the three fields in similar facies, suggesting that the primary feldspar content was likewise similar. The latter point is examined more fully in a latter section.

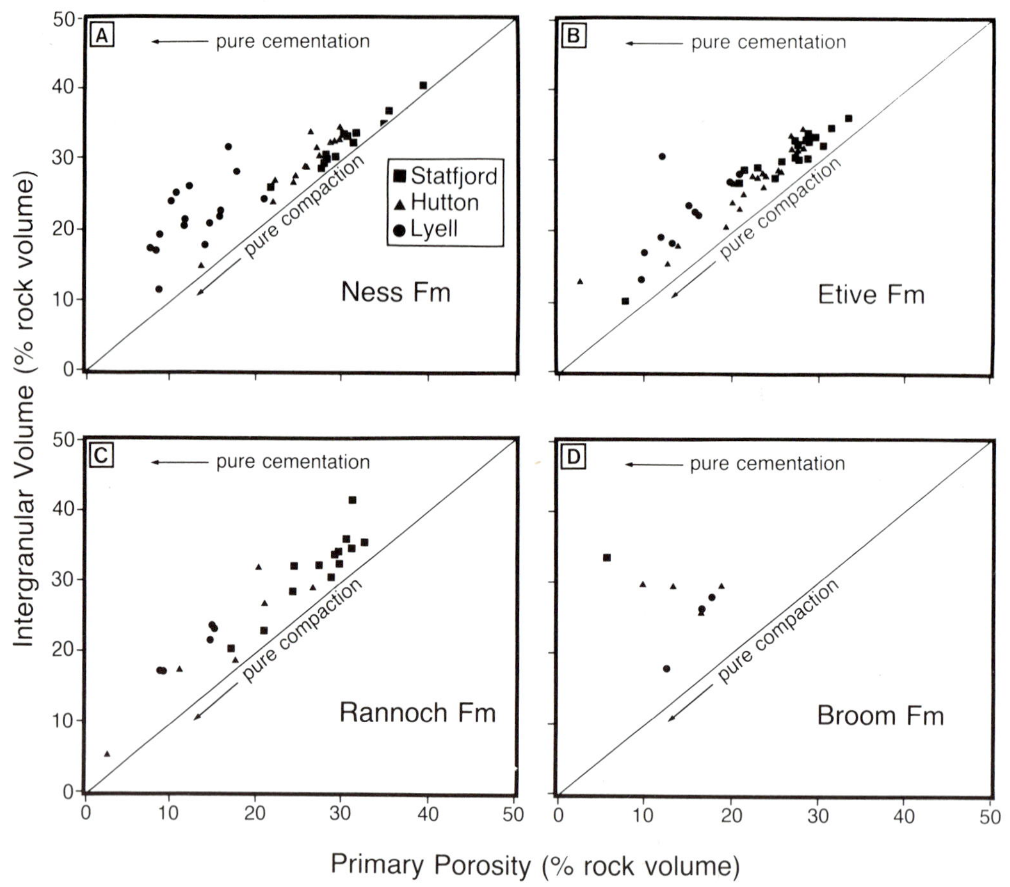


**Fig. 12.** Plot of primary porosity versus intergranular volume (volume of primary porosity plus cement): (**A**) Ness Formation; (**B**) Etive Formation; (**C**) Rannoch Formation; (**D**) Broom Formation. Data are based on thin section point counts. Sandstones with low porosity and high intergranular volume have undergone early cementation (e.g. doggers). Sandstones with low porosities and low intergranular volume are highly compacted. Cement includes kaolinite.

In spite of pervasive feldspar dissolution, the volume of secondary porosity due to grain dissolution remains constant at relatively low volumes, averaging 3% (Fig. 13B). Elsewhere (Harris 1989), I have suggested that the volume of secondary porosity is limited by the mechanical strength of a sandstone. As a result, most of the large secondary pores produced by feldspar dissolution are crushed by compaction (for example, Fig. 10A; see also Harris 1990*b*). Dip-oriented profiles of secondary porosity, based on petrographic point counts from Statfjord and Hutton fields (Harris 1990*a*), show that the volume of secondary porosity does not vary with respect to the subcrop of the Brent on the Cimmerian unconformity. Scotchman *et al.* (1989) reached a similar conclusion at NW Hutton field.

At shallow depths (Statfjord) and in more plagioclase-rich rocks at intermediate depth (Hutton), plagioclase is preferentially dissolved. As a result of plagioclase dissolution, the feldspar component of the sandstones rapidly becomes enriched in K-feldspar. At Lyell, the feldspar component is dominated by K-feldspar, similar to other deep Brent fields in the East Shetland Basin, such as NW Hutton (Scotchman *et al.* 1989) and Heather (Glasmann *et al.* 1989*b*). This contrasts with well-documented trends of K-feldspar dissolution and albitization noted in the Tertiary section of the Gulf of Mexico (Boles 1982; Gold 1987; Milliken *et al.* 1989). Interestingly, sandstones in the Bergen High area may not show preferential dissolution of plagioclase. Petrographic data from Glasmann *et al.* (1989*a*) show that the dominant

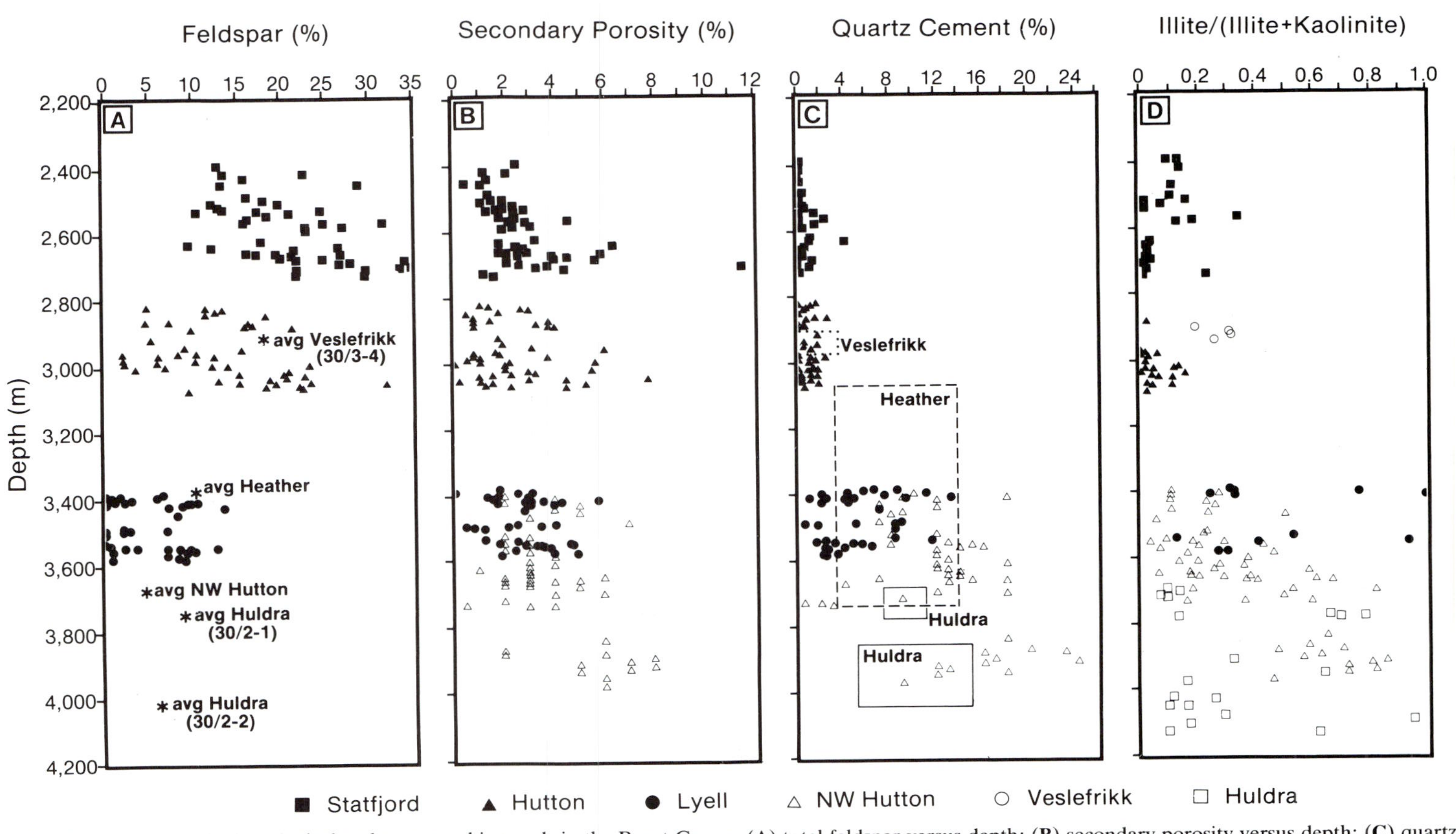


**Fig. 13.** Summary of mineralogical and petrographic trends in the Brent Group: (**A**) total feldspar versus depth; (**B**) secondary porosity versus depth; (**C**) quartz cement versus depth; (**D**) clay content versus depth. Data are from Statfjord, Hutton and Lyell fields (this study), with additional data summarized from NW Hutton (Scotchman *et al.* 1989), Heather (Glasmann *et al.* 1989*b*), Hild, Huldra and Veslefrikk fields (Glasmann *et al.* 1989*a*). Fields from Bergen High fields (Hild, Huldra & Veslefrikk) appear to contain slightly more feldspar at a given depth than fields from East Shetland Basin.

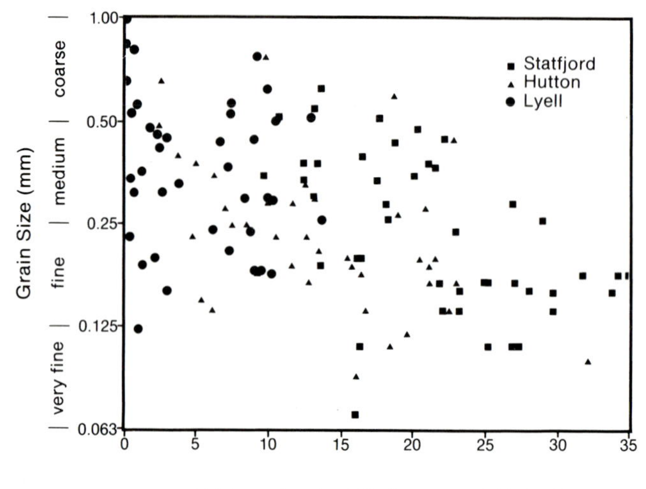


**Fig. 14.** Plot of feldspar content versus grain size at Statfjord, Hutton and Lyell fields.

feldspar at Huldra field (3700–4130 m) is plagioclase, albeit present only in small amounts.

## *Quartz cement*

Quartz cement formed in two stages: an early stage, probably associated with meteoric water leaching, that formed on average 0.5% cement by volume at Statfjord field; and a late stage associated with stylolite and microstylolite development and dissolution of feldspars. Stylolites and microstylolites form in micaceous rocks (e.g. Rannoch Formation) at depths greater than 2700 m and temperatures higher than *c.* 100°C. As a result, quartz cement volume increases at this level and deeper, averaging 1.0% at Hutton field and 5.2% at Lyell field (Fig. 13C). At NW Hutton (average depth of 3660 m), slightly deeper than Lyell, quartz cement is even more abundant, averaging 12% and ranging up to 24% by volume (Scotchman *et al.* 1989). At Heather field, where depths range from *c.* 3050 m to *c.* 3890 m, average quartz cement contents range from 3.5% to 13.4% among the different reservoir zones (Glasmann *et al.* 1989*b*). The systematic regional increase in quartz cement below a threshold depth and the association of chemically-catalysed pressure solution with quartz cement both suggest that the dissolution of detrital quartz provides much of the silica cement. The association of stylolization and quartz cement has also been noted in mid-Norway (Bjørlykke *et al.* 1986).

The dissolution of feldspar may also have been an important source of silica. The reaction of feldspar to kaolinite produces excess silica, possibly resulting in quartz precipitation. Such a mechanism was probably responsible for the small amount of quartz cement at Statfjord, where feldspar was dissolved by meteoric leaching. A similar process may have operated during burial diagenesis; the amount of feldspar dissolved from the Brent could have formed a considerable amount of the quartz cement. If most of the feldspar in a sandstone consisting of 75% quartz were altered to kaolinite, the resulting quartz cement would constitute as much as 8–10% of the solid rock volume. Somewhat lesser volumes would result from alteration of K-feldspar to illite.

The mechanisms proposed here for quartz cementation differ from those proposed by some authors. For example, Scotchman *et al.* (1989) at NW Hutton, Glasmann *et al.* (1989*b*) at Heather and Jourdan *et al.* (1987) in the Alwyn area attribute precipitation of quartz cement to an influx of hot formation waters, probably driven by compaction in deeper parts of the basin. These arguments are largely based on fluid inclusion data, which in the case of Heather field include inclusion filling temperatures hotter than present-day temperatures. However, Bjørlykke *et al.* (1989) point out that silica solubility is too low at these temperatures and pressures for this form of silica transport in deep formation water to be a viable mechanism for forming such large volumes of cement.

## Clays

Kaolinite is virtually the only clay present at shallow depths (Statfjord) with small amounts of wispy illite present in micaceous sections of the Brent (Fig. 13D). Illite first forms in significant amounts at intermediate depths (Hutton field), coincident with the onset of microstylolite development and quartz cementation, and dominates the clay assemblage in many samples from deep Brent fields (Fig. 13D) such as Lyell, NW Hutton (Scotchman *et al.* 1989) and Huldra (Glasmann 1989*a*).

Various workers, for example Bjørlykke & Brensdal (1986), have argued that kaolinite at Statfjord field formed during meteoric leaching. Glasmann *et al.* (1989*b*) concluded that kaolinite at Heather field precipitated from warm meteoric water (45–60°C) during early- to mid-Cretaceous, based on isotopic composition of the kaolinite. Other workers believe that most kaolinite formed during burial diagenesis, for example Scotchman *et al.* (1989) and Jourdan *et al.* (1987). Petrographic analyses reported here do not provide absolute constraints on kaolinite timing, though the extensive dissolution of plagioclase between Statfjord and Hutton suggests that at least some of the kaolinite formed during burial diagenesis.

The pattern of late, deep-formed illite is reported in many other Brent fields, including Heather (Glasmann *et al.* 1989*b*), Murchison (Morton & Humphreys 1983), Veslefrikk (Glasmann *et al.* 1989*a*) and the Greater Alwyn area (Jourdan *et al.* 1987) and has also been reported in mid-Norway (Bjørlykke *et al.* 1986). Occurrences of both fibrous authigenic illite and illite formed by alteration of older kaolinite are described at Alwyn North (Jourdan *et al.* 1987) and Heather field (Glasmann *et al.* 1989*b*). The increase in illite parallels the increase in quartz cement and pressure solution, a coincidence also noted by Glasmann *et al.* (1989*a*, *b*) at Huldra and Heather fields. A number of workers have associated the formation of illite with the flux of hot, presumably saline formation waters (for example Sommer 1978; Jourdan *et al.* 1987). Glasmann *et al.* (1989*b*) concluded that at Heather field, illite precipitated from a mixture of meteoric pore water and saline compaction water, based on isotopic composition of illite.

Recent studies of illite stability (Aagard & Helgeson 1983; Sass *et al.* 1987; Bjørkum & Gjelsvik 1988) demonstrate that illite forms at higher $[K^+/H^+]$ and $[H_4SiO_4]$ than kaolinite. In part, then, the formation of late illite may be attributable to enrichment of formation waters in potassium. The coincidence of illite formation and K-feldspar dissolution in the Brent supports this model. K-feldspar dissolution may be due to influx of organic acids from the overlying, maturing Heather and Kimmeridge Clay Formations, a process which reaches its peak in the temperature range 80 to 120°C (Surdam *et al.* 1989). If so, acid production in the shales must be balanced by acid consumption in the sandstones, presumably by feldspar dissolution; otherwise, $[K^+/H^+]$ would decrease, shifting formation water composition away from the illite stability field. Alternatively, kaolinite and K-feldspar react to form illite at temperatures above 100°C, though all three may coexist stably at lower temperatures (Bjørkum & Gjelsvik 1988). This suggests that K-feldspar dissolution and illite formation in the Brent may be caused simply by higher temperature.

Illite formation may also result from increased silica activity in formation waters. The coincident formation of quartz cement and illite suggests that dissolution of quartz on stylolites and microstylolites raises silica activity and drives formation water compositions from the kaolinite stability field into the illite field. This model requires that quartz precipitates at a rate slower than at which it dissolves on stylolites. That this is feasible is shown in an experimental study by Rimstidt & Barnes (1980), who demonstrated that at temperatures of 100°C, precipitation of quartz is extremely slow even under conditions of very high silica supersaturation.

## Chemical composition of Brent Sandstones

Sandstones from Statfjord, Hutton, and Lyell fields were chemically analyzed to assess the effects of feldspar and clay diagenesis (Table 3 and Figs 15 and 16). This part of the study was undertaken in part to test the model of pervasive feldspar dissolution. If sandstone compositions upon deposition at Lyell were similar to compositions at Hutton and Statfjord, and if aluminium were immobile during diagenesis, then the ratio of silica to alumina should remain constant during feldspar dissolution. Alternatively, if the Lyell sandstones were originally extremely quartzose, that should be reflected in a low aluminium to silica ratio. The model will be more complicated if aluminium remains in solution during feldspar dissolution, for example in the presence of organic acids (Surdam *et al.* 1989). This would have the effect of enriching sandstones in silica during diagenesis. In addition, if silica cements are in part derived from

**Table 3.** *Average chemical compositions of Brent sandstones from Statfjord, Hutton and Lyell fields*

| | Statfjord | | | | Hutton | | | |
|---|---|---|---|---|---|---|---|---|
| | Broom ($n$ = 2) | Rannoch ($n$ = 9) | Etive ($n$ = 9) | Ness ($n$ = 8) | Broom ($n$ = 4) | Rannoch ($n$ = 8) | Etive ($n$ = 9) | Ness ($n$ = 12) |
| $SiO_2$ | 59.7% (16.4) | 58.99% (11.47) | 79.6% (5.56) | 76.9% (6.31) | 50.9% (19.5) | 58.5% (21.0) | 75.7% (16.1) | 84.9% (8.08) |
| $TiO_2$ | 0.82% (1.00) | 1.58% (0.77) | 0.04% (0.49) | 0.87% (0.59) | 0.13% (0.03) | 1.03% (0.39) | 0.93% (0.83) | 0.33% (0.23) |
| $Al_2O_3$ | 15.27% (10.76) | 17.82% (4.40) | 8.91% (2.15) | 9.74% (3.58) | 6.29% (4.13) | 14.05% (4.66) | 7.89% (1.80) | 5.64% (2.28) |
| $Fe_2O_3$ | 6.25% (3.32) | 5.36% (4.12) | 0.79% (0.36) | 1.33% (0.78) | 8.68% (8.07) | 8.28% (13.27) | 2.34% (3.98) | 0.49% (0.82) |
| MgO | 1.39% (1.32) | 1.35% (0.90) | 0.18% (0.11) | 0.29% (0.21) | 0.44% (0.13) | 1.17% (0.87) | 0.21% (0.19) | 0.18% (0.20) |
| MnO | 0.17% (0.06) | 0.15% (0.09) | 0.03% (0.01) | 0.05% (0.03) | 0.09% (0.08) | 0.11% (0.17) | 0.04% (0.08) | 0.02% (0.03) |
| CaO | 3.93% (3.95) | 0.67% (0.35) | 0.29% (0.09) | 0.38% (0.20) | 19.8% (13.9) | 7.14% (12.43) | 3.34% (9.31) | 1.10% (3.05) |
| $K_2O$ | 3.12% (1.24) | 3.68% (0.72) | 2.17% (0.77) | 2.65% (0.82) | 1.34% (0.30) | 2.87% (0.97) | 1.74% (0.41) | 1.05% (0.36) |
| $Na_2O$ | 0.67% (0.13) | 0.71% (0.15) | 0.65% (0.24) | 0.50% (0.13) | 0.39% (0.08) | 0.58% (0.18) | 0.50% (0.13) | 0.28% (0.14) |
| $P_2O_5$ | 0.08% (0.06) | 0.11% (0.05) | 0.04% (0.02) | 0.05% (0.04) | 0.17% (0.08) | 0.12% (0.06) | 0.08% (0.07) | 0.04% (0.02) |
| S | 0.16% (0.08) | 0.10% (0.04) | 0.145% (0.14) | 0.27% (0.19) | 0.05% (0.03) | 0.09% (0.06) | 0.05% (0.02) | 0.06% (0.02) |
| Cl | 0.18% (0.01) | 0.11% (0.08) | 0.08% (0.09) | 0.05% (0.06) | 0.03% (0.03) | 0.02% (0.04) | 0.01% (0.02) | 0.01% (0.02) |
| Zr (ppm) | 126 (31) | 456 (693) | 595 (497) | 431 (360) | 77 (21) | 287 (287) | 481 (602) | 311 (448) |
| Cr (ppm) | 298 (47) | 199 (175) | 228 (160) | 200 (166) | | 52 (144) | 156 (175) | 62 (115) |

Table 3 (*Continued*)

| | Lyell | | | | |
|---|---|---|---|---|---|
| | Broom ($n$ = 6) | Rannoch ($n$ = 9) | Etive ($n$ = 9) | Ness ($n$ = 10) | Tarbert ($n$ = 1) |
| $SiO_2$ | 75.6% (13.2) | 51.9% (16.0) | 79.4% (7.59) | 84.9% (7.73) | 80.0% |
| $TiO_2$ | 0.18% (0.17) | 1.56% (1.01) | 0.66% (0.81) | 0.24% (0.21) | 0.19% |
| $Al_2O_3$ | 10.63% (4.37) | 20.24% (7.99) | 9.78% (2.76) | 4.62% (2.31) | 9.41% |
| $Fe_2O_3$ | 2.05% (2.33) | 4.85% (4.35) | 0.85% (0.86) | 0.63% (0.30) | 0.70% |
| MgO | 0.15% (0.10) | 0.75% (0.26) | 0.18% (0.05) | 0.10% (0.03) | 0.08% |
| MnO | 0.03% (0.03) | 0.10% (0.13) | 0.02% (0.03) | 0.01% (0.01) | 0.01% |
| CaO | 9.14% (12.9) | 9.23% (17.51) | 1.34% (2.84) | 0.39% (0.54) | 0.24% |
| $K_2O$ | 1.20% (0.33) | 3.72% (1.20) | 2.00% (0.66) | 0.69% (0.58) | 1.05% |
| $Na_2O$ | 0.21% (0.04) | 0.24% (0.03) | 0.19% (0.03) | 0.16% (0.05) | 0.14% |
| $P_2O_5$ | 0.09 | 0.05% (0.01) | 0.03% (0.03) | 0.04% (0.07) | 0.07% |
| S | 0.47% (0.67) | 0.53% (0.38) | 0.19% (0.21) | 0.13% (0.10) | 0.07% |
| Cl | 0.05% (0.04) | 0.03% (0.03) | 0.05% (0.03) | 0.05% (0.05) | 0.09% |
| Zr (ppm) | 95 (75) | 423 (505) | 305 (499) | 217 (154) | 86 |
| Cr (ppm) | 178 (148) | 155 (143) | 195 (146) | 304 (29) | 294 |

Standard deviation in parentheses.

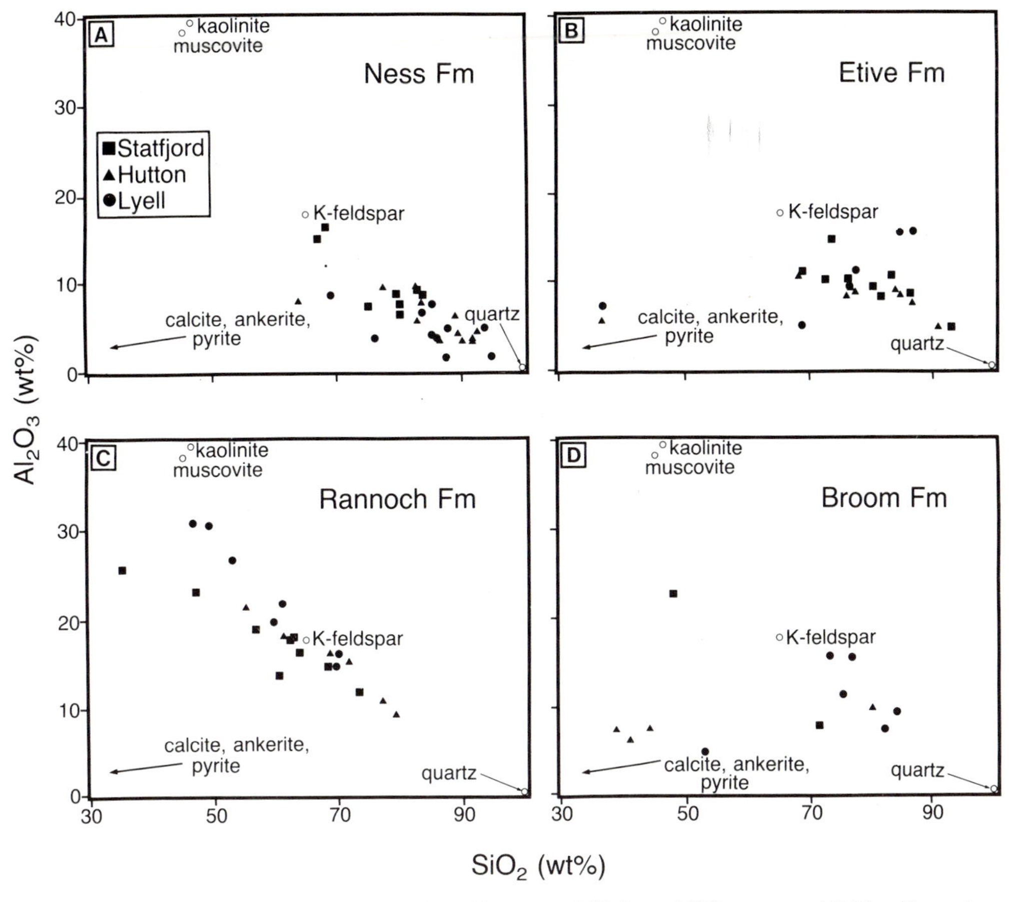


**Fig. 15.** Compositions of Brent sandstones plotted in terms of $Al_2O_3$ and $SiO_2$ content: (**A**) Ness Formation; (**B**) Etive Formation; (**C**) Rannoch Formation; (**D**) Broom Formation. Broom data exhibit considerable scatter, largely due to presence of carbonate cement. In other formations, data from Hutton and Lyell fields are very similar; Statfjord data are somewhat enriched in $Al_2O_3$ in the Ness; other formations are similar.

quartz dissolution on stylolites, that would also enrich the sandstones in silica.

The silica and alumina contents of the Brent sandstones are depicted in Fig. 15. The data indicate that most sandstone compositions can be treated as mixtures of quartz, feldspar, mica, or clay; the Rannoch, by virtue of its high alumina and low silica content must contain considerable mica or clay (confirmed by petrographic analysis). Deviations in composition to the low silica-low alumina side of quartz-feldspar and quartz-mica mixing lines reflect the presence of carbonate or pyrite cements. The data show considerable overlap among the three fields. In particular, Hutton and Lyell fields cannot be distinguished on the basis of bulk silica and alumina contents in either the Ness, Etive, or Rannoch Formations. Data from all three fields overlap particularly closely in the Etive Formation. Statfjord sandstones are somewhat enriched in aluminium relative to Hutton and Lyell fields in the Ness and Rannoch Formations, which may be taken to reflect: (1) difference in primary feldspar and/or mica composition of the Brent; (2) that aluminium was leached from the formations at Hutton and Lyell fields; or (3) that aluminium was diluted by addition of quartz cement at Hutton and Lyell. While the latter hypothesis may bear upon the Lyell data, there is too little quartz cement at Hutton to account for the apparent differences. In summary, the chemical data suggest that detrital compositions in the Ness, Etive, and Rannoch Formations were similar at Hutton and Lyell and that the Etive was similar in all three fields.

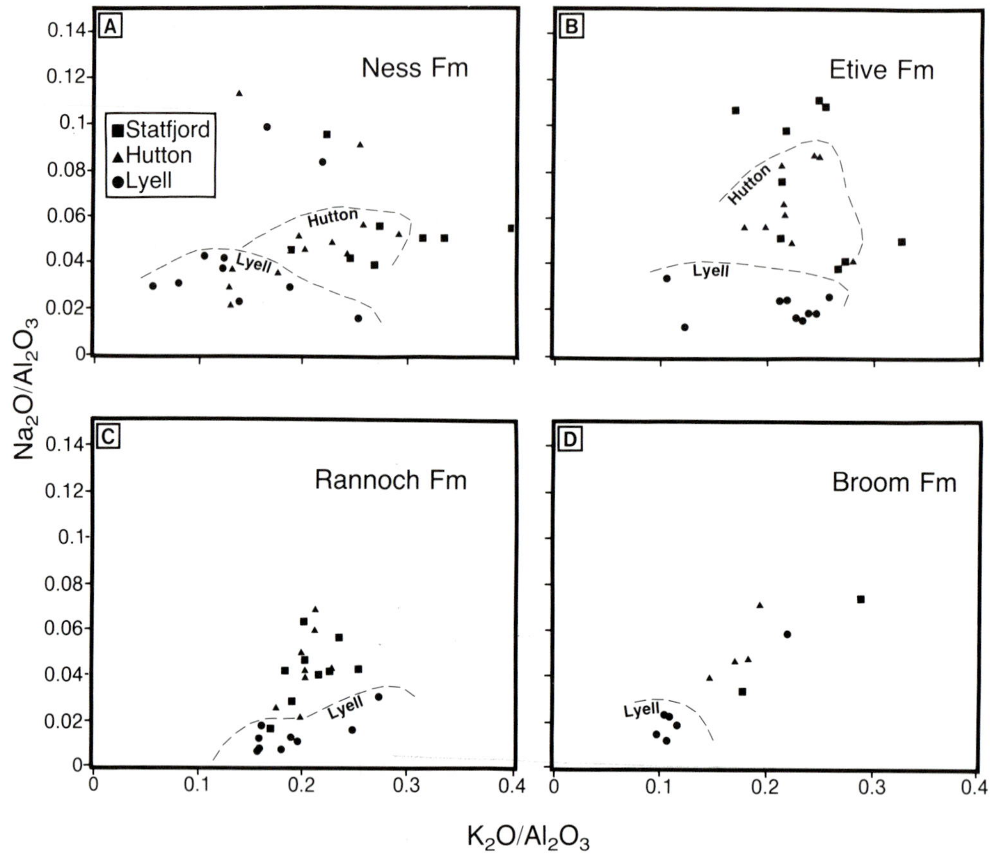


**Fig. 16.** Compositions of Brent sandstones from Statfjord, Hutton and Lyell fields, plotted in terms of $K_2O/Al_2O_3$ versus $Na_2O/Al_2O_3$: (**A**) Ness Formation; (**B**) Etive Formation; (**C**) Rannoch Formation; (**D**) Broom Formation. $Na_2O/Al_2O_3$ decreases with increasing depth, reflecting dissolution of plagioclase and loss of sodium from rock. Ness shows decrease in $K_2O/Al_2O_3$, reflecting loss in potassium during K-feldspar dissolution and illitization.

Chemical data also provide the means to test whether potassium in illite is derived from an external source such as a deep basinal brine or whether it is internally sourced from alteration of mica or dissolution of K-feldspar. If the potassium is externally derived, then the potassium content of the rock should increase during illite formation (Fig. 16). Alternatively, if potassium is internally derived, the potassium content of the rocks might remain constant, (if the potassium freed by feldspar dissolution is balanced by illite precipitation), or it might decrease if the potassium freed by feldspar dissolution exceeds that consumed by illite precipitation. Potassium and sodium contents of the sandstones are shown graphically in Fig. 16. Here, the concentrations of these elements are normalized to the aluminium content to minimize the effect of original variation in feldspar and mica content. Therefore, when a feldspar, for example K-feldspar, dissolves and aluminum and silica precipitate as kaolinite and quartz, the $K_2O/Al_2O_3$ ratio will decrease.

The sodium content of the Brent decreases in all formations from the Statfjord to Hutton to Lyell fields (Fig. 16). The contrast between Lyell and Hutton composition is more striking than that between Hutton and Statfjord. In fact, most of the Lyell sandstones contain little sodium, consistent with the petrographic observation that they contain little or no plagioclase. Clearly, then, to the extent that any albitization took place in the three Brent fields studied here, it involved redistribution of sodium within the Brent and not addition of sodium from some external source such as brines or shales with mixed layer illite/smectite clays. Figure 16 also shows that potassium content remains constant or decreases somewhat, particularly between Hutton and Lyell field. This suggests that potassium in illite was derived internally and not from a basinal brine. The Brent sandstones therefore act as net exporters of sodium and possibly potassium. The fates of these components are not clear. The formation waters may become more concentrated in sodium and potassium with increasing depth. In addition, some potassium and sodium may move upsection with the overall flow of formation waters.

## Conclusions

The Brent sandstones undergo a series of linked burial diagenetic reactions between 80° and 120°C, reactions which: (1) transform the detrital composition of the sandstones from arkose or subarkose to quartzarenite; (2) cement the sandstone with quartz; and (3) transform the clay minerals assemblage from one dominated by kaolinite to one dominated by illite.

The compositional changes result from massive dissolution of feldspar, almost exclusively plagioclase at lower temperatures and increasing amounts of K-feldspar at temperatures above 100°C. In spite of the feldspar dissolution, secondary porosity volume remains consistently low, averaging approximately 3% of rock volume. The mechanical strength of the sandstones limits the amount of secondary porosity that can be preserved.

The formation of large volumes of quartz cement coincides with the first occurrence of stylolites and microstylolites. Stylolites and microstylolites first form at temperatures of 90–100°C. More deeply buried sandstones have been more intensely pressure-solved and consequently are more cemented. Dissolution of feldspar may provide additional quartz cement.

Illite formation probably results from several factors. Dissolution of K-feldspar provides a source of potassium, with the dissolution caused either by the thermodynamic instability of the assemblage illite, K-feldspar, and kaolinite at temperatures greater than 100°C or by an influx of acid from the overlying organic-rich shales. Increased silica activity due to quartz dissolution on stylolites and microstylolites may also serve to stabilize illite.

Chemical analyses of sandstones from Statfjord, Hutton and Lyell fields show that the Brent systematically loses sodium and, to a lesser extent, potassium during burial diagenesis. Sodium loss is almost complete, supporting petrographic observations that plagioclase is preferentially dissolved, in contrast to the well-known phenomenon of albitization in the Tertiary sandstones of the Gulf of Mexico. Potassium loss indicates that potassium in illite is derived from the Brent itself, not from an external source such as a deep basinal brine. The Brent is thus potentially a net exporter of both sodium and potassium, though the fate of these components is not presently known.

I thank Conoco Inc., Conoco (UK) Ltd., Conoco Norway Inc., and the Statfjord, Hutton, and Lyell field partners for providing access to core material and for permission to publish this study. I am grateful to my colleagues R. W. Lahann, R. W. Mitchell, and R. M. Siebert and three anonymous reviewers for critical reviews of the manuscript. Lastly, I thank M. Skaggs for drafting the figures and A. Randol and C. Larmer for typing the manuscript.

## References

Aagaard, P. & Helgeson, H. C. 1983. Activity/composition relations among silicates and aqueous solutions: II. Chemical and thermodynamic consequences of ideal mixing of atoms on homological sites in montmorillonites, illites, and mixed-layer clays. *Clays and Clay Minerals*, **31**, 207–217.

Beard, D. C. & Weyl, P. K. 1973. Influence of texture on porosity and permeability of unconsolidated sand. *American Association of Petroleum Geologists Bulletin*, **57**, 349–369.

Bjørkum, P. A. & Gjelsvik, N. 1988. An isochemical model for formation of authigenic kaolinite, K-feldspar and illite in sediments. *Journal of Sedimentary Petrology*, **58**, 506–511.

Bjørlykke, K. & Brensdal, A. 1986. Diagenesis of the Brent sandstone in the Statfjord field, North Sea. *In*: Gautier, D. L. (ed.) *Roles of organic matter in sediment diagenesis*. Society of Economic Paleontologists and Mineralogists Special Publication, **38**, 157–167.

——, Aagaard, P., Dypvik, H., Hastings, D. S. & Harper, A. S. 1986. Diagenesis and reservoir properties of Jurassic sandstones from the Haltenbanken area, offshore mid Norway. *In*: Spencer, A. M., Holter, E., Campbell, C. J., Hanslein, S. H., Nelson, P. H. H., Nysæther, E. & Ormaasen, E. G. (eds) *Habitat of hydrocarbons on the Norwegian continental shelf*. Norwegian Petroleum Society, Graham & Trotman, London, 275–286.

——, Ramm, M. & Saigal, G. C. 1989. Sandstone diagenesis and porosity modification during basin evolution. *Geologische Rundschau*, **78**, 243–268.

Blanche, J. B. & Whitaker, J. H. McD. 1978. Diagenesis of part of the Brent Sand Formation (Middle Jurassic) of the northern North Sea basin. *Journal of the Geological Society, London*, **135**, 73–82.

Boles, J. R. 1982. Active albitization of plagioclase,

Gulf Coast Tertiary. *American Journal of Science*, **282**, 165–180.

BROWN, S. & RICHARDS, P. C. 1990. The development of stratigraphy and depositional models for the Brent Group (Middle Jurassic): A review [abst.]. *Geological Society of London, Newsletter*, **19** (2), 29.

——, —— & THOMSON, A. R. 1987. Patterns in the deposition of the Brent Group (Middle Jurassic) UK North Sea. *In*: BROOKS, J. & GLENNIE, K. (eds) *Petroleum Geology of northwest Europe*. Graham & Trotman, London, 899–913.

BUDDING, M. C. & INGLIN, H. F. 1981. A reservoir geological model of the Brent sands in Southern Cormorant. *In*: ILLING, L. V. & HOBSON, G. D. (eds) *Petroleum Geology of the continental shelf of northwest Europe*. Heyden & Son, London, 326–334.

CANNON, S. J. C., GILES, M. R., WHITAKER, M. J., PLEASE, P. M., & MARTIN, S. V. 1992. A regional reassessment of the Brent Group, UK sector, North Sea. *This volume*.

COUSTAU, J., LEE, P. J., DU PUY, J. & JUNCA, J. 1988. The Jurassic oil resources of the East Shetland Basin. *Bulletin of Canadian Petroleum Geology*, **36**, 177–185.

ERICHSEN, T., HELLE, M., HENDEN, J. & ROGNEBAKKE, A. 1987. Gullfaks. *In*: SPENCER, A. M., HOLTER, E., CAMPBELL, C. J., HANSLEIN, S. H., NELSON, P. H. H., NYSÆTHER, E. & ORMAASEN, E. G. (eds) *Geology of the Norwegian oil and gas fields*. Graham & Trotman, London, 273–286.

EYNON, G. 1981. Basin development and sedimentation in the Middle Jurassic of the northern North Sea. *In*: ILLING, L. V. & HOBSON, G. D. (eds) *Petroleum Geology of the continental shelf of northwest Europe*. Heyden & Son, London, 196–204.

FRASER, H. J. 1935. Experimental study of the porosity and permeability of clastic sediments. *Journal of Geology*, **43**, 910–1010.

GLASMANN, J. R., CLARK, R. A., LARTER, S., BRIEDIS, N. A. & LUNDEGARD, P. D. 1989*a*. Diagenesis and hydrocarbon accumulation, Brent sandstone (Jurassic), Bergen High area, North Sea. *American Association of Petroleum Geologists Bulletin*, **73**, 1341–1360.

——, LUNDEGARD, P. D., CLARK, R. A., PENNY, B. K. & COLLINS, I. D. 1989*b*. Geochemical evidence for the history of diagenesis and fluid migration: Brent sandstone, Heather field, North Sea. *Clay Minerals*, **24**, 255–284.

GOLD, P. B. 1987. Textures and geochemistry of authigenic albite from Miocene sands, Louisiana Gulf Coast. *Journal of Sedimentary Petrology*, **57**, 353–362.

GRATON, L. C. & FRASER, H. J. 1935. Systematic packing of spheres, with particular relation to porosity and permeability. *Journal of Geology*, **43**, 785–909.

GRAUE, E., HELLAND-HANSEN, W., JOHNSEN, J., LOMO, L., NØTTVEDT, A., RONNING, K., RYSETH, A. & STEEL, R. 1987. Advance and retreat of Brent delta system, Norwegian North Sea. *In*: BROOKS, J., & GLENNIE, K. (eds) *Petroleum Geology of northwest Europe*. Graham & Trotman, London, 915–937.

HARRIS, N. B. 1989. Diagenetic quartzarenite and destruction of secondary porosity: An example from the Middle Jurassic Brent sandstone of northwest Europe. *Geology*, **17**, 361–364.

—— 1990*a*. Diagenetic quartzarenite and destruction of secondary porosity: An example from the Middle Jurassic Brent sandstone of northwest Europe – Reply. *Geology*, **18**, 288.

—— 1990*b*. Diagenetic quartzarenite and destruction of secondary porosity: An example from the Middle Jurassic Brent sandstone of northwest Europe – Reply: *Geology*, **18**, 799–800.

HOUSEKNECHT, D. W. 1987. Assessing the relative importance of compaction processes and cementation to reduction of porosity in sandstones. *American Association of Petroleum Geologists Bulletin*, **71**, 633–642.

JOURDAN, A., THOMAS, M., BREVART, O., ROBSON, P., SOMMER, F. & SULLIVAN, M. 1987. Diagenesis as the control of the Brent sandstone reservoir properties in the Greater Alwyn area (East Shetland Basin). *In*: BROOKS, J. & GLENNIE, K. (eds) *Petroleum Geology of northwest Europe*. Graham & Trotman, London, 951–961.

KIRK, R. H. 1980. Statfjord field: A North Sea giant. *In*: HALBOUTY, M. T. (ed.) *Giant oil and gas fields of the decade: 1968–1978*. American Association of Petroleum Geologists Memoir **30**, 95–116.

MCBRIDE, E. F. 1963. A classification of common sandstones. *Journal of Sedimentary Petrology*, **33**, 664–668.

MILLIKEN, K. L., MCBRIDE, E. F. & LAND, L. S. 1989. Numerical assessment of dissolution versus replacement in the subsurface destruction of detrital feldspars, Oligocene Frio Formation, south Texas. *Journal of Sedimentary Petrology*, **59**, 740–757.

MORTON, A. C. 1985. A new approach to provenance studies: Electron microprobe analysis of detrital garnets from Middle Jurassic sandstones of the northern North Sea. *Sedimentology*, **32**, 553–566.

—— 1992. Provenance of Brent Group sandstones: heavy mineral constraints. *This volume*.

—— & HUMPHREYS, B. 1983. The petrology of the Middle Jurassic sandstones from the Murchison field, North Sea. *Journal of Petroleum Geology*, **5**, 245–260.

NAGY, J., DYPVIK, H. & BJÆRKE, T. 1984. Sedimentological and paleontological analyses of Jurassic North Sea deposits from deltaic environments. *Journal of Petroleum Geology*, **7**, 169–188.

ODOM, I. E., DOE, T. W. & DOTT, R. H., JR. 1976. Nature of feldspar-grain-size relations in some quartz-rich sandstones. *Journal of Sedimentary Petrology*, **46**, 862–870.

RIMSTIDT, J. D. & BARNES, H. L. 1980. The kinetics of silica-water reactions. *Geochimica et Cosmochimica Acta*, **44**, 1683–1699.

ROBERTS, J. D., MATHIESON, A. S. & HAMPSON, J. M. 1987. Statfjord. *In*: SPENCER, A. M., HOLTER, E., CAMPBELL, C. J., HANSLIEN, S. H., NELSON, P. H. H., NYSÆTHER, E. & ORMAASEN, E. G. (eds) *Geology of the Norwegian oil and gas fields*. Graham & Trotman, London, 319–340.

SASS, B. M., ROSENBERG, P. E. & KITTRICK, J. A. 1987. The stability of illite/smectite during diagenesis: An experimental study. *Geochimica et Cosmochimica Acta*, **51**, 2103–2115.

SCHERER, M. 1987. Parameters influencing porosity in sandstones: A model for sandstone porosity prediction. *American Association of Petroleum Geologists Bulletin*, **71**, 485–491.

SCHMIDT, V. & MCDONALD, D. A. 1979. The role of secondary porosity in the course of sandstone diagenesis. *Society of Economic Paleontologists and Mineralogists Special Publication*, **26**, 157–207.

SCOTCHMAN, I. C., JOHNES, L. H. & MILLER, R. S. 1989. Clay diagenesis and oil migration in Brent Group sandstones of NW Hutton field, UK North Sea. *Clay Minerals*, **24**, 339–374.

SOMMER, F. 1978. Diagenesis of Jurassic sandstones in the Viking Graben. *Journal of the Geological Society, London*, **135**, 63–67.

SURDAM, R. C., CROSSEY, L. J., HAGEN, E. S. & HEASLER, H. P. 1989. Organic–inorganic interactions and sandstone diagenesis. *American Association of Petroleum Geologists Bulletin*, **73**, 1–23.

# K-Ar dating of illites in Brent Group reservoirs: a regional perspective

P. J. HAMILTON[1,2], M. R. GILES[3] & P. AINSWORTH[1,4]

[1] *Isotope Geology Unit, SURRC, East Kilbride, Glasgow G75 OQU, UK*

[2] *Present address: Division of Exploration Geoscience, CSIRO, 51 Delhi Road, North Ryde, NSW, Australia*

[3] *Shell UK Exploration and Production Shell-Mex House, Strand, London WC2R ODX, UK*

*Present address: RR/21, KSEPL, Shell Research, Volmerlaan 6, 2281GD Rijswijk, Netherlands*

[4] *Present address: Department of Geology, University of Glasgow, Glasgow G12 8QQ, UK*

**Abstract:** This paper reviews existing K-Ar dates from diagenetic illites from the Brent Group and presents new age data from 11 wells from across the East Shetland Basin.

K-Ar ages from diagenetic illites record illite growth in various parts of the Brent Group of the East Shetland Basin from about 60 Ma until 17 Ma. Diagenetic illites from the Cormorant field yield ages, which when combined with burial history simulation, suggest substantial illite growth from 75°C. This temperature is believed to be close to the temperature for the start of illite growth in the Brent Group. Other illite dates, when combined with burial history simulations, demonstrate continued illite growth to temperatures in excess of 110°C. Etive Formation samples of wells in which the overlying Kimmeridge Clay Formation source rocks are mature for oil generation, show a strong correlation between the K-Ar age of the finest size fraction of diagenetic illite and the calculated time at which the overlying source rocks reached a vitrinite reflectance of 0.62%. Data from other formations do not follow this trend and frequently give ages which are considerably older than those from the Etive Formation. It is considered likely that many of these older dates may arise as a result of contamination by detrital illites which are likely to be less common in the highly reworked sands of the Etive Formation barrier complex.

No recent (<10 Ma) K-Ar dates have been obtained from diagenetic illites of the water zone of any wells so far studied in the Brent province. It is also apparent from many wells where diagenetic illites from both the hydrocarbon and water zones have been dated, that illite growth apparently stops in the water zone relatively soon after hydrocarbon emplacement. Furthermore, the time span for illitization, suggested by dating different illite size fractions of the same sample, is much shorter than that suggested by comparison of the range of temperatures illite apparently grew at and burial histories of Brent Group wells. Consideration of these points has led the authors to conclude that illite begins to form at a temperature of about 75°C probably at quite a low reaction rate. The process of oil migration at somewhat higher temperature perturbs the static aqueous pore fluid medium and increases effective water-to-rock ratio by fluid transport. This results in increased rates of illite growth. Subsequent to the peak of oil generation, fluid flow rates decrease, with a concomitant decrease in illite-forming reaction rates. Any attempt to isolate illite formed at lower or higher temperatures will inevitably also sample illite from the most rapid period of growth and so yield a false and reduced time span of illite formation and apparently old ages for most recently formed illite in the water zone.

It is commonly observed from regional petrographic studies of sediment burial that specific diagenetic reactions are depth-related (e.g. Hower *et al.* 1976; Irwin *et al.* 1977; Dypvik 1983; Green *et al.* 1989; Giles & de Boer 1990; Giles *et al.* this volume). Furthermore, in hydrocarbon bearing basins, it is also often remarked that diagenetic modification of porosity and permeability is more advanced in the water saturated zone than in the overlying hydrocarbon zone (e.g. Lee *et al.* 1985). These observations have significant implications for radiometric dating of the timing and duration of diagenetic events. This is especially so within

*From* MORTON, A. C., HASZELDINE, R. S., GILES, M. R. & BROWN, S. (eds), 1992, *Geology of the Brent Group*, Geological Society Special Publication No. 61, pp. 377–400.

the context of predictions of reservoir quality at the time of hydrocarbon migration. Whilst there are a considerable number of naturally occurring radioactive elements that are used for geochronology, scope for their application to dating diagenetic events is severely limited. The Rb-Sr system has been successfully applied in dating early diagenetic glauconitization (Smalley *et al.* 1986), illite-smectite transformation in shales (Morton 1985) and sedimentation ages (e.g. Rundberg & Smalley 1986). Model ages may also be calculated for single minerals with high Rb/Sr ratios, although this requires assumption of the initial Sr isotope compositions which are often only poorly constrained. However, this particular dating scheme, and most others, are generally of very limited applicability with regard to age determinations of diagenetic minerals formed during sandstone burial.

A notable exception, now gaining increasing usage, is application of the $^{40}$K-$^{40}$Ar decay scheme to dating of diagenetic illite, a mica-like clay mineral containing essential potassium and of ubiquitous occurrence in deeply buried Brent Group sandstones.

The aims of this paper are to (i) present new K-Ar ages for illites from Brent Group sandstones, (ii) review these data, and other published ages, for quantitative constraints on the timing of the beginning and cessation of illite growth relative to hydrocarbon migration, and (iii) attempt to relate these age data to an understanding of illite formation in the diagenetic environment. The filamentous and pore-bridging habit of diagenetic illite grains has a markedly deleterious effect on reservoir permeability, hence an understanding of illite genesis on a regional basis is of great value to both exploration and production.

## Systematics of the K-Ar geochronometer

Several texts exist giving the systematics of the K-Ar geochronometer in substantial detail (e.g. Faure 1987). For the sake of completeness a brief summary is given below.

Radioactive decay is described statistically by the number of atoms of a radioactive isotope undergoing decay per unit time, -d*N*/d*t*, which is proportional to the number of atoms present, *N*, i.e.

$$-\frac{dN}{dt} = \lambda N \tag{1}$$

where $\lambda$ is the decay constant, and is the probability of decay within unit time. Integration of equation (1) leads to:

$$N = N_0 e^{-\lambda t} \tag{2}$$

where, $N_0$ is the initial number of radioactive atoms and $N$ is the number present after time $t$. The number of daughter or radiogenic atoms, $D^*$, formed in time t is given by:

$$D^* = N_0 - N \tag{3}$$

$$= N(e^{\lambda t} - 1). \tag{4}$$

Equation (4) may be solved for $t$ as:

$$t = \frac{1}{\lambda} \ln\left(\frac{D^*}{N} + 1\right). \tag{5}$$

Thus if the number of atoms (or mass) of parent and daughter isotopes are known at present then the time elapsed, $t$, since the mineral or rock analysed was formed, can be determined.

Potassium has three naturally occurring isotopes; $^{39}$K, $^{40}$K and $^{41}$K. Of these, only $^{40}$K is radioactive and has a present day abundance of 0.01167% of total K. Argon has three naturally occurring isotopes; $^{36}$Ar, $^{38}$Ar and $^{40}$Ar. The relative abundance of $^{40}$Ar is variable in nature because of radioactivity. $^{40}$K decay is of a branched type and produces $^{40}$Ca by $\beta^-$ decay of 88.8% of $^{40}$K atoms with the remainder decaying to $^{40}$Ar by K-shell electron capture or positron emission.

In order to calculate a K-Ar age a number of constants are assumed and those in current use are as recommended by Steiger & Jaeger (1977), viz:

$$\lambda_\beta = 4.962 \times 10^{-10} a^{-1}$$

$$\lambda_\varepsilon = 0.581 \times 10^{-10} a^{-1}$$

$$\lambda\ (\text{total}) = \lambda_\varepsilon + \lambda_\beta = 5.543 \times 10^{-10} a^{-1}$$

$$^{40}\text{Ar}/^{36}\text{Ar(atmosphere)} = 295.5.$$

For application of the $^{40}$K-$^{40}$Ar branch of decay the age equation, (5), is

$$t = \frac{1}{\lambda} \ln\left(\frac{^{40}\text{Ar}\ \lambda}{^{40}\text{K}\ \lambda_\varepsilon} + 1\right) \tag{6}$$

$$t = 1.804 \times 10^9 \ln\left(9.540\frac{^{40}\text{Ar}}{^{40}\text{K}} + 1\right).$$

Thus in order to determine an age for an illite separate, potassium and radiogenic $^{40}$Ar abundances are measured. Potassium is usually determined by flame photometry and in duplicate or triplicate. Some atmospheric argon is desorbed from a separate illite aliquot by pre-heating at *c.* 120°C for *c.* 24 hours. Argon is then released by sample fusion in vacuum, purified of other released gases (e.g. $N_2$, $CO_2$, $H_2O$) and added to a known amount of $^{38}$Ar spike, prior to mass spectrometer analyses of isotope ratios. From the measured $^{40}Ar/^{38}Ar$ and $^{40}Ar/^{36}Ar$ and the known amount of spike added and its composition, a correction is made for residual atmospheric argon and the $^{40}Ar^*$ abundance determined. The underlying assumptions of the method in its particular application to illite dating are critically assessed in Hamilton *et al.* (1989).

Eleven wells were sampled from the East Shetland Basin for this study (Fig. 1). The core pieces used came from the water zones (legs) (five Etive Formation samples) and from the oil zones (legs) (four Ness, six Etive and three Tarbert Formation samples) and spanned a depth range from 9942 ft to 13 532 ft (3030.36 to 4124.60 m). Illite separation procedures are detailed in Hamilton *et al.* (1989) and were adapted from Jackson (1979).

## Interpretation of K-Ar illite ages

Prior to presentation and discussion of the previously published and new K-Ar illite ages from Brent sandstones some brief consideration is given here both to problems of contamination and to the timing and nature of illite growth in sandstones. Detailed discussion of hypothetical age-depth profiles is given in Hamilton *et al.* (1989).

Diagenesis results in the dissolution/transformation of detrital minerals and the formation of new minerals. An aqueous pore-fluid medium is required for these reactions to occur, with water acting both as a mineral solvent and as a nutrient source for newly forming minerals. It is generally assumed that once the aqueous pore-fluid is displaced by hydrocarbons, mineral diagenesis will cease. Thus if the last formed portion can be isolated from a diagenetic mineral that is amenable to dating, then a maximum age constraint can be established for the timing of hydrocarbon charge (see Fig. 2). If growth of that mineral ceases due to the presence of hydrocarbons then, theoretically at least, the exact time of hydrocarbon accumulation may be realized. Diagenetic illite, occurring as a late cement with a flakey and/or filamentous habit extending into and across free pore space (Fig. 3B, D) is the best candidate for this geochronological application.

Lee *et al.* (1985, 1989), in particular, have extolled the principle illustrated in Fig. 5 that during illite growth the last formed (youngest) illite will always be at the ends of filaments extending into pore space. Conversely, coarser grained illite occurs closer to the points of attachment to substrate grains (see Fig. 3D). Analogously, final growth of flakey illite grains would have been at flake edges and older growth more towards the flake interior. A series of K-Ar ages for size fractions of a pure illite should thus constrain the time span of illite diagenesis.

Two major problems potentially exist with this approach of equating different grain sizes with different ages of illite growth. A range of size fractions for an illite separate do frequently exhibit increases in K-Ar age with increase in grain size (e.g. see Glasmann *et al.* 1989*a*, *b*; Lee *et al.* 1989; Hamilton *et al.* 1987). However, it is exceedingly difficult to monitor small degrees of detrital contaminant that may equally well explain such correlations (see Hamilton *et al.* 1989; Ehrenberg & Nadeau 1989). Though tedious, TEM observations on illite fractions (Fig. 4) are the most effective means of monitoring for such contamination effects (Hamilton *et al.* 1989; Glasmann *et al.* 1989*b*).

Worthy of mention here in the context of contamination is the general assumption that an illite age is unaffected by admixture of kaolinite in the analyzed illite 'separate'. This has yet to be carefully documented in published literature. Consider for example, a replacive vermicular kaolinite that has a petrogenetic origin in splayed out lath ends of altered muscovite booklets as illustrated in Fig. 3A. Electron microprobe traverses measuring the K-content from unaltered muscovite into the kaolinite indicate a transition zone with kaolinite morphology but bearing essential K at a level of 1–2% (P. J. Hamilton, unpublished data). This could present a contaminant of the illite K-Ar system if included in the analyzed material.

The second negative aspect of this sort of analytical approach lies in the uncertainty that only one illite forming event has been sampled. A single morphology does not in itself constitute evidence of synchronous growth. Figure 3A, B and C show examples of fine-size filamentous illite which could all occur in the same sandstone sample and in the same illite size separate. Without exceptionally good petrographic control an assumption that these occurrences of filamentous illite are contemporaneous would be unwarranted. This serves to illustrate the important point that each size fraction age is in fact an average age for a range of grain sizes formed over an interval of time that could be of significant duration. The possibility is examined later that even illite of only one petrogenetic origin may have fine sized particles of very different ages (see Fig. 5) as a consequence of the nature of nucleation and growth.

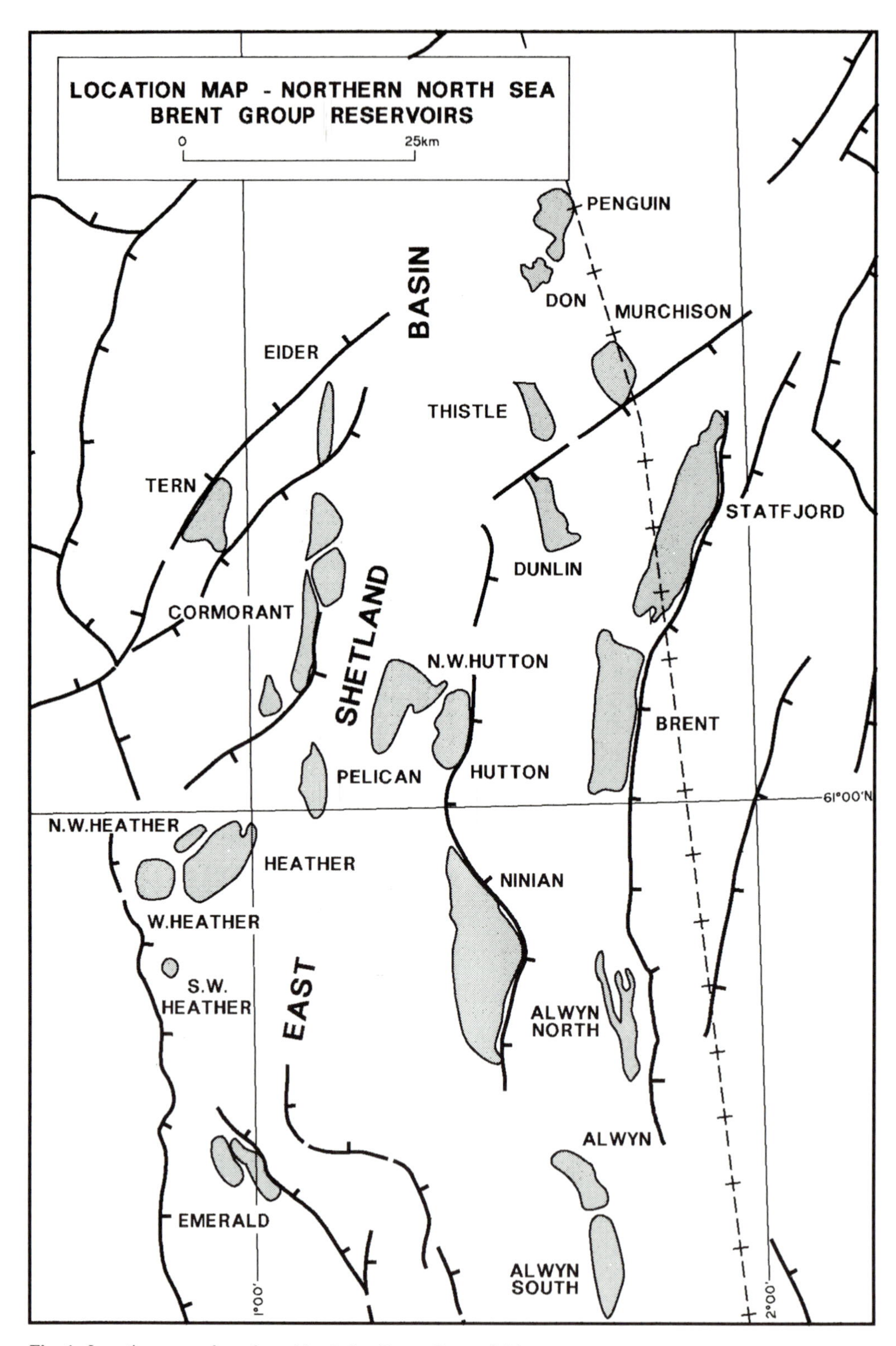


**Fig. 1.** Location map of northern North Sea Brent Group fields.

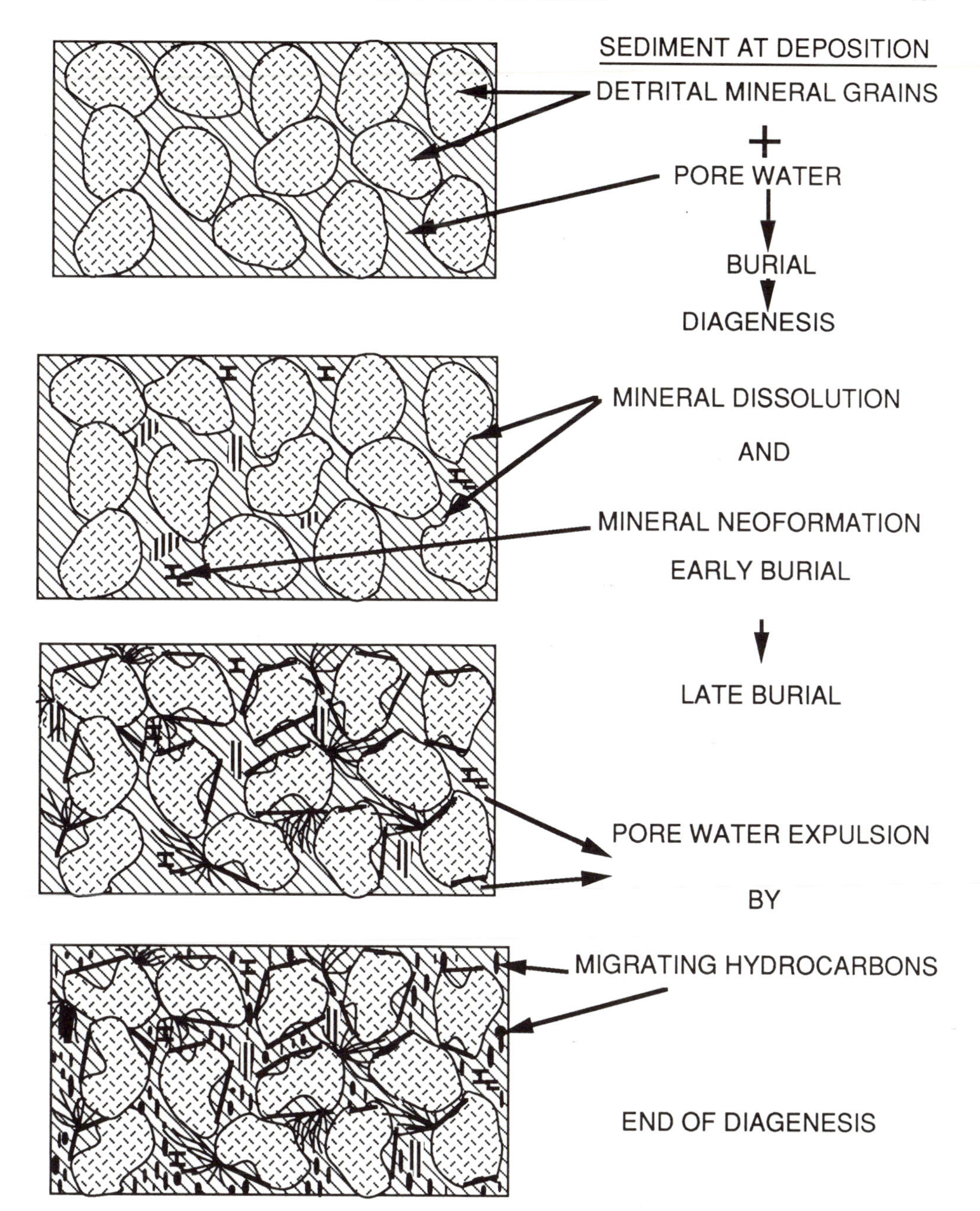


**Fig. 2.** Schematic representation of burial diagenesis in relation to hydrocarbon charge to a reservoir sandstone. Most detrital mineral grains are out of thermodynamic equilibrium with their pore waters. This results in detritus/pore-water chemical reactions that modify porosity and permeability through dissolution of old mineral grains and the formation of new minerals. Once the aqueous fluid medium is replaced by hydrocarbons these reactions will cease. Any mineral, such as illite, growing at the time of hydrocarbon charge and that is amenable to radiometric dating can be used to constrain the emplacement age of the oil or gas.

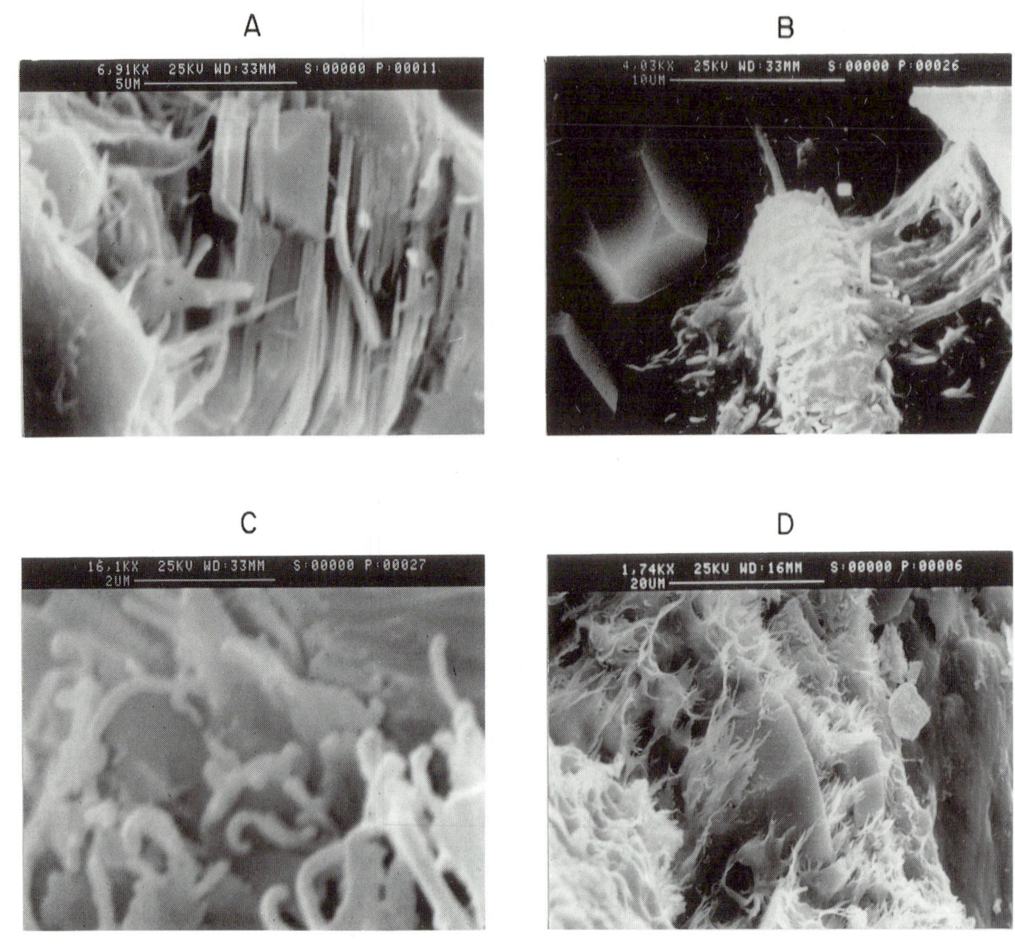


**Fig. 3.** Filamentous illite developed on different kaolinite generations: (**A**) on early kaolinite developed at splayed out end of muscovite lath, (**B**) on late blocky porosity-occluding stack of kaolinite plates which appear also to be being replaced by the illite. The relative timing of these two generations of illite is unconstrained yet both could contribute to the same size fractions of illite separate. (**C**) Filamentous illites emanating from detrital illitic clay grain coating. (**D**) Diagenetic illite developed on late ankerite and showing older, larger, platey morphology of illite near substrate, and younger, smaller, filamentous morphology extending into pore space.

## Interpretation of K-Ar diagenetic illite ages: Insights provided by considerations of illite nucleation & growth rates

The available K-Ar dates from diagenetic illites raise a number of important questions which suggest that their interpretation is far from simple.

(1) Why are there no K-Ar dates in the North Sea Basin which indicate recent growth of illite (i.e. near zero ages)?

(2) Why should illite growth in the water leg frequently appear to stop soon after illite growth in the oil leg?

(3) Why is the time span for illite growth, indicated by sampling various illite size fractions, so narrow?

(4) In a basin with an approximately uniform geothermal gradient, the oldest oil zone illite ages should occur in the deepest structures i.e. those which were charged first. On a basin-wide scale, then, the deepest structures should have the oldest dates and the shallowest illite bearing structures the youngest. Thus there should be a correlation of increasing illite age with depth on a basin wide scale. This simple trend is not obvious in illite age data from the Brent Group (see Fig. 6).

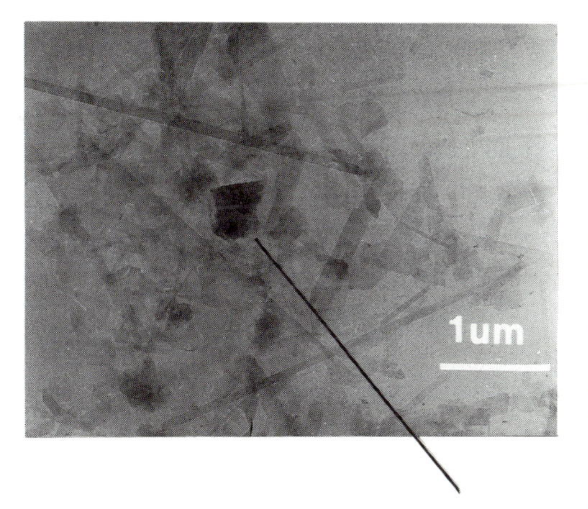


**Fig. 4.** Transmission electron photomicrograph of 0.1 $\mu$m illite fraction from Cormorant well at TVD 9966 ft after high gradient magnetic separation. This has been highly successful in reducing the amount of kaolinite but scattered grains remain of an electron dense 2M mica contaminant.

The answer to some of these questions might be found by considering the nucleation and growth rates of illite crystals.

## *Illite nucleation and growth*

Consider a number of illite crystals growing from a substrate (Fig. 5). These crystals must grow by adding new sections to older crystals. This may occur by preferential growth of the ends of the illite fibres. Thus any single crystal does not have a unique age but is an aggregate of sections of different ages. It follows that the older larger crystals do not date the start of illite growth but contain sections of a wide variety of ages.

Returning to the crystals shown in Fig. 5, after an initial phase of growth, some crystals will be larger than others. The larger crystals, by virtue of their larger specific surfaces, will start to grow even faster and deplete the 'nutrient' supply from the adjacent pore fluid. This will have the effect of stunting the size of nearby crystals. At another point in the pore network, new crystals may start to form, possibly because of increased K supply caused by the decomposition of a nearby K-feldspar. This too will result in the growth of larger and smaller crystals as shown in Fig. 5.

If the finest illite fraction is sampled, it will comprise old 'stunted' crystals and younger crystals. Thus the K-Ar date obtained will be an average weighted toward the last phase of growth but by no means an exact measure of it. Therefore, K-Ar dates from the finest illite fraction of hydrocarbon bearing sequences only give dates which approximate to the time of hydrocarbon emplacement. It also follows that even if illite had been forming in the rock right up to the present day, then ages of near zero would still not be found from diagenetic illites of the finest size fraction because these will be composed of crystals of a range of ages. Near zero ages might be encountered where illite formation had only just begun. However, in such circumstances, as the initial growth rates are not likely to be high, there would be the practical difficulty of obtaining enough illite to date.

## *Illite growth and reaction rates*

Chermak & Rimstidt (1990) have shown that the rate of conversion of kaolinite to illite may be written:

$$R = \frac{\mathrm{d}m_{\mathrm{illite}}}{\mathrm{d}t} = \frac{A}{M}\,\kappa_{\leftarrow}\,m_{\mathrm{K}} + \left(K - \frac{m_{\mathrm{H^+}}}{m_{\mathrm{K^+}}}\right) \quad (7)$$

Where, $R$ is the rate of illite growth in moles per unit time, $\kappa_{\leftarrow}$ the reverse rate constant, $A$ is the total surface area of illite, $M$ is the mass of the solution in kg, $m_{\mathrm{K^+}}$ concentration of potassium ions, $m_{\mathrm{H^+}}$ is the concentration of hydrogen ions and $K$ is the equilibrium constant for the reaction. As an aqueous pore fluid is necessary for the reaction, the presence of such a pore fluid is implicit in the equations that follow.

The rate constant ($\kappa_{\leftarrow}$) depends upon temperature, via the Arrhenius equation i.e.

$$\kappa_{\leftarrow} = A_{\leftarrow}\mathrm{e}^{-\Delta E_{\mathrm{a}}/RT} \quad (8)$$

Where $A_{\leftarrow}$ is the so-called pre-exponential constant, $\Delta\ E_{\mathrm{a}}$ the activation energy for the reaction, $R$ the gas constant and $T$ the absolute temperature. Combining these two equations gives amount of illite ($\Delta m_{\mathrm{illite}}$) grown in a solution of constant composition over some time interval, $t_1$ to $t_2$.

$$\Delta m_{\mathrm{illite}} = A_{\leftarrow}\mathrm{e}^{-\Delta E_{\mathrm{a}}/RT} \times \frac{A}{M} \times m_{\mathrm{K}} + \left(K - \frac{m_{\mathrm{H^+}}}{m_{\mathrm{K^+}}}\right) \int_{t=t_1}^{t=t_2} \mathrm{d}t \quad (9)$$

The equation demonstrates that the amount of illite produced is dependent on both time ($t$) and temperature ($T$). Consequently, the temperature at which significant quantities of illite may be found in one basin will not necessarily

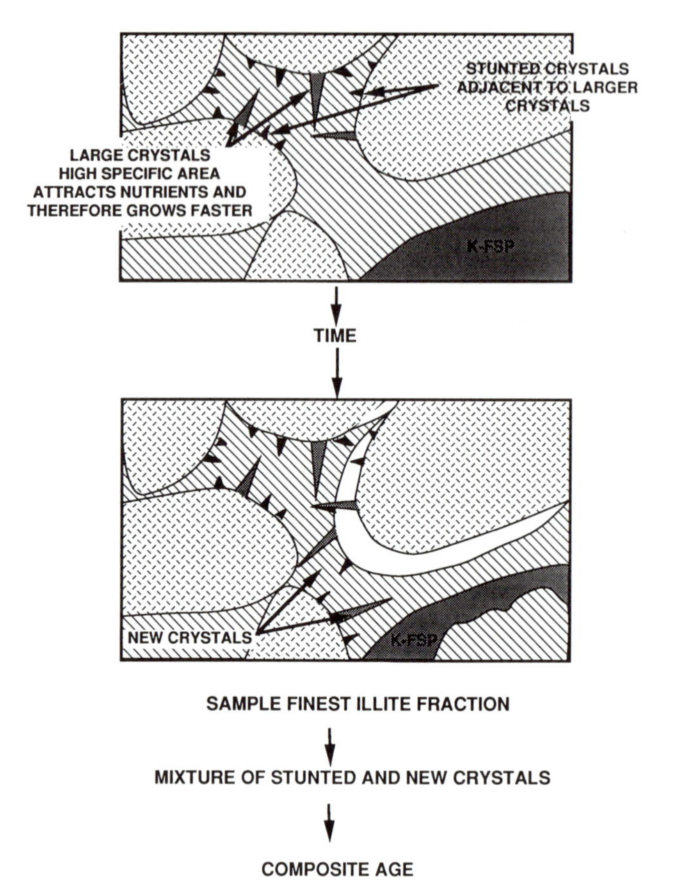


**Fig. 5.** Schematic illustration of illite crystal growth showing (**a**) accompanying increase in grain size, and (**b**) the ends of illite fibres which formed at different times but which would be incorporated into the same size fraction.

be the same as in another basin with a different burial history.

As the concentration of the reactants is unlikely to remain constant, then variations in these will also effect the reaction rate. The rate of supply of ions to the growing crystals will be particularly important. The link between reaction rates and diffusion and/or advection, is provided by the mass transport equation (Bear 1972).

The 1-dimensional form of this equation is:

$$\frac{\partial C_i}{\partial t} = D_h \frac{\partial^2 C_i}{\partial X^2} - V \frac{\partial C_i}{\partial X} + \sum_{i=1}^{i=n} R_i \quad (10)$$

Here $D_h$ is the hydrodynamic dispersion coefficient which takes into account the combined effects of diffusion and mechanical dispersion.

$V$ is the ground water velocity.

$\frac{\partial^2 C_i}{\partial X^2}$ and $\frac{\partial C_i}{\partial X}$ are the second and first derivatives of the rate of change of concentration of species i in the pore fluid with distance.

$\frac{\partial C_i}{\partial t}$ is the rate of change of species i in the pore fluid at a given point with time.

$\sum_{i=1}^{i=n} R_i$ is the rate of production of species i by $n$ reactions at the solid/liquid interface.

This equation relates the rate of change of concentration of a given species in time at a given point, to the rate of change of its concentration with distance and allowing for all the reactions between minerals and the pore fluid. One of these equations is needed for each species involved in the reaction. The equation

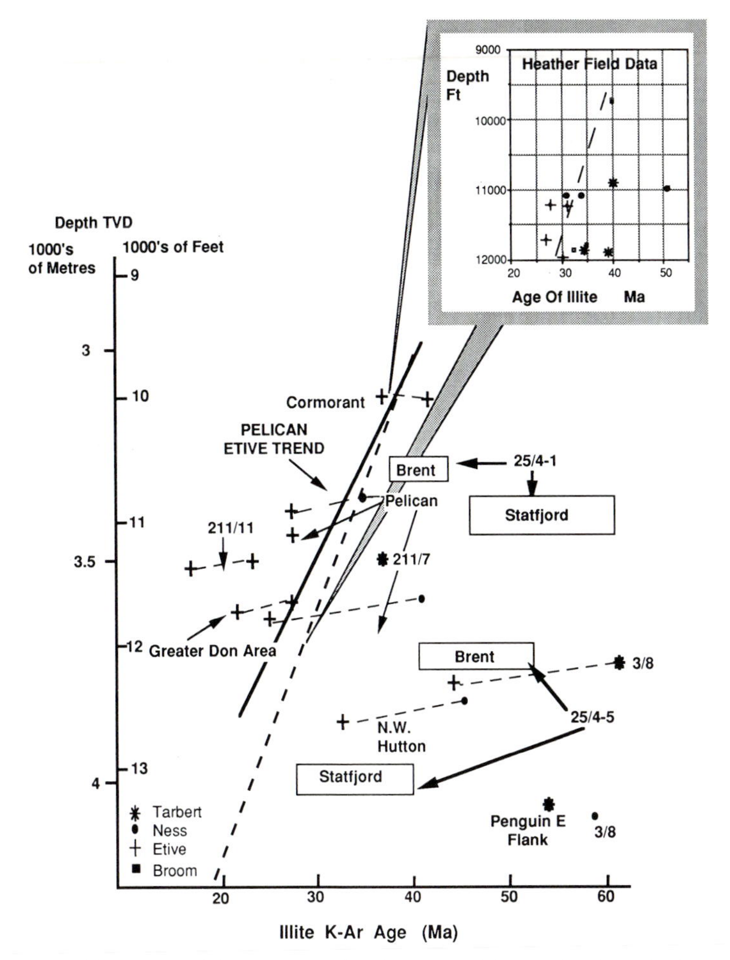


**Fig. 6.** Comparison of K-Ar ages of fine size illite fractions from different depths and fields of the Brent oil province.

can be used to assess the relative importance of advective transport as opposed to diffusive transport of the reactants necessary for illite growth (for more details see Paliauskas & Domenico 1976; Giles 1987; Giles & de Boer 1990).

The variables which control the rate of illite growth are therefore:

(1) temperature
(2) concentration of reactants
(3) rate of supply of reactants
  (3)a. pore fluid velocity
  (3)b. rate of diffusion/mechanical dispersion
(4) time
(5) presence of an aqueous pore fluid.

The hydrothermal experiments of Huang *et al.* (1986), thermodynamic considerations of Bjørkum & Gjelsvik (1988), and thermodynamic and kinetic considerations of Giles and de Boer (1990) suggest closed system authigenic formation of illite is possible via reaction between silica in solution, feldspar and kaolinite. Giles & de Boer (1990) further note that acidity is not a prerequisite for feldspar dissolution, which can in fact occur under neutral and alkaline conditions. Bjørkum & Gjelsvik (1988) suggest a temperature for the onset of illite growth of less than 100°C and possibly as low as 50°C for initiation of this reaction. This is essentially in agreement with the idea of a critical threshold temperature of about 75°C suggested by the K-Ar illite ages from Cormorant Field and from the depth trends of illite abundance and illite/(illite + kaolinite) ratio reported in Giles *et al.* (this volume). Illite stability is favoured by: the presence of silica at levels equal to or exceeding that required for saturation with alpha quartz, high $K^+$ to $H^+$ activity ratio, and the unstable coexistence of feldspar and kaolinite. If the younger K-Ar ages obtainable for illite from water-wet samples truly represent cessation of illite forming processes, then one or more of the parameters that favour illite formation must have changed.

Consider a rock which is undergoing steady burial as in Fig. 7a. In such a system the temperature is increasing steadily in response to increased burial. If illite starts to form once the critical temperature is exceeded, then the rate of illite growth will increase exponentially (i.e. in line with equation 8) with increased temperature and therefore with depth, as shown in Fig. 7b. This will be true provided that the supply of reactants is unlimited by mass transport constraints (i.e. rate at which ions/molecules are transported through the pore fluid to the growing crystals). In the initial period of growth ($\Delta t_1$) of Fig. 7b very little illite will be formed (see hatched area for this time interval in Fig. 7b). In each subsequent time step ($\Delta t$), progressively more illite will form, as indicated by the hatched areas under the curve shown in Fig. 7b which is equivalent to the integral given in equation 9 over a time difference equal to $\Delta t$. Therefore, the vast majority of the illite is formed long after growth began. Because reaction rates are slow at the onset of illite growth, it will probably not be possible to isolate illite crystals from this initial phase of growth and therefore use K-Ar techniques to date it. It also follows from the form of Fig. 7b, that the diagenetic illite population is dominated by those crystals that formed at the highest temperature. Thus sampling a range of illite size fractions for K-Ar dating will give an artificially short time span for illite growth. The time span obtained will be heavily biased toward illite formed at the highest temperatures.

If illite growth continued until the present

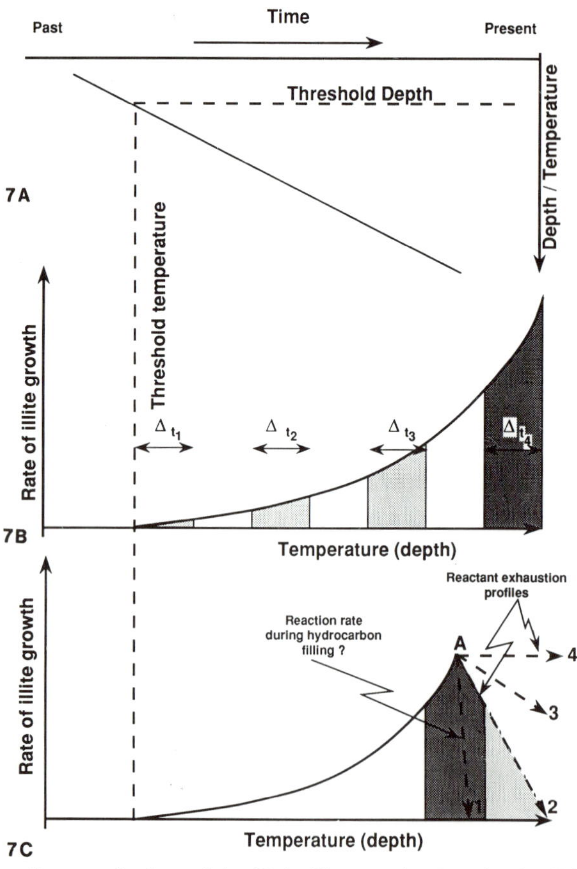


**Fig. 7.** Importance of reaction rates in determining K-Ar illite ages (see text for details).

day along a curve such as Fig. 7b, then the K-Ar dates will show an average age even though recently formed illite will dominate the population. It will be impossible to obtain a diagenetic K-Ar date of near zero as a result of the aggregate nature of the sample. The only exception to this will be where illite growth began relatively recently. However, the problem here will be obtaining sufficient illite to date.

Exponential growth rates are probably only sustainable for a limited time. Eventually the rate of reaction will be constrained by the rate of transport of reactants to the growing crystals. Once this occurs, reaction rates will no longer increase exponentially with increasing temperature, but will be determined by the rate of supply of reactants. At this point, providing the rate of supply of reactants is approximately constant, the reaction rate will remain constant too, giving a rate profile such as A4 of Fig. 7c. Illite formed during this phase will still dominate that formed by the initial exponential phase of growth.

If illite precipitation was to become blocked in some way, then it is likely that precipitation would not begin until very high supersaturations are achieved. In such circumstances rapid precipitation of large amounts of illite might occur in a very short time span and thus K-Ar ages would yield the true age of illite precipitation. Illite which formed from such a solution would be expected to be morphologically distinct, and would probably give rise to small crystals or fibres.

Sooner or later the supply of nutrients must run out. $K^+$ is probably the most critical variable in this respect as it can only be derived from the decomposition of K-feldspars or micas. Curves A-2 and A-3 of Fig. 7c represent schematically progressively falling reaction rates as a result of the exhaustion of reactants. In each equal time step, despite increased temperatures, the reaction rate falls because of falling quantities of reactants. Consequently smaller quantities of illite are produced, as indicated by the hatched area below the curve A-2 in Fig. 7c. This will mean that the population of illites to be dated will still be dominated by those grown during the peak of illite growth. Thus the most recent period of illite growth (that along the segment of curve labelled A-2 in Fig. 7c) will not be apparent from the K-Ar dates which will represent an average of the ages of diagenetic illites weighted toward the period of maximum growth, which was prior to the decay of reaction rates. Regional data presented in Giles *et al.* (this volume) shows that feldspars disappear with depth in the Brent Province, which suggests that the illite forming reaction is eventually limited by the availability of reactant.

Charging a reservoir with hydrocarbons is generally assumed to stop illite growth. However, there is no reason to assume that this will be an instantaneous process, as the pore fluid has to go from a 100% aqueous fluid to some level of hydrocarbon saturation. The decay of reaction rates might follow curve A-1 of Fig. 7c. As relatively little illite will be formed during hydrocarbon charging, as opposed to the period preceding charging, then sampling the diagenetic illites will yield K-Ar dates which are generally biased toward the period prior to hydrocarbon charging. In other words, K-Ar dates from the finest illite fraction will not date hydrocarbon emplacement but will give dates which should at best approximate to the timing of charge. There is of course the open question of whether illite growth will cease in an hydrocarbon bearing reservoir. As most hydrocarbon reservoirs are water wet, i.e. the pore walls are coated with an aqueous fluid and the hydrocarbons occupy the centre of the pore, then there is no reason why illite growth should cease. However, solute transport can only occur by diffusion through the aqueous fluid, which will be slow and limit the rate of illite growth. One would expect that the higher the irreducible water saturation in the reservoir, the greater the scope and amount of closed system diagenesis that will be possible, such as the growth of illite or the decomposition of feldspars. There is, therefore, a probability that some limited level of closed system diagenesis will continue in water wet rocks once they are charged with oil. It is difficult to assess the impact this may have on K-Ar dates from diagenetic illites from oil zones as it is impossible to quantify the rate of illite growth under such conditions.

The dissolution of feldspars is a key factor in the growth of illite. Feldspar decomposition rates are a function of pH. Reaction rates are highest in either alkaline or acid solutions, but are also finite and measurable in neutral solutions (see Giles & de Boer 1990). K-feldspars might be rapidly used up if pore waters were to become notably more acid or alkaline. Such a situation may arise by the introduction of carboxylic acids generated as part of decarboxylation reactions in organic matter, which occur between 80 and 100°C (see Surdam & Crossey 1987). Such an event would promote the rapid removal of feldspars in this temperature interval and the rapid growth of illite. Thus all the illite size fractions would have formed during a narrow timespan encompassing the most rapid illite growth rates. This hypothesis explains both why very young ages are not recovered from the water zone samples and why the size fraction age differences are small. However, the importance of carboxylic acids in the dissolution of feldspars has been questioned by Giles & Marshall (1986) and Giles & de Boer (1990).

As a structure is buried, illite growth begins in the deeper parts first as the critical temperature for illite formation is exceeded. If the supply of reactants is limited, then illite will cease to form in the deeper parts of the structure first. Therefore the exhaustion of reactants will lead to illite age/depth profiles which show increasing illite ages with depth, as observed for instances, by Hogg (1989) in a study of the Alwyn Field.

Uplift will cause progressively smaller quantities of illite to be produced in each time interval until the temperature falls below the critical temperature for illite growth. The averaging, inherent in sampling diagenetic illites, will result in a K-Ar date which does not date uplift, nor the peak of illite generation, but some time in between.

## Review of published K-Ar data for Brent Sandstones

The first published references to illite K-Ar ages in Brent sandstones are from Sommer (1975, 1978). Youngest ages of 45–55 Ma from three wells drilled by TOTAL–SNEAP were interpreted as dating the time of hydrocarbon charge to the reservoir units. The field was not identified and no sample depths are given.

Jourdan *et al.* (1987) report some general features of K-Ar ages for <0.2 μm illites from the Alwyn area but give no details of burial history or sample depths. They note that illite formation is only minor in Alwyn North (3/4–8, 3/9a–1 and 3/9a–4). K-Ar ages of 75 Ma reflect a maximum age of oil accumulation in this structure. On the south flank of the Alwyn North structure, younger ages of 35–45 Ma (3/9–6) reflect a later phase of hydrocarbon charging in sandstones that did not receive the initial oil pulse.

In the southern part of the field (3/14–3, –4, –6, –7, –8 and –11, and 3/15–2 and –4) K-Ar illite ages of 45–55 Ma record a generally more sustained period of diagenesis prior to hydrocarbon accumulation. In the western part of the Alwyn South structure, most wells (3/8–4, 3/14–9, –10, –12) yield even younger ages of 35–45 Ma. This geographic progression of decreasing illite age is paralleled by a progressive increase in quartz overgrowth fluid inclusion temperatures (pressure corrected, $T_p$) from <120°C to >140°C. The authors conclude from these data that all late diagenesis, from 75 to 35 Ma, occurred under the influence of hot fluids derived from depth, initially from the Viking Graben trough, and later from deeper parts of the East Shetland Basin.

Studies by Hogg *et al.* (1987) and Hogg (1989) on Alwyn South wells confirmed the K-Ar illite ages of 45–56 Ma for 1.0–0.5 μm and 0.5–0.1 μm fractions. With a few exceptions the ages increase with increasing depth over the intervals sampled. This is not a profile that would result from a downward migration of the oil–water contact. This may reflect later migration into shallower horizons. In fact, Leythaeuser & Ruckheim (1989) have observed a relationship between oil composition and porosity–permeability that suggests oil charge will occur first to more porous and permeable sandstones, and that accumulation will not necessarily occur in a regular manner from the top downwards. Burial history curves indicate depths of 1.8–2.2 km during illite precipitation (e.g. see Liewig *et al.* 1987). Normal geothermal gradients would imply temperatures of 50–80°C. However, quartz overgrowth fluid inclusions have $T_p$ values that indicate much higher temperatures and are attributed to migration of hot aqueous fluids through the reservoir prior to hydrocarbon accumulation.

Interpretation of K-Ar illite data reported by Liewig *et al.* (1987) from Alwyn Brent sands is again made difficult by lack of information on burial histories and well localities. Illite ages are given from three samples, each of the <0.6 μm size fraction and from within the oil zone, but close to the oil–water contact (OWC). Sample depths of 10 636 ft (3241.89 m) 10 625 ft (3238.54 m) TVD (true vertical depth) yielded ages of 35 ± 1 and 44 ± 2 Ma. All other samples were variably contaminated with feldspar, and gave apparent ages of 35–129 Ma that broadly correlate with illite to feldspar ratio. Extrapolation of this contamination trend to zero feldspar yields an age of *c.* 40 Ma.

Overall, these studies of the illite and quartz diagenesis of the Alwyn Brent area suggest periods of late diagenesis that ranged from short to protracted, prior to oil accumulation, and were a response to interaction of reservoir sands with fluids that were much hotter than recorded from elsewhere during the Eocene in the Brent province. The significance of hot, exotic fluids in Brent diagenesis is addressed in Giles *et al.* (this volume).

In block 25/4, Thomas (1986) reported K-Ar illite ages (plotted on Fig. 6) for <0.2 μm separates in a comparative diagenetic study of

Brent and Statfjord Group sandstones. Both formations are in the oil zone of 25/4–1, a crestal well, and both are in the water zone in well 25/4–5 located in a downfaulted block. In the former, between *c.* 10 430 ft (3179.10 m) to 10 600 ft (3230.91 m), three illite ages in the range 44–38 Ma were obtained from Brent sands. Statfjord samples yielded generally older ages of 62–44 Ma for illites at depths of between *c.* 10 780 ft (3285.78 m) to 11 070 ft (3374.18 m).

The deeper, water-bearing well, 25/4–5 shows an illite age profile between the Brent and Statfjord Groups that is the opposite from that in the oil-bearing 25/4–1 well (Fig. 6). Brent sand at approx. 12 120–12 220 ft (3694.22–3724.70 m) yielded two illite ages of *c.* 52 Ma and two of *c.* 40 Ma. For the Statfjord Group (*c.* 12 980–13 170 ft, 3956.35–4014.26 m, sampling interval) one *c.* 40 Ma and three 28 Ma ages were obtained. It would seem that illite diagenesis was brought to a premature end in sands of the Statfjord Group by oil migration but was still active 20–30 Ma later in the water zone. To account for these differences in terms of hydrocarbon emplacement, then different fluid dynamic regimes must exist between formations, thus enabling oil accumulation to occur earlier at deeper levels. Within the constraints of the limited detail supplied in Thomas (1986) no significant age profile can be inferred.

From the Rannoch Formation of the northern Brent province (no TVD or well localities available) Hamilton *et al.* (1987) reported an age decrease with decrease in illite grain size from 60 to 45 Ma. These authors also found a trend of increasing illite age ($<0.05$ $\mu$m size fraction) with depth from 59 Ma to 46 Ma over a 109 ft (33.22 m) depth range. Haszeldine *et al.* (this volume) reported a K-Ar age of 59 Ma for $<0.1$ $\mu$m illite from a Rannoch Formation sample (10 202 ft, 3109.61 m) in well 211/19–4 of the Murchison Field.

Glasmann *et al.* (1989*a*) reported on diagenetic studies of Heather Field samples and included K-Ar ages for illite clay fractions. These authors also note, with some exceptions, decreases in age with depth and decreasing grain size (Fig. 6). Less than 0.1 $\mu$m fractions are typically 4–12 Ma younger than coarser fractions and provide some estimate of the minimum time span of illitization. Reconstructed burial and thermal histories (no detail given), quartz overgrowth fluid inclusion homogenization temperatures and the K-Ar illite ages suggest a temperature range of 80–130°C for illite precipitation.

The K-Ar illite ages from Heather Field (Glasmann *et al.* 1989*a*) were interpreted as documenting progressive oil filling of reservoirs from shallower to deeper levels through mid Eocene to late Oligocene times. Where reversals in the general age–depth trend occur (e.g. in 2/5–9) it is suggested that poroperm controls have locally led to some preferential focusing of fluid flow during earliest illite formation. Overall, however, the age–depth trend is viewed as sufficiently established to be incorporated into reservoir modelling.

Illites from NW Hutton Field have K-Ar ages from 39 Ma to 49 Ma (J. Cocker, pers. comm. 1988 quoted in Scotchman *et al.* 1989). Water zone ages average 41 Ma and oil zone ages average 43.1 Ma. Oil charge to the reservoir is suggested to have occurred over a relatively short time interval of 5 Ma with water zone illite diagenesis also ceasing shortly thereafter.

## Results from the present study

The new K-Ar age results from Brent Group samples are given in Table 1. They are discussed first on an individual basis with highlighted examples of interpretation in relation to hydrocarbon migration and the timing of illite growth.

*NW Hutton Field.* A water zone Etive Formation sandstone and an oil zone Ness Formation sandstone were selected from a well in this field. Insufficient illite of $<0.1$ $\mu$m was recovered in each case necessitating combination of this fraction with a fine sized (0.1–0.3 $\mu$m) aliquot from the next size fraction (0.1–0.5 $\mu$m). This finer fraction ($<0.3$ $\mu$m) for the Ness Formation sample yielded an older age (56 Ma) than that from the coarser fraction (45 Ma), the reverse of what might be expected. XRD analyses did not show any K-bearing contaminants, but this is insufficient proof that a low level of contamination was not present. The younger age may therefore be considered as providing a constraint on the maximum age of oil accumulation. Geohistory analysis for this well shows that this age is coincident with the time at which the overlying Kimmeridge Clay Formation source rocks reached a vitrinite reflectance value ($R_0$) of 0.62% and thus was generating hydrocarbons (Fig. 8A). Two size fractions from the water zone sample gave ages, within error, of 34 Ma. These data can be interpreted as recording the persistence of illite forming processes in the water zone for *c.* 10 Ma subsequent to oil emplacement at 45 Ma (Fig. 8A) and document a longer period of water zone illite formation than suggested in Scotchman *et al.* (1989) but are otherwise in broad agreement.

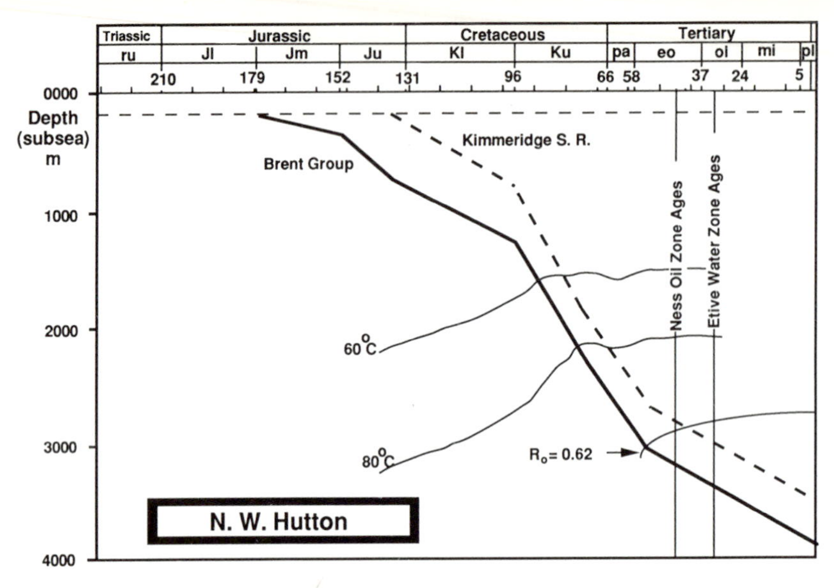

Triassic
Jurassic
Cretaceous
Tertiary
ru
Jl
Jm
Ju
Kl
Ku
pa
eo
ol
mi
pl
210
179
152
131
96
66
58
37
24
5
0000
Depth (subsea) m
1000
2000
3000
4000
Kimmeridge S. R.
Brent Group
Ness Oil Zone Ages
Etive Water Zone Ages
60°C
80°C
R<sub>o</sub>= 0.62
N. W. Hutton


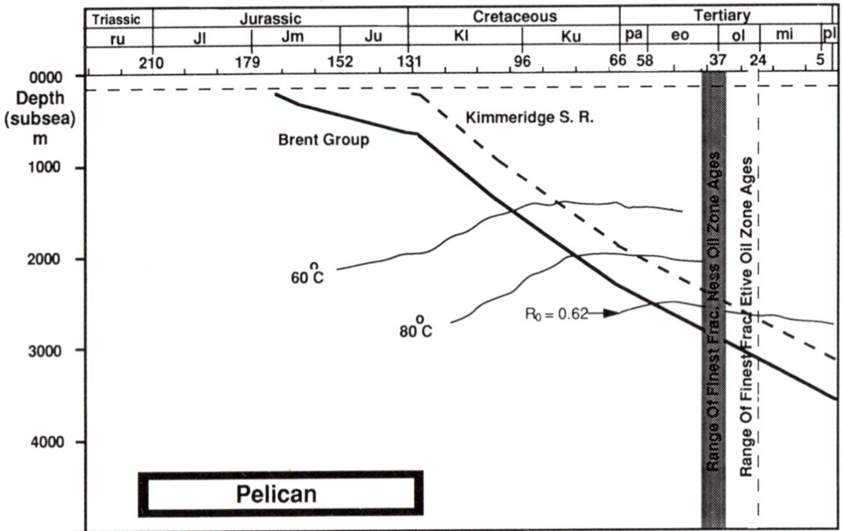

Triassic
Jurassic
Cretaceous
Tertiary
ru
Jl
Jm
Ju
Kl
Ku
pa
eo
ol
mi
pl
210
179
152
131
96
66
58
37
24
5
0000
Depth (subsea) m
1000
2000
3000
4000
Kimmeridge S. R.
Brent Group
Range Of Finest Frac. Ness Oil Zone Ages
Range Of Finest Frac. Etive Oil Zone Ages
60°C
80°C
R0 = 0.62
Pelican


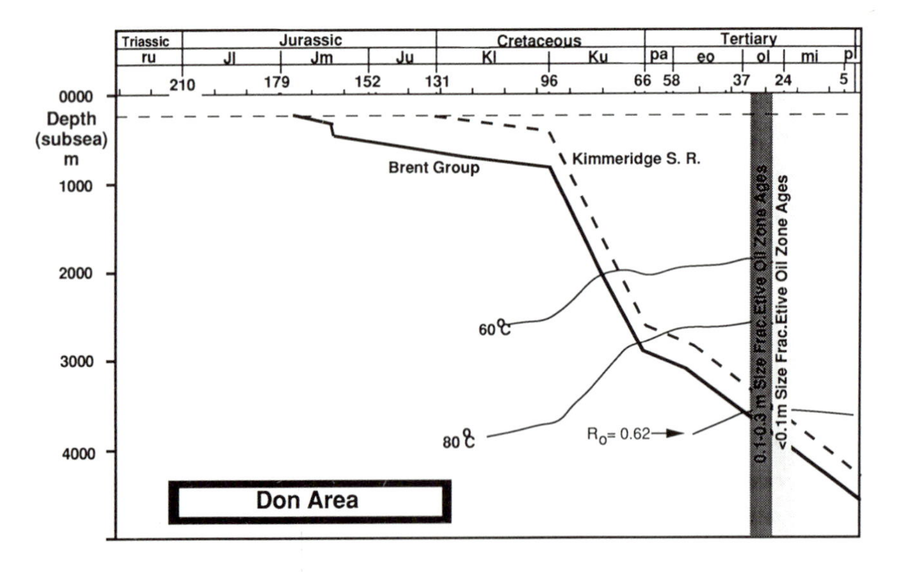

Triassic
Jurassic
Cretaceous
Tertiary
ru
Jl
Jm
Ju
Kl
Ku
pa
eo
ol
mi
pl
210
179
152
131
96
66
58
37
24
5
0000
Depth (subsea) m
1000
2000
3000
4000
Kimmeridge S. R.
Brent Group
0.1-0.3 m Size Frac. Etive Oil Zone Ages
<0.1m Size Frac. Etive Oil Zone Ages
60°C
80°C
Ro= 0.62
Don Area

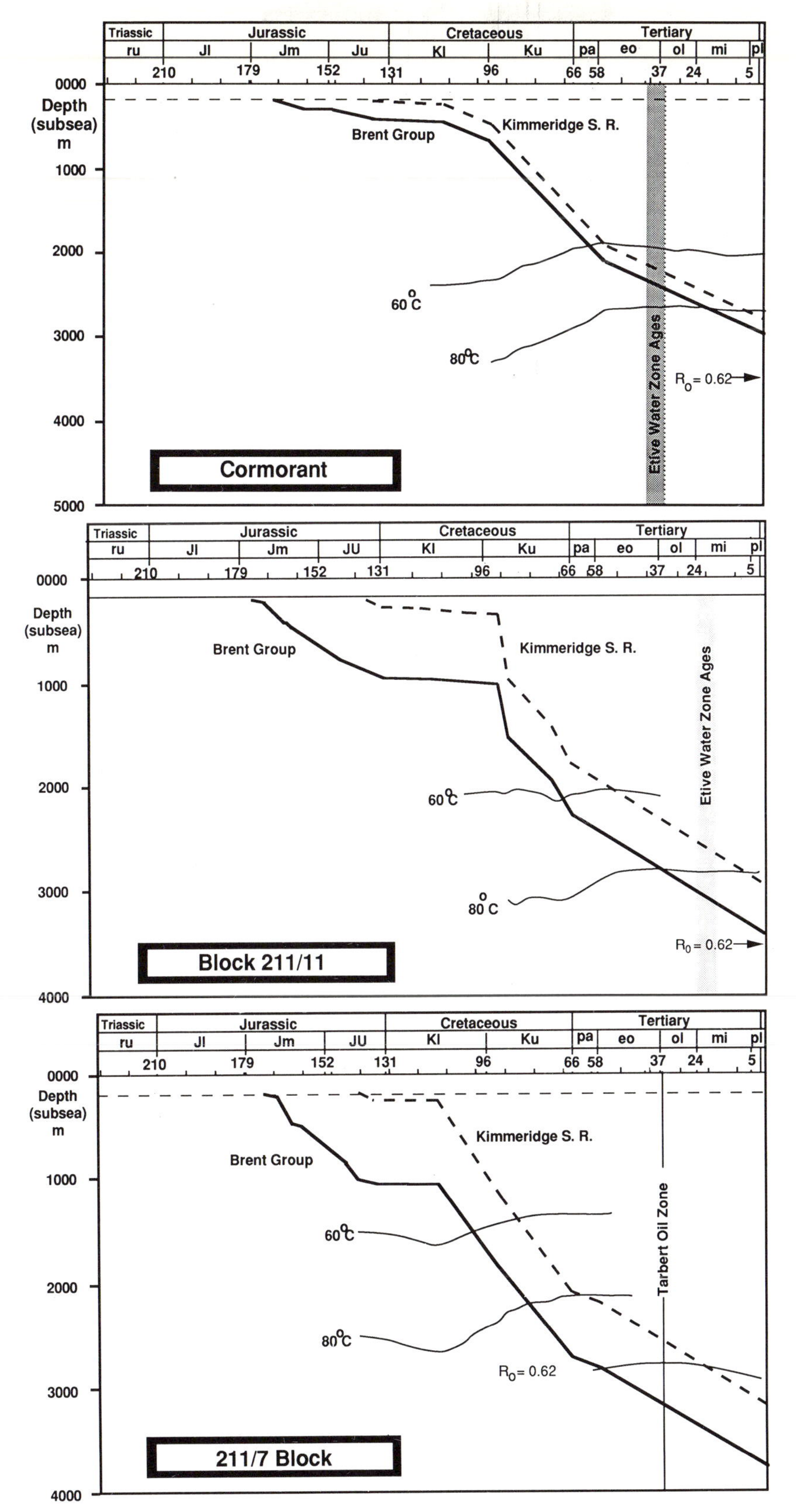


**Fig. 8.** (**a**) NW Hutton. Geohistory curves showing 80°C isotherm, isoreflectance curve for $R_0$ = 0.62% and illite ages. (**b**) Pelican. Geohistory curves showing 80°C isotherm, isoreflectance curve for $R_0$ = 0.62% and illite ages. (**c**) Don Area. Geohistory curves showing 80°C isotherm, isoreflectance curve for $R_0$ = 0.62% and illite ages. (**d**) Cormorant. Geohistory curves showing 80°C isotherm, isoreflectance curve for $R_0$ = 0.62% and illite ages. (**e**) Block 211/11. Geohistory curves showing 80°C isotherm, isoreflectance curve for $R_0$ = 0.62% and illite ages. (**f**) 211/7 Block. Geohistory curves showing 80°C isotherm, isoreflectance curve for $R_0$ = 0.62% and illite ages.

**Table 1.** *K-Ar results*

| Area (see Fig. 1) | Depth (TVD, ft) | Formation /leg | Size fraction (μm) | K (wt%) | $*Ar^{40}$ ($10^{-10}$ mole $gm^{-1}$) | Age (Ma) ± σ |
|---|---|---|---|---|---|---|
| 211/7 | 11243.00 | Tarbert Fm. /residual oil | 0.1–0.5 | 3.98 | 2.5428 | 36.5 ± 0.9 |
| 211/11 | 11325.00 | Etive Fm. /water | <0.1 | 5.26 | 1.5925 | 17.4 ± 0.7 |
| | 11285.00 | Etive Fm. /water | <0.1 | 4.80 | 1.9982 | 23.8 ± 0.7 |
| Penguin E. Flank | 13321.5 | Tarbert /Oil? | <0.1 | 4.46 | 4.2575 | 54.2 ± 1.4 |
| Greater Don Area | 11709.5 | Etive Fm. /oil | <0.1 | 5.29 | 2.0325 | 22.0 ± 0.7 |
| | | | 0.1–0.3 | 5.9 | 2.9339 | 28.4 ± 0.7 |
| | 11620.5 | Etive Fm. /oil | <0.1 | 4.80 | 2.3157 | 27.6 ± 1.1 |
| | | | 0.1–0.3 | 6.23 | 3.7195 | 34.1 ± 0.9 |
| Cormorant | 9966.00 | Etive Fm. /water | <0.1 | 3.72 | 2.6841 | 41.3 ± 0.9 |
| | | | 0.1–0.3 | 1.37 | 3.6113 | 147.0 ± 3.0 |
| | 9942.00 | Etive Fm. /water | <0.1 | 4.02 | 2.5283 | 35.9 ± 0.8 |
| Pelican Well A | 11760.9 | Etive Fm. /oil | 0.1–0.5 | 5.95 | 2.5716 | 24.8 ± 0.7 |
| | 11601.50 | Ness Fm. /oil | <0.1 | 3.98 | 2.8453 | 40.8 ± 1.5 |
| | | | 0.1–0.5 | 4.54 | 4.4068 | 55.1 ± 1.39 |
| Well B | 11070.7 | Etive Fm. /oil | 0.1–0.5 | 5.40 | 2.5754 | 27.3 ± 0.6 |
| Well C | 10905.50 | Etive Fm. /oil | 0.1–0.3 | 6.03 | 2.8544 | 27.1 ± 0.6 |
| | 10779.5 | Ness Fm. /oil | <0.1 | 4.55 | 2.7476 | 34.8 ± 0.9 |
| | | | 0.1–0.5 | 4.64 | 3.5702 | 43.9 ± 1.0 |
| N.W. Hutton | 12617.95 | Etive Fm. /water | <0.3 | 4.61 | 2.6718 | 33.1 ± 0.7 |
| | | | 0.1–0.5 | 5.2 | 3.1426 | 34.5 ± 0.8 |
| | 12466.80 | Ness Fm. /oil | <0.3 | 3.72 | 3.6693 | 56.0 ± 1.4 |
| | | | 0.1–0.5 | 5.23 | 4.1588 | 45.3 ± 1.0 |
| 3/8 Well 1 | 13531.80 | Ness Fm. /oil | 0.1–0.5 | 4.32 | 4.4986 | 59.1 ± 1.4 |
| Well 2 | 12308.00 | Etive Fm. /oil | <0.1 | 4.74 | 3.7032 | 44.5 ± 1.1 |
| | 12176.00 | Tarbert Fm. /oil | <0.1 | 3.97 | 4.324 | 61.8 ± 1.4 |

*Pelican.* Five oil zone samples, three from the Etive Formation and two from the Ness Formation, were selected from three wells in this field (Table 1). The following observations are made.

(i) From one sample of the Ness Formation of well A, an age difference of 14 Ma was found between size fractions (<0.1 μm, 40.8 ± 1.5 Ma; 0.1–0.5 μm, 55.1 ± 1.4 Ma). This represents the largest time difference between size fractions of the same sample found in this study. However, the XRD pattern shows a suggestion that 2M mica may be present in the coarser (0.1–0.5 μm) fraction. Larger size fractions (e.g. 2–5 μm) contain significant muscovite. Only about 6% of muscovite of Caledonian provenance (i.e. K-Ar age about 300–430 Ma, see

Glasmann *et al.* 1989*b*; Hamilton *et al.* 1987) is required in the coarser fraction to generate this age difference.

(ii) Age differences between Ness Formation size fraction pairs of Well C are about 9 Ma (Table 1). These differences are not as obviously due to possible contaminant phases and conventionally would be taken to indicate minimum time spans for illite growth.

(iii) The Etive Formation oil zone samples show a decrease in illite age of 2.5 Ma between the shallowest and deepest samples. In contrast, the two Ness Formation samples have finest size fraction ages which show the reverse relationship. The Pelican Etive Formation oil zone age–depth trend is also indicated on Fig. 6 for comparison with other data.

(iv) The Etive Formation ages range from 24.8 to 27.3 Ma and are approximately coincident with the estimated time when the overlying Kimmeridge Clay source rock achieved a vitrinite reflectance value of 0.62% (Fig. 8B) and were consequently generating oil. The Ness Formation illite ages are older than this by *c*. 8–14 Ma.

*Greater Don area (Block 211/13)*. Two size fractions were dated from each of two Etive Formation oil zone samples from one well in this area. The finer fractions gave ages about 6.4–6.5 Ma younger than the coarser fractions and indicate a minimum time span of illitization similar to that inferred from the Pelican data. Accepting the <0.1 μm ages as dating hydrocarbon emplacement suggests an interval of 6 Ma for the oil–water contact to migrate downwards from 11 620 ft (3541.82 m) at 28 Ma to 11 710 ft (3569.25 m) at 22 Ma. If oil accumulation had been more rapid than this at 28 Ma, the 22 Ma age suggests the possibility of sustained illite growth subsequent to oil charge. This point is discussed further later in the text. These ages are significantly younger than those for other fields in the area such as the 59 Ma age for Rannoch Formation illite in the Murchison oil zone (Haszeldine *et al.* this volume) and the 60–45 Ma ages for Rannoch Formation illite fractions for another northern Brent province field (Hamilton *et al.* 1987) and suggests that in these areas, illite precipitation ceased before oil migration. In the Greater Don well discussed here hydrocarbon generation from the Kimmeridge Clay source rocks would have been underway by an $R_0$ value of 0.62%, which from geohistory curves (Fig. 8C) is estimated to have been reached at 28 Ma. This age is broadly coincident with the time suggested for cessation of illite diagenesis.

*Cormorant*. Two Etive Formation samples from the water zone, which differed in depth by only 24 ft (7.32 m), yielded similar ages (41.3 and 35.9 Ma) for finest fraction (<0.1 μm) illite. The 9966 ft (3037.67 m) sample may have a minor detrital mica contaminant so resulting in the high apparent age of 41.3 Ma. Considerable difficulty was experienced with this particular sample in isolating any size fraction for which XRD analysis did not indicate the presence of 2M mica. Whilst the finest size fraction analyzed gave no indication from XRD of such a contaminant, detrital illite and also kaolinite are significant components of the next size fraction at 0.1–0.3 μm. This yielded an age of 147 Ma as would be consistent with the presence of a detrital contaminant. A small amount of sample remained after K-Ar analysis of the <0.1 μm fraction and was used to assess the efficacy of high grade magnetic separation to obtain purer illites. The TEM image of Fig. 4 shows this procedure to have been successful in removing the kaolinite. However, in addition to the thin elongate particles typical of diagenetic illite, it is evident that rare thick particles of detrital illite still remain, and confirms that even this fine size fraction was contaminated.

The 36 Ma age for the shallower sample thus provides the best estimate of cessation of illite diagenesis in water wet Etive Formation sandstones from this Cormorant Well. At the depth of burial (*c*. 8040 ft, 2450.62 m) corresponding to this time the overlying Kimmeridge Clay source rocks were certainly immature, and burial temperature of the Etive Formation is estimated at about 75°C (Fig. 8D). This value is the lowest temperature recorded for significant illite growth as noted in Giles *et al.* (this volume).

The age differences between the two samples (41.3 ± 0.9, 35.9 ± 0.8 Ma) taken together suggest that the period of illite diagenesis lasted less than 10 Ma. If 'nutrient' supply for such illite growth is now present, and as these samples are still in the water zone, then a longer period of illitization should have occurred and preponderance of younger illite would be expected.

*Block 211/11*. Two Etive Formation samples from one well (same block as Eider) yielded K-Ar ages for <0.1 μm illites of 17 and 24 Ma from depths of 11 325 ft (3451.90 m) and 11 285 ft (3439.71 m) respectively. Thus the oft-repeated pattern of decrease in age with increase in depth in oil zone samples exists here, but in a water zone (Fig. 8E). The former age is the youngest obtained in this study. It is surprising that there is not a predominance of recent illite precipitation as the structure is not hydrocarbon filled and therefore there is no clear reason why illite diagenesis should have ceased.

*211/7.* One well was sampled in this block (about 10 km NW of Penguin) for a residual oil zone sample from the Tarbert Formation. The 0.1–0.5 μm illite separate yielded an age of 36 Ma. Geohistory analysis indicates palaeotemperatures at this time of *c.* 105°C for this sample and 95°C for the overlying Kimmeridge Clay Formation source rocks. The overlying Kimmeridge Clay Formation source rocks did not become mature for oil generation ($R_0$ = 0.62%) until *c.* 22 Ma (Fig. 8F).

*Penguin E, flank.* A Tarbert Formation oil zone sample yielded an <0.1 μm illite fraction with a K-Ar age of 54 ± 1.4 Ma. This is within two standard errors of the 56 Ma age, at which time the Kimmeridge Clay Formation source rocks in this well are estimated to have begun to generate significant volumes of oil (i.e. when $R_0$ = 0.62%.

*Block 3/8 (south of Ninian).* The <0.1 μm fractions from two oil zone samples, one from the Etive Formation and one from the Tarbert Formation, yielded significantly different ages of 45 Ma and 62 Ma respectively. Again the older age derives from the shallower sample. The overlying Kimmeridge Clay Formation source rocks did not reach $R_0$ = 0.62% and expel significant oil until much later (*c.* 45 Ma, see Fig. 8I). In contrast the Etive illite age of 45 Ma is coincident with the probable time of hydrocarbon generation.

A second well from this block provided the deepest sample of this study from a depth of 13 532 ft (4124.60 m) located within the Ness Formation and in the oil zone. The 0.1–0.5 μm fraction yielded an age of 59 Ma. Geohistory analysis indicates this is the time at which Kimmeridge Clay Formation source rocks would have reached $R_0$ = 0.5% and significant oil generation would only have begun at about 42 Ma when $R_0$ = 0.62%. The point will be explored later that 'dirtier' Brent sands such as from the Ness Formation tend to yield ages in excess of the time at which significant oil migration begins.

### Illite diagenesis in oil zone samples

Pore fluid chemistry (pH, potassium activity) plays a major role in defining the illite stability field, and consequently, when an aqueous pore-fluid is displaced by chemically 'inert' hydrocarbons it has been suggested that illite may no longer precipitate. If hydrocarbon charging of a reservoir occurs from the top downwards, the result should be a negative correlation between the ages of finest size (last-formed) illite, and depth. Much of the age–depth data for the Heather Field together with the Pelican and Greater Don Etive Formation data (Fig. 6) can be interpreted as resulting from downward migration of the oil–water contact with time. From Fig. 9 it is evident that the finest illite size fraction ages from the Etive Formation of these fields correlates very well with the modelled timing of when a vitrinite reflectance of 0.62% was reached in the overlying Kimmeridge Clay Formation source rocks. One datum, however, from the deeper sample from the Greater Don well, indicates illite diagenesis ceased about 6–8 Ma later than the calculated time of oil generation/migration. However, difficulties in assessing errors in estimating the timing of maturity are such that it is not possible to determine if this 6–8 Ma time difference is real.

A possible explanation of the age difference between the two Etive Formation samples of the Greater Don well is that the oil fill occurred earlier to the shallower sample depth at 11 620 ft (3541.82 m) but did not reach the 90 ft (27.43 m) deeper level until 6–8 Ma later. Another possibility is that illite diagenesis persisted at the deeper level subsequent to oil charge at 28 Ma. This would require water saturations within the oil zone sufficient to permit diffusive or fluid flow transport of requisite 'nutrients' for illite growth to continue. In fact, Ehrenberg & Nadeau (1989) have tentatively suggested that, in the Garn field, illite diagenesis persisted subsequent to hydrocarbon charging with residual water saturations of 10–50%. In general terms however it may be concluded that finest size illites from oil legs in the Etive Formation have K-Ar ages that accurately reflect cessation of diagenesis at the time of oil migration.

Observations of illite abundance variations with depth, together with geohistory analysis, suggest a temperature interval for illite growth extending from about 75°C until oil charging typically at about 110°C, and extending in time from at least 17 Ma to in excess of 50 Ma. The inferred time spans for illitization in these Etive Formation samples is thus tens of millions of years. Age spreads for different size fractions of illite are considerably less than this, typically 5–10 Ma, and may reflect a limitation of the method imposed by a requirement for purity not met by coarse, and older, illite fractions. Further discussion of the time span of illitization is presented in consideration of water zone illite growth.

Most of the oil zone illite ages, other than for the Etive Formation samples, show older values than calculated times of oil generation in the overlying Kimmeridge Clay Formation source rocks (Fig. 9). It would be expected that as Etive Formation sands have a much higher degree of connectivity, as would be expected from a barrier sand complex (see Cannon *et al.* this volume), oil charging would preferentially occur via this formation (as illustrated in Fig. 10). Geochemical evidence (Miles 1990) confirms the important role played by the Etive Formation as a migration route. Younger estimates of the end of illite formation would therefore be expected for these other formations — the reverse of what is observed (see Fig. 9).

If these older ages represent the end of the illite forming processes then it must be unrelated to oil charge. This was suggested earlier in the context of data from the Rannoch Formation (Hamilton *et al.* 1987 and Haszeldine *et al.* this volume) and from a residual oil zone sample from the Tarbert Formation of a 211/7 well. This hypothesis would also serve well to explain the anomalous data of Glasmann *et al.* (1989) from West Heather. Arguing against this idea is the observation that general trends of variations in illite abundance, and in distribution of different morphologies with depth, pertain to all Brent formations (see Giles *et al.* this volume). It would not therefore be expected that Etive Formation sands would experience illite forming processes for tens of millions of years longer, and to considerably greater burial depths, than for juxtaposed formations.

A major difference between the Etive Formation sands and other Brent Group sands is that the latter at deposition are 'dirtier', that is they contain higher proportions of detrital clay in the form of grain-coating mud and mud-clasts, or detrital mica. This clay includes a detrital illite component and an illite-smectite mixed-layer clay component. Much of the latter may have had an origin by neoformation in Jurassic weathering horizons (Burley & Flisch 1989). If not subjected to burial illitization and smectite destruction, K-Ar ages should lie close to sedimentation ages of (*c.* 175–160 Ma). K/Ar ages similar to or slightly younger than sedimentation ages are in fact recorded for some illite/smectite clay separates from shallow Piper Formation mudrocks and oil zone sandstones in the Tartan and Piper fields in the Outer Moray Firth (Burley & Flisch 1989). The persistence of old illite ages such as these, together with the apparent age differences between different illite size fractions, strongly suggests that the illitization event cannot easily reset the K-Ar dates of the older illites. With onset of illitization of mixed-layer clays during burial, K-Ar ages become increasingly younger. Clay separate ages will reflect variable mixtures of old detrital illite, variably transformed mixed-layer clay, and authigenic illite neo-formed independently of the mixed-layer clay reactions. Morphological or size distinction of the latter two is difficult for fine-size illite fractions, as it is difficult to determine unambiguously the paragenesis of illite filaments or thin flakes emanating from grain coating clay or mud pellets. It may therefore be concluded that significantly older K-Ar ages for Brent Group sandstones may result from samples which are rich in clay matrix as a consequence of sampling fine size illite of more than one mode of formation. Though illitization may cease as a consequence of oil charge, the ages obtained may not reflect this event. For samples, such as from West Heather, with ages very much greater than the time of oil migration, the possibility exists that there was no relationship between an early period of illite formation and the time of oil migration.

A few non-Etive data points also fall close to the Etive correlation line in Fig. 9 and this suggests that the youngest fraction of authigenic illite may occasionally be uncontaminated. These few data are interpreted in the same way as for the Etive oil zone illites and therefore also reflect the time of oil charge to these sandstones.

### *Illite diagenesis in water zone samples*

A total of five water zone samples, all from the Etive Formation, have fine-fraction K-Ar illite ages of 41–17 Ma. Surprisingly none of these reflect very recent diagenesis ($\leq$10 Ma), not even those at depths where ample potassium feldspars are available to supply the requisite K by dissolution (e.g. in Cormorant). The absence of very young illite ages is also evident from published studies. However, in a few cases such as Alwyn, the absence of young dates may reflect exhaustion of $K^+$ required for illite growth from the feldspar detritus.

A water-wet Ness Formation sandstone in well 2/5–9 (depth 10973 ft) from the Heather field yielded illite size fractions of 0.5–0.2, 0.2–0.1 and <0.1 $\mu$m with K-Ar ages of 56.9, 50.3 and 50.9 Ma respectively (Glasmann *et al.* 1989*a*). These data would not support a hypothesis of very young illite diagenesis either, and their similarity across the size fraction range suggests a lack of contaminant phases. Rather,

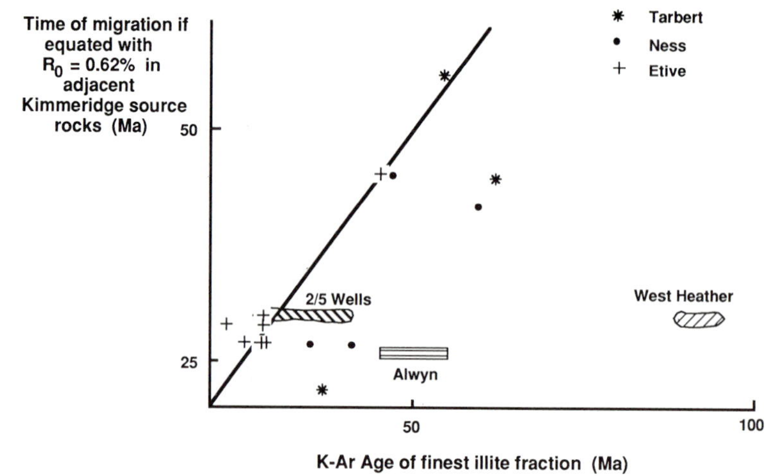


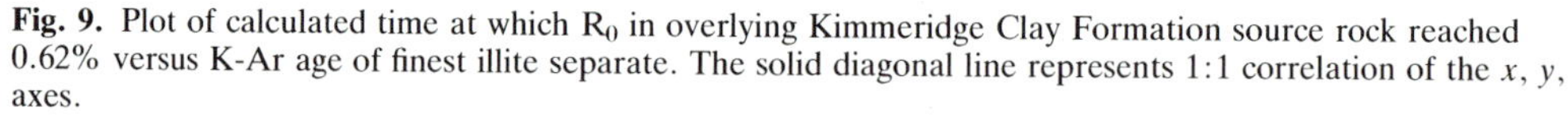

**Fig. 9.** Plot of calculated time at which $R_0$ in overlying Kimmeridge Clay Formation source rock reached 0.62% versus K-Ar age of finest illite separate. The solid diagonal line represents 1:1 correlation of the *x*, *y*, axes.

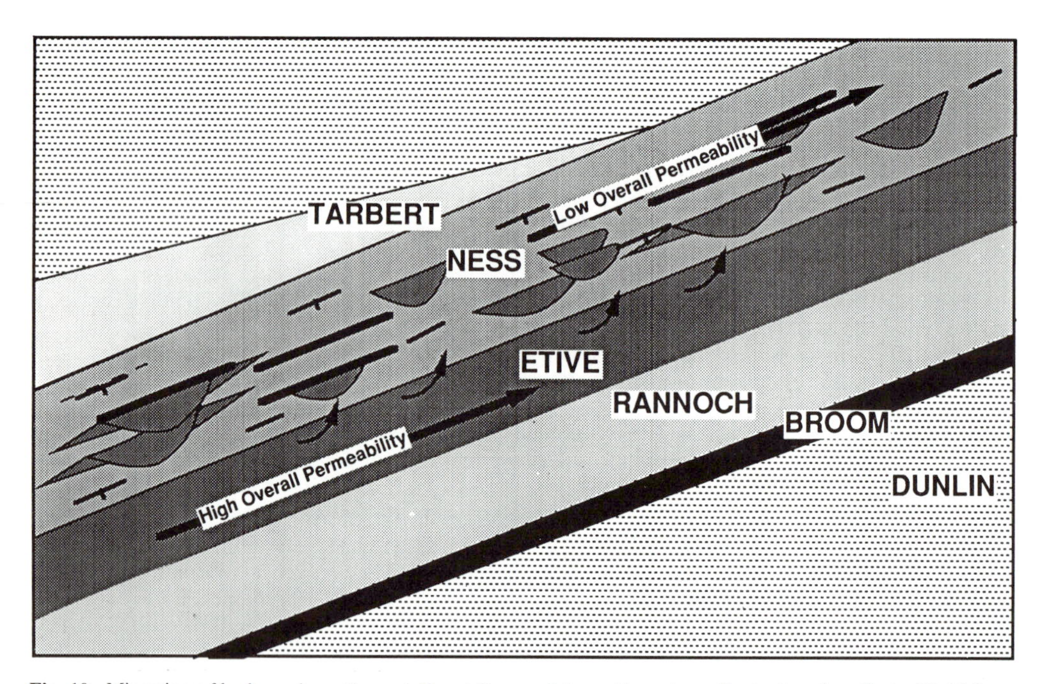


**Fig. 10.** Migration of hydrocarbons through Brent Group. Schematic cartoon illustrating the effect of fluid flow being preferentially focused at the time of hydrocarbon migration by overall permeability contrast between the 'clean' Etive sands and the 'dirtier' sands of other Brent Formations.

a brief period of Eocene illite formation was halted for some unknown reason.

Ehrenberg & Nadeau (1989) have suggested the existence of a critical thermal threshold for illitization in Garn Formation sandstones which corresponds to the change from incipient to extensive illitization. It coincides with a present-day burial depth of 3.7 km and temperature of 140°C. They observe no difference of illite abundance in extensively illitized sandstones within hydrocarbon legs or across oil–water contacts. Furthermore, K-Ar dates for illites from water filled portions are not consistently younger than those from hydrocarbon-filled portions. Hydrocarbon charge to the reservoirs studied occurred during earliest Eocene time and the K-Ar ages for illites (55–31 Ma) are mostly younger than this. Ehrenberg & Nadeau (1989) suggest that a water saturation as low as 10% may be sufficient for the closed system illite-forming reaction to proceed. Illitization here would seem to be a simple function of temperature or temperature and time. Geohistory analysis suggests that for several wells extensive illitization should have occurred in the last 8 million years. The actual illite ages are considerably older than this and again raise the question of possible contamination.

## Summary of observations of K-Ar dates from the Brent Group

The K-Ar dates from diagenetic illites of this and previous studies of the Brent Group may be summarized as follows.

(1) The ages are broadly coincident with the expected timing of hydrocarbon migration.

(2) The expected decrease in age of the finest illite size fraction with depth, a feature which is expected as the result of the progressive filling of a structure by hydrocarbons is visible in the data from a few fields e.g. Heather (Glasmann *et al.* 1989*a*), Greater Don area and the Etive Formation dates from the Pelican field.

(3) Such a pattern of decreasing K-Ar illite ages with depth is also visible in the data from the water filled 211/11 structure.

(4) Ages from the finest illite size fraction of Etive Formation samples from wells where the overlying Kimmeridge Clay Formation source rocks are mature for oil generation show a remarkable correlation between K-Ar illite age and the calculated time when the overlying source rocks reached a vitrinite maturity of 0.62%.

(5) Dates from the Ness, Tarbert, Rannoch and Broom Formations are often older than the dates obtained from the Etive Formation. Furthermore they frequently do not fall into any clear pattern when compared to other samples (as in the Pelican Field).

(6) Water zone ages are usually only slightly younger than oil zone ages, suggesting that illite formation stopped in the water zone soon after oil emplacement.

(7) Data from the Cormorant Field indicates that illite may start to form from about 75°C and data from deeper wells indicates that it can continue to temperatures in excess of 110°C. Similarly age ranges for the basin as a whole show a protracted period of illite growth from *c.* 60 Ma to 17 Ma. This is in sharp contradiction to the period of illitization deduced from dating illites of different size fractions.

## The significance of K-Ar diagenetic illite ages from Brent Province

What then are the possible causes of such a change in the physio-chemical conditions that would account for the cessation of illite growth in water zone samples? Modelling water movement in subsiding basins, in the absence of large scale meteoric systems, by de Caritat (1990) has shown that sediments "fall" slowly through a water column whose composition is initially determined by the pore fluids in the sediments at their time of deposition and subsequently modified by diagenetic reactions between the sediment and its pore fluid. Such a system is not one in which drastic changes in pore water chemistry are likely. Furthermore, the East Shetland Basin and Viking Graben are generally overpressured (Chiarelli & Duffaud 1980), a situation which is believed to have existed since Late Cretaceous times (Doligez *et al.* 1987). The peak of oil generation in the Viking Graben is believed to have occurred during the Palaeocene, while in the East Shetland Basin it was not reached until the late Eocene (Goff 1983). Thus it coincided with early development of overpressures. The presence of long lived overpressures such as these imply a largely closed system with respect to water and consequently no large scale changes in pore water composition would be expected. As shown by the burial histories presented in this study (Fig. 8a–f) the sediments are presently at their maximum burial depth and maximum temperature; thus, a drop in temperature is not the likely cause either. As shown by Giles *et al.* (this volume), feldspars, the likely source of the K for illite growth, also persist down to at least 13 000 ft (3962.45 m). The above discussion strongly suggests that there is no obvious reason why illite diagenesis should have ceased at the times indicated by the K-Ar ages of finest size fractions from the water leg.

The mass transport equation (Eqn. 8) discussed earlier indicates the importance of the rate of supply of reactants as a result of fluid flow and diffusion. As discussed, the rate of supply of reactants may limit the rate of illite formation. The experiments of Huang *et al.* (1986) also indicate that effective water-to-rock ratio is an important control on the amount of illite that may form. This ratio is higher for a dynamic, as opposed to static, fluid regime, i.e. when nutrient supply by fluid flow transport dominates over diffusive transport. There are two major periods when fluid flow was active in the Brent depositional basin.

The earliest occurred when Cimmerian uplift promoted meteoric water flushing (Bjørlykke & Brendsdal 1986). At the low prevailing temperatures and shallow depths (<1.5 km) of active flow (e.g. Hamilton *et al.* 1987) this would not have affected illite diagenesis.

The second period of aqueous fluid transport must occur in parallel with hydrocarbon migration. Large scale water movements are likely to accompany oil migration as hydrocarbons migrate through open structures into closed structures and stratigraphic traps, thus forcing out aqueous pore fluids which interact with rock units below the lengthening hydrocarbon/water contact. The increase in effective water-to-rock ratio in the water leg at higher temperatures would promote more rapid illite growth rates. As the peak of hydrocarbon migration passed, the aqueous fluid flow regime in the water leg would approach a static condition. Water-to-rock ratio would decrease and illite growth rates would decline. Higher rates of illite growth are therefore promoted at the onset of hydrocarbon migration. This higher rate of illite growth would continue in the water leg for longer than in the hydrocarbon zone. The static pore fluid regime prior to, and post hydrocarbon generation, would favour solute transport by diffusional processes, and thus low rates of illite growth. K-Ar dating of the illites would therefore reflect dates which approximately coincide with this period of rapid growth and would display slightly younger dates from the water leg as from the oil leg.

## Conclusions

The major points to emerge from this regional review of K-Ar illite ages are considered to be as follows.

(1) A model is proposed for illite growth that is in concert with observed patterns of K-Ar ages, petrographic data and thermodynamic considerations. Illite begins to form at a temperature of about 75°C probably at quite a low reaction rate. The process of oil migration perturbs the static pore fluid medium and increases the effective water-to-rock ratio by fluid transport. This effects an increased rate of illite growth. Subsequent to the peak of oil generation, flow rates decrease with a concomitant decrease in illite forming reaction rates. Any attempt to isolate illite formed at lower or higher temperatures will inevitably also sample illite from the most rapid period of growth and so yield a falsely reduced estimate of time span of illite formation and apparently old ages for most recently formed illite in the water zone.

(2) Illite ages of finest size fractions from the 'clean' Etive sandstones are significantly younger than those from juxtaposed 'dirty' (i.e. bearing detrital illite and mixed-layer illite–smectite clays) sandstones of other formations. The former correspond closely with calculated times of oil generation and migration from overlying Kimmeridge Clay source rocks, and approximate the timing of the end of illite diagenesis as a consequence of the oil accumulation. It is however possible that silicate diagenesis does persist in water wet portions of oil legs though at much reduced rates. The illite ages from the dirty sandstones of the Ness, Tarbert and Broom Formations generally reflect contamination of neoformed authigenic illite with older parageneses such as detrital and diagenetically illitized mixed layer illite-smectite clay.

(3) Measurement of the duration of illite growth by direct K-Ar dating of different size fractions is obfuscated by the need for freedom from K-bearing contaminates and by the aggregate nature of the fine size fractions analyzed. Thus water wet sandstones do not yield very young (<15 Ma) illite ages that would indicate recent illite formation.

The authors wish to thank Shell UK Exploration & Production and Esso UK Exploration & Production for permission to publish this paper. In addition the authors would like to thank B. Hanncock and A. Morris for their draughting efforts and patience.

The analytical work was undertaken at the Scottish Universities Research and Reactor Centre, which is funded by the Universities and the NERC.

## References

Bear, J. 1972. *Dynamics of Fluids in Porous Media.* Elsevier, New York.

Bjørkum, P. A. & Gjelsvik, N. 1988. An isochemical model for formation of authigenic kaolinite, K-feldspar and illite in sediments. *Journal of Sedimentary Petrology*, **58**, 506–511.

Bjørlykke, K. & Brendsdal, A. 1986. Diagenesis of the Brent Sandstone in the Statfjord Field, North Sea. *In*: Gautier, D. L. (ed.). *Roles of organic matter in sediment diagenesis.* SEPM Special Publication, **38**, 157–167.

Burley, S. D. & Flisch, M. 1989. K-Ar chronology and the origin of illite in the Piper and Tartan Fields, Outer Moray Firth, U.K. North Sea. *Clay Minerals*, **24**, 285–315.

Cannon, S. J. C., Giles, M. R., Whittaker, M. J.

PLEASE, P. M. & MARTIN, S. V. 1991 Regional review of the Brent Group, UK Sector, North Sea. *This volume*.

CARITAT, P. DE 1990. *Aspects of sediment diagenesis: Empirical investigation (Denison Trough, Queensland) and theoretical modelling*. PhD thesis, The Australian National University.

CHERMAK, J. A. & RIMSTIDT, J. D. 1990. The hydrothermal transformation rate of kaolinite to muscovite/illite. *Geochimica et Cosmochimica Acta*, **54**, 2979–2990.

CHIARELLI, A. & DUFFAUD, F. 1980. Pressure origin and distribution in Jurassic of Viking basin (United Kingdom-Norway). *Bulletin of the American Association of Petroleum Geologists*, **64**, 1245–1266.

DOLIGEZ, B., UNGERER, P., CHENET, P. Y., BURRUS, J., BESSIS, F. & BESSEREAU, G. 1987. Numerical modelling of sedimentation, heat transfer, hydrocarbon formation and fluid migration in the Viking Graben, North Sea. *In*: BROOKS, J. & GLENNIE, K. W. (eds) *Petroleum Geology of North West Europe*. Graham & Trotman, London, 1039–1048.

DYPVIK, H. 1983. Clay mineral transformations in Tertiary and Mesozoic sediments from the North Sea. *Bulletin of the American Association of Petroleum Geologists*, **67**, 160–165.

EHRENBERG, S. N. & NADEAU, P. H. 1989. Formation of diagenetic illite in sandstones of the Garn Formation, Haltenbanken area, mid-Norwegian continental shelf. *Clay Minerals*, **24**, 233–253.

FAURE, G. 1987. *Principles of isotope geology* 2nd Ed., John Wiley and Sons.

GILES, M. R. 1987. Mass transfer and the problems of secondary porosity creation in deeply buried hydrocarbon reservoirs. *Marine and Petroleum Geology*, **4**, 188–204.

—— & DE BOER, R. B. 1990. Origin and significance of redistributional secondary porosity. *Marine and Petroleum Geology*, **7**, 378–397.

—— & MARSHALL, J. D. 1986. Constraints on the development of secondary porosity in the subsurface: Re-evaluation of processes. *Marine and Petroleum Geology*, **3**, 243–255.

——, STEVENSON, S., MARTIN, S. V., CANNON, S. J. C., SAMWAYS, G. M. & HAMILTON, P. J. 1992. The diagenesis and reservoir properties of the Brent Group: A regional perspective. *This volume*.

GLASMANN, J. R., LUNDEGARD, P. D., CLARK, R. A., PENNY, B. & COLLINS, I. 1989*a*. Geochemical evidence for the history of diagenesis and fluid migration: Brent Sandstone, Heather Field, North Sea. *Clay Minerals*, **24**, 255–284.

——, CLARK, R. A., LARTER, R., BRIEDIS, N. A. & LUNDEGARD, R. D. 1989*b*. Diagenesis and hydrocarbon accumulation, Brent Sandstone (Jurassic), Bergen High area, North Sea. *Bulletin of the American Association of Petroleum Geologists*, **73**, 1341–1360.

GOFF, J. C. 1983. Hydrocarbon generation and migration from Jurassic source rocks in the E. Shetland Basin and Viking Graben of the Northern North Sea. *Journal of the Geological Society, London*, **140**, 445–474.

GREEN, P. M., EADINGTON, P. J., HAMILTON, P. J. & CARMICHAEL, D. C. 1989. Regional diagenesis – an important influence in porosity development and hydrocarbon accumulations within the Hutton Sandstone, Eromanga Basin. *In*: O'NEILL, B. J. (ed.) *The Cooper and Eromanga Basins Australia*. Proceedings of PESA Conference, Adelaide, 619–628.

HAMILTON, P. J., FALLICK, A. E., MACINTYRE, R. M. & ELLIOTT, S. 1987. Isotopic tracing of the provenance and diagenesis of Lower Brent Group Sands, North Sea. *In*: BROOKS, J. & GLENNIE, K. W. (eds) *Petroleum Geology of N.W. Europe* Graham & Trotman, London, 939–949.

——, KELLEY, S. and FALLICK, A. E. 1989. K-Ar dating of illite in hydrocarbon reservoirs. *Clay Minerals*, **24**, 215–231.

HASZELDINE, R. S., BRINT, J. F., FALLICK, A. E., HAMILTON, P. J. & BROWN, S. 1991. Open and restricted hydrologies in Brent Group diagenesis, North Sea. *This volume*.

HOGG, A. J. C., PEARSON, M. J., FALLICK, A. E., HAMILTON, P. J. & MACINTYRE, R. 1987. Clay mineral and isotope evidence for controls on reservoir properties of Brent Group sandstones, British North Sea. *Terra Cognita*, **7**, 342.

HOGG, A. J. C. 1989. *Petrographic and isotopic constraints on the diagenesis and reservoir properties of the Brent Group sandstones, Alwyn South, Northern North Sea*. PhD Dissertation, University of Aberdeen, UK. University Microfilms International Publication, No. DX89573.

HOWER, J., ESLINGER, E. V., HOWER, M. E. & PERRY, E. A. 1976. Mechanism of burial metamorphism of argillaceous sediment. 1. Mineralogical and geochemical evidence. *Geological Society of America Bulletin*, **87**, 725–737.

HUANG, W. L., BISHOP, A. M. & BROWN, R. W. 1986. The effect of fluid/rock ratio on feldspar dissolution and illite formation under reservoir conditions. *Clay Minerals*, **21**, 585–601.

IRWIN, H., CURTIS, C. D. & COLEMAN, M. L. 1977. Isotopic evidence for source of diagenetic carbonates formed during burial of organic-rich sediments. *Nature*, **269**, 209–213.

JACKSON, M. L. 1979. *Soil chemical analysis – advanced course*, 2nd Ed. Published by the author, Madison, W.S., USA.

JOURDAN, A., THOMAS, M., BREVART, O., ROBSON, P., SOMMER, F. & SULLIVAN, M. 1987. Diagenesis as the control of the Brent sandstone reservoir properties in the Greater Alwyn area (East Shetland Basin). *In*: BROOKS, J. & GLENNIE, K. W. (eds) *Petroleum Geology of N.W. Europe*. Graham & Trotman, London, 951–961.

LEE, M., ARONSON, J. L. & SAVIN, S. M. 1985. K/Ar dating of Rotliegendes Sandstone, Netherlands. *Bulletin of the American Association of Petroleum Geologists*, **68**, 1381–1385.

——, —— & —— 1989. Timing and conditions of Permian Rotliegende Sandstone diagenesis, southern North Sea: K-Ar and oxygen isotope data. *Bulletin of the American Association of Petroleum Geologists*, **73**, 195–215.

LEYTHAEUSER, D. & RUCKHEIM, J. 1989. Heterogeneity of oil composition within a reservoir as a reflection of accumulation history. *Geochimica et Cosmochimica Acta*, **53**, 219–223.

LIEWIG, N., CLAUSER, N. & SOMMER, F. 1987. Rb-Sr and K-Ar dating of clay diagenesis in a Jurassic sandstone oil reservoir, North Sea. *Bulletin of the American Association of Petroleum Geologists*, **71**, 1467–1474.

MILES, J. A. 1990. Secondary migration routes in the Brent Sandstones of the Viking Graben and East Shetland Basin: Evidence from oil residues and subsurface pressure data. *Bulletin of the American Association of Petroleum Geologists*, **74**, 1718–1735.

MORTON, J. P. 1985. Rb-Sr dating of diagenesis and source age of clays in Upper Devonian Black Shales of Texas. *Geological Society of American Bulletin*, **96**, 1043–1049.

PALIAUSKAS, V. V. & DOMENICO, P. A. 1976. Solution chemistry, mass transfer and the approach to chemical equilibrium in porous carbonate rocks and sediments. *Geological Society of American Bulletin*, **87**, 207–214.

RUNDBERG, Y. & SMALLEY, P. C. 1989. High-resolution dating of Cenozoic sediments from northern North Sea using $^{87}Sr/^{86}Sr$ stratigraphy. *Bulletin of the American Association of Petroleum Geologists*, **73**, 298–308.

SCOTCHMAN, I. C., JOHNES, L. H. & MILLER, R. S. 1989. Clay diagenesis and oil migration in Brent Group sandstones of the NW Hutton field, U.K. North Sea. *Clay Minerals*, **24**, 339–374.

SMALLEY, P. C., NORDAA, A. & RAHEIM, A. 1986. Geochronology and palaeothermometry of Neogene sediments from the Voring Plateau using Sr, C and O isotopes. *Earth and Planetary Science Letters*, **78**, 368–378.

SOMMER, F. 1975. Histoire diagenetique d'une series gresuese de Mer du Nord. Datation de l'introduction des hydrocarbures. *Revue de l'Institut de Français du Petrole*, **30**, 729–741.

SOMMER, F. 1978. Diagenesis of Jurassic sandstones in the Viking Graben. *Journal of the Geological Society London*, **135**, 63–67.

STEIGER, R. H. & JAEGER, E. 1977. Subcommission on geochronology: Convention on the use of decay constants in geo- and cosmochronology. *Earth and Planetary Science Letters*, **36**, 359–362.

SURDAM, R. C. & CROSSEY, L. J. 1987. Integrated diagenetic modelling: A process-oriented approach for clastic systems. *Annual Reviews in Earth and Planetary Science*, **15**, 141–170.

THOMAS, M. 1986. Diagenetic sequences and K-Ar dating in Jurassic sandstones, Central Viking Graben effects on reservoir properties. *Clay Minerals*, **21**, 695–710.

# Open and restricted hydrologies in Brent Group diagenesis: North Sea

R. STUART HASZELDINE[1], J. F. BRINT[1], A. E. FALLICK[2], P. J. HAMILTON[2], S. BROWN[3]

[1] *Department of Geology and Applied Geology, University of Glasgow, Glasgow G12 8QQ, UK*

[2] *Isotope Geology Unit, Scottish Universities' Research and Reactor Centre, East Kilbride G75 OQU, UK*

*Present address: CSIRO, 51 Delhi Rd., North Ryde, Sydney, Australia*

[3] *Hydrocarbons Research Programme, British Geological Survey, Grange Terrace, Edinburgh, UK*

*Present address: Petroleum Science and Technology Institute, Riccarton, Edinburgh, UK*

**Abstract:** Diagenetic cements in the mid Jurassic Brent Group sandstones of the North Sea have been studied in the Thistle, Murchison, Dunlin and Alwyn South (3/13a-1) oilfields. Volumetrically important cements start with early siderite, kaolinite and calcite, and continue to later kaolinite, ankerite, quartz and illite. The mineralogies of early siderites and calcites are homogeneous in formations with different depositional porewaters. Carbonate concretions nucleated on biotite grains, and their carbon supply was very local and facies controlled. Strontium was derived from dissolution of silicates and shell debris and $^{87}Sr/^{86}Sr$ rapidly became homogeneous in different formations. Oxygen was derived from meteoric water and was homogeneous throughout, even in marine deposits. These data suggest an open-flow system with uniform meteorically-derived pore fluids and local ion supply throughout the sandstones.

Diagenetically later ankerite formed below 1.1 km and shows mineralogies which are less homogeneous, with $^{87}Sr/^{86}Sr$ variable. Cement ions could all have been very locally supplied from within the formation. Late kaolinite could also have been formed locally from feldspars within the formation. The exception to this was quartz cement, its large volumes indicating import across formations or from outside the Brent. Fluid inclusions in quartz show abnormally hot temperatures equating to 80–100°C at 2.2 km. These can be interpreted in two ways. First to indicate a large circulation of hot evolved meteoric water within the basin. This would have been induced by tensional fracturing of the crust related to the final phase of North Atlantic rupture in the Palaeocene and Eocene. However, fluid inclusion temperatures within quartz overgrowths show a systematic increase of temperature with present-day burial depth, suggesting that these temperature records could have been reset, and are now too hot. If these temperature data are rejected, a second interpretation is possible. The $\delta^{18}O$ isotopic signatures show that late diagenetic silicates and carbonates could have grown in meteoric water, unmodified in its $\delta^{18}O$ value, in a normal geothermal gradient. Illite grew in a slowly flowing fluid, which rapidly evolved to todays $\delta^{18}O$ values, as individual groups of pores attempted to equilibrate with the rock around them Oil accumulation halted diagenesis and was permitted by buoyant trapping as this porewater system became stagnant.

The Brent Group of the Northern North Sea is one of the major oil reservoirs of the UKCS (Brennand *et al.* 1990), sited on the western shelf of the Viking Graben Rift. The first diagenetic studies were texturally based (Blanche & Whittaker 1978; Hancock & Taylor 1978), and since then only a few workers have addressed the processes of diagenesis in these sediments in terms of fluid flow (Sommer 1978; Glasmann *et al.* 1989; Jourdan *et al.* 1987) or fluid stagnation (Bjørlykke & Brendsal 1986; Giles 1987). This paper, based on the work of Brint (1989), records a textural and isotopic study, based on samples from the Thistle, Dunlin, Murchison and Alwyn South fields (Fig. 1). Samples span the Tarbert, Ness, Etive and Rannoch Formations, ranging in depth from 8624 ft (2628 m) to 12459 ft (3797 m) TVD.

*From* Morton, A. C., Haszeldine, R. S., Giles, M. R. & Brown, S. (eds), 1992, *Geology of the Brent Group*, Geological Society Special Publication No. 61, pp. 401–419.

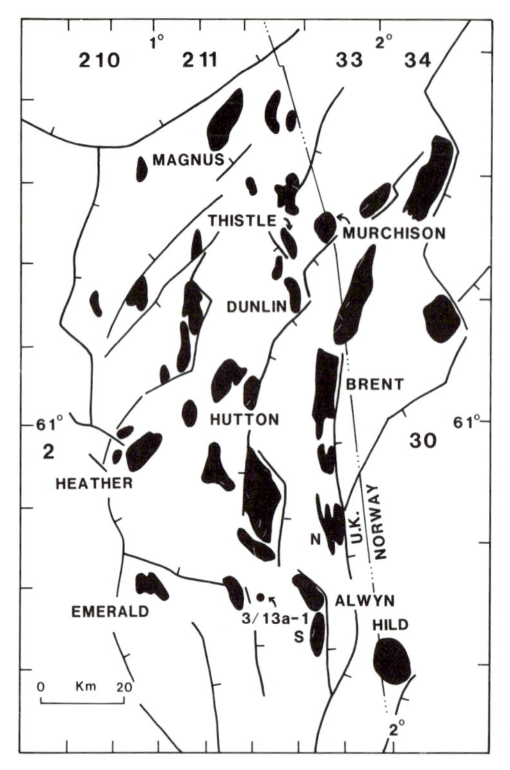


**Fig. 1.** Location map of oilfields mentioned in text, sited in the East Shetland Basin, on the western flank of the Viking Graben, large NNW line is the UK–Norway international boundary.

Methods of mineral separation, isotopic analysis and point counting are conventional, and can be found on open file in Brint (1989), in Brint *et al.* (1991) for quartz overgrowth analysis, or by writing to the authors. Tabulations of the isotopic and other analytical data have been deposited as Supplementary Publication No. SUP 18074 (6pp) with the British Library Document Supply Centre, Boston Spa, Wetherby, W. Yorks LS23 7BQ and with the Society Library.

As with any study outside a major oil company, it is time consuming to amass very large data sets, covering geographically large areas. Therefore we have accurately measured, and present here, a small number of very carefully obtained data sets. Our data are similar to those published by Glasmann *et al.* (1989) working on the Heather field and to Jourdan *et al.* (1987) working on the Alwyn field. This enables us to assert that the isotopic pattern of our detailed information is representative of the Brent province as a whole.

Our objective is to infer the first-order hydrological controls which operated during diagenesis of these Brent sandstones. Therefore we are concerned to distinguish between the end-member hypotheses: either the pore fluid was static and/or very slow moving, so that it did not physically transport large quantities of diagenetic ions, except by diffusion; or the pore fluid was moving, did transport ions during diagenesis, and was inherently out of equilibrium with the rock immediately surrounding it. A few moments' thought will convince you that these two alternatives make very different predictions. In th first case, diagenetic mineralogies, timings and quantities can be modelled by 'simple' equilibrium and kinetic geochemistry, and are 'theoretically predictable' by considering the original mineralogy and organic mix of the sandstone and adjacent mudrocks. In the second case, diagenetic mineral quantities depend upon the volume and flow path of the fluid, so that consideration must be given to sediments outwith the Brent Group, combined with fluid volumes and flow paths across or within the whole basin. Our purpose here is not to construct a precisely quantified model, but to point a qualitative finger at the more likely of the two alternatives.

## Petrography

Using thin section, cathode-luminescence and scanning electron microscope petrography, the diagenetic sequence of common cements in the Brent Group can be determined (Fig. 2). These can conveniently be considered as 'early dia-

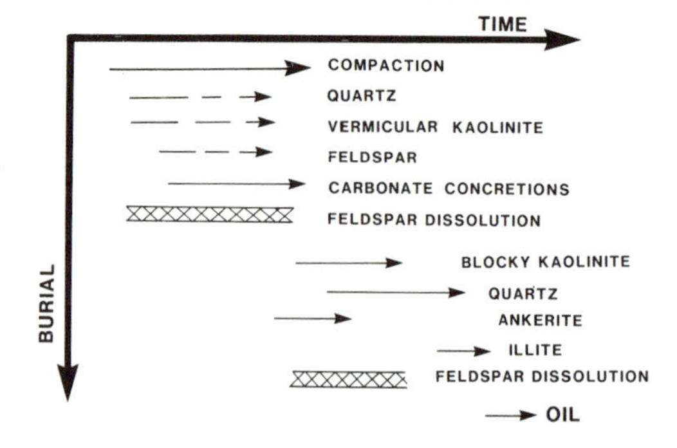


**Fig. 2.** Paragenetic sequence of the volumetrically most abundant diagenetic minerals. This has been derived from study of more than 75 thin sections from the Tarbert, Ness, Etive and Rannoch Formations, cement areas were estimated by 500 point counts per thin section.

genetic' cements up to and including growth of carbonate concretions, and 'late diagenetic' cements which post-date concretion growth.

## *Early cements*

The 'early' cements will be discussed first; they can be safely examined only within concretions, for outwith concretions the early cements are subject to physical mixing with later cements or even to dissolution. Carbonate cemented zones 1.3–2.6 m thick occur in three places in two wells from the Dunlin Field. These zones do not correlate between wells spaced km apart and are inferred to be concretions. We have examined the Tarbert and Ness Formations; Hamilton *et al.* (1987) examined concretions from the Rannoch Formation in the Thistle Field. No concretions occur in the most porous and permeable Etive Formation. Small feldspar and quartz overgrowths and large vermicular kaolinite crystals (20 $\mu$m wide) predate calcite. Siderite rhombs and aggregates also occur inside concretions, and have often nucleated on or around biotite mica. The depth at which these concretions grew can be estimated by point-counting the cement volumes between detrital sand grains, assuming that this represented primary porosity, and plotting this in terms of solidity (Baldwin & Butler 1985). These solidity values are superimposed onto a general compaction curve for North Sea sandstones (Sclater & Christie 1982). Although this compaction curve is only an approximation for the different facies in the Dunlin Field, nevertheless we can still use it to demonstrate from simple textural information alone that the siderites and calcites formed above 1.1 km and maybe within the first hundreds of metres of burial (Fig. 3). By contrast the ankerite rims to concretions formed in more compacted sediment, at depths down to 1.5 km.

## *Late cements*

The later cements are dominated by kaolinite, ankerite, quartz and illite. Kaolinite occurs in a blocky habit, consisting of crystals less than 10 $\mu$m wide. Ankerite occurs as rims to the concretions, and as rare small (10–20 $\mu$m) rhombs dispersed throughout the sandstones. Quartz forms syntaxial overgrowths on detrital grains, which sometimes grow around kaolinite, and so partly post-date it. Illite grows on top of kaolin-

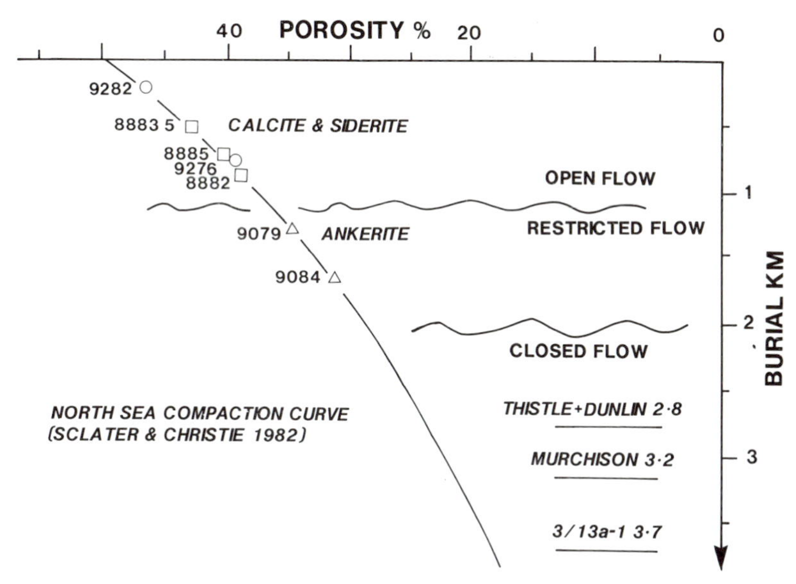


**Fig. 3.** The growth depth of carbonate concretions can be estimated by assuming that diagenetic cements filled all inter-grain porosity, and that no major replacement by carbonate has occurred. Thus point count data of all cements can be used as a porosity estimate, to superimpose on the general porosity curve for the North Sea (Sclater & Christie 1982). This simple textural information shows that most concretions formed at very shallow depths. Numbers and symbols on the curve refer to the present day depth of samples in feet, but symbols are plotted at their inferred depths of carbonate growth. The present-day oilfield depths are shown in kilometres by horizontal lines.

ite and quartz, and sometimes intergrows with the final stages of quartz overgrowths. By contrast with the early carbonate cements, these later cements are most abundant in today's most porous and permeable Formations, and are not abundant in the Rannoch Formation (Fig. 4). We infer that permeability in the Rannoch was reduced by physical compaction early in the sediment's diagenetic history, preventing any easy through-flow of fluid. The Rannoch today typically has low permeabilities of 10 mD with streaks up to 500 mD. This tempts us to speculate that the supply of these cements was, in some way, linked to the ease of fluid movement transporting ions through the Formations. We also suggest that quartz has been imported into the Etive, Ness and Tarbert Formations, and exported from the Rannoch Formation (Fig. 5). These estimates are derived from a modified version of Houseknecht's (1987) method, where we have simply point counted the amount of quartz cement in a section (later cross-checks with cathodeluminescence show that this is likely to be a 20–30% under estimate) and have point counted and estimated the amount of grain-to-grain dissolution in the same thin section (likely to be an over estimate).

## Early carbonates: compositions and Sr isotopes

### *Siderite geochemistry*

Siderite is observed to have three growth zones on back-scatter electron microscope (BSEM) images (Fig. 6), and to grow together with pyrite inside detrital biotite flakes (Fig. 7). The three siderite zones are common to three concretions studied in detail in the Tarbert and Ness Formations in the Dunlin Field. Microprobe analyses of the three zones shows siderites from all three concretions, in both formations from these two wells, show identical compositions for each growth zone. These define a looped trend (Fig. 8). Even though this is a small set of data, its interpretation is highly significant (cf. Mozley & Hoernle 1990, who showed that siderite composition accurately reflects porewater chemistry). If these sands had been isolated aquifers during diagenetic siderite growth, why should the compositions be similar? The Tarbert and Ness Formations are so different in original sedimentary facies (marine transgressive sand complexes, and river channels) that an exact coincidence of detrital chemistry to generate an identical closed dia-

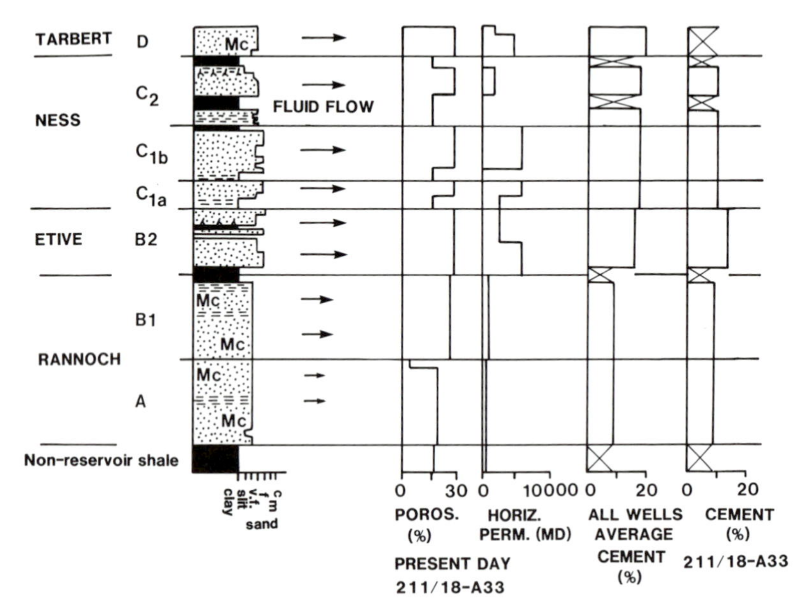


**Fig. 4.** Thin section observations show that the greatest percentages of diagenetic minerals by area (predominantly quartz) occur in the most porous and permeable reservoir sandstones of the Brent Group. The average values for all thin sections studied are schematically shown here, with actual present-day poroperm values plotted from 211/18-A33 to illustrate a typical well. We infer more fluid flow through porous and permeable sandstone in the past, just as today.

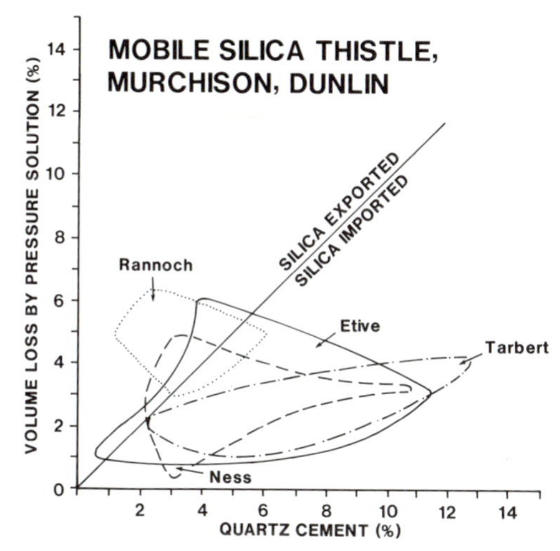


**Fig. 5.** Thin section counts of the area of quartz cement versus the area lost by pressure solution between quartz grains. This suggests that the Rannoch Formation exported quartz during diagenesis, whilst the Tarbert, Ness and Etive imported quartz cements. When the likely contribution of 10% feldspar dissolution producing 4% quartz cement is added in, and the similar thicknesses of Rannoch = Etive + Ness + Tarbert considered, then very little, if any, quartz need have been imported from outside the Brent. However, quartz has certainly been unevenly re-distributed into more permeable Formations.

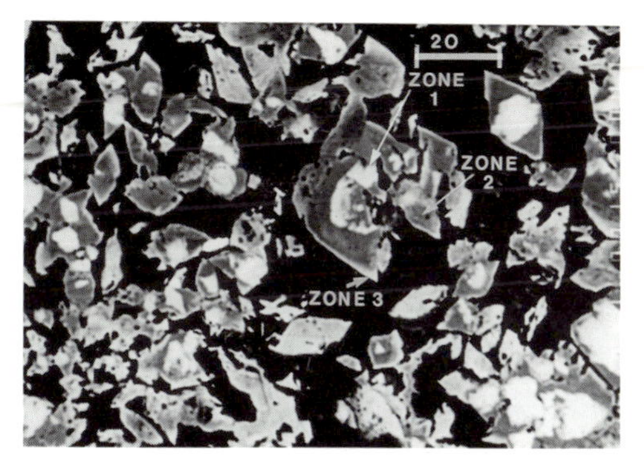


**Fig. 6.** In Backscatter Electron imaging of a polished thin section, early diagenetic siderite is observed to have three growth zones. These occur in the Tarbert *and* Ness formations of the Dunlin field. The innermost zone has a ragged outline, suggesting dissolution prior to formation of the second zone. Each zone has the same composition in the two different Formations (Fig. 8). 211/23–4, 2829.0 m Ness Formation, scale bar 20 $\mu$m.

genetic system is improbable. Our preferred explanation is that the aquifers were connected, and so the aqueous chemistry was homogenized. By contrast, aquifers in the Magnus oilfield, inferred from independent evidence to have been stratified with different compositions, show inhomogeneous siderite chemistries (Macaulay 1990). We can make a conceptual analogy to the uniform growth zones of diagenetic calcite and of fluorite found throughout the same aquifer in the Illinois basin. Tnese authors interpret such zones to indicate aquifer continuity during cementation.

Strontium isotopes are capable of providing more detail on this water homogenisation. We have physically separated the siderites and measured the C, O and Sr isotopes on batches of whole siderite grains from within these same three concretions. This technique inevitably measures all three siderite growth zones together. We find that siderites show a trend from high $^{87}Sr/^{86}Sr$ to a lower value characteristic of the texturally later calcite (Fig. 9). The four samples from the Ness Formation concretion show an excellent linear correlation ($R$ = 1.0, significant at the 99% level). Unmodified depositional pore water in the two Formations cannot account for these values, for it was meteoric in one case (with a very low content of radiogenic Sr), and marine in the other case (with a moderate content of non-radiogenic Sr). Our interpretation of these values is that Sr mixing occurred between small amounts of a very radiogenic strontium supplied by micas (muscovite $^{87}Sr/^{86}Sr$ = 0.79; Hamilton *et al.* 1987), or other feldspars and clays ($^{87}Sr/^{86}Sr$ = 0.72; Sullivan *et al.* 1990) and dominated by larger quantities of less radiogenic strontium supplied by dissolution of marine shell debris. During the growth of siderites, the component of ¨‚solved shell Sr gradually increased, to become 'balanced' at the value of 0.711 found in the concretions. Inside the concretions we observe siderite associated with reacting biotite (Boles & Johnson 1983), and so we infer that radiogenic Sr (and Fe and Mg) to form siderites was supplied by biotite breakdown (Fig. 7). We speculate that the $^{87}Sr/^{86}Sr$ 'fingerprint' of carbonates was initially very local, depending upon the composition of adjacent biotites. However, by the time that calcite formed, $^{87}Sr/^{86}Sr$ had become uniform throughout the Ness and Tarbert Formations. This is compatible with a flow of fluid, which moved the Sr from its local sources, causing values to become homogenized within the same formation and between formations. The Sr trend in siderite could also be compatible with a local diffusive supply of $^{87}Sr/^{86}Sr$ becoming gradually swamped by an

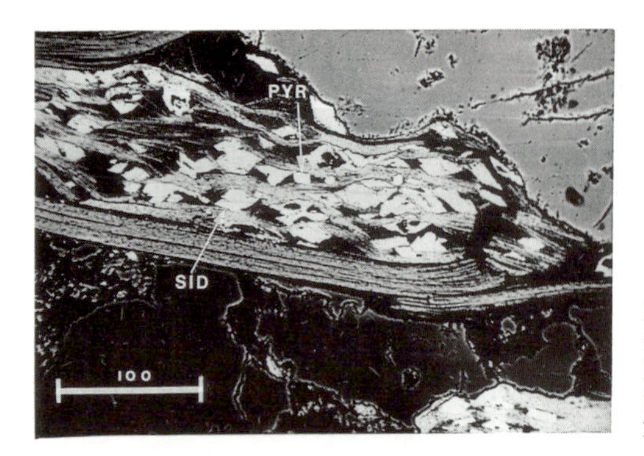


**Fig. 7.** Backscatter photograph of siderite and pyrite rhombs growing within expanding biotite cleavage layers. This abundance of pyrite suggests growth in the shallow sulphate reduction zone. 211/23–4, 2829.9 m Ness Formation, scale bar 100 $\mu$m.

increasing supply of homogenous $^{87}Sr/^{86}Sr$ derived from dissolution of local detrital marine shells. However, this is considered less plausible, as marine Sr to form siderite and calcite in the non-marine Ness Formation would need to be imported.

## *Calcite geochemistry*

All the three concretions show a very uniform calcite mineralogy (Fig. 8), without any ferroan or magnesian calcite solid solution. The $^{87}Sr/^{86}Sr$ isotopic signatures are identical in the Ness and Tarbert Formations (Fig. 9). This Sr is highly concentrated, and the $^{87}Sr/^{86}Sr$ value is above that of Middle Jurassic seawater (Burke *et al.* 1982). We interpret a mixed source of Sr, from dissolving detrital marine shell debris, combined with a minor radiogenic contribution from dissolving silicates such as biotite or feldspar. We note that these concretions occur in Tarbert marine depositional facies (where detrital shell debris could be anticipated), and in Ness non-marine depositional facies (where detrital marine shell debris would not be expected). Thus calcite recycled from marine fossils, with its characteristic $^{87}Sr/^{86}Sr$ ratio must have been transported between formations to precipitate in the Ness Formation. We also note that Wilkinson & Dampier (1990) studied analogues of these Brent concretions exposed in the Middle Jurassic onshore on the West coast of Scotland. They found that concretions more than 1 m in diameter grew in 9.5 Ma by diffusive processes, and in only 4 Ma in moving pore water. Extrapolating their graphs implies that Brent concretions 1.3–2.6 m in diameter would need water flowing past them at 5 m per year to reduce their growth times from 15–40 Ma to 6–16 Ma. Thus it seems that water flow is required to form Brent concretions for reasons of: enabling calcite supply to reach non-marine sediment; uniform mineralogy (rather than compositions controlled by local diffusion); uniform $^{87}Sr/^{86}Sr$ ratios; and rapid rates of ion supply. In this light, we can now look back to siderite growth, and the well established mixing line of $^{87}Sr/^{86}Sr$ between calcite and the most radiogenic siderite in the Ness Formation (Fig. 9). If marine $^{87}Sr/^{86}Sr$ and Ca ions were being gradually supplied to these siderites by moving water as a precursor to growth of calcite, then it is unreasonable to suggest that siderites grew in static water, and quite consistent to interpret them as growing in moving water, which was consequently homogeneous between formations.

## *Ankerite geochemistry*

The last growth phases of concretions are recorded by poikilotopic ankerite. These concretion ankerites show variable composition, even on opposite sides of the same concretion (Fig. 8). These compositions are Ca-rich compared to 'ideal' ankerite (Deer *et al.* 1962), and petrographic examination shows that some calcite crystals were dissolved at their margins before being overgrown by ankerite. Our three ankerite isotopic analyses show high Sr concentrations, but less radiogenic $^{87}Sr/^{86}Sr$ values than the calcites. These values are much more variable than those of the calcites, but do not show the extreme variability of the siderites (Fig. 9). We interpret these effects to represent ankerite forming partly by dissolution of calcite concretions, and partly by continued supply of dissolved shell debris. The non-radiogenic $^{87}Sr/^{86}Sr$ values are unusual, for we would expect texturally later carbonate cements to show increasingly radiogenic $^{87}Sr/^{86}Sr$ values (Sullivan *et al.* 1990). Our less-radiogenic values here imply a minimal contribution of radiogenic Sr from biotite and feldspar dissolution, and a continued dominance of marine Sr, either by dissolution of earlier diagenetic calcite, or con-

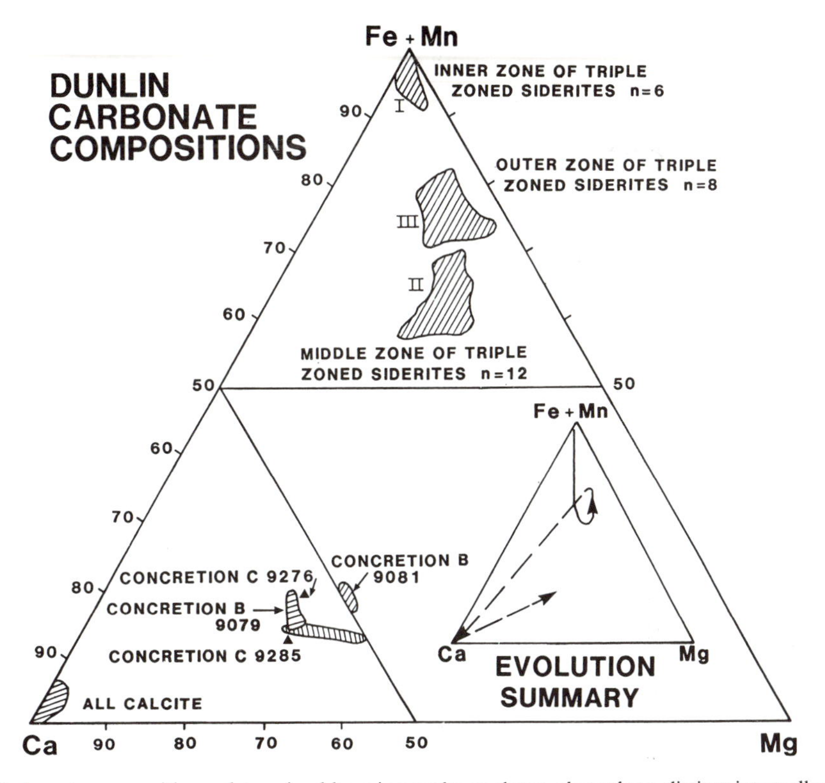


**Fig. 8.** Carbonate compositions, determined by microprobe analyses, show three distinct internally uniform zones in early siderite, very uniform calcite, and less homogeneous ankerite. Inset summarizes the pore fluid geochemical path during carbonate growth. Calcite analyses, Dunlin field 2701.1–2829 m, $n$ = 29. Ankerite analyses, Dunlin field 2767.3–2768.8 m, 2827.3–2830.0 m. Ankerite margins of concretions ($n$ = 17) have calcium excess, petrographic observations show some dissolution of calcite prior to ankerite, and $^{87}Sr/^{86}Sr$ data suggest inheritance of Sr, maybe Ca was inherited too?

tinued swamping of $^{87}Sr/^{86}Sr$ by dissolution of marine shells. We suggest that the lack of uniformity in both composition and Sr fingerprints represents poorer fluid communication between the different concretion sites, each using local ion supplies. If the siderites and calcites (above) were homogenized in a moving pore fluid, then conversely, these ankerites were not so homogenized, and could have grown in a slower-moving or static pore fluid. We note that ankerites formed deeper than siderites and calcites (Fig. 3), where we might intuitively expect fluid flows to be slower than near the surface.

### *Summary*

All ions to form carbonate cements could have been supplied from within the Brent Group, siderite Sr was initially very local, but gradually mixed with Sr from dissolving shell debris to form a homogeneous fluid throughout the different formations; homogenizing effects are inferred to have occurred in a moving pore fluid. Ankerite Sr was less homogeneous.

## Early carbonates: C and O isotopes

Analysis of the O and C isotopes in pure separated samples gives us additional information about ion sources and water origins in these early cements. A plot of the $\delta^{13}C$ and $\delta^{18}O$ results for siderite, calcite and ankerite shows (Fig. 10) that $\delta^{13}C$ was very variable. The changes in $\delta^{13}C$ between the various siderite, calcite and ankerite cements reflect local variations in $\delta^{13}C$ supply to the cements. The values around $\delta^{13}C = O$ reflect a carbon supply from marine shells and porewater in the marine Tarbert Formation. The negative $\delta^{13}C$ values reflect a local supply of bicarbonate produced from plant debris in the Ness Formation, which has lagoonal organic rich mudrocks and coaly debris only 3 m below from the concretion concerned. The Rannoch Formation con-

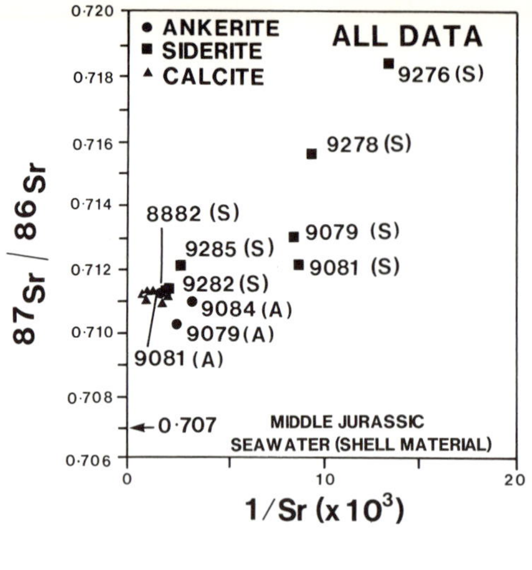


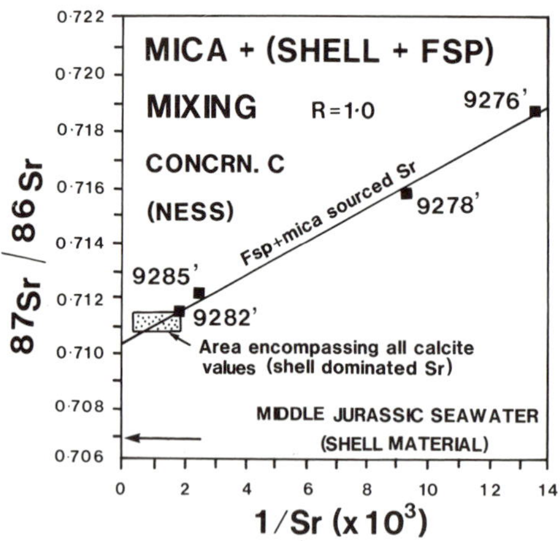


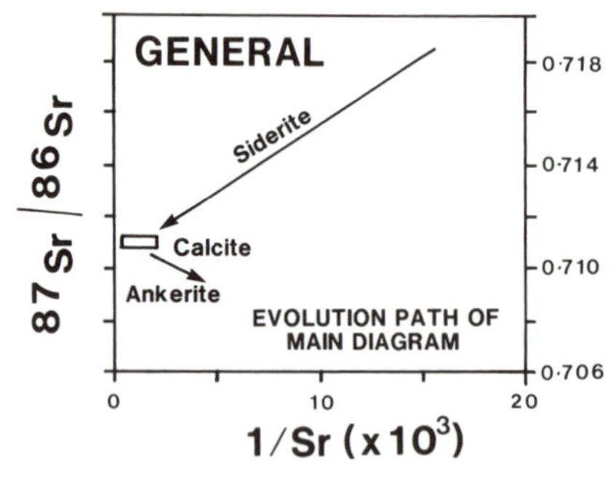


**Fig. 9.** Plots of all $^{87}Sr/^{86}Sr$ from Dunlin concretions, and (middle) a two-component mixing line illustrated by the concretion centred on 211/23–4, 2827 m (9279 ft). This illustrates fingerprints of $^{87}Sr/^{86}Sr$ versus 1/Sr ($\times 10^3$) during diagenesis. Early Sr ratios were dominated by local radiogenic ?biotite sources (Fig. 7) and low Sr concentrations. These mixed gradationally with the 0.711 ratio, high Sr concentration end member characteristic of homogeneous calcite concretions. This constant end member was presumably itself a product of a constant mixing ratio dominated by abundant shell dissolution (seawater values) and minor concentrations of radiogenic (0.72?) ratios from silicate dissolution. Three ankerite data points suggest reduced Sr concentrations, with ratios inherited from calcite (excess calcium on Fig. 8).

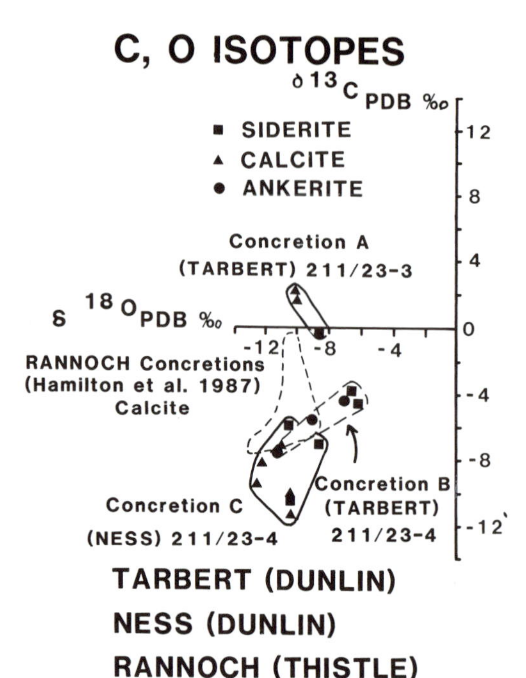


**Fig. 10.** $\delta^{13}C$, $\delta^{18}O$ plot for 3 concretions sampled from the Dunlin field 211/23–3, 211/23–4, and Rannoch Formation from the Thistle field (Hamilton *et al.* 1987). This shows that $\delta^{13}C$ was locally controlled by sedimentary facies, with marine values in one Tarbert concretion, merging to negative values derived from organic carbon oxidation in the Rannoch and Ness Formations. $\delta^{18}O$ shows a range of values, but these can be interpreted as growth from a uniform meteoric water through a range of temperatures (Fig. 11).

cretions lie in an intermediate position, as does its depositional shoreface environment with mixed marine and terrestrial carbon sources. Thus $\delta^{13}C$ shows a very local facies control of ion supply, and cannot help us to distinguish if the diagenetic system contained moving or static water.

Oxygen isotopic signatures in these siderite and calcite cements show a range of $\delta^{18}O$ values. We can use these analyses to estimate the isotopic composition of the palaeo-fluid. As will be seen below, this fluid can be deduced to have been constant in its $\delta^{18}O$ composition during growth of all these cements. To do this, we need to use the isotope fractionation equation to make a visual plot of the three inter-dependent variables of: present isotope value of the mineral; ancient water isotope composition during growth; ancient temperature of growth (Fig. 11). In this case, we know our present mineral value- measured in our mass spectro-

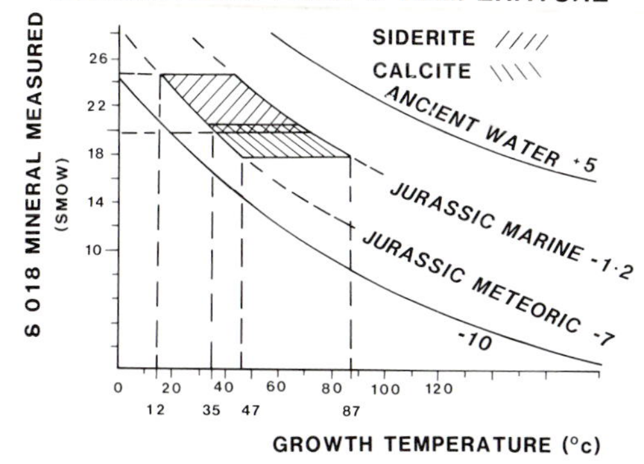


**Fig. 11.** Temperature versus $\delta^{18}$O mineral, and curves for different porewaters for siderite and calcite. This shows that water values of −7 must be used to give realistically low temperatures of siderite and calcite growth. This meteoric value is similar to that used by Hamilton *et al.* (1987) and Hudson & Andrews (1987) for early diagenesis in mid-Jurassic sandstones or paralic limestones from western Scotland. Fractionation equations appropriate for our analyses are $1000\ \ln\alpha_{siderite-water} = 3.13 \times 10^6 \times T^{-2} - 3.5$ (Carothers *et al.* 1988); $1000\ \ln\alpha_{calcite-water} = 2.78 \times 10^6 \times T^{-2} - 2.89$ (Friedman & O'Neil 1977).

meter, and we also infer from the thin-section textures that these cements grew early in diagenesis, during the first 1.1 km of burial (Fig. 3). Therefore we can assume that growth temperatures reflected this, and were in the 0–40°C range.

Two extreme possibilities exist for the diagenetic waters, firstly that they were marine. Mid-Jurassic marine waters were not affected by an icecap and so had a $\delta^{18}O = -1.2$ (Hudson & Andrews 1987; Hamilton *et al.* 1987). Using the curve for this water value on Fig. 11 yields a temperature range of 42–72°C for siderite. This is too hot to be realistic for such an early cement, and so is an unlikely possibility. The second possibility is that diagenetic porewater was meteoric, as we know from many present-day and ancient examples (for example Bjørlykke *et al.* 1989 and others) that meteoric waters are likely to immediately displace depositional waters in coastal facies assemblages. In the Mid-Jurassic of the North Sea, we can infer a −6 or $-7\delta^{18}$O value for meteoric water (Hudson & Andrews 1987; Hamilton *et al.* 1987). Using a −7 curve on Fig. 11, we obtain a temperature range of 16–34°C for siderite, and 31–45°C for calcite. These are credible temperatures for cements growing during the first 1.1 km of burial. So, the implication is that porewater $\delta^{18}$O was meteoric and insignificantly modified during siderite and calcite growth. An identical conclusion can be drawn from the very much larger data compilation of C and O isotopes in the Brent Sandstones made by Samways (*in* Giles *et al.* this volume). Their data, drawn from eleven different oilfields across the East Shetland Basin show that $\delta^{18}$O from the earliest carbonate cements is compatible with growth from meteoric water. This suggests that the water: rock ratio was large, enabling water, rather than rock, to dominate the diagenetic system. The range of $\delta^{18}$O values measured on the minerals is compatible with growth from the same $\delta^{18}$O water at different temperatures. This deduction implies that meteoric water found easy access to all Brent Formations during shallow burial. Enhanced porosities have also been observed at fault block crests beneath the Upper Jurassic unconformity (Harris 1989), although Bjorkum *et al.* (1990) argue that such a zone is thin. Nevertheless, such evidence suggests that fault block crests were emergent as islands above sea level during 5–10 Ma of the late Jurassic and allowed rainwater to penetrate directly into the Brent. This isotopic and textural evidence is in direct conflict with tectonophysical reconstructions (Yielding *et al.* this volume), which suggest that unconsolidated sands were eroded at wavebase, and fault block footwalls remained submarine.

## Late cements: petrography and fluid inclusions

Let us first look at some of the basic petrographic information for these deeper-formed cements. By means of point counting thin-sections, we can determine that more diagenetic cement occurs in the sandstones which are the most permeable (Fig. 5). We can also point count specifically for diagenetic quartz, and compare that with the quartz lost in the same sample by pressure solution between grains (Fig. 6). Consequently, we can infer that the more permeable Brent Formations have imported quartz, whereas the less permeable Rannoch has exported quartz. Mullis (1991) finds from the theoretical calculations that it is very difficult to envisage diffusion of quartz cement for more than 2 or 3 metres. Consequently, the simplest interpretation of these observations is that some sort of mass-transfer process has occurred in these sandstones, and may have been influenced by the permeability, suggesting flow of porewater.

Fluid inclusions a few microns in size are sometimes trapped within diagenetic quartz overgrowths in these oilfields, although it is much more common to find larger inclusions up to 15 μm trapped at the grain/overgrowth boundary. Such inclusions have been used by many workers to infer the salinity and palaeotemperature of cementing fluids (Hollister & Crawford 1981; Haszeldine *et al.* 1984; Burley *et al.* 1989). The homogenization temperatures in quartz overgrowths from Thistle, Murchison and South Alwyn are shown in Fig. 12. No pressure correction has been made to these data, as the aqueous fluid is assumed to be saturated in methane (Burley *et al.* 1989). The modal and maximum trapping temperatures are very warm- ranging from 80° up to 135°C. Figure 13 presents our salinity data for these same fields, and suggests that a fluid of twice seawater salinity was present, together with a low salinity fluid similar to the present day reservoir fluid. It is tempting to use these temperatures to estimate the temperatures at which diagenetic quartz grew. However Osborne & Haszeldine (in prep.) have pointed out that North Sea fluid inclusion temperatures from several different Formations show a good correlation of temperature with present-day reservoir depth. Inspection of Fig. 12, where the present-day burial depths are also listed, shows an obvious trend of fluid inclusion homogenization temperature increasing with burial depth. If a geothermal gradient of 35°C/km is assumed, then the modal temperatures lie close to present day reservoir temperatures. We also note that the great majority of our readings were obtained from inclusions at the grain/overgrowth boundary. Thus we must now regard the interpretation of ancient temperatures from these fluid inclusion data as unreliable, although the salinity data may still be good. Note that the higher salinity data are only abundant in part of the Thistle oilfield (Fig. 13): the major part of the study area records fresh water.

## Late cements: K-Ar date

Authigenic illite has been separated as pure samples from 10 202 ft (3109 m) in well 211/19–4 (Rannoch Fmn, Murchison). K-Ar dating of the less than 0.1 μm size fraction gave $4.60 \times 10^{-10}$ moles/gram radiogenic argon, and an age of 58.5 ± 1.4 Ma. When plotted onto a burial curve (Fig. 14), this age equates to a burial depth of approximately 2.2 km, which is equivalent to a temperature of some 80°C, if the geothermal gradient was 35°C/km.

## Late cements: oxygen isotopes

We have argued above that the $\delta^{18}O$ values from siderite and calcite can be used to constrain

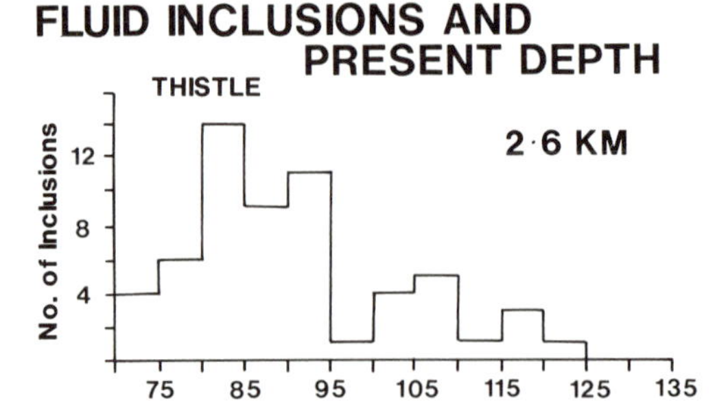


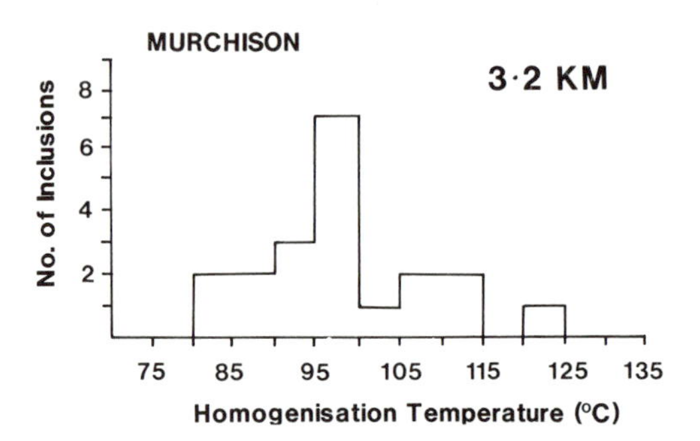


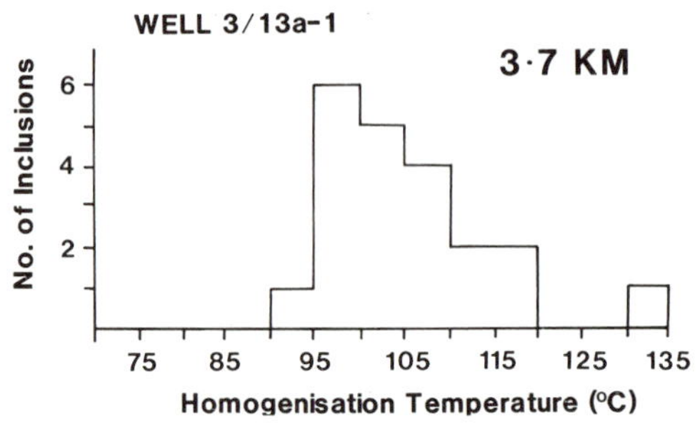


**Fig. 12.** Fluid inclusion homogenisation temperatures measured for Thistle, Murchison, and 3/13a-1. Each histogram shows a modal spike, usually interpreted as the temperature at which most of the mineral grew. If the histograms are combined (not shown), small modes at 80, 90, and 105°C are seen. Total number of measurements 102. Most of these inclusions occur at the boundary between the detrital grain and the overgrowth, the position most susceptible to leakage. The approximate present day depths of these oilfields are also shown. Notice that the modal spike of each field increases in temperature with depth, suggesting that these inclusion temperatures may have been reset, and are not true fossils of their growth temperatures.

the pore fluid present during their growth, because we had an independent estimate, from petrography, of the temperature at which those cements grew. For these deeper-formed cements, a similar approach can be used, for we

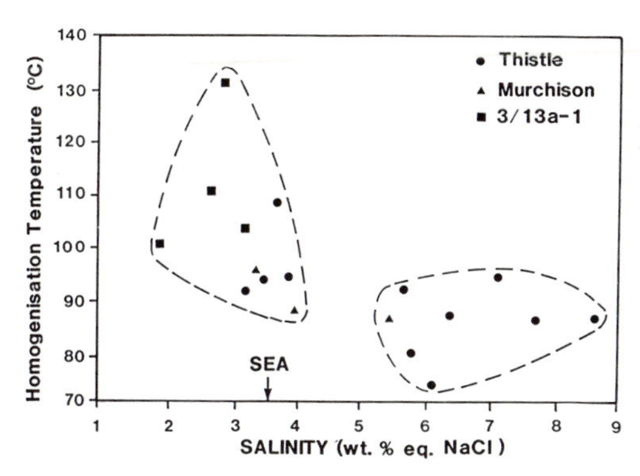


**Fig. 13.** Cross-plot of homogenization temperature versus salinity measured from fluid inclusions shows a low salinity and high salinity group (sea water = 3.5 wt% eq.NaCl). Comparison with fluid inclusion data from Heather (Glasmann *et al.* 1989) and from Alwyn (Malley *et al.* 1986) implies that most Brent diagenetic waters were seawater or less in salinity. The more saline waters in Thistle may represent a contribution from deeper waters, or a fluid compartment connected to Triassic evaporites.

have deduced above that the $\delta^{18}O$ value of the early diagenetic fluid was −7, and we are fortunate to know the present-day reservoir temperatures (110°C in Murchison, 107°C in Thistle, and 115°C in the area of Alwyn). The present-day Thistle pore fluid is $\delta^{18}O$ +2 (British Petroleum pers. comm. 1989), which is in good accord with the positive $\delta^{18}O$ values reported from present day waters in oilfields from the Norwegian sector of the North Sea (Egeberg & Aagaard 1989).

Consequently, we have separated pure samples of each of the late cements and measured the $\delta^{18}O$ values of each. Brint (1989) contains details of methods, which are standard techniques. The acid leaching method of Milliken *et al.* (1981) was used for quartz overgrowth analysis (Brint *et al.* 1990). Figure 15 shows the measured mineral isotope values plotted as curves on axes representing palaeotemperature, and palaeo-water $\delta^{18}O$. We observe no significant differences in the mineral $\delta^{18}O$ values measured for the same cements in the four different oilfields studied (Thistle, Murchison, Dunlin and 3/13a-1 Alwyn South). Thus ranges of pore fluid $\delta^{18}O$ and temperature conditions were uniform in all fields during mineral growth. Using a similar logic to that for the carbonates, we can assume a water $\delta^{18}O$ value, read in from the $\delta^{18}O$ water axis to the measured mineral curves, and obtain a range of temperatures compatible with the measured readings. Note that none of these minerals give unique 'answers', rather a combination of possibilities is possible within the 'area' between the curves of measured values for each mineral. We must now model the porewater evolution by starting at a meteoric value for siderite and calcite, passing through the whole permitted area of $\delta^{18}O$ for each mineral, in the order

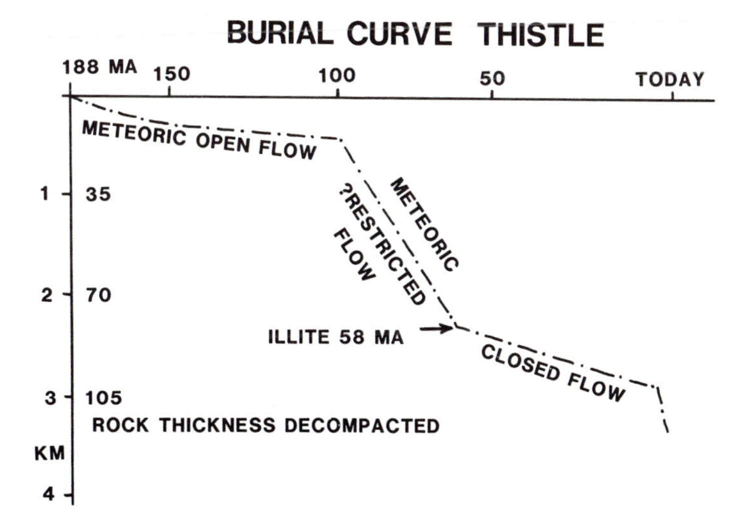


**Fig. 14.** Schematic burial curve for the Thistle oilfield, using sediment thicknesses incrementally decompacted from present values. Illite is the last mineral to form in these sandstones (Fig. 2), so that K-Ar dating of illite at 58 Ma implies that all diagenetic minerals had grown before then. This is particularly important in constraining the depth at which quartz cements grew, to 2.2 km or less, as quartz predates illite. The curve is annotated with the general hydrological regime we infer for different time periods.

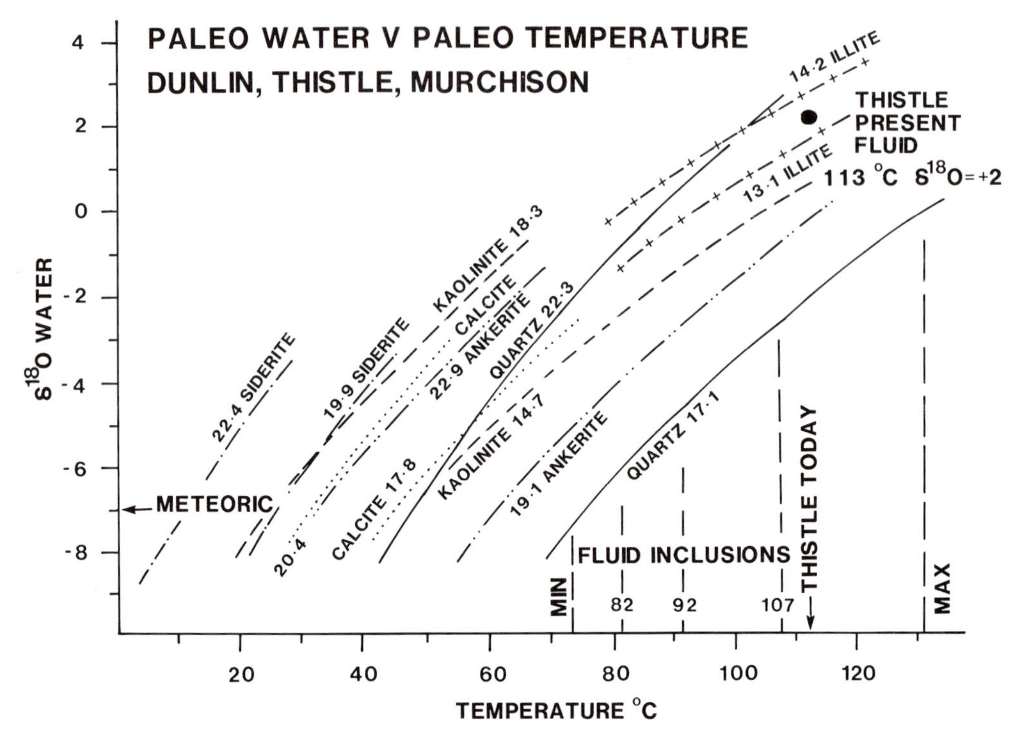


**Fig. 15.** Composite diagram, of temperature against palaeo-pore fluid composition, showing fluid inclusion temperature modes and maximum, as well as curves of the different $\delta^{18}O$ mineral values measured. Each mineral has a pair of curves, being the maximum and minimum values measured. A range of plausible water and temperature compositions of growth for each mineral is possible between the curves. To obtain unique values, we must either assume, or otherwise constrain a porewater $\delta^{18}O$, or independently constrain temperature by fluid-inclusions. Curve equations (cf. Fig. 11):
$1000 \ln\alpha_{kaolinite-water} = 2.5 \times 10^6 \times T^{-2} - 2.87$ (Land & Dutton 1978);
$1000 \ln\alpha_{ankerite-water} = 2.78 \times 10^6 \times T^{-2} + 0.32$ (Dutton & Land 1985);
$1000 \ln\alpha_{quartz-water} = 3.34 \times 10^6 \times T^{-2} - 3.31$ (Matsuhisa *et al.* 1979);
$1000 \ln\alpha_{illite-water} = 2.43 \times 10^6 \times T^{-2} - 4.82$ (Yeh & Savin 1977)

dictated by the paragenetic sequence, and finishing at the present day temperature and water $\delta^{18}O$ conditions for the reservoir. Alternative routes for this evolution are discussed below.

## Late cements: hydrogen isotopes

Hydrogen isotope compositions have been measured on pure kaolinite and illite clay samples, with a reproducability of ±2‰. We also know the present day water in Thistle to be $\delta D$ −24 to −27 (BP pers. comm. 1989). As $\delta D$ is not significantly affected by temperature, we can estimate palaeo-water values from mineral values by adding 24‰ to illite (Yeh 1980) and 20‰ to kaolinite (Liu & Epstein 1984). Thus we obtain palaeo-water ranges of $\delta D$ −28 to −48‰ for kaolinite, and $\delta D$ −31 to −36‰ for illite. These values are compatible with a meteoric water of $\delta^{18}O$ −6 to −7‰ (Craig 1961).

## Discussion of late diagenetic porewater evolution

There are three basic ways of modelling the porewater evolution. First (Fig. 16a) by drawing a horizontal meteoric water line on the data of Fig. 15, representing water precipitating the siderite and calcite cements. Then drawing an approximately diagonal line to link the finish of meteoric calcite cementation with todays reservoir water. Such gradual and continuous water evolution during late diagenesis could be envisaged as characteristic of a diffusion dominated system (Fig. 17) where the same

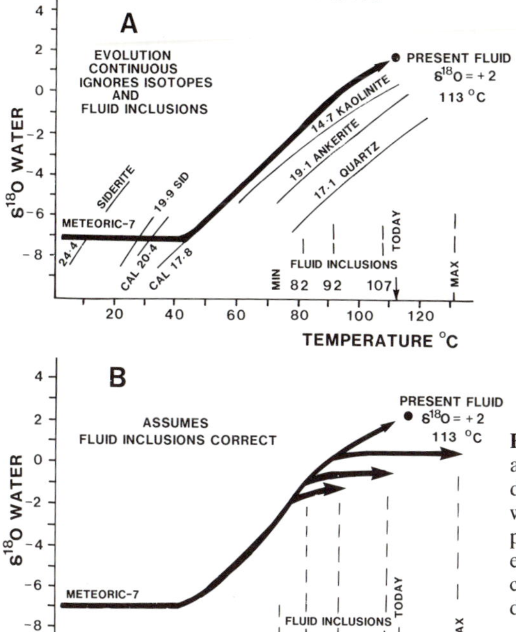


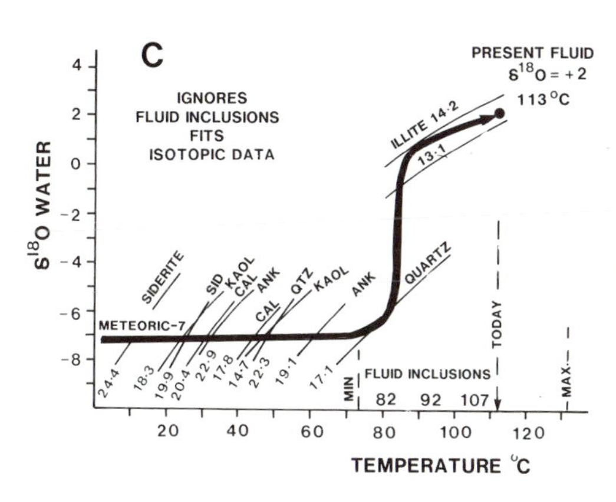


**Fig. 16.** (**a**) Interpretation of pore water evolution, assuming initial meteoric water flow during shallow diagenesis, followed by a closed diffusive system in which the porewater gradually moves towards the present-day oilfield value, attempting to reach equilibrium with the rock. This hypothetical curve cannot satisfy either the isotopic or the fluid inclusion data, and so is rejected.

(**b**) Interpretation of porewater evolution, assuming gradual equilibrium evolution of pore water, similar to that shown in (a), but with diagenesis controlled by pulses of hot fluid. These enable the porewater to accommodate the least positive ankerite and quartz $\delta^{18}O$ values, and the hottest fluid-inclusion palaeo-temperatures. Our fluid inclusion measurements provide only equivocal support for this model, for the temperatures show a systematic increase with present day burial depth.

(**c**) Interpretation of pore water evolution, assuming that meteoric flow unmodified in $\delta^{18}O$ continued through permeable aquifers to depths of at least 2.2 km. The sequence of mineral growths are produced in their correct order, and cementation ceases at 80°C, which fits nicely with the equilibrium temperature expected at 2.2 km when illite formed (Fig. 14). This model neglects fluid-inclusion data, and no extraneous heat is required. Cementing ions could all be supplied within the Brent, but re-distributed by the water flow, according to matrix permeability.

pore water remained static throughout deep diagenesis. This line does not pass through all of the areas enclosed by the measured isotope values on the different minerals, and so this alternative can be rejected.

The second alternative (Fig. 16b) makes the assumption that fluid inclusion homogenization temperatures are correct. To satisfy all the isotopic measurements, we then need to identify areas from Fig. 15 where both fluid inclusion temperatures and the least positive values of $\delta^{18}O$ quartz overlap. Consequently, we are forced to postulate that quartz grew from a pore fluid which had evolved its $\delta^{18}O$ value from −7 to at least 0‰. In addition, we are forced to postulate that quartz formed at less than 2.2 km, predating illite on Fig. 14. Consequently, the equilibrium reservoir temperature at this depth should have been about 80°C, so that fluids of this 0‰ $\delta^{18}O$ signature and 80–130°C fluid inclusion temperature were much hotter than equilibrium. These constraints can

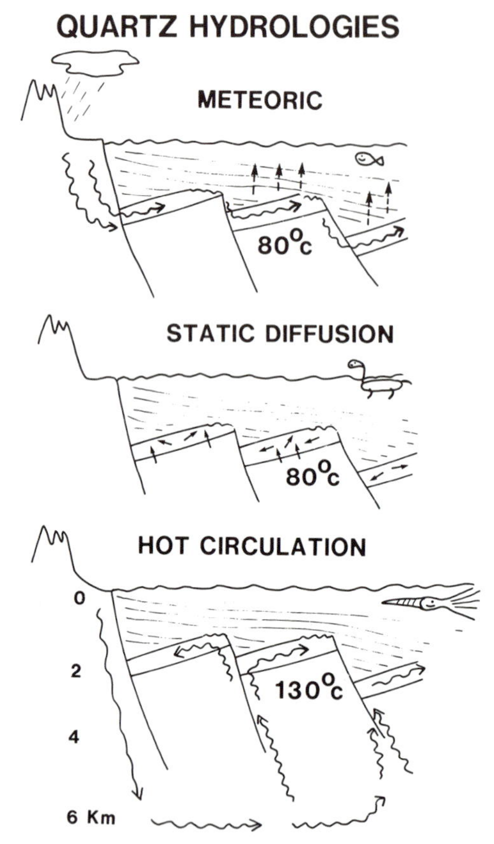


**Fig. 17.** Cartoons illustrating the pore-fluid flow options considered for the creation of quartz overgrowths in the early Tertiary, with the temperatures inferred for each. Tilted fault blocks of mid-Jurassic sediment are shown, overlain by late Jurassic to late Cretaceous sediments, and a western land area elevated to perhaps 1000 m which supplied the rapidly deposited Palaeocene sandstones. Wavy arrows indicate large flows, shorter arrows indicate lesser flows, and short arrows indicate little or no flow of pore-water.

be reconciled by envisaging a system which was spasmodically or continually open to flows of hot water ascending (Fig. 17). Pulses of such flows may have given rise to the individual modes of fluid inclusion temperatures observed in our data (Fig. 12) and in the data of Malley *et al.* (1986) and Jourdan *et al.* (1987) from Alwyn South. In addition, Hogg (1990, this volume) has observed zoned quartz overgrowths using cathode-luminescence (CL) microscopy, suggesting that quartz chemistries and/or growth rates varied with time. By contrast, CL observations by Macaulay (1990) where a pore fluid is inferred to have been static do not show zoned overgrowths, suggesting that the Alwyn zones record a moving fluid. Advection of such hot fluids could have transported the required extra silica into or within the Brent Formations, although well known problems with this include the $10^6$ pore volumes of fluid needed to cement each pore (Haszeldine *et al.* 1984), and the lack of evidence for silica transport, such as filled veins. One preliminary fission track study from Alwyn (Meyer *et al.* 1990) suggests that elevated temperatures of 115–125°C lasted for 40 Ma until the late Eocene. The reliability, or not, of fission track and/or fluid inclusion palaeotemperature measurement is pivotal evidence in accepting or rejecting this hypothesis.

The third hypothesis (Fig. 16c) is one which ignores the fluid inclusion temperatures. In this approach, the starting composition of meteoric water is maintained, so that a pore fluid with a constant $\delta^{18}O$ value simply becomes hotter. When such an effect is plotted graphically (Fig. 16c), we observe that the paragenetic sequence of cements is followed exactly, that the entire range of $\delta^{18}O$ mineral areas is covered, and that quartz must finish its growth at about 80°C. This is a remarkable coincidence of temperature with that deduced above for the growth of illite via its K-Ar age date and depth–temperature equivalence. If we propose that illite grew from the same pore fluid as the other minerals, then illite growth does not fit into the correct paragenetic position, consequently it must have grown in its correct paragenetic sequence from a pore fluid of different $\delta^{18}O$ composition. It is a notable observation that the permitted pore water for illite contains the present day pore water within its bounds. We see no requirement from the measured or observed data why this third pathway should be false, and so accept it as a probable fluid evolution path. Thus we must envisage a meteoric flow system (Fig. 17), where water flow was sufficiently voluminous to preserve its isotopic signature. The CL evidence in favour of water flow (discussed above) applies equally well to this system, and quartz would be transported preferentially into the most permeable zones. When fluid movement eventually ceased, then growth of quartz and illite in equilibrium with a few pore fluid volumes forced the fluid to partly equilibrate with the surrounding rock and then evolve to its more positive present day value by diffusion from surrounding mudrocks. A potential objection to this hypothesis, is that Palaeocene meteoric waters deduced from lava alteration on Skye, some 550 km distant to the SW, record $\delta^{18}O$ −10‰ and $\delta D$ −80‰ (Forester & Taylor 1977). The values recorded in Brent diagenetic kaolinites and illites are not so depleted ($\delta$D −28 to −48‰, $\delta$D −31 to −36‰). Thus to support a meteoric hypothesis, we need to suggest a less depleted meteoric water, perhaps due to rainfall at sea level rather than on a 3 km high volcano.

## Flow models

We suggest that broad types of palaeo-hydrology can be reconstructed through the diagenetic history of the Brent sandstones. In the earliest burial stages down to 1.1 km, open-system flow occurred (Fig. 18a). Ions for cements were all supplied from within the Brent Group sandstones, with some C from mudstones. The homogeneous $\delta^{18}O$ porewater values (Figs 11 & 15) required to form these minerals at low temperatures suggest that un-evolved meteoric waters flowed through these sediments in a water-dominated open-flow system. Different

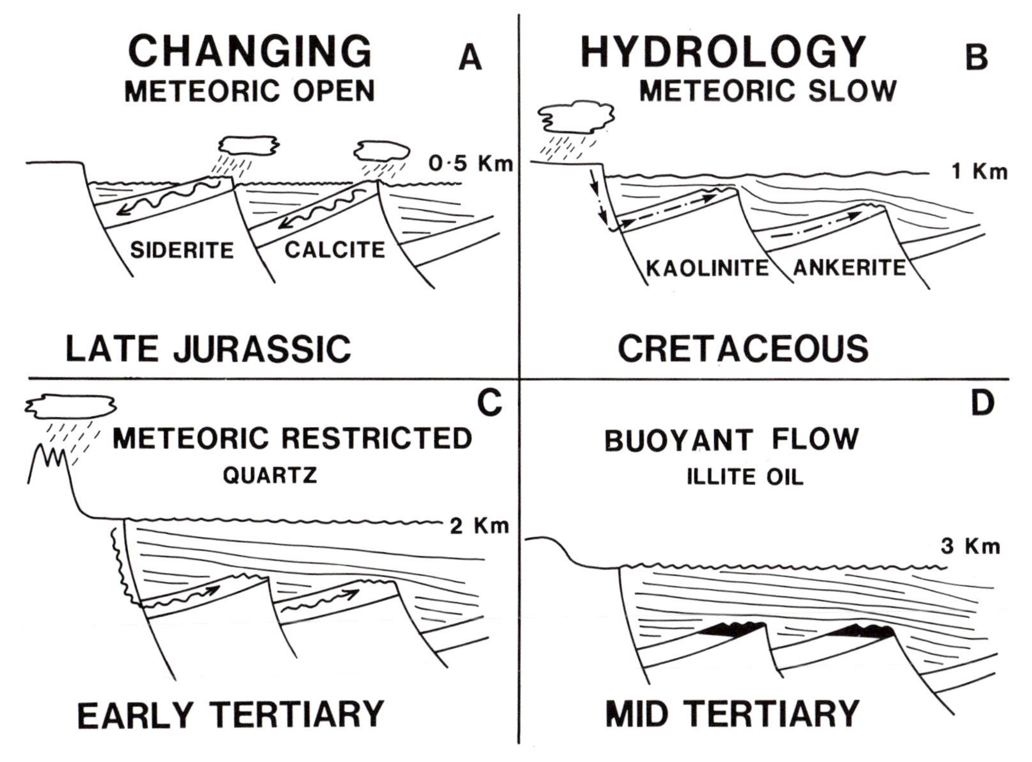


**Fig. 18.** Summary of diagenetic hydrologies inferred for the Brent sandstones. (**A**) Open system inflow and throughput of meteoric waters. This must have first occurred immediately after Ness river deposition, before Tarbert deposition. The strong and unevolved meteoric $\delta^{18}O$ signatures of carbonate concretions imply that fault block crests were exposed during the late Jurassic 'Cimmerian' rifting event, allowing throughflow of meteoric water. Siderite, calcite, minor quartz, feldspar, some pyrite and vermiform kaolinite grew.

(**B**) When the Brent aquifer became partly sealed beneath Lower Cretaceous mudrocks, a restricted system of slow flow developed. Ankerite mineralogies indicate local inhomogeneous pore-water, but $\delta^{18}O$ values of ankerite, kaolinite and quartz are compatible with an unevolved meteoric water.

(**C**) Open system rapid flow of deeply penetrating meteoric waters. These were driven by a high topographic head from the epeiric uplift at the margin of the East Shetland Basin and on the west of Scotland. Note that hot water circulation (Fig. 17) cannot be disproved with our information. In either case, quartz could have been supplied by pressure solution from within the sands and muds of the Brent Group. Such flow regimes may have been induced during fracturing of the sediment pile by externally controlled events, such as crustal tension accompanying rupture of the Greenland-European crust, and loading of the basin by Palaeocene sands producing rapid subsidence and overburden stress.

(**D**) Rapid erosion and subsidence of the western land area, combined with deposition of thick sequences of Palaeocene and Eocene muds drastically reduced water circulation. Illite grew in this closing system when fluid in individual pores became high enough in alkalinity; pore water $\delta^{18}O$ evolved rapidly, as small water volumes attempted to equilibrate with surrounding rock. Oil migrated in buoyantly and waterproofed the rock, so that rapid diagenesis ceased.

atomic or ionic species could travel different distances from their sources, before precipitating in their mineral 'sink'. Thus carbon could not travel far at all, and $\delta^{13}C$ signatures are controlled by local facies (Fig. 10). Strontium was initially very locally supplied from degrading micas but as diagenesis proceeded, Sr from dissolving marine shell debris became mixed in and was transported across-formations to produce homogeneous values in marine and non-marine formations. Eventually shell debris saturated the Sr supply, with only minor radiogenic supply from silicates, so that diagenetic calcite $^{87}Sr/^{86}Sr$ was nearly marine in its ratio. Iron, calcium and magnesium were unreactive enough and abundant enough to homogenize in siderite through different formations at an early stage (Fig. 8). The Fe and Mn rich chemistry of siderite attests to a meteorically supplied porewater, rather than being Mg rich from seawater (Mozley & Hoernle 1990). Oxygen in the pore fluid was much more abundant than the oxygen exchanging during mineral growth, so that $\delta^{18}O$ of the pore fluid remained homogeneous. Meteoric waters must have recharged this flow system through the erosional unconformity on the footwall of many Brent Group fault blocks. Harris (1989) notes additional porosity beneath the unconformity, which he ascribes to subaerial exposure, whereas Bjorkum *et al.* (1990) calculate that the sediment erosion rate during the late Jurassic should not have permitted such a zone to be preserved. Sequence stratigraphical studies indicate a subaerial unconformity beneath the Tarbert Formation in many areas (Helland Hansen *et al.* and Mitchener *et al.* this volume), so that a meteoric flush could have displaced and homogenized the original porewaters at that time. However concretions in the depositionally marine Tarbert show $\delta^{18}O$ isotopic signatures similar to those of the Ness or Rannoch (Fig. 10), suggesting that water in all Formations was meteoric. Structurally-based models of footwall uplift (Yielding *et al.* this volume) interpret the flat unconformity as indicating a sub-marine origin; however BP (N. Milton pers. comm. 1990) prefer a subaerial interpretation. Nevertheless, the isotopic evidence is very strong in favour of a uniform fluid of meteoric origin in all formations during early diagenesis, so that many of these footwall unconformities must have been sub-aerial.

Deeper than 1.1 km, the rate of flow in the system became slower (Fig. 18b), so that ankerite compositions are different in different formations (Fig. 8). Was the supply of these ions also slower, so that kinetically rapid local reactions 'mopped up' ions before they could travel far and homogenize compositions through the different formations? Strontium for the ankerites remained very un-radiogenic. This may have been due to local dissolution of calcite recycling Sr from diagenetic cement or shell debris, or to a throughput of meteoric water with a very low content of radiogenic Sr. A similar flow regime must have affected kaolinite, which overlaps with early calcites and late ankerites in its paragenesis.

Diagenetic quartz could also have grown in two distinct settings. (1) If fluid inclusion temperatures are correct in recording an apparent geothermal gradient of 40–60°C/km then circulation of hot water with an evolved $\delta^{18}O = 0$ is required from depths of at least 4 km up to quartz precipitation depths of 2 km (Fig. 17). (2) Alternatively, if fluid inclusion temperatures are discredited as potentially unreliable records, then we can model the entire sequence of cements as having grown in a normal geothermal gradient from a meteoric water unmodified in $\delta^{18}O$ (Figs 17 & 18c). There is no requirement from our data that the water started to evolve to more positive values. This suggests that water and rock were not in equilibrium, and so the system must have been dominated by moving pore water in order to continually replenish the pore fluid which precipitated these diagenetic minerals. Most kaolinite, most ankerite, and all quartz, grew in such a flow system which would have been restricted to the more permeable aquifers (Fig. 18c). Ions for cements could have been locally supplied from sands and muds, but transported in the circulating fluid, just as in the early meteoric diagenetic system.

In either of these scenarios for quartz cementation, large scale movement of porewater was induced by the external geological conditions, for rapid subsidence was occurring in the North Sea, the continental crust was experiencing extreme E–W tension together with dyke intrusion during rupture between Hatton Bank and Greenland. Tertiary volcanoes on the west of Britain were elevated at least 3 km, and the edge of a mantle plume may have extended beneath the northern North Sea (White & McKenzie 1989). In short, tectonic conditions were combining to produce tectonic and overpressure fracturing of the sedimentary pile, and a large hydrostatic head was ready to force artesian water flow down through the sedimentary pile, for it to return upwards carrying its heat. The problem of identifying hot fluid influxes from deeper in the basin remains, and revolves around the reliability of fluid-inclusion palaeotemperature information. If Osborne & Haszeldine (in prep.) are correct, and most inclusion temperatures published to date are unreliable, then there is, unfortunately, no compelling evidence in favour of additional heat. Thus the lack of extensive veining in core, the good coincidence of isotopic temperatures with equilibrium burial temperatures, and the potential uncertainty of fluid inclusion interpretation leads us to favour the open system 'cool meteoric' model, although we cannot refute the hot circulation model.

Illite can easily be envisaged to have formed in a slowly flowing system (Fig. 18d) just before oil migration (Jourdan *et al.* 1987; Hogg 1989). In such a low water:rock ratio, the pore fluid would rapidly become dominated by rock composition as minerals (illite and the last skin of quartz) crystallised, so that water composition became heavier in $\delta^{18}O$ and the reactants available in solution became exhausted. Buoyant migration of oil into the pore system halted rapid diagenesis. The present-day water saturations of good quality sandstone reservoirs in Brent Sandstones are only 10–20% (Moss, this volume), so that only extremely slow and isolated diffusive processes are feasible in continuing any diagenetic reactions.

## Conclusions

(1) Diagenesis in the first 1.1 km of burial occurred under open-flow conditions dominated by unevolved meteoric water (Fig. 18a). Ions were supplied from within the Brent sands and muds, and travelled either locally or for longer distances, depending on their reactivity and their concentration. Distances of travel were O>Ca>Sr>Fe>C. Early siderite, calcite and kaolinite cements record homogeneous porewater in depositionally distinct formations, and different ions became 'closed systems' at different burial depths.

(2) Ankerite grew as rims to concretions below 1.1 km, and shows a variable mineralogy, suggesting more local ion supplies in slower restricted flows of water. Most kaolinite growth overlapped with this, and $\delta^{18}O$ of the porewater can be modelled as meteoric (Fig. 18b).

(3) Quartz diagenesis cannot be definitively modelled. One alternative is that fluid inclusions record the temperatures of hot fluids circulating to at least 4 km depth in a fractured crust (Fig. 17), and bringing isotopically evolved porewater, heat and quartz to the Brent Group as they rose to 2 km. However, if fluid inclusion temperatures have been reset around present day values, a second alternative is that quartz and all the 'late' burial cements could equally well have grown under a normal geothermal gradient in a flow of unevolved meteoric water restricted to permeable aquifers, and driven by rapid elevation of the landmass west of the East Shetland Basin (Fig. 18c). In either case the unusual external geological rifting and uplift events during the Palaeocene and Eocene were the fundamental driving causes of an unusual event of porewater circulation.

(4) Illite grew in a rock dominated system as water flows declined towards being stationary (Fig. 18d). Porewater isotopes rapidly evolved to present day values. Oil migrated buoyantly and accumulated after porewater had ceased moving.

We thank our Glasgow colleagues for discussions, and help, sometimes unknowingly, especially C. Macaulay, M. Sullivan, M. Russell, A. Hall and G. Bowes. J. F. B. was supported by a NERC, studentship. SURRC is supported by NERC and the Scottish Universities. We are grateful to Britoil and the British Geological Survey for supplying Core. Last, but not forgotten, J. Kantorowicz (Shell) and another anonymous referee helped an earlier and rather different typescript to mature. Editorial handling M. R. Giles.

## References

Baldwin, B. & Butler, C. O. 1985. Compaction curves. *Bulletin of the American Association of Petroleum Geologists*, **69**, 622–626.

Bjorkum, P. A., Rune, M., Walderhaug, O. & Hurst, A. 1990. The role of the late Cimmerian unconformity for the distribution of kaolinite in the Gullfaks field, northern North Sea. *Sedimentology*, **37**, 395–406.

Bjørlykke, K. & Brendsdal, A. 1986. Diagenesis of the Brent sandstone in the Statfjord field, North Sea. *In*: Gautier, D. L. (ed.) *Roles of Organic Matter in Sediment Diagenesis*. SEPM Special Publication, **38**, 157–167.

——, Ramm, M. & Saigal, G. C. 1989. Sandstone diagenesis and porosity modification during basin evolution. *Geologische Rundschau*, **78**, 243–268.

Blanche, J. B. & Whitaker, J. H. McD. 1978. Diagenesis of part of the Brent Sand Formation (Middle Jurassic) of the northern North Sea Basin. *Journal of the Geological Society, London*, **135**, 73–82.

Boles, J. R. & Johnson, K. S. 1983. Influence of mica surfaces on pore water pH. *Chemical Geology*, **43**, 303–317.

Brennand, T. P. Van Hoorn, B. & James, K. H. 1990. Historical review of North Sea exploration. *In*: Glennie, K. W. (ed.) *Petroleum geology of the North Sea 3rd edn*. Blackwell Scientific, Oxford, 1–33.

Brint, J. F. 1989. *Isotope diagenesis and palaeofluid movement: Middle Jurassic Brent sandstones, North Sea*. PhD thesis, University of Strathclyde, U.K.

——, HAMILTON, P. J., HASZELDINE, R. S. & FALLICK, A. E. 1991. Oxygen isotopic analysis of diagenetic quartz overgrowths from the Brent sands: a comparison of two preparation methods. *Journal of Sedimentary Petrology*, **61**, 527–533.

BURLEY, S. D., MULLIS, J. & MATTER, A. 1989. Timing diagenesis in the Tartan reservoir (North Sea): constraints from combined cathodoluminescence microscopy and fluid inclusion studies. *Marine and Petroleum Geology*, **6**, 98–120.

BURKE, W. H., DENISON, R. E., HETHERINGTON, E. A., KOEPNICK, R. B., NELSON, H. F. & OTTO, J. B. 1982. Variation of seawater $^{87}Sr/^{86}Sr$ throughout Phanerozoic time. *Geology*, **10**, 516–519.

CAROTHERS, W. L., ADAMI, L. H., ROSENBAUER, R. J. 1988. Experimental oxygen isotope fractionation between siderite-water and phosphoric acid liberated $CO_2$- siderite. *Geochimica et Cosmochimica Acta*, **52**, 2445–2450.

CRAIG, H. 1961. Isotopic variations in meteoric waters. *Science*, **133**, 1702–1703.

DEER, W. A., HOWIE, R. A. & ZUSSMAN, J. 1962. *Rock forming minerals vol. 5, Non-silicates*. Wiley, London.

DUTTON, S. P. & LAND, L. S. 1985. Meteoric burial diagenesis of Pennsylvanian arkosic sandstones, southwestern Anadarko Basin, Texas. *Bulletin of the American Association of Petroleum Geologists*, **69**, 22–38.

EGBERG, P. K. & AARGAARD, P. 1989. Origin and evolution of formation waters from oil fields on the Norwegian shelf. *Applied Geochemistry*, **4**, 131–142.

FORESTER, R. W. & TAYLOR, H. P. 1977. $^{18}O/^{16}O$, D/H and $^{13}C/^{12}C$ studies of the Tertiary igneous complex of Skye, Scotland. *American Journal of Science*, **277**, 136–177.

FRIEDMAN, I. & O'NEIL, J. R. 1977. Compilation of stable isotope fractionation factors of geochemical interest. *In*: FLEISCHER, M. (ed.) *Data of Geochemistry* (6th edn). United States Geological Survey Professional Paper 440-KK.

GILES, M. R. 1987. Mass transfer problems of secondary porosity creation in deeply buried hydrocarbon reservoirs. *Marine and Petroleum Geology*, **4**, 188–204.

——, STEVENSON, S., MARTIN, S. V., CANNON, S. J. C., HAMILTON, P. J., MARSHALL, J. D. & SAMWAYS, G. R. 1992. The diagenesis and reservoir properties of the Brent Group, a regional perspective. *This volume*.

GLASMANN, J. R., LUNDEGARD, P. D., CLARK, R. A., PENNY, B. K. & COLLINS, I. D. 1989. Geochemical evidence for the history of diagenesis and fluid migration: Brent sandstone, Heather Field, North Sea. *Clay Minerals*, **24**, 255–284.

HAMILTON, P. J., FALLICK, A. E., MACINTYRE, R. M. & ELLIOTT, S. 1987. Isotopic tracing of the provenance and diagenesis of Lower Brent Group Sands, North Sea. *In*: BROOKS, J. & GLENNIE, K. (eds) *Petroleum Geology of North West Europe*. Graham & Trotman, London, 936–949.

HANCOCK, N. J. & TAYLOR, A. M. 1978. Clay mineral diagenesis and oil migration in the middle Jurassic Brent Sand Formation *Journal of the Geological Society, London*, **135**, 69–72.

HARRIS, N. B. 1989. Diagenetic quartzarenite and destruction of secondary porosity; an example from the middle Jurassic Brent sandstone *Geology*, **17**, 361–364.

HASZELDINE, R. S., SAMSON, I. M., & CORNFORD, C. 1984. Quartz diagenesis and convective fluid movement: Beatrice oilfield UK North Sea. *Clay Minerals*, **19**, 391–402.

HELLAND-HANSEN, W., ASHTON, M., LOMO, L. & STEEL, R. 1992. Advance and retreat of the Brent delta: recent contributions to the depositional model. *This volume*.

HOGG, A. J. C. 1989. *Petrographic and isotopic constraints on the diagenesis and reservoir properties of the Brent Group sandstones, Alwyn South*. PhD thesis, University of Aberdeen, UK.

HOGG, A. J. C. 1992. Cathodoluminescence of quartz cements in Brent Group sandstones, Alwyn South. *This volume*.

HOLLISTER, L. S. & CRAWFORD, M. L. 1981. *Short course in fluid inclusions: applications to petrology*. Mineralogical Association of Canada. Short course handbook vol. **6**.

HOUSEKNECHT, D. W. 1987. Assessing the relative importance of compaction processes and cementation to reduction of porosity in sandstones. *American Association of Petroleum Geologists Bulletin*, **71**, 633–642.

HUDSON, J. D. & ANDREWS, J. E. 1987. The diagenesis of the Great Estuarine Group, Middle Jurassic, Scotland. *In*: MARSHALL, J. D. (ed) *Diagenesis of Sedimentary Sequences*. Geological Society, London, Special Publication, **36**, 259–276.

JOURDAN, A., THOMAS, M., BREVART, O., ROBSON, P., SOMMER, F. & SULLIVAN, M. 1987. Diagenesis as the control of the Brent sandstone reservoir properties in the Greater Alwyn area (East Shetland Basin). *In*: BROOKS, J. & GLENNIE, K. (eds) *Petroleum Geology of North West Europe*. Graham & Trotman, London, 951–961.

LAND, L. S. & DUTTON, S. P. 1978. Cementation of a Pennsylvanian deltaic sandstone: isotopic data. *Journal of Sedimentary Petrology*, **48**, 1167–1176.

LIU, K. K. & EPSTEIN, S. 1984. The hydrogen isotope fractionation between kaolinite and water. *Isotope Geoscience*, **2**, 335–350.

MACAULAY, C. I. 1990. *Classic diagenesis and porefluid evolution: an isotopic study, Magnus Oilfield, North Sea*. PhD thesis, University of Strathclyde.

MALLEY, P., JOURDAN, A. & WEBER, F. 1986. Etude des inclusions fluides dans les nourrissages siliceux des gres reservoirs de la Mer du Nord. *Acadamie des Sciences*, **302**, 653–658.

MATSUHISA, Y., GOLDSMITH, J. R. & CLAYTON, R. N. 1979. Oxygen isotopic fractionation in the system quartz-albite-anorthite-water. *Geochimica et Cosmochimica Acta*, **43**, 1131–1140.

MEYER, A. J., PIRONON, J. & PAGEL, M. 1990. Fluid

inclusion and fission track thermochronology from Brent sandstone reservoir (Alwyn) North Sea. *In*: *Geochemistry of earth's surface and mineral formation 2nd Symposium, Aix en Provence*, 241–242.

Milliken, K. L., Land, L. S. & Loucks, R. G. 1981. History of burial diagenesis determined from isotopic geochemistry, Frio Formation, Texas. *American Association of Petroleum Geologists Bulletin*, **65**, 1397–1413.

Mitchener, B. C., Lawrence, D. A., Partington, M. A., Bowman, M. B. J. & Gluyas, J. 1992. Brent Group: sequence stratigraphy and regional implications. *This volume*.

Moss, B. P. 1992. Petrophysical characteristics of the Brent sandstone. *This volume*.

Mozley, P. S. & Hoernle, K. 1990. Geochemistry of carbonate cements in the Sag River and Shublick Formations, Alaska. *Sedimentology*, **37**, 817–836.

Mullis, A. 1991. The role of silica precipitation kinetics in determining the rate of quartz pressure solution. *Journal of Geophysical Research*, **96B**, 10 007–10 023.

Sclater, J. G. & Christie, P. A. F. 1982. Continental stretching: an explanation of the post-mid-Cretaceous subsidence of the Central North Sea Basin. *Journal of Geophysical Research*, **85**, 3711–3739.

Sommer, F. 1978. Diagenesis of Jurassic sandstones in the Viking Graben. *Journal of the Geological Society, London*, **135**, 63–67.

Sullivan, M., Haszeldine, R. S. & Fallick, A. E. 1990. Linear coupling of strontium and carbon isotopes, evidence for cross-formational fluid flow. *Geology*, **18**, 1215–1218.

White, R. S. & McKenzie, D. P. 1989. Volcanism at rifts. *Scientific American*, **260**, July, 44–45.

Wilkinson, M. & Dampier, M. D. 1990. The rate of growth of sandstone-hosted calcite concretions. *Geochimica et Cosmochimica Acta*, **54**, 3391–3399.

Yeh, H. W. 1980. D/H ratios and late stage dehydration of shales during burial. *Geochimica et Cosmochimica Acta*, **44**, 341–352.

Yeh, H. W. & Savin, S. M. 1976. The extent of oxygen isotope exchange between clay minerals and seawater: *Geochimica et Cosmochimica Acta*, **40**, 743–748.

Yielding, G., Badley, M. & Roberts, A. 1992. The structural evolution of the Brent Province. *This volume*.

# Cathodoluminescence of quartz cements in Brent Group sandstones, Alwyn South, UK North Sea

A. J. C. HOGG[1], E. SELLIER[2] & A. J. JOURDAN[2, 3]

[1] *Department of Geology & Petroleum Geology, University of Aberdeen, Marischal College, Aberdeen AB9 1AS, UK*

*Present Address: BP Research, Chertsey Road, Sunbury-on-Thames, Middlesex TW16 7LN, UK*

[2] *Total – Compagnie Francaise des Petroles, Laboratoires Exploration, 218–228 Avenue du Haut-Leveque, 33605 Pessac Cedex, France*

[3] *Present address: Total Oil Marine plc, Crawpeel Road, Altens, Aberdeen AB9 2AG, UK*

**Abstract:** Quartz cements can, at best, enhance sandstone reservoir quality through strengthening the clastic framework against compaction or, at worst, degrade poroperm properties by occluding pore spaces and constricting fluid pathways. An understanding of the controls on distribution and timing of quartz cementation can therefore aid exploration and production strategy.

In the Middle Jurassic Brent Group sandstones in Alwyn South authigenic quartz, kaolinite and illite development has resulted in severe deterioration of reservoir properties. Further resolution is obtained using a scanning electron microscope (SEM) adapted for cathodoluminescence (CL) mode observations. CL images show a clear distinction between detrital and authigenic quartz. In detail quartz overgrowths are complex and multiply zoned and a recurrent sequence of three major and two minor CL zones is distinguished. Zones become increasingly euhedral from core to rim and reflect variation in crystal growth rates and silica supply during several pulses of silicification. Heterogeneous zones and dissolution-boundary development shows that quartz precipitation is flux-controlled. Correlation of CL zones with clay mineral inclusions allow cement paragenesis to be refined and correlations with fluid inclusion data allow the conditions of quartz precipitation to be constrained.

Sandstone reservoir quality is dependent on original sedimentary facies and diagenetic overprinting. In the upper formations of the Brent Group sandstones in the Alwyn South area. facies variation is relatively limited but there is considerable variation in reservoir quality. Although compaction may reduce porosities and permeabilities in deeper wells, the deterioration of normally good reservoir properties is attributed to authigenic quartz, kaolinite and illite development.

The extent of silica precipitation is critical because quartz cements may, at best, retain permeability through strengthening the framework against further compaction or, at worst, severely reduce porosity and permeability by constricting fluid pathways or occluding pore spaces. Quartz cements exist throughout the area with an apparent increase in abundance with depth.

In their model of the diagenesis of the Greater Alwyn Area, Jourdan *et al.* (1987) recognized three distinct hydrodynamic systems with related diagenetic episodes.

(i) Neoformation of vermicular kaolinite, feldspar alteration and minor quartz cementation resulting from compaction-induced meteoric pore water flow soon after deposition.

(ii) Often intense development of vermicular kaolinite, feldspar alteration, slight quartz cementation and authigenic carbonate dissolution. This phase, related to meteoric water influx, is best developed where the Brent Group subcrops the basal Cretaceous unconformity.

(iii) K-feldspar dissolution overlapped by synchronous or later intense quartz-overgrowth formation, blocky kaolinite formation and, in Alwyn South, intense illitization.

Sources contributing silica to the pore waters were chemical compaction and pressure solution during intense Cretaceous subsidence, dissolution of framework silicates, and water expelled on increasing burial from argillaceous source rocks. In Alwyn South, data from fluid

*From* Morton, A. C., Haszeldine, R. S., Giles, M. R. & Brown, S. (eds), 1992, *Geology of the Brent Group*. Geological Society Special Publication No. 61, pp. 421–440.

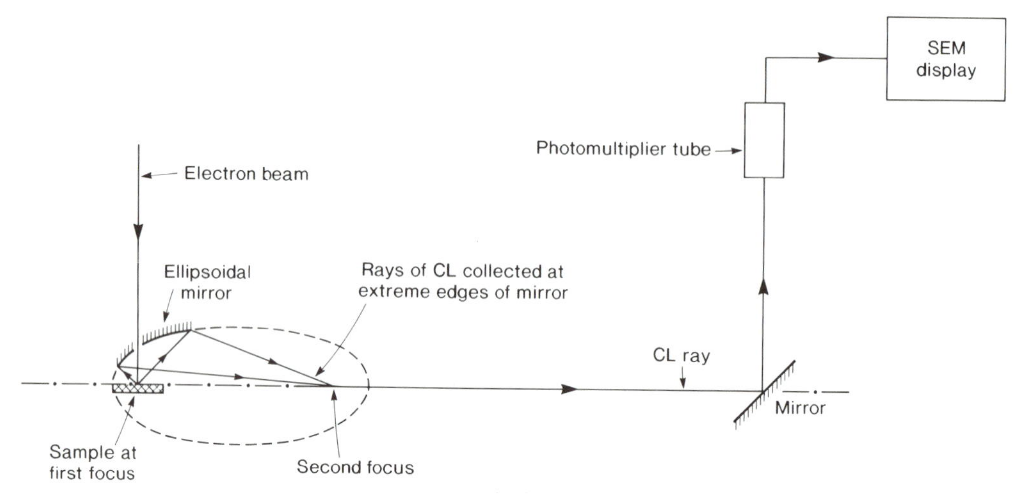


**Fig. 1.** Schematic diagram showing the arrangement of optical components in the Oxford Instruments ellipsoidal-mirror light collector system and beam pathway to JEOL JSM 840 scanning electron microscope.

inclusions in overgrowths indicate mean trapping temperatures of 107–120°C (Malley *et al.* 1986; Jourdan *et al.* 1987; Hogg 1989). In the intensely silicified, deep 3/14a–8 well trapping temperatures as great as 140°C suggest a high geothermal gradient in the vicinity of that well.

## Objectives

This model generates several questions which we attempt to address here.

(i) Can the quartz cement be resolved petrographically into its component early, middle and late stages?

(ii) If the major phase of quartz precipitation occurred over a long period, was overgrowth development continuous and in continuity with the original cements? Did quartz precipitation occur in several pulses?

(iii) How does fluid inclusion data relate to the different cementation phases?

(iv) How does the cementation history of tightly cemented sands, as in 3/14a–8, differ from moderately cemented sands?

Because quartz cements are typically clear, and in optical continuity with the host grain these problems are not easily solved using conventional petrography but may be resolved using cathodoluminescence (CL) petrography.

CL microscopy, which has for the past two decades been successfully applied in studies of cement textures and zonation in carbonate diagenesis, has until recently seen limited use in clastic diagenesis. This may be because quartz is only weakly luminescent using commercially available cold cathode luminoscopes (Ramseyer *et al.* 1989). Good results have however been achieved using a hot-cathode luminoscope which allows weak and short-lived quartz luminescence to be recorded with small losses in luminescence quality during observation (Zinkernagel 1978; Ramseyer *et al.* 1988, 1989; Burley *et al.* 1989).

The problem of detecting weak quartz luminescence can also be overcome using the scanning electron microscope (SEM) which is widely available and can be adapted for CL work. This paper addresses the questions outlined above and demonstrates the potential of SEM CL as a powerful tool in assessing and understanding sandstone reservoir properties.

## SEM cathodoluminescence

Models for the origin of CL have been outlined by Nickel (1978), Walker (1985), Marshall (1988) and Miller (1988). Cathodoluminescence of quartz cements has been discussed by Zinkernagel (1978); Burley *et al.* (1985, 1989) and Ramseyer *et al.* (1988, 1989).

An SEM equipped with an efficient CL detector can be suitable for luminescence studies (e.g. Grant 1978). It is particularly useful for diagenetic quartz which generally luminesces poorly compared with feldspar and carbonates. Although prolonged electron bombardment is

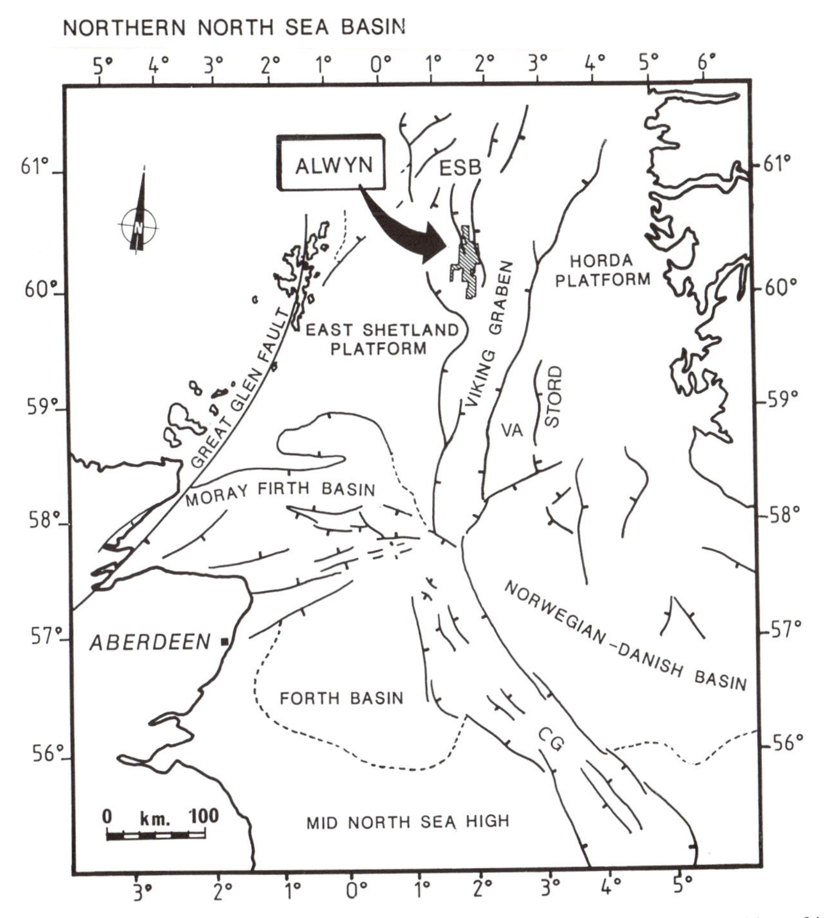


**Fig. 2.** Mesozoic structural elements of the Northern North Sea Basin with location of the Alwyn 3/14 licence area. CG, Central Graben; ESB, East Shetland basin; VA, Vestland Arch.

known to induce alterations in quartz CL quality (e.g. Zinkernagel 1978; Ramseyer *et al.* 1988, 1989), we have found that under the operating conditions used for this work (next section), beam damage through observation can be minimized.

The rastering imaging system on the SEM offers a greater range of magnification than optical CL systems. However, in the absence of an optics system an important drawback of SEM CL is that luminescence colours are not readily observed. This may be compensated for if the system is fitted with a monochromator which allows wavelength (colour) and intensity differences to be resolved.

To date, few SEM CL studies applied to sedimentary petrology have been reported. This may be due to the early availability of simpler cold cathode luminoscopes, problems with detector efficiency and resolution or to smearing of the CL image due to the rastering of the SEM beam. Remond *et al.* (1970) and Remond (1977), reported good results but with minerals such as cassiterite and sphalerite which have high luminescence efficiencies. Grant & White (1978), using an SEM in CL mode recognized a poorly resolved blocky texture in overgrowths from aeolian sandstones. Using an electron microprobe Henry & Toney (1987) obtained high resolution images of CL textures in quartz sandstone by combining backscatter electron (BSE) and CL images. Similar to the SEM, the images were obtained by rastering an electron microprobe beam across the area of interest. Combining the BSE with CL images enhanced grain boundaries and fractures. Suchecki & Bloch (1988), using a similar system, reported overgrowths distinguished by luminescent and non-luminescent zones with ‘flamboyant and irregular textures’ in a variety of sandstones from North America and the Jurassic of offshore Norway. More recently, using SEM CL Meunier *et al.* (1989) observed radiation damage to quartz in uranium–vanadium-bearing sands from the Jurassic of Colorado.

## Experimental and instrumental

Twenty samples were selected, using composite logs and a thin section survey, on the basis of lithological similarity and the presence of well developed silica cements. To assess the range and continuity of CL character two pairs of samples 0.5 m apart were also included. Observations were made on 55 $\mu$m doubly-polished impregnated thin sections. The sections were first examined under transmitted light and modal mineralogies were determined by point-counting. Samples with fluid inclusions were cut into fragments and used for microthermometric measurements prior to CL work. The sections were then given a *light* carbon coating.

All CL observations were made with a JEOL JSM 840 scanning electron-microscope fitted with an Oxford Instruments CL detector (Fig. 1). In this system the CL signal is collected by an ellipsoidal mirror situated several millimetres above the sample surface. The signal is transmitted via an inclined mirror to a photomultiplier and from there to an amplifier which outputs to the SEM cathode ray tube display. Accelerating voltages were 20 kV with a beam current intensity of between $3 \times 10^{-9}$ to $1 \times 10^{-6}$ amperes. The operating vacuum was $2 \times 10^{-6}$ Torr. with an argon microflow.

For this work images were observed at magnifications of between ×30–×700 with resolutions of the order of several microns (*c.* 3–7 $\mu$m). Observation times were $\leq$1 minute and typically 40 seconds. Using this system we were able to retrieve the same CL images after one or even several periods of observation. Images were recorded using a slow scanning beam raster onto polaroid print film. For the present work over 300 CL views were examined.

In this work the CL images are in negative format and, in general, dark areas are more strongly luminescent than bright areas. This method alleviates the problem of light from the otherwise brightly luminescing detrital quartz cores swamping the dully luminescing overgrowths during long exposure times. For the moment no direct correlations are made between the CL intensity in the photographs and the actual luminescence intensity of the sample. This is because the quality of the CL signal collected, for example intensity or signal to noise ratio, can vary with machine working conditions. Nevertheless, valuable structural and textural information can be derived from these SEM CL images.

## Geology

The Alwyn South area lies between 60°31′–60°40′N and 1°36′–1°56′E and centres on the Total UK 3/14 licence area. This work focuses on wells 3/14a–7, 3/14a–8, 3/14b–9 and 3/14a–11 drilled between 1981 and 1985 (Fig. 2).

The regional structure is dominated by Middle Jurassic rifting which resulted in the now well defined East Shetland Basin and Viking Graben structures (Fig. 2). Alwyn South, which lies on the western margin of the graben is divided by a series of major N–S normal faults into three N–S-trending structural units (Fig. 3). These comprise a Western Flank, which has a westerly structural dip, and two zones of rotated fault blocks dipping westwards. These fault blocks, designated the Central and Frontal Panels, are progressively downthrown eastwards.

In this area the generalized Brent Group succession (the Broom, Rannoch, Etive, Ness and Tarbert Formations; Deegan & Scull 1977) represents an extensive Bajocian–Bathonian regressive sequence followed by a transition to more marine conditions.

In Alwyn South, the Ness Fm represents a Bajocian alluvial/deltaic aggregational sequence. It is argillaceous but with both sandy and coaly horizons. In 3/14a–7 and 3/14a–8 the Upper Ness Fm is predominantly sandy with 60–80% coarse to fine grained river channel fill sequences. Shales and coals make up the remaining 40–20%. An unconformable (?) transgressive surface separates the Ness Fm from the overlying Tarbert Fm. The Tarbert Fm is essentially sandy, comprising basal littoral sandstones grading to micaceous shore-face sands. This represents the transition between the alluvial or deltaic Ness and the marine Heather Fm above.

In the Frontal Panel wells 3/14a–7 and 3/14a–8 and 3/14a–11 a further unit distinct from the underlying Tarbert Fm is found. The Upper Massive Sands (UMS) is a progradational fan delta sequence deposited with a coarse detrital input from the Ness and Tarbert Fms eroded at fault block crests to the west (Johnson & Eyssautier 1987). The unit is dominated by >70 m of medium to coarse grained massive sandstones overlain by a transitional sequence of shaly sand and silty shales. A transgressive surface separates the UMS from the overlying Heather Fm.

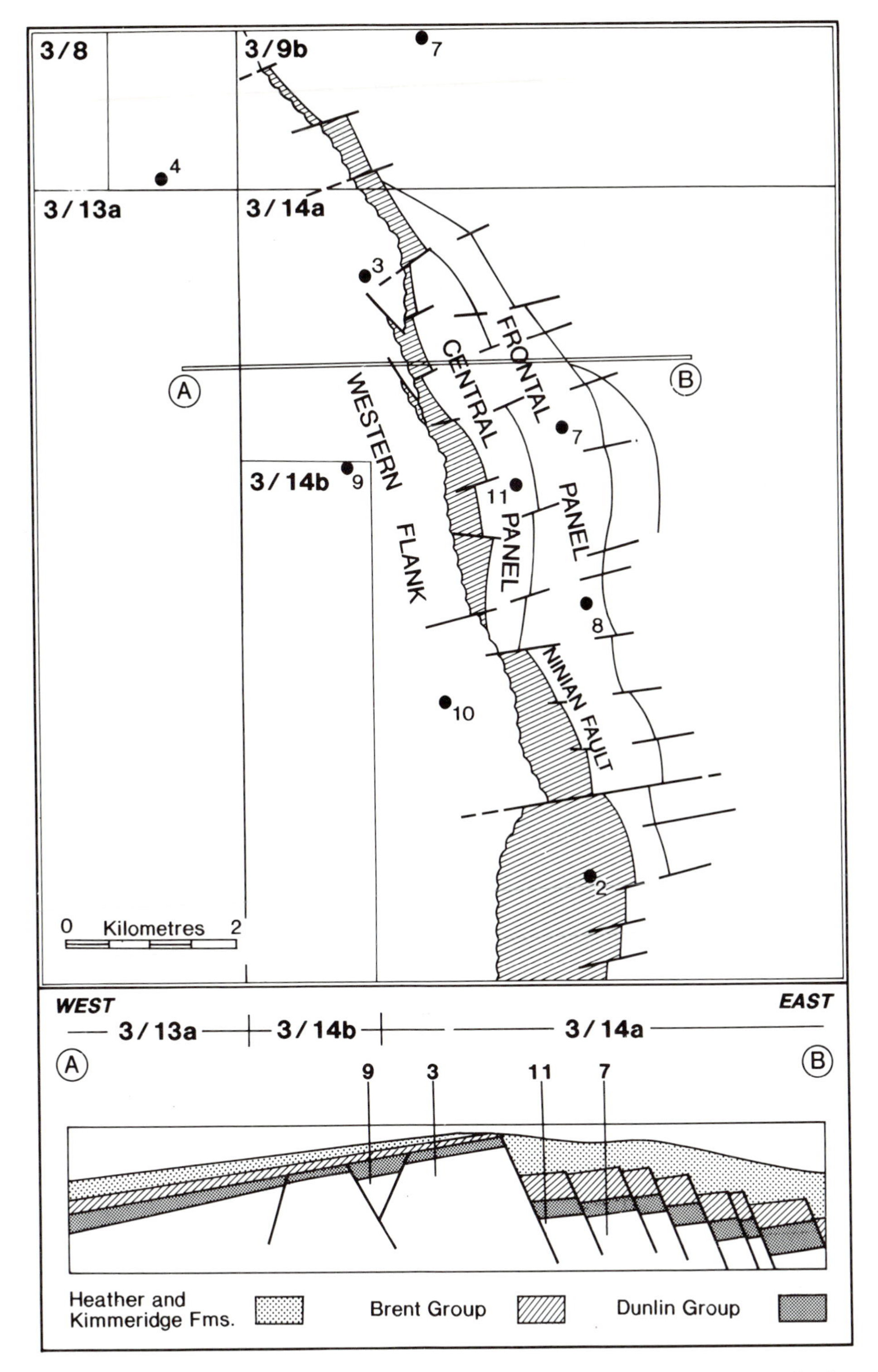


**Fig. 3.** Structural elements at base Cretaceous and locations of wells in Alwyn South. Top Brent Gp absent in hachured areas.

## Petrography

In Alwyn South the Brent Group sands form well to moderately sorted, upper fine to lower coarse grain sized sandstones. Originally deposited as quartz arenites and subarkoses the principal detrital components are now quartz and rock fragments (schists, cherts and quartzites). Detrital feldspar, initially more common, has undergone dissolution and now forms <1% of the mode.

The sandstones have undergone both mechanical and chemical compaction. Some mechanical compaction is indicated by deformation of micas but, in general, this was halted by quartz cementation.

Because compaction and cementation obscure original grain shapes, rounding is difficult to assess using optical petrography. However, where dust films allow detrital surfaces to be distinguished, grains are subangular to subrounded.

The sandstones are well cemented, in order of abundance, by illite, quartz and kaolinite. Pyrite, albite and ankerite occur as minor authigenic phases. Porosity includes both primary (interparticle) and secondary porosity after feldspar dissolution. This ranges from 3% in the UMS at 3731 m in 3/14a–7 to 24% in the Tarbert Fm at 3644 m in 3/14a–11.

### *Quartz overgrowths*

Syntaxial quartz overgrowths (Fig. 4) form 3–12% of the mode and are up to 150 μm in thickness. In these sandstones the overgrowths are generally distinguished by euhedral faces where they grow into voids and by thin lines of impurities (detrital clays, iron oxides and fluid inclusions) at overgrowth/detrital grain interfaces. In compacted zones, or where these impurities do not exist, the proportions of cement are likely to have been underestimated. Euhedral faces are separated by lamellar gaps in less well cemented sandstones while in others, faces are in contact and the pore-spaces are completely occluded.

In general, overgrowths are best developed in the Frontal Panel wells with abundances highest in 3/14a–8. The best reservoir properties occur in clean, moderately cemented sands with *c.* 5% authigenic silica e.g. Tarbert Fm and UMS of 3/14a–7 or the Upper Ness in 3/14a–8. In 3/14a–8, 36% estimated porosity at deposition has been reduced to 4% by: authigenic clay cement (8%), silica (8%) and compaction (16%) (from optical petrography).

The relatively open structure of the detrital grain framework, point contacts, and the absence of grain fracturing suggests that quartz cementation began relatively early in the burial history of the sands. Taking relatively clean sands in the Tarbert Fm and UMS in all four wells, and using the sandstone compaction curve of Baldwin & Butler (1985), average minus-cement porosities of 30% (assuming all clays are authigenic) suggest that quartz cementation began at about 1850 m burial. Geohistory curves indicate that the Brent in this area was at these depths at about the early Palaeocene.

Chemical compaction operated when mechanical compaction ceased and microstylolites are found where these sandstones contain clay laminae. In clean sandstones, pressure solution features observed include grain dissolution contacts, concave–convex grain contacts and microstylolites. Both detrital and authigenic quartz are affected. Sutures on relict feldspars and uncollapsed secondary pores after feldspar indicate that chemical compaction preceded their dissolution.

The processes may have liberated silica for transport and precipitation elsewhere in the system but, since much of the quartz cementation occurred prior to chemical compaction, they are not considered to be the major source of silica.

### *Quartz crystallites*

Small (>50 μm) euhedral quartz crystallites are found in secondary pores after feldspar dissolution (Fig. 4C), in authigenic clay masses (SEM evidence) or on overgrowths in primary pores. Modal proportions are low (≪1%) and do not influence reservoir properties. This phase is considered to be a late stage feature.

## Cathodoluminescence petrography

CL observations on Brent sandstones from the 3/14 area show a clear contrast between detrital and authigenic quartz (Figs 5–8). Detrital quartz is dark, whilst authigenic quartz generally ranges from white/light grey to dark grey.

Cemented grains have euhedral to subhedral grain boundaries whereas, under CL, the detrital grains are angular to subangular or subrounded. The original detrital framework was relatively open and in CL point-grain contacts are common. However some compaction precementation is indicated by interpenetrating point contacts. Some grains with darker angular fragments in a medium light grey structureless matrix suggest reworked quartz breccia or in situ annealing after compactional fracturing

**Fig. 4.** (**A**) 3/14a–11, 3638.25 m, Tarbert Fm. Moderately sorted, medium-grained sandstone. Moderate quartz cementation supports detrital quartz framework partially reducing well-connected primary porosity (a). Compare (B). Clay impurities and fluid inclusions (b) define the overgrowth/detrital grain boundaries. 19% optical porosity.

(**B**) 3/14a–8, 3731 m, UMS. Moderately sorted, medium-grained sandstone. Quartz cement partially (c) or completely (d) occludes primary pores. 3% optical porosity.

(**C**) 3/14a–7, 3753.5 m, Upper Ness Fm. Secondary pore (e) after feldspar (?) with clay film (f) at original outline. Compromise boundary (g) terminating against relict grain implies quartz cementation predates dissolution. Quartz crystallites (h) indicate post dissolution, cementation also post-dating enclosed clay.

(**D**) 3/14a–8, 3732 m, UMS. Liquid–vapour fluid inclusions (i) and impurities at or close to detrital quartz (j) boundary. Clean overgrowths occlude porespace and enclose authigenic clay (at trilete junction).

(Fig. 7E,F). Compaction continued post-cementation and, in CL, stylolites cross-cut both detrital grains and cement (e.g. 3/14a–11, 3644 m).

More compaction in the deepest well, 3/14a–8, is indicated under CL by a less open detrital grain framework with fewer point contacts and more planar and concavo-convex sutured grain boundaries. However, stylolites cross-cutting both detrital grain and overgrowth indicate chemical compaction continued after cementation.

A.

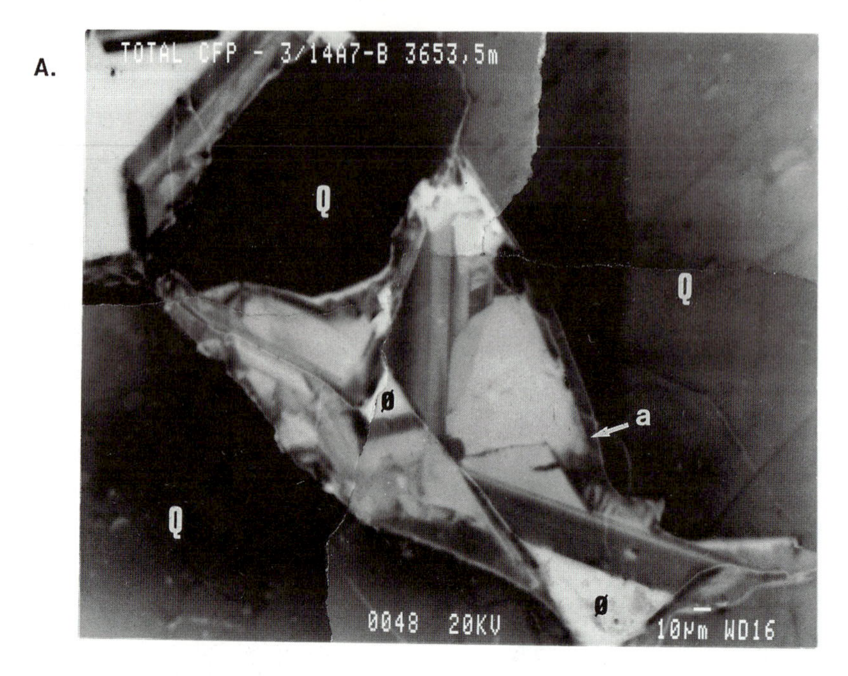


B.

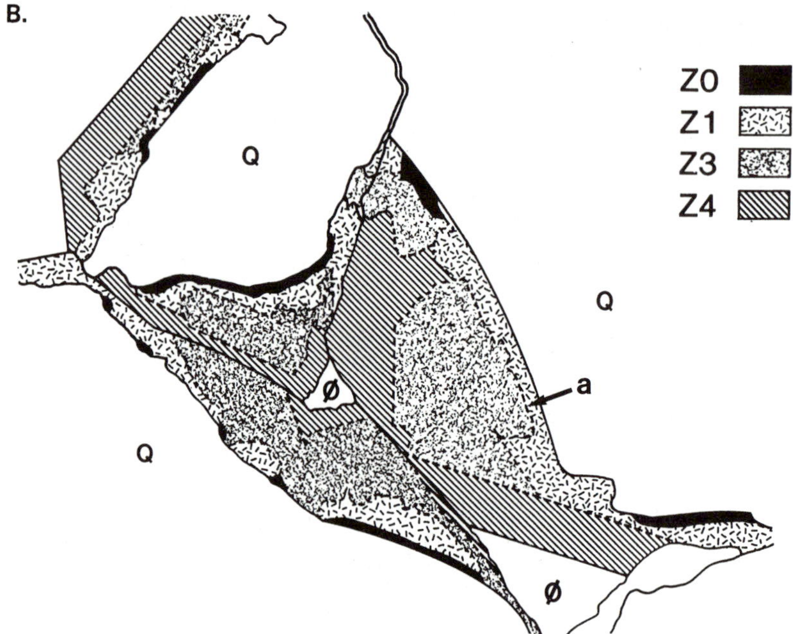


**Fig. 5.** 3/14a–7, 3653.5 m, UMS. Photomosaic of CL images (**A**) with diagram of same area showing the sequence of zones (**B**). This *negative* CL image shows dark to bright multiphase quartz overgrowths on dark subangular detrital quartz (Q). Four CL zones are defined; Z0, a thin bright zone adjacent to the detrital grain surface; Z1, thin with a dark heterogeneous or turbid luminescence with an irregular dissolution boundary at a; Z3, a thick bright/moderate zone with euhedral edges; Z4, multiphase zone with oscillatory dark/medium grey subzones with euhedral terminations. The CL zones become progressively well ordered and euhedral from core to rim. $\phi$ = porosity.

A

B

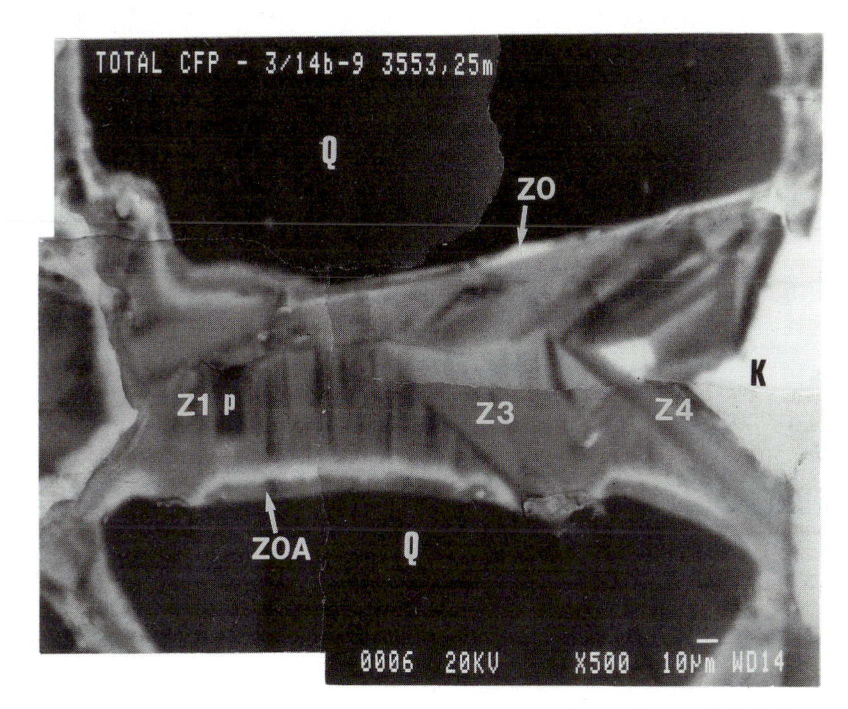


**Fig. 6.** (**A**) 3/14b–9, 3553.25, Tarbert Fm. Photomosaic backscatter electron image. Euhedral overgrowths meet at a compromise boundary (a). Arrow b shows evacuated fluid inclusion positioned at the grain-overgrowth boundary.

(**B**) Photomosaic CL image of same area as (A). Z0, a thin bright zone is adjacent to the detrital grain (Q, dark grey) or encloses an extra zone Z0A, which is medium grey and occurs between Z0 and the host grain. Z0 does not necessarily correspond with impurities on the detrital grain surface (compare backscatter image). Z1 at p has a perpendicular oscillatory (palisade) luminescence texture. This is succeeded by Z3, medium grey, poorly structured and Z4, a multiphase zone with oscillatory dark/medium grey sub-zones with euhedral faces. K: kaolinite.

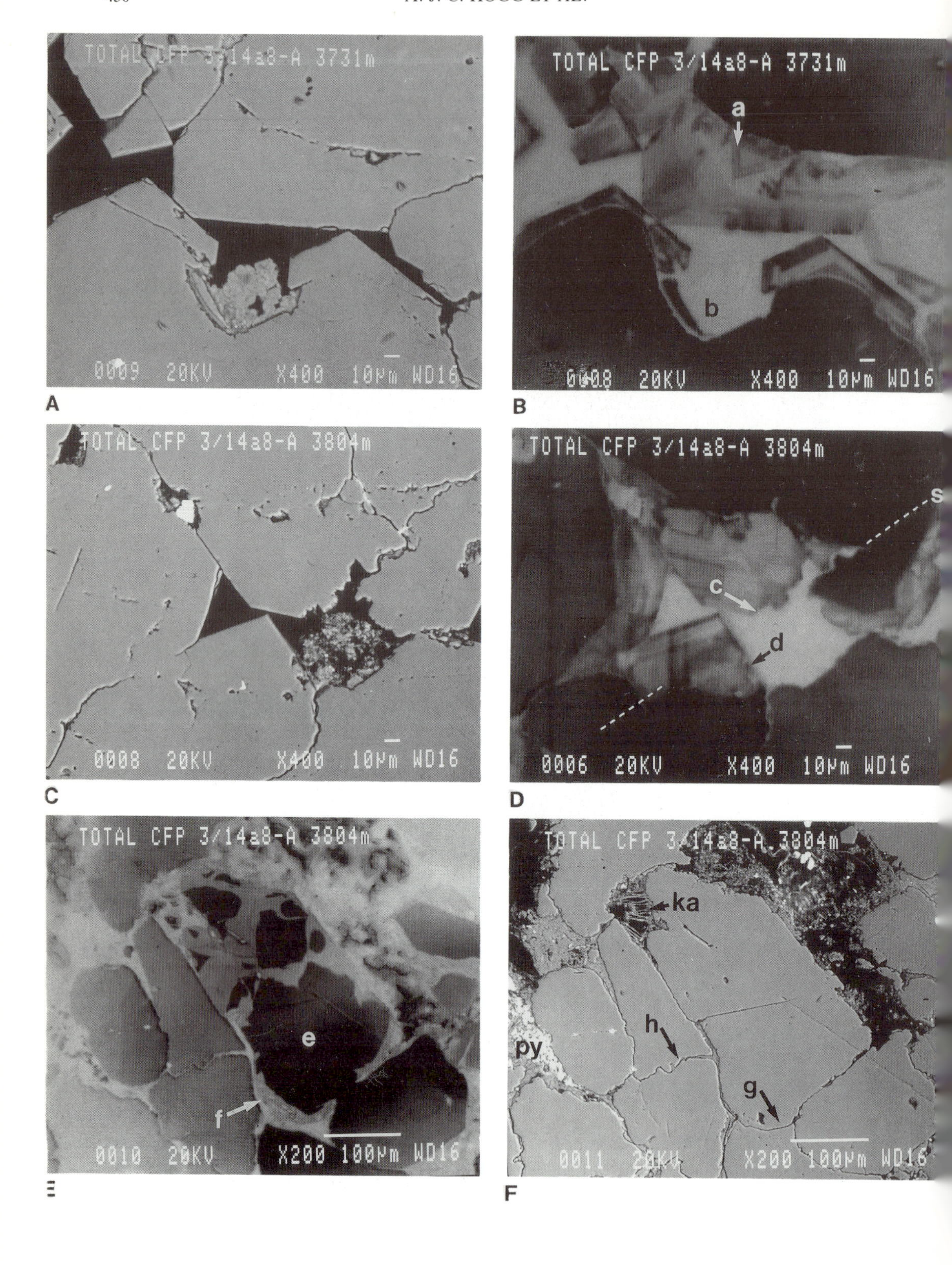

TOTAL CFP 3/14a8-A 3731m
0009 20KV X400 10µm WD16
A
TOTAL CFP 3/14a8-A 3731m
a
b
0008 20KV X400 10µm WD16
B
TOTAL CFP 3/14a8-A 3804m
0008 20KV X400 10µm WD16
C
TOTAL CFP 3/14a8-A 3804m
s
c
d
0006 20KV X400 10µm WD16
D
TOTAL CFP 3/14a8-A 3804m
e
f
0010 20KV X200 100µm WD16
E
TOTAL CFP 3/14a8-A 3804m
ka
h
py
g
0011 20KV X200 100µm WD16
F

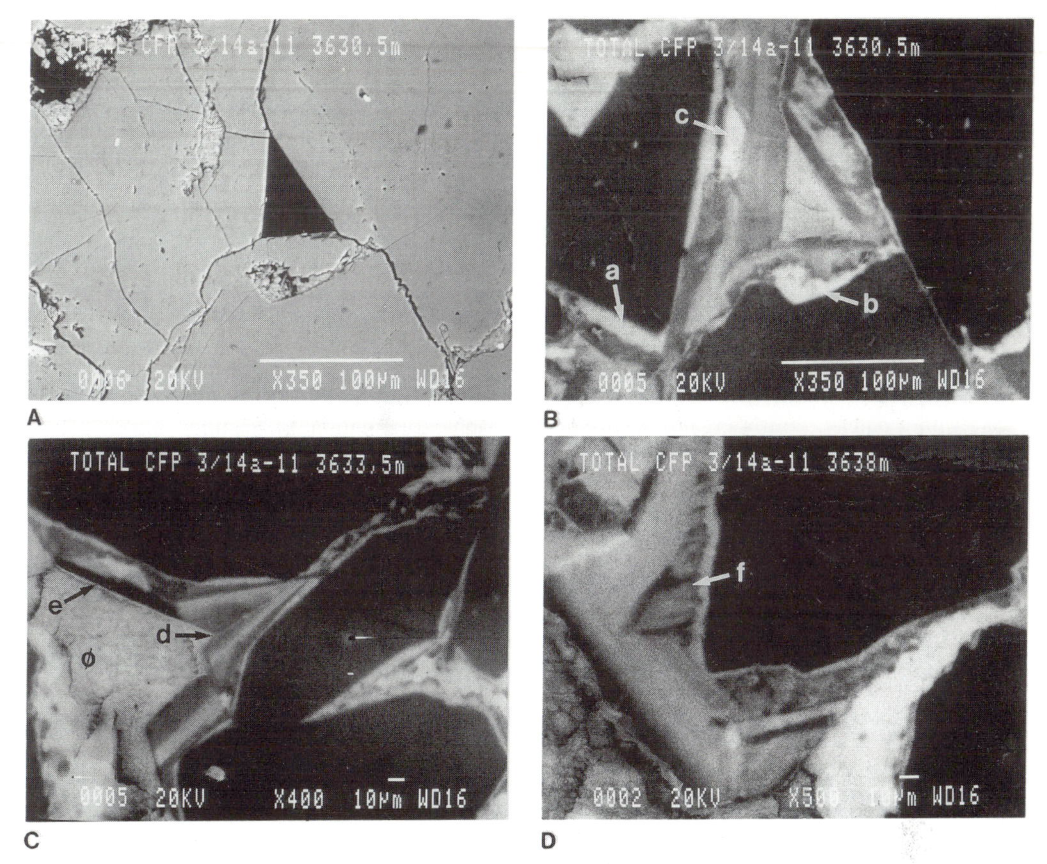


**Fig. 8.** (**A**) and (**B**) 3630.5 m, 3/14a–11, Tarbert Fm. Backscatter and CL image pair. Polyphase overgrowths partially occlude primary pore space. Zones become increasingly euhedral from core to rim. Z0, is well developed (a) and does not always correspond with clay inclusions (compare backscatter). Authigenic/detrital grains included next to host grain (b). Clays within overgrowths (c) (between Z1 and Z4) are authigenic.
(**C**) 3/14a–11, 3633.5 m, Tarbert Fm. CL image. Discrete, zoned overgrowths form a mirror pair along compromise boundary (d). Z0 is relatively continuous. Z5 forms dark zone (e).
(**D**) 3/14a–11, 3638 m, Tarbert Fm. CL image. Z1 forms fibrous or palisade texture (f).

**Fig. 7.**(**A**) and (**B**) 3/14a–8, 3731, UMS. Backscatter and CL image pair. Euhedral multiphase quartz overgrowths on rounded detrital quartz grains. Zone sequence suggests a mode of development for quartz overgrowths. An early turbid, poorly crystallized zone precipitated adjacent to the host grain (Z1). This was succeeded by small (*c.* 15 $\mu$m) quartz crystallites (Z2) (a) and enveloped by a medium grey euhedral zone, Z3, which presumably nucleated on precursor crystallites growing in favourable crystallographic orientation. Finally a thicker euhedral sequence of euhedral oscillatory zoned quartz was deposited (Z4). Clay impeded overgrowth development at b.
(**C**) and (**D**) 3/14a–8, 3804, Tarbert Fm. Backscatter and CL image pair. Euhedral multiphase quartz overgrowths on dark detrital quartz. Z2 crystallites occur at grain boundary of grain at top field of view. Pressure solution plane (c) cross cuts both overgrowth and detrital planes and zoning in overgrowth is disrupted. Disruption of overgrowth zones by incipient pressure solution at d.
(**E**) and (**F**) 3/14a–8, 3804, Tarbert Fm. Backscatter and CL image pair. Rock fragment (e) with structureless, lighter, quartz cement from first cycle lithification. Clast in fragment hosts overgrowth at (f,g). Concavo-convex boundaries with interstitial mica (f) and sutured grain contacts (h) indicate compaction post-cementation. ka, kaolinite; py, pyrite.

Under CL, overgrowths range from several microns to 125 μm in thickness. In 3/14b–9 and 3/14a–11, they range from 40–70 μm with individual overgrowths up to 105 μm; in 3/14a–7 they are up to 115 μm in thickness. Overgrowths are thickest in 3/14a–8 where they range from a few microns to 125 μm but are generally of the order of 90 μm thickness.

Quartz overgrowths have a complex, polyphase internal structure. A synthesis diagram of observed CL features is shown in Fig. 9. Four to five distinctive recurrent zones or groups of subzones, can be identified (Table 1). These include a thin white zone at the grain surface (designated Z0), followed by a medium grey zone with a patchy, poorly organized, internal structure (Z1), then white-intermediate grey zones, which are homogeneous (Z3) or with well ordered oscillatory subzones (Z2 and Z4/Z5) (Figs 5–8).

The CL sequences and zone thicknesses are shown for each well in Table 2. While slight variation in individual zones is expressed from well to well the zonal sequences are similar, both on the small scale, as in samples taken 0.5 m apart, and on the well scale. Except in very thin overgrowths, three major zones, Z1, Z3 and Z4/Z5, are present throughout. Minor zones, Z0 and Z2, are more sporadic. Zone boundaries may be euhedral, subhedral, or irregular and fretted.

No strong correlation is observed between facies and CL zone sequence. However, Z0 appears more likely to occur in the UMS (fan delta facies) of 3/14a–7 and 3/14a–8 and in the Tarbert Fm (fan delta facies) of 3/14a–11.

Authigenic clays (illite/kaolinite) are commonly intergrown with the silica cements. Under CL these clays are usually found next to the host grain or within, and between, zones Z0 and Z1 (Fig. 8A & B).

Complex intergrowth of zones, competition for growth space (compromise boundaries), zone dissolution and variation in crystallographic orientation of sections across zones make quantitative assessment of zones difficult. Some observations can however be made. Zone Z0 is the thinnest zone occurring against the host grain boundary. This zone is best developed in 3/14a–11 where it is up to 12 μm thick. Zone Z1 is irregular with fretted, irregular and anhedral to subhedral boundaries. These irregular boundaries may indicate dissolution events (Fig. 5). Z1 is never the final zone. Z2 forms quartz crystallites (*c*. 15 μm) with oscillatory subzones. Z3 is thick and well developed in 3/14a–7, 3/14a–11 and in 3/14a–8. Z4 is thickest in 3/14a–8.

**Table 1.** *Summary of CL zone characteristics*

| Zone | Sub-zone | Mean max.* thickness (μm) | Description | Observations |
|---|---|---|---|---|
| Z0 | Z0A | 7 | Thin, patchy uniform grey, adjacent to host grain | Rare, 3/14b–9, 3553.25, (69.6) |
| | Z0 | 6 | Thin, bright irregular selvage, | Usually in UMS/Tarbert Fm |
| Z1 | Z1A | 16 | Thin, intermediate/dark, diffuse oscillatory subzones, euhedral | Only in 3/14a–8, 3830 |
| | Z1 | 30 | Thick, intermediate/dark heterogeneous, turbid/patchy, irregular/subhedral boundaries | Fluid inclusions, kaolinite, dissolution extensive, palisade texture |
| Z2 | | 18 | Well-organized euhedral crystallites, dark/moderate diffuse subzones | Minor zone, 3/14a–7, 3/14a–8 |
| Z3 | | 31 | Thick, bright/moderate, homogeneous/gradational bright to dark, euhedral | Usually present |
| Z4 | | 29 | Multiphase oscillatory dark/moderate or bright thick, euhedral | Usually present, more diffuse in 3/14a–11 sector zones |
| Z5 | | 10 | Thin, dark, euhedral | Rare, dissolution may grade from Z4 to Z5 in 3/14a–11 |

* Average maximum thickness of the zone in grains where it is observed.

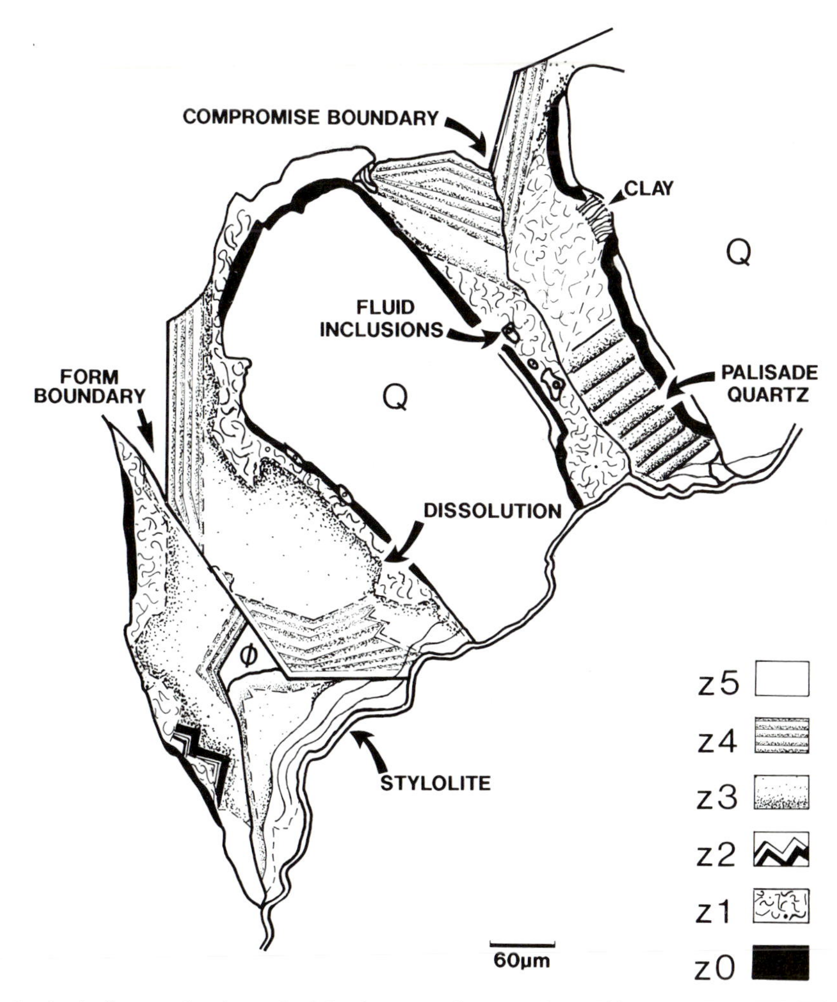


**Fig. 9.** Synthetic diagram showing cathodoluminescence features observed in quartz overgrowths from Alwyn South. Z0–Z5 represents the sequence of CL zones observed in the quartz cements. These are defined from the CL images, which are in negative format, by a combination of tonal and textural characteristics. Z0 is a thin white zone adjacent to the detrital quartz (Q). Z1 is turbid, poorly crystallized with generally irregular, anhedral, upper boundaries. Occasionally this zone has a fibrous oscillatory luminescence texture perpendicular to the growth direction of the other zones (palisade quartz). Z2 forms quartz crystallites with oscillatory subzones. Overlying in Z3, a thick, generally light, homogeneous or gradational CL zone. Z4 is a euhedral multiphase zone with oscillatory dark/medium grey subzones. The final zone, Z5 cross-cuts Z4 and is thin, dark and euhedral. Zones are disrupted where they are cross-cut by stylolites.

Some variation in overgrowth thickness (measured normal to the detrital grain surface) is recognized. In 3/14a–7 overgrowths thicken slightly down well. Overall the overgrowths are thickest in the deep 3/14a–8 well, where thickness variations correlate with thickening in zones Z3 and Z4. Where overgrowths are thin several of the zones may be absent. Whether this represents non-deposition or dissolution of the interval is not clear.

Internal structures vary between wells. In 3/14b–9 Z1 cements are well developed, up to 50 $\mu$m thick, moderate grey and euhedral to subhedral. They are poorly organized, with indistinct oscillatory and thick uniform subzones. However, in one sample palisade quartz (a zonal sequence comprising thin (4 $\mu$m) alternating dark and medium grey luminescent subzones perpendicular to the detrital grain) has been observed (Fig. 6). Palisade quartz has also been

**Table 2.** *CL zone sequences: Alwyn South wells*

| Depth (m) | Zone sequence | Mean max. thickness ($\mu$m) | Remarks | Fluid inclusions $x\ T_{L\text{-}V}$ (°C)* |
|---|---|---|---|---|
| *13/4a–7* | | | | |
| 3635.50 | Z0 | 6 | | 115.1 ± 5.9 ($n$ = 15) |
| | Z1 | 26 | | |
| | Dissolution | | | |
| | Z2 | – | | |
| | Z3 | 29 | | |
| | Z4 | 26 | | |
| | Dissolution | | | |
| 3653.50 | Z0 | 4 | | |
| | Z1 | 30 | | |
| | Dissolution | | | |
| | Z3 | 47 | | |
| | Dissolution | | | |
| | Z4 | 45 | | |
| 3697 | Z0 | 9 | Rare | 104.7 ± 1.8 ($n$ = 8) |
| | Z1 | 38 | | |
| | Z3 | 49 | | |
| | Z4 | 27 | | |
| 3697.25 | Z0 | 2.5 | Rare | |
| | Z1 | 31 | | |
| | Z2 | 20 | Rare | |
| | Z3 | 34 | | |
| | Z4 | 18 | Usually present | |
| 3753.50 | Z1 | 23 | | |
| | Dissolution? | | | |
| | Z2 | 20 | Sometimes present | |
| | Dissolution? | | | |
| | Z3 | 24 | | |
| | Z4 | 23 | | |
| | Dissolution | | | |
| | Z5 | 16 | Rare | |
| 3731 | Z0 | 3 | Sometimes absent | |
| | Kaolinite | | | |
| | Z1 | 41 | | |
| | Dissolution | | | |
| | Z2 | 23 | Zones not well differentiated | |
| | Dissolution | | | |
| | Z3 | 37 | | |
| | Z4 | 48 | | |
| | Z5 | 21 | | |
| *3/14a–8* | | | | |
| 3731.25 | Z0A | – | Thin grey zone | |
| | Z0 | 5 | | |
| | Kaolinite | | | |
| | Z1 | 26 | | |
| | Z2 | 10 | Rare | |
| | Z3 | 28 | | |
| | Z4 | 24 | | |
| | Dissolution | | | |
| | Z5 | 6 | Rare | |
| 3759 | Kaolinite | | | |
| | Z0 | 3 | Rare | |
| | Z1 | 28 | Euhedral | |
| | Dissolution | | | |
| | Z3 | 39 | | |
| | Z4 | 40 | | |
| | Z5 | 11 | Rare | |

| | | | | |
|---|---|---|---|---|
| 3759.25 | Z0 | 4 | | |
| | Z1 | 37 | | |
| | Kaolinite | | | |
| | Z3 | 21 | | |
| | Z4 | 43 | | |
| | Dissolution | | | |
| 3804 | Z1 | 21 | | |
| | Z3 (?) | 25 | Not clear | |
| | Z4 | 36 | | |
| 3830 | Z0 | 2 | Rare | |
| | Z1A | 16 | | |
| | Z1 | 22 | | |
| | Kaolinite | | | |
| | Z3 | 25 | | |
| | Z4 | 14 | | |
| | Z5 | 6 | Rare | |
| *3/14b–9* | | | | |
| 3553.25 | Z0A | 7 | Grey | 103.2 ± 6.0 ($n$ = 25) |
| | Z0 | 9 | | |
| | Z1 | 54 | Palisadal | |
| | Z4 | 45 | | |
| *3/14a–11* | | | | |
| 3630.50 | Z0 | 7 | | |
| | Kaolinite | | Prominent | |
| | Z1 | 27 | | |
| | Kaolinite | | | |
| | Z3 | 31 | | |
| | Z4 | 27 | | |
| | Z5 (?) | 5 | Poorly zoned<br>Grades from Z4 | |
| 3632 | Kaolinite | | | |
| | Z0 | 8 | Prominent | |
| | Z1 | 32 | | |
| | Kaolinite | | | |
| | Z3 | 26 | | |
| | Z4/Z5 | 20 | | |
| | Z5 | 4 | Rare | |
| 3633.50 | Z0 | 7 | | |
| | Z1 | 22 | | |
| | Kaolinite | | | |
| | Z3 | 20 | | |
| | Z4/Z5 | 24 | | |
| | Z5 (?) | 10 | | |
| 3638 | Z0 | 5 | | |
| | Kaolinite | | | |
| | Z1 | 27 | | |
| | Kaolinite | | | |
| | Z3 | 41 | | |
| | Z4/Z5 | 16 | | |
| 3644 | Z1 | 23 | | |
| | Z2 | 28 | Rare or may | |
| | Z3 | 24 | be absent | |
| | Z4/Z5 | 15 | | |

* Mean homogenization temperatures for aqueous liquid–vapour fluid inclusions

observed in Z1 in 3/14a–11 (Fig. 8D). Sector zoning, not previously reported in quartz overgrowths, has been distinguished in several samples.

With some minor differences, the zone sequences are similar in each of the wells. Zonal sequences in 3/14a–8 are in general more complex than in the two other wells. Z4 in 3/14a–7 and 3/14a–8 is subzoned with oscillatory subzones usually terminating in a dark zone which in some cases is distinct enough to be assigned as zone Z5. In contrast, subzoning in Z4 in 3/14a–11 is indistinct and the zone is homogeneous or gradational becoming darker towards the outer boundary. Zone sequences are, however, broadly similar with some variations in CL texture or zone thickness.

In general, the CL sequence shows a progression towards a more organised growth pattern from anhedral towards increasingly euhedral crystal forms in the outermost zones.

## Fluid inclusions

Liquid–vapour fluid inclusions are commonly trapped within quartz overgrowths. In these samples inclusions are confined exclusively to early zones at, or close to, the detrital grain boundary. Lines of pits or cavities after evacuated fluid inclusions are, in CL, next to the detrital grain boundary, and correlate with Z0. However, in some cases large inclusions appear to have impeded Z0 development and are capped by later cements (Fig. 6). In addition, inclusions in 3/14b–9 are up to 20 $\mu$m diameter and are too large to be trapped by a 0–16 $\mu$m Z0, but are sealed by Z1. Thus, in general fluid inclusions were trapped syn-Z0 but also in early Z1, interstitial to or against Z0.

Homogenization temperatures (Table 2) of fluid inclusions for those samples examined under CL range from 103–115.1°C with a mean of 107.2 ± 7.7 (1 $\sigma$ standard deviation) for 48 inclusions. Temperatures of fusion of ice during freezing experiments indicate aqueous fluid salinities of between 2–4 wt% equivalent NaCl.

## Discussion

The CL images are in negative format and indicate that, in these Brent Group sandstones, the detrital grain cores are more brightly luminescent under SEM CL that their overgrowths. This is consistent with hot-cathode CL work by Zinkernagel (1978) and Houseknecht (1987) showing that quartz cements are non- or poorly luminescent compared with brighter luminescing detrital quartz.

The origin of quartz CL has been discussed by Walker (1985), Marshall (1988) and Ramseyer *et al.* (1988, 1989). Although the controls on its luminescence are not fully resolved, it appears that quartz CL originates in either, or some combination of, intrinsic lattice defect centres or lattice impurities such as transition metals or $OH^-$ groups.

Grant & White (1978) suggested that variation in Al content may be important in controlling quartz CL. Microprobe analysis of quartz overgrowths displaying CL zonations showed that Al concentrations were of the order of 700 ppm in the overgrowth as compared with 200 ppm in the host grain (Henry & Toney 1987) and CL intensity was found to vary inversely with Al content in the overgrowth. Where $Al^{+3}$ substitution for $Si^{+4}$ occurs charge compensating $H^+$ or monovalent alkali elements may act as activation centres (Mitchell & Denure 1973; Walker 1985). Luminescence character may also originate from lattice distortions due to the difference in the ionic radii of Al and Si. The disrupted appearance of CL zones adjacent to stylolites (e.g. Fig. 7) may also suggest that lattice ordering is a control on luminescence texture.

Zoning in overgrowths is therefore likely to depend on some combination of growth rate (controlled by temperature, kinetics and pore fluid composition) and the uptake of lattice impurities.

### *Zone Z0*

Zone 0 is a thin (3–9 $\mu$m), structureless bright (non-luminescent) zone adjacent to the host grain. Some bright areas within overgrowths clearly originate from authigenic and detrital clays included at, or close to, the detrital grain boundaries (Fig. 8A & B).

Iron and other impurities, common on the surfaces of detrital quartz grains may, if they are incorporated into the lattice in this early zone, inhibit luminescence and account for the bright appearance of Z0 in these CL negatives. Chemical analysis of the walls of fluid inclusions from the Alwyn area using laser mass spectrometry show that silica in adjacent quartz overgrowths is hydrated and, in addition to hydrocarbons, contains trace potassium and phosphorus (Malley *et al.* 1986). These latter two may also affect Z0 CL behaviour.

Fluid inclusions, which range from 1–25 $\mu$m diameter, are trapped syn-Z0 but also by the early Z1 zone. Trapping temperatures for fluid inclusions in those samples examined in CL range from between 103–115°C (mean 107°C).

These diagenetically high temperatures indicate that, unless recrystallization has occurred, these Z0 cements are not of early diagenetic origin and are unlikely to be associated with influxes of meteoric waters (salinities are between 2–4% eq. NaCl). Therefore, unless Z0 cements have been recrystallized since formation, fluid inclusion evidence precludes a synsedimentary or meteoric origin for this zone. Z0 cements therefore represent the initial stages of the latest, major, silicification phase described by Jourdan *et al.* (1987).

### *Zone Z1*

Fluids inclusions are trapped in Z0 but also in early Z1. Following the discussion above this suggests that Z1 and later cements are also related to last major phase of quartz cementation.

Fluid inclusion data shows that quartz precipitation took place from fluids at *c.* 110°C. If the cement is the product of the influx of hot (as great as 140°C in 3/14a–8) compactional waters via faults into the cooler sandstone body (Jourdan *et al.* 1987) the consequent drop in silica solubility will result in rapid, poorly organized, crystallization. This will integrate CL activators and quenchers and entrap fluid inclusions into the cement structure, which could account for Z1 CL texture seen in Figs 5, 7B, 8B & D. This poorly organized structure will be prone to dissolution and recrystallization. Evidence for dissolution (fretted and irregular grain boundaries) is observed in Z1 (e.g. Fig. 5).

Palisade quartz has been observed in two samples forming in Z1 at high angles to the grain surface. This form of quartz is likely to form parallel to the crystallographic *c.* axis; however the mechanism by which it forms is uncertain. The Z1 texture shown in Fig. 8D is similar (though at different scales) to that observed by Ramseyer *et al.* (1988) in slowly grown fissure quartz. We would however prefer an alternative interpretation that this texture represents rapid precipitation, or was developed during recrystallization of rapidly deposited silica.

### *Zone Z2*

These cements are infrequent and reflect subhedral to euhedral crystallites or outgrowths seeded onto suitable nucleation sites on the host grains or Z1 surfaces (Fig. 7B, D). Indistinct, oscillatory subzones suggest that these cements were precipitated in several phases.

### *Zone Z3*

Relatively thick and homogeneous, this CL zone may indicate a steady state precipitation of silica. The gradation in tone from light at the Z1 contact through medium grey at the upper surface suggests that rates of precipitation changed towards the outer euhedral rim of the zone.

### *Zone Z4, Z5*

Oscillatory subzones in Z4 may originate from precipitation in several pulses from silica-rich fluids. Each subzone grades from light to darker tones at the upper surface. The changing luminescence may indicate changing rates of precipitation. A hiatus is then followed by renewed precipitation. Finally, overgrowth formation is terminated by Z5, a variably thick, well-organized euhedral zone probably precipitated slowly as pore fluids are depleted in silica or flux rate slows.

### *General discussion*

Whether CL zonations represent fluid pulses introduced from outside the reservoir or closed-system circulation is unclear. However, the homogeneity of Z4/Z5 in 3/14a–11 (Central Panel) compared with the complexity of Z4 zoning in 3/14a–7 and 3/14a–8 suggests a more dynamic fluid regime in the two Frontal Panel wells. The 3/14a–8 well which is faulted, intensely cemented and more complexly zoned was considered by Jourdan *et al.* (1987) to be point of entry of the hot compactional fluids (Fig. 10). Multiple zones within quartz cements may represent repeated injections of silica bearing fluids from depth along faults through some mechanism such as seismic pumping described by Sibson (1981). Burley *et al.* (1989) also favoured this process for cyclical cementations and multi-phase CL zonation in the upper Jurassic Piper Sandstone of the Tartan field. In Alwyn South, the relative complexity of CL textures in 3/14a–8 relative to shallower up-flank and crestal wells may be due to thermal and compositional homogenization of the incoming hot basinal fluids with existing reservoir waters. In this case we would expect homogenization to be more advanced resulting in slower and more continuous cement growth in panels furthest from the point of entry of the cementing fluids.

Overall, the zonal sequence shows a progressive ordering of the silica cements. Z0 and Z1 are anhedral and irregular, Z3 subhedral to euhedral and Z4/Z5 are euhedral. Increasing

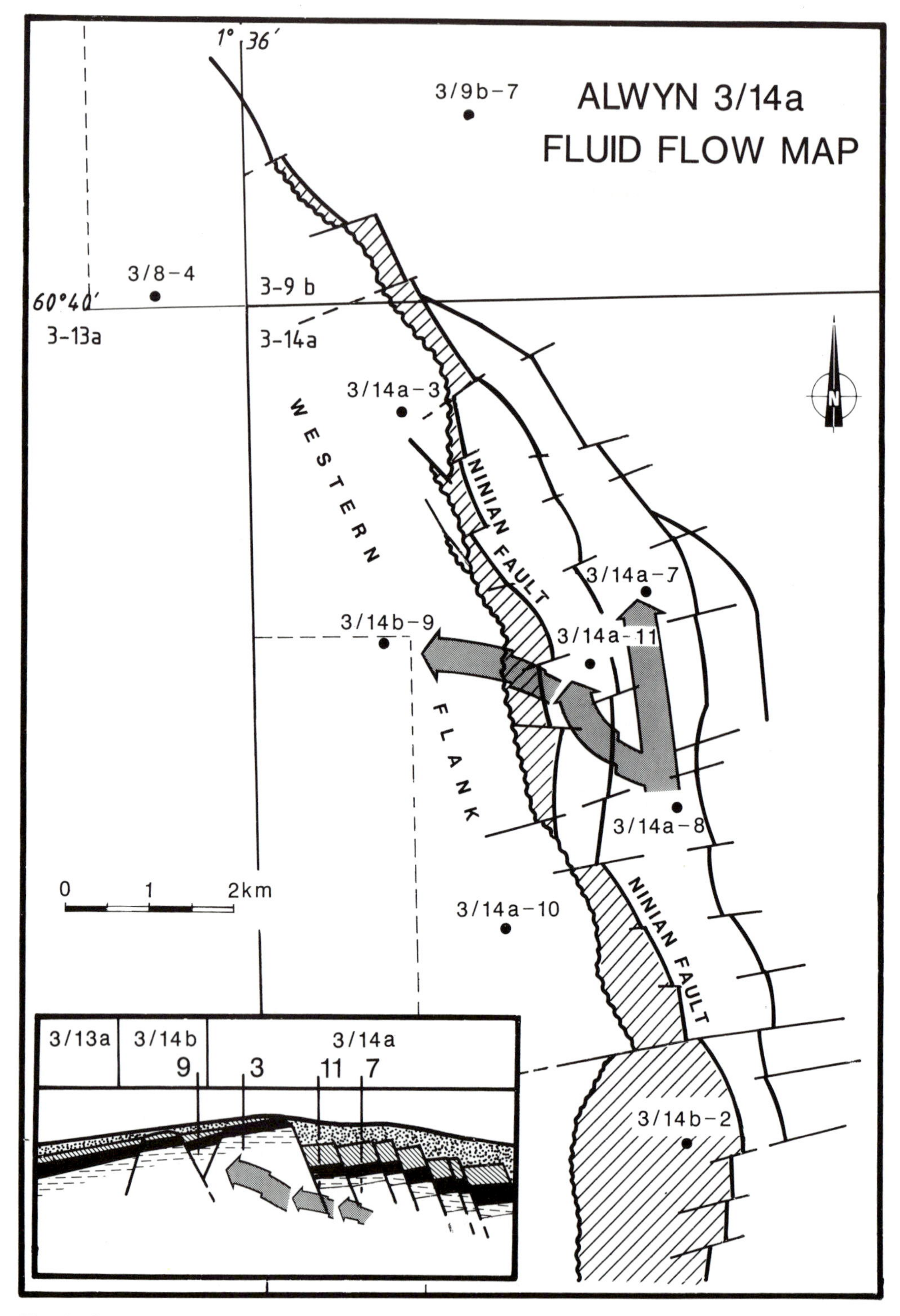


**Fig. 10.** Structural map showing direction of fluid-flow during quartz cementation of the Alwyn South reservoirs. The flow pattern is supported by K-Ar dating of illite coexisting with the quartz cements (Hogg 1989).

organization may record a genetic evolution during overgrowth formation. This increasing organization may reflect either or both of: (i) nucleation problems at the start of overgrowth development or (ii) competition with other minerals such as illite for growth space. Problems with nucleation will be exacerbated where precipitation has taken place rapidly. We have recognized similar patterns of SEM CL zone development in quartz overgrowth in other reservoir sandstones both from the North Sea and elsewhere.

On the pore scale, heterogeneous development of cement zones and sporadic dissolution zones suggest that quartz overgrowth development may be a flux-controlled process.

## Conclusions

In the quartz arenites/subarkoses of the Alwyn South Brent Group severe deterioration of normally good reservoir properties is primarily the result of quartz cement precipitation overlapped by later illite formation. Sandstone from the Upper Ness, Tarbert Fms and the UMS in wells 3/14a–7, 3/14a–8, 3/14b–9 and 3/14a–11 illustrate the potential of CL in unravelling the reservoir cementation history.

Under CL, detrital and authigenic quartz are clearly distinct and the detrital framework of these sandstones is relatively open with some point contacts. An average minus cement porosity of 30% (from optical data) suggests that cementation took place at approximately 1850 m depth; that is about the early Palaeocene. Compaction continued post-cementation and in both transmitted light and CL stylolites cross-cut detrital grains and cements.

In SEM CL the quartz overgrowths display complex multiphase zonations. Three recurring major and two minor CL zones have been recognized in the four wells. A thin early (Z0) irregular silica phase associated with detrital clay and authigenic kaolinite and containing fluid inclusions corresponds to an early phase of cementation. Fluid inclusion homogenization temperatures of *c*. 107°C suggest that Z0 may represent an early phase of the main cementation event. Z1 is a thick irregular/subhedral microcrystalline cement overlying the detrital grain or Z0. This poorly organized cement may be related to rapid silica precipitation during the initial influx of hot silica bearing fluids into cooler reservoir sands, trapping fluid inclusions at relatively high temperatures. Steady-state precipitation of homogeneous Z3 cements followed. This was succeeded by Z4, a thick euhedral zone with oscillatory subzones indicating several pulses of cementation but together forming an important later event. Silica deposition gradually slowed with the precipitation of final, well-organized, Z5 cements.

Cumulatively, zones Z1 to Z5 show that quartz cementation occurred in several pulses of varying importance interrupted by periods of non-deposition. Zonal character is similar throughout the four wells but sequences are thicker and more complex in 3/14a–8 suggesting a more dynamic cementation regime in this deep Frontal Panel well. This well is cross-cut by a fault and may have been the point of entry of hot, silica-bearing fluids from deeper in the graben. Sequence variations suggest that quartz cementation is, on the pore scale, a flux-controlled process. The progression towards increasing organization from overgrowth core to rim may reflect nucleation effects and/or rapid crystallization during early overgrowth development. This might represent a general mode of formation for quartz overgrowths.

Cathodoluminescence microscopy has hitherto had limited application in clastic sedimentology because of the low intensity of quartz luminescence. However, where technical difficulties have been overcome SEM CL is a powerful tool both in general sandstone petrography and in understanding the evolution of sandstone reservoir properties during diagenesis.

This research, financially and materially supported by Total Oil Marine plc, was undertaken at Total-CFP Exploration Laboratories, Bordeaux and at the University of Aberdeen. Technical assistance from A. Houtmann, C. Palus, and V. Pradet is gratefully acknowledged. PhD work by A. J. C. H. at the University of Aberdeen was supervised by M. J. Pearson. The paper is published with the kind permission of Total Oil Marine plc and Elf (UK) plc.

## References

Baldwin, B. & Butler, C. O. 1985. Compaction curves. *American Association of Geologists Bulletin*, **69**, 622–266.

Burley, S. D., Kantorowicz, J. D. & Waugh, B. 1985. Clastic Diagenesis *In*: Brenchley, P. J. & Williams, B. P. J. (eds) *Sedimentology: Recent development and applied aspects*. Geological Society, London, Special Publication, **18**, 189–226.

Burley, S. D., Mullis, J. & Matter, A. 1989. Timing diagenesis in the Tartan Reservoir (UK North Sea): Constraints from combined cathodoluminescence microscopy and fluid inclusion studies. *Marine and Petroleum Geology*, **6**, 98–120.

Deegan, C. E. & Scull, B. J. 1977. *A standard*

*lithostratigraphic nomenclature for the central and northern North Sea*. Report of the Institute of Geological Sciences, **77/25**.

GRANT, P. 1978. The role of the scanning electron microscope in cathodoluminescence petrology. *In*: WHALLEY, W. B. (ed.) *Scanning Electron Microscopy in the study of sediments*. Geo Abstracts, Norwich, 1–12.

GRANT, P. R. & WHITE, S. H. 1978. Cathodoluminescence and microstructure of quartz overgrowths and quartz. *In*: JOHARI, O. (ed.) *Scanning Electron Microscopy/1978*, **1**, Scanning Electron Microscopy Inc., Illinois, 789–794.

HENRY, D. J. & TONEY, J. B. 1987. Combined cathodoluminescence/backscattered electron imaging and trace element analysis with the electron microprobe: Applications to geological materials. *In*: GEISS, R. H. (ed.) *Microbeam Analysis – 1987*. San Francisco Press Inc., San Francisco, 339–342.

HOGG, A. J. C. 1989. *Petrographic and isotopic constraints on the diagenesis and reservoir properties of the Brent Group Sandstones, Alwyn South, Northern U.K. North Sea*. PhD dissertation, University of Aberdeen, UK. University Microfilms International Publication No. DX 89573.

HOUSEKNECHT, D. W. 1987. Assessing the relative importance of compaction processes and cementation to reduction of porosity in sandstones. *American Association of Petroleum Geologists Bulletin*, **71**, 633–642.

JOHNSON, A. & EYSSAUTIER, M. 1987. Alwyn North Field and its regional geological context. *In*: BROOKS, J. & GLENNIE, K. W. (eds) *Petroleum geology of North West Europe*. Graham & Trotman, London, 963–977.

JOURDAN, A., THOMAS, M., BREVART, O., ROBSON, P., SOMMER, F. & SULLIVAN, M. 1987. Diagenesis as the control of the Brent sandstone reservoir properties in the Greater Alwyn Area (East Shetland Basin). *In*: BROOKS, J. & GLENNIE, K. (eds), *Petroleum geology of North West Europe*, Graham & Trotman, London, 951–961.

MALLEY, P., JOURDAN, A. & WEBER, F. 1986. Etude des inclusions fluides dans les nourrissages siliceux des gres reservoirs de Mer du Nord: une nouvelle lecture possible de l'histoire diagenetique du Brent de la region d'Alwyn. *Comptes Rendus de L' Academie des Sciences de Paris*, **302**, Serie II, 653–658.

MARSHALL, D. J. 1988. *Cathodoluminescence of geological materials*. Unwin Hyman, Boston.

MEUNIER, J. D., SELLIER, E. & PAGEL, M. 1989. Radiation damage rims in quartz from uranium-bearing sandstones. *Journal of Sedimentary Petrology*, **60**, 53–58.

MILLER, J. 1988. Cathodoluminescence microscopy. *In*: TUCKER, M. (ed.) *Techniques in sedimentology*. Blackwell Scientific Publications, Oxford, 174–190.

MITCHELL, J. P. & DENURE, D. G. 1973. A study of Si layers on Si using cathodoluminescence spectra. *Solid-State Electronics*, **16**, 825–839.

NICKEL, E., 1978. The present status of cathodoluminescence as a tool in sedimentology. *Minerals Science and Engineering*, **10**, 73–100.

RAMSEYER, K., BAUMANN, J., MATTER, A. & MULLIS, J. 1988. Cathodoluminescence colours of α-quartz. *Mineralogical Magazine*, **52**, 669–677.

——, FISCHER, J. MATTER, A., EBERHARDT, P. & GEISS, J. 1989. A cathodoluminescence microscope for low intensity luminescence. *Journal of Sedimentary Petrology*, **59**, 619–622.

REMOND, G. 1977. Application of cathodoluminescence in mineralogy. *Journal of Luminescence*, **15**, 121–155.

——, KIMOTO, S. & OKUZUMI, H. 1970. Use of the SEM in cathodoluminescence observations in natural samples. *In*: O'HARE (ed.) *Scanning electron microscopy/1970, Proc. 3rd Annual Scanning Electron Microscope Symposium*. I.T.T. Research Institute, Chicago, 33–39.

SIBSON, R. H. 1981. Fluid flow accompanying faulting: field evidence and models. *In*: SIMPSON, D. W. & RICHARDS, P. G. (eds) *Earthquake prediction: An international review*. American Geophysical Union, Maurice Ewing Series, **4**, 593–603.

SUCHECKI, R. K. & BLOCH, S. 1988. Complex quartz overgrowths as revealed by microprobe cathodoluminescence (abstract). *American Association of Petroleum Geologists Bulletin*, **72**, 252.

WALKER, G. 1985. Mineralogical applications of luminescence techniques. *In*: BERRY, F. J. & VAUGHAN, D. J. (eds) *Chemical bonding and spectroscopy in mineral chemistry*. Chapman and Hall, London, 103–140.

ZINKERNAGEL, U. 1978. *Cathodoluminescence of quartz and its application to sandstone petrology*. Contributions to Sedimentology, **8**.

# Migration of petroleum into Brent Group reservoirs: some observations from the Gullfaks field, Tampen Spur area North Sea

STEPHEN LARTER[1] & IDAR HORSTAD[2]

[1] *Newcastle Research Group in Fossil Fuels & Environmental Geochemistry (NRG), Drummond Building, University of Newcastle, Newcastle upon Tyne NE1 7RU, UK*
[2] *Saga Petroleum a.s., Kjørboveien 16, Postboks 490, 1301 Sandvika, Norway*

**Abstract:** Whereas the processes of petroleum generation and primary migration are beginning to be understood from a quantitative viewpoint, the process of secondary migration is still incompletely quantitatively understood. We estimate that typically the carrier systems feeding Brent Group reservoirs contain an average of around 3% residual oil saturation, the petroleum flowing in high saturation petroleum rivers through the most permeable carrier bed zones. Most of the carrier porosity contains no petroleum. The filling of the Brent Group reservoir in the Gullfaks field (Tampen Spur) is elucidated by petroleum geochemistry and this study indicates that the field has filled from more than one source basin and has been biodegraded post oil-emplacement. The complex filling and degradation history suggests that geochemical study of reservoir petroleum columns is a useful exercise providing information on the fillpoints of these complex reservoir systems and potentially information on the palaeohydrogeology of the area.

It is generally agreed that the major source rocks for the medium and high gravity black oil accumulations found in Brent Group sandstones in the rotated fault block structures of the Viking Graben and East Shetland basin are the organic rich marine shales of the U. Jurassic Draupne or Kimmeridge Clay Fms., (Goff 1983; Cornford *et al.* 1986; Thomas *et al.* 1985). While of variable organic facies (Huc *et al.* 1985; Larter 1985), at maturities of around 1% $R_o$ vitrinite reflectance equivalent, the locally very rich (initial potential as high as 30 kg/tonne or more, Thompson *et al.* 1985) marine organic matter dominated clastic source rocks in the basinal areas may have expelled up to 80% of their generated oil (Cooles *et al.* 1986; Mackenzie *et al.* 1987) which may represent as much as 30%+ of the initial organic content of the sediment (Larter 1988). This efficient expulsion of oil occurs as a single phase (Mackenzie *et al.* 1987) from organic rich sediments such as the Draupne Fm (up to 10% TOC) driven by the compaction resulting from volumetrically significant conversion of solid phase kerogen to fluid petroleum and overburden driven expulsion. The volume lost from the sediment due to petroleum expulsion may be up to 7% in the oil window and may account for most of the volume loss during deep burial of source rocks (Larter 1988).

In addition to gross chemical changes and reduction of hydrogen index for the source rock kerogens with increased burial, maturation and petroleum generation, molecular indicators of source maturity also show systematic changes. For example Fig. 1 shows the evolution of the isomerisation at $C_{20}$ of the $C_{29}$ regular sterane (Mackenzie 1984) in extracts of Draupne Fm cuttings samples from the N. Viking Graben. This is one of a great many properties used by petroleum geochemists as indicators of source maturity (cf. Mackenzie 1984). While the mechanism of this reaction (20R → 20R + 20S) is unclear and simple isomerisation can be eliminated (Requejo 1989; Abbott *et al.* 1990), published 'kinetic' models (Mackenzie & McKenzie 1983) can be used in an empirical prediction engineering manner (Larter 1989) to provide viable predictions of the evolution of this parameter with maturation. Thus Fig. 1 shows data from analysis of the Draupne Fm. from several wells in the N. Viking Graben area and while the data shows much scatter (a result of caving problems primarily) a general agreement is seen between the source rock data and predicted profiles based on computer models using a variety of linear (constant heat flow with respect to time) and rift basin heat flow models (rifting event at 120 Ma, $\beta$ = 1.8; after Mackenzie & McKenzie 1983).

The principal zone of hydrocarbon generation and expulsion covers the depth range 2.5–3.5 km in this area and in this range the sterane parameter 20S/20S + 20R) changes from about 30–55% 20S. The data on Fig. 1 from a cored petroleum reservoir shows values around 55% 20S and suggests this oil was sourced from source rocks around 3500 m. As

*From* Morton, A. C., Haszeldine, R. S., Giles, M. R. & Brown, S. (eds), 1992, *Geology of the Brent Group*. Geological Special Publication No. 61, pp. 441–452.

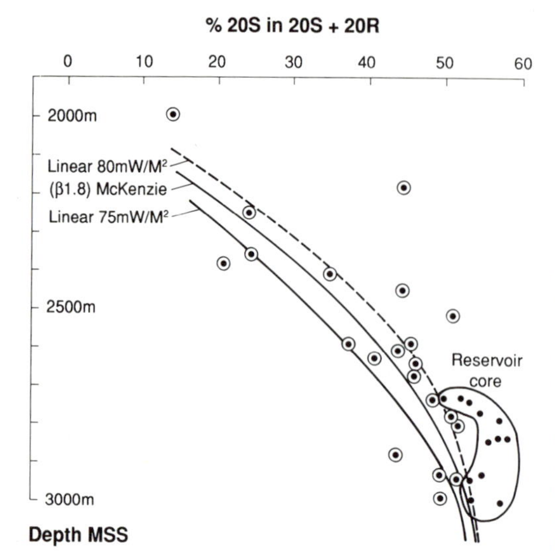


**Fig. 1.** Depth distribution of $C_{29}$ sterane 20S/(20S + 20R) for Draupne Formation samples, N. Viking Graben. Computed profiles for this parameter are also shown based on generalised burial histories and the kinetics published by Mackenzie & McKenzie (1983). Data from analysis of reservoired petroleums are also shown.

secondary migration predominantly involves buoyant upward migration to cooler regions it is reasonable to assume that in many situations, temperature driven reactions involved in maturity parameters are effectively frozen once the petroleum has left the source rock. Thus the molecular markers carry information on the source rock temperature (maturity) and organic facies (cf. Mackenzie 1984 for a review). As we see later the use of such molecular markers as remote source rock probes may provide us with information on the actual processes involved in the filling of individual Brent Group reservoirs and we demonstrate their use in a study of the Gullfaks field from the Tampen Spur area.

## Brent Reservoir geochemistry: examples from the Gullfaks field

Large-scale compositional heterogeneities in petroleum accumulations have been known for many years but it is only in the last few years that chemical heterogeneities in petroleum columns have been interpreted geochemically (England *et al.* 1987; England & Mackenzie 1989; Karlsen & Larter 1989; Leythaeuser & Ruckheim 1989; Larter *et al.* 1990; Horstad *et al.* 1990). It seems from these recent studies that, in addition to biodegradation and water washing effects, variations in petroleum column composition may additionally be interpretable as being due to source facies and/or maturity variations in the petroleum charges feeding the oil accumulation. The variable composition of the petroleum charges feeding oil fields is integrated and may be preserved as compositional variation in the petroleum column due to the slow rate of diffusive or density driven mixing processes in the reservoir (England & Mackenzie 1989; Larter *et al.* 1990) and the absence of thermal convection in most liquid petroleum columns.

In this paper we summarize a reservoir geochemical study of the Gullfaks field, paying attention to the clues provided on the filling of the Brent Group reservoir there. Petroleum reservoir geochemistry may prove to be one of the most powerful applications of petroleum geochemistry to geological basin evaluation problems. Even in well-drilled provinces such as the Brent Province, North Sea area, the vast majority of well locations are aimed at evaluation of structural highs. Mature petroleum source rocks relevant to known petroleum accumulations are therefore only rarely and often incompletely sampled. A systematic study of petroleum column heterogeneity in this province is therefore not only of academic interest but is also directly relevant to determining, using molecular marker approaches discussed briefly above, the location off-structure, type and maturity of the actual mature source rocks responsible for the accumulation. Further we have suggested before (Horstad *et al.* 1990) that by directly determining the location of field fill points for these fields it may prove possible to place more accurate estimates on the volumes of oil stained carrier beds involved in trap filling and thus better define relevant petroleum carrier systems and petroleum losses associated with the carrier system. We expand on this below.

Figure 2 shows a summary calculation of the volumetric aspects of petroleum entrapment assuming a perfect seal on the trap. The volume of petroleum trapped is equal to the volume expelled from the source bed minus the volume lost in the carrier as residual oil stain (oil shows). Corrections can be made for PVT effects on the migrating charge using empirical density, pressure, temperature relationships (cf. Glasø 1980; England *et al.* 1987). Mature source rock volumes (ST × SA) can be fairly accurately estimated using modelling and seismic definition and the mass of petroleum expelled ($M_e$) can usually be estimated to much better than an order of magnitude using mass balance ap-

Volume Petroleum in Trap ($V_T$)

= Volume Petroleum (trap P.T.) expelled from source $V_S$

minus

Volume Petroleum (trap P.T.) lost in carrier VL

$$V_T = V_S - V_L$$

$$V_S = Me \times ST \times SA / \rho_T \text{ (trap P.T.)}$$

$$V_L = CT \times CA \times \varnothing_{av} \times RS \times \rho_C \text{ (carrier P.T.)} / \rho_T \text{ (trap P.T.)}$$

| | | |
|---|---|---|
| Me | = | Mass Petroleum expelled / CuM Source Rock |
| ST | = | Mature Source Rock Thickness |
| SA | = | Mature Source Rock Area |
| $\rho_T$ | = | Total Petroleum density at trap P.T. |
| $\rho_C$ | = | Total Petroleum density at carrier P.T. |
| CT | = | Carrier Thickness |
| CA | = | Carrier Area |
| $\varnothing_{av}$ | = | Average Carrier Porosity |
| RS | = | Average Carrier Residuai Petroleum Saturation |

**Fig. 2.** Petroleum migration volumetrics expressed in terms of the volume of petroleum in trap ($V_T$), the volume expelled from the source ($V_S$) and the volume lost in the carrier ($V_L$).

proaches (Goff 1983; Larter 1985; Cooles *et al.* 1986) calibrated where possible using geochemical data from wells (cf. Mackenzie & Quigley, 1988). In contrast the petroleum loss, $V_L$, is much less easily estimated.

The volume of total possible carrier bed connecting source rock basin to trap can usually easily be estimated from seismic studies and simplistic concepts of up-dip migration. Average carrier porosities can be estimated empirically or by calibration from well studies. However a major uncertainty exists on the magnitude of the effective residual petroleum saturation (RS) existing in most carrier systems (England *et al.* 1987; Macgregor & Mackenzie 1986). As petroleum will only flow as a discrete phase through a water bearing rock when the petroleum saturation reaches high values (greater than 50–80%; England *et al.* 1987) and as there is rarely enough petroleum to completely saturate an entire carrier system, petroleum will flow in a few zones of high saturation, most of the carrier being water saturated and not exposed to petroleum. As it is not practical to assess remotely where petroleum is flowing in the gross carrier volume, an accountancy parameter, the effective residual saturation, RS, is needed for the calculation. This parameter has no physical significance but merely represents an averaged petroleum saturation for the carrier as defined in terms of gross geometry.

England *et al.* (1987) estimate that the parameter RS varies from about 1–10% in most carrier systems and Macgregor & Mackenzie (1986) estimated using case histories and volume balances that for fluvial sand carriers in the Central Sumatra basin RS was about 2%. Such volume balances for Brent Group reservoirs, using estimates for petroleum expelled and subtracting in place petroleum, can be used to place broad estimates of RS for the East Shetland basin (using Goff 1983 data) and for other Brent Group reservoirs. While most Brent Group-reservoired fields are filled to trap spill-point by petroleum (in that regard Gullfaks is exceptional being underfilled) we can make crude maximum estimates that the average carrier residual petroleum saturations for Brent + Statfjord carriers are around 3% of the porous volume of the potential carrier volume for oil reservoirs in the N. Viking Graben and E. Shetland basin areas. What is concerning, is that doubling RS often results in a 'disappearance' of the accumulations i.e. $V_L$ exceeds $V_S$. Clearly this empirically and crudely determined parameter is the critical parameter in a volumetric migration analysis yet is not accessible by any realistic deterministic approach that we know of. It is not yet clear to us whether the large uncertainties in $V_S$, but particularly in $V_L$ (Fig. 2), render the estimation of $V_T$ as anything more useful than a purely academic exercise.

While large uncertainties exist on our estimates of the fraction of a carrier system actively carrying petroleum we can be certain that only a small fraction of the carrier is active and that petroleum must flow into the traps through a few small 'keyhole' conduits. Analysis of published reservoir geochemical data from the Draugen field (Trøndelag Platform, Norway) certainly suggest that large oil fields do not fill on broad fronts (the classic text book drainage area) but fill through rather restricted zones (Karlsen & Larter 1989) from petroleum rivers, the location of which is perhaps not easily predictable from analysis of the carrier system alone. This strengthens our view that reservoir geochemical study may prove to have value in defining field fill points or keyholes.

We can now illustrate the field fill point concept with a geochemical study of Brent Fm. and Cook/Statfjord Fms. reservoirs in the Gullfaks field.

The Gullfaks field is located in Block 34/10, 175 km from the Norwegian coast (Fig. 3). A

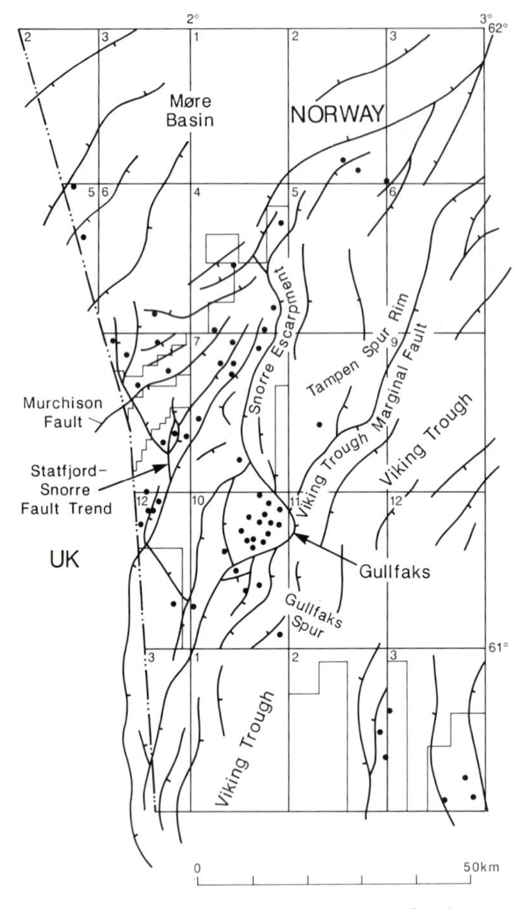


**Fig. 3.** Geographic location of the main Gullfaks field in block 34/10, 175 km off the Norwegian coast. (Modified after Karlsson 1986).

total of 500 million $Sm^3$ oil is present in the Middle Jurassic (Bajocian–Bathonian 183–169 Ma) Brent Group and the Triassic to Early Jurassic Statfjord formation (Rhaetian–Sinemurian 218–196 Ma) and the Jurassic Cook formation (Pliensbachian–Toarcian 196–198 Ma) sediments (Erichsen *et al.* 1987). In this study we concentrate on the Brent Group reservoir, but to complete the picture some analyses of oil samples from the Cook and Statfjord formations are included. The sediments of the Brent Group in the Gullfaks field area are buried at 1800–2000 meters, and are located on the structurally high Tampen Spur areas, between the East Shetland basin, East Shetland platform and the Viking Graben (Karlsson 1986). The Brent Group sediments (Bajocian–Bathonian age) represent a regressive–transgressive megasequence showing a complete delta cycle (Eynon 1981), and consist of five different formations with large variations in lithology and reservoir properties (papers in this volume). After deposition of the Brent Group the Gullfaks field area was eroded and uplifted in the Late Jurassic, resulting in several westerly tilted fault blocks (Fig. 4). The erosion of the Brent Group sediments was most severe on the crests of the rotated fault blocks, and on the eastern most fault blocks the whole Brent Group sedimentary section has been eroded away (Karlsson 1986). In this part of the field, reservoired petroleums are therefore found only in the Triassic Lunde and Statfjord formations and the Lower Jurassic Cook formation. Most of the Upper Jurassic and Lower Cretaceous sediments are missing on the main Gullfaks structure, only 1 m of the Draupne Formation is present in the deepest and most westerly well

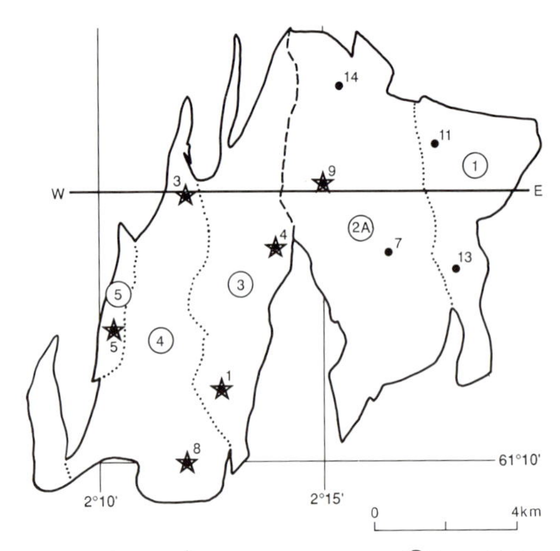


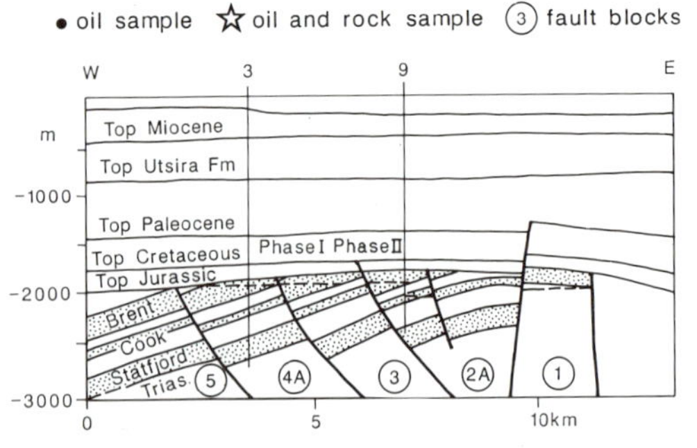


**Fig. 4.** Detailed map of the Gullfaks field showing the location of each well, and the type of samples available. An E–W geological cross section of the field shows the orientation of the rotated fault blocks in the area, and the erosion of the Brent Group sediments on the crest of the eastern fault blocks. (After Erichsen *et al.* 1987).

(34/10–5) and the total thicknesses of Cretaceous sediments is between 100–180 m, the greatest thicknesses being present in the northern part of the field (wells 34/10–3, 9 and 14). Since Upper Cretaceous time the whole area has been under general subsidence (Graue *et al.* 1987).

The Jurassic and Triassic Gullfaks petroleum reservoirs are sealed by Upper Cretaceous fine grained sediments, but leakage from the at present undersaturated and overpressured reservoirs to the Paleocene sediments above has been reported (Irwin 1989). This petroleum leakage from the Gullfaks reservoirs means that the system cannot be treated as a simple trap but may represent a system in which more than one current reservoir volume has been swept by petroleum during the trap's lifetime.

Figure 4 shows the locations of reservoir cores and reservoir fluid samples studied. For complete details of the study the reader is referred to Horstad (1989) and Horstad *et al.* (1990). The reservoir was studied using the general procedure outlined by Larter *et al.* (1990) the objective being to define a filling history for the Gullfaks field.

## Petroleum geochemical approach

A model for petroleum accumulation development should among other features comprise the following information:

(1) definition of the organic facies, maturity level, expulsion efficiency, volume and location of source rocks charging petroleums to the reservoir;

(2) definition of the location, orientation, connectivity and residual petroleum saturation of the regional carrier system transporting petroleum from source basins to trap;

(3) determination of the times at which various portions of the field reached significant petroleum saturation levels and how this distribution of petroleum is related to the sedimentology and diagenesis of the reservoir;

(4) determination of the extent of post-filling mixing or transport of petroleum within the field by diffusive or advective processes such as mechanical stirring or reservoir tilting;

(5) evaluation of the mechanisms and extent of alteration of the petroleum column by biodegradation or other in-reservoir processes such as tar mat precipitation.

Our approach initially involves the 3-D compositional mapping of the petroleum accumulation (cf. Karlsen & Larter 1989) and the definition of the maturity structure of the field (the spatial relationships between petroleums of differing source maturities). The maturity structure of an oil accumulation is determined using one of the multitude of biomarker (Mackenzie 1984; Cornford *et al.* 1986) or alkylaromatic hydrocarbon based maturity parameters (Radke 1988) or in terms of gross maturity related parameters such as solution gas-oil ratio or saturation pressure/bubble point (England & Mackenzie 1989; Karlsen & Larter 1989). A filling direction for a single reservoir field may then potentially be ascertained in terms of a gradient of maturity within the field, the most mature petroleums being nearest the charging points of the reservoir (cf. Karlsen & Larter 1989; Leythaeuseur & Ruckheim 1989; England & Mackenzie 1989).

Mapping of the petroleum column must be based on a high resolution petroleum population distribution map of the reservoir (the term petroleum population refers to a petroleum zone of definable bulk composition with similar maturation and/or source characteristics). An approach to petroleum population mapping which permits rapid mapping of petroleum reservoirs on a metre scale for several wells within a field was described in Karlsen & Larter (1989) and Horstad (1989) using combinations of a thin layer chromatography-flame ionization detector system (TLC-FID) and Rock-Eval screening of reservoir cores to provide gross compositional information (i.e. concentrations of total petroleum, saturated hydrocarbons; saturated/aromatic hydrocarbon ratios etc.). This approach was applied to reservoir cores from the Gullfaks field and the screening formed the basis for selection of samples for further detailed molecular analysis.

The vertical composition of the petroleum column in the Brent Group reservoir at the Gullfaks field, as determined by the screening procedure is generally very homogeneous at any location (Fig. 5). The only exceptions are samples selected from the Ness Formation, where the bulk composition of core extracts varies due to local contamination from in situ coals (Horstad 1989). The screening procedure in which several hundred analyses were performed indicated that for this reservoir, DST samples provided an adequate description of the petroleum column. In other reservoirs this may not be so (Larter *et al.* 1990).

While at any location the vertical composition of the petroleum is quite constant, systematic variations in the chemical composition of the petroleum within the Brent Group reservoir are recognized laterally across the field. Based on

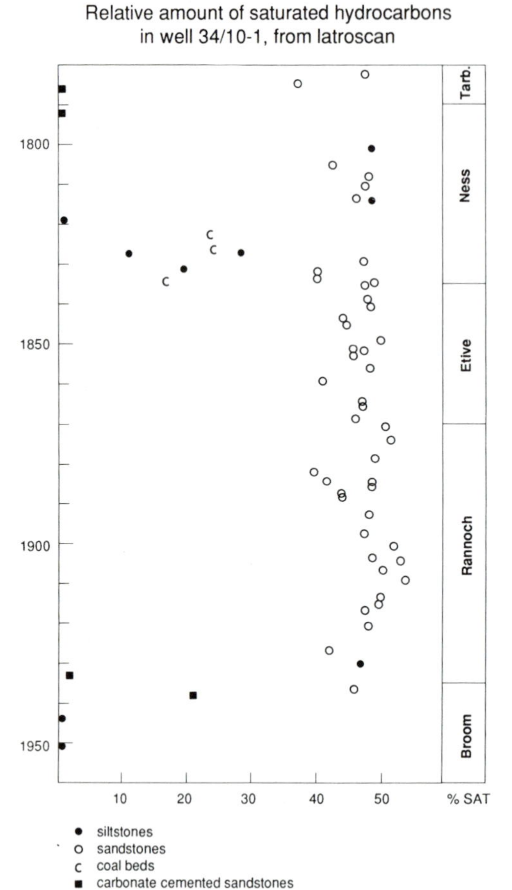


**Fig. 5.** The vertical composition of the petroleum column in the Brent Group reservoir is very homogeneous, shown here by the relative amount of saturated hydrocarbons as determined by Iatroscan TLC-FID.

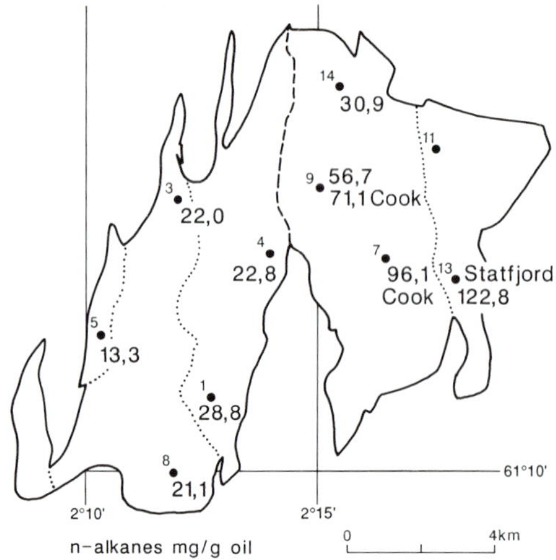


**Fig. 6.** Total yield of n-alkanes ($C_9-C_{34}$) in all the wells. The data is the average of all available Drill Stem Test samples in each well. Note the easterly increase in the absolute amount of n-alkanes due to more severe degradation in the western part of the field. It is stressed that the samples within the Cook and Statfjord formations are not in communication with the Brent Group. Based on G.C. analysis of whole oil samples with internal standard quantitation.

analysis of 12 different DST oil samples from the Brent Group reservoir in seven wells, a systematic gradient in the absolute amount of $C_9$ to $C_{34}$ n-alkanes (determined by internal standard quantitation) is observed. The n-alkane concentration ($\Sigma nC_9-nC_{34}$) varies systematically from 56.7 mg/g oil in the eastern part of the reservoir to 13.3 mg/g oil in the western part of the reservoir (Fig. 6). The absolute amount of $C_9$ to $C_{34}$ n-alkanes is even higher in petroleum reservoired within the Cook and Statfjord formations further east, but this petroleum belongs to another slightly more mature petroleum population, although of related source facies. This systematic removal of n-alkanes in the west of the Gullfaks structure can also be seen by isoprenoid alkane/n-alkane ratios (i.e. pristane/n-$C_{17}$, phytane/n-$C_{18}$ ratios) demonstrating the same systematic gradient across the field. In the most degraded oils in the western Brent reservoir removal of pristane and phytane is also observed (Horstad 1989). By comparison with several studies of biological degradation reported in the literature (Winters & Williams 1969; Bailey *et al.* 1973; Connan 1984; Jobson *et al.* 1979; Milner *et al.* 1977) it is concluded that the decrease in the absolute amount of n-alkanes across the field is due to biological degradation of the petroleum. Large scale convection in petroleum columns would result in a uniform petroleum composition both vertically and laterally across at least each fault block in the field. This of course assumes that vertical and lateral reservoir barriers are essentially absent. While we cannot demonstrate this, a common oil–water contact in most wells does suggest that large scale communication on a geological time scale is good across the field. As lateral gradients are observed in the n-alkane concentration gradients (Fig. 6) large scale pervasive thermal convection over a long time can be eliminated as a mixing mechanism and this is confirmed by calculation of Rayleigh numbers for the petroleum accumulation (Horstad 1989) which indicate a stable thermally non-convective state for the reservoir.

It is concluded based on estimates of diffusive equilibration times that diffusion is the main mixing process eliminating chemical concentration gradients in the Brent Group reservoir in the Gullfaks field (Horstad *et al.* 1990).

Although anaerobic bacterial in-reservoir degradation of petroleum cannot be ruled out by laboratory study as slow laboratory degradation rates may be significant over geological time, it is generally agreed that aerobic bacterial activity is the principal initiator of biological degradation of crude oils in the geosphere (Connan 1984). Although fermentative processes have occurred in the Gullfaks reservoir at some time as indicated by the geochemistry of carbonate benches in the reservoir (Saigal & Bjorlykke 1987), fermentation is not linked to the petroleum degradation at this time. We therefore consider that the degradation of the Gullfaks petroleum was caused by bacterial degradation involving the interaction of molecular oxygen and petroleum mediated by bacteria.

An oxygen mass balance suggests that the observed petroleum degradation could be carried out in less than 10 Ma with a 1 m $a^{-1}$ meteoric water flow through the Brent reservoir below the oil column (Horstad 1989). The gradient in partial degradation observable in Fig. 6 is interpreted to reflect an input of meteoric water, molecular oxygen and perhaps nutrients at least locally from the west of Gullfaks. Bacteria may have been supplied from the recharge area with the water flow, by particulate diffusion from the recharge area or may represent ancient dormant bacteria deposited with the Brent sediments and reactivated by oil and nutrient flow into the reservoir. While estimates of bacterial diffusion coefficients in water using Stokes-Einstein equations suggest that bacterial diffusion may be adequate to import bacteria into a reservoir from recharge areas this process most likely occurs by flow of bacteria with the oxygenated water from the recharge area which is relatively rapid (thousands of years). A gradient in n-alkane concentrations suggests that degradation post-dated reservoir filling. While we cannot conclusively eliminate a later phase of non-degraded oil mixing into the degraded reservoir the interpreted maturity gradients discussed below, are inconsistent with this hypothesis.

## Facies and maturity differences among the two petroleum populations in the Gullfaks field

Karlsen & Larter (1989) defined a petroleum population as a contiguous charge of petroleum in a field, which is chemically definable in terms of a source facies or maturation level. Based on GC/MS analysis of the saturated hydrocarbon fractions of the Gullfaks oils, it is possible to distinguish two different petroleum populations within the sample set from the Gullfaks field: one early to mid-mature population present in the Brent Group in the western part of the field, and a slightly more mature population within the Cook, Statfjord and Lunde formations in the eastern part of the field. This is reflected in slight differences in the relative distribution of $C_{27}$, $C_{28}$, $C_{29}$ $5\alpha(H),14\beta(H),17\beta(H)$ cholestanes in the petroleums. Although there is a considerable scatter in the data set, samples from the Brent Group reservoir generally contain relatively less $5\alpha(H),14\beta(H),17\beta(H)$-methylcholestane ($C_{28}$ sterane) compared to the $C_{27}$, $C_{29}$ homologues than samples from the Cook, Statfjord and Lunde formations. The presence of two petroleum populations is also supported by the $17\alpha(H),21\beta(H)$-28,30-bisnorhopane/$17\alpha(H),21\beta(H)$-30-norhopane ratio (Fig. 7).

The statistical significance of these conclusions was performed by applying Student's t-Test for two sample analysis (Davis 1973), which confirmed discrimination of the two petroleum populations, at a significance level of 0.05 (5%), both for the distribution of $5\alpha(H),14\beta(H),17\beta(H)$-methyl-cholestane (as a % of 27–29 homologues) and the $17\alpha(H),21\beta(H)$-28-30-bisnorhopane/$17\alpha(H),21\beta$ (H)-30-norhopane ratio (Horstad 1989). In-field differences in petroleum composition generally require the statistical testing of hypotheses as compositional differences are often small. We conclude that the Brent Group reservoir was filled from a related but slightly different source to those filling the Cook/Statfjord Fm reservoir. Though we have no direct basin oil-source rock correlations to fall back-on, regional considerations favour an Upper Jurassic source for both petroleum populations.

## Petroleum maturity gradients across the field

Since the generation and expulsion of liquid petroleum from a source rock occurs over a certain depth and temperature range (typically 50°C at any location corresponding to times of 5–50 Ma with usual geological heating rates), a petroleum reservoir should receive more and more mature petroleum as it is filled from a continuously subsiding source rock basin, providing it remains in communication with the source rock. Thus, if a reservoir is filled from

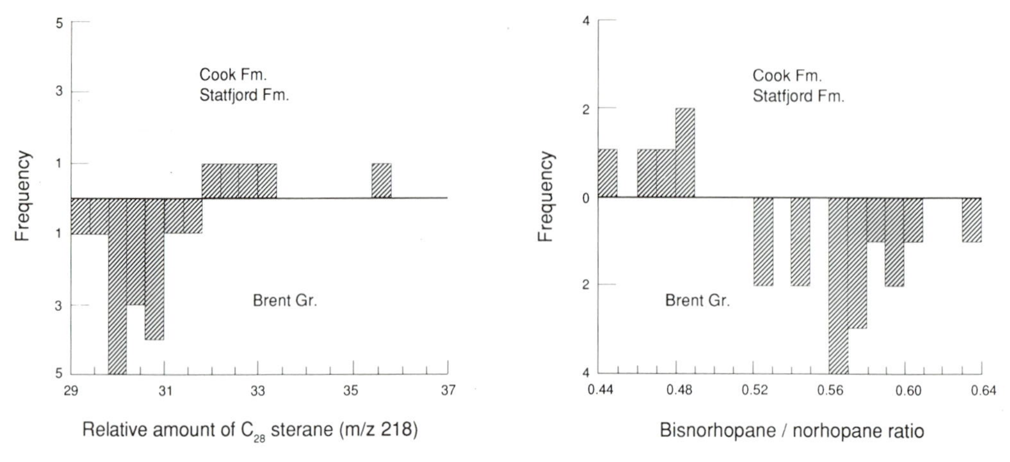


**Fig. 7.** Statistical testing of the relative amount of $C_{28}$ $5\alpha$(H),$14\beta$(H),$17\beta$(H)-methyl-cholestane and the $17\alpha$(H),$21\beta$(H)-28-30-bisnorhopane/$17\alpha$(H), $21\beta$(H)-30-norhopane ratio, suggest that the Brent Group oils and the Cook/Statfjord formations oils belongs to two statistically different petroleum populations, at a 5% level of significance (based on a students-T test analysis of 21 samples).

one end in a sequential manner, the most mature petroleum in the reservoir should be located nearest the source rock basin or probable filling point (England *et al.* 1987; England & Mackenzie 1989; Karlsen & Larter 1989; Larter *et al.* 1990). We envisage the reservoir filling initially as suggested by England *et al.* (1987), by a progressive local petroleum saturation increase near the fill point. As filling continues we envisage that once extensive oil water contacts were locally established filling took place largely by density driven flow (England & Mackenzie 1989) driving petroleum in a rotational manner from fill point to reservoir extremity, thus maintaining a maturity gradient. Since the temperature range over which liquid petroleum is expelled from a high quality source rock such as the Draupne Formation/Kimmeridge Clay is rather limited, the corresponding shift in the different maturity indicators will be small, typically in the range equivalent to 0.6–1.0% vitrinite reflectance. This approach therefore requires a very precise determination of the source maturity indicators carried by the petroleum charge.

When the hopane X/(hopane X + $17\alpha$(H),21 $\beta$(H) hopane) ratio (Cornford *et al.* 1986), the proposed isomerization at C-20 in $5\alpha$(H),$14\alpha$(H),$17\alpha$(H)-ethylcholestanes S/(S + R) (Fig. 8), the apparent isomerization from $14\alpha$(H), $17\alpha$(H) to $14\beta$(H),$21\beta$(H) in ethylcholestane ($\beta\beta/(\alpha\alpha + \beta\beta)$ ratio) and the $18\alpha$ (H)-22,29,30-trisnorneohopane/$17\alpha$(H)-22,29,30-trisnorhopane (Ts/Tm) ratio are plotted on a map of the field, related very tentative gradients in the interpreted petroleum maturities are observed (Horstad *et al.* 1990). The most mature petroleum in the Brent Group reservoir appears to be present in the western part of the field, while the most mature petroleum in the Cook/Statfjord reservoir system is present in the eastern part of the field. The relatively small maturity gradients across the field (3–4 standard deviations) observed here may reflect the observation made earlier that the Gullfaks system may have been partially open during filling, the present gradients reflecting a relatively late stage maturity window in the evolution of the source rock basins. However it is also possible that a Late Cretaceous/Early Tertiary age for filling of the reservoir will have allowed for a significant diffusion affected lateral mixing of the petroleum columns resulting in reduced gradients.

On the basis of statistically tested differences in source rock facies indicators and maturity differences between the two main petroleum populations and the more tentative maturity gradients within each population shown in Fig. 8, it is very tentatively concluded that the Gullfaks structure probably was filled from two different source rock basins, one to the west and one slightly more mature to the east or northeast of the Gullfaks structure. We are currently testing this hypothesis with maturation modelling methods. The conclusion based on maturity gradients is of necessity weaker based on few data points but a typical gradient of several standard deviations is observed across the field for most of the maturity parameters. It is further noted that the standard deviations are

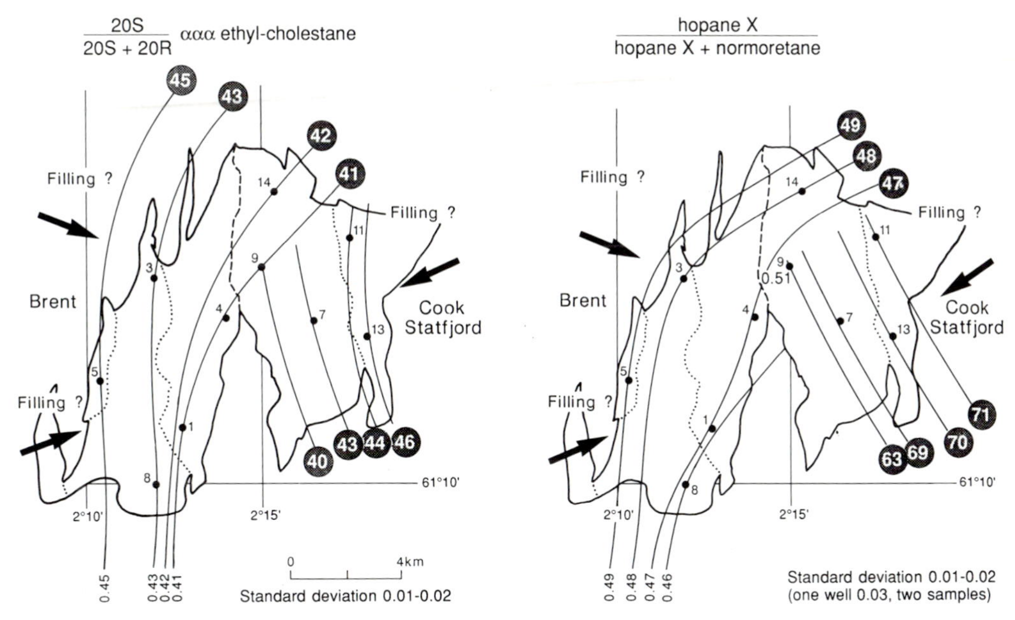


**Fig. 8.** A tentative interpretation of two different maturity gradients in the field based on the % hopane X/(hopane X + hopane) ratio and isomerization at C-20 in $5\alpha$(H),$14\alpha$(H),$17\alpha$(H)-ethyl-cholesterane, (%20S/(20S + 20R)), one within the Brent Group in the western part of the field and one within the Cook/Statfjord formations in the eastern part of the field. The contours are highly schematic and are for illustrative purposes only. These two maturity gradients might reflect the filling of the field from two continuously subsiding source rock basins. Based on up to 12 samples in each well.

calculated from analysis of several different petroleum samples (up to 12 in well 34/10–1) at each location, not reanalysis of the same sample. All the analyses were obtained in sequence by the same operator on the same mass spectrometer set up. While we cannot exclude biodegradation as a factor in influencing these maturity gradients we feel the bulk of the evidence supports the conclusion that these are maturity gradients. Recent oil correlation studies from oils in Tordis field (west of Gullfaks) and from block 34/8 are consistent with our conclusions.

To date, one of the major problems in this reservoir study, has been determination of the timing of the different events related to filling and degradation.

Since the rotated fault blocks in the western part in the field are less eroded on the crests than the fault blocks to the east, it is likely that this part of the field has always been deeper, after the fault block rotation in Jurassic time. Thus, since the most severely degraded petroleum is located where the water first reached the reservoir, the input of oxygen-rich meteoric water to degrade the petroleum must have been derived from water flow at least locally from the west.

To date we have not been able to reconcile all our observations of the Gullfaks field with the geology of the area. We can conclude with some certainty that the trap has been filled from at least two different basins and carrier systems but we cannot adequately define the fill time. The most probable time of degradation of the oils in the western part of the field is prior to the rapid subsidence of the whole area in early Tertiary time (Frost 1989) when, presumably, overpressuring of the reservoir and removal of recharge areas would have stopped meteoric water flow. While a Tertiary degradation event cannot be ruled out from our data, we have no geological information to support this concept and suggestions that oxygenated water could have been driven into the reservoirs by the up to 3 km ice sheets present in the North Sea area water columns during glacial times (Cornford, pers. comm.) are not physically convincing. It seems probable therefore that the oils were emplaced in late Cretaceous/early Tertiary time. This in itself is somewhat problematic in that preliminary maturation modelling suggests that only the Viking Graben source rocks would be mature at this time (Glasmann *et al.* 1989.) The major problem now concerns the explanation of the westerly degradation gradient observed in the Brent reservoir. Our early attempts to explain the inflow of water from the west of the Brent Group reservoir by recharge from elevated areas along the Statfjord/Snorre fault trend (Horstad *et al.* 1990), in the manner described by Glasmann *et al.* (1988) for the Heather field, seem complicated by the absence, to our knowledge, of any indication of an exposed recharge area there in late Cretaceous/early Tertiary time. Simple calculations suggest that a hydraulic head of 30 m in a recharge area feeding a 1 Darcy permeability Brent sandstone could deliver enough water to degrade the Gullfaks reservoir in less than 4 Ma (Horstad, 1989), but we cannot define such an area at present. We cannot rule out much more complex filling/degradation scenarios that might explain the observed maturity/degradation gradients but these are not discussed here. We are continuing to study this problem by looking at the petroleum geology of the area and by continuing this study as part of a longer study of petroleum accumulations in the whole Tampen Spur area. Additionally we are reviewing the geochemistry of subsurface petroleum biodegradation which we feel is far from understood.

## Conclusions

On the basis of a petroleum column geochemical study of the petroleum in the Brent Group and Cook/Statfjord Fm reservoirs at the Gullfaks field, the petroleum in the reservoir can be divided into two main petroleum populations on the basis of statistically testable biological marker analysis. Further, filling directions into the reservoir are possibly reflected in two more tentative gradients in the maturity of the petroleum suggesting the Brent Group reservoir filled from the west, the Cook and Statfjord Fms reservoirs filling from the north or north east.

After emplacement, the petroleum in the Brent Group, and to a lesser extent the petroleum in the Cook and Statfjord formations, were degraded by aerobic bacteria. The oxygen and nutrient supply was charged from unknown elevated areas driving water flow from the west causing a gradient in the extent of biological degradation of the Brent Group oil columns from west to east.

The compositional variations in the field are not connected with convective mixing, and diffusion, the major mixing process, is only able to homogenize the petroleum composition on a relatively small scale. Lateral compositional heterogeneities may therefore persist for millions of years, but a relatively long time since filling (*c.* 60 Ma) may have enabled diffusion to markedly attenuate compositional gradients across the field. A further complication may be that since the Gullfaks seal may not have been efficient throughout the fields history, the current reservoir charge may therefore only represent a fraction of the petroleum that has migrated into the trap (Irwin 1989). This reservoir study confirms the view that petroleum traps fill in a complex manner, often from several source basins, reservoir geochemistry helping to better define the location of active fill points or 'keyholes' into the reservoirs.

We wish to thank the Norwegian Petroleum Directorate for supporting the Gullfaks reservoir study both financially and for providing rock samples. We also want to thank Statoil who made all the DST oil samples available and BP, Norway for financial support of the Organic Geochemistry laboratory at the University of Oslo. We thank Saga Petroleum for supporting a follow up study of the whole Tampen Spur area. We thank UNOCAL for permission to publish the data shown in Fig. 1. The manuscript was prepared by Y. Hall (NRG) and B. Brown (NRG) assisted with drafting. The paper benefitted from reviews by W. England (BP) and C. Cornford (Integrated Geochemistry Services).

## References

Abbott, G. D., Wang, G. Y., Eglinton, T. I., Home, A. K. & Petch, G. S. 1990. The kinetics of sterane biological marker release and degradation processes during the hydrous pyrolysis of vitrinite kerogen. *Geochimica et Cosmochimica Acta*, **54**, 2451–2461.

Bailey, N. J. L., Jobson, A. M. & Rogers, M. A. 1973. Bacterial degradation of crude oil: comparison of field and experimental data. *Chemical Geology*, **11**, 203–211.

Connan, J. 1984. Biodegradation of crude oils in reservoirs. *In*: Brooks, J. & Welte, D. H. (eds) *Advances in Petroleum Geochemistry*. Academic Press, London, **1**, 299–235.

Cooles, G. P., Mackenzie, A. S. & Quigley, T. M. 1986. Calculation of petroleum masses generated and expelled from source rocks. *Organic Geochemistry*, **10**, 235–246.

Cornford, C., Needham, C. E. J. & De Walque, L. 1986. Geochemical habitat of North Sea oils. *In*: Spencer, A. M. *et al.* (eds) *Habitat of Hydrocarbons on the Norwegian Continental Shelf*. Norwegian Petroleum Society, Graham & Trotman, 39–54.

Davis, J. C. 1973. *Statistics and Data Analysis in Geology*. John Wiley & Sons, Inc., New York.

England, W. A. & Mackenzie, A. S. 1989. Geochemistry of petroleum reservoirs. *Geologische Rundschau*, **78**, 214–237.

——, Mackenzie, A. S., Mann, D. M. & Quigley,

T. M. 1987. The movement and entrapment of petroleum fluids in the subsurface. *Journal of the Geological Society, London*, **144**, 327–347.

ERICHSEN, T., HELLE, M. & ROGNEBAKKE, A. 1987. Gullfaks. *In*: SPENCER *et al.* (eds) *Geology of Norwegian Oil and Gas Fields*. Graham & Trotman, 273–286.

EYNON, G. 1981. Basin development and sedimentation in the Middle Jurassic of the Northern North Sea. *In*: ILLING, L. V. & HOBSON, G. D. (eds) *Petroleum Geology of the Continental Shelf of North West Europe*. Heyden & Sons, London, 196–204.

GLASØ, O. 1980. Generalized pressure-volume-temperature correlations. *Journal of Petroleum Technology*, **32**, 789–795.

GLASMANN, J. R., LUNDEGARD, P. D., CLARK, R. A., PENNY, B. K. & COLLINS, I. D. 1988. Isotopic evidence for the history of fluid migration and diagenesis: Brent Sandstone, Heather Filed, North Sea. *Clay Minerals*, **24**, 255–264.

—— CLARK, R. A., LARTER, S. R., BRIEDIS, N. A. & LUNDEGARD, P. D. 1989. Diagenesis and hydrocarbon accumulation, Brent Sandstone (Jurassic), Bergen High Area, N. Sea. *Bulletin of the American Association of Petroleum Geologists*, **73**, 1341–1360.

GOFF, J. C. 1983. Hydrocarbon generation and migration from Jurassic source rocks in the East Shetland Basin and Viking Graben of the Northern North Sea. *Journal of the Geological Society, London*, **40**, 445–474.

GRAUE, E., HELLAND-HANSEN, W., JOHNSON, J., LOMO, L., NOTTVEDT, A., RØNNING, K., RYSETH, A. & STEEL, R. 1987. Advance and retreat of Brent Delta system, Norwegian North Sea. *In*: BROOKS, J. & GLENNIE, K. (eds) *Petroleum Geology of North West Europe*. Graham & Trotman, London, 915–938.

HORSTAD, I. 1989. *Petroleum composition and heterogeneity within the Middle Jurassic reservoirs in the Gullfaks field area, Norwegian North Sea*. Cand. Scient thesis, Department of Geology, University of Oslo.

——, LARTER, S. R., DYPVIK, H., AAGAARD, P., BJORNVIK, A. M., JOHANSEN, P. E. & ERIKSEN, S. 1990. Degradation and maturity controls on oil field petroleum column heterogeneity in the Gullfaks field, Norwegian N. Sea. *Organic Geochemistry*, **16**, 1, 497–510.

HUC, A. Y., IRWIN, H. & SCHOELL, M. 1985. Organic matter quantity changes in an U. Jurassic shale sequence from the Viking Graben. *In*: THOMAS, B. M. *et al.* (eds) *Petroleum Geochemistry in Exploration of the Norwegian Shelf*. Graham & Trotman, 179–183.

IRWIN, H. 1989. *Hydrocarbon leakage, biodegradation and the occurrence of shallow gas and carbonate cement*. Norwegian Petroleum Society, 7053. Stavanger 10–11 April 1989.

JOBSON, A. M., COOK, F. D. & WESTLAKE, D. W. S. 1979. Interaction of aerobic and anaerobic bacteria in petroleum biodegradation. *Chemical Geology*, **24**, 355–365.

KARLSEN, D. A. & LARTER, S. R. 1989. A rapid correlation method for petroleum population mapping within individual petroleum reservoirs: Applications to petroleum reservoir description. *In*: COLLINSON, J. D. (ed.) *Correlation in Hydrocarbon Exploration*. Graham & Trotman, London, 77–85.

KARLSSON, W. 1986. The Snorre, Statfjord and Gullfaks oilfields and the habitat of hydrocarbons on the Tampen Spur, offshore Norway. *In*: SPENCER, A. M. *et al.* (eds) *Habitat of Hydrocarbons on the Norwegian Continental Shelf*. Norwegian Petroleum Society, Graham & Trotman, 181–197.

LARTER, S. R. 1985. Integrated kerogen typing in the recognition and quantitative assessment of petroleum source rocks. *In*: THOMAS, B. M. *et al.* (eds) *Petroleum Geochemistry in Exploration of the Norwegian Shelf*. Graham & Trotman, p. 269–286.

—— 1988. Some pragmatic perspectives in source rock geochemistry. *Marine and Petroleum Geology*, **5**, 192–204.

—— 1989. Chemical models of vitrinite reflectance evolution. *Geologisches Rundschau*, **78**, 349–359.

——, BJØRLYKKE, O., KARLSEN, D. A., NEDKVITNE, T., EGLINTON, T., JOHANSEN, P. E. & LEYTHAEUSER, D. 1990. Determination of petroleum accumulation histories: Examples from the Ula field, Central Graben, Norwegian N. Sea. *In*: BULLER, A. (ed.) *N. Sea Oil and Gas Fields II*. Graham & Trotman, London, p. 319–330.

LEYTHAEUSER, D. & RUCKHEIM, J. 1989. Heterogeneity of oil composition within a reservoir as a reflectance of accumulation history. *Geochimica et Cosmochimica Acta*, **53**, 2119–2123.

MACGREGOR, D. S. & MACKENZIE, A. S. 1986. Quantification of oil generation and migration in the Malacca Strait Region. *Proc. 15th Annual Convention of the Indonesian Petroleum Association*, 7–9 October, 1986, Jakarta.

MACKENZIE, A. S. 1984. Application of biological markers in petroleum geochemistry. *In*: BROOKS, J. & WELTE, D. (eds) *Advances in Petroleum Geochemistry 1983*. Academic Press, London, VI, 115–214.

—— & MCKENZIE, D. P. 1983. Isomerization and aromatization of steroid hydrocarbons in sedimentary basins formed by extension. *Geological Magazine*, **120**, 417–470.

—— & QUIGLEY, T. M. 1988. Principles of geological prospect appraisal. *Bulletin of the American Association of Petroleum Geologists*, **72**, 399–415.

——, PRICE, I., LEYTHAUSER, D., MULLER, P., RADKE, M. & SCHAEFFER, R. G. 1987. The expulsion of petroleum from Kimmeridge clay source rocks in the area of the Brae oilfields, U.K. continental shelf. *In*: BROOKS, J. & GLENNIE, K. (eds) *Petroleum Geology of N.W. Europe*. Graham & Trotman, London, 865–877.

MILNER, C. W. D., ROGERS, M. A. & EVANS, C. R.

1977. Petroleum transformation in reservoirs. *Journal of Geochemical Exploration*, **7**, 101–153.

RADKE, M. 1988. Applications of aromatic compounds as maturity indicators in source rock and crude oils. *Marine and Petroleum Geology*, **5**, 224–236.

RECQUEJO, A. G. 1989. Quantitative analysis of triterpane and sterane biomarkers, methodology and applications in petroleum geochemistry in biological markers in sediments and petroleum. *Proceedings of the Siefert Memorial Symposium*. American Chemical Society. (In press).

SAIGAL, G. C. & BJØRLYKKE, K. 1987. Carbonate cements in clastic reservoirs from offshore Norway – relationships between isotopic composition, textural development and burial depth. *In*: MARSHALL, J. D. (ed.) *Diagenesis of Sedimentary Sequences*. Geological Society, Special Publication, London, **36**, 313–324.

THOMAS, B. M., MØLLER-PEDERSEN, P., WHITTAKER, M. F. & SHAW, N. D. 1985. Organic facies and hydrocarbon distributions in the Norwegian N. Sea. *In*: THOMAS, B. M. *et al*. (eds) *Petroleum Geochemistry in Exploration of the Norwegian Shelf*. Graham & Trotman, 3–26.

THOMPSON, S., COOPER, B. S., MORLEY, R. J. & BARNARD, P. C. 1985. Oil-generating coals. *In*: THOMAS, B. M. *et al*. (eds) *Petroleum Geochemistry in Exploration of the Norwegian Shelf*. Graham & Trotman, 59–73.

WINTERS, J. C. & WILLIAMS, J. A. 1969. Microbial alteration crude oil in the reservoir. Symposium on Petroleum Transformations in Geologic Environments. *American Chemical Society Meeting, New York, September 7–12, 1969*, E22–E31.

# Integration of petroleum engineering studies of producing Brent Group fields to predict reservoir properties in the Pelican Field, UK North Sea

J. D. KANTOROWICZ[1,2], M. R. P. EIGNER[1,3], S. E. LIVERA[1,4], F. S. VAN SCHIJNDEL-GOESTER[1], AND P. J. HAMILTON[5]

[1] *Koninklijke/Shell Exploratie en Produktie Laboratorium, Volmerlaan 6, 2280 AB Rijswijk, Netherlands*

[2] *Present address: Conoco (UK) Ltd, Park House 116 Park Street, London W1Y 4NN, UK*

[3] *Present address: Shell Training Centre, Leeuwenhorst, Noordwijkerhout, Netherlands*

[4] *Present address: Petroleum Development Oman, PO Box 81, Muscat, Oman*

[5] *SURRC, East Kilbride, Glasgow, UK; Present address: CSIRO, Mineral Resources Labs, 51, Delhi Road, North Ryde, New South Wales, Australia*

**Abstract:** Diagenetic models and the results of laboratory experiments on cores from producing Brent Group fields (Cormorant and Tern) have been combined to aid prediction of reservoir properties in a deeply buried prospect (Pelican).

In Cormorant and Tern illite particles are thin (20–40 Å), lath-shaped and liable to damage during drying before air permeability measurement, leading to erroneously high permeabilities. The magnitude of this drying effect ($K_{air}$ conventional drying/$K_{air}$ critical point drying) varies with illite morphology and the timing of hydrocarbon emplacement. In Tern pore-filling illites occur above and below the oil–water contact, and there is no significant difference in drying effect. In Cormorant, the drying effect is higher below the oil–water contact where a diversity of late diagenetic pore-filling illite morphologies occur.

Injectivity tests on cores from Cormorant indicate that sea water injection will not lead to impairment of reservoir quality as a result of clay dispersal or other rock-related phenomena. Impairment occurred when flooding cores with produced water and was avoided by introducing divalent cations, saline brines or approximating in situ equilibrium conditions (pH 6.4). This equilibrium point is strongly influenced by the surface chemistry of feldspar and illite, rather than the more abundant kaolinite and quartz. Burial diagenetic replacement of kaolinite by illite will increase this equilibrium towards pH 7 but should not cause impairment during sea water injection into deeper reservoirs.

Pelican contains abundant pore-filling illite. K–Ar age dates (44–25 Ma) suggest illite growth started after burial to over 8000 ft and palaeoburial temperatures of over 80–90°C. Compared with nearby reservoirs, variations in the nature and abundance of illite result from illitisation being arrested by hydrocarbon charge at different times. Optimal reservoir and aquifer properties occur in shallower structures charged from more deeply buried source rocks. Good reservoir properties occur with charge after short-lived illite growth, although later burial may reduce aquifer properties. The more prolonged illite growth experienced by deeply buried reservoirs such as Pelican, charged late in the Tertiary, has resulted in poor reservoir properties and poor aquifer permeabilities may be expected.

The initial phases of oilfield development in the UK sector of the Brent province focussed on larger and shallower structures such as Brent, Cormorant and Ninian. Recent development has turned towards smaller and more deeply buried structures such as Pelican (Fig. 1). In view of the more marginal economics of developing these new fields, intensive efforts should be made to predict both reservoir properties and the properties of sandstones below the oil–water contact (in the aquifer). In general, reservoir properties in Brent Group sandstones are lithofacies controlled and so will reflect the depositional model (Johnson & Stewart 1985). It can also be shown that reservoir properties deteriorate with increasing depth as a result of compaction and the growth of cements such as illite (Giles *et al.* this volume). The situation becomes more complicated as variations in the relative timing of illite growth and hydrocarbon

*From* MORTON, A. C., HASZELDINE, R. S., GILES, M. R. & BROWN, S. (eds), 1992, *Geology of the Brent Group*. Geological Society Special Publication No. 61, pp. 453–469.

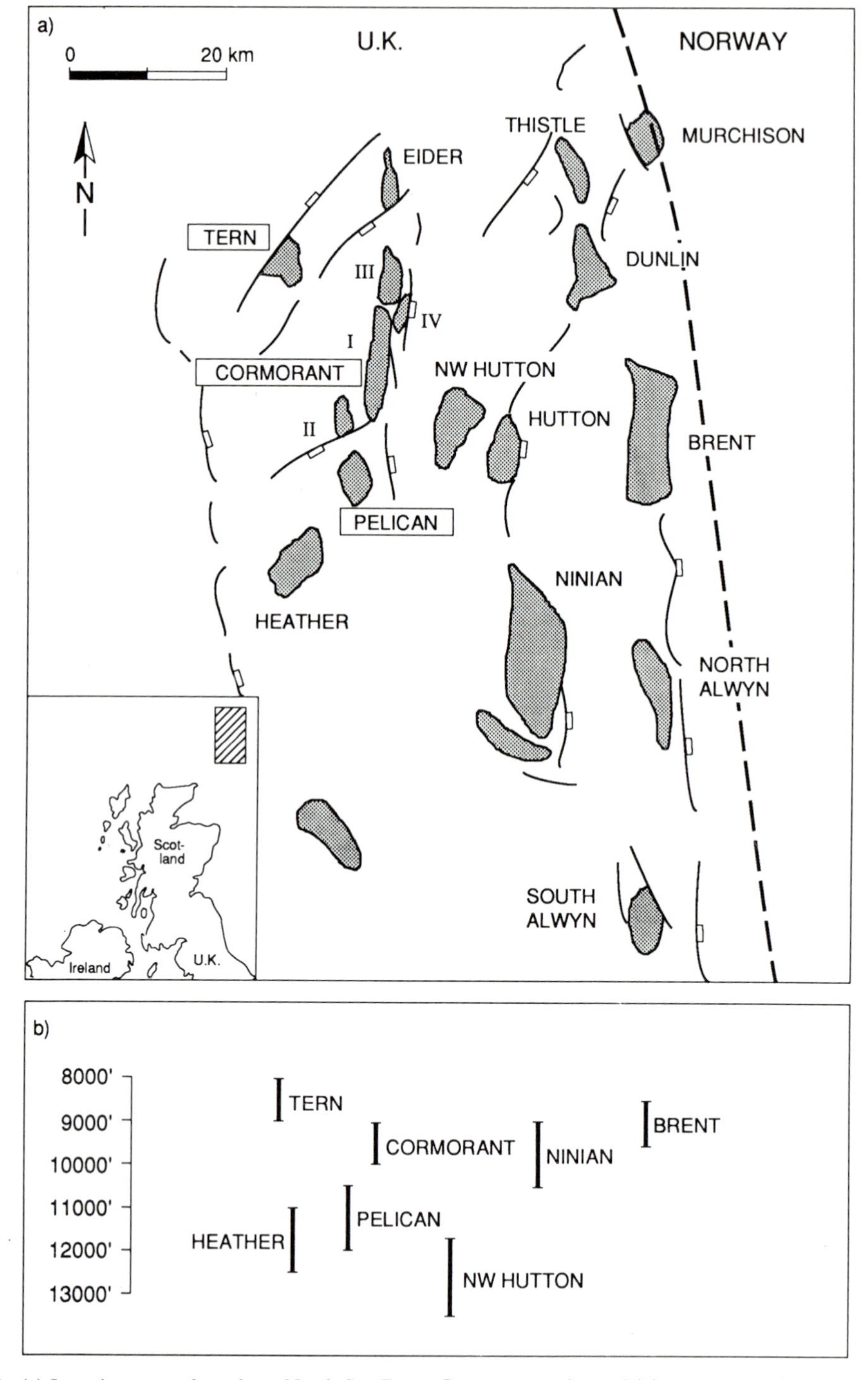


**Fig. 1.** (**a**) Location map of northern North Sea Brent Group reservoirs and (**b**) east–west schematic cross section showing burial depths (in feet) of fields mentioned in the text.

emplacement (charge) into the reservoir will determine whether illite is ubiquitous or restricted to aquifer sandstones (Thomas 1986, Hamilton *et al.* 1989). Recognizing the potential for late diagenetic illite growth in aquifer sandstones is particularly important at an early stage of oil field appraisal and development planning as preferential illitization of the aquifer may have an adverse affect on injection rates (e.g. Heather field, Gray & Barnes 1981). Injection may even become uneconomic, necessitating injection into the oil column itself, with a potential reduction in ultimate recovery (e.g. Drymond & Spurr 1988).

The purpose of this paper is to show how data and previous experience from the Brent province can be used to improve development of new fields such as Pelican, where at present core materials has only been recovered from above the oil–water contact, and where early water injection may be needed. Specifically, this paper describes two laboratory investigations into permeability in cores from the Cormorant and Tern Fields (Fig. 1). These assess the effect of core preparation for laboratory analysis, and will show how the subsequent results may be misleading. Subsequently, K-Ar age dates from the Pelican Field (Fig. 1) are presented and combined with data from Tern, Cormorant and other fields in the area in order to try and predict the likely reservoir properties of sandstones locally. The approach described here should be seen as complementing the work of Giles *et al.* and Hamilton *et al.* (this volume), and highlighting the production geological aspects of diagenetic studies.

Of course, attempts to model the impact of one or two minerals in a reservoir depend upon selecting the most appropriate diagenetic model as the basis for any predictions about mineral distribution. Broadly comparable sequences of diagenetic modifications are reported from most Brent Group sandstone reservoirs (Fig. 2), although it should be appreciated that this over-simplifies local diagenetic complexity. For example, lithofacies controls on mineral authigenesis exist in places (e.g. Sommer 1978; Kantorowicz 1984; Scotchman *et al.* 1989),

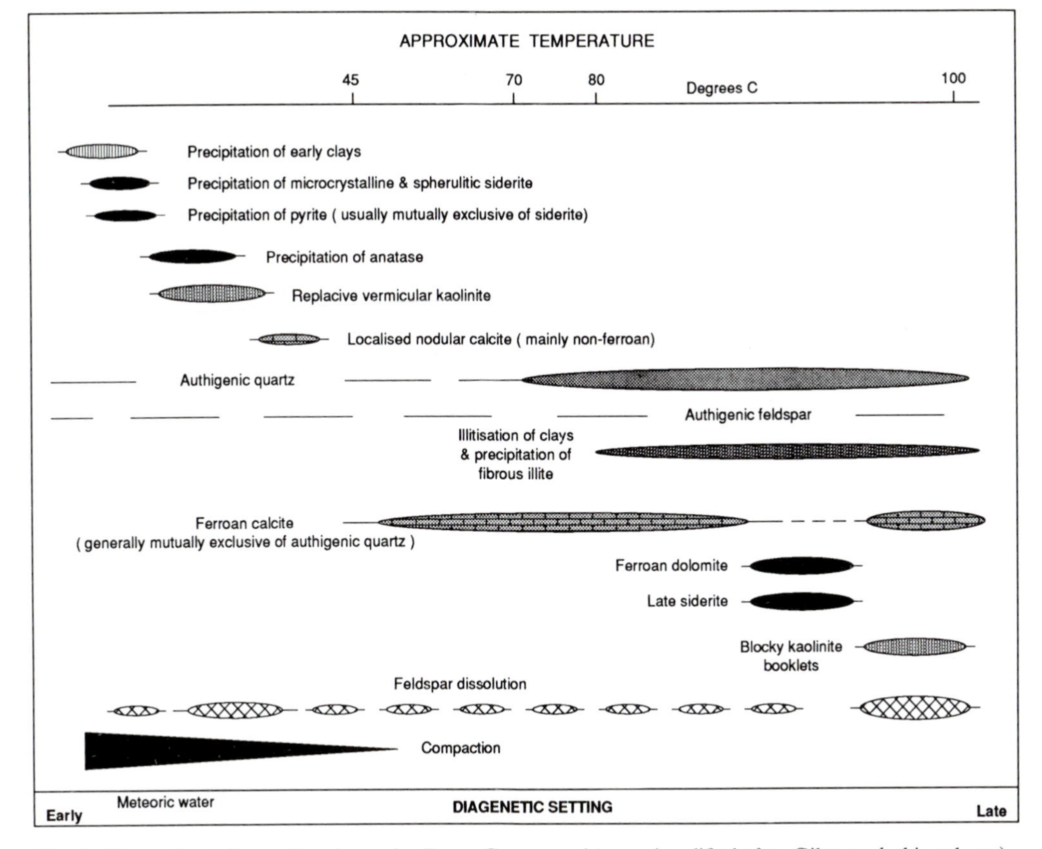


**Fig. 2.** Generalized diagenetic scheme for Brent Group sandstones (modified after Giles *et al.* this volume).

whilst many of the late diagenetic cements are restricted to more deeply buried reservoirs (e.g. Kantorowicz 1985; Jourdan *et al.* 1987). There is, however, little consensus as to the precise nature and timing of the processes responsible. Considering the clays specifically, kaolinite is the most volumetrically abundant, forming up to 20% of some sandstones, and is widely distributed. There appear to be at least two generations of kaolinite that are variously attributed to syn- and early post-depositional meteoric water penetration (Kantorowicz 1984; Morton & Humphreys 1983); meteoric water flushing during Cimmerian exposure (Bjørlykke & Brendsdal 1986; Glasmann *et al.* 1989); and to burial diagenetic processes such as the introduction of acidic fluids and alteration of feldspars (Blackbourn 1984; Kantorowicz 1984; Scotchman *et al.* 1989). Illite typically forms 1–2% of the sandstones, and is commonly attributed to increasing alteration of feldspars or kaolinite with depth (Jourdan *et al.* 1987; Bjørkum & Gjelsvik 1989; Glasmann *et al.* 1989), although its abundance is often facies controlled and so may be related to either illite nucleation on, or replacement of detrital or earlier authigenic clays (Sommer 1978; Kantorowicz 1984). The distribution of these clays is likely to vary as a result of their different origins. The impact of any reservoir quality predictions should be calibrated to an appropriate diagenetic model.

## The impact of illitic clays on permeability measurement

Recent studies have shown that the observed morphology of authigenic illites (or illite-smectites with minor mixed layering) varies depending upon the way in which the core samples are prepared for analysis (Fig. 3; McHardy *et al.* 1982; Pallatt *et al.* 1984; de Waal *et al.* 1988; Purvis 1991). It has also been shown that illite morphology can have a direct impact on air permeability measurements. Specifically, permeabilities measured after conventional or air drying, during which core plugs are dried in an oven or a humidifier, may be significantly higher than permeabilities measured after critical point drying. This is because air drying involves passing an air/water interface through the pores, and critical point drying does not. It is inferred from the matted and clumped morphology of illitic particles in many air dried samples that the interfacial tension exerted at this interface is sufficient to pull the clay particles together and so reduce tortuosity. (This damage should not be confused with the collapse that can result from fibres charging and discharging in the electron beam within the scanning electron microscope; see Purvis 1991). The increase in permeability that can result may be quantified as the drying effect:

$$K_{air} \text{ air dried}/K_{air} \text{ critical point dried}$$

The critical point drying method is discussed in detail by de Waal *et al.* (1988).

### *Cormorant Field*

In the Cormorant Field, the drying effect is different when comparing sandstone cores from above and below the oil–water contact, and can be correlated with the morphology and relative abundance of illite (Table 1, Fig. 4, Kantorowicz 1990). Pore-lining (grain-coating) illites occur above and below the oil–water contact, whilst pore-filling morphologies occur around and below the oil–water contact (in the aquifer). As these illites comprise relatively low percentages in these samples (0–1% from modal analyses) any quantitative contrasts are likely to be within analytical error. However, qualitative examination of scanning electron micrographs suggests that not only is the morphology of illite below the oil–water contact more diverse, but also that illites are more abundant here. Damage to these more abundant pore-filling clays leads to a higher drying effect being measured on core samples from the aquifer (Table 1). From the difference in the types of illite, it is likely that oil emplacement arrested short-lived pore-lining illite growth above the oil–water contact, whilst growth of a variety of pore-filling morphologies continued for sometime in the water-bearing sands below.

### *Tern Field*

In the Tern Field, well 210/25A-5 was cored above and below the oil-water contact. Petrographical analyses reveal that broadly comparable amounts of illite (1–2% from bulk and clay fraction X-Ray diffraction), and similar illite morphologies occur throughout the cores. Illite lines pores, replaces kaolinite, and has a blooming morphology (Fig. 5) similar to that observed in Cormorant. Transmission electron microscope examination of separated clay fractions reveals that most illite particles in Tern are 20–40 Å thick, and lath shaped (Nadeau & McHardy pers. comm. 1985; see Nadeau 1985 for a discussion of the method). The studied samples also contain quartz and feldspar overgrowths and kaolinite. Critical point drying data from Tern reveal that drying effects differ between formations but above and below the oil–water contact the effects are comparable (Table 2). The average drying effects, 1.9 above the

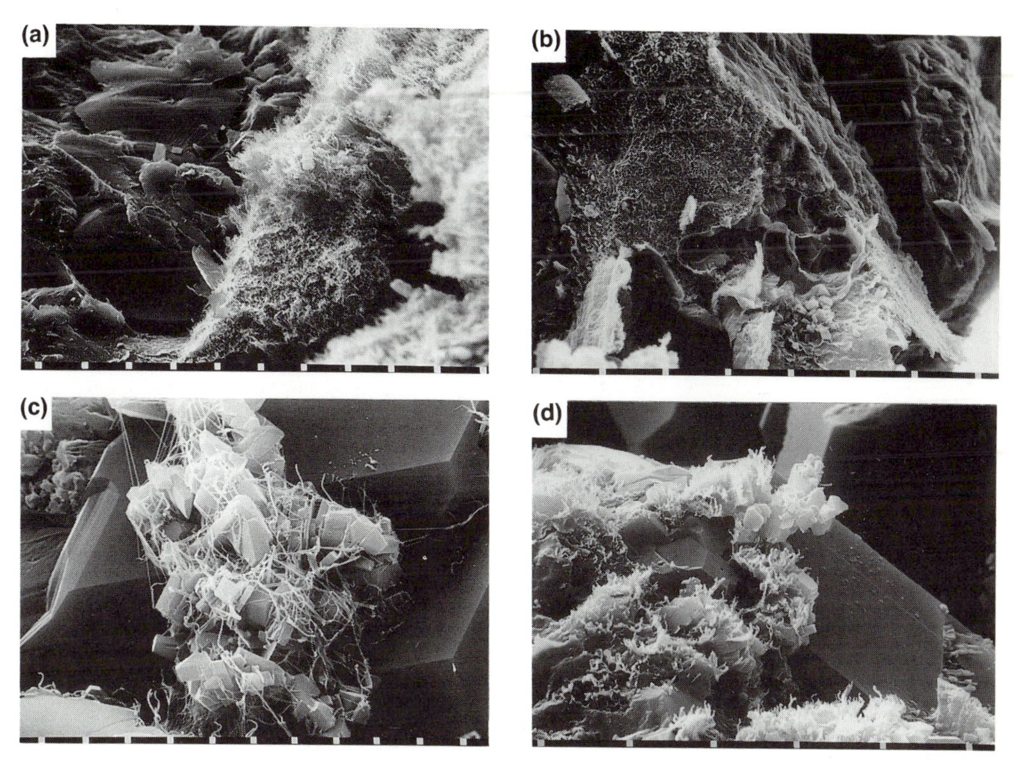


**Fig. 3.** Scanning electron micrographs of illite-bearing sandstones (a) and (c) after critical point drying and (b) and (d) after air drying. Depths are drillers' depths in feet. (a) and (b), 211/26-CA30, 13409 ft, (c) and (d) 211/26-CAUW2, 10044 ft. Scale bars: (a) 10 $\mu$m, (b) 30 $\mu$m, (c) 3 $\mu$m, and (d) 10 $\mu$m.
In the critical point dried sample (**a**) fibrous illite occurs coating the entire grain on the right-hand side of the field of view. The bright surface in places reflects 'charging' in the SEM due to the delicate illite particles not being thoroughly coated with carbon. After air drying (**b**) much of the illite has been pulled together to present ridges of illite to the open pore space. The delicate fibres seen projecting from the surface of the critical point dried sample are also not evident. Critical point drying also reveals thin delicate illite fibres growing on and between kaolinite particles (**c**), that do not survive air drying (**d**).

**Table 1.** *Summary of illite distributions and drying effects, Cormorant Field*

| Sample details | | Pore fluid | Illite morphology | | | | Drying effect |
|---|---|---|---|---|---|---|---|
| Formation | No. | | Pore lining | Blooming | Pore bridging | Replacing kaolinite | |
| *Well 211/26-CA35* | | | | | | | |
| Ness | 2 | Oil | x | | | | 1.08 |
| Etive | 3 | Oil | x | | | | 0.84 |
| Rannoch | 3 | Oil | | | | x? | 1.11 |
| *Well 211/26-CA30* | | | | | | | |
| Ness | 3 | Oil | x | x | | | 2.46 |
| Ness | 2 | Water | x | x | x | x | 2.12 |
| Etive | 10 | Water | x | x | x | x | 3.67 |
| *Well 211/26-CAUW2* | | | | | | | |
| Ness | 5 | Water | x | x | x | x | 3.44 |
| Etive | 2 | Water | x | x | | x | 2.78 |
| Rannoch | 1 | Water | x | | | | 2.77 |

Note: the average drying effects quoted within the text include additional samples that were not examined petrographically.

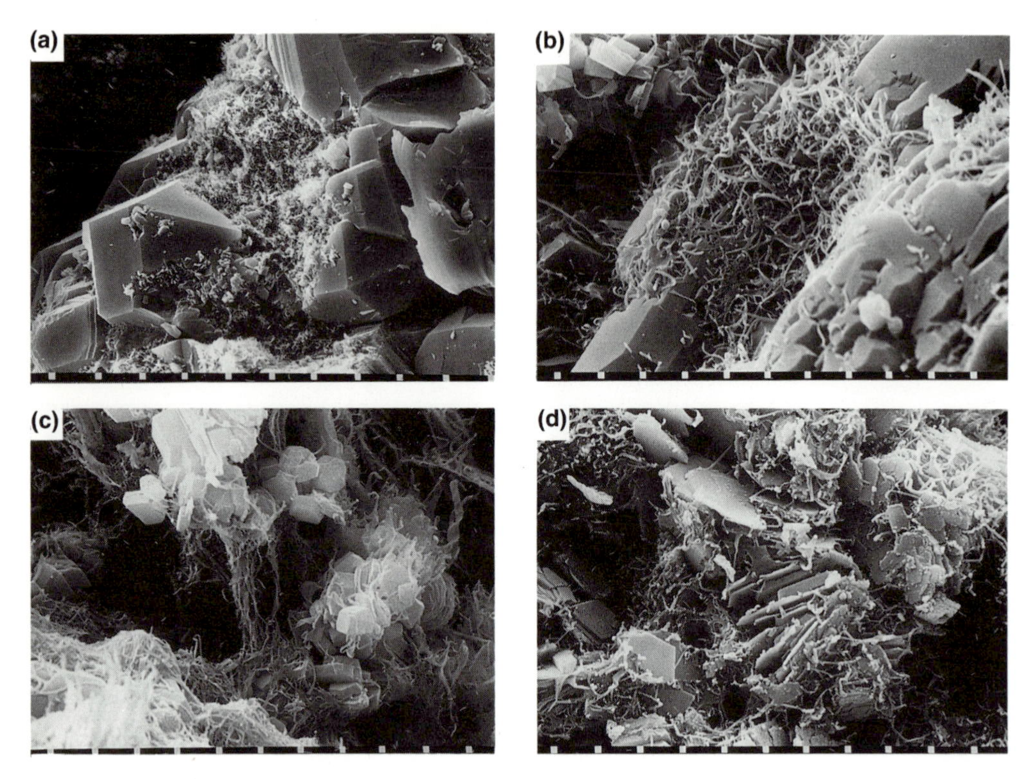


**Fig. 4.** Scanning electron micrographs of the four types of illite present in the Cormorant Field in critical point dried samples. Depths are drillers' depths in feet. (**a**) Pore-lining or grain-coating illite on a detrital grain and partially covered by quartz overgrowths, 211/26-CAUW2, 9927 ft. Scale bar 10 μm. (**b**) Blooming illite that forms broad laths of illite growing out between partially developed quartz overgrowths. Blooming illite is considered to represent continued or renewed growth of the grain-coating illite seen in (a), 211/26-CAUW2, 9981 ft. Scale bar 3 μm. (**c**) Pore-bridging illite that occurs growing between grain-coating illites on one side of a pore and the kaolinites and illite that replaces kaolinites within the pore, 211/26-CA30, 13724 ft. Scale bar 3 μm. (**d**) Illite that replaces kaolinite, 211/26-CAUW2, 10024′. Scale bar 3 μm.

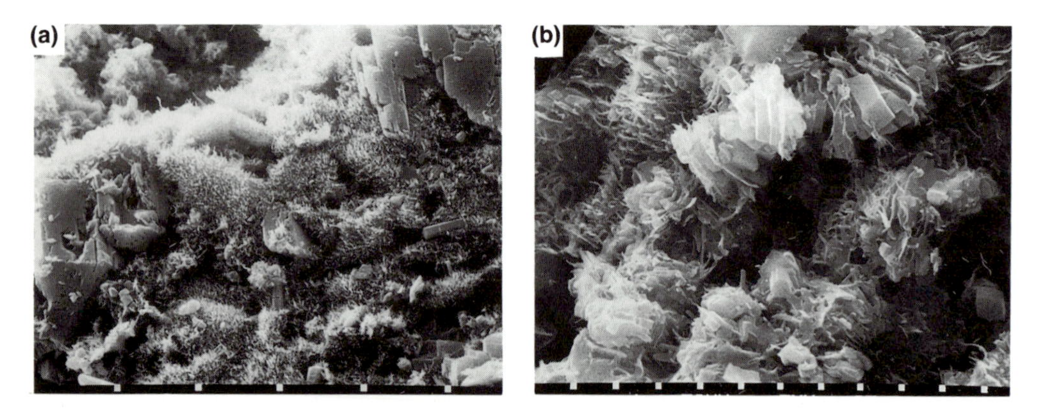


**Fig. 5.** Scanning electron photomicrographs of Tern Field illites after critical point drying (depths are driller's depths in feet). (**a**) Pore-lining and blooming (pore-filling) illites partially coating a feldspar grain and overgrowths. 210/25A-5, 8347.6 ft, scale bar 10 μm. (**b**) Illite intimately intergrown with and replacing late blocky kaolinite. 210/25A-5, 8314.0 ft, scale bar 3 μm.

**Table 2.** *Permeabilities and drying effects, Brent Group sandstones, Well 210/25A-5, Tern Field*

| Formation | Pore fluid | $K_{air}$ CPD (mD) | Drying effect | ($K_{air}$ AD/$K_{air}$ CPD) |
|---|---|---|---|---|
| Upper Ness | Oil | 52.8 | 1.46 | |
| Upper Ness | Oil | 2.04 | 1.23 | |
| Lower Ness | Oil | 3600.0 | 1.71 | |
| Lower Ness | Oil | 45.1 | 2.59 | |
| Etive | Oil | 442.0 | 1.80 | |
| Etive | Oil | 836.0 | 2.55 | |
| Etive | Oil | 569.0 | 1.99 | Av. 1.9 |
| Rannoch | Water | 140.0 | 2.39 | |
| Rannoch | Water | 595.0 | 1.73 | |
| Broom | Water | 0.018 | 0.94 | Av. 1.7 |

oil–water contact, and 1.7 below (2.1 excluding the low permeability Broom Formation sample), indicate that the conventional core analysis measurements in this well are almost double as a result of damage to illites during drying. The similarity in illite morphology and resulting average drying effect above and below the oil–water contact indicate that in Tern, hydrocarbon emplacement occurred after illite had grown throughout the reservoir.

### *Discussion*

The damage to illites observed here increases measured permeability by a factor of 2 or 3 when pore-filling illites are present. The role of illite particle strength has also been highlighted by Pallatt *et al.* (1984) and Kantorowicz (1990): particle thicknesses are generally inversely proportional to the drying effect. The particle thickness data and drying effects for Tern are comparable to those for the Cormorant aquifer sandstones. Additional data are required to enable prediction of drying effects in deeper Brent Group reservoirs with more prolonged illite growth (see below). It is possible that increased particle size and an increasing proportion of pore-bridging illites would lead to a decrease in the drying effect (*cf.* data from the Rotliegend; Kantorowicz 1990). The fact that variations in the morphology of a mineral that represents 1–2% of a sandstone can have such an effect highlights the importance of being able to predict where abundant and diverse illite morphologies may be present, and of collecting and suitably preserving cores from which to obtain realistic data.

## Cormorant injectivity tests

The development of Cormorant Block II involves water injection to provide aquifer support (Stiles & McKee 1986; Stiles & Hutfilz 1988). Core flooding experiments were undertaken to establish if sea water could be injected without causing interaction with the reservoir leading to a decline in permeability and consequently in injection rates.

### *Injection of produced water*

The first experiments used water similar to that produced from the field (Table 3) as this was expected to be in equilibrium with cores from the reservoir. Injection caused an initial permeability drop, irrespective of the rate (Fig. 6). As the reduction in permeability (impairment) was of a similar magnitude at high and low flow rates, it is more likely to have been caused by chemical processes such as clay dispersal rather than the physical mobilization of fines. Chemical analysis indicated that cation exchange between the injection waters and the core occurred at the onset of injection, after which the effluent water compositions stabilised.

The behaviour of clay minerals can be explained by their surface chemistry. Clays have a net-negative surface charge due to low- for high-

**Table 3.** *Chemical composition of produced water, Cormorant Field*

| | mg/l |
|---|---|
| $Na^+$ | 5860 |
| $K^+$ | 129 |
| $Ca^{++}$ | 100 |
| $Mg^{++}$ | 23 |
| $Ba^{++}$ | 20 |
| $Sr^{++}$ | 25 |
| $Cl^-$ | 9300 |
| $SO_4^{2-}$ | 17 |
| $HCO_3^-$ | 170 |
| pH | 8.5 |

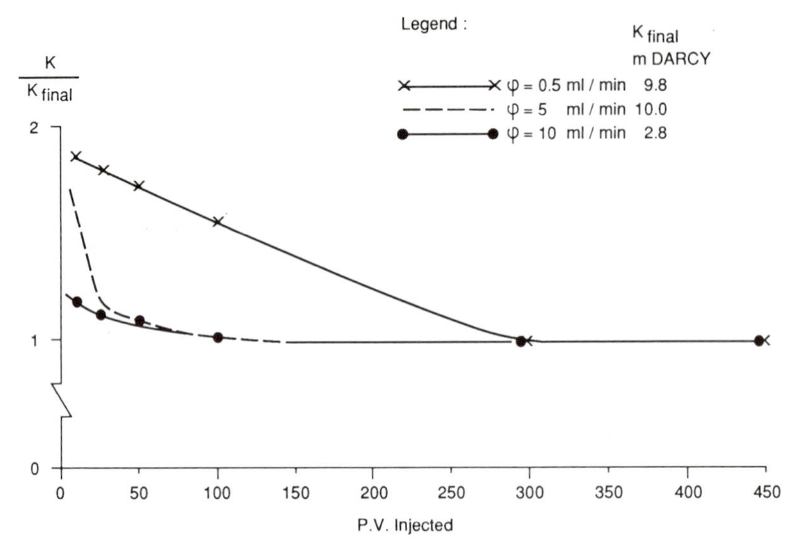


**Fig. 6.** Permeability response curves for Etive Formation samples, 211/26-CA30, during core flooding with simulated produced waters. At low flow rates ($\varphi$ = 0.5 ml/min) permeability ($k$) dropped gradually, and stabilized after around 300 pore volumes (p.v.) of water injection. At higher rates (5 ml/min) permeability dropped rapidly during the first 20 pore volumes of injection and then slowly stabilised. At very high rates (10 ml/min) an apparently stable permeability is measured throughout. This may reflect an initial drop before the first measurements were taken.

valence cation substitution in the lattice (van Olphen 1977). This surface charge is balanced by a 'double-layer' of cations, on or near the clay surface that are exchangeable with those in the pore water, giving rise to the clay's cation exchange capacity (van Olphen 1977). Clay behaviour reflects attractive van der Waal's forces between the particles and repulsive (osmotic) forces between the surface cations. Saline pore waters cause the double layer to contract, and particles become attracted to each other and start to flocculate. When pore water salinity is lowered the layer expands, separating clay particles until van der Waal's forces are overcome by osmotic forces and the particles disperse, reducing permeability.

Analysis of the effluent water compositions following produced water flooding of Etive Formation cores indicates that impairment resulted from cation exchange and hence incompatibility between the introduced fluids and the core.

## *Injection of stabilizing brines*

The impairment that results from changes in pore-water chemistry can be avoided by stabilizing the clays present. This can be achieved by injecting brines with a ratio of monovalent to divalent cations of 10:1 or less, and by injecting waters with salinity equivalent to or greater than the original formation water (Khilar & Fogler 1983; Lever & Dawe 1984; Scheuerman & Bergersen 1989).

During flooding experiments with a variety of brine compositions, clays are stabilised and impairment avoided with $NaCl:CaCl_2$ ratios of 10:1. In addition, permeability increases when 10:1 $NaCl:MgCl_2$ solutions or greater salinity solutions are injected (Fig. 7). Effluent water compositions measured during brine flooding revealed the extent of cation exchange. Effluent pH also varied (Fig. 8), with a sharp initial rise caused by hydrogen behaving in the same way as other pore water cations and being exchanged with the clay surface cations.

## *Sea water injection*

After stabilizing the clays, sea water was injected into the cores (Fig. 8). The initial and equilibrium pH varied, but no impairment occurred. Although the clays were stabilized with respect to the injection waters, it is still unclear whether this reflects in situ conditions, or 'stabilisation' due to clay flocculation in the injection brines. To assess this question, additional experiments were performed to try and establish a pH that would not vary during flooding (cf. Sharma *et al.* 1985). Figure 9 presents the results

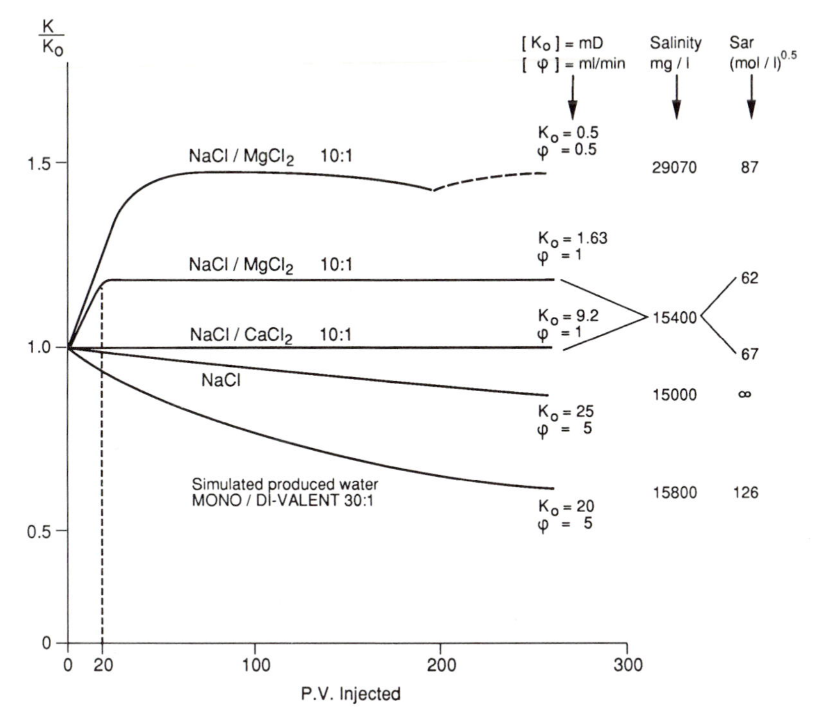


**Fig. 7.** Core flooding tests to show the sensitivity of 211/26-CA30, Etive Formation core, to variations in injection water chemistry. Experiments were undertaken with varying flow rates, and as a result of the differing injection water chemistries, varying sodium adsorption rates (SAR). The SAR is defined as $Na/(Ca + Mg)^{0.5}$ (Reeve & Tamaddoni 1965; Naghshineh Pour *et al.* 1970). $K_0$ is the original permeability.

for kaolinite and illite standards and Etive Formation core (quartz 58%, kaolinite 7%, illite 1%, feldspar 9% and 25% porosity). The results show that for kaolinite the point at which pH remains constant during flooding is 4.1. This may be termed the point of zero net proton change (PZNPC) and corresponds to the minerals' Zeta Potential. For illites this was found to be 7.2 and for Etive cores 6.4.

Having established that no proton exchange took place during flooding with pH 6.4 solutions, a simulated formation water composition was prepared with a pH of 6.4, rather than the 8.3 measured from the produced water. No impairment occurred when this water was injected into the cores, or during subsequent sea water injection, indicating that sea water injection in the field should not lead to clay dispersal and impairment. By increasing the injection rate it was possible to achieve stable brine permeabilities that approached the air permeabilities measured after critical point drying (Fig. 10). This suggests that realistic permeabilities can be measured on cores that contain chemically stable clays.

## *Discussion*

These results demonstrate that a cautionary approach should be taken when interpreting experimental data. It is well known that cores may be unrepresentative of reservoir conditions. However, the introduction of brines and divalent cations can stabilise the clays to such an extent that erroneously optimistic results are obtained.

In the previous section it was shown that permeability measurements are sensitive to the way in which the cores are prepared. These flooding experiments show how the response of a sandstone to water flooding is more closely related to the surface chemistry of a clay that represents 1–2% of the reservoir than to the properties of more abundant clays. Predicting the likely behaviour of a deeply buried sandstone during water injection is therefore dependant upon being able to predict the likely abundance of the more sensitive minerals. Figure 9 shows that the progressive alteration of kaolinite to illite during burial will alter the equilibrium pH of the rock towards 7. This

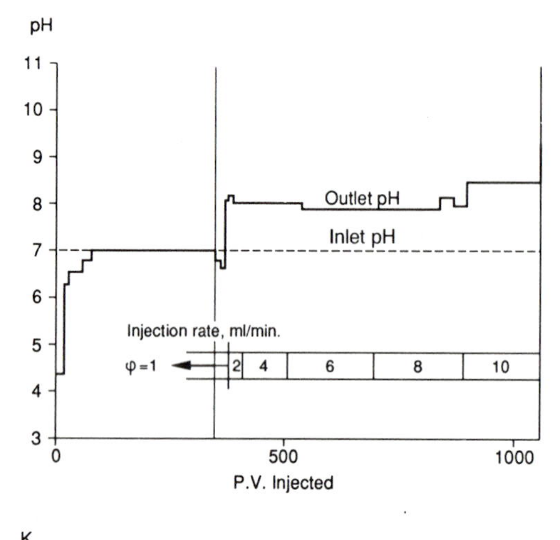


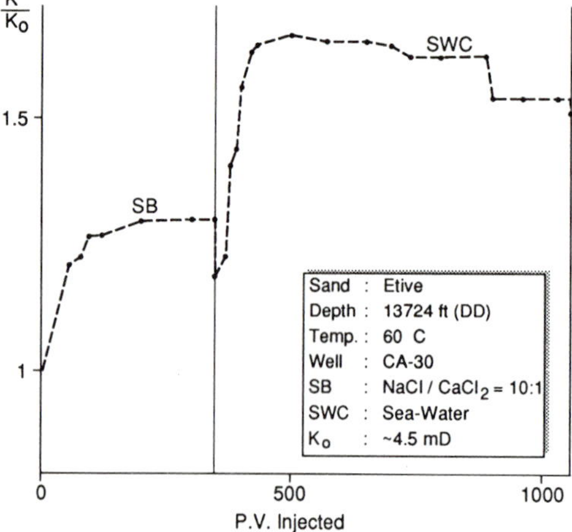


**Fig. 8.** Permeability and pH response during core flooding with stabilizing brine and sea water. Permeability initially increases following hydrogen ion exchange with the stabilizing brine, and then increases further as sea water injection occurs. $K_0$ is the original permeability.

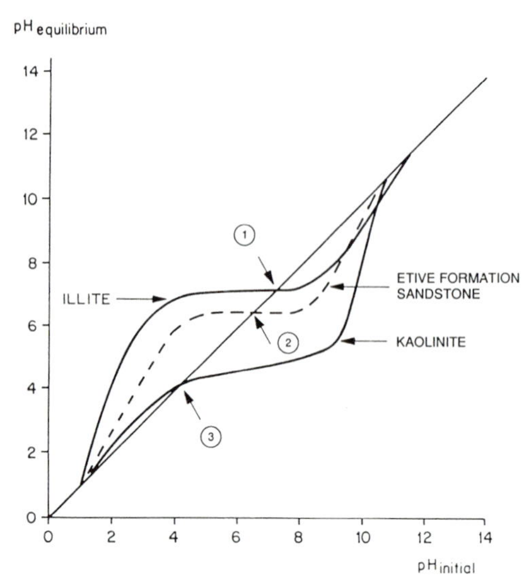


**Fig. 9.** Results of core flooding experiments on proton exchange reactions. Curves for illite, kaolinite and Etive formation sandstone cores are shown. The curves show the variations in initial and equilibrium pH for each mineral assemblage, and the point at which no proton exchange occurs (PZNPC): point 1, pH 7.2 for illite; point 2, pH 6.4 for the Etive core; and point 3, pH 4.1 for kaolinite. These are the points at which no change in pH and thus no cation exchange occurs during flooding.

should not cause impairment as a result of sea water injection into deeper Brent Group reservoirs.

## Diagenesis of the Pelican Field

The history of diagenetic modifications to the Brent Group sandstones in Pelican is broadly comparable to the diagenesis of other Brent Group reservoirs in the area (Fig. 2). Early clays (predominantly vermiform kaolinites) and carbonates (calcite and siderite) occur in places. These are followed by late blocky kaolinite (0–25%, average 5%), quartz overgrowths (0–20%, average 10%) and illite (0–4%, average 1%). Illite replaces kaolinite and occurs as a well developed pore lining from which blooming and pore-bridging illites project (Fig. 11). Late ferroan dolomite also occurs locally (cf. Kantorowicz 1985). All the core recovered to date from Pelican is from oil-bearing sandstones and, assuming oil emplacement has arrested diagenesis, the observed modifications predate oil emplacement. The porosity and permeability of the sandstones now averages between 10–20% and 0.1–50 mD for the various reservoir units.

The timing of events within the paragenetic sequence can be constrained in two ways. Firstly, quartz overgrowths become increasingly abundant below 6000–7000 ft in most Brent Group sandstones (Giles *et al.* this volume), corresponding to precipitation at burial temperatures of 70–90°C. Assuming a relatively constant geothermal gradient quartz probably began to form abundantly in the late Cretaceous and after burial to over 2 km (Fig. 12). Secondly, K-Ar age dates from the authigenic illites in Pelican range from around 35–44 million years ago for Ness Formation samples, to 25–

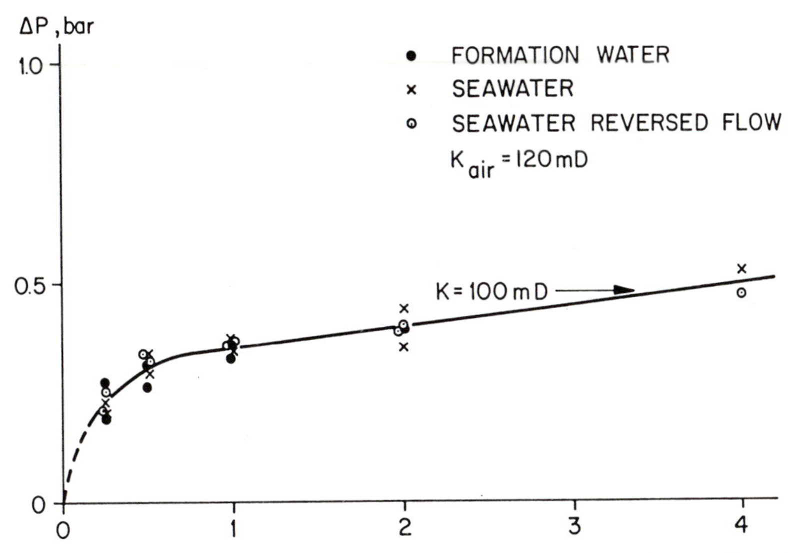


**Fig. 10.** Results of Etive Formation core flooding tests at pH 6.4 (equal to PZNPC or Zeta Potential of zero). The derivative of this curve is a measure of the permeability. In this case the permeability to brine, 100 mD, approaches that to air measured on a critical point dried twin plug. $\Delta p$ is the pressure drop measured across the plug during core flooding.

39 Ma for the Etive Formation (Table 4; the methodology and applications of K-Ar age dating are described by Hamilton *et al.* 1989). Coarse fractions from the illite separates have K-Ar ages up to about 9 million years older than the finer fractions (cf. Glasmann *et al.* 1989). This indicates that illite growth began at about 50 Ma or older, at a burial depth of around 8000 ft and continued during the Eocene and Oligocene (Fig. 12).

The Etive dates are younger than the Ness as direct illite precipitation in these cleaner Etive sandstones may have required higher temperatures than, for example, nucleation on detrital clays in the Ness. The Ness dates may reflect lower temperature nucleation on such clays, or lower temperature illitisation of precursor clays such as detrital illite or earlier authigenic mixed-layer illite-smectite (or contamination).

These dates are consistent with thermodynamic suggestions that illite may form at lower temperatures, but forms most readily at temperatures approaching 90–100°C (Bjørkum & Gjelsvik 1988). In Pelican, oil emplacement post dates illitization and, from the dates of the finer fractions in the Etive sandstones, is constrained to within the last 28–25 million years.

## Prediction of aquifer properties

Reservoir development may be complicated by the presence of extensive pore-filling illites

**Table 4.** *Summary of illite age dates from Pelican Field*

| Core depth | Depth (tvss ft) | Formation | Size fraction (μm) | Age (Ma)* |
|---|---|---|---|---|
| *Well 211/26–4* | | | | |
| 11759 | 11635 | Etive | 0.1–0.5 | 24.8 ± 0.7 |
| 11601 | 11490 | Ness | <0.1 | 40.8 ± 1.5 |
| *Well 211/26A-12* | | | | |
| 10999 | 10905.5 | Etive | 0.1–0.3 | 27.1 ± 0.6 |
| 10871 | 10799.5 | Ness | <0.1 | 34.5 ± 0.9 |
| 10871 | 10799.5 | Ness | 0.1–0.5 | 43.9 ± 1.0 |
| *Well 211/26A-13* | | | | |
| 12470 | 11079 | Etive | 0.1–0.5 | 27.3 ± 0.6 |

* Age dates include an error bar reflecting the quality of the age determination

(a)

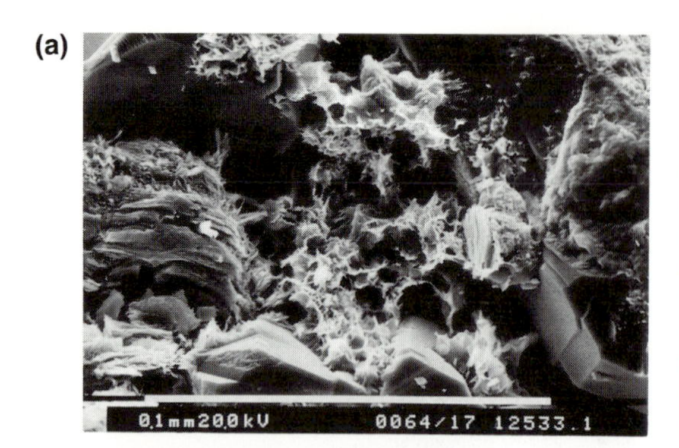


(b)

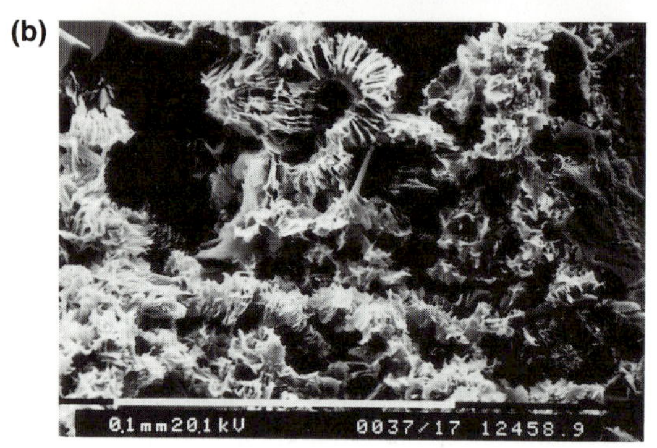


**Fig. 11.** Scanning electron micrographs of authigenic illites, Pelican Field. Depths are driller's depths in feet. (**a**) Pore-filling illite, growing from the grains around the edges of the pore and perhaps replacing a grain in the centre of the pore. 211/26A-13, 12533.1 ft, scale bar 100 $\mu$m. (**b**) Illite replacing kaolinite, 211/26A-13, 12458.9 ft, scale bar 100 $\mu$m.

around and below the oil–water contact. For example, in Heather (Fig. 1) the 'positioning of down-dip injection wells' was 'seriously affected' by reservoir properties (Gray & Barnes 1981). In NW Hutton (Fig. 1), up to 450 ft of oil-bearing sandstones occur below a productive limit of 12500 ft subsea (Scotchman *et al.* 1989). The impact of abundant illites is seen in the Upper Jurassic Magnus reservoir, where reduced aquifer permeabilities necessitated direct water injection into the oil column, and even this was only possible after novel stimulation (Drymond & Spurr 1988).

The amount of illite present in a reservoir is related to several factors (Macchi 1987; Bjørkum & Gjelsvik 1988), whilst with an unlimited supply of its constituents, its growth rate is probably related to both the rate and depth of burial (Bjørkum & Gjelsvik 1988; Kantorowicz 1990). In the absence of hydrocarbons, illite might be expected to grow uniformly throughout a sandstone, with local factors such as nucleation points or sources of potassium influencing its timing and abundance. Differences in the amount of illite above and below hydrocarbon-water contacts are generally taken to indicate that hydrocarbon migration into a reservoir has arrested illitization (Lee *et al.* 1985; Hamilton *et al.* 1989). If the timing of illitization is known, this information can be used in a variety of ways. Firstly, differences in age dates above and below an oil–water contact may constrain the timing of charge. Secondly, differences in the amounts or types of illite present may help to constrain and elucidate any general models predicting increased illite growth with depth.

Illite age dates are available for several fields in the area around Pelican. Data from NW Hutton are quoted in Scotchman *et al.* (1989) although no details of sample numbers, size fractions, depths and error ranges are given. The data quoted suggest the onset of illitisation at 49 Ma (Fig. 13), arrested on average in the oil-leg at 43 Ma (Scotchman *et al.* 1989) whilst deeper (water leg) dates are only 1–2 million years younger on average. Data from NW Hutton quoted by Hamilton *et al.* (this volume) of 45 Ma in the oil-leg and 33 Ma in the water extend these ranges but do not alter interpretation of the timing of oil emplacement as around 43–40 Ma. In Heather illitization began in the oil-bearing sands around 55 Ma and continued until 27 Ma when the reservoir was charged (Glasmann *et al.* 1989, their fig. 18). No aquifer dates are available. Data from the finest fractions are interpreted by Glasmann *et al.* (1989) to show that illitization was arrested by slow downward migration of the oil-water contact from 45 Ma until 27 Ma. In Pelican, the onset of illitisation is around 44 Ma or older and continued until 25 Ma (Fig. 13, Table 4).

In these Brent Group sandstones the onset of illitization may be constrained to burial through a window of approximately 80–90°C and depths of 8–9000 ft. Comparing Heather, NW Hutton and Pelican, the differences in the onset of illitisation reflect these reservoirs being buried into an 'illite window' at different times. The variations could reflect differences in the burial histories of the individual reservoirs, or variations in local geothermal gradient histories causing the position of the illite window to vary. (Obviously, the picture could be modified by the publication of more details concerning NW Hutton). Illite formation began relatively early in both Pelican and Heather and continued for up to 30 Ma before charge. This prolonged illite growth produced relatively poor reservoirs in which a diversity of illite morphologies occur above the oil-water contact (Fig. 11, see also

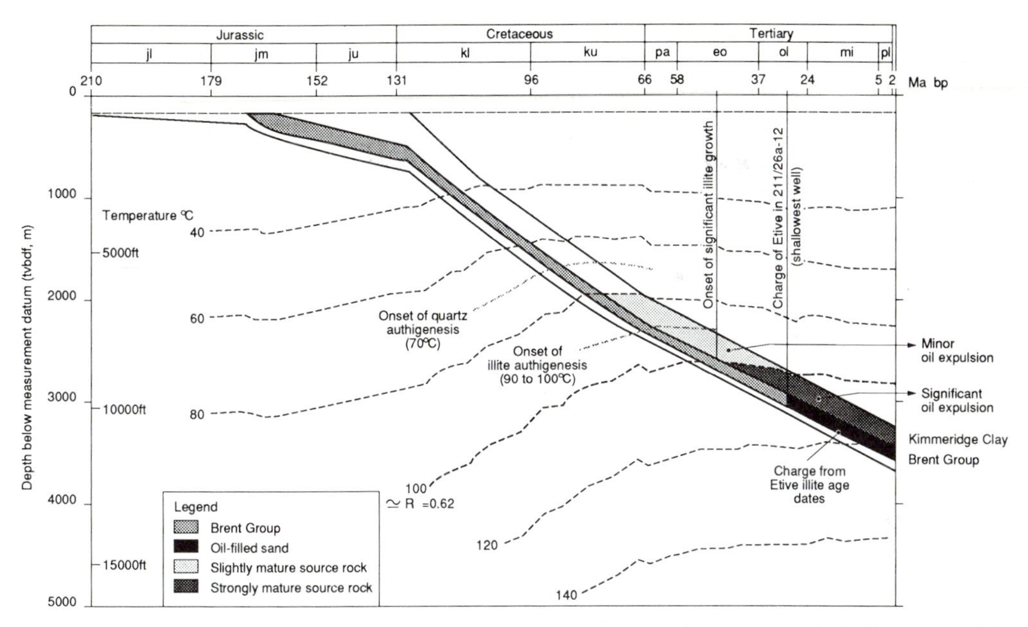


**Fig. 12.** Burial history of Brent Group sandstones, Pelican Field. The burial history of the Pelican reservoir is shown together with isotherms to constrain the timing of quartz and illite authigenesis. The timing of maturation of adjacent Kimmeridge Clay Formation source rocks is indicated. Peak maturation in the source rocks ($R_0$ 0.62) overlaps the timing of the youngest illitization prior to oil emplacement.

Glasmann *et al.* 1989, fig. 11). By contrast, charge appears to have reached NW Hutton relatively soon after the onset of illite growth and so better reservoir properties have been preserved in shallower sandstones. More abundant illite growth occurred subsequently in deeper sandstones and in those below the oil–water contact (see Scotchman *et al.* 1989, their figs 7 and 9). In Cormorant, no data are available to constrain the timing of charge, although it is observed that more illite occurs in the aquifer (Kantorowicz 1990). In Heather, NW Hutton and Cormorant the small quantities of illite present have reduced aquifer permeabilities (Gray & Barnes 1981; Scotchman *et al.* 1989; Kantorowicz 1990). There are no data to explain why much younger dates are not recorded in these sandstones although considering that illite precipitation is likely to involve 'growth' of existing crystals rather than new crystal formation (Nadeau 1985), the dates recorded from aquifer sands probably reflect averages for the duration of each illite particle's growth. It is not unreasonable that no ages approaching 0 Ma are recorded. See also discussion in Hamilton *et al.* (this volume).

Combining the data for the different reservoirs suggests that the structures were charged with oil from different areas of deeply buried Kimmeridge Clay source rock ('kitchens'): NW Hutton around 40 Ma, Heather from around 45 Ma until 27 Ma and Pelican from around 28 Ma onwards. Pelican is one of the deepest reservoirs, contains a diversity of pore-filling illites, and has amongst the poorest reservoir properties. The field was charged relatively late. Rather than being charged during early burial with oil from a deep source, Pelican was charged from a nearby source rock that matured after the reservoir entered the illite window (Fig. 14a). Hamilton *et al.* (this volume) stress that many fields in the area record illitization until local source rocks are matured to a level of $R_0$ 0.62, suggesting charge from the adjacent Kimmeridge Clay. Cormorant, by contrast, may be seen as charged prior to significant illite growth but then buried to a depth at which illitization has reduced aquifer properties significantly (Fig. 14b). Optimal reservoir and aquifer properties can be expected where the reservoir remains above the illite window, and is charged from a more deeply buried source (Fig. 14c).

Water injection into Brent Group sandstones will be easiest in shallow fields unaffected by illite growth. As permeability declines with increasing depth, achievable injection rates are likely to decline to a point where water injection is uneconomic. Nevertheless, the injectivity

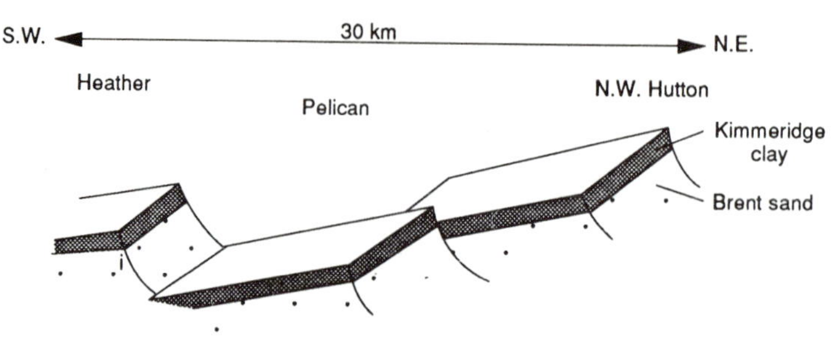


(a) 60-50 million years ago. Onset of illite growth in Heather (57 Ma)

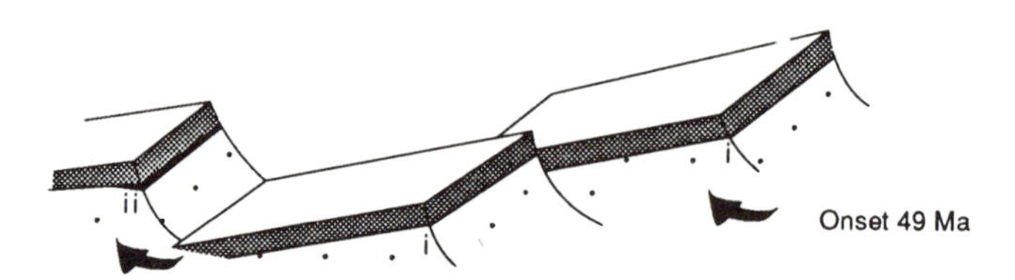


(b) 50-40 million years ago. Continued illite growth in Heather with possible onset of charge(45-40 Ma). Onset of illite growth (49 Ma) and subsequent arrest by charge (43 Ma) in NW Hutton. Onset of significant illite growth in Pelican (44 Ma)

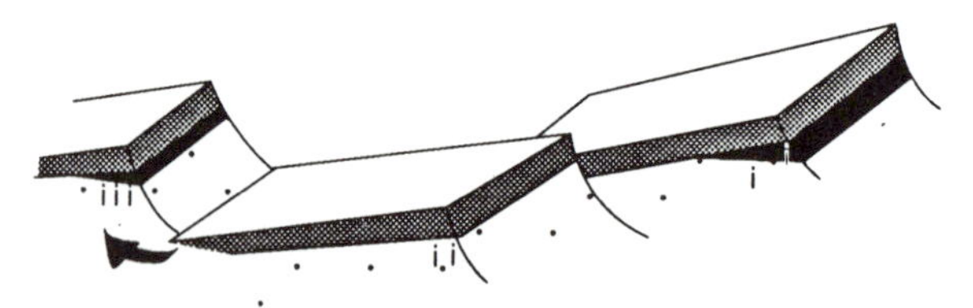

(c) 40-30 million years ago. Oil charge continued to fill Heather. Illite growth continued in Pelican and Heather, and in Hutton aquifer.

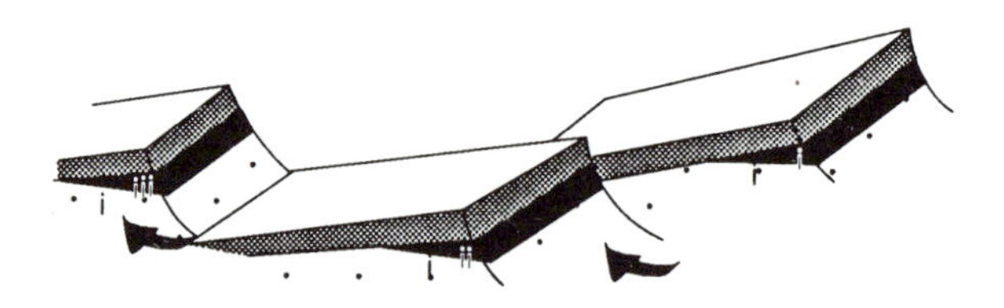

(d) 30-20 million years ago. Illite growth arrested by charge in Heather (27 Ma) and in Pelican (25 Ma).

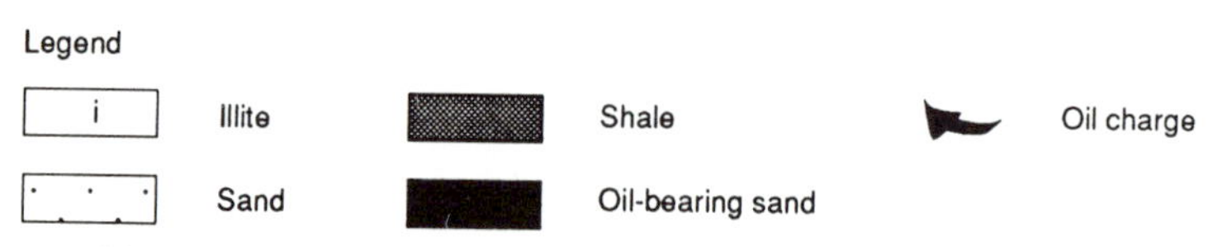


**Fig. 13.** Schematic evolution of Heather, Pelican and N.W. Hutton structures through the Tertiary illustrating the relative timing of illite growth and oil emplacement. See Fig. 1 for field locations.

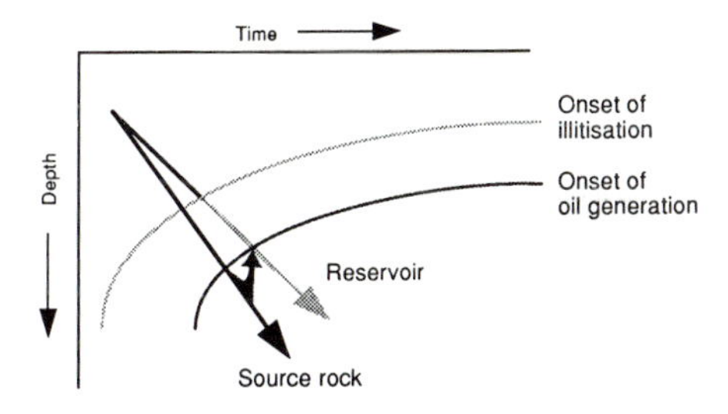


a) Charge into poor reservoir with poor aquifer properties

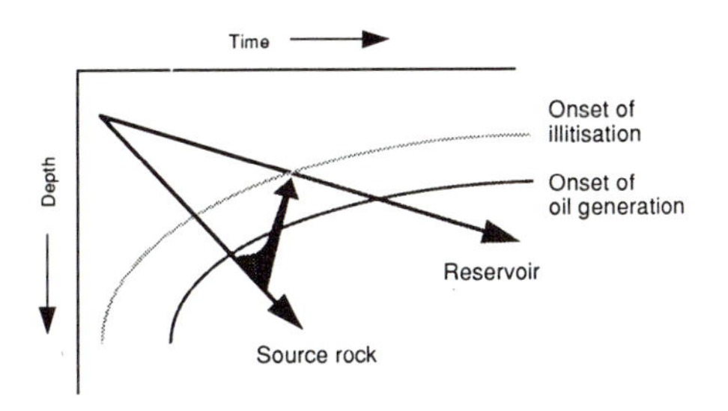


b) Charge into good reservoir with potential illitisation of aquifer

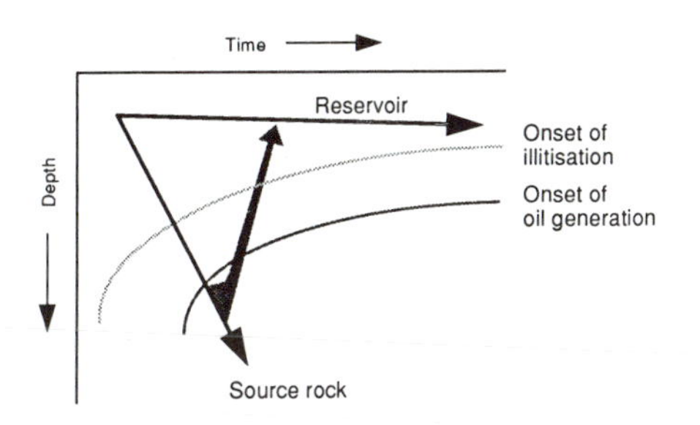


c) Charge into good reservoir with good aquifer properties

**Fig. 14.** Comparison of burial curves for reservoirs and source rocks illustrating the impact of relative burial rates through the illite and oil windows respectively. (**a**) Rapid burial of a reservoir and source rock leads to the reservoir passing through the illite window before the source rock passes through the oil window. Illite therefore grows throughout the reservoir prior to hydrocarbon charge. (**b**) Slower reservoir burial rates lead to the source rock passing through the oil window and charging of the reservoir before the onset of significant illitization. Illite may grow in the aquifer subsequently. (**c**) A more deeply buried source liberates oil that migrates up structure into a reservoir that does not enter the illite window. Illite will not occur in the oil reservoir and may not grow in the aquifer either.

studies from the Cormorant field suggest that as long as fines mobilization is avoided, sea water injection into these reservoirs should not lead to formation damage and impairment.

## Conclusions

Brent Group cores require careful handling and preparation before laboratory analysis to avoid gathering spurious data. Permeability measurements may be factors of ×2 to ×3 too high as a result of damage that can be related to the morphology of the illitic clays present. This in turn can be related to the duration of illite growth. Differences between oil-bearing and aquifer sandstones reflect the relative timing of illitization and charge.

Sea water injection into Brent Group sandstone reservoirs containing small amounts of illite should not lead to clay dispersal and impairment problems. Where the amount of illite present has decreased permeabilities to levels at which injection is unlikely to achieve economic rates, then direct injection into the oil column should be considered. Whether reservoir properties will differ significantly from aquifer properties will depend upon the burial history of both the reservoir and the hydrocarbon source rock. Only reservoirs above the illite window, or charged very late after burial through the illite window can be expected to have similar properties above and below the oil–water contact. Reservoirs charged above and then buried below the illite window may have relatively poor aquifer properties and reduced injectivity in which case injection into the oil column may be favourable.

The reservoirs around Pelican reveal a gradual increase in illite abundance and corresponding decrease in reservoir quality with depth. The diversity of illite morphologies that develop also increases with depth. Pore-lining clays form first, followed by a variety of pore-filling morphologies. Pore-lining clays are present above the oil-water contact in Cormorant whilst pore-filling morphologies occur below. In Pelican, the prolonged illitization recorded by the K-Ar data is reflected by an extensive and diverse development of pore-filling clays within oil-bearing sandstones. It is likely that the aquifer will contain at least the same amount of illite and the same diversity of morphologies. As illite growth in Pelican is around the youngest recorded by any of these studies and charge is relatively late, aquifer properties are likely to be poor, but are unlikely to be significantly worse.

We should like to thank Shell International Research Maatschappij, Shell International Petroleum Maatschappij, Shell UK Exploration and Production Ltd., and Esso Exploration and Production UK Ltd., for permission to publish this paper. K. J. Bil, J. F. Brint, W. van Driel, P. J. Ealey, A. P. Heward, J. D.

Marshall, R. B. Mattern, M. E. van der Stadt, and J. A. de Waal, for assistance and advice during the separate studies that are discussed herein. L. van Eijk and H. Nieuwstraten for drawing the diagrams, and reviewers R. Garden, D. Emery and T. J. Primmer for their comments.

## References

BJØRKUM, P. A. & GJELSVIK, N. 1988. An isochemical model for formation of authigenic kaolinite, K-feldspar and illite in sediments. *Journal of Sedimentary Petrology*, **58**, 506–511.

BJØRLYKKE, K. & BRENDSDAL, A. 1986. Diagenesis of the Brent sandstone in the Statfjord field. *In*: GAUTIER, D. L. (ed.) *Roles of organic matter in sediment diagenesis*. Society of Economic Paleontologists and Mineralogists, Special Publication, Tulsa, Oklahoma, **38**, 157–167.

BLACKBOURN, G. A. 1984. Diagenetic history and reservoir quality of a Brent sand sequence. *Clay Minerals*, **19**, 377–389.

DRYMOND, P. F. & SPURR, P. R. 1988. Magnus Field: surfactant stimulation of water injection wells. *Society of Petroleum Engineers Reservoir Engineering*, February, 165–174 (paper first presented as SPE Paper 13980, *Society of Petroleum Engineers Offshore Europe Conference*, Aberdeen, 1985).

GILES, M. R., STEVENSON, S., MARTIN, S. V., CANNON, S. J. C., HAMILTON, P. J., MARSHALL, J. D., & SAMWAYS, G. R. 1992. The reservoir properties and diagenesis of the Brent Group: a regional perspective. *This volume*.

GLASMANN, J. R., LUNDEGARD, P. D., CLARK, R. A., PENNY, B. & COLLINS, I. 1989. Isotopic evidence for the history of fluid migration and diagenesis: Brent sandstone, Heather Field, North Sea. *Clay Minerals*, **24**, 255–284.

GRAY, W. D. T. & BARNES, G. 1981. The Heather oil field. In: ILLING, L. V. & HOBSON, G. D. (eds) *Petroleum Geology of the Continental Shelf of North West Europe*. Heyden and Sons, London, 335–341.

HAMILTON, P. J., KELLEY, S. & FALLICK, A. E. 1989. K-Ar dating of illite in hydrocarbon reservoirs. *Clay Minerals*, **24**, 215–231.

——, GILES, M. R. & AINSWORTH, P. K-Ar dating of illites in Brent Group reservoirs: a regional perspective. *This volume*.

JOHNSON, H. D. & STEWART, D. J. 1985. Role of clastic sedimentology in the exploration and production of oil and gas in the North Sea. *In*: BRENCHLEY, P. J. & WILLIAMS, B. P. J. (eds) *Sedimentology: Recent developments and applied aspects*. Geological Society, London, Special Publication, **18**, 249–310.

JOURDAN, A., THOMAS, M., BREVART, O., ROBSON, P., SOMMER, F. & SULLIVAN, M. 1987. Diagenesis as the control of the Brent sandstone reservoir properties in the Greater Alwyn area (East Shetland Basin). *In*: BROOKS, J. & GLENNIE, K. W. (eds) *Petroleum geology of North West Europe*, Graham and Trotman Ltd. London, 951–961.

KANTOROWICZ, J. D. 1984. The nature, origin and distribution of authigenic clay minerals from the Middle Jurassic Ravenscar and Brent Group sandstones. *Clay Minerals*, **19**, 359–375.

—— 1985. The origin of authigenic ankerite from the Ninian Field, U.K. North Sea. *Nature*, **315**, 214–216.

—— 1990. The influence of variations in illite morphology on the permeability of Brent Group sandstones, Cormorant Field, U.K. North Sea. *Marine and Petroleum Geology*, **7**, 66–74.

KHILAR, K. C. & FOGLER, H. S. 1983. Water sensitivity of sandstones. *Society of Petroleum Engineers Journal*, 55–64.

LEVER, A. & DAWE, R. A. 1984. Water-sensitivity and migration of fines in the Hopeman sandstones. *Journal of Petroleum Geology*, **7**, 97–108.

LEE, M., ARONSON, J. L. & SAVIN, S. M. 1985. K/Ar dating of time of gas emplacement in Rotliegendes sandstones, Netherlands. *American Association of Petroleum Geologists Bulletin*, **69**, 1381–1385.

MACCHI, L. 1987. A review of sandstone illite cements and aspects of their significance to hydrocarbon exploration and development. *Geological Journal*, **22**, 333–345.

MCHARDY, W. J., WILSON, M. J. & TAIT, J. M. 1982. Electron microscope and X-ray diffraction studies of filamentous illitic clay from sandstones of the Magnus Field. *Clay Minerals*, **17**, 23–29.

MORTON, A. C. & HUMPHREYS, B. 1983. The petrology of the Middle Jurassic sandstones from the Murchison field, North Sea. *Journal of Petroleum Geology*, **5**, 245–260.

NADEAU, P. H. 1985. The physical dimensions of fundamental clay particles. *Clay Minerals*, **20**, 499–515.

NAGHSHINEH-POUR, B., KUNZE, G. W. & CARSON, C. D. 1970. The effect of electrolyte composition on hydraulic conductivity of certain Texas soils. *Soil Science*, **110**, 124–127.

OLPHEN, H., VAN 1977. *An introduction to clay colloid chemistry*, 2nd edition. John Wiley and Sons.

PALLATT, N., WILSON, M. J. & MCHARDY, W. J. 1984. The relationship between permeability and the morphology of diagenetic illite in reservoir rocks. *Journal of Petroleum Technology*, **36**, 2225–2227.

PURVIS, K. 1991. Fibrous clay mineral collapse. *Clay Minerals*, **26**, 141–145.

REEVE, R. C. & TAMADDONI, G. H. 1965. Effect of electrolyte concentration on laboratory permeability and field intake rate on sodic soil. *Soil Science*, **99**, 216–266.

SCHEUERMAN, R. F. & BERGERSEN, B. M. 1989. Injection water salinity, formation pretreatment and well operations fluid selection guidelines. *Society of Petroleum Engineers International Symposium on oilfield chemistry*, Paper 18641, Houston, Texas, 33–49.

SCOTCHMAN, I. C., JOHNES, L. H. & MILLER, R. S. 1989. Clay diagenesis and oil migration in Brent Group Sandstones of NW Hutton Field, U.K.

North Sea. *Clay Minerals*, **24**, 339–374.

Sharma, M. M., Yortos, Y. C. & Handy, L. L. 1985. Release and deposition of clays in sandstones. *Society of Petroleum Engineers 59th Annual Conference and Exhibition*. Paper 13562, Houston, Texas.

Sommer, F. 1978. Diagenesis of Jurassic sandstones in the Viking Graben. *Journal of the Geological Society, London*, **135**, 63–67.

Stiles, J. H. & Hutfilz, J. M. 1988. The use of routine and special core analysis in characterising Brent Group reservoirs, U. K. North Sea. *Society of Petroleum Engineers European Petroleum Conference*, SPE Paper 18386, London, October 1988.

Stiles, J. H. & McKee, J. W. 1986. Cormorant: development of a complex field, *Society of Petroleum Engineers 61st Annual Technical Conference and Exhibition*, New Orleans, SPE Paper 15504, October 5–8.

Thomas, M. 1986. Diagenetic sequences and K/Ar dating in Jurassic sandstones, central Viking Graben: effects on reservoir properties. *Clay Minerals*, **21**, 695–710.

Waal, J. A. De, Bil, K. J., Kantorowicz, J. D. & Dicker, A. I. M. 1988. Petrophysical core analysis of sandstones containing delicate illite. *Log Analyst*, **29**, 317–331.

# The petrophysical characteristics of the Brent sandstones

BRIAN MOSS

*Moss Petrophysical Ltd, 1 Swaynes Lane, Merrow, Guildford, Surrey, UK*

**Abstract:** The commonly occurring ranges and the principal controlling factors of the porosity, permeability, capillary characteristics and fluid saturations of the Brent sandstones are presented. Whilst there is insufficient space to discuss these parameters in specific detail, their variability and the degree to which they are influenced by the geological characteristics of the Brent sandstones is reviewed. The paper shows, through the use of examples, that a thorough understanding of the petrophysical characteristics of a reservoir rock requires the integration of several types of data into a single model. The wireline log and laboratory core analyses provide a quantitative input, whilst geological factors such as mineralogy, diagenetic history, and sedimentary fabric provide a most necessary framework within which the pore geometry variability, detected by log and core data, can be examined. An appendix summarizes the principal quantitative techniques involved.

Petrophysical studies of reservoir rocks have three broad objectives:

(1) to establish the reserves present in a field;
(2) to derive a quantitative description of reservoir quality and its variability in three dimensions within the formation or formations comprising the field;
(3) to establish the extent of hydraulic continuity within the formations in order to allow the modelling of the behaviour of the contained fluids under the dynamic conditions of production and/or injection.

In order to attain these aims the petrophysicist works towards an understanding of the field in terms of attributes such as its pore volume and type, the geometry of that pore system, the type of fluids present and their saturations, the permeability distribution and the capillary pressure characteristics. A complete understanding requires knowledge of several geological attributes such as grain size, sorting, mineralogy and depositional environment. An intermediate objective is to describe the formation in terms of one or more 'rock types', within each of which the response or behaviour of the fluids to imposed pressure gradients is consistent wherever that rock type is developed within the field. Often, a single lithostratigraphic unit is subdivided into several rock types, but it is also possible that more than one lithostratigraphic unit can be described within a single rock type.

Johnson & Stewart (1985) stress the necessity of conducting careful well-log-to-core calibrations and interpretations in order to identify and then predict, from the widest possible data set (usually the log data), the distribution of particular facies associations. This is particularly important for the Brent Group because the fundamental rock-type characteristics of the various sand bodies are generally controlled by their primary sedimentary fabric. Subsequent diagenetic overprint varies in importance from minor (e.g. Brent Field) to highly significant (e.g. North Alwyn and Heather), but in terms of the Brent Group as a whole the dominant control on reservoir quality and extent is the depositional environment(s) in which the particular sand was deposited.

Therefore, there is, or should be, an intimate relationship between petrophysical and geological studies because the controls on petrophysical parameters are inherently geological, and the resulting petrophysical model will be used to 'translate' the geological model into terms that the reservoir and production engineer can employ in their dynamic representations of the reservoir.

## Geological setting

The Brent Group comprises five formations that characterize an overall progradation of a river delta system into a shallow sea in Mid-Jurassic times (Budding & Inglin 1981). The river sediment forming the delta was reworked into a coastal barrier system which separated the back barrier lagoons and delta plains from the higher energy marine environment.

At the base, the Broom Formation comprises what is thought to be an offshore sand sheet, genetically distinct from the rest of the Brent Group (Brown *et al.* 1987). It consists of poorly sorted, thoroughly bioturbated sands, interbedded with both micaceous and pebbly sands in a regionally continuous layer some 9–18 feet thick. Overlying the Broom, the Rannoch For-

*From* Morton, A. C., Haszeldine, R. S., Giles, M. R. & Brown, S. (eds), 1992, *Geology of the Brent Group*. Geological Society Special Publication No. 61, pp. 471–496.

mation comprises muds and fine-grained laminated micaceous sands, thought to be the product of lower shoreface environments whilst the Etive Formation comprises typically medium-grained, rather massive sands considered to have been deposited in the upper shoreface and beach environments of a barrier complex, with tidal channels cutting through the barrier sediments.

Overlying the Etive, the Ness Formation comprises the many different environmental settings found on the delta plain, e.g. interbedded sandstones, siltstones, mudstones and coals. The sandstones can be either upward-fining (channels) or upward-coarsening (bars) and the mudstones can be massive beds with rootlets or heterolithic units with silt or very fine sand streaks.

At the top of the sequence, the Tarbert Formation is often characterized by a sharp erosional boundary overlain by bioturbated coarse sands, in turn succeeded by fine-grained micaceous sands resembling the storm-laid Rannoch shoreface sands lower in the sequence. The facies associations represent a complex of shoreline sands, back-barrier and lagoon-fill sands, and lagoonal silts, mudstones and coals in some areas. Most authors (e.g. Budding & Inglin 1981; Ronning & Steel 1987; Brown *et al.* 1987) consider the Tarbert to represent a transgressive phase in which the delta and barrier sediments have been reworked.

On a regional basis the five major formations in the Brent Group are easily identified and correlated (Brown *et al.* 1987), but within these major subdivisions considerable variability is seen.

## Principal geological influences upon the petrophysical characteristics of the Brent sands

Figure 1 shows a standard log interpretation of a Brent sequence, using parameters derived from crossplots (scatter diagrams) and published tables of the data (Serra 1984 p. 3). The standard log interpretation methodologies are outlined in the Appendix. As discussed later, this standard approach to reservoir evaluation can be improved upon, by identifying different rock types in the section, if the core data and geological considerations are combined with the wireline log interpretation. Despite this, the data in Fig. 1 are a reasonable approximation of the pore volume and saturation distribution that may be expected in this field (Ninian).

In petrophysical terms, the most important geological factors to consider are mineralogy and rock fabric. In the following, each of the parameters that the petrophysicist seeks to measure will be examined in the light of these factors as they affect the Brent sands.

### *Mineralogical influences on log-derived porosity derivation*

While very accurate porosities and grain density measurements may be obtained from core plugs using helium porosimeters and careful volumetric assessment, the principal logging devices used to determine porosity are the bulk density, the compensated neutron, and the acoustic travel time tools. There are at least two major uncertainties in the estimation of porosity from wireline log data in the Brent Group. These are, firstly, whether the assumption that the matrix (i.e. the grains) can be adequately represented by a single endpoint number for each logging tool (such as 2.65 gm/cc for the density of quartz), and, secondly, whether all the porosity that is measured contains fluids that are free to contribute to the production/injection stream.

Pevararo & Russell (1984) published the results of log interpretation for mineralogical content compared with core measurements for a Brent sequence and amply demonstrated the extent to which the constituent formations varied from the simple case (Fig. 2).

The Rannoch Formation contains significant quantities of micas which increase average matrix density to about 2.75–2.8 gm/cc. Therefore, the use of 2.65 gm/cc as an average matrix density produces erroneously low porosities. This can be seen in the Rannoch section of Fig. 1 (10540 ft to 10587 ft) where log porosity understates core porosity by a consistent margin of 4 to 5 porosity units. The millimetre-scale laminations of mica-rich and mica-poor sediments in the Rannoch is beyond the resolution of the principal logging tools, which therefore produce a smoothed average of the bulk volume of mica. Commonly, log derived volumes of mica are in the range 10–20% (Suau & Spurlin 1982). The influence of mica in the Rannoch interval is also seen on the density curve, (Rhob, Track 2, Fig. 1) which shifts towards higher densities (lower apparent porosities). The neutron curve is hardly affected, reflecting the relatively high porosity of these sands despite the presence of the mica.

The Etive tends to be a more simple system (Fig. 2). The major components of the sands are quartz and feldspar, with some heavy

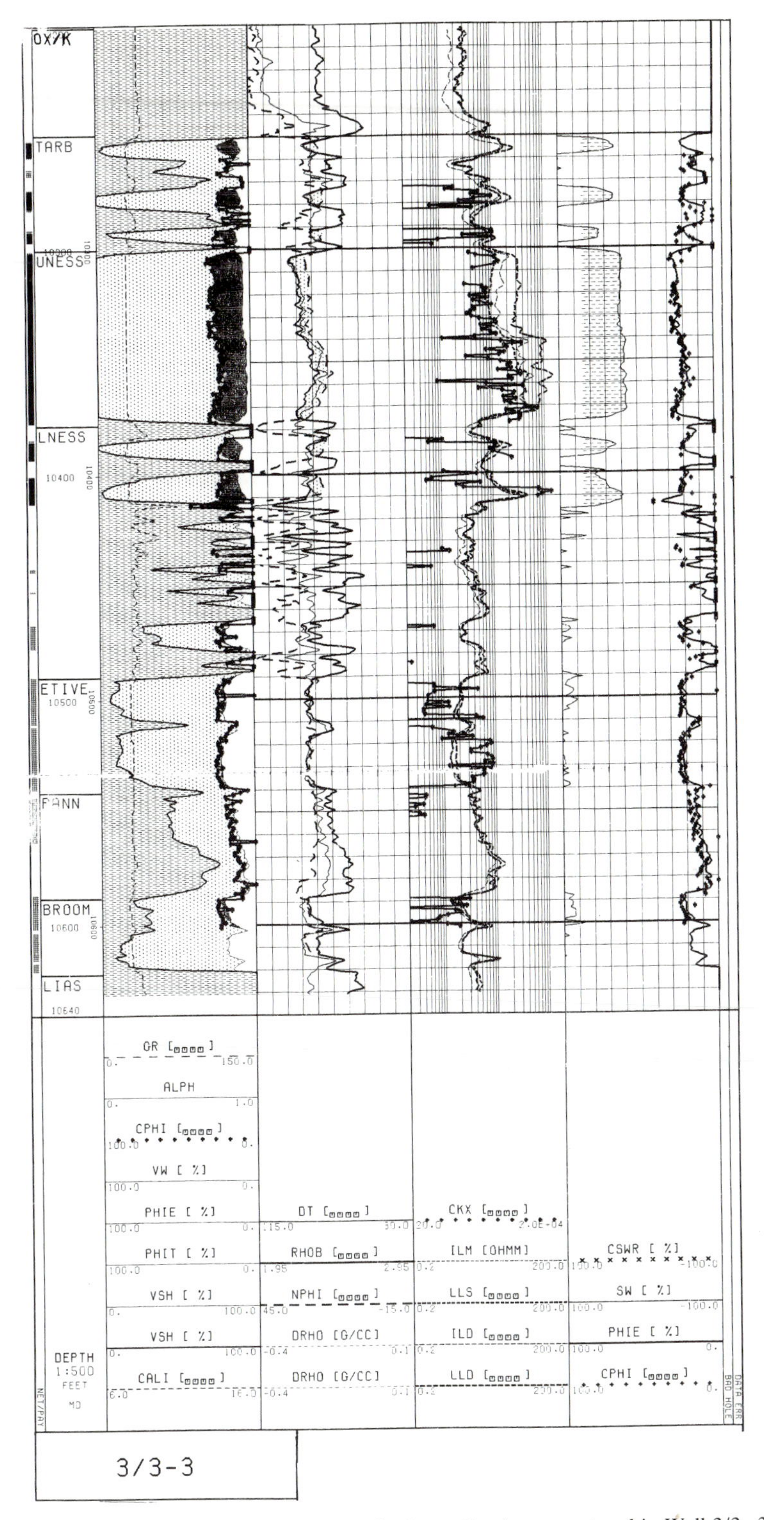


**Fig. 1.** Standard wireline log interpretation through the Brent Sands encountered in Well 3/3–3. Track 1 is computed data with core porosity. Track 2 is density, neutron and sonic log data. Track 3 is induction resistivity, and core permeability. Track 4 is computed saturation and porosity, with core porosity.

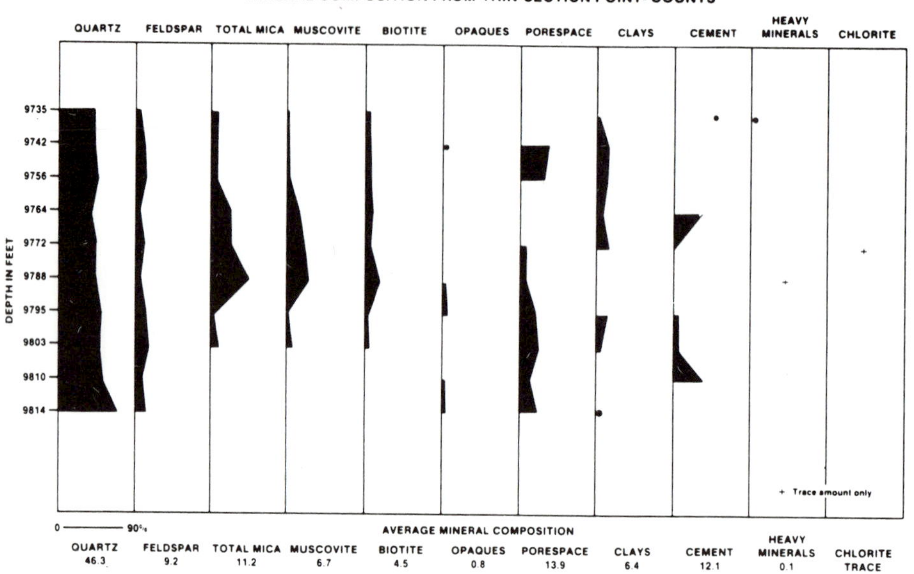


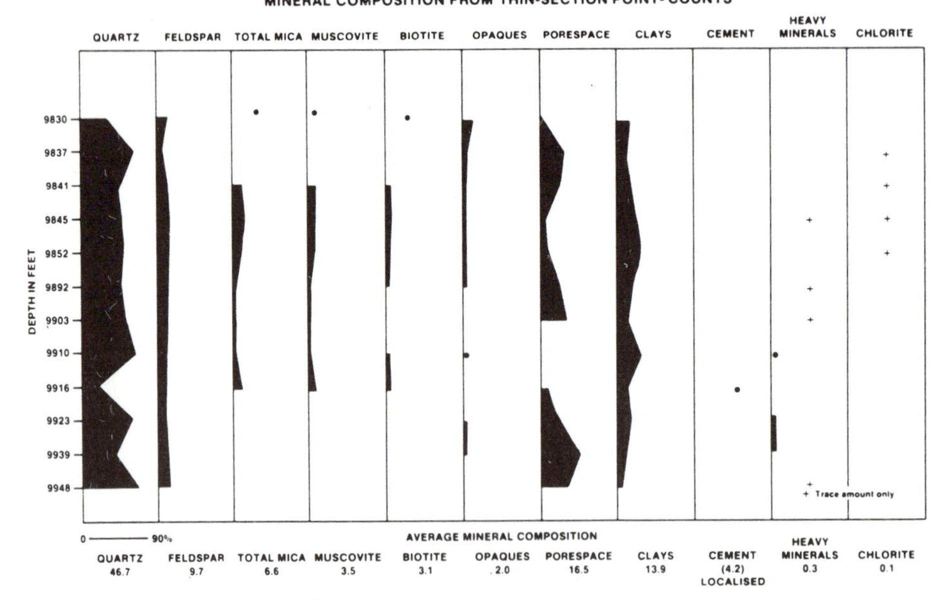


detrital mineral laminae. Whilst potassium feldspar profoundly affects clay mineral determinations using the natural gamma-ray log, its influence on the calculation of porosity is slight. Its presence reduces the actual matrix density slightly, but it is virtually invisible on the sonic and neutron devices. In general, therefore, potassium feldspar is not a problem for the bulk

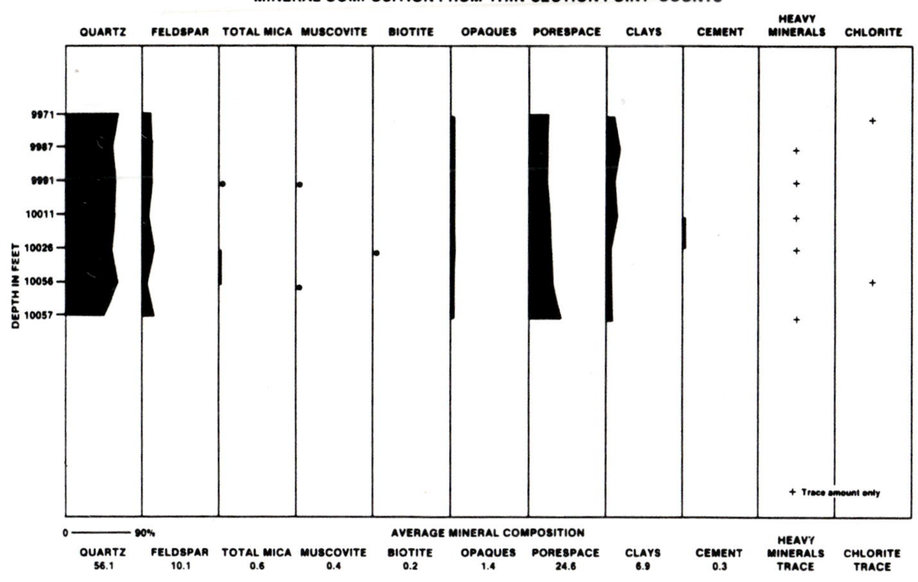


RANNOCH FORMATION

MINERAL COMPOSITION FROM THIN-SECTION POINT-COUNTS

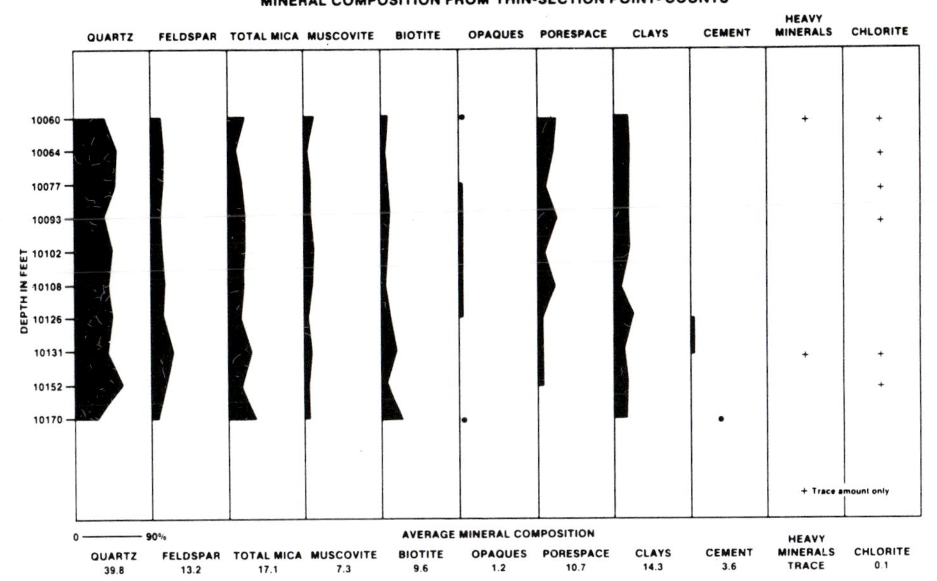


**Fig. 2.** Point-count data from a Brent sequence (no Broom data available). From Pevararo & Russell (1984).

estimation of total porosity. The thin carbonate cemented zones and concentrations of heavy minerals in the Etive increase the matrix density and thus counteract the effect of the feldspar presence. Consequently, the Etive often has the normal range of sandstone matrix densities (2.63–2.7 gm/cc).

The Ness sands are far more variable. Some

consist of virtually pure quartz, whilst others contain varying quantities of feldspar, mica, cements (both clay minerals and carbonates), and heavy minerals such as pyrite and siderite. Very few are simple systems amenable to being characterised by a single end-point number for each log. Nevertheless, Pevararo & Russell (1984) show the Ness nett sand characteristics as essentially similar to those of the Etive (mainly ranging between 2.63 and 2.68 gm/cc) but they note that the logs are affected by the thin-bedded nature of parts of the Ness Formation and that gradational zones between the sands (simple or complex) and the interbedded mudstones and silts can complicate interpretation. The impact of such gradational boundaries is greatest in the selection of nett sand for inclusion as meaningful reservoir volume (Hurst 1987).

The Tarbert Formation shows a range of composition from pure quartz to complexes of minerals, similar to those of the Ness sands. Some intervals contain more mica than is found in the Rannoch Formation (Fig. 2). Matrix density values for these sands lie in the range 2.65–2.8 gm/cc.

The effect of mineralogical variation on logging tools is most obvious on the density device, which is also the predominant tool for porosity assessment. The sonic tool is largely unable to distinguish between minerals other than detecting the presence of clays in general, and the actual observed response is a function of clay morphology, amongst other factors. The neutron tools have the least predictable responses to clay minerals, in general. These tools respond principally to the hydrogen content of the formation/fluid system, and thus the mineralogical variety and varied compaction history of clays, leads to a wide range in neutron responses (Patchett & Coalson 1982).

To achieve the highest accuracy, it is essential throughout much of the Brent sands that evidence for the perturbing effects of mineralogy is sought and that appropriate adjustments are made before making automatic calculations of porosity from log data. Fortunately, the minerals present do not pose any particular problems for core analysis, and so it should be possible to obtain accurate values for total porosity and bulk grain densities for cored intervals, which can then be used to calibrate log data in similar facies associations. However, partial coring of the Ness sands, in particular, could generate incomplete or unrepresentative data because of the high degree of variability of this formation. Crossplots of two logging tools may be used to derive porosity (see Appendix), but care must be taken in assigning the matrix and 'shale' end-points for the interpretations, or unrepresentative results will occur. Confusing the high gamma-ray, high density signature of the Rannoch Formation, caused principally by the presence of detrital micas, with a clay effect, for instance, results in too much 'shale' volume and too little porosity being computed (Fig. 1). Spectral gamma-ray data may help distinguish the minerals in such cases (Marett *et al.* 1976; Suau & Spurlin 1982).

Another source of uncertainty in porosity measurements emanates from the fact that not all the calculated porosity is able to contribute effectively to the reservoir system. In this regard feldspars and clays pose the principal mineralogical problems.

Diagenetic processes affecting the Brent sands have had a variable impact on the destruction or preservation of reservoir quality. Some parts of the province have suffered far less from diagenetic alteration than have others. Pevararo & Russell's (1984) data came from an area only lightly affected by diagenesis, as has Johnson & Stewart's (1985) and, indeed, many other authors'. However, studies in Heather (Glasmann *et al.* 1989) and Alwyn (Jourdan *et al.* 1987) show that considerable reduction in reservoir quality can result from more extensive diagenetic activity.

One result of diagenesis in the Brent sandstones, however severe, is to introduce complexity to the pore geometry by the creation of secondary porosity and microporosity, neither of which commonly contributes to the effective movement of fluids in a reservoir. One early phase in the diagenetic history of the Brent Sands, an episode of feldspar dissolution under conditions of acidic pore water, affects reservoir quality by the creation of pore space through leaching and etching of feldspar grains, and by the reduction in porosity and pore throat dimensions caused by the reprecipitation of the dissolution products as kaolinite.

In the Brent sandstones this leaching of the feldspar grains can be complete or partial. When leaching is complete the resultant void space either remains open or acts as the precipitation site for later phases. If it remains open, such secondary porosity can give rise to highly porous but poorly permeable sands, by virtue of the fact that the relatively large void left by the feldspar grain is often only poorly connected to its neighbours. This is the case in the North West Hutton Field (Scotchman *et al.* 1989) where the secondary porosity can reach 8%. However, Scotchman *et al.* report that only in the Broom Formation does the secondary porosity connect sufficiently frequently to the intergranular pore system for the overall reservoir quality to be enhanced.

When the leaching is incomplete, the grains are etched and thus contain voids that are too small to allow the entry of hydrocarbons against the high capillary forces associated with such microporosity. Under these circumstances the total porosity measured by the density tool is greater than the effective porosity available for hydrocarbon storage and movement. Ineffective microporosity has also been noted in association with clay minerals within Brent sandstones (Keir & Brown 1978; Scotchman *et al.* 1989). Whilst microporosity associated with etched feldspars rarely attains volumetric significance, that associated with clay minerals may be of the order of several porosity units and therefore be of some importance to the correct characterization of the rocks. Fine-grained sands such as the Rannoch Formation and in parts of the Ness, particularly those intervals with a high clay content as well, can be dominated by microporosity to the extent that moderately porous sand intervals have very low permeabilities and are almost completely non-effective reservoirs. (See Fig. 1 between 10540 and 10587 ft, where the core permeabilities in track 3 are less than 0.01 mD.

An overall decline in porosity with depth is commonly observed in the Brent sandstones. Scotchman *et al.* (1989) quantify the effect in the North West Hutton Field at an average decline of 0.46 porosity units (PU) per 100 ft, from an average porosity of 20.7 PU in the crestal area at 11 130 ft subsea to 12.7 PU downflank at 12 870 ft subsea. In general, the decline is caused by an increasing abundance of quartz overgrowths with depth.

There is some evidence that differences in the porosity values calculated at each depth increment from the single-tool equations using the density and neutron log data (see Appendix) are indicative of changing geological factors in a given section. Figure 3 from an early paper on the Brent Sands (Lavers *et al.* 1975) shows that significant variation in this differential porosity perhaps has geological meaning, as the variation seems to correlate with the natural gamma-ray data. It has already been stated how the presence of micas may be detected from the density-neutron separation turning negative (e.g. the Rannoch interval, Fig. 1). Fluid type in the pores must be taken into account in such exercises, and, always, inferring too much from such qualitative indications is to be avoided. Hard evidence from core examination is needed before one can be objective about the causes of any observed variation, but a varying differential porosity is common and the fact that it is related to subtle changes in pore geometry can be used to improve dramatically the prediction of permeability from porosity data in the Brent sandstones. Patterns in the sonic-density data may also be used to highlight subtle changes in pore geometry.

### *Clay mineral influences on log-derived water saturation*

A major consequence of the diagenetic processes that were active in the Brent sands is the precipitation of authigenic clay minerals within the sand bodies. Their petrophysical significance commonly is profound, for the following reasons: (i) clays are associated with non-effective microporosity, (ii) clays form in a number of different crystalline habits, some of which are particularly prone to blocking intergranular pore throats and thus drastically reducing permeability, e.g. illite (Pallatt *et al.* 1984), and (iii) clays minerals have an associated conductivity which must be explicitly accounted for in the accurate determination of water saturation from electrical logging tools (see Appendix).

The amount of conductivity associated with the clay mineral is measured by the mineral's cation-exchange capacity (CEC). CEC is related to the specific surface of the clay and is thus dependent upon the clay type. The different clay minerals exhibit markedly different CEC values. Kaolinites have the lowest CEC at 3.25 meq/100 gm, illites record CEC values in the range 30–40 meq/100 gm whilst smectites and vermiculites exhibit CEC values in the range 80–260 meq/100 gm (Serra 1986, p. 463). Therefore, in the presence of clay minerals dispersed through the pore system, the formation may exhibit an excess of conductivity over what would be expected for an equivalent volume of electrolyte in a clay-free rock. Where clays are present as laminae, or discrete grains, this effect is reduced since the total surface area of such clays is smaller.

As for the magnitude of the excess conductivity within the Brent sands, Aaboe (1984) urges caution in the choice and/or measurement of parameters for use in either saturation model (see Appendix). His work suggests that the actual excess conductivity due to clay content in the sand bodies is insignificant for much of the Brent sands, in particular when the 'shale' volume is less than 40% of the total rock volume. In other words, his results indicate that

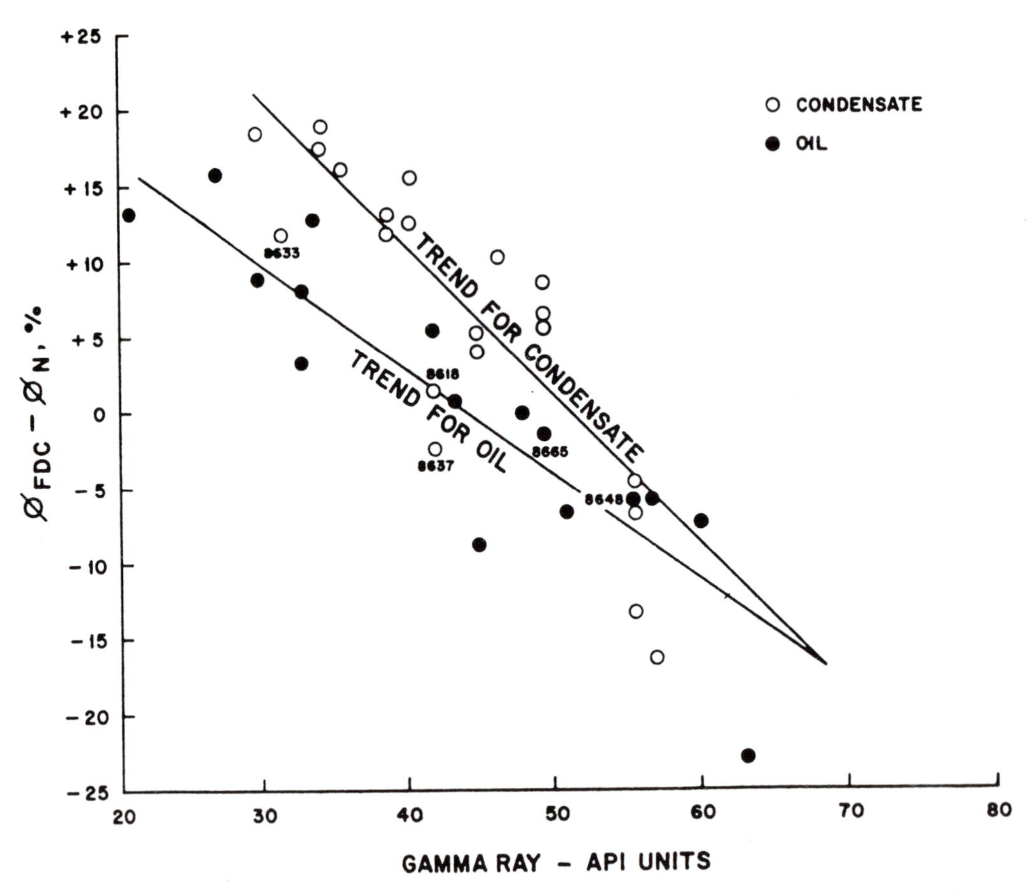


**Fig. 3.** Variation in density-neutron separation, expressed as the differences in their calculated porosities, plotted against natural gamma-ray data for different fluid types. From Lavers *et al.* (1976).

the nett reservoir sands of the Brent Formations are largely unaffected by clay conductivity. During the course of project work the author has corroborated this view, at least as regards part of the Norwegian Sector. On Fig. 4 the scatter at low shale volumes in the vicinity of the actual formation water resistivity of 0.08 ohm-metres is due to uncertainty in the choice of cementation exponent as discussed in a later section. The cluster of data plotting at 0.08 ohm-metres and 45% 'shale' volume is actually the micaceous Rannoch interval and is therefore not a simple clay-related phenomenon in this well. Consequently, the graph shows a stepped relationship, with the excess conductivity of the true clays not being effective until more than about 50% 'shale' volume is exceeded; there is essentially no excess conductivity in the cleaner sand intervals in this well.

The determination of the volume of clay present is problematical. Commonly, a volume of 'shale' is calculated from crossplots of two or more logging curves, or by making transformations directly from a single curve such as the natural gamma-ray (see Appendix). This 'shale' volume is necessarily, therefore, a composite of all other minerals present in the formation that, for each particular logging tool, can not be grouped together with the matrix. As a consequence, different logging tools will register different volumes of 'shale'. The difficulties in interpreting such data become severe when the formation is not a simple sand–clay system. The preceding discussion of the mineralogical complexity of the Brent Group sands made it clear that most Brent sands are not simple two component systems. Figures 5 and 6 illustrate the variability in the data away from the areas of pure sand responses. To what extent the non-sand components are actually clay minerals rather than micas, carbonates and other, heavier, minerals requires careful study of the

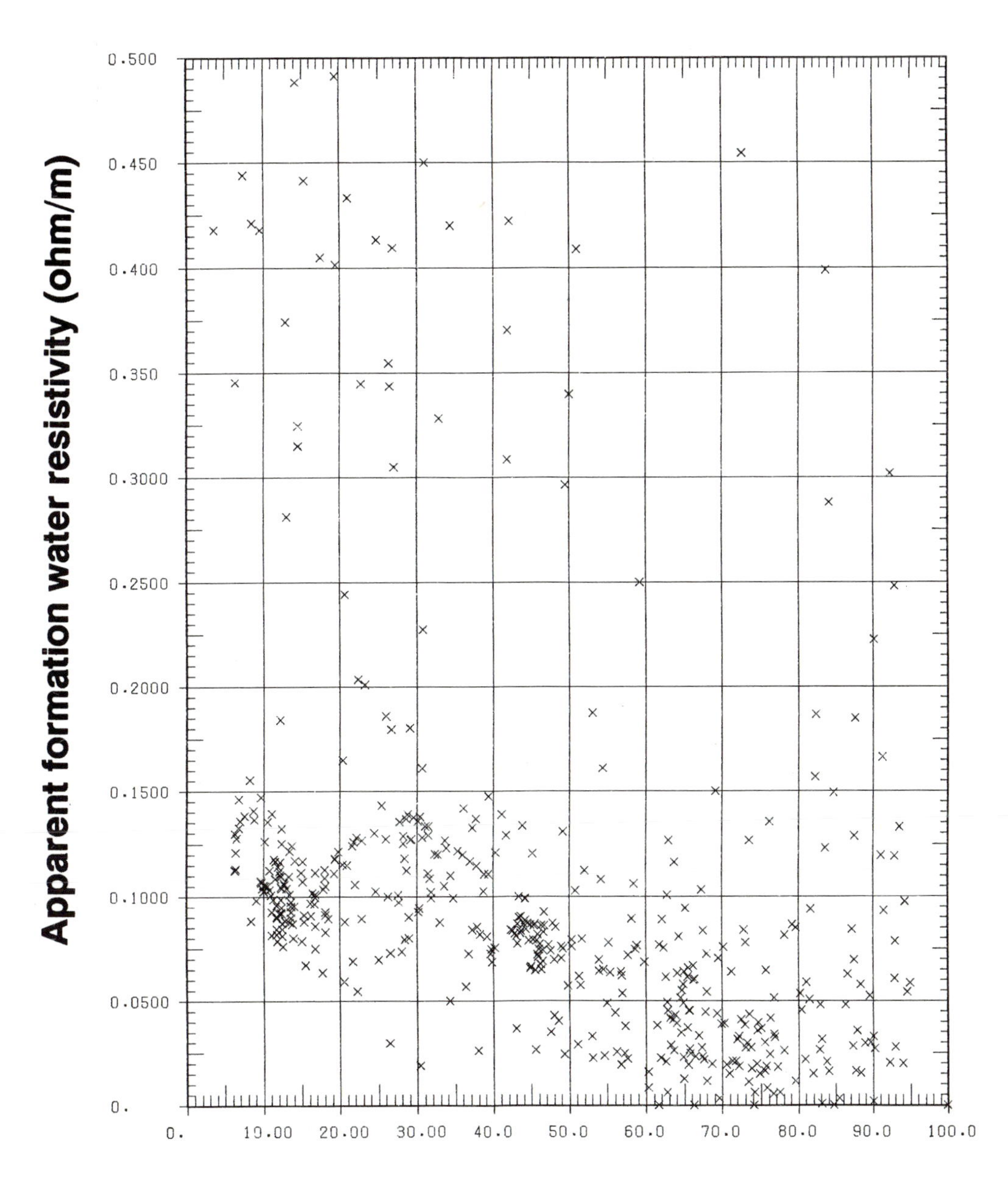


**Fig. 4.** Apparent log-derived water resistivity versus volume of 'shale' Well 3/3–3. Formation water resistivity is 0.08 ohm/m. The scatter at low volume of shale is attributed, in large part, to uncertainty in saturation computation parameters *m* and *n*.

data, and additional geological input in most instances. Figures 5 and 6 show the effect of mica on the logs, as core analysis indicates that the Rannoch in this well contains very little clay, and the data clouds for the micaceous interval are almost invisible on the crossplots of the entire section. Thus, assigning all of the non-quartz rock volume in this formation as

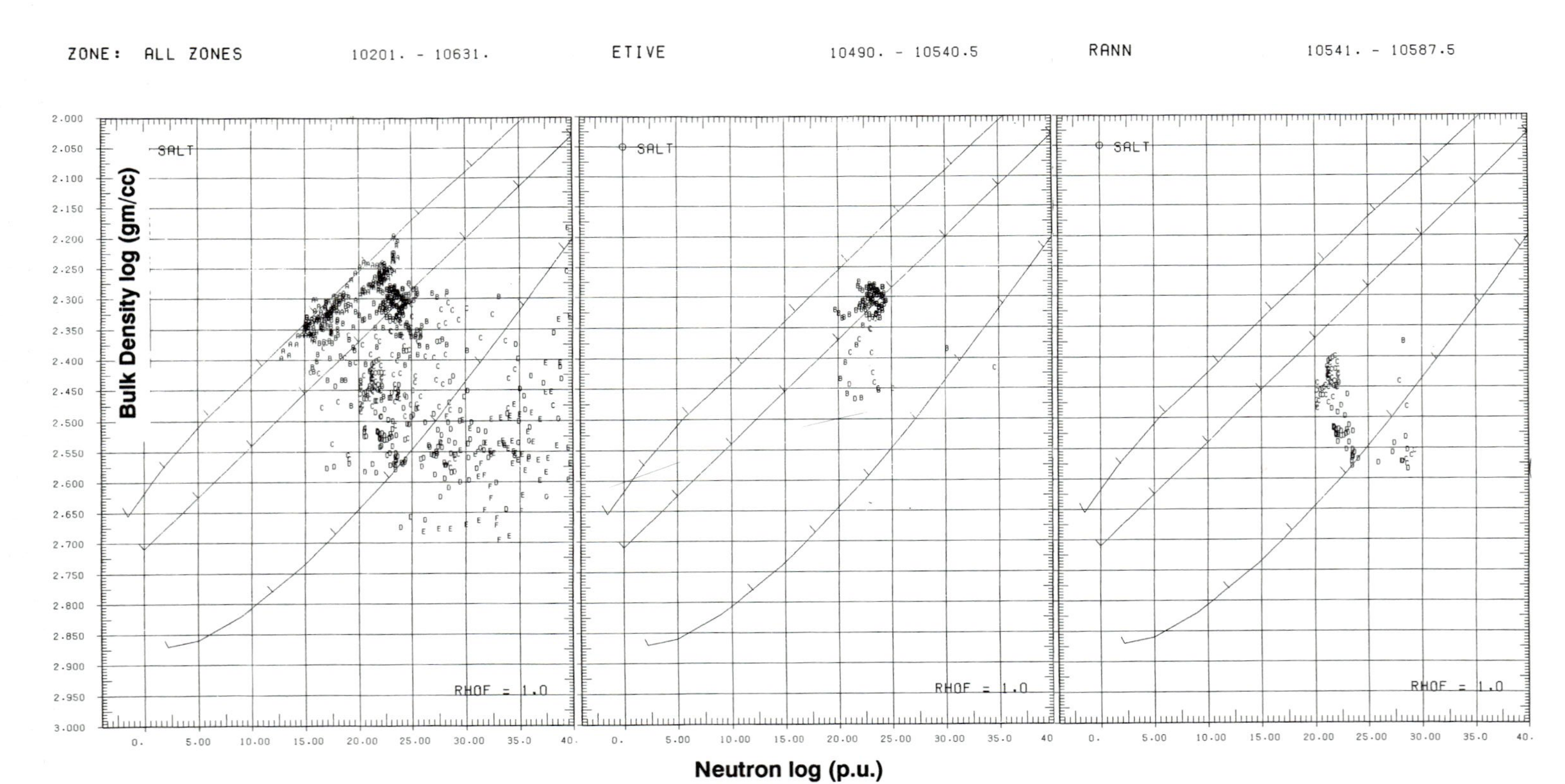


**Fig. 5.** Bulk density versus Neutron porosity data, Well 3/3–3. Pure sandstone response line is shown as the most northwesterly line on the plots. The plots show the entire Brent sequence on a single, merged, plot and the individual lithostratigraphic units plotted separately.

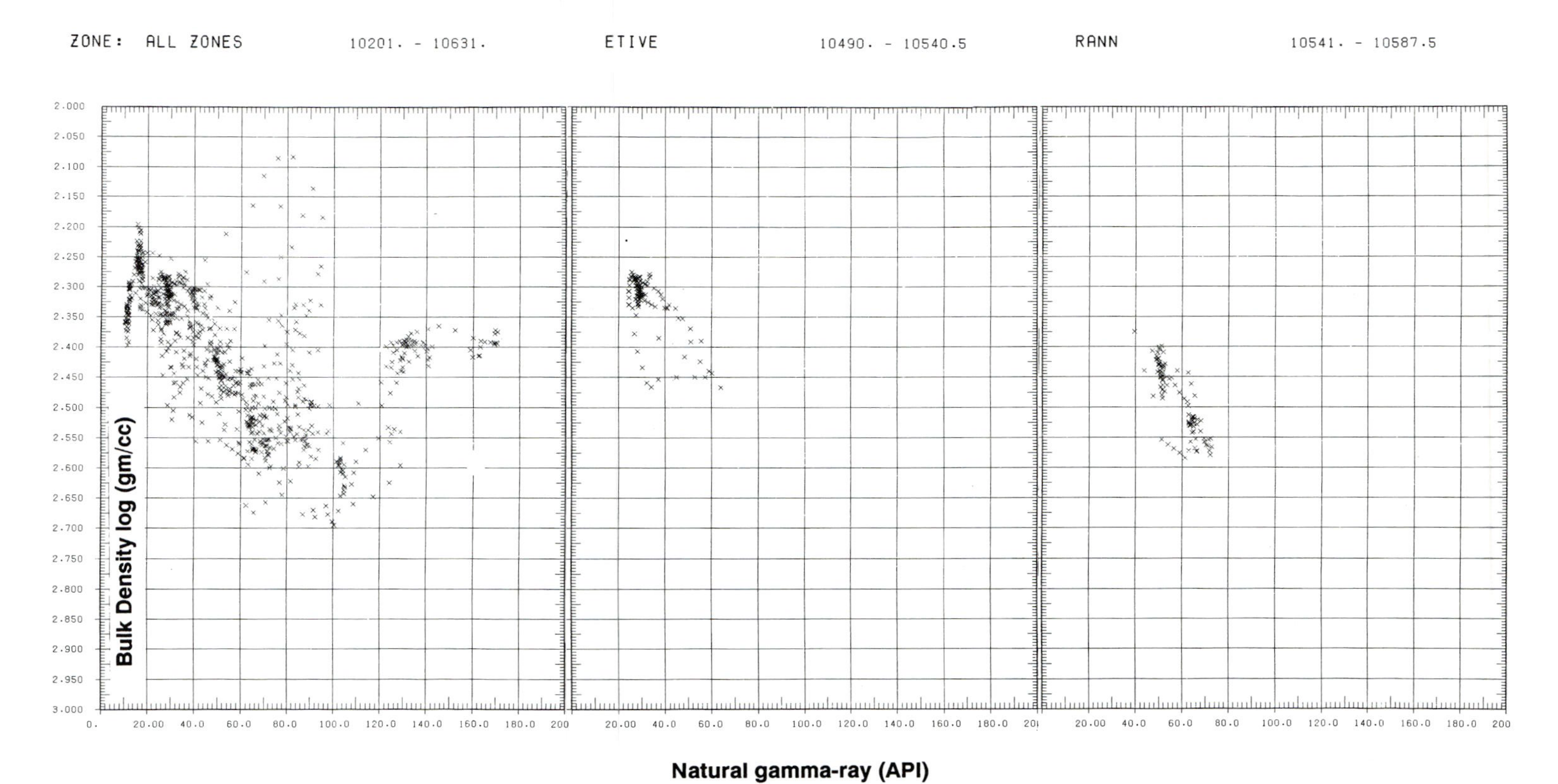


**Fig. 6.** Bulk density versus natural gamma-ray data, Well 3/3–3. The plots show the entire Brent sequence on a single, merged, plot and the individual lithostratigraphic units plotted separately. Gamma-ray activity in the Etive is due to heavy mineral content.

'clay' can result in misleading answers for hydrocarbon saturation, and porosity is also understated. The mica requires explicit identification.

Unfortunately for the petrophysicist nearly all of the non-quartz components of these sands, at least in the volumes in which they are usually encountered, appear as 'shales' on crossplots such as Figures 5 and 6. The Brent sands are sufficiently complex that no single determinant of 'V-shale' is applicable throughout. Rather, several different estimates derived from different tools and tool combinations must be evaluated at each depth in the section. That which provides the most consistent model with respect to all the data is the one that should be selected. Detailed discussion of the effectiveness of each techniques is beyond the scope of this paper. The interested reader is referred to Suau & Spurlin (1982) and to Fertl (1987).

## Pore geometry influence on permeability, capillarity, and electrical properties

Differentiation of a sedimentary sequence into hydraulic flow units is a principal goal for petrophysicists. This is important when translating the reservoir description from the geological model to the engineering model, because it is necessary to recognize which parts of the formation have the same hydrocarbon storage and transmissibility characteristics, as measured by the properties of absolute and relative permeability and capillarity. Pore geometry is the fundamental controlling factor of these properties, so the task is to identify those areas of the formation with similar pore geometries. The electrical properties of a formation are also dependent upon pore geometry, and the values of such terms as the cementation exponent and the saturation exponent, as used in the log analysts' determination of fluid saturation, are often different in radically different pore geometries. Failure to recognize the existence of distinct 'rock types' within the petrophysical model for the formation can lead to erroneous results for fluid saturation volumes and an oversimplification of the flow units present.

Although multiple rock types occur in the Brent sequence because of variations in depositional environment, the most important factor is diagenesis. This is because the principal effects of diagenesis are manifested at pore level and it is the pore geometry that ultimately controls the ability of a rock to transmit fluids under a pressure potential. The rock types defined on the basis of capillary and relative permeability properties can then be examined in terms of the sedimentary architecture to ascertain their connectivity and hence the production/injection characteristics of the Brent sequence as a whole.

Nevertheless, many studies have indicated the importance of original depositional fabric in controlling the diagenetic history of sediments (Fuchtbauer 1983; Bjorlykke 1983) and this is apparently true for the Brent sequence (e.g. Hancock 1978; Hurst & Irwin 1982; Blackbourn 1984; Kantorowicz 1984; Jourdan *et al.* 1987; Scotchman *et al.* 1989). Grain-size distributions are generally considered to be the principal controlling influence. The subsequent course of diagenetic activity and its effects are obviously driven and modified by a complex of different factors. In the case of the Brent sequence, the model derived by Blackbourn (1984) wherein finer-grained less well-sorted material is preserved with higher permeabilities than adjacent coarse-grained beds apparently conflicts with the findings of Scotchman *et al.* (1989) who report that fine-grained sediments are much more severely affected by diagenesis than are coarser-grained sediments at similar depths. Presumably, factors such as mineralogical content, heat flow and burial rate, among others, are the key to understanding the apparent conflict, although the specific diagenetic trends common in the Brent Sands are beyond the scope of this paper.

### *Permeability trends in the Brent Sands*

The permeability of a formation may be defined as the ease with which fluids move through the rock. It is a dynamic attribute of the formation and cannot be observed without the presence of a pressure gradient to instigate flow. The definition implies that permeability is a measure both of the amount of connectedness of the void space containing the fluid(s), and of the sizes of the connections. Although permeability and porosity frequently correlate, permeability has no direct relationship with porosity: for example, large pores give rise to high porosity, but if they are connected by narrow, constricted or clogged pore throats the permeability is low or non-existent. Thus, diagenetic processes are generally deleterious to permeability, as they commonly result in pore throat restrictions.

In Brent sequences that have not been severely affected by diagenesis, it is possible to formulate predictors of permeability magnitude based on porosity trends, because the primary sediment texture is the dominant factor controlling pore geometry and connectivity. In contrast, in areas where diagenetic alteration has been severe, permeability is no longer a simple function of porosity because the pore geometry has been rendered very complex by the various phases of mineral dissolution and precipitation.

The Ninian Field in the UK Sector is an example of the former case. Figure 7a shows the porosity–permeability (to air at laboratory conditions) data for the whole of the Brent Sands of well 3/3–3. This plot suggests that above about 17% porosity there is no relationship between porosity and permeability in these sands. However, when the data are subdivided into individual lithostratigraphic units, distinct trends are discernable within the overall data cloud (Fig. 7a, b and c) and two or three well-defined groups or trends can be used to characterize these data. Moreover, in this example, the same trend occurs in different formations, particularly with the Rannoch and Broom data, implying that the actual 'rock types' for which the trends are appropriate are present in different lithostratigraphic units. These 'rock types' need to be identified early in the evaluation of a field, so that representative sampling for advanced core analysis can be achieved.

More severe diagenetic alteration is found in the Alwyn area (Jourdan *et al.* 1987; Fig. 8). In this area the breakdown into lithostratigraphic units does not improve the correlation between porosity and permeability. In these rocks the pore geometry is sufficiently complex that for essentially invariate porosities the permeability can range over two to three orders of magnitude (albeit with low absolute permeability values). Jourdan *et al.* (1987) attribute the deterioration in permeability to late stage diagenetic events associated with the influx of hot acidic pore waters along nearby fault planes. These waters generated abundant quartz overgrowths, feldspar dissolution, and both kaolinite and illite precipitation in different parts of the fields. The poorest permeabilities are associated with neoformed illite precipitation in the Alwyn South Field, whereas in Alwyn North the precipitation of dickite (blocky kaolinite) has lead to a reduction of porosity rather than loss of permeability. This verifies the conclusions of Pallatt *et al.* (1984) that clay mineral morphology is of paramount importance in the extent of any diminution in permeability. Blocky kaolinite forms relatively large crystals of low surface area that reside in the centres of pore spaces (Jourdan *et al.* 1987), whereas the neoformed illite takes the habit of fibrous crystals of relatively low volume but high surface area that bridge the pores and pore throats and therefore block them. A small increase in the volume of such pore-bridging illite therefore leads to drastic reductions in permeability without altering porosity significantly, so destroying porosity–permeability trends.

Scotchman *et al.* (1989) observed an overall decline in permeability with depth in diagenetically affected Brent sandstones in the North-West Hutton field. Of special note is a much more rapid decline in permeability with depth for about two hundred feet below the oil–water contact in the central and eastern fault blocks, attributed to the increase in authigenic illite in the water leg. The overall permeability trend was attributed to the progressive increase in quartz overgrowths with depth.

Keir & Brown (1978) showed how micro-porosity within particular minerals can confuse the permeability–porosity trend (Fig. 9). They studied the Cormorant Field and reported a notable loss of permeability with increased kaolinite volume. They measured a reduction in permeability of a factor of 40 in the most kaolinitic sands, although the same trend existed in both kaolinitic and non-kaolinitic sands. The most likely explanation is the presence of significant micro-porosity associated with the kaolinite crystals. Although this did not contribute to the available volume through which fluids could move, and hence did not contribute to formation permeability, it was included in the measure of total porosity that was plotted against permeability. This has the effect of shifting the poro-perm data to the right, giving the appearance of low permeabilities for the given porosities (Fig. 9). In fact, had they recognized the micro-porosity and discounted it, the data would have shifted to the left by the amount of such micro-porosity and would probably have fallen on the normal trend line, but at much lower porosities. Such an effect may also result from isolated porosity arising from the dissolution of feldspar grains: this will increase bulk porosity but not permeability. It is clear, however, that this is not the case in the Cormorant Field.

Anisotropy within formations causing directional preferences in permeabilities will vary as the internal structure of the sand bodies varies. Weber (1986) has shown the potential significance of accretion surface development, channel cross-cutting and other internal sediment structure in modifying the permeability fields of a sediment. Currently engineers and petrophysicists measure permeability on core plugs cut parallel to and orthogonal to the principal bedding surfaces identified in the core sections. These are compared in a simple fashion to indicate anisotropy, but more complex representations may be required in order to capture

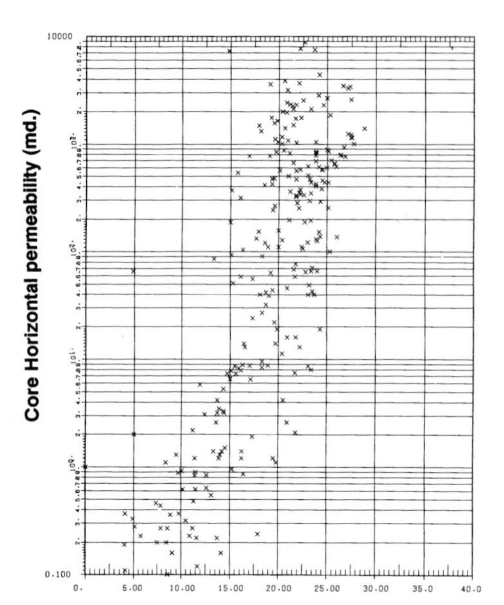


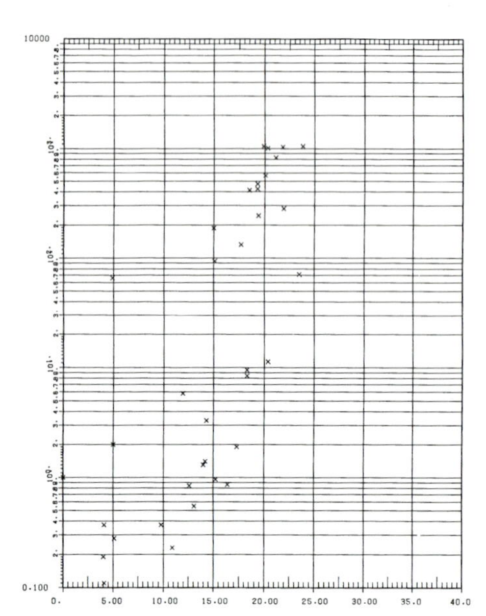


Core porosity (p.u.)

(a)

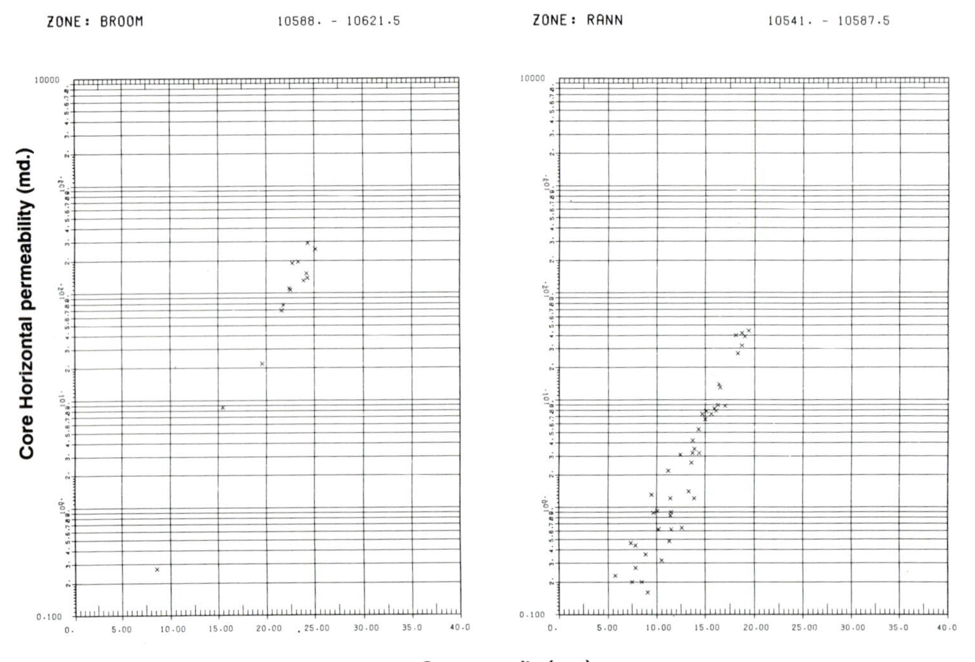


Core porosity (p.u.)

(c)

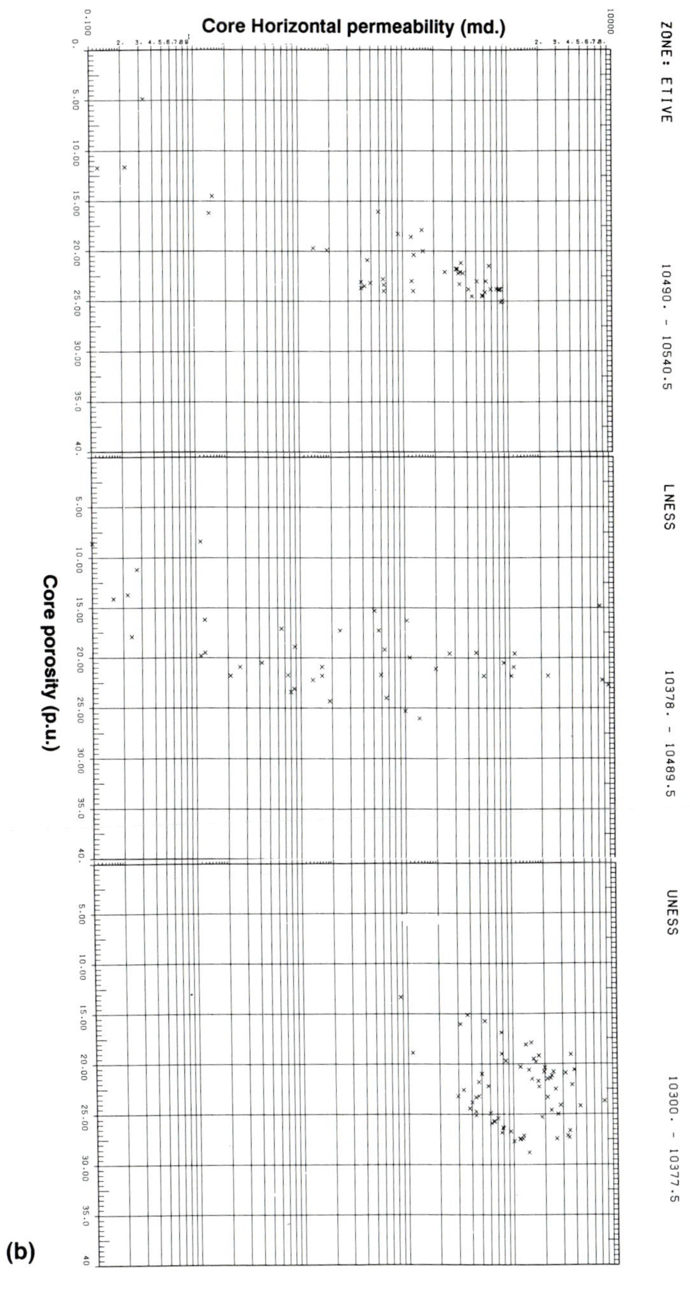


**Fig. 7.** Core horizontal permeability to air versus core porosity, Well 3/3–3. (**a**) Total well set data and Tarbert Formation, (**b**) Ness and Etive Formation; (**c**) Rannoch and Broom Formation. These data are ambient condition measurements.

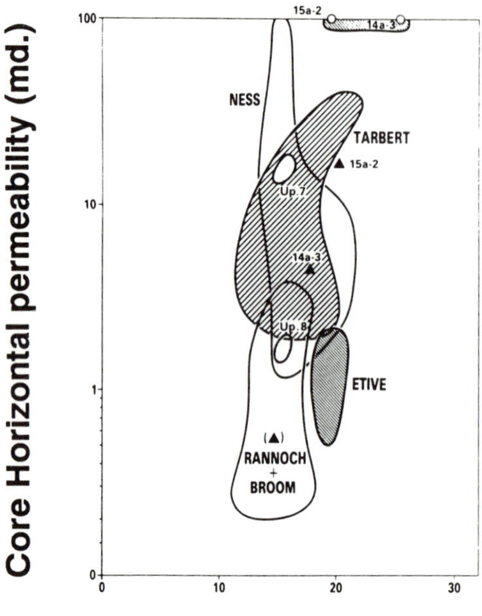


**Fig. 8.** Core porosity versus core permeability sketch for the different reservoirs of the Brent Group as encountered in the Alwyn South and South-East structures. From Jourdan *et al.* (1987).

the significance of other permeability directions in strongly cross-bedded sands (Weber 1982) such as occur in the Rannoch and Ness Formations.

Laboratory measurements may be optimistic unless taken under simulated conditions of effective overburden stress in most of the rock types present in the Brent Sands. Where the sandstones are severely affected by diagenetic processes and have been very well cemented by one mineral or another, dilatant failure of the grain fabric by the development of micro-cracks during drilling and core recovery may occur (Kaye 1990). This effect is probably not significant volumetrically but could be significant for the measurement of capillary and electrical properties. If the pore throats have been altered by mineral precipitation to become flat-sided rather than rounded, small amounts of grain slip during application of the simulated overburden may be sufficient to close off completely some of the throats and cause drastic reductions in permeability to occur.

The Brent Sequence commonly exhibits at least five per cent pore volume expansion (i.e. one porosity unit in twenty) when measured at surface conditions, compared with measurements made under simulated effective overburden conditions. Laboratory procedures, it should be stated, can have a serious effect on these measurements (Nieto *et al.* 1990). Correction for this 'overburden' effect should be made in order to relate core porosities to in-situ porosities. However, the impact of core relaxation on permeability measurements may be more severe if there is any directional dependency in either the permeability trends or in the in-situ formation effective stress field. If either such dependency exists, it may not be possible to recreate it accurately in the laboratory. Thus, the simulated conditions may not reflect the in-situ conditions properly, and the results obtained may be misleading. Such directional dependencies may be caused by sedimentary structures or by tectonic activity in the region of the well.

### *Relative permeability trends*

When a formation is filled with more than a single fluid phase the ability for any one of the phases to move through the pore system is hampered by the presence of the other phase(s). Hence one of the most important attributes of the formation for the reservoir engineer is the 'relative permeability' to each fluid phase, and the saturation level at which they become immobile. There are many influences on the relative permeability of a particular formation but principal among them are the wettability of the pore surfaces and the pore geometry.

For the most part the Brent Sands are preferentially water-wet systems. Calcite precipitation or high concentrations of carbonaceous material may cause localized development of mixed-wettability or preferentially oil-wet conditions, but these are not dominant. Initial and residual saturations of the fluids are directly controlled by the pore geometry of the system.

Complex pore geometries that comprise small irregular pore throats or that have blind pores with only a single entry throat to them lead to a high immobile saturation of the wetting phase (Honarpour *et al.* 1986; Mohanty *et al.* 1987). The residual saturation of the non-wetting phase is also high because of the frequent isolation of large oil ganglia by pore bridging of the water phase. Such geometries are associated with the fine grained sands (e.g. the Rannoch) and with the areas that have undergone severe diagenesis. In more open pore geometries capillary forces are less dominant, resulting in the wetting phase immobile saturation being lower and oil being more efficiently displaced from the pores, hence a lower residual oil saturation. Such pore ge-

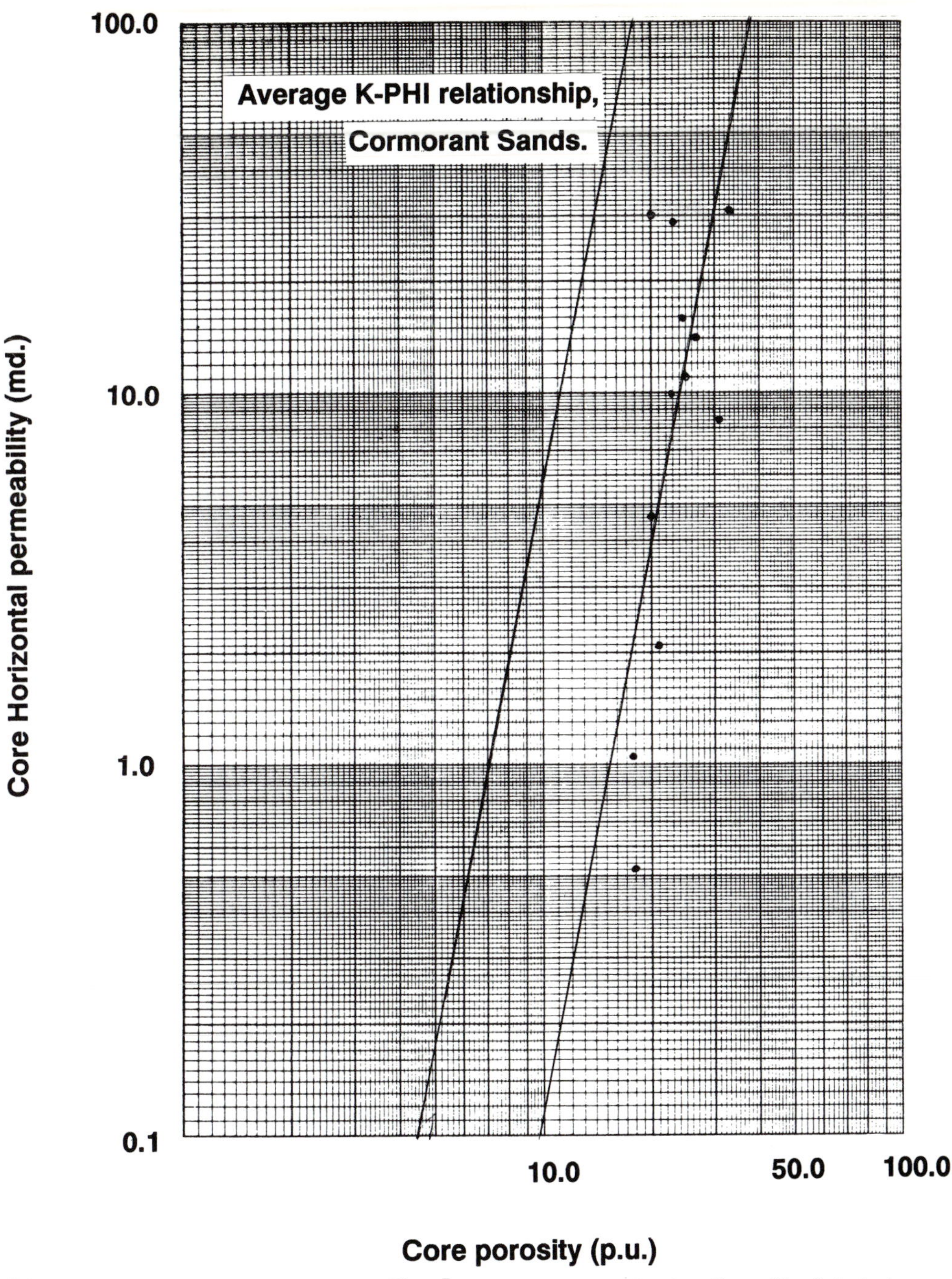


**Fig. 9.** Permeability versus porosity crossplot from the Cormorant Field showing effects of kaolinite in the pore system, compared to the normal trend for the Field. After Keir & Brown (1978).

ometries are typical of many of the better quality sands throughout the Brent sequence.

Figures 10 and 11 illustrate typical relative permeability curves for some Brent sandstones. Figure 10 represents data normalized to a constant irreducible water saturation. The similarity in curve shape and end-points possibly indicates that there is little real variation between the lithostratigraphic units sampled. Figure 10 should be compared to Fig. 11 that shows considerable variation in curve shape and end-points in unnormalised data measured on

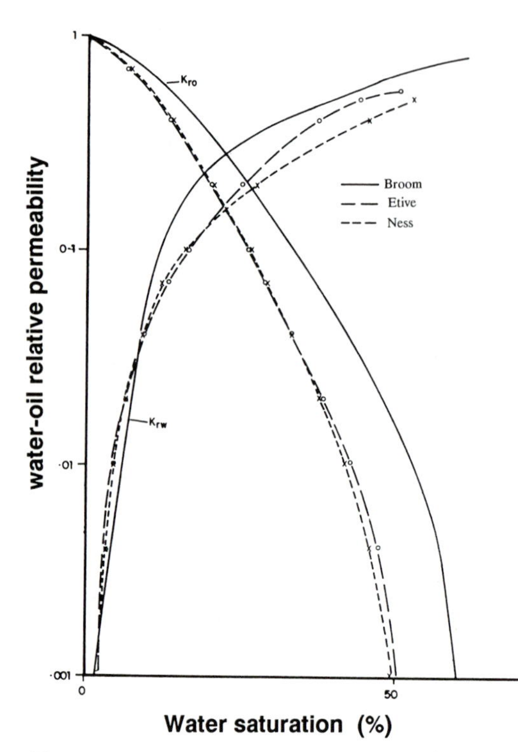


**Fig. 10.** Sketch of normalized water–oil relative permeability curves from three Formations of the Brent Group.

core plugs with widely different absolute permeabilities. The range in irreducible water saturation for the different absolute permeabilities sampled is entirely expected; the shape of the curves may reflect variable wettability conditions in the plug samples used for the experiments (Honarpour *et al.* 1986) among other factors, while the end-points are largely governed by pore geometry.

The evidence from these curves (Fig. 11) suggests that for the highest permeability systems the water phase will flow more readily than the oil phase once the water saturation has increased to approximately 35%. However, in the tightest sample, the permeability to water does not exceed that to oil until 50% water saturation has been achieved. This phenomenon is a reflection of differences in pore geometry between the two samples. The high permeability sample has a near-unimodal pore size distribution (cf. Fig. 12 and the next section of text), whilst the low permeability sample has a much more tortuous and complex pore system. In the former, the wetting phase starts to move, under the imposed pressure gradient, when its saturation rises above about 18%. In the latter case, the wetting phase remains immobile until it reaches a saturation of about 40%. This difference reflects the presence of an abundance of single-entry dead-end pores and pores with very tight throats between them in the low per-

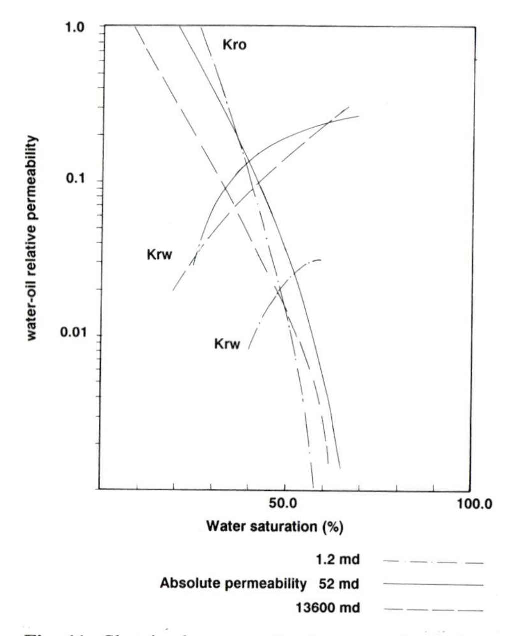


**Fig. 11.** Sketch of unnormalized water–oil relative permeability curves for three samples of Brent Sands with markedly different absolute permeailities.

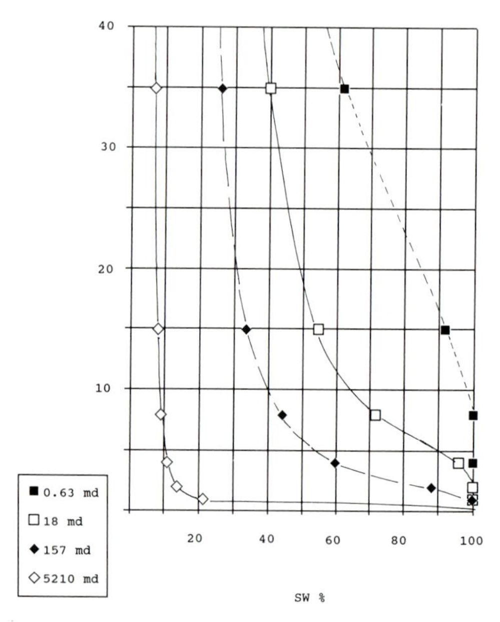


**Fig. 12.** Representative capillary pressure curves from Brent Sands.

meability sample, both of which phenomena give rise to a higher capacity for retaining the wetting phase. Figure 11 also shows that the relative permeability to oil ($K_{ro}$) in the low permeable sample declines much more rapidly (has a steeper slope) than is the case for the higher permeability sample. These are 'relative' permeabilities, the cross-over point of the two curves in the lowest permeability sample has an actual permeability to each phase of the order of 0.02 mD, whereas the highest permeability sample has the cross-over of the two curves occurring at an actual phase permeability of about 900 mD.

### *Trends in formation capillarity*

Another important attribute of the formation that is dependent upon the pore geometry, and is of crucial importance in the derivation of distinct rock types within a sand unit, is the formation capillarity. The standard presentation of capillary curves plots the volume of non-wetting fluid that is introduced into a sample originally fully saturated with wetting fluid (when conducting a 'drainage' test) against cumulative applied pressure (Fig. 12). Their shape can be informative of the pore throat size distribution and the accessibility or connectedness of the pore system as a whole. In a pore system with uniform pore throat dimensions (for example, the higher permeability samples in Fig. 12), as soon as the non-wetting phase pressure exceeds the threshold for the size of throat present, all of the available pore space is occupied by the non-wetting fluid, and the irreducible saturation of the wetting fluid is very quickly reached. Where the pore system has a wide range of pore throat sizes the volume of non-wetting fluid injected at each pressure increment is small as the cumulative pressure reached only exceeds the threshold for a small fraction of the pore throats. The lowest permeability sample of Fig. 12 has a pore system that exhibits this type of behaviour. These curves also may be used to calculate the mean effective radius of the pore throats in each sample. For the most permeable sample on Fig. 12, this is in the range 12.0–15.0 $\mu$m.

Capillary characteristics correlate well with permeability and have a fundamental control over the relative permeability properties of a formation, as they directly control both initial and residual saturation conditions. It is therefore most important in the Brent Sequence to sample each separate cluster of points that are seen on a permeability–porosity crossplot (Fig. 7), in order to characterize the formation properly and to identify all the rock types that may be present. The poro-perm plot offers direct evidence for the existence of significantly different rock types in the formation of interest. This evidence may be confirmed by correlating the capillary curves from within a poro-perm cluster using a 'J-function' correlation analysis, and comparing them to similarly treated curves from other data clusters (Archer & Wall 1986). Note that a J-function analysis requires there to be a degree of correspondence between porosity and permeability in a rock type, and so is unlikely to be effective in heavily digenetically affected sediments.

### *Saturation equation exponents*

The tortuosity of the pore system governs the values of cementation and saturation exponents that the log analyst uses in determining water saturation. In regular intergranular pore systems Archie (1942) demonstrated that these exponents should take the value of 2.0. For more tortuous pore systems these exponents are higher, and, conversely, are lower in less tortuous pore systems. Open fractures have the theoretical minimum value of 1.0, whilst highly tortuous microporosity in oil-wet carbonates have values in excess of 5.0. In the Brent sandstones the range generally lies between 1.75 and 2.1.

As Fig. 1 shows, there is a slight oil saturation computed for the upper Etive and the upper Broom intervals. As these sands are well below the oil–water contact, it may be supposed that these oil saturations are artefacts of the interpretation. In this particular interpretation, constant cementation and saturation exponents (2.076 and 1.908 respectively) were used throughout the well. In the Etive interval (from 10[illegible] ft to 10 540 ft), the lower sand has similar porosities but lower resistivities (on track three of Fig. 1) compared to the upper part. Variable resistivity for the same porosity indicates that the value of the saturation equation constants should be different in the two zones. Since the lower interval computes as 100% water, it is the upper interval that is inferred to have the wrong values and either or both the cementation and the saturation exponents should be increased. The same is implied for the upper Broom sand. On the plot of apparent formation water resistivity against volume of shale (Fig. 6) for this well, the scatter seen at low shale volumes is due in large part to the upper intervals in the Etive and Broom Formations having been processed with the wrong parameters.

The need to alter the saturation equation

parameters in the two zones strongly indicates that these intervals will also exhibit different capillary pressure and relative permeability characteristics. Both intervals form discrete clusters on the poro-perm plot (Fig. 7). Figure 13 shows the range in permeability values for a typical Etive section in another location, and shows the same patterns as can be seen in the Etive of well 3/3–3 (Fig. 1). The data on Fig. 13 are from an oil leg and are rather

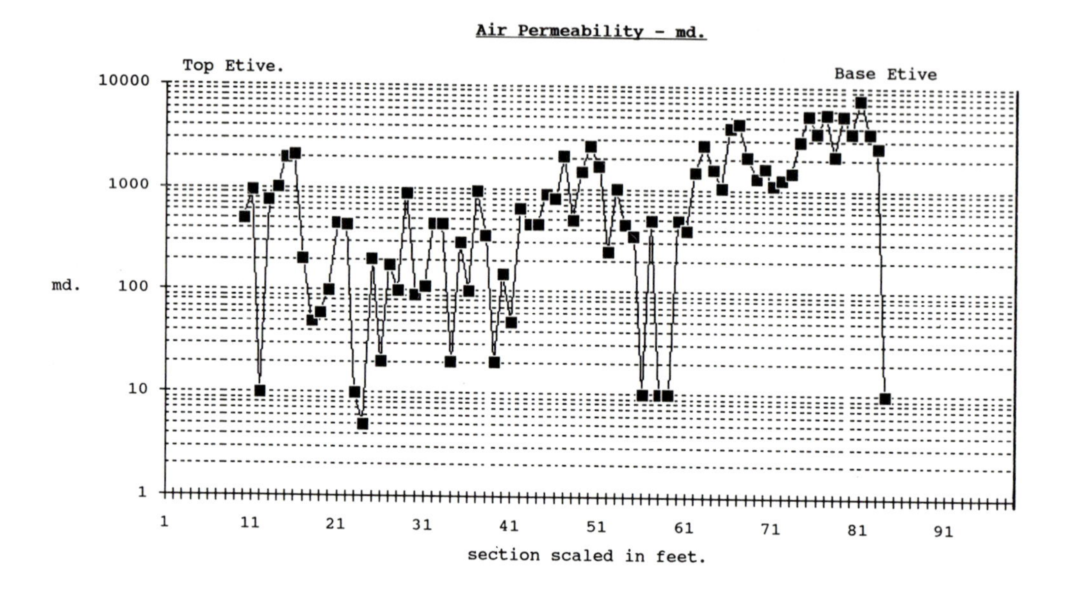


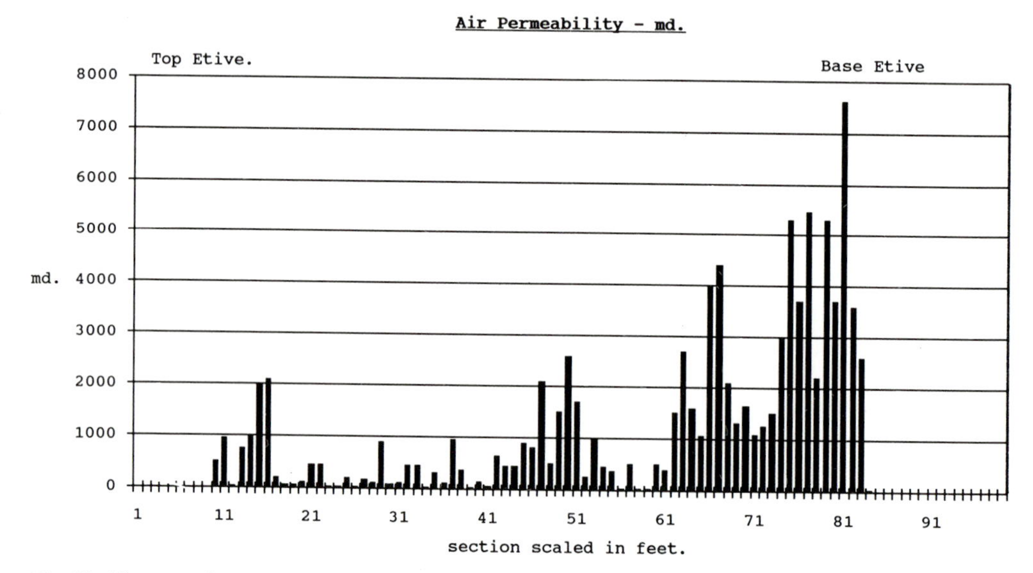


**Fig. 13.** Upper portion: conventional logarithmic display of horizontal permeability data from core through an Etive section. Permeability is generally in excess of 100 md, with an apparently smooth increase in permeability with depth (To the right).

Lower portion: same data plotted on linear scale; magnitude of permeability contrast at the base is more apparent; this pattern of permeability distribution, so common in the Etive, is not conducive to stable waterflood.

higher values than those of Fig. 1. The increased permeabilities towards the base of the formation in each case are attributed to the presence of a tidal channel cutting in to the barrier beach sediments at the location of the well. The display on the linear scale serves to emphasize the range in values, and provides a better visual representation of the relative way in which the various parts of the section will behave under dynamic conditions during production/ injection.

Thus, petrophysical evidence from saturation exponents and poro-perm clusters indicates that these tidal channels constitute a significantly different rock type that may be expected to transmit fluids differently from other parts of the Etive Formation. As a result of their very high permeabilities and their near-unimodal pore size distribution, they tend to act as 'thief' zones through which injected water can move more rapidly than through surrounding sands. Plotting the lateral extent of these tidal channels is, therefore, of the utmost importance in designing the most effective secondary recovery strategy. The effect is exacerbated by their occurring directly on top of the low permeability Rannoch, and it is the contrast in permeability between the rock types that is the problem. Figure 14 shows that there is commonly good pressure communication between the Rannoch and the Etive formations, but this only serves to worsen the situation. The indication from this, and similar, plots is that the high residual oil saturation in the Rannoch, remaining after the more permeable intervals have been completely flushed to high water-cut production, will nevertheless be at the lower overall reservoir pressure and so be all the harder to recover by tertiary recovery strategies.

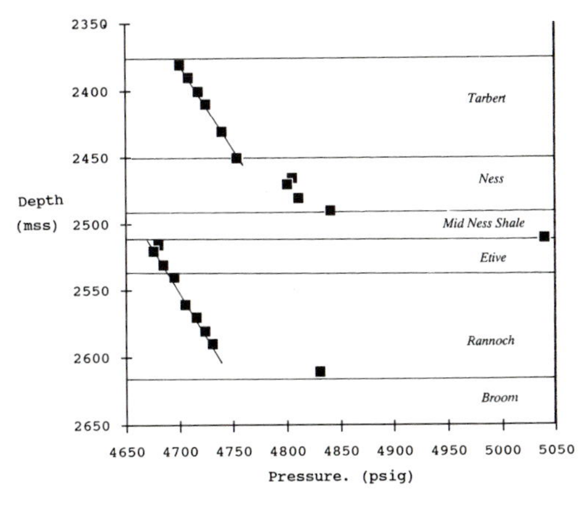


**Fig. 14.** Repeat Formation Test profile from the Brent Group of the Statfjord Field, taken in late 1986.

## Summary

In petrophysical terms, the Brent sands comprise very fine to very coarse sands, interbedded with siltstones, mudstones and some conglomerates and varying widely in interstitial clay content from essentially absent to very 'shaly'. Mineralogical complexity exists in that mica, feldspar, siderite, calcite, pyrite, and clay minerals are variably present within a predominantly quartz matrix. The resultant variability in the sediment fabric gives rise to a very wide range of pore size distributions and hence considerable range in permeability and saturation characteristics. Almost every Brent Group reservoir contains a different subset of the possible variations, giving rise to a different petrophysical model. A thorough understanding of the actual petrophysical characteristics encountered in any given reservoir requires the integration of data from several sources and disciplines.

Table 1 summarizes the broad ranges of porosity, net to gross, saturation and permeability characteristics of the major lithostratigraphic units of the Brent Sands. It is hoped that this paper has indicated the considerable variation that is hidden behind such bland averaging of the data. Nevertheless, the table serves to illustrate the impact of diagenetic variability in the Brent sandstones by summarising the properties of the mildly affected Statfjord Field as compared with the more strongly affected North West Hutton Field.

As production experience builds up for the Brent sandstones the realization is growing that

**Table 1.** *Formation properties*

| | N/G | Phi(pu) | Perm (md) |
|---|---|---|---|
| Tarbert | | | |
| Statfjord | 0.9 | 30 | 3000 |
| NW Hutton | | 13 | 0.52 |
| Ness | | | |
| | 0.5 | 26 | 1000 |
| | | 10–22 | 0–2600 |
| Etive | | | |
| | 1.0 | 31 | 5000 |
| | | 14–22 | 1–600 |
| Rannoch | | | |
| | 0.85 | 27 | 10–1000 |
| | | 13–20 | 0.5–8.0 |
| Broom | | | |
| | | 2–22 | 0–630 |

having a more precise reservoir description implemented in dynamic models provides considerable benefits in understanding field performance. Implementing that degree of geological reality will require continued and increasing overlap between the disciplines of geology, geophysics, petrophysics, reservoir engineering and production engineering.

The author gratefully acknowledges the editorial assistance provided by A. C. Morton (BGS), and the helpful comments made by A. Corrigan (Corrigan Associates) and an anonymous reviewer on an earlier draft of the manuscript.

## Appendix

Useful references include Fertl (1987), Serra (1986, ch. 9), Suau & Spurlin (1982) and Worthington (1985) and their extensive reference lists.

### *Volume of shale models*

The term 'shale' comprises the material in a formation other than the dominant mineral component: the 'matrix'. Hence, in a sandstone the dominant mineral is quartz and the term 'shale' normally comprises a composite of all the other components present. A major portion of the shale volume in a sand will comprise clay minerals, by virtue of their relative abundance in sandstone formations, and it is the clay minerals that are of primary interest to the petrophysicist for reasons that are stated in the text. Note that, unless particular precautions are taken, a mixture of feldspar and quartz will compute as a very 'shaly' sandstone by most of the methods listed.

Many tools are examined for their 'shale' influence by rescaling their observed dynamic measurement range across a sand-claystone sequence such that one extreme is deemed '100% shale', whilst the other extreme is considered to be 'shale-free'; such 'shale' parameters, therefore, are not picked from the sand beds to which they are applied, but from adjacent fine-grained intervals having potentially different clay mineralogy, depositional environment and diagenetic history. Equations of the type:

$$V_{sh} = \frac{(\mathrm{Obs} - \mathrm{Min})}{(\mathrm{Max} - \mathrm{Min})} \qquad \mathrm{A1}$$

Where: Obs = Log reading;
Max = Reading in 100% 'shale' and
Min = Reading in 0% 'shale'
$V_{sh}$ = Fractional volume of 'shale',

can be calculated for tools such as the gamma-ray, spectral gamma-ray, spontaneous potential, neutron porosity and pulsed neutron. These methods suffice when the single tool's response with shale volume is clear and linear, although there remains the problem of the parameters being derived from neighbouring beds.

For more complex sandstones where the 'shale' component is a composite of several minerals, the effects of carbonates, micas and feldspars might be quantifiable by comparing the above methods with a crossplot method. In the latter, two curves are simultaneously processed such that the position of an observed data pair is scaled against its distance from the matrix (0% 'shale') and the 'shale' (100% 'shale') points, as they fall on a crossplot of one curve against the other. Equations of the type:

$$V_{sh} = \frac{Y_{obs}(X_{ma} - X_{fl}) - X_{obs}(Y_{ma} - Y_{fl}) - Y_{fl}X_{ma} - Y_{ma}}{(Y_{sh} - Y_{fl})(X_{ma} - X_{fl}) - (X_{sh} - X_{fl})(Y_{ma} - Y_{fl})} \qquad \mathrm{A2}$$

Where: $Y$ = density log
$X$ = either neutron or acoustic log
obs = log reading
sh = log value at 100% shale point
ma = log value at 0% shale point
fl = log value at 100% fluid point

are used in these instances.

Each shale indicator tends to over-estimate the shale volume (Poupon & Gaymard 1971), hence, selecting the minimum volume computed from among several indicators at each depth increment will minimize this over-statement.

### *Porosity models*

Porosity is computed from each of the single tool indicators by equations that consider the tool signal as comprising a linear mixture of components (minerals, fluids) multiplied by their relative effect on the tool, calculated as follows:

$$\text{Tool signal} = V_{fl}\mathrm{Res}_{fl} + (1 - V_{fl})\mathrm{Res}_{ma} \qquad \mathrm{A3}$$

Where: Res = long response in 100% fluid or 100% matrix.

Clearly, the matrix term can be expanded to include several different components within the capabilities of tool resolution. Commonly, this will include at least a 'shale' term. Similarly, the fluid term is expanded to account for both hydrocarbons and water.

When solved for porosity these equations revert to:

$$\text{Porosity} = \frac{(\text{Res}_{ma} - \text{Res}_{obs})}{(\text{Res}_{ma} - \text{Res}_{fl})} \quad \text{A4}$$

where the subscripts are as before.

These equations are closely similar to the single tool shale volume calculations.

The clay minerals contain water in their structure that is immobile either through chemical, electrical or capillary forces. This water is seen by the porosity tools as bulk porosity, but it is not capable of contributing to the storage capacity or flow transmissivity of the formation. It may also have significantly different electrical properties that require explicit recognition. The observed tool readings require explicit corrections to remove the effects of clay minerals. Thus was born the concept of 'effective' porosity as distinct from 'total' porosity.

$$\text{Effective porosity} = \text{Total porosity} - (V_{sh} \cdot \text{PHI}_{sh}) \quad \text{A5}$$

The 'shale' porosity ($\text{PHI}_{sh}$) is computed from each single tool equation by substituting the tool reading in 100% 'shale' with the observed tool reading in equations of type A4 given above.

Crossplot solutions can be made for porosity by an analogous means to that used in equations of type A2 for shale, except that for porosity the data are first corrected for shale effects, if necessary, then the ratio of the distance between the matrix point and the corrected datum point, over the distance between the matrix point and the fluid point, is computed as 'effective' porosity.

### *Water saturation models*

The essence of the clay problem in respect of the calculation of fluid saturations in petrophysics lies in the fact that clays have a net negative charge because of substitution of lower-valence cations in the clay lattice structure, e.g. $Al^{+3}$ for $Si^{+4}$ and $Mg^{+2}$ for $Al^{+3}$. To maintain electrical neutrality, the resulting negative lattice charge is balanced by positive cations (counter-ions) located on and near the clay surface. The spatial distribution of counter-ions is determined by the opposing forces of electrostatic attraction to the negative clay lattice and diffusion into the surrounding bulk solution. The resultant counter-ion cloud surrounding the clay particle is capable of transmitting an electric current and thus clay minerals in the hydrated state exhibit a definite conductivity (Vinegar & Waxman 1984).

Petrophysicists have developed two approaches to the problem of accounting for the magnitude of the clay conductivity effects. The earliest methods were empirical, but nevertheless continue in widespread use today because all the parameters required for their evaluation are at least notionally available through the interpretation of wireline log data. These methods require the assumption that the 'shale' volume ($V_{sh}$) is equal to the volume of the clay minerals plus the volume of the water that is chemically, or otherwise, bound to their crystals. 'Effective' porosity is calculated as in A5, and fluid saturations are then calculated on the basis of a parallel resistor model with one component representing the contribution of the clay-bound water and the other component representing the volume of the water within the rest of the pore system. The saturation of hydrocarbons is taken to be the difference between the volume of the pores and the volume of the water in those pores. The equation of this type in commonest use in the Brent Group, although there is no theoretical basis for its applicability in these sands, is the so-called 'Indonesian Equation' (Poupon & Leveaux 1970):

$$\frac{1}{(R_t)^{0.5}} = \left[\frac{V_{sh}^{(1-V_{sh}/2)}}{(R_{sh})^{0.5}} + \frac{\text{PHI}_{eff}^{m/2}}{(a \times R_w)^{0.5}}\right] Sw^{n/2} \quad \text{A6}$$

where:
$R_t$ = true formation resistivity (ohm/m)
$R_{sh}$ = resistivity of 'shale' (ohm/m)
$R_w$ = formation water resistivity (ohm/m)
$V_{sh}$ = volume of 'shale' (fraction)
$\text{PHI}_{eff}$ = 'effective' porosity (fraction)
$a$ = cementation coefficient
$m$ = cementation exponent
$n$ = saturation exponent
$S_w$ = water saturation as fraction of $\text{PHI}_{eff}$.

The other type of saturation model that has been developed is soundly based upon the double-layer phenomenon of the clay conductivity. These approaches treat the water volume in the rocks as comprising two different electrolytes, one associated with the clay surfaces, and the other in the open pore system (Waxman & Smits 1967). Therefore, the need to compute an 'effective porosity' disappears, as does the requirement for the determination of a V-shale term. This model calculates water saturation as follows:

$$S_w = \frac{(F^* \cdot R_w)}{(R_t \cdot (1 + B \times Q_v \times (R_w/S_w))^{1/n*}} \quad \text{A7}$$

where:
$F^*$ = clean rock formation factor ($\text{PHI}_t^{-m}$)

$R_w$ = formation water resistivity (ohm/m)
$R_t$ = true formation resistivity (ohm/m)
$B$ = equivalent conductivity of clay ions
$Q_v$ = cation exchange capacity per unit volume
$n^*$ = clean rock saturation exponent

There are drawbacks to the double-layer approach, however. The CEC parameter (i.e. the measure of the clay conductivity) either requires direct measurement in the laboratory on core samples, or requires estimation from other log data, which can mean that the approach reverts to an empirical V-shale method unless extreme care is taken in such estimation of CEC. However, an alternative double-layer model devised by one of the logging companies (Clavier *et al.* 1977) to overcome the need for core data requires that the clay-bound water parameters are read from an adjacent mudstone bed and may thus bear little relation to the properties that pertain to the dispersed clays in the pores of the sand bodies.

### *Capillarity*

A pressure differential is required before non-wetting fluid may enter pores that already are fully saturated with a wetting fluid (Archer & Wall 1986). For each pore the pressure required is equivalent to a minimum threshold capillary pressure which is a function of several factors: the physical characteristics of the two fluids in question, the wettability of the surfaces of the pore, and of the radius of the pore throat controlling entry to the pore space. Unless this minimum capillary pressure is exceeded the non-wetting fluid will not enter the pore. Larger pore throats will have lower capillary thresholds than will smaller pore throats. The amount of pressure exerted by the non-wetting fluid at a particular pore throat is dependent upon the density contrast between the two fluids and the height of the pore throat above the zero capillary pressure level or 'free-water-level' (FWL). The FWL is a property of the reservoir and fluid system. The non-wetting fluid will enter progressively more of the pore throats, filling more of the available pore space, with height above the FWL until the wetting fluid has been reduced to its immobile saturation. This irreducible saturation is again a function of the capillarity of the formation.

### *Permeability*

The permeability of a formation can be defined as the ease with which fluids can flow through the rock. It is used as a constant of proportionality in an equation relating the amount of fluid flow with the type of fluid and the potential of that fluid, the so-called 'Darcy Equation':

$$\frac{Q}{A} = -\frac{k}{u} \cdot \frac{dP}{dL} \qquad \text{A8}$$

where:
$k$ = permeability in horizontal direction (Darcies)
$u$ = fluid viscosity (centipoise)
Q = flow rate (cc/sec)
$A$ = cross-sectional area of flow (sq cm)
$dP$ = change in pressure (atm)
$dL$ = length over which dP measured (cm)

Apart from the fact that a non-zero permeability requires a non-zero porosity, there is no general relation between permeability and porosity: permeability measures the continuity of the pore space. A good correlation between these two parameters is frequently observed in clastic sediments, however, and this results in porosity often being used to predict permeability in the formation, based on core-log calibrations. Enhancements to the relationship may be used that exploit the fact that permeability is related to the surface area of the pore system. This can be input into the prediction as the irreducible water saturation of a formation, but this then limits the prediction to use data measured in unswept reservoir zones only. Marked improvements in porosity:permeability models can accompany the subdivision of the formation into constituent rock types.

The fact that a fluid gradient must be applied before the permeability of a formation can be observed, means that logging data acquired in an essentially static environment cannot directly measure permeability. The formation-sampler equipment can investigate a fluid gradient in the immediate vicinity of the well-bore, but its interpretation in terms of permeability is very difficult owing to the shallow depth of the pressure perturbation induced, amongst other factors (Serra 1986).

### *Wettability*

Wettability has been defined as the 'relative preference of a surface to be covered by one of the fluids under consideration' (Cuiec 1987); quantitative definitions are possible based upon thermodynamic considerations. The entire surface of a porous medium may have uniform wettability or may have heterogeneous or 'dalmatian' wettability (leading to the notion of 'fractional' wettability). In the latter case the overall wettability depends on whether the heterogeneity exists throughout the pore system or whether some pores are preferentially water-wet and others are preferentially oil-wet.

Mineralogical variation will effect wettability, as will oil chemistry. Silica is strongly water wet, whilst calcite is strongly oil wet. Clay mineralogy may lead to the existence of sites on the clay crystal surfaces to which polar compounds within the oil phase are attracted. Carbonaceous material is preferentially oil-wet. High gas–oil ratio crudes are more likely to have their heavier ends 'plating-out' on the grain surfaces to create oil-wet conditions. Mixed wettability leads to lower residual oil saturations (i.e. that saturation of oil at which it ceases to move through the pores) through the reduction of capillary forces which enables better displacement of the larger pores.

## References

AABOE, E. 1984. Influence of shaliness upon conductivity in shaly sandstones in the northern North Sea area. *Transactions of SPWLA 25th Annual Logging Symposium*, June 10–13, Paper LL.

ARCHER, J. S. & WALL, C. G. 1986. *Petroleum Engineering, Principles and Practice*. Graham & Trotman, London.

ARCHIE, G. E. 1942. The electrical resistivity log as an aid in determining some reservoir characteristics. *Petroleum Transactions of AIME*, **126**, 54–62.

BLACKBOURN, G. A. 1984. Diagenetic history and reservoir quality of a Brent Sand sequence. *Clay Minerals*, **19**, 377–390.

BJORLYKKE, K. 1983. Diagenetic reactions in sandstones. *In*: PARKER, A. & SELLWOOD, B. W. (eds) *Sediment Diagenesis*. D. Reidel Publishing Company, Dordrecht, Holland, 169–213.

BROWN, S., RICHARDS, P. C. & THOMPSON, A. R. 1987. Patterns in the deposition of the Brent Group (Middle Jurassic) UK North Sea. *In*: BROOKS, J. & GLENNIE, K. (eds) *Petroleum Geology of North-West Europe*. Graham & Trotman, 899–913.

BUDDING, M. C. & INGLIN, H. F. 1981. A reservoir geological model of the Brent Sands in southern Cormorant. *In*: ILLING, L. V. & HOBSON, G. D. (eds) *Petroleum Geology of the Continental Shelf of North-West Europe*. Heyden, London, 326–334.

CLAVIER, C., COATES, G. & DUMANOIR, J. 1977. The theoretical and experimental bases for the 'Dual Water' model for the interpretation of shaly sands. SPE 6859. *52nd Annual Fall Technology Conference, SPE, AIME, Denver, Colorado October 1977*.

CUIEC, L. 1987. Wettability and oil reservoirs. *In*: KLEPPE, J. *et al.* (eds) *North Sea Oil and Gas Reservoirs*. Graham & Trotman, 193–207.

FERTL, W. H. 1987. Log-derived evaluation of shaly clastic reservoirs. *Journal of Petroleum Technology*, Feb 1987, 175–194.

FUCHTBAUER, H. 1983. Facies controls on sandstone diagenesis. *In*: PARKER, A. & SELLWOOD, B. W. (eds) *Sediment Diagenesis*. D. Reidel Publishing Company, Dordrecht, Holland, 269–288.

GLASMANN, J. R., LUNDEGARD, P. D., CLARK, R. A., PENNY, B. K. & COLLINS, I. D. 1989. Geochemical evidence for the history of diagenesis and fluid migration: Brent Sandstone, Heather Field, North Sea. *Clay Minerals*, **24**, 255–284.

HANCOCK, N. J. 1978. Diagenetic modelling of the Middle Jurassic Brent Sand of the Northern North Sea. *Transactions of the European Offshore Petroleum Conference and Exhibition*, Paper EUR 92.

HAUGEN, S. A., LUND, O. & HOYLAND, L. A. 1988. Statfjord Field: Development Strategy and Reservoir Management. SPE 16961. *Journal of Petroleum Technology*, July 1988, 863–873.

HONARPOUR, M., KOEDERITZ, L. & HARVEY, A. H. 1986. *Relative Permeability of Petroleum Reservoirs*. CRC Press Inc., Boca Raton, Florida.

HURST, A. 1987. Problems of reservoir characterization in some North Sea sandstone reservoirs solved by the application of microscale geological data. *In*: KLEPPE, J. *et al.* (eds) *North Sea Oil and Gas Reservoirs*. Graham & Trotman, 153–167.

—— & IRWIN, H. 1982. Geological modelling of clay diagenesis in sandstones. *Clay Minerals*, **17**, 5–22.

KAYE, L. 1990. A discussion of the influence of mechanical disturbance on the quality of log and core data obtained from stress-sensitive reservoir materials. *2nd EAPG Conference*, Copenhagen, May 28–June 1, 1990, Paper E042.

JOHNSON, H. D. & STEWART, D. J. 1985. Role of clastic sedimentology in the exploration and production of oil and gas in the North Sea. *In*: BRENCHLEY, P. J. & WILLIAMS, B. P. J. (eds). *Sedimentology: Recent Developments and Applied Aspects*. Geological Society, London, Special Publications, **18**, 249–310.

JOURDAN, A., THOMAS, M., BREVART, O., ROBSON, P., SOMMER, F. & SULLIVAN, M. 1987. Diagenesis as the control of the Brent sandstone reservoir properties in the Greater Alwyn area East Shetland Basin). *In*: BROOKS, J. & GLENNIE, K. (eds) *Petroleum Geology of North-West Europe*. Graham & Trotman, London, 951–961.

KANTOROWICZ, J. 1984. Nature, origin and distribution of authigenic clay minerals from Middle Jurassic Ravenscar and Brent Group sandstones. *Clay Minerals*, **19**, 359–376.

KEIR, C. A. & BROWN, J. 1978. Permeability impairment due to the presence of kaolinite in the matrix of reservoir rock. *Transactions of the European Offshore Petroleum Conference and Exhibition*, Paper EUR 94.

LAVERS, B. A., SMITS, L. J. M. & van BAAREN, C. 1975. Some fundamental problems of formation evaluation in the North Sea. *The Log Analyst*, May–June, 1975.

Marett, G., Chevalier, P., Souhaite, P. & Suau, J. 1976. Shaly sand evaluation using gamma-ray spectroscopy applied to the North Sea Jurassic. *Transactions of SPWLA 17th Annual Logging Symposium*, paper DD.

Mohanty, K. K., Davis, H. T. & Scriven, L. E. 1987. Physics of oil entrapment in water-wet rock. SPE 9406. *SPE Reservoir Engineering*, February 1987, 113–128.

Nieto, J. A., Yale, D. P. & Evans, R. J. 1990. Core compaction correction – a different approach. *Proceedings of the First Society of Core Analysts European Core Analysis Symposium*, London, 21–23 May 1990, 139–156.

Pallat, N., Wilson, J. & McHardy, W. 1984. The relationship between permeability and the morphology of diagenetic illite in reservoir rocks. SPE 12798. *Journal of Petroleum Technology*, December 1984, 2225–2227.

Patchett, J. G. & Coalson, E. B. 1982. The determination of porosity in sandstone and shaly sandstone, part two: effects of complex mineralogy and hydrocarbons. *Transactions of SPWLA 23rd Annual Logging Symposium*, 6–9 July 1982, paper T.

Poupon, A. & Gaymard, R. 1970. The evaluation of clay content from logs. *Transactions of SPWLA 11th Annual Logging Symposium*, 1970, paper G. (Reprinted in full in *SPWLA Reprint Volume: Shaly Sand*, 1982, II 83–II 103.)

—— & Leveaux, J. 1971. The evaluation of water saturations in shaly formations. *Transactions of SPWLA 12th Annual Logging Symposium*, 1971, paper O. (Reprinted in full in *SPWLA Reprint Volume: Shaly Sand*, 1982, IV 81–IV 95.)

Peveraro, R. C. A. & Russell, K. J. 1984. Interpretation of wireline log and core data from a mid-Jurassic sand/shale sequence. *Clay Minerals*, **19**, 483–506.

Richards, P. C. & Brown, S. 1986. Shoreface storm deposits in the Rannoch Formation (Middle Jurassic), North West Hutton Oilfield. *Scottish Journal of Geology*, **22**, 367–375.

Ronning, K. & Steel, R. J. 1987. Depositional sequences within a 'transgressive' reservoir sandstone unit: the Middle Jurassic Tarbert Formation, Hild area, northern North Sea. *In*: Kleppe J. *et al.* (eds) *North Sea Oil and Gas Reservoirs*. Graham & Trotman, 169–176.

Scotchman, I. C., Johnes, L. H. & Miller, R. S. 1989. Clay diagenesis and oil migration in Brent Group sandstones of NW Hutton Field, UK North Sea. *Clay Minerals*, **24**, 339–374.

Serra, O. 1984. *Fundamentals of Well-Log Interpretation. 1. The acquisition of logging data*. Developments in Petroleum Science 15A. Elsevier, Amsterdam.

—— 1986. *Fundamentals of Well-Log Interpretation. 2. The interpretation of logging data*. Developments in Petroleum Science, 15B. Elsevier, Amsterdam.

Suau, J. & Spurlin, J. 1982. Interpretation of Micaceous Sandstones in the North Sea. *Transactions SPWLA 23rd Annual Logging Symposium*, July 6–9 1982, Paper G.

Vinegar, H. J. & Waxman, M. S. 1984. Induced polarization of shaly sands – the effect of clay counter ion type. *Transactions of SPWLA 25th Annual Logging Symposium*, June 10–13, Paper LLL.

Waxman, M. S. & Smits, L. J. M. 1968. Electrical conductivities in oil-bearing shaly sands. *Journal of the Society of Petroleum Engineers*, **8**, 107–122.

Weber, K. J. 1982. The influence of common sedimentary structures on fluid flow in reservoir models. SPE 9247. *Journal of Petroleum Technology*, March 1982, 665–672.

—— 1986. How heterogeneity affects oil recovery. *In*: Lake, L. W. & Carroll, H. G. Jr. (eds) *Reservoir Characterization*. Academic Press, New York, 487–544.

Worthington, P. F. 1985. The evolution of shaly-sand concepts in reservoir evaluation. *The Log Analyst*, Jan–Feb 1985, 23–40.

# Index